MW00449416

Digital Signal and Image Processing

TAMAL BOSE
Utah State University

with
Francois Meyer contributing Chapter 10 and
Mei-Qin Chen contributing the Appendix

JOHN WILEY & SONS, INC.

EXECUTIVE EDITOR *William Zobrist*
MARKETING MANAGER *Ilse Wolfe*
SENIOR PRODUCTION EDITOR *Norine M. Pigliucci*
SENIOR DESIGNER *Dawn Stanley*
PRODUCTION MANAGEMENT SERVICES *Hermitage Publishing Services*
SENIOR EDITORIAL ASSISTANT *Angie Vennerstrom*

The cover is a photograph courtesy of Gerhard Lang. It shows the peaks of Everest (left) and Lhotse (right). The 3-D graph is the magnitude spectrum of Everest, the highest peak in the world. This picture is meant to convey a sense of grandeur and challenge in the study of signal processing which is shared by many of us in the community.

This book was set in New Caledonia by Hermitage Publishing Services and printed and bound by Malloy, Inc. The cover was printed by Phoenix Color Corp.

This book is printed on acid-free paper. ♾

Library of Congress Cataloging-in-Publication Data
Bose, Tamal.
Essentials of digital signal and image processing / Tamal Bose ; with Francois Meyer contributing Chapter 10 and Mei-Qin Chen contributing the Appendix.
p. cm.
Includes bibliographical references.
ISBN 0-471-32727-1 (cloth)
WIE ISBN 0-471-45230-0
1. Image processing—Doigoitsal techniques. 2. Holography—Data processing. 3. Signal processing—Digital techniques. I. Title.
TA1637.B67 2003
621.3—dc21 2003047903

Printed in the United States of America

10 9 8 7 6 5 4 3 2 1

To my wife Tammy
And our children Cameron and Megana

Preface

This text is unique because it presents the fundamentals of Digital Signal Processing (DSP), Adaptive Signal Processing (ASP), and Digital Image Processing (DIP) under one cover. With this text, the reader can learn several different combinations of these important subjects.

The subject of Digital Signal Processing (DSP) deals with the theory of processing any type of digital signal. Digital signals can be processed with a lot more ease, flexibility, and speed than analog signals. This is precisely the reason that DSP has become such an important field over the last three decades. Parallel advancements in microelectronics and computational hardware have also led to an increased importance and widespread use of DSP. The broad area of DSP encompasses all the tools, techniques, and algorithms that are used for processing. DSP algorithms are ubiquitous in modern technology. In consumer electronics, they are found in mobile phones, home security systems, digital watches, CD players, modems, High Definition Television (HDTV), portable digital assistants, electronic toys and GPS systems, among many others. In aerospace and military applications, DSP algorithms are used almost everywhere from aircraft controls and communications to sophisticated missile guidance systems.

An important subset of the broad area of DSP is Adaptive Signal Processing (ASP). ASP deals with the design and implementation of algorithms that adapt to changing environments. For example, consider a mobile phone in a car. The communication signal received by the phone is distorted as it propagates through air and is reflected many times by buildings, other vehicles, and hills. In addition, the distortion characteristics change as the car moves from one location to another. The telephone receiver uses algorithms to compensate for this distortion. These are adaptive filtering algorithms. There are many such applications where adaptive algorithms are imperative. The theory of ASP has therefore become very important.

Another subset of DSP is Digital Image Processing (DIP). Many of the techniques of DSP can be extended to two dimensions in order to process images. The field of DIP has become increasingly important in the past two decades due to its numerous applications. The algorithms of DIP are used for image enhancement, restoration, compression and many other functions. These algorithms can be found in image transmission and reception over the Internet and satellite links. Digital cameras and camcorders use many DIP algorithms. Photo editing programs found on most computers are nothing but implementations of image processing algorithms.

This textbook is written primarily for use in a traditional DSP course. The main strengths of this book as a DSP text are the following: (a) Clarity and rigor in the presentation of mathematical concepts; (b) Algorithms are described in such a way that students can code them in MATLAB with little difficulty; (c) The breadth of DSP topics is just enough for an introductory DSP course; (d) Basic MATLAB commands are described at the end of most chapters to help students with the assigned computer projects; (e) Extension of many one-dimensional concepts to two dimensions are presented in most chapters. This allows the use of examples with images in addition to 1-D signals, thereby making it more interesting; and (f) Comprehensive computer projects at the end of each chapter.

Key Features

- The essentials of DSP, ASP, and DIP are presented in a single text.
- The topics are presented with sufficient mathematical rigor, which is not compromised for simplicity. However, mathematical derivations are presented in the simplest possible way.
- Several computer projects are included at the end of each chapter. As opposed to short computer problems, these are comprehensive projects requiring knowledge of several topics in the chapter.
- At the end of most chapters, a section on MATLAB is included. This section presents a discussion on some of the MATLAB commands that may be useful for the projects in that chapter.
- Topics on 2-D signal processing are given in the same chapter as their 1-D counterparts. This makes for a smooth transition to the 2-D concepts.
- The material is organized in such a way that the subjects of DSP, ASP, and DIP can be taught independently.
- Some well known government-adopted standards are described, such as the JPEG lossless and lossy standards, and some speech coding standards. The standards are described in enough detail that computer projects may be assigned to implement these algorithms.
- An appendix on Linear Algebra serves as a reference for readers who need to brush up on some useful mathematical tools.

Overview of Contents

The book is divided into 11 chapters plus an appendix. The appendix is a reference for the fundamental concepts in linear algebra. It begins with basic matrices, operations, and properties. Many special matrices, norms and their properties are then discussed. The material is limited to the concepts that are useful in signal and image processing.

Chapter 1 presents the fundamental concepts of signals and systems. It covers a variety of topics including classification of signals, classification of systems, convolution, impulse response, and different types of filters. A section is also presented that extends these topics to 2-D systems.

Chapter 2 is devoted to Fourier analysis. Different types of Fourier techniques are described for analyzing and designing signal processing systems in the frequency domain. We begin with a brief review of the Continuous-Time Fourier Transform (CTFT). The Discrete-Time Fourier Transform (DTFT) and its properties are then presented in detail. This sets the stage for discussions of the sampling theorem, frequency response, and decimation/interpolation. The Discrete Fourier Transform (DFT) and the basic radix-2 Fast Fourier Transforms (FFT) are also presented. The above concepts are generalized to 2-D systems for applications in image processing.

Chapter 3 is primarily devoted to Z-transform. It also presents the concept of linear phase in digital filters and the state-space representation of filters. Algorithms for performing linear filtering using FFT, namely, the overlap-add and overlap-save methods, are discussed. Algorithms are also described for computing the DFT using linear filtering. Some of these topics are then generalized for 2-D systems.

In Chapter 4, we present several different methods for designing FIR and IIR filters. Various filter structures are then derived for implementation and their relative advantages and disadvantages are discussed. In the 2-D sections, the design of the FIR filter is achieved by straightforward extensions of the 1-D methods. Different methods for designing 2-D IIR filters are finally described.

In Chapter 5, we first review the concepts of decimation and interpolation and present the effects of these operations in the frequency domain and the Z-domain. Computationally efficient filter structures are derived for these operations. These are called polyphase structures. The concept of subband coding is described with motivation in speech and image compression. This leads to the theory of filter banks. The simple DFT filter bank is presented. Quadrature Mirror Filter (QMF) banks are then described in detail. Design methodologies for these filters are also developed.

Chapter 6 begins with a discussion of fixed-point and floating-point number representations. It then describes the various sources of quantization noise and its effect on digital filters. Methods to analyze and minimize noise are described. Different design constraints are derived so that different filter structures are stable and have minimum noise due to quantization effects.

Chapter 7 is devoted to the fundamental concepts in Adaptive Signal Processing. Concepts such as the correlation matrix and Wiener filtering are described in detail. Simple gradient search algorithms for optimization such as the steepest descent method and Newton's method are presented. The celebrated filtering method called the Least Mean Square (LMS) algorithm is derived and analyzed. Then we move into the frequency domain adaptive filtering method, which gives rise to several block adaptive algorithms. In particular, we derive the Block LMS (BLMS) algorithm, the frequency domain adaptive filter (FDAF), and the Fast-LMS (FLMS) algorithm.

In Chapter 8, we present least squares algorithms for adaptive filtering. The Recursive Least Squares (RLS) algorithm and the Euclidean Direction Search (EDS) algorithm are described in detail. Applications of these algorithms are illustrated with an example of a channel equalizer. The topic of Adaptive Pulse Code Modulation (ADPCM) is then presented. This is important for data compression. The last section of this chapter deals with 2-D adaptive filters. The LMS and EDS algorithms are extended for 2-D filters. Examples on image processing are given to illustrate the applications of these algorithms.

Chapter 9 deals with the theory and applications of linear prediction. We begin with the concepts of forward and backward prediction. The Levinson-Durbin algorithm is then presented. The Gradient Adaptive Lattice (GAL) is then derived. The advantage of this algorithm and its computational complexity are also discussed. Speech coding is an important application of linear prediction. Several techniques of speech coding are discussion including the waveform coder, transform coder, subband coder, and linear predictive coder. The applications of these techniques are discussed in relation to some established speech coding standards.

Chapter 10 is devoted to the fundamentals of image processing. We first describe several methods for image enhancement. Image interpolation is then discussed including linear and polynomial techniques. Edge detection is presented and several different methods are discussed in detail. Topics in image restoration including inverse filtering, lowpass filtering, and median filtering are covered in detail. Finally, the basics of color image processing are presented.

Chapter 11 presents the fundamentals of image compression and coding. Several different types of image transforms are presented with mathematical derivations and properties. The technique of 2-D DPCM is then presented in detail. In image coding, we present two popular algorithms, Huffman coding and Lempel-Ziv-Welch (LZW) coding. The JPEG standards for both lossy and lossless compression are then briefly discussed.

Possible Course Guidelines

This textbook can be used in many courses depending on the selection of material. I have used the material from this book to teach courses on Digital Signal Processing (DSP), Adaptive Signal Processing (ASP), and Digital Image Processing (DIP) at the University of Colorado at Denver, University of Colorado at Boulder, and at Utah State University.

The prerequisites for this text are (a) basic calculus, (b) differential equations, and (c) linear system theory. These courses are typical requirements for most electrical and computer engineering juniors. No prior knowledge of discrete-time system theory is necessary. Some course suggestions are given below.

Undergraduate courses
(No DSP prerequisite)
✓ Entire chapter

Course	Ch 1	Ch 2	Ch 3	Ch 4	Ch 5	Ch 6	Ch 7	Ch 8	Ch 9	Ch 10	Ch 11
One-semester course on DSP:	✓ Omit 2-D	✓ Omit 2-D	✓ Omit 2-D	✓ Omit 2-D	Select Topics	Select Topics					
One-semester course on DSP+DIP:	✓	✓	✓	Select Topics						Select Topics	
One-semester course on DSP+ASP:	✓ Omit 2-D	✓ Omit 2-D	✓ Omit 2-D	Select Topics			Select Topics	Select Topics			
One-semester course on DSP+DIP+ ASP:	✓ Omit 2-D	✓ Omit 2-D	✓ Omit 2-D				✓			✓	
Two-quarter sequence on DSP + DIP (course 1):	✓ Omit 2-D	✓ Omit 2-D	✓ Omit 2-D	✓ Omit 2-D							
Two-quarter sequence on DSP + DIP (course 2):	2-D sections	2-D sections	2-D sections	2-D sections						✓	✓
Two-quarter sequence on DSP + ASP (Course 1):	✓ Omit 2-D	✓ Omit 2-D	✓ Omit 2-D	✓ Omit 2-D							
Two-quarter sequence on DSP + ASP (Course 2):							✓	✓	✓		

Graduate courses
(One DSP course as prerequisite)

Course	Ch 1	Ch 2	Ch 3	Ch 4	Ch 5	Ch 6	Ch 7	Ch 8	Ch 9	Ch 10	Ch 11
Graduate course on DSP + DIP:	2-D sections	2-D sections	2-D sections	2-D sections	Select Topics	Select Topics				✓	✓
Graduate course on DSP + ASP:					✓	✓	✓	✓	✓		
Graduate course on DIP + ASP:	2-D sections	2-D sections	2-D sections	2-D sections			✓	✓	Select Topics	✓	Select Topics

Supplements

The text website is http://www.wiley.com/college/bose. This site contains the MATLAB code for most of the computer-based examples in the text. It also contains the Powerpoint slides for the figures and graphs in the text. A solution manual with fully worked solutions can be found at the text website. Adopting instructors should register for access to download an electronic copy.

I have made every possible effort to make this book as error-free as possible. However, I am sure that some errors went undetected. An erratum can be found on the text website for any such occurences. I would appreciate it if readers bring to my attention any errors that they find. This can be done electronically through the text website.

Acknowledgments

Many people have contributed to make this book a reality. First of all, I would like to thank the contributing authors. Dr. Francois Meyer of the University of Colorado at Boulder contributed Chapter 10. Dr. Mei-Qin Chen of The Citadel contributed the appendix on linear algebra.

Several experts in the field of signal and image processing reviewed the manuscript. I sincerely thank the reviewers for their suggestions, comments, and corrections. The reviewers include Drs. Andreas Spanias of Arizona State University, Jim Schroeder of University of Adelaide, Rob Nowak of University of Wisconsin at Madison, Panos Agathoklis of University of Victoria, Erchin Serpedin of Texas A & M University, Artice Davis of San Jose State University, Charles Owen of Michigan State University, James Kang of Cal Poly San Luis Obispo, Stan Kesler of Drexel University, and Stanley Reeves of Auburn University.

I would also like to acknowledge the help of several former students. Dr. Guo Fang Xu of Data Play proofread Chapters 1 through 3 and also wrote some problems for these chapters. Dr. Ratchaneekorn (Kay) Thamvichai of St. Cloud State University proofread several chapters and is the primary author of the solution manual. Mike Larsen, Leah Roberts and John Bushman proofread and typed several parts of the manuscript and solution manual.

I acknowledge Bill Zobrist of John Wiley for doing a superb job as editor. I want to also thank Hermitage Publishing Services for help with production. In particular, I would like to thank the copy editor Betty Pessagno and the production manager Larry Meyer for a job well done.

Finally, I want to thank my family for their support during this long project. They have sacrificed long hours of family time with me while I worked on the manuscript. I dedicate this book to my wife Tammy and our children Cameron and Megana.

TAMAL BOSE

Contents

CHAPTER 1

Fundamental Concepts

Signals or waveforms are commonplace occurrences in everyday life. Our lights and appliances are powered by electrical signals. From our circuits course, we know that these are usually sinusoidal current and voltage waveforms. The computer uses, processes, and stores information that is actually composed of electrical signals. When we turn the radio on, electrical signals are received from the air and converted into sound. On the other hand, sound is converted into electrical signals and transmitted through air for our radios to receive them. A similar process occurs for television. Images and audio are converted to electrical signals and transmitted through air. Our TV sets receive this information and convert the electrical signals into audio and video.

The subject of digital signal processing (DSP) encompasses the tools and techniques required for processing a variety of signals in engineering. The rapid advancement of microprocessor technology over the last two decades has made DSP a popular and attractive discipline. The omnipresence of DSP techniques in applications can be attributed to the speed, flexibility, and economy of processing digital signals as opposed to analog. New and improved signal processing algorithms are being constantly developed for many applications. Often these algorithms are computationally demanding. Parallel advancements in microprocessor technology provide the speed and power for implementing these new algorithms. To study DSP, we must understand the basic theory of signals and systems.

This chapter presents the fundamental concepts of signals and systems. It begins with a classification of signals and briefly discusses a typical signal processing system. Then it discusses simple two-dimensional (2-D) signals followed by an overview of digital image processing. The subsequent sections cover a variety of topics from discrete-time system theory, such as system classification, convolution, impulse response, and types of filters. A section is also presented that extends these topics to 2-D systems.

1.1 SIGNAL PROCESSING OVERVIEW

A digital signal processing (DSP) system operates on a digital signal and converts it to another signal that has certain desirable characteristics. In other words, a DSP system processes a digital signal. We begin this section with the definition and classification of signals.

Signal Classification

A *signal* is defined as a physical quantity that is a function of some independent variable such as time or space. A signal can be represented as $x(t)$, where t stands for the independent variable time. An example is the familiar sinusoid

$$y(t) = A\sin(\Omega t - \theta) \tag{1.1}$$

where A is the amplitude, Ω is the frequency in radians per second, and θ is the phase in radians. This signal is shown in Fig. 1.1. Let T and F denote the time-period (in seconds) and frequency (in cycles/second or Hertz) of the signal. The angular frequency is related to these parameters by

$$\Omega = 2\pi F = \frac{2\pi}{T}\text{rad/sec.} \tag{1.2}$$

The normalized time-period is 2π radians; that is, this is the time-period when the angular frequency is unity. If the sinusoid is shifted by t_θ seconds, then the phase θ is given by

$$\theta = 2\pi\frac{t_\theta}{T}\text{ radians.} \tag{1.3}$$

Another example is the exponential signal given by

$$x(t) = e^{-\alpha t}, \quad \alpha > 0. \tag{1.4}$$

This signal is depicted in Fig.1.2. These examples present signals from a class called continuous-time signals. Some other types of signals also occur in engineering. In the following, we present definitions and discussions of the different types of signals.

Continuous-Time Signal

A *continuous-time* signal is a function of time that is defined for every time instant within a given range of time. This range may be infinite. The sinusoidal and exponential functions given above are examples of continuous-time signals. A continuous-time signal is different from a *continuous* function, which is defined as a function differentiable everywhere in a given range. For example, a square-wave is a continuous-time signal but is not a continuous function.

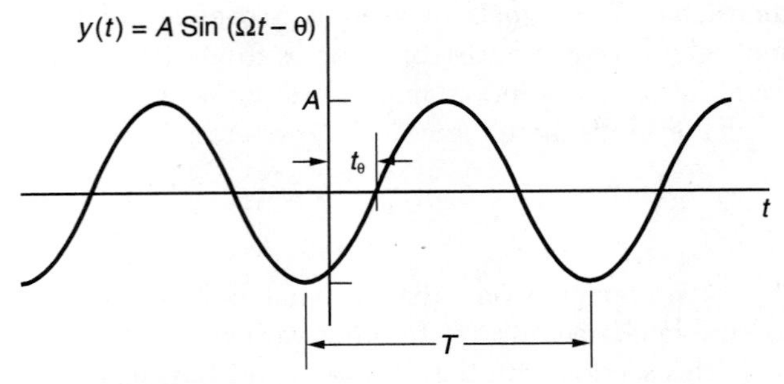

Figure 1.1 Sinusoid signal.

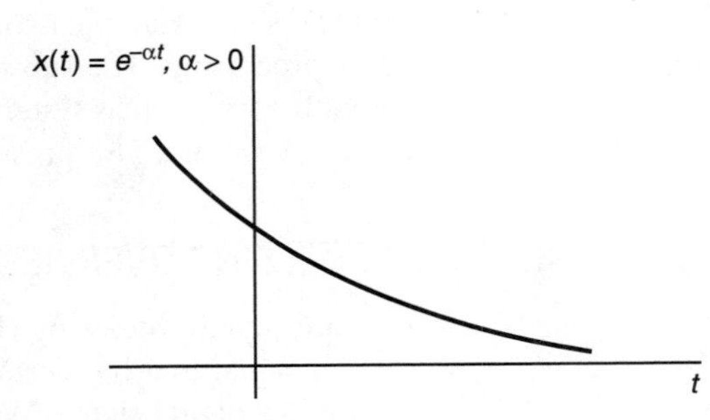

Figure 1.2 Exponential signal.

Discrete-Time Signal and Sequence

We first explain the concept of quantization, which is necessary for defining this class of signals. A variable t is said to be *quantized* if it can only assume values from a specified set of numbers. For example, if this set is the set of integers, then t can only be an integer; that is, it is quantized to the nearest integer. Consider the quantization of t as

$$t = nT_s \tag{1.5}$$

where n is an integer and T_s is some constant. In this case, t assumes the values $\ldots, -2T_s, -T_s, 0, T_s, 2T_s, \ldots$ The continuous-time variable t is quantized in steps of T_s. Here the quantization step size is constant, and this is referred to as uniform quantization. In practice, quantization does not have to be uniform, but it usually is.

A *discrete-time signal* is one that is quantized in time. Typically, a discrete-time signal is obtained from a continuous-time signal by uniformly quantizing the time variable as in (1.5). This process is also called *sampling*. Therefore, we define a discrete-time signal or a sampled signal as[1]

$$x(nT_s) \triangleq x(t)\,|_{t=nT_s}$$

where $x(t)$ is the continuous-time signal. A sampler is symbolically denoted as shown in Fig. 1.3. It can be thought of as a switch that periodically opens and closes. It stays open for a period given by T_s and is closed for an infinitesimally small amount of time. For example, the continuous-time sinusoid and exponential can be sampled. The discrete-time sinusoid and exponential are shown in Fig. 1.4(a) and (b), respectively. In these graphs we see that the time axis is still present, but it is quantized. The sampled (or discretized) signal has the same amplitude as the continuous-time signal at integer multiples of T_s and is zero elsewhere. By virtue of this last statement, we see that the discrete-time signal is defined for every instant of time and therefore fits the definition of the continuous-time signal. Thus, a discrete-time signal is a special case of a continuous-time signal.

When we remove the reference of time from a discrete-time signal, we get a discrete-time *sequence*, which is defined as

$$\{x(n)\} \triangleq x(nT_s) = x(t)|_{t=nT_s}, \quad n = \ldots, -2, -1, 0, 1, 2, \ldots \tag{1.6}$$

The resulting function $\{x(n)\}$ is simply a sequence of numbers with no reference to time between samples of data. An example of a sequence is

$$\{x(n)\} = \{\ldots, 2, -3, 7, 8, \mathbf{3}, 9, 2, 1, \ldots\}$$

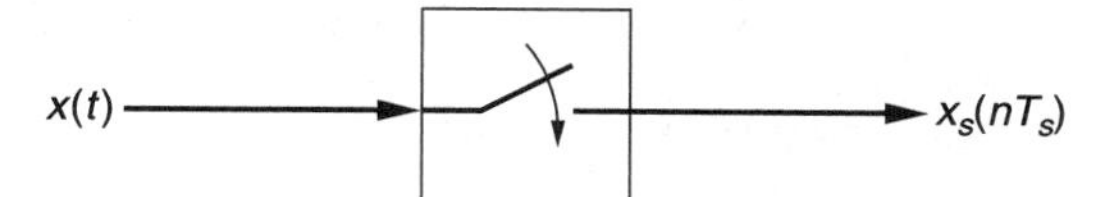

Figure 1.3 A sampler.

[1] The symbol $\triangleq$ will be used throughout this text to denote a definition.

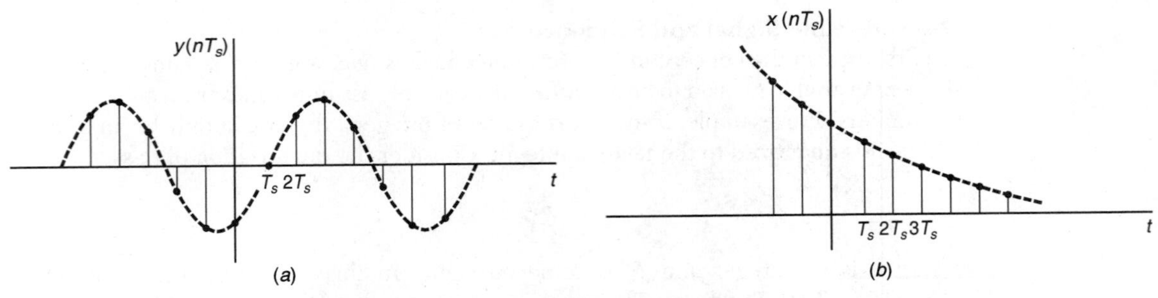

Figure 1.4 Sampled signals: (a) Sinusoid, (b) exponential.

where the value in boldface represents $x(0)$. For notational simplicity, the brackets will be dropped and $x(n)$ will be used to denote a sequence. It will also denote the value of the sequence at sample index n. The actual meaning of $x(n)$ should be clear from the context.

Digital Signal and Sequence

When the amplitude of a discrete-time *signal* is quantized, it is called a *quantized signal*. Most signals are coded in binary representation; these are called *digital signals*. Quantization occurs when this coding is performed. If a discrete-time *sequence* is quantized in amplitude, it is called a *digital sequence*. As the name suggests, amplitude quantization is done to a certain number of bits of binary. The amplitude of a digital signal or sequence can therefore assume values from a specified set of numbers. The process of converting an analog signal to digital is performed by the analog-to-digital (A/D) converter. Amplitude quantization of a sequence $x(n)$ will be denoted by $Q\{x(n)\}$. There are two basic types of quantization: magnitude-truncation and rounding. The input–output characteristics of these quantizers are given in Fig. 1.5(a) and (b), respectively. The quantization step size is the distance between two quantization levels and is denoted by q. In the magnitude-truncation scheme, the magnitude of the signal is always reduced to the lower quantization level. On the other hand, in the rounding scheme, the magnitude of the value of the signal is assigned to the nearest quantization level. If the signal has a value that is exactly halfway between two levels, then the quantized value becomes the quantization level which is greater in magnitude.

A Typical DSP System

Now we present a typical DSP scheme and briefly describe its components. A basic block diagram of a DSP system is shown in Fig. 1.6. This system will be explained with an example that uses a sum of two sinusoids as the continuous-time signal and is given by

$$x(t) = \sin(25\pi t) + \sin(100\pi t).$$

Clearly, the sinusoids have frequencies of $F_1 = 12.5$ Hz and $F_2 = 50$ Hz. The signal $x(t)$ is shown in Fig. 1.7(a). It is then sampled with a sampling frequency of $F_s = 200$ Hz that is, $T_s = \frac{1}{F_s} = 0.005$ second. The corresponding discrete-time signal $x(nT_s)$ is shown in Fig. 1.7(b). The sampler has therefore performed time quantization as explained earlier.

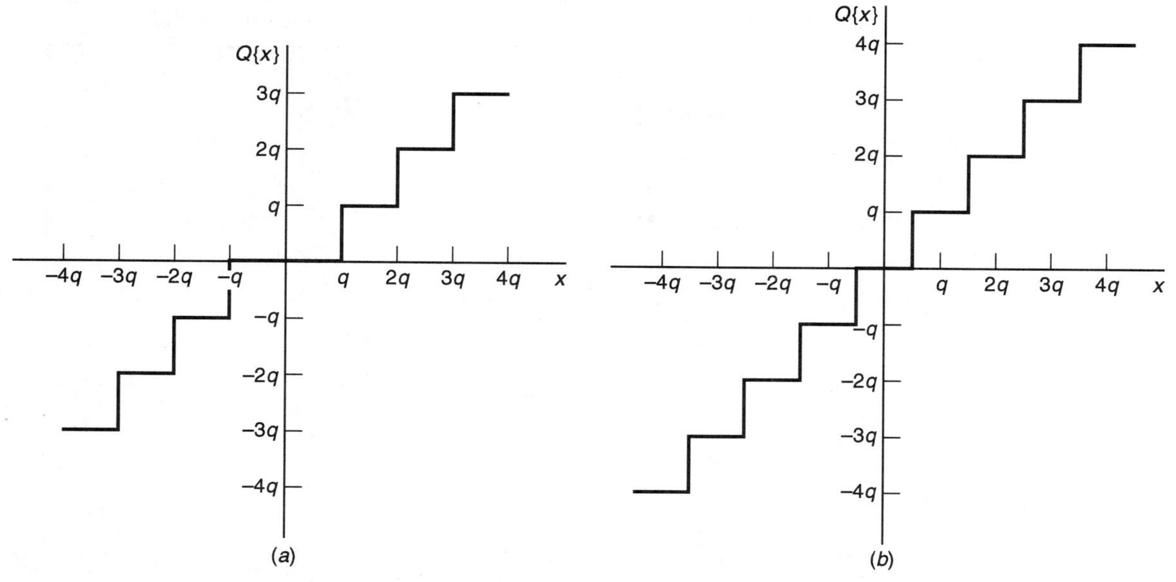

Figure 1.5 Quantization schemes: (*a*) Magnitude truncation, (*b*) rounding.

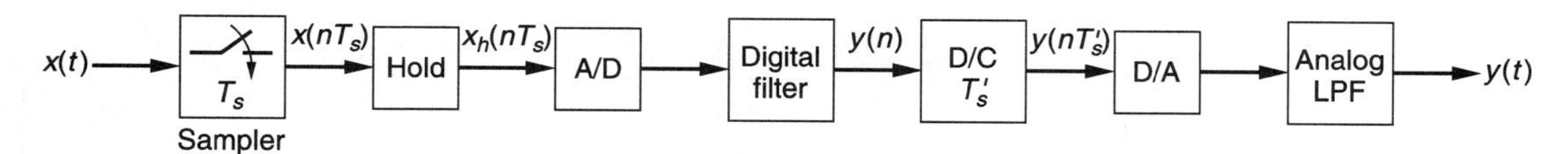

Figure 1.6 Typical DSP system.

As can be seen from $x(nT_s)$, each sample appears for an infinitesimally small period of time shown by the dots. This is not sufficient for the A/D converter to perform the conversion to digital. This is why the "hold" block is needed. The simplest of these circuits is called the Zero-Order-Hold (ZOH). It simply holds the current sample for one full sampling time-period, T_s. Other types of holding circuits perform some type of interpolation using the last few samples. The reader is referred to [1] for further reading on this topic. After the ZOH, we get the signal $x_h(nT_s)$, which is shown in Fig. 1.7(*c*). The A/D converter then quantizes this signal to the appropriate quantization levels and converts to digital. In this example, we use a 4-bit magnitude-truncation quantizer. That is, there are $2^4 = 16$ levels. These levels are shown as horizontal broken lines in Fig. 1.7(*c*) and (*d*). In (*c*), the levels of $x_h(nT_s)$ are not the same as the horizontal lines since we have not performed quantization yet. The quantized signal is labeled $x_q(nT_s)$ and is shown in Fig. 1.7(*d*). Note that in (*d*), the levels of the plot of $x_q(nT_s)$ mesh with the horizontal lines. These levels are converted to digital and processed by the digital signal processor (DSP). The processing algorithm in the DSP is normally referred to as a *digital filter*. More on digital filters will be given later in this chapter. In this example, we use a simple digital filter described by the difference equation

$$y(n) = x(n) + x(n-2), \quad n = 0, 1, 2, \ldots.$$

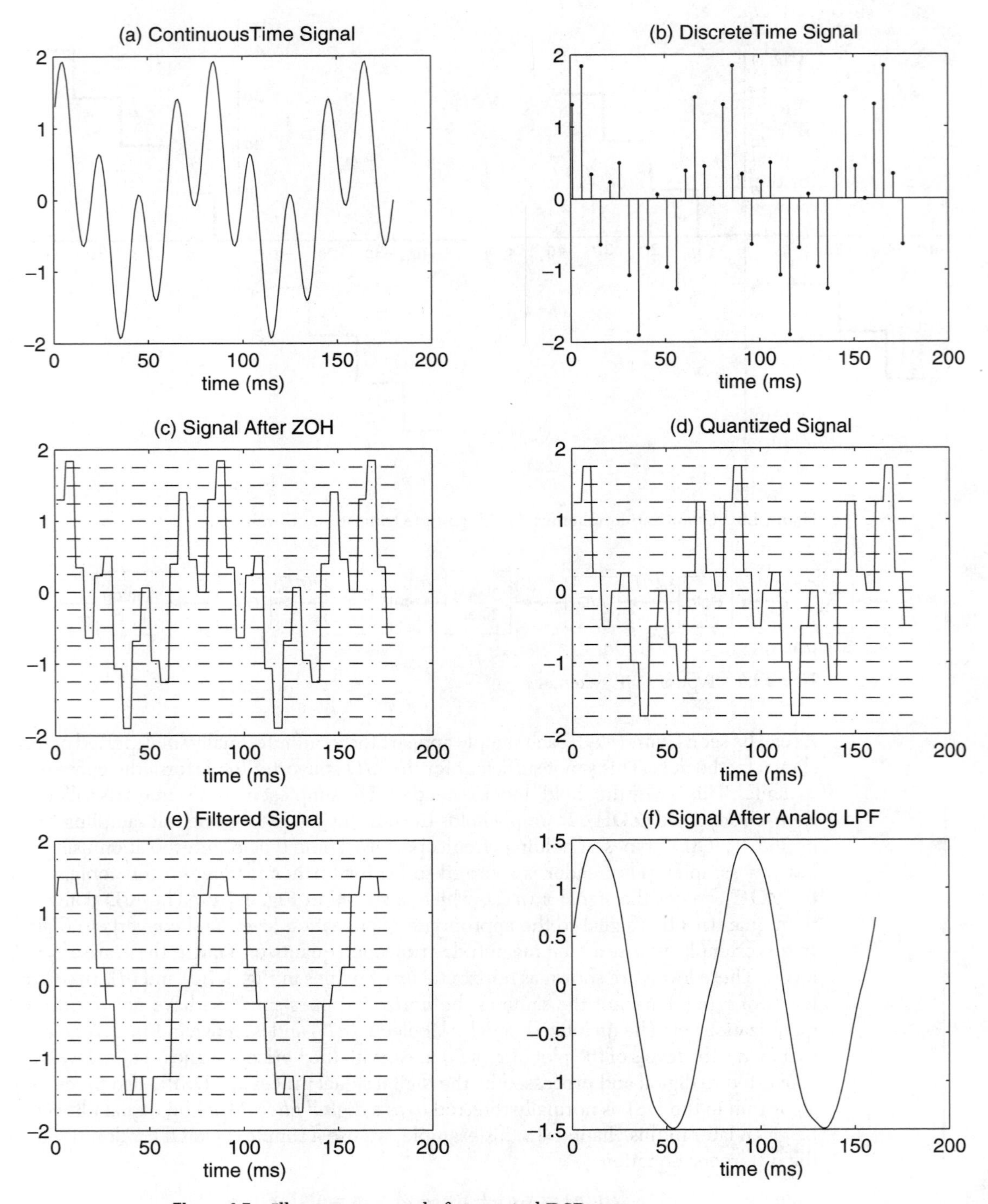

Figure 1.7 Illustration example for a typical DSP system.

The coefficients of $x(n)$ and $x(n-2)$, which are unity in this case, are called the filter coefficients. The coefficients are chosen to be unity so that the filter becomes a so-called notch filter. This filter adds the input data from two samples ago, $x(n-2)$ to the current input sample, $x(n)$ in order to find the output. The filter is also assumed to have 4 bits of precision. Note that as far as the filter is concerned, the time between samples (T_s) is immaterial. The filter simply operates on the samples of data as they arrive. The output of the filter is then passed on to the discrete-to-continuous (D/C) converter where the samples are spaced appropriately in time. This is the opposite of sampling. The samples can be spaced by any desired amount of time T'_s, which becomes the new sampling frequency. Of course, T_s and T'_s can be the same, which is the case in this example. Next, the D/A converter converts the signal from digital to the analog levels. It is important to note that the output of the D/A converter is still quantized. This is because the D/A converter simply converts from a digital representation (binary) to an analog format. Since the digital representation is quantized (to 4 bits), so is its analog counterpart. The output of the D/A converter is labeled as $y(nT'_s)$ and shown in Fig. 1.7(e). Note that this signal resembles a single sinusoid. Also, it is very "edgy" and contains many high-frequency components. The reason for this will be clear in Chapter 2. To smooth out the signal, an analog lowpass filter (LPF) is used. The output of this filter is shown in Fig. 1.7(f). This signal is approximately a sinusoid with a frequency of 12.5 Hz, which is one of the components of the original signal. So, what did the digital filter do? It notched out one of the sinusoids and passed the rest. This is called a notch filter. Analysis and design of this type of filter will be presented in a subsequent chapter.

Now consider the above system with the digital filter as a simple unit gain, that is, $y(n) = x(n)$. In other words, the digital filter does nothing. In the absence of the quantizer, the A/D and D/A converters, we expect the final output of the system $y(t)$ to be exactly the same as the the original input $x(t)$. The quantizer is not included because it is a nonlinear operation, and once it is performed, we introduce an error that cannot be reversed. If there is no amplitude quantization, then there is no need for the A/D and D/A converters. The question that arises now is the following: How fast do we have to sample the input signal so that it can be reconstructed perfectly? The answer is given by the famous *sampling theorem* which is stated below. Analysis and proof of this theorem require a background in Fourier analysis and will therefore be postponed until Chapter 2.

Theorem 1.1 (Sampling Theorem) *Let F_H represent the highest frequency component of a continuous-time signal $x(t)$. If this signal is sampled at a frequency F_s that satisfies $F_s > 2F_H$, then the original signal can be perfectly reconstructed from its sample values.*

Theorem 1.1 is credited to Nyquist [2]. A rigorous proof and a reconstruction formula for the theorem were given by Shannon [3]. The highest frequency component of the signal, F_H is called the *Nyquist frequency*. The lower limit of the sampling frequency, $F_N = 2F_H$ is called the *Nyquist rate*. If the sampling rate is not higher than the Nyquist rate, then a phenomenon called *aliasing* occurs, and the input signal cannot be reconstructed. The sampling theorem assumes a knowledge of the frequency content of the input signal so that the sampling rate can be established. For example, we know that all the frequency components of audio signals are less than 20 kHz. Therefore, audio signals are typically sampled at 44.1 kHz. Setting the sampling frequency a little above the Nyquist rate is common practice in order to allow some safety margin. Sometimes, this a priori information about the frequency content of the signal is not available. In such cases, the input signal is processed by an analog lowpass filter before sampling. This establishes the Nyquist frequency and hence the sampling rate. This filter is called an *anti-aliasing filter*.

1.2 IMAGE PROCESSING BASICS

In this section, we extend some of the above concepts to two dimensions (2-D) and then present some general discussion on image processing.

2-D Signals

A two-dimensional continuous-time signal can be denoted by $x(t_1, t_2)$ where t_1 and t_2 are the independent variables and do not necessarily represent time. As in the 1-D case, this 2-D continuous-time signal must be defined for every value of (t_1, t_2) in some given range. A 2-D signal is actually plotted in three dimensions, where t_1 and t_2 are two of the axes and the amplitude $x(t_1, t_2)$ is plotted on the third axis. An example of a 2-D signal is shown in Fig. 1.8(a). If the amplitude represents intensity on a black-white scale, then this 2-D signal is an image. When the amplitude is plotted as intensity, the result is an image and is shown in Fig. 1.8(b).

A 2-D continuous-time signal can be sampled to obtain the discrete-time signal $x(n_1T_{s1}, n_2T_{s2})$, where n_1 and n_2 are sample indices, and T_{s1} and T_{s2} are the respective sampling periods. When we drop the sampling periods from the notation, we obtain the 2-D sequence $x(n_1, n_2)$. A 2-D sequence can be represented in four different ways as follows: (1) It can be represented as shown in Fig. 1.9, where the numbers in the parentheses denote the amplitudes. If there is no number in parentheses, then the amplitude is defined as unity. (2) If the image such as the one shown in Fig. 1.8(b) is sampled in both directions, we get a finite dimensional matrix, each element of which is referred to

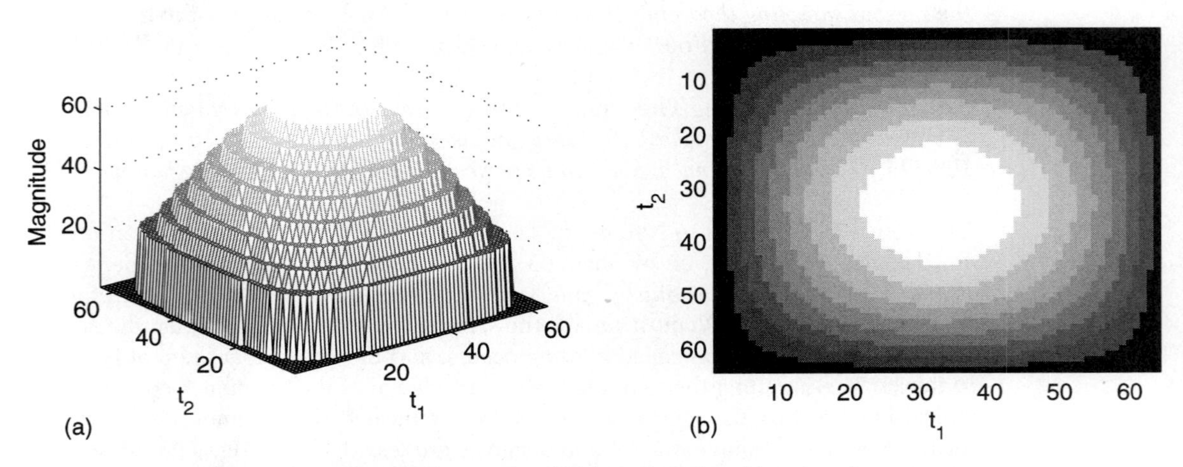

Figure 1.8 Example of a 2-D signal: (a) Surface plot, (b) image plot.

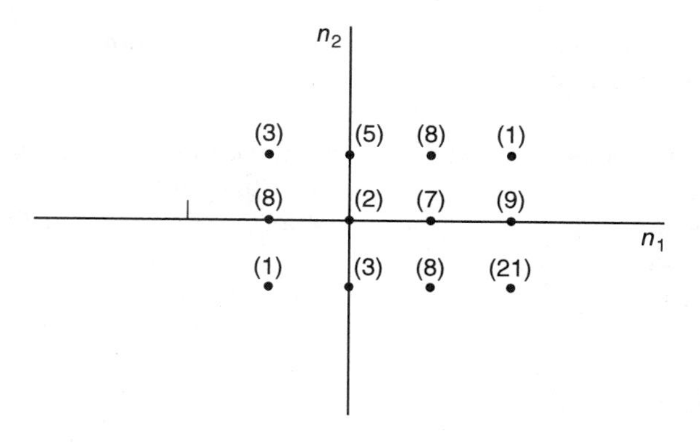

Figure 1.9 2-D signal representation.

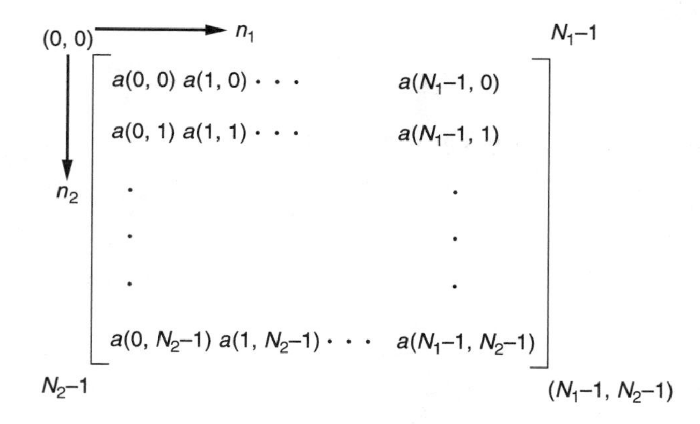

Figure 1.10 Matrix representation of an image.

as a *pixel*. An example of a $N_1 \times N_2$ image represented as a matrix is shown in Fig. 1.10. (3) This 2-D signal matrix may be plotted with the amplitudes as intensity resulting in an image. (4) The 2-D signal may also be represented as a surface plot as before.

Just as in the 1-D case, a 2-D discrete-time sequence can be quantized in amplitude to obtain a digital sequence. For images, the amplitude represents intensity and is therefore nonnegative in value. For quantization, we first decide on the number of bits, say B. Then the number of intensity levels are from 0 through $2^B - 1$. These are called *gray* levels. The amplitudes of the 2-D sequence can be scaled appropriately and quantized to these gray levels. The digital image is then referred to as a B-bit image.

Overview

Many of the 1-D techniques of digital signal processing can be extended and used for processing images. Of course, many techniques are special for image processing alone. In this book, we will cover several methods for both 1-D and 2-D signal processing. The typical DSP system depicted in Fig. 1.6 is also applicable to 2-D signal processing. In this case, we have two indices (n_1, n_2) and two sampling periods (T_{s1}, T_{s2}). The subject of image processing can be subdivided into the main categories of (a) image restoration,

(b) image enhancement, and (c) image compression and coding. In the following we briefly discuss each category.

Image Restoration. Image restoration refers to the process of restoring a corrupted image to its original state. Therefore, it is assumed that an original image was available and that it got corrupted during transmission, storage, or some other process. The goal is to process the corrupted image so that it resembles the original as much as possible. The success of the restoration process must be evaluated by some qualitative or quantitative measure. One of the commonly used quantitative measures is called the *signal to noise ratio improvement* (SNRI), which will be defined below. Several other measures will be presented in later chapters. Let $x(n_1, n_2)$ denote an $N_1 \times N_2$ original image, $c(n_1, n_2)$ the corrupted image, and $\widehat{x}(n_1, n_2)$ the restored image. The signal to noise ratio (SNR) of any signal is defined as

$$SNR \triangleq 10 \log_{10} \frac{\text{var}\{\text{signal}\}}{\text{var}\{\text{noise}\}} \quad \text{decibels(dB)} \tag{1.7}$$

where var denotes the variance. The variance of a $N_1 \times N_2$ sequence $y(n_1, n_2)$ is defined as

$$\text{var}\{y(n_1, n_2)\} \triangleq \frac{\sum_{n_1=0}^{N_1-1} \sum_{n_2=0}^{N_2-1} [y(n_1, n_2) - \bar{y}(n_1, n_2)]^2}{N_1 N_2} \tag{1.8}$$

where $\bar{y}(n_1, n_2)$ represents the mean of the sequence. The mean of this sequence is given by

$$\bar{y}(n_1, n_2) \triangleq \frac{\sum_{n_1=0}^{N_1-1} \sum_{n_2=0}^{N_2-1} y(n_1, n_2)}{N_1 N_2}. \tag{1.9}$$

The corrupted image $c(n_1, n_2)$ consists of two parts, namely, the signal part given by $x(n_1, n_2)$ and the noise part which is $c(n_1, n_2) - x(n_1, n_2)$. The SNR of the corrupted signal is therefore

$$SNR_c = 10 \log_{10} \frac{\text{var}\{x(n_1, n_2)\}}{\text{var}\ \{c(n_1, n_2) - x(n_1, n_2)\}} \text{ dB}. \tag{1.10}$$

In the restored image, the remaining noise is $\widehat{x}(n_1, n_2) - x(n_1, n_2)$. The SNR of this image is therefore

$$SNR_{\widehat{x}} = 10 \log_{10} \frac{\text{var}\{x(n_1, n_2)\}}{\text{var}\{\widehat{x}(n_1, n_2) - x(n_1, n_2)\}} \text{ dB}. \tag{1.11}$$

The performance of a restoration algorithm can be measured by the *SNR Improvement* (SNRI), which is defined as

$$\begin{aligned} SNRI &= SNR_{\widehat{x}} - SNR_c \text{ dB} \\ &= 10 \log_{10} \frac{\text{var}\{c(n_1, n_2) - x(x_1, n_2)\}}{\text{var}\{\widehat{x}(n_1, n_2) - x(n_1, n_2)\}} \text{dB}. \end{aligned} \tag{1.12}$$

In equation (1.12), the numerator represents the amount of noise in the corrupted image, and the denominator gives the noise in the restored image. Therefore, the ratio appropriately represents the meaning of SNRI. If the above ratio is greater than unity, then the logarithm is positive and the SNRI shows a positive improvement. If the ratio is unity or less, then the SNRI is zero or negative, respectively; that is, the restoration process yielded no improvement or image deterioration, respectively. Please note that the above definitions are also valid for 1-D signals. Of course, the mean and variance definitions given in (1.9) and (1.8) need to be appropriately adjusted for 1-D signals.

Image Enhancement and Analysis. This refers to enhancing some particular characteristic of the image. Some examples are discussed in the following. Detailed algorithms and images are given in Chapter 10. The reader is encouraged to refer to the relevant images in that chapter while reading the following discussion.

(a) *Contrast improvement*: Contrast refers to the overall brightness or darkness of the objects within the image. In many cases, the contrast is not acceptable and needs to be corrected. This can be done by digital signal processing and is often referred to as histogram modification.

(b) *Edge detection*: In some images, the edge information is of vital importance. An example is the X-ray image of a blocked artery. In this image, the cardiologist is interested in finding the percentage of blockage. The image is usually blurry around the edges, and the physician estimates the percentage from experience. Wouldn't it be nice if we could detect the edges of the artery? Another algorithm could then measure the exact percentage of blockage. Edge detection is also the first stage of many object recognition algorithms.

(c) *Image sharpening*: In many cases, the sharpness of the image is not quite acceptable. Several linear and nonlinear techniques are available for improving the sharpness. Also, in some cases the image is too blurry to even recognize any of the objects in it. There are existing techniques that work very well at deblurring images.

(d) *Image segmentation*: This is often one of the stages in image analysis or automatic object recognition. This technique subdivides the image into segments according to some desired characteristics. Segmentation may be object-oriented, texture-oriented, frequency spectrum-oriented, and so on. These algorithms are usually iterative, and the processing should be stopped when the desired level of segmentation is achieved. This is usually problem dependent. Segmentation is one of the most difficult tasks in image processing.

Image Compression. Image representations require a lot of data. For example, a typical 8-bit 500×500 image requires 2 Mb of data. Millions of images are transmitted daily over satellite links, the Internet, local area networks, and the like. In addition, just as many images are stored in various types of data storage devices such as magnetic disks, optical devices, and electronic circuitry. To increase the rates of image transmission and storage, these images must be compressed and coded. Compression is performed by an algorithm that removes the redundancy in images so that it only requires a fraction of the original data to represent the image. Several components of compression algorithms can be used, such as image transformer, pulse code modulator, quantizer, and entropy coder. Detailed analysis of these components will be presented in Chapter 11.

An Image Processing Example

In this subsection, we present a simple example of image restoration. The goal is to reduce *impulse noise* using a *median filter*. Two new terminologies have just been introduced, and their explanations are now in order. While we are at it, we will also define a *mean filter*.

Median and Mean Filters. A 2-D median filtering operation is performed by sliding a $P \times P$ window over the entire image and replacing the center pixel by the median of the pixels in that window. Obviously, P must be an odd integer in order to define a center pixel. Consider the following 3×3 window of pixels:

$$\begin{bmatrix} 12 & 20 & 5 \\ 11 & 19 & 33 \\ 25 & 9 & 15 \end{bmatrix}$$

The center pixel has the value of 19, and the median of the above numbers is 15. Thus, the filtered version of the center pixel has the value 15. This window slides over the entire image centering on each pixel at a time. The filtered pixel values are stored in another matrix, which forms the processed image. What about the pixels in the boundary of the image? At this time, we choose to leave them alone. There are several ways to handle the bordering pixels, and these will be discussed in Chapter 10. A mean filter works the same way as above except that the filtered version of the center pixel is the mean of the window. Mean and median filters are effective for reducing different types of noise. Computer Project 3 addresses this topic in a little more detail.

Impulse Noise. Sometimes a few of the pixels in a B-bit image are destroyed and replaced by the highest possible pixel level $(2^B - 1)$. When this happens, the image is said to be corrupted by impulse noise. The amount of noise is specified by the percentage of pixels that are corrupted.

Getting back to the example at hand, consider the image shown in Fig. 1.11(a). We add 10% of impulse noise to this image. That is, 10% of the pixels are randomly selected and replaced by the highest pixel level. The resulting corrupted image is shown in Fig. 1.11(b). The SNR of this image is calculated by (1.7) and found to be 2.3 dB. A 3×3 median filter is then applied to this image. The filtered image is shown in Fig. 1.11(c), and the SNRI given by (1.12) is found to be 16 dB. The median filter has significantly reduced the impulse noise. As we can see from the image, there is still some impulse noise remaining. So, we try a 5×5 median filter. The resulting image is shown in Fig. 1.11(d). The SNRI in this case is 14.8 dB. The impulse noise is not noticeable at all. However, the SNRI in this case is not as high as the other. The reason for this is that with a larger window size, the image tends to get slightly blurred, which in turn lowers the SNRI. This blurring is visible in the checkered floor of the image.

▶ 1.3 FUNDAMENTAL SIGNALS

Certain elementary functions are essential to an understanding of the theory of signal processing. These functions can be either continuous-time or discrete-time. In this

Figure 1.11 Examples of median filtering.

section, the definitions are presented with some examples. These functions may be shifted and combined to generate numerous other functions.

Continuous-Time Signals

Rectangular Function

The rectangular function is defined by

$$\Pi(t) \triangleq \begin{cases} 1, & -\frac{1}{2} \leq t \leq \frac{1}{2} \\ 0, & \text{otherwise.} \end{cases} \tag{1.13}$$

This function is depicted in Fig. 1.12(a). The width and height of the rectangular function are unity. Now consider the rectangular function $\frac{1}{\beta} \prod(\frac{t}{\beta})$, $\beta > 0$. By the above definition,

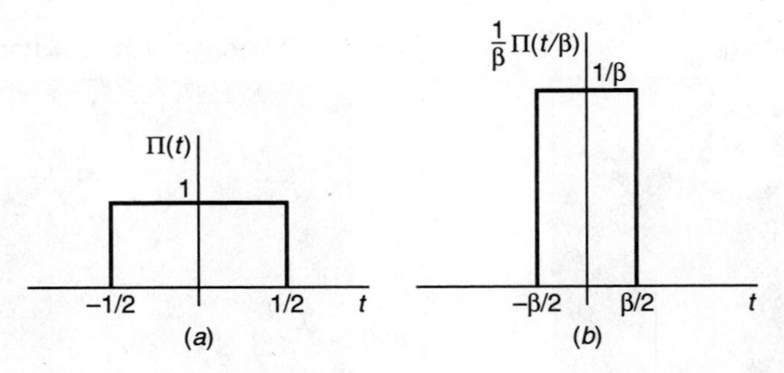

Figure 1.12 (a) Rectangular function; (b) $\frac{1}{\beta}\pi(1/\beta)$.

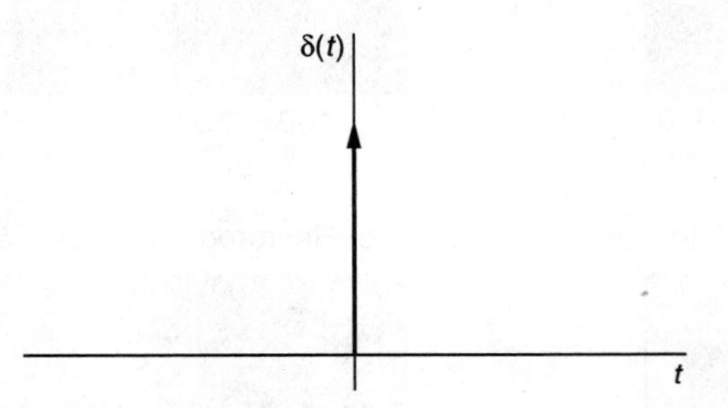

Figure 1.13 Continuous-time unit impulse function.

the amplitude of this function is $\frac{1}{\beta}$, for $-\frac{1}{2} \leq \frac{t}{\beta} \leq \frac{1}{2}$, or $-\frac{\beta}{2} \leq t \leq \frac{\beta}{2}$. Thus, the function has a width of β and a height of $\frac{1}{\beta}$ and is shown in Fig. 1.12(b). This function is the basis for the definition of the continuous-time unit impulse, which is also called the *Dirac delta function*.

Unit Impulse

The continuous-time unit impulse function is defined as

$$\delta(t) \triangleq \lim_{\beta \to 0} \frac{1}{\beta} \Pi\left(\frac{t}{\beta}\right). \tag{1.14}$$

The unit impulse is therefore a narrow and tall rectangular function with unit area. It is graphically depicted in Fig. 1.13. Based on the above definition, the unit impulse has the property

$$\int_{-\infty}^{+\infty} \delta(t)dt = 1, \quad \delta(t) = \begin{cases} 0, & t < 0 \\ 0, & t > 0 \end{cases}. \tag{1.15}$$

Note that the limits of the integration do not have to be $(-\infty, \infty)$ but can be (a, b), where $a < 0$ and $b > 0$; that is, the limits must cover the impulse. Now consider a function $f(t)$ multiplied by an unit impulse $\delta(t - t_0)$ appearing at $t = t_0$. By (1.15), the product $f(t)\delta(t - t_0)$ is zero everywhere except at $t = t_0$. This leads to the following properties:

Prop. 1: $f(t)\delta(t-t_0) = f(t_0)\delta(t-t_0)$.

Prop. 2: $\int_a^b f(t)\delta(t-t_0)dt = \int_a^b f(t_0)\delta(t-t_0)dt = f(t_0)$ where $a < t_0$ and $b > t_0$.

Property 2 is commonly referred to as the *sampling property* or the *sifting property* of the impulse. The derivative of an impulse is also conceptually an interesting function. The *nth* derivative of an impulse is denoted by $\delta^{(n)}(t)$. For even n, this function is defined to be an even function,[2] and for odd n, it is an odd function. Now we list some other properties of the impulse.

Prop. 3: $\int_a^b f(t)\delta^{(n)}(t-t_0)dt = (-1)^n f^{(n)}(t_0)$ where $a < t_0$ and $b > t_0$.

Prop. 4: $\delta(at) = \frac{1}{|a|}\delta(t)$.

Prop. 5: $f(t) * \delta(t) = f(t)$.

Prop. 6: $f(t) * \delta^{(n)}(t) = f^{(n)}(t)$.

Properties 5 and 6 involve the continuous-time convolution operation, which is assumed as a prerequisite. However, for completeness, we now define the convolution of two functions as

$$\begin{aligned} x(t) * h(t) &= \int_{-\infty}^{\infty} x(\tau)h(t-\tau)d\tau \\ &= \int_{-\infty}^{\infty} h(\tau)x(t-\tau)d\tau. \end{aligned}$$

Using the above convolutions formula, properties 5 and 6 can be proved. The proofs are assigned as problems.

Unit Step. The continuous-time unit step function is defined as

$$u(t) \triangleq \begin{cases} 1, & t > 0 \\ 0, & t < 0. \end{cases} \tag{1.16}$$

This function is shown in Fig. 1.14. From (1.15), it can be seen that $\frac{d}{dt}u(t) = \delta(t)$.

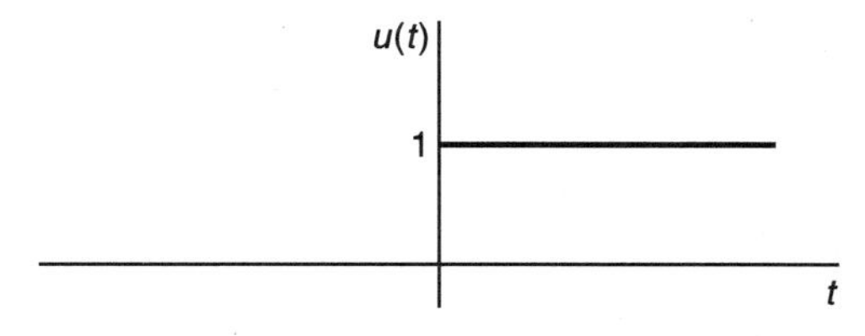

Figure 1.14 Continuous-time unit step function.

[2]A function $x(t)$ is even if it satisfies the condition $x(t) = x(-t)$. It is said to be an odd function if it satisfies $x(t) = -x(-t)$.

Discrete-Time Signals

Unit Impulse

The discrete-time unit impulse is defined as

$$\delta(n) \triangleq \begin{cases} 1, & n = 0 \\ 0, & n \neq 0 \end{cases}. \tag{1.17}$$

This signal is also referred to as the *unit sample* function and is shown in Fig. 1.15(a). The discrete-time impulse is not obtained by sampling the continuous-time impulse function. The function $\delta(n - n_0)$ is a shifted impulse that has a unit amplitude at $n = n_0$ and zero elsewhere. If some discrete-time function $f(n)$ is multiplied by $\delta(n - n_0)$, the resulting function $f(n)\delta(n - n_0)$ is zero everywhere except at $n = n_0$, where it has a value of $f(n_0)$. This leads to the following properties that are analogous to those of the continuous-time impulse.

$$f(n)\delta(n - n_0) = f(n_0)\delta(n - n_0) = \begin{cases} f(n_0), & n = n_0 \\ 0, & n \neq n_0 \end{cases} \tag{1.18}$$

$$\sum_{n=a}^{b} f(n)\delta(n - n_0) = \sum_{n=a}^{b} f(n_0)\delta(n - n_0) = f(n_0) \tag{1.19}$$

where $a < n_0$ and $b > n_0$.

Using (1.18), we can show that any sequence $x(n)$ can be represented in terms of impulses. The product $x(0)\delta(n)$ gives a function with amplitude $x(0)$ appearing at $n = 0$. Similarly, the product $x(1)\delta(n - 1)$ gives a function with amplitude $x(1)$ appearing at $n = 1$, and so on. Thus, we can represent $x(n)$ as

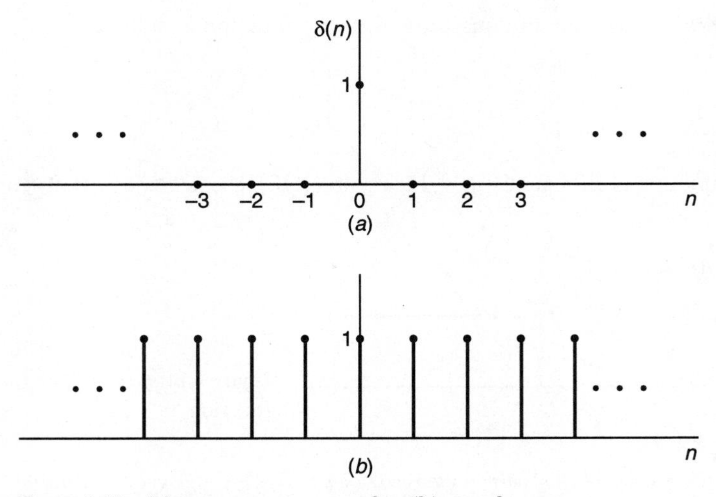

Figure 1.15 (a) Discrete-time impulse; (b) impulse train.

$$x(n) = \ldots + x(-1)\delta(n+1) + x(0)\delta(n) + x(1)\delta(n-1) + \cdots$$
$$= \sum_{m=-\infty}^{\infty} x(m)\delta(n-m). \tag{1.20}$$

The function $\sum_{m=-\infty}^{\infty} \delta(n-i)$ is called the impulse train and is shown in Fig. 1.15(b). The above representation of any sequence in terms of impulses is called the impulse train representation.

Unit Step

The discrete-time unit step function is defined as

$$u(n) \triangleq \begin{cases} 1, & n \geq 0 \\ 0, & n \leq -1 \end{cases}. \tag{1.21}$$

This function is shown in Fig. 1.16(a). By (1.20), the unit step can be represented in terms of impulse functions as

$$u(n) = \sum_{i=0}^{\infty} \delta(n-i). \tag{1.22}$$

The sequence $u(n-n_0)$ represents a shifted version of the step. If n_0 is positive, the step shifts to the right as in Fig. 1.16(b); otherwise it shifts to the left. Now consider the function $u(n_0-n)$. To figure this out, we can go back to the definition in (1.21). The function $u(n_0-n)$ is unity for $n_0 - n \geq 0$, that is, $n \leq n_0$. This is shown in Fig. 1.16(c)

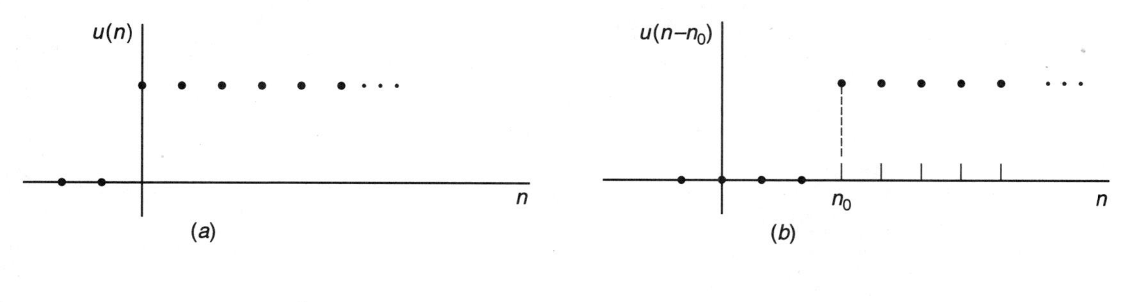

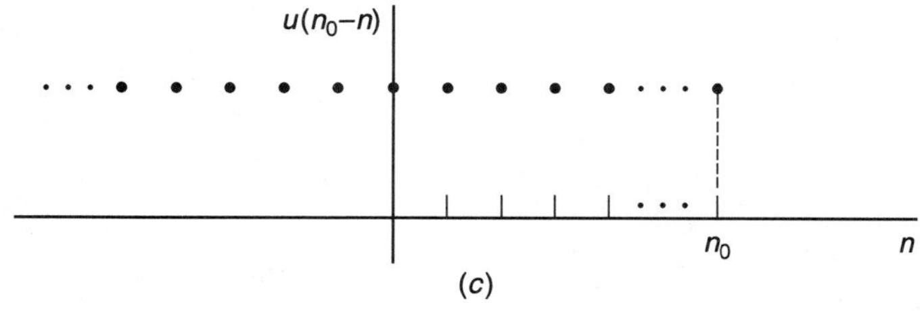

Figure 1.16 Discrete-time step functions: (a) $u(n)$, (b) $u(n-n_0)$, and (c) $u(n_0-n)$.

for positive n_0. The discrete-time unit impulse can be represented in terms of shifted steps as

$$\delta(n) = u(n) - u(n-1). \tag{1.23}$$

Exponential

The discrete-time exponential signal is defined as

$$f(n) \triangleq a^n, \quad n = 0, 1, 2 \ldots \tag{1.24}$$

where a can be real or complex. If $|a| < 1$, this function decays to zero as $n \to \infty$. If $0 < a < 1$, the function decays monotonically. However, if $-1 < a < 0$, the function oscillates between positive and negative values as it decays to zero. Figure 1.17 shows typical plots for these cases.

If a is complex, it can be written as

$$a = re^{j\omega_0} = r\measuredangle\omega_0$$

where r is the magnitude and ω is the argument or angular frequency. The exponential signal then becomes

$$f(n) = r^n e^{j\omega_0 n} = r^n(\cos\omega_0 n + j\sin\omega_0 n).$$

The real and imaginary parts of the function can be plotted separately. Another option is to plot the magnitude, $|f(n)|$ and the phase, $\measuredangle f(n)$ of the function given by

$$|f(n)| = r^n$$

$$\measuredangle f(n) = \omega_0 n.$$

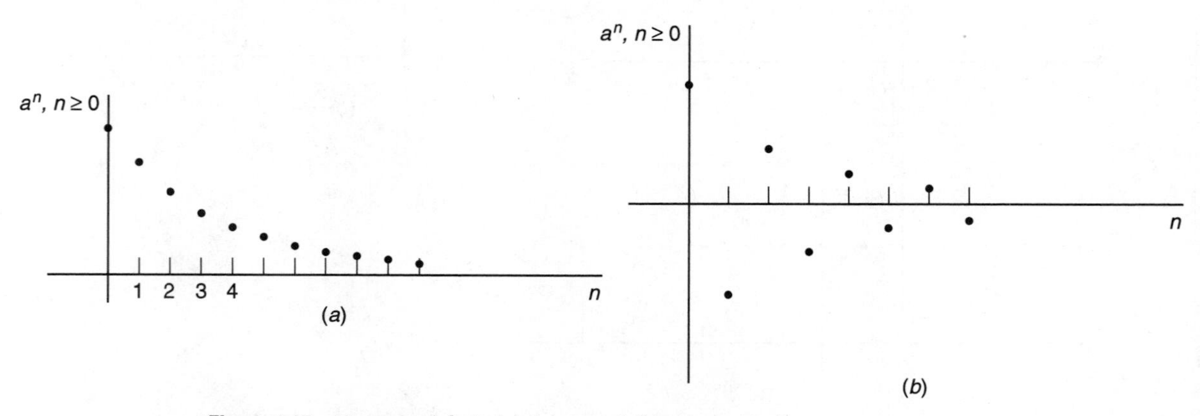

Figure 1.17 Exponential a^n: (a) $0 < a < 1$ (b) $-1 < a < 0$.

It is conventional to plot the phase in the range $(-\pi, \pi)$. This can be done by subtracting an integer multiple of 2π from the phase. In other words, the phase is plotted as $\measuredangle f(n)$ modulo 2π.

▶ **EXAMPLE 1.1** *Consider the signal*

$$x(n) = (0.65 + j0.5)^n u(n).$$

The real and imaginary parts of the signal are shown in Fig. 1.18(a) and (b), respectively. The magnitude and phase are plotted in Fig. 1.18 (c) and (d).

Periodic Signals

Periodic signals are important because they occur often in many application areas of signal processing. As we know, periodic functions can be represented as a Fourier series, which is an indispensable analysis tool. Most of the signals of interest encountered in practice are finite length. A finite-length nonperiodic signal can be duplicated and concatenated to

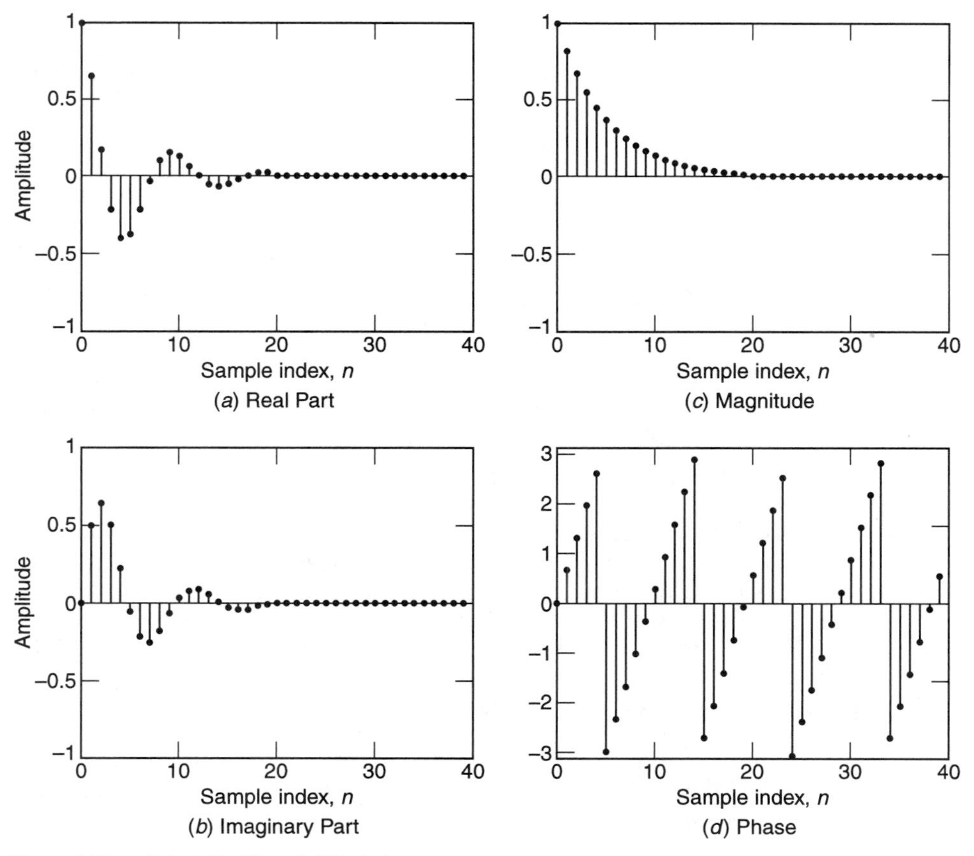

Figure 1.18 $x(n) = (0.65 + j0.5)^n u(n)$.

produce a periodic function with period equal to the length of the original signal. Fourier domain techniques can then be applied to this periodic signal for analysis. Periodic signals can of course be continuous-time or discrete-time.

A continuous-time signal $x(t)$ is said to be periodic with time-period T_0 if it satisfies

$$x(t) = x(t + T_0), \quad \text{for all } t. \tag{1.25}$$

A sinusoidal signal $\sin \Omega_0 t$ is obviously periodic with a time-period of $T_0 = \frac{2\pi}{\Omega_0}$ secs, where Ω_0 is the angular frequency in rads/sec.

A discrete-time sequence $x(n), n = 0, 1, 2, \ldots$ is said to be periodic with period P if it satisfies

$$x(n) = x(n + P), \quad \text{for all } n \tag{1.26}$$

where P is a positive integer. Let the continuous-time sinusoid $\sin(\Omega_0 t)$ be sampled to obtain the discrete-time sequence $\sin(w_0 n)$. This sequence is not necessarily periodic in the sense of (1.26). Consider the equation

$$\begin{aligned} x(n + P) &= \sin(w_0(n + P)) \\ &= \sin(w_0 n)\cos(w_0 P) + \cos(w_0 n)\sin(w_0 P). \end{aligned}$$

The right-hand side of the above is equal to $x(n) = \sin(w_0 n)$ if and only if

$$\cos(w_0 P) = 1 \quad \text{and } \sin(w_0 P) = 0.$$

The above hold if and only if there exists an integer m such that

$$w_0 P = 2\pi m$$

That is

$$m = \frac{w_0 P}{2\pi} = \frac{P}{T_0}. \tag{1.27}$$

Thus, the discrete-time sinusoidal sequence $\sin(w_0 n)$ is periodic with period P if and only if there exists an integer m such that $P = mT_0$ is an integer. If a sequence is periodic with period P, then it is also periodic with a period that is an integer multiple of P. The smallest period of the sequence is called the *fundamental period*.

▶ **EXAMPLE 1.2** *Consider the sequence $\sin(3n)$. The angular frequency and the time-period are $\omega_0 = 3$ and $T_0 = \frac{2\pi}{3}$, respectively. In this case, there exists no integer such that $P = mT_0$ is also an integer. Thus, this sequence is nonperiodic. Now consider the sequence $\sin(3\pi n)$. In this case, $T_0 = \frac{2}{3}$, and there exists an integer such that $P = \frac{2}{3}m$ is also an integer. The smallest integer value of P is obtained when we choose $m = 3$. Thus, this sequence is periodic with a fundamental period of $P = 2$.*

Frequency in Continuous versus Discrete-Time

Now we examine the relationship between the frequency of a continuous-time sinusoid and that of its discrete-time counterpart. Reconsider the continuous-time sinusoid as given earlier by

$$y(t) = A\sin(\Omega t - \theta). \tag{1.28}$$

Discretize the above to obtain the discrete-time signal

$$y_s(nT_s) = A\sin(\Omega nT_s - \theta). \tag{1.29}$$

Now consider a discrete-time sinusoidal sequence given by

$$y(n) = A\sin(\omega n - \theta) \tag{1.30}$$

where ω is the frequency of the discrete-time sequence. Comparing (1.29) and (1.30), we have the following relationship:

$$\begin{aligned}\omega &= \Omega T_s \\ &= \frac{2\pi\Omega}{\Omega_s}\end{aligned} \tag{1.31}$$

where Ω_s is the sampling frequency in radians/sec. Now, what is the unit of the frequency of the discrete sequence? We know that Ω is in radians/sec; therefore, ω is in radians. Also, by substituting $\omega = 2\pi f$ in the above, we get

$$f = \frac{F}{F_s} \tag{1.32}$$

where $F_s = \frac{1}{T_s}$ is the sampling frequency. The unit of F is Hz or cycles/sec and that of F_s is samples/sec. Therefore, the unit of the discrete frequency f is cycles/sample.

Let's make another important observation based on (1.31). If we set $\Omega = \Omega_s$, then we have $\omega = 2\pi$. That is, the sampling frequency corresponds to the digital frequency of 2π. This also implies that frequency, $\frac{\Omega_s}{2}$ corresponds to the digital frequency of π. It will be shown in Chapter 2 that the frequency response[3] of a discrete sequence is periodic with period 2π. Therefore, it is customary to plot the frequency response of a discrete sequence in the range $(-\pi, \pi)$ or $(0, 2\pi)$.

1.4 CLASSIFICATION OF SYSTEMS

Systems can be classified into different types according to their input-output relationships. These classifications hold for both continuous-time and discrete-time systems. The focus here will be on discrete-time systems. Let $x(n)$ be the input to a system and $y(n)$ be the corresponding output. We will represent the output as

$$y(n) = \Re\{x(n)\}$$

where $\Re$ denotes the operator that operates on the input signal. This representation will be used throughout this section.

[3] From linear systems, we know that the frequency response of a signal or system is defined as its gain and phase as a function of frequency.

Linearity

A system is said to be *linear* if it obeys the law of superposition. For inputs $x_1(n)$ and $x_2(n)$, let the system outputs be $y_1(n)$ and $y_2(n)$, respectively. For constants α_1 and α_2, a linear system satisfies

$$\begin{aligned}\Re\{\alpha_1 x_1(n) + \alpha_2 x_2(n)\} &= \alpha_1 \Re\{x_1(n)\} + \alpha_2 \Re\{x_2(n)\} \\ &= \alpha_1 y_1(n) + \alpha_2 y_2(n).\end{aligned} \tag{1.33}$$

The above definition can be generalized to any number of inputs. Another way of stating the definition is the following: If a sum of different inputs is applied to a linear system, then the output is the same as that obtained by adding all the outputs of the system when the inputs are applied individually. The definition is graphically illustrated in Fig. 1.19. An example is now in order.

▶ EXAMPLE 1.3 *The input-output relationship of a system is given by*

$$y(n) = 2x(n) + c. \tag{1.34}$$

where $x(n)$ is the input, $y(n)$ is the output and c is a constant. Determine whether or not this system is linear.

SOLUTION If an input $x_1(n)$ is applied to the system, the output is

$$y_1(n) = 2x_1(n) + c.$$

If another input $x_2(n)$ is applied to the system, the response is

$$y_2(n) = 2x_2(n) + c.$$

Now let the sum of the above inputs be applied to the system. By the definition of the system given in (1.34), the resulting output is

$$y_s(n) = 2[x_1(n) + x_2(n)] + c.$$

Since $y_s(n) \neq y_1(n) + y_2(n)$, the system is nonlinear. Note that although this system equation does not have a nonlinear term involving the input, it still does not obey superposition. The reason is that the system is not initially at rest; that is, $\Re\{0\} \neq 0$. ◀

Shift-Invariance

A system is said to be *shift-invariant* if

$$\Re\{x(n - M)\} = y(n - M)$$

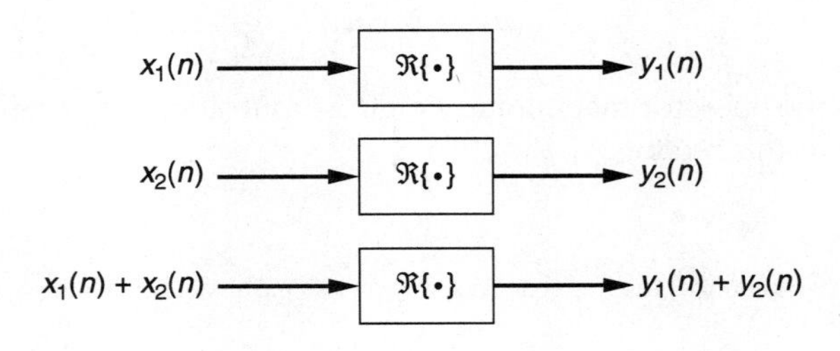

Figure 1.19 Illustration of linearity.

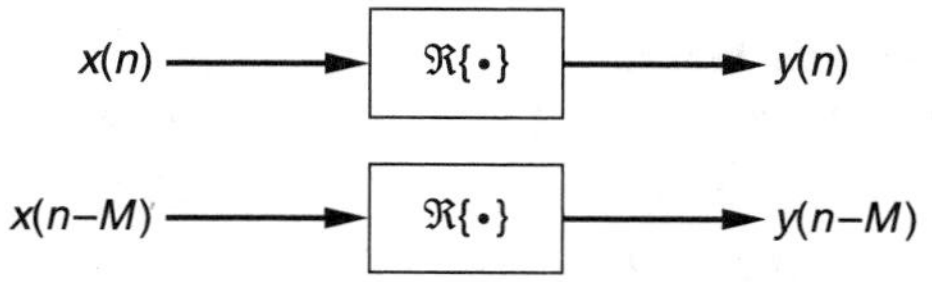

$x(n) \longrightarrow \boxed{\Re\{\bullet\}} \longrightarrow y(n)$

$x(n-M) \longrightarrow \boxed{\Re\{\bullet\}} \longrightarrow y(n-M)$

Figure 1.20 Illustration of Shift-invariance.

where $x(n)$ and $y(n)$ are the input and output, respectively and M is an integer. In other words, if the result of shifting the input to a system is simply a shift in the output by the same amount, then the system is said to be shift-invariant. Otherwise, the system is shift-variant. A graphical illustration is given in Fig. 1.20. Continuous-time systems that satisfy this definition are called time-invariant.

EXAMPLE 1.4 *Consider the system in the last example, that is, $y(n) = 2x(n) + c$. Determine whether or not the system is shift-invariant.*

SOLUTION If the input is shifted by M and applied to the system, then the output is

$$\Re\{x(n-M)\} = 2x(n-M) + c.$$

If the original output is shifted, the result is

$$y(n-M) = 2x(n-M) + c.$$

Since $\Re\{x(n-M)\} = y(n-M)$, the system is shift-invariant. This system is therefore referred to as Nonlinear Shift Invariant (NLSIV).

Causal

A system is said to be *causal* if the output at n_0 does not depend on the input at $n > n_0$. In other words, a causal system does not predict the input. A real-time signal processing system operates on each incoming input sample (and possibly past input and output samples) to produce the output. Therefore, a real-time DSP system must be causal. The system used in the last two examples is causal.

EXAMPLE 1.5 *Consider the system given by*

$$y(n) = x(Pn)$$

where P is a positive integer. This operation is called downsampling or decimation and will be discussed in Chapter 2. Determine whether this system is linear, shift-invariant, and causal.

SOLUTION First we determine linearity. If an input $x_1(n)$ is applied, the output is

$$y_1(n) = x_1(Pn).$$

If another input $x_2(n)$ is applied to the system, the response is

$$y_2(n) = x_2(Pn).$$

Now let the sum of the above inputs be applied to the system. By the definition of the system given in (1.34), the resulting output is

$$y_s(n) = x_1(Pn) + x_2(Pn).$$

Since $y_s(n) = y_1(n) + y_2(n)$, the system is linear.

Now let us test for shift-invariance. If the input is shifted by M and applied to the system, then the output is

$$\Re\{x(n-M)\} = x(nP - M).$$

If the original output is shifted, the result is

$$y(n-M) = x(P(n-M)) = x(nP - PM).$$

Since $\Re\{x(n-M)\} \neq y(n-M)$, the system is shift-variant. This system is therefore referred to as Linear Shift Variant (LSV).

In this system, the output at n depends on the input at Pn. If $n > 0$, then $Pn > n$, and the system is noncausal. However, for $n < 0$, $Pn < n$, and the system is causal. Overall, the system is said to be noncausal. ◀

▶ 1.5 CHARACTERIZATION OF LSI SYSTEMS

There are several different ways to characterize discrete-time Linear Shift Invariant (LSI) systems, including difference equation, impulse response, transfer function, and frequency response. This section covers difference equation representation and impulse response. Transfer function and frequency response will be presented in subsequent chapters.

Difference Equations

A discrete-time LSI system may be characterized by a difference equation, a general form of which is

$$y(n) + a_1 y(n-1) + a_2 y(n-2) + \cdots + a_N y(n-N) = b_0 x(n) + b_1 x(n-1) + \cdots b_M x(n-M) \quad (1.35)$$

where $x(n)$ and $y(n)$ are the input and output, respectively. Finite integers M and N represent the maximum delays in the input and output, respectively. As mentioned earlier, a discrete-time system is commonly referred to as a digital filter. The constants a_i and b_i are called the filter coefficients. Note that the coefficient of $y(n)$ is $a_0 = 1$. If the coefficient of $y(n)$ in a given difference equation is not unity, then we can divide the entire equation by this coefficient so that we get the form in (1.35), where $a_0 = 1$. Equation (1.35) can be written in a more compact form as

$$\sum_{i=0}^{N} a_i y(n-i) = \sum_{j=0}^{M} b_j x(n-j), \quad a_0 = 1. \quad (1.36)$$

A system characterized by the above equation is said to be a *recursive* filter. If the output $y(n)$ does not depend on the past output samples, then the system is called a *nonrecursive* filter. The equation of a nonrecursive filter is

$$y(n) = \sum_{j=0}^{M} b_j x(n-j). \quad (1.37)$$

Difference equations (1.36) and (1.37) can be solved either in the time domain or in the transform domain such as Z-transform or discrete-time Fourier transform (DTFT). Transform-domain methods will be presented in Chapters 2 and 3. Time-domain techniques usually do not provide any further insight into the system dynamics and therefore will not be covered. At this time, it would be a good exercise to show that the nonrecursive system of (1.37) is linear, shift-invariant, and causal.

Impulse Response

The *impulse response* of a system is defined as its output when the input is a unit impulse (see Fig. 1.21). The impulse response is usually denoted by $h(n)$. Consider the nonrecursive filter of (1.37). Its impulse response is obtained by substituting $\delta(n)$ for $x(n)$ as given by

$$\begin{aligned} h(n) &= \sum_{j=0}^{M} b_j \delta(n-j) \\ &= b_0\delta(n) + b_1\delta(n-1) + \cdots + b_M\delta(n-M). \end{aligned} \tag{1.38}$$

This impulse response is plotted in Fig. 1.22. Clearly, the impulse response of this nonrecursive filter is of finite length. Therefore, nonrecursive systems are commonly referred to as *Finite Impulse Response (FIR)* filters. On the other hand, the impulse response of the recursive system given in (1.36) is infinite. This can be shown by a simple example. Consider the simple first-order recursive system

$$y(n) = ay(n-1) + x(n), n = 1, 2, \ldots$$

with the initial condition $y(n) = 0, n \leq -1$. The impulse response can be found by solving the above with $x(n) = \delta(n)$. By simple recursion, we get

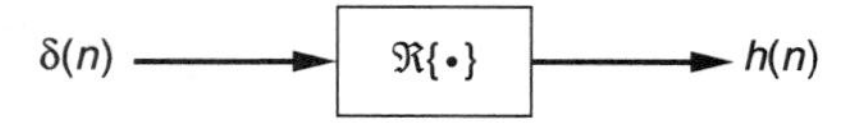

Figure 1.21 Definition of impulse response.

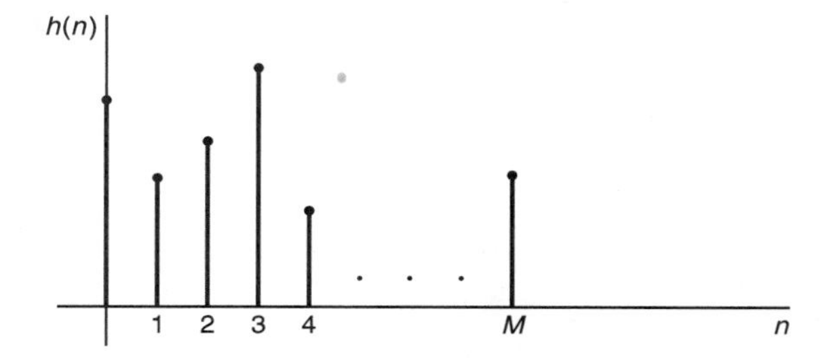

Figure 1.22 Impulse response of a nonrecursive system.

$$y(0) = 1$$
$$y(1) = a$$
$$y(2) = a^2$$
$$\vdots$$
$$h(n) \triangleq y(n) = a^n.$$

The above impulse response is of infinite length for any nonzero value of a. Therefore, recursive systems are in general commonly referred to as *Infinite Impulse Response (IIR)* filters. Special cases can be found where a recursive system has a finite impulse response. Try finding an example!

The impulse response is also used to characterize a system or filter. This is because if the impulse response of a system is specified, then the output can be determined for any given input. That is, the impulse response completely describes the system. The following theorem states this relationship.

Theorem 1.2 *If $x(n)$ and $h(n)$ are the input and impulse response of a LSI system denoted by the operator $\Re\{\cdot\}$, the output is obtained by the convolution of these functions given by*

$$y(n) = x(n) * h(n) \tag{1.39}$$

$$\triangleq \sum_{m=-\infty}^{\infty} x(m)h(n-m) \tag{1.40}$$

$$= \sum_{m=-\infty}^{\infty} h(m)x(n-m) \tag{1.41}$$

where $*$ denotes the convolution operation.

Proof. By using the impulse train representation in (1.20), $x(n)$ can be written as

$$x(n) = \sum_{m=-\infty}^{\infty} x(m)\delta(n-m). \tag{1.42}$$

The output of the system is

$$\begin{aligned} y(n) &= \Re\{x(n)\} \\ &= \Re\left\{\sum_{m=-\infty}^{\infty} x(m)\delta(n-m)\right\} \\ &= \sum_{m=-\infty}^{\infty} x(m)\Re\{\delta(n-m)\}. \end{aligned}$$

The operator $\Re$ does not operate on $x(m)$ because it is not a function of n. From the definition of impulse response, $h(n) = \Re\{\delta(n)\}$. The above therefore becomes

$$y(n) = \sum_{m=-\infty}^{\infty} x(m)h(n-m)$$

and the proof is complete for (1.40). Equation (1.41) can be obtained by a simple substitution of variables in the above equation. ■

Now that the convolution formula is established, we proceed to explain how to evaluate it for two given sequences. Either one of the formulas (1.40) or (1.41) can be used. Consider (1.40) for explanation purposes. Given $x(n)$ and $h(n)$, we perform the sequence of operations "*Switch-Shift-Sum.*" That is, the independent variable for $x(n)$ is switched from n to m to form $x(m)$. The other sequence $h(n)$ is shifted by this new independent variable to form $h(n-m)$. Then the product of $x(m)$ and $h(n-m)$ is summed from $m = -\infty$ to ∞. For each value of n, the sum is calculated to yield $y(n)$. Another way of explaining the above is as follows. After switching variables, we have a graph of the function $x(m)$ versus m, which is really the same as the graph $x(n)$ versus n. To obtain the graph of $h(n-m)$ versus m, consider the following steps. We first graph $h(m)$ versus m, which is again the same as the graph of $h(n)$ versus n. Then the graph of $h(-m)$ versus m is obtained by flipping the graph of $h(m)$ over the vertical axis. The graph of $h(n-m)$ versus m is then obtained by sliding $h(-m)$ by n. If n is positive, the graph slides to the right. Otherwise, it slides to the left. That is, as n varies from $-\infty$ to ∞, the graph of $h(n-m)$ slides discretely from left to right. For each position of the graph of $h(n-m)$, it is multiplied by the graph of $x(m)$. The product is summed from $-\infty$ to ∞ to yield the value of $y(n)$ for that particular n. The process is now illustrated with an example.

▶ EXAMPLE 1.6 *Consider an LSI system with an impulse response*

$$h(n) = \begin{cases} 1 & N_1 \le n \le N_2, \\ 0 & \text{otherwise.} \end{cases}$$

Find the output of the system when the input sequence is

$$x(n) = a^n u(n) \quad \text{with } |a| < 1.$$

SOLUTION The output $y(n)$ can be calculated by using either of the convolution formulas given by (1.40) or (1.41). In (1.40), we flip and slide $h(n)$, whereas in (1.41) we flip and slide $x(n)$. Let us use (1.41) in this case. Recall that the operations "Switch-Shift-Sum" must be performed. In the "switch" operation, the independent variable is switched from n to m. The impulse response $h(m)$ and the input sequence $x(m)$ are plotted versus m in Fig. 1.23(b) and (a), respectively. The "shift" operation $h(n-m)$ is performed in two steps. First, we obtain $h(-m)$ by flipping $h(m)$ over the vertical axis, that is, a simple reflection of $h(m)$ about $m = 0$. The function $h(-m)$ is shown in Fig. 1.23(c). Replacing $-m$ by $n-m$ leads to shift the sequence $h(-m)$ to $h(n-m)$. $n > 0$ corresponds to a shift to the right, while $n < 0$ corresponds to a shift to the left. For n, a large negative number, the graphs of $h(n-m)$ and $x(m)$ are shown in Fig. 1.23(d). This is the case of no overlap. As n increases, the sequence $h(n-m)$ slides from left to right, and we have the two additional cases of partial overlap and full overlap as shown in Fig. 1.23(e) and (f). ◀

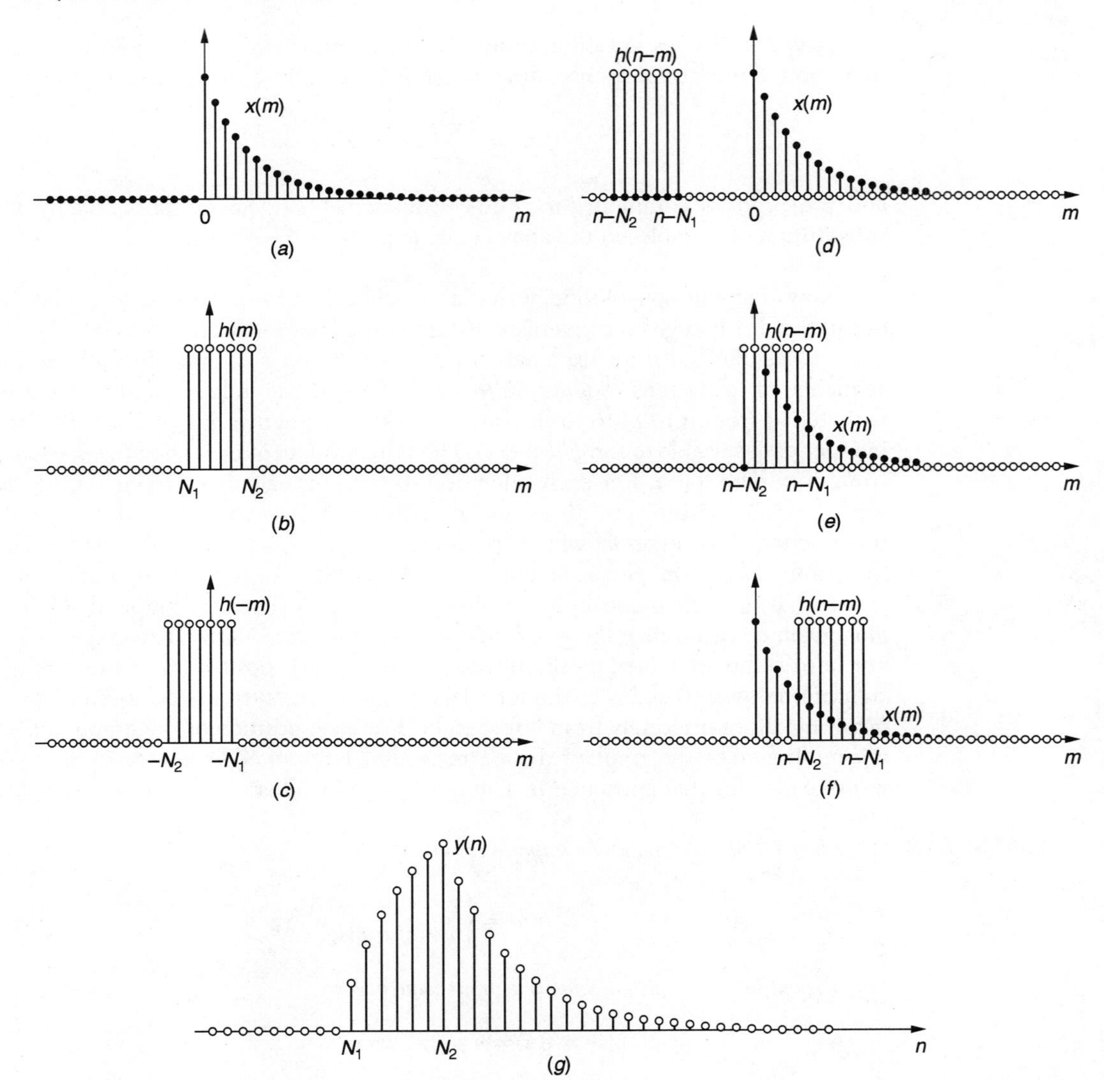

Figure 1.23 Illustration of convolution.

Case 1: The no overlap case occurs when $h(n-N_1)$ is to the left of $x(0)$; that is, $(n-N_1) < 0$ or $n < N_1$. In this case the output is

$$y(n) = 0. \tag{1.43}$$

Case 2: Partial overlap occurs when $(n - N_1) \geq 0$ and $(n - N_2) < 0$. The number of overlapped samples is a function of n; that is,

$$x(m)h(n-m) = \begin{cases} a^m & \text{if } x(m) \text{ and } h(n-m) \text{ overlap,} \\ 0 & \text{otherwise.} \end{cases}$$

In this case, the output is the sum

$$y(n) = \sum_{m=0}^{n-N_1} a^m. \tag{1.44}$$

Case 3: Clearly, when $n - N_2 \geq 0$, that is, $n \geq N_2$ there is complete overlap. The output is

$$y(n) = \sum_{m=n-N_2}^{n-N_1} a^m. \tag{1.45}$$

Recall that the sum of a geometric progression is

$$\sum_{k=K_1}^{K_2} \gamma^k = \frac{\gamma^{K_1} - \gamma^{K_2+1}}{1-\gamma}, \quad K_2 \geq K_1. \tag{1.46}$$

Applying this to (1.44) and (1.45), we obtain the closed-form expression for $y[n]$ as

$$y(n) = \begin{cases} 0 & n < N_1; \\ \frac{1-a^{n-N_1+1}}{1-a} & N_1 \leq n < N_2; \\ a^{n-N_2}\frac{1-a^{N_2-N_1+1}}{1-a} & n \geq N_2. \end{cases} \tag{1.47}$$

The output sequence $y(n)$ is plotted in Fig. 1.23(g).

For computing the convolution of two finite sequences, it is often easier to perform the operations on the vectors rather than graphically as above. This is illustrated in the following example.

EXAMPLE 1.7 *Perform convolution of the following sequences:*

$$x(n) = \begin{bmatrix} 5 & 2 & \mathbf{7} & -6 & 10 \end{bmatrix}, \quad h(n) = \begin{bmatrix} \mathbf{1} & -4 & 3 \end{bmatrix}$$

Recall that the digits in bold represent the signal value at $n = 0$.

SOLUTION

To perform the convolution of the above sequences, we perform the "Switch-Shift-Sum" operation as before. First, we switch the independent variable from n to m. But this does not change the vectors. Next, we perform the shift operation $h(n - m)$. To do this, we flip the sequence $h(m)$ to get $h(-m)$. This flipped sequence can then be shifted, multiplied by $x(m)$ and summed as below.

$$\begin{bmatrix} x(m) \\ h(-m) \end{bmatrix} = \begin{bmatrix} 5 & 2 & \mathbf{7} & -6 & 10 \\ 3 & -4 & \mathbf{1} & & \end{bmatrix} \Rightarrow y(0) = 5 \cdot 3 + 2 \cdot (-4) + 7 \cdot 1 = 14$$

Negative shifts

$$\begin{bmatrix} x(m) \\ h(-1-m) \end{bmatrix} = \begin{bmatrix} & 5 & 2 & \mathbf{7} & -6 & 10 \\ 3 & -4 & 1 & & & \end{bmatrix} \Rightarrow y(-1) = 5 \cdot (-4) + 2 \cdot (1) = -18$$

$$\begin{bmatrix} x(m) \\ h(-2-m) \end{bmatrix} = \begin{bmatrix} & & 5 & 2 & \mathbf{7} & -6 & 10 \\ 3 & -4 & 1 & & & & \end{bmatrix} \Rightarrow y(-2) = 5 \cdot 1 = 5$$

$$\begin{bmatrix} x(m) \\ h(-3-m) \end{bmatrix} = \begin{bmatrix} & & & 5 & 2 & \mathbf{7} & -6 & 10 \\ 3 & -4 & 1 & & & & & \end{bmatrix} \Rightarrow y(n) = 0, \quad n \leq -3$$

Positive shifts

$$\begin{bmatrix} x(m) \\ h(1-m) \end{bmatrix} = \begin{bmatrix} 5 & 2 & \mathbf{7} & -6 & 10 \\ & 3 & -4 & 1 & \end{bmatrix} \Rightarrow y(1) = 2 \cdot 3 + 7 \cdot (-4) + (-6)(1) = -28$$

$$\begin{bmatrix} x(m) \\ h(2-m) \end{bmatrix} = \begin{bmatrix} 5 & 2 & \mathbf{7} & -6 & 10 \\ & & 3 & -4 & 1 \end{bmatrix} \Rightarrow y(2) = 7 \cdot 3 + (-6) \cdot (-4) + 10.1 = 55$$

$$\begin{bmatrix} x(m) \\ h(3-m) \end{bmatrix} = \begin{bmatrix} 5 & 2 & \mathbf{7} & -6 & 10 & \\ & & & 3 & -4 & 1 \end{bmatrix} \Rightarrow y(3) = (-6) \cdot 3 + 10 \cdot (-4) = -58$$

$$\begin{bmatrix} x(m) \\ h(4-m) \end{bmatrix} = \begin{bmatrix} 5 & 2 & 7 & \mathbf{-6} & 10 & & \\ & & & & 3 & -4 & 1 \end{bmatrix} \Rightarrow y(4) = (10) \cdot 3 = 30$$

$$\begin{bmatrix} x(m) \\ h(5-m) \end{bmatrix} = \begin{bmatrix} 5 & 2 & 7 & \mathbf{6} & 10 & & & \\ & & & & & 3 & -4 & 1 \end{bmatrix} \Rightarrow y(n) = 0, \quad n \geq 5.$$

The final convolution result is

$$y(n) = \begin{bmatrix} \ldots & 0 & 5 & -18 & \mathbf{14} & -28 & 55 & -58 & 30 & 0 & \ldots \end{bmatrix}.$$

Properties of Convolution

Some properties of discrete-time convolution are presented here. These properties will be useful in the section on filter interconnections, which will follow shortly in this chapter.

Theorem 1.3 *Convolution of two discrete-time sequences $h(n)$ and $g(n)$ is commutative; that is,*

$$h(n) * g(n) = g(n) * h(n). \tag{1.48}$$

Proof. The proof follows immediately by an inspection of the convolution formulas given in (1.40) and (1.41). ■

The following theorems can also be proved quite easily by using the convolution formulas. The proofs are assigned as problems at the end of the chapter.

Theorem 1.4 *Convolution of discrete-time sequences is associative; that is,*

$$h(n) * \{g(n) * p(n)\} = \{h(n) * g(n)\} * p(n). \tag{1.49}$$

Theorem 1.5 *Convolution of discrete-time sequences is distributive; that is,*

$$h(n) * \{g(n) + p(n)\} = h(n) * g(n) + h(n) * p(n). \tag{1.50}$$

Stability

Stability is an important criterion in the design of digital filters. Filters must be designed so that the output is bounded for any given bounded input. In this section, we formally define stability of LSI systems and establish a condition for it.

Definition: A digital filter is said to be *Bounded Input Bounded Output (BIBO)* stable, if for any bounded input the output of the filter is also bounded.

EXAMPLE 1.8 *Consider the system with input-output description given by*

$$y(n) = e^{ax(n)+bx(n-1)}$$

where $x(n)$ and $y(n)$ are the input and output, respectively, and a and b are constants. Determine whether this system is BIBO stable.

SOLUTION Taking the absolute value of both sides gives

$$\begin{aligned} |y(n)| &= |e^{ax(n)+bx(n-1)}| \\ &\le e^{|ax(n)+bx(n-1)|} \\ &\le e^{|a||x(n)|+|b||x(n-1)|}. \end{aligned}$$

It is assumed that the input $x(n)$ is bounded, that is, $|x(n)| \le \gamma < \infty$, for all n. The above therefore becomes

$$|y(n)| \le e^{\gamma(|a|+|b|)} < \infty.$$

The output is thus bounded, and the given system is BIBO stable.

EXAMPLE 1.9 *Consider the digital filter with input-output relationship*

$$y(n) = \sum_{i=0}^{n} x(n-i)$$

where $x(n)$ and $y(n)$ are the input and output, respectively, and b_i are the filter coefficients. Determine whether the system is BIBO stable.

SOLUTION Note that this is a nonrecursive digital filter with a filter order (maximum delay) that increases with every sample of incoming data. The length of its impulse response is $(n+1)$, which is finite only for finite n. So, this filter is nonrecursive but not FIR. Clearly, as $n \to \infty$, we have an infinite number of terms in the expression for $y(n)$. To show that this system is unstable, we need to find some bounded input for which the output is unbounded. Let the input be simply $x(n) = 1$. As $n \to \infty$, the output is $y(n) = 1 + 1 + \cdots \to \infty$, and the system is therefore unstable.

The following theorem gives a necessary and sufficient condition for the stability of a LSI system in terms of its impulse response $h(n)$.

Theorem 1.6 *An LSI system is BIBO stable if and only if its impulse response is absolutely summable; that is,*

$$\sum_{n=-\infty}^{\infty} |h(n)| < \infty. \tag{1.51}$$

Proof. Let the system input and output be represented by $x(n)$ and $y(n)$, respectively. We now consider the two parts of the theorem separately.

If Part (sufficient condition). Suppose that (1.51) holds. The output of the system is the convolution of the input and the impulse response given by

$$y(n) = \sum_{m=-\infty}^{\infty} h(m)x(n-m).$$

Taking the absolute value of both sides, we have

$$\begin{aligned} |y(n)| &= \left| \sum_{m=-\infty}^{\infty} h(m)x(n-m) \right| \\ &\leq \sum_{m=-\infty}^{\infty} |h(m)||x(n-m)|. \end{aligned}$$

Since the input is assumed to be bounded, we have $|x(n)| \leq \gamma < \infty$. Then the above becomes

$$|y(n)| \leq \gamma \sum_{m=-\infty}^{\infty} |h(m)| < \infty.$$

Thus, the output is bounded and the system is BIBO stable. This completes the proof of the sufficient condition.

Only If Part (necessary condition). Here we need to show that if the system is BIBO stable, then the condition of (1.51) is satisfied. If we reverse the logic, we need to show that if the condition of (1.51) is not satisfied, then the system is not stable. So, suppose that the condition is not satisfied; that is, $\sum_{n=-\infty}^{\infty} |h(n)| = \infty$. We need to prove that the system is not stable; that is, there exists an input for which the output is not bounded. Now let the input be the following bounded sequence: $x(n) = \frac{h^*(-n)}{|h(-n)|}$. The output at $n = 0$ is

$$\begin{aligned} y(0) &= \sum_{m=-\infty}^{\infty} h(m)x(-m) \\ &= \sum_{m=-\infty}^{\infty} h(m)\frac{h^*(m)}{|h(m)|} \\ &= \sum_{m=-\infty}^{\infty} |h(m)| = \infty. \end{aligned}$$

Thus, we have found a bounded input for which $y(0) = \infty$; that is, the system is unstable. This completes the proof of the necessary condition. ■

EXAMPLE 1.10 *Consider the FIR filter given by (1.37). The impulse response of this filter is of finite length, and therefore the sufficient condition of Theorem 1.6 is satisfied. FIR filters are therefore always stable.*

EXAMPLE 1.11 *Consider the first-order IIR filter described by*

$$y(n) - ay(n-1) = x(n), \quad y(-1) = 0, \quad n = 0, 1, 2, \ldots$$

Determine whether or not this system is BIBO stable.

SOLUTION The impulse response was found to be $h(n) = a^n, n \geq 0$. Now we check if this impulse response satisfies the condition of absolute summability. That is,

$$\sum_{n=-\infty}^{\infty} |h(n)| = \sum_{n=0}^{\infty} |a|^n$$
$$= \frac{1}{1-|a|}, \quad |a| < 1. \tag{1.52}$$

Equation (1.52) is the formula for the infinite geometric progression and converges under the given condition. Thus, the impulse response of this filter is absolutely summable if $|a| < 1$, and the system is BIBO stable under this constraint.

1.6 IMPLEMENTATION OF DIGITAL FILTERS

There are two basic types of digital filters: FIR and IIR. In this section, we present simple implementations (or structures) of these filters. We begin with some fundamental building blocks for filter implementation. The three basic building blocks are adder, multiplier and delay (see Fig. 1.24). The output of an adder is simply the sum of all the input sequences. The multiplier is a gain that is multiplied to the input. The delay element takes the sequence $x(n)$ and shifts it by one unit to produce the sequence $x(n-1)$. Before discussing the implementation of filters with these basic blocks, it is important to study the effects of interconnecting LSI systems in different ways.

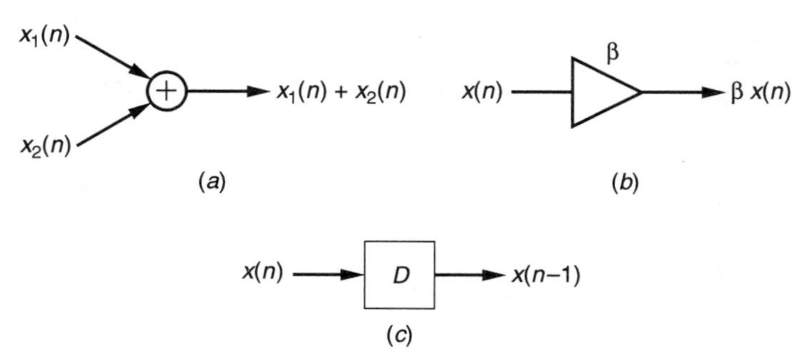

Figure 1.24 (a) Adder, (b) multiplier, (c) delay.

Filter Interconnections

Consider the connection of two filters in series as shown is Fig. 1.25. This is called a cascade connection. The output of the first filter is $h(n) * x(n)$. The final output is

$$y(n) = g(n) * [h(n) * x(n)].$$

Using the associative property of convolution given by (1.49), we see that above becomes

$$y(n) = [g(n) * h(n)] * x(n).$$

This equation implies that the impulse response of a cascade to two filters is the convolution of the impulse responses of the individual filters. Also, since by (1.48) the convolution operation is commutative, the order of the filters in the cascade connection does not matter. That is, if the filters $h(n)$ and $g(n)$ are interchanged in Fig. 1.25, the output will remain unchanged.

Now consider the parallel connection of two filters as shown in Fig. 1.26. The output of each filter is simply the convolution of the input with its impulse response. The adder sums up these outputs to produce the signal

$$y(n) = h(n) * x(n) + g(n) * x(n).$$

By the distributive property of convolution given in (1.50), the above becomes

$$y(n) = [h(n) + g(n)] * x(n).$$

This equation establishes the fact that the impulse response of the parallel connection of two filters is obtained by adding the individual impulse responses.

▶ **EXAMPLE 1.12** *Find the impulse response of the overall system shown in Fig. 1.27.*

SOLUTION From the above discussion on cascade and parallel connection of filters, the impulse response of the overall system, $h(n)$ can easily be found to be

$$h(n) = h_4(n) + h_1(n) * [h_2(n) + h_3(n)].$$

◀

In the following subsections, some basic structures are presented for implementing FIR and IIR filters. For now, this will be sufficient to give a feel for how digital filters are implemented. A comprehensive treatment of filter structures is given in Chapter 4.

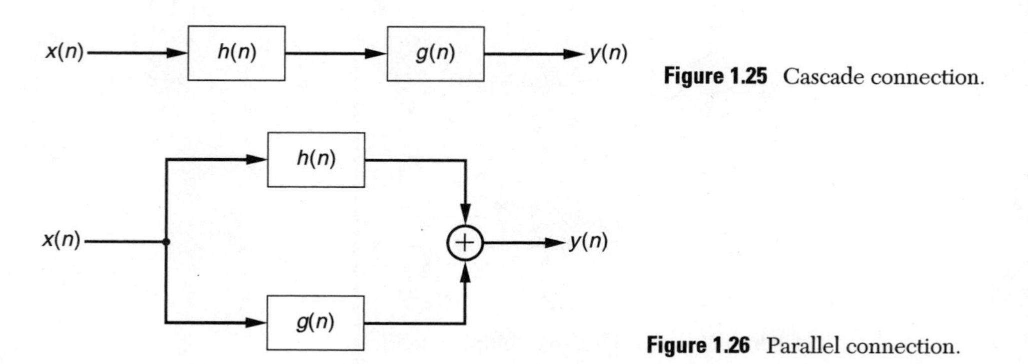

Figure 1.25 Cascade connection.

Figure 1.26 Parallel connection.

FIR Filters

FIR filter structures are straightforward and are implemented directly from the difference equation, which is repeated here for convenience:

$$y(n) = b_0 x(n) + b_1 x(n-1) + \cdots + b_M x(n-M). \tag{1.53}$$

From equation (1.53), it is clear that all we need is a series of delays, some multipliers, and one adder for implementing this digital filter. How about initial conditions? In order to calculate the first output $y(0)$, we need the input data $x(0), x(-1), x(-2), \ldots, x(-M)$. That is, we need M initial conditions. Note that $x(0)$ is the current data sample and not an initial condition. Unless otherwise specified, the initial conditions are assumed to be zero. The flow diagram for this filter is shown in Fig. 1.28, where **D** denotes a unit delay. Filter structures such as the one shown in this figure can be implemented in hardware or in software. For hardware implementation, the flow diagram represents the actual circuit diagram in terms of delays, multipliers, and adders, except that the addition is performed with two signals at a time using many adders. For software implementation, the flow diagram depicts the algorithm for calculating all the signals. Let us rewrite the above equation as

$$y(n) = \sum_{j=0}^{M} b_j x(n-j).$$

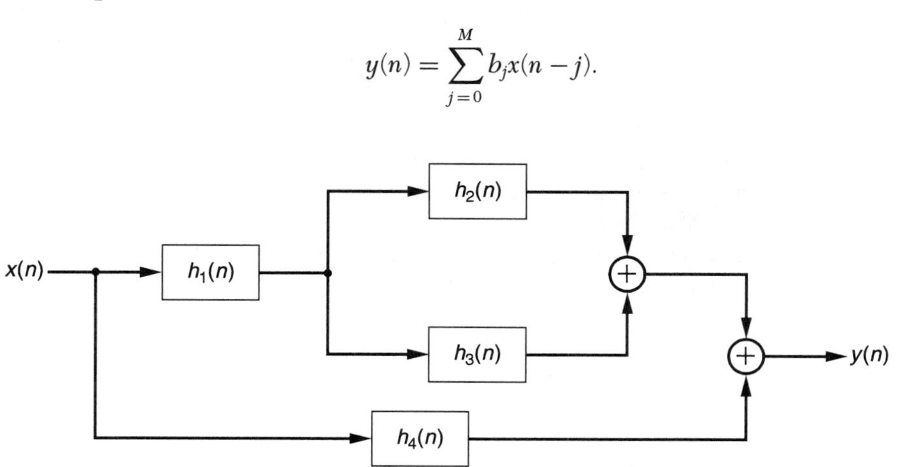

Figure 1.27 Example on filter interconnections.

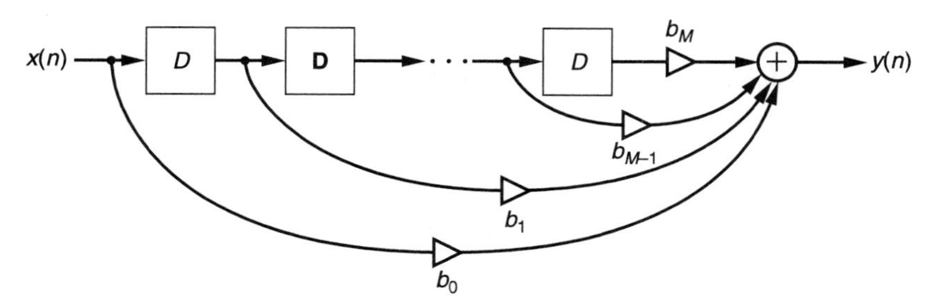

Figure 1.28 FIR filter structure.

This fits the definition of convolution, and so we have

$$y(n) = b_n * x(n). \tag{1.54}$$

The output of an FIR filter can therefore be calculated by convolving the input sequence with the sequence of filter coefficients. We also know from the previous section that the output of a linear system is the convolution of its input with the impulse response. Therefore, the sequence of filter coefficients must be the impulse response. Indeed, this is what we saw in equation (1.38) and Fig. 1.22! The impulse response has a finite number of terms and is therefore absolutely summable. Therefore, the FIR filter is always stable. This is one advantage of the FIR filter.

There is another way of representing the output of an FIR filter, which uses matrix notation. Define the following:

$$\mathbf{X}(n) = \left[x(n)\, x(n-1) \ldots x(n-M)\right]^T \tag{1.55}$$

$$\mathbf{w} = \left[b_0\, b_1\, \ldots b_M\right]^T \tag{1.56}$$

where T denotes the transpose. Then (1.53) can be written as

$$\begin{aligned} y(n) &= \mathbf{w}^T\mathbf{X}(n) \\ &= \mathbf{X}^T(n)\mathbf{w}. \end{aligned} \tag{1.57}$$

Note that for the FIR filter implementation, we have an input vector and a weight vector setup. The weight vector is constant whereas the input vector is dynamic. For every new value of n, the elements of the vector (1.55) slide to the right. The rightmost or the oldest input data goes out of the vector, and the current input data slides in at the leftmost element. At every sample instant, (1.57) is computed to find the output.

IIR Filters

The IIR filter is described by the recursive difference equation

$$\sum_{i=0}^{N} a_i y(n-i) = \sum_{j=0}^{M} b_j x(n-j), \quad a_0 = 1 \tag{1.58}$$

where $x(n)$ are $y(n)$ are the input and output, respectively. There are several different filter structures for IIR filters. Only the *Direct Form* structures are presented here; the rest will be given in Chapter 4. We begin with the first-order filter given by

$$y(n) + a_1 y(n-1) = b_0 x(n) + b_1 x(n-1). \tag{1.59}$$

The order of an IIR filter is defined as the maximum delay in the output. Let $\mathbf{D}\{.\}$ denote the unit delay operation. The above equation can be rearranged as

$$\begin{aligned} y(n) &= b_0 x(n) + b_1 x(n-1) - a_1 y(n-1) \\ &= b_0 x(n) + \mathbf{D}\{b_1 x(n) - a_1 y(n)\}. \end{aligned} \tag{1.60}$$

This equation can be implemented as shown in Fig. 1.29. Now we proceed to second-order filters. These can be easily generalized to higher order structures as we shall see shortly. The second-order difference equation is given by

$$y(n) + a_1 y(n-1) + a_2 y(n-2) = b_0 x(n) + b_1 x(n-1) + b_2 x(n-2). \quad (1.61)$$

Transposing all terms to the right except for $y(n)$ and using the delay operator gives

$$\begin{aligned} y(n) &= b_0 x(n) + \{b_1 x(n-1) - a_1 y(n-1)\} + \{b_2 x(n-2) - a_2 y(n-2)\} \\ &= b_0 x(n) + \mathbf{D}\{b_1 x(n) - a_1 y(n)\} + \mathbf{D}\{\mathbf{D}\{b_2 x(n) - a_2 y(n)\}\}. \end{aligned} \quad (1.62)$$

This structure is shown in Fig. 1.30(a) and is called Direct Form I Transposed, or in short *Direct Form* I_T. This implementation requires three delay elements. However, we can set up a pipeline of two delays and thereby save one delay. The term $\{b_2 x(n) - a_2 y(n)\}$ is delayed twice and therefore must enter the beginning of the pipeline. The other term, $\{b_1 x(n) - a_1 y(n)\}$, is delayed only once and therefore enters the middle of the pipeline. The flow diagram is shown in Fig. 1.30(b); and this structure is called Direct Form II Transposed, or *Direct Form* II_T.

Now we find the standard direct form structure. Consider the second-order IIR filter equation (1.61) again. Define the right-hand side as

$$f(n) = b_0 x(n) + b_1 x(n-1) + b_2 x(n-2) \quad (1.63)$$

which is an FIR filter. The IIR filter can now be written as

$$y(n) + a_1 y(n-1) + a_2 y(n-2) = f(n). \quad (1.64)$$

The implementation of equations (1.63) and (1.64) is given in Fig. 1.31(a), where the left side is the FIR part and the right side is the IIR. This structure is called *Direct Form I*, and it has more delays than are really necessary. Note that this structure consists of two filters in cascade. From the earlier discussion on filter interconnections, we know that owing to the commutative property of convolution, the two filters in the cascade connection can be interchanged without affecting the output. The resulting structure is shown in Fig. 1.31(b). The output of the first filter (IIR part) is labeled as $g(n)$. From this figure, we have the following important equations:

$$g(n) + a_1 g(n-1) + a_2 g(n-2) = x(n) \quad (1.65)$$

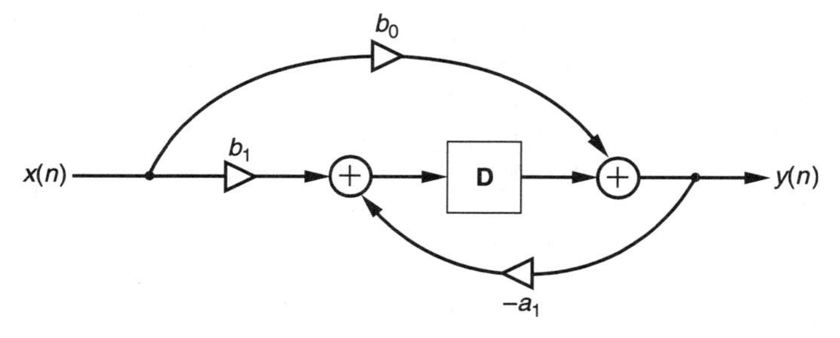

Figure 1.29 First-order IIR filter.

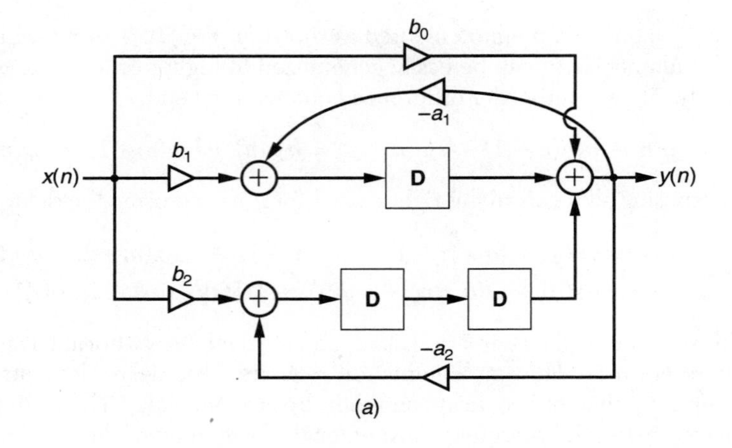

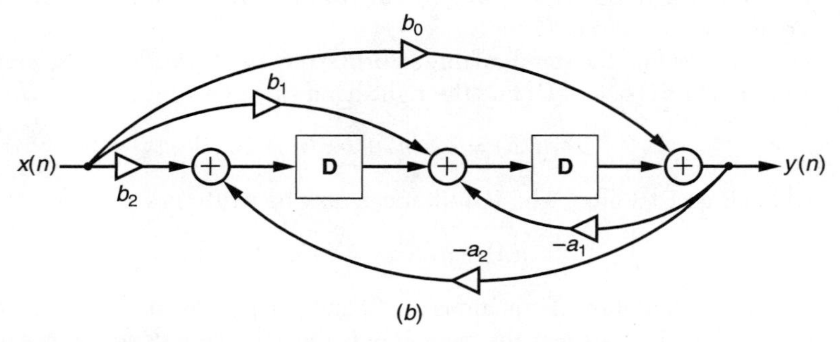

Figure 1.30 (a) Direct Form I_T; (b) Direct Form II_T.

$$y(n) = b_0 g(n) + b_1 g(n-1) + b_2 g(n-2). \tag{1.66}$$

Observe that in the second filter of Fig. 1.31(b), that is, in equation (1.66), the output is obtained by using $g(n)$, $g(n-1)$ and $g(n-2)$. All of these signals are available from the first filter and can be tapped off. The two delays in the second filter are therefore unnecessary. The flow diagram with only two delays is given in Fig. 1.31(c), and this structure is called *Direct Form II*. This is the most popular of all the Direct Form structures discussed. If we compare the original difference equation in (1.61) to (1.65) and (1.66), we see that a new variable $g(n)$ has been introduced which relates to the original equation in a special way. This observation actually yields an algorithm for obtaining the Direct Form II structure, which is easily extended to any order. The algorithm is given in the following.

Given the IIR filter equation

$$y(n) + a_1 y(n-1) + \cdots + a_N y(n-N) = b_0 x(n) + b_1 x(n-1) + \cdots + b_M x(n-M), \tag{1.67}$$

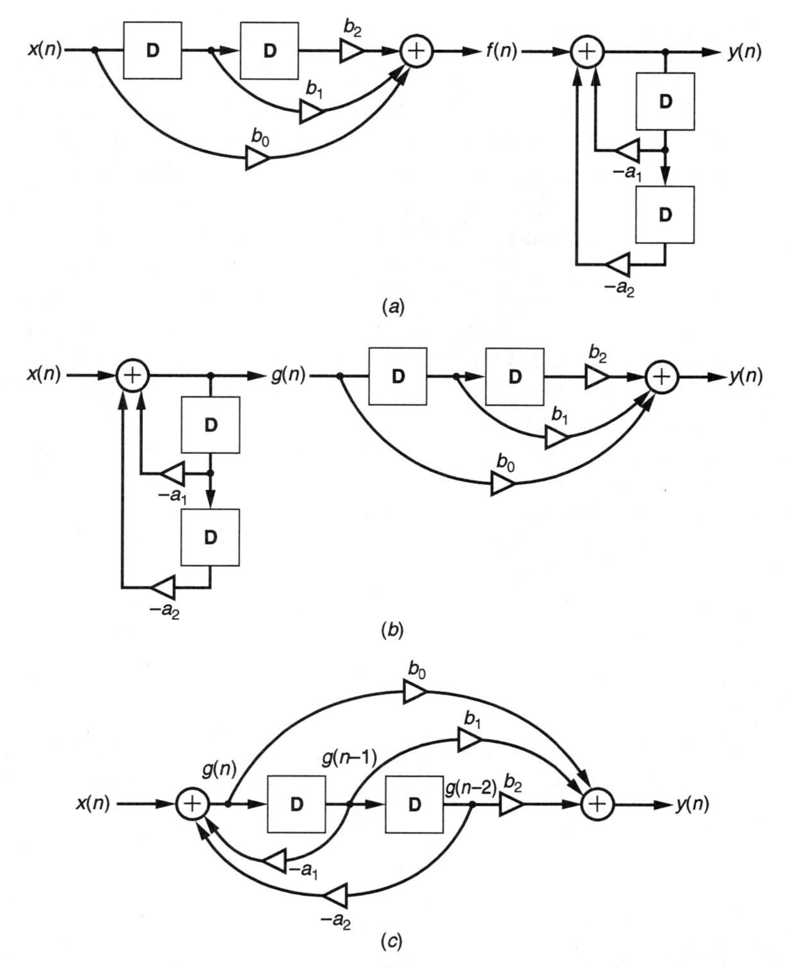

Figure 1.31 (*a*) Direct Form I; (*b*) Direct Form I with the cascaded filters interchanged; (*c*) Direct Form II.

perform the following steps.

1. Replace the output signal variables $y(.)$ by a new variable $g(.)$ and set the resulting equation equal to the input $x(n)$. This gives the equation

$$g(n) + a_1 g(n-1) + \cdots + a_N g(n-N) = x(n). \tag{1.68}$$

2. Replace the input signal variables $x(.)$ by the new variable $g(.)$ and set the resulting equation equal to the output $y(n)$. This gives the equation

$$y(n) = b_0 g(n) + b_1 g(n-1) + \cdots + b_M g(n-M). \tag{1.69}$$

3. Use (1.68) and (1.69) to sketch the flow diagram for the Direct Form II structure.

1.7 BASICS OF RANDOM PROCESSES

Discrete-time signals that can be mathematically described, such as the sin() function or the step function, are called *deterministic* signals. The value of such a signal can therefore be determined at every sample instant provided all the parameters of the function are known. Discrete-time signals that cannot be predicted ahead of time are called *random* signals or processes. The value of such a process at a given sample instant is generated in a random fashion. Such signals are common phenomena in many areas of engineering. Speech signals, radar signals, video signals, biomedical signals, and a variety of noise sources are some examples. A random process is not really a single sequence but a collection of sequences. To explain this, consider the following example. Let us assume that we want to measure the temperature in a given city by thermometers placed at a variety of different locations within the city limits. Let $x_1(n), n = 0, 1, 2, \ldots$ denote a sequence of temperature measurements every second for a given duration measured by thermometer no. 1. The sequence $x_2(n)$ denotes the measurements made by thermometer no. 2, and so on. The sequences are shown in Fig. 1.32. The set of all such sequences is called an *ensemble* of sequences or simply a random process. One particular sequence in this process is called a *realization* or a *time series*. Such a realization is denoted simply by $x(n)$, where the subscript is dropped for simplicity.

Consider a realization $x(n)$ of a possibly complex random process given by $\{x(n), x(n-1), x(n-2), \ldots, x(n-N)\}$. The *mean sequence* of the process is given by

$$\mu(n) = E\{x(n)\} \tag{1.70}$$

where $E\{.\}$ denotes the mathematical expectation operation. The above definition means that for each value of n, we take the average across the ensemble of sequences at that

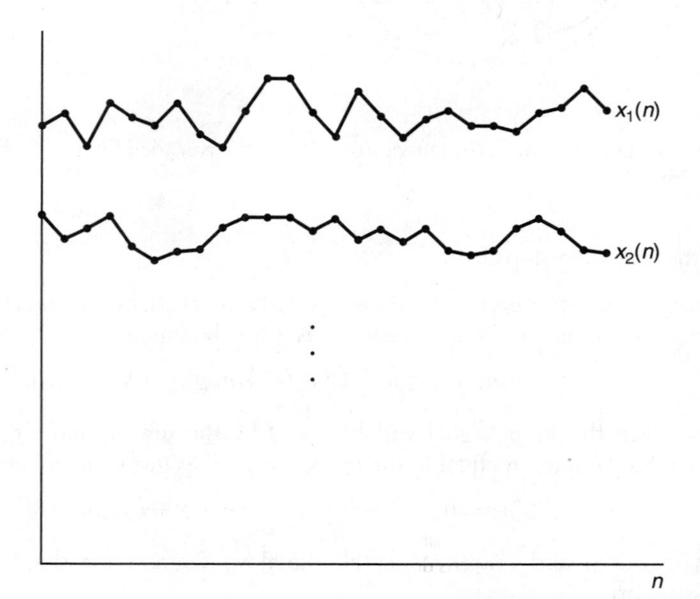

Figure 1.32 An example of a random process.

particular n. Thus, the mean value is clearly a sequence given by $\{\mu(n), \mu(n-1), \mu(n-2), \ldots, \mu(n-N)\}$.

The *autocorrelation function* of the process is defined by

$$r(n, n-k) = E\{x(n)x^*(n-k)\}, \quad k = 0, \pm 1, \pm 2, \ldots \tag{1.71}$$

where $*$ denotes complex conjugation. Equation (1.71) is therefore a 2-D sequence. To explain this definition, recall that the random process is the ensemble of sequences given by $\{x_1(n)\}, \{x_2(n)\}, \ldots$ as shown in Fig. 1.32. For a fixed value of n, say n_0, we vary k and find the sequences

$$x_1(n_0)x_1^*(n_0-k), \quad k = 0, \pm 1, \pm 2, \ldots$$

$$x_2(n_0)x_2^*(n_0-k), \quad k = 0, \pm 1, \pm 2, \ldots$$

$$\vdots$$

The above is again an ensemble of sequences. The mean sequence of this ensemble is then found and written as $r(n_0, n_0-k), k = 0, \pm 1, \pm 2, \ldots$. When we repeat this process for all possible values of n_0, we get the autocorrelation function, which is an ensemble of sequences. This definition can be extended to two random processes $x(n)$ and $y(n)$. The *cross-correlation function* for these processes is defined as

$$\begin{aligned} r_{xy}(n, n-k) &= E\{x(n)y^*(n-k)\}, \quad k = 0, \pm 1, \pm 2, \ldots \\ &= E\{y(n)x^*(n-k)\}, \quad k = 0, \pm 1, \pm 2, \ldots \end{aligned} \tag{1.72}$$

In the above definitions, if we subtract the mean before multiplication, then we have another definition. The *autocovariance function* of a random process $x(n)$ is defined as

$$c(n, n-k) = E\{[x(n) - \mu(n)][x(n-k) - \mu(n-k)]^*\}, \quad k = 0, \pm 1, \pm 2, \ldots \tag{1.73}$$

The cross-covariance function is defined in a similar fashion.

A random process is said to be *wide-sense stationary* [4] or in short, simply *stationary* if the mean, autocorrelation, and autocovariance are shift-invariant, that is,

$$\mu(n) = \mu$$

$$r(n, n-k) = r(k)$$

$$c(n, n-k) = c(k).$$

The above means that the mean is constant across the ensemble of sequences. The correlation and the covariance are independent of the position n and simply depend on the shift k. This shift is simply the difference between the arguments, that is, $n-(n-k) = k$. For a stationary process with $k = 0$, we have

$$\begin{aligned} r(0) &= E\{x(n)x^*(n)\} \\ &= E\{|x(n)|^2\}. \end{aligned} \tag{1.74}$$

[4] The wide-sense stationary process is most commonly used in engineering. The notion of a *strictly stationary* process is too restrictive that most signals do not satisfy. A random process is strictly stationary if its probability density function is time independent.

The above is called the mean square value. The variance of a stationary random process is defined as

$$\begin{aligned}\text{var}\{x(n)\} &= \sigma^2 \\ &= E\{|x(n) - \mu|^2\}\end{aligned} \tag{1.75}$$

Thus, if the process is zero mean, then $r(0) = \sigma^2$. Now we give a property of the autocorrelation sequence.

Fact 1: *The autocorrelation sequence of a stationary random process satisfies*

$$r(-k) = r^*(k). \tag{1.76}$$

Proof. From the definition of autocorrelation, we have $r(k) = E\{x(n)x^*(n-k)\}$. Taking the complex conjugate of both sides

$$\begin{aligned}r^*(k) &= E\{x(n)x^*(n-k)\}^* \\ &= E\{x(n-k)x^*(n)\} \\ &= r(-k). \quad \clubsuit\end{aligned} \tag{1.77}$$

Fact 2: *The autocovariance of a random process can be written as*

$$c(n, n-k) = r(n, n-k) - \mu(n)\mu^*(n-k). \tag{1.78}$$

Proof. The proof is straightforward and is assigned as a problem.

For a stationary random process, Fact 2 simplifies to the following:

$$c(k) = r(k) - |\mu|^2. \tag{1.79}$$

In the above discussion, we assumed the availability of an ensemble of sequences or a random process. However, in practice we often have a single realization of the random process, that is, a single sequence. In this situation, it is convenient to estimate the mean and the correlation of the random process from the single sequence. Consider a sequence $x(n), n = 0, 1, 2, \ldots, N$. The *time-average* of this sequence is defined as

$$\widehat{\mu} = \frac{1}{N+1} \sum_{n=0}^{N} x(n). \tag{1.80}$$

The *time-correlation* of the sequence is defined as

$$\widehat{r}(k) = \frac{1}{N+1} \sum_{n=0}^{N} x(n)x^*(n-k). \tag{1.81}$$

A stationary random process is said to be *mean-ergodic* if the ensemble average is equal to the time-average in the limit as $N \to \infty$; that is,

$$\mu = \lim_{N \to \infty} \widehat{\mu}. \tag{1.82}$$

A stationary random process is said to be *correlation-ergodic* if the autocorrelation is equal to the time-correlation in the limit as $N \to \infty$; that is,

$$r(k) = \lim_{N\to\infty} \widehat{r}(k). \tag{1.83}$$

Based on the above definitions, we now state the following theorem, which gives sufficient conditions for mean and correlation ergodicity.

Theorem 1.7 *A sufficient condition for a stationary process to be mean ergodic is*

$$\sum_{k=0}^{\infty} |c(k)| < \infty \tag{1.84}$$

where $c(k)$ is the autocovariance of the process. The sufficient condition for correlation-ergodicity is

$$\text{var}\{\widehat{r}(k)\} \to 0 \quad \text{as} \quad N \to \infty. \tag{1.85}$$

The above conditions are not difficult to prove and are usually covered in a course on random processes. Refer to [21] for the proof. The conditions in the above theorem hold for most zero-mean processes encountered in engineering.

If we modify the definition of time-correlation given in (1.81) by removing the $(N+1)$ factor in the denominator, then the new definition is called *unscaled time-correlation* given below.

Unscaled time-autocorrelation

$$\tilde{r}_{xx}(k) = \sum_{n=0}^{N} x(n)x^*(n-k). \tag{1.86}$$

Unscaled time-cross-correlation

$$\begin{aligned} \tilde{r}_{xy}(k) &= \sum_{n=0}^{N} x(n)y^*(n-k) \\ &= \sum_{n=0}^{N} y(n)x^*(n-k). \end{aligned} \tag{1.87}$$

EXAMPLE 1.13 *Find the unscaled time-cross-correlation of the sequences*

$$x(n) = \begin{bmatrix} 1 & \mathbf{3} & -1 & 5 \end{bmatrix} \text{ and } \quad y(n) = \begin{bmatrix} \mathbf{7} & -2 & 9 & -4 \end{bmatrix}$$

Recall that the numbers in boldface denote the signal value at $n = 0$.

SOLUTION In accordance with equation (1.87), for a given value of k, we shift one of the sequences (say $y(n)$) by k units, multiply by the other sequence, and sum up all the samples in the product. For $k = 0$, we have the following.

$$\begin{bmatrix} x(n) \\ y(n-0) \end{bmatrix} = \begin{bmatrix} 1 & \mathbf{3} & -1 & 5 & \\ & \mathbf{7} & -2 & 9 & -4 \end{bmatrix} \Rightarrow \tilde{r}_{xy}(0) = 3 \cdot 7 + (-1) \cdot (-2) + 5 \cdot 9 = 68$$

For $k = 1$, we have

$$\begin{bmatrix} x(n) \\ y(n-1) \end{bmatrix} = \begin{bmatrix} 1 & \mathbf{3} & -1 & 5 & & \\ & & 7 & -2 & 9 & -4 \end{bmatrix} \Rightarrow \tilde{r}_{xy}(1) = (-1) \cdot 7 + 5 \cdot (-2) = -17.$$

For $k = -1$, we have

$$\begin{bmatrix} x(n) \\ y(n+1) \end{bmatrix} = \begin{bmatrix} 1 & \mathbf{3} & -1 & 5 \\ 7 & \mathbf{-2} & 9 & -4 \end{bmatrix} \Rightarrow \tilde{r}_{xy}(-1) = 1 \cdot 7 + 3 \cdot (-2) + (-1) \cdot 9 + 5 \cdot (-4) = -28,$$

and so on. The final answer is

$$\tilde{r}_{xy}(n) = \begin{bmatrix} 0 & -4 & -3 & 29 & -28 & 68 & -17 & 35 & 0 & 0 \end{bmatrix}.$$

◀

Now we present some results on the effects of random signals on a linear system. Consider a stationary random process, $x(n)$, as an input to a linear system with impulse response $h(n)$. This is depicted in Fig. 1.33.

Fact 3: *If the input to a shift-invariant real linear system is a zero-mean stationary random process, the output is also a zero-mean stationary random process.*

Proof. Consider the system of Fig. 1.33 where $x(n)$ is a zero-mean stationary random process. Let the impulse response be real and be denoted by $h(n)$. The output is given by the convolution formula

$$\begin{aligned} y(n) &= \sum_{m=-\infty}^{\infty} x(m)h(n-m) \\ &= \sum_{m=-\infty}^{\infty} h(m)x(n-m) \end{aligned} \tag{1.88}$$

x(n) → h(n) → y(n)

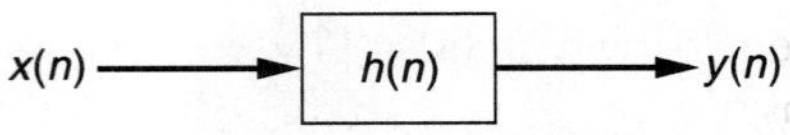

Figure 1.33 A linear system with impulse response $h(n)$.

Since $x(n)$ is zero-mean and stationary, its mean is constant (zero) and its autocorrelation is independent of n; that is,

$$\mu_x = E\{x(n)\} = 0$$
$$r_x(k) = E\{x(n)x^*(n-k)\}.$$

Now we take the expectation of both sides of the convolution equation to get

$$E\{y(n)\} = E\left\{\sum_{m=-\infty}^{\infty} x(m)h(n-m)\right\}. \tag{1.89}$$

Since $x(n)$ is a random process and $h(n)$ is not, we have

$$\begin{aligned} E\{y(n)\} &= \sum_{m=-\infty}^{\infty} E\{x(m)\}h(n-m) \\ &= \mu_x \sum_{m=-\infty}^{\infty} h(n-m) \\ &= 0. \end{aligned} \tag{1.90}$$

Thus, the output has zero-mean and is independent of the time index n. Now let's find the autocorrelation function of the output as follows. The autocorrelation of the output is defined by $E\{y(n)y^*(n-k)\}$. If $y(n)$ is given by (1.88), then $y^*(n-k)$ is given by

$$y^*(n-k) = \sum_{l=-\infty}^{\infty} h(l)x^*(n-l-k). \tag{1.91}$$

The autocorrelation of the output now becomes

$$\begin{aligned} E\{y(n)y^*(n-k)\} &= E\left\{\sum_{m=-\infty}^{\infty} h(m)x(n-m) \sum_{l=-\infty}^{\infty} h(l)x^*(n-l-k)\right\} \\ &= \sum_{m=-\infty}^{\infty}\sum_{l=-\infty}^{\infty} E\{x(n-m)x^*(n-l-k)\}h(m)h(l) \\ &= \sum_{m=-\infty}^{\infty}\sum_{l=-\infty}^{\infty} h(m)h(l)r_x(l+k-m). \end{aligned} \tag{1.92}$$

In the above formula, for some value of n, the right-hand side is independent of n and dependent only on k. It can therefore be written as $r_y(k)$. We have thus proved that the mean of the output is constant (zero) and that the correlation is independent of the time index n, which implies that the output is stationary. ♣

1.8 2-D EXTENSIONS

In this section, some of the concepts from the foregoing sections are extended to two dimensions. Most of the extensions are straightforward. The derivations or proofs of the results are similar to 1-D and will not be given. Some of these proofs are assigned as problems at the end of the chapter. We begin this section with the fundamental 2-D discrete sequences.

2-D Sequences

The *2-D unit impulse* is defined as

$$\delta(n_1, n_2) \triangleq \begin{cases} 1, & n_1 = n_2 = 0 \\ 0, & \text{otherwise.} \end{cases} \tag{1.93}$$

This function is depicted in Fig. 1.34(a). Recall from an earlier section that since there is no number in parentheses next to the dot, the amplitude is unity. This function is a direct extension of the 1-D unit impulse. There are certain 2-D function definitions that do not have a 1-D counterpart [5]. We will denote these with 2d as a suffix. For the impulse function, we define

$$\delta_{2d}(n_1) \triangleq \begin{cases} 1, & n_1 = 0 \\ 0, & \text{otherwise.} \end{cases} \tag{1.94}$$

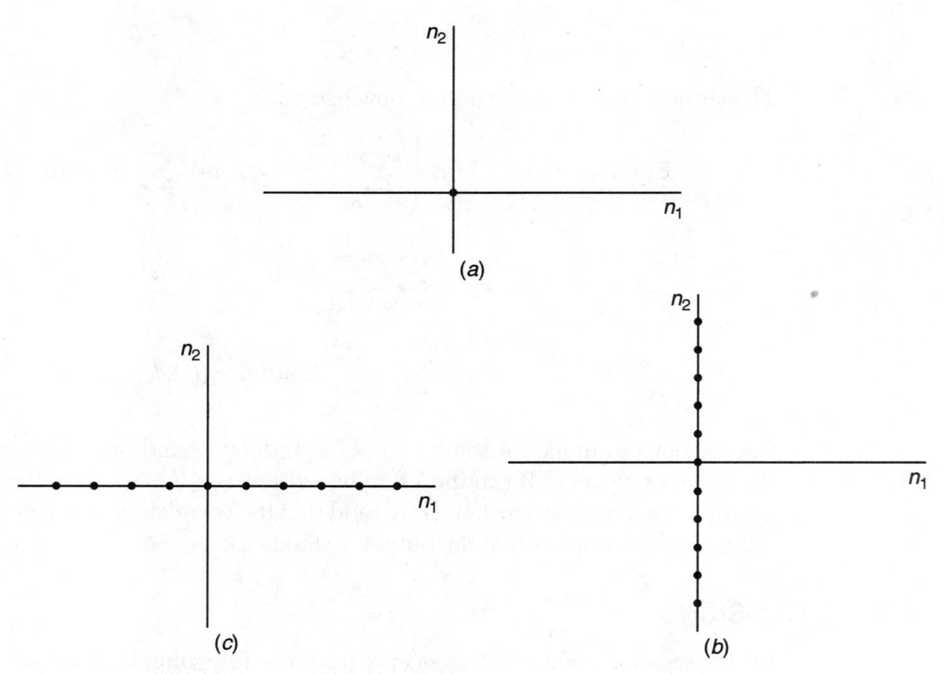

Figure 1.34 (a) $\delta(n_1, n_2)$; (b) $\delta_{2d}(n_1)$; (c) $\delta_{2d}(n_2)$.

and

$$\delta_{2d}(n_2) \triangleq \begin{cases} 1, & n_2 = 0 \\ 0, & \text{otherwise.} \end{cases} \tag{1.95}$$

These functions are shown in Fig. 1.34(*b*) and (*c*), respectively.

The *2-D unit step* is defined as

$$u(n_1, n_2) \triangleq \begin{cases} 1, & n_1 \geq 0, \quad n_2 \geq 0 \\ 0, & \text{otherwise.} \end{cases} \tag{1.96}$$

This function is depicted in Fig. 1.35(*a*). We also define the following 2-D specific functions:

$$u_{2d}(n_1) \triangleq \begin{cases} 1, & n_1 \geq 0 \\ 0, & \text{otherwise.} \end{cases} \tag{1.97}$$

$$u_{2d}(n_2) \triangleq \begin{cases} 1, & n_2 \geq 0 \\ 0, & \text{otherwise.} \end{cases} \tag{1.98}$$

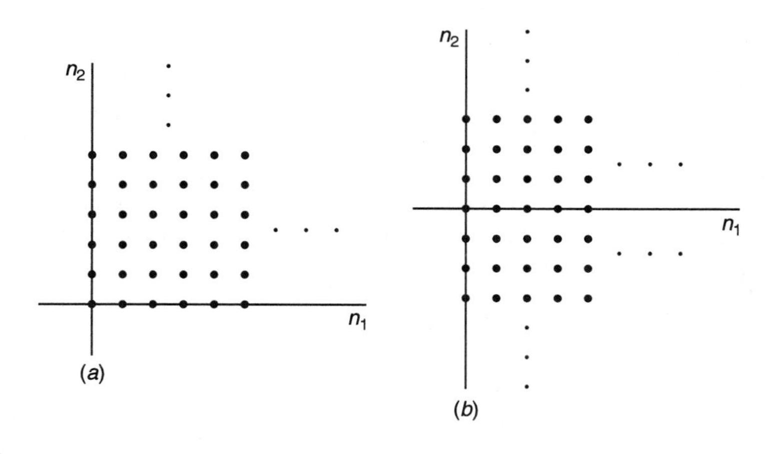

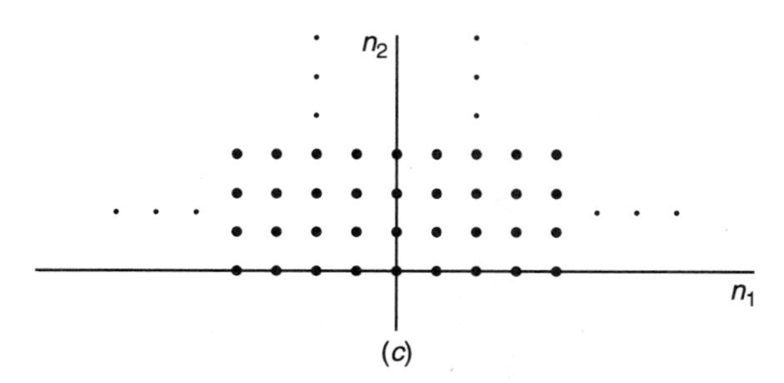

Figure 1.35 (*a*) $u(n_1, n_2)$; (*b*) $u_{2d}(n_1)$; and (*c*) $u_{2d}(n_2)$.

The above functions are depicted in Fig. 1.35(b) and (c), respectively.

The *2-D exponential* is defined as

$$x(n_1, n_2) = a^{n_1} b^{n_2}.$$

Notice that this sequence is really a product of two 1-D exponentials — one as a function of n_1 and the other of n_2. This brings us to the definition of separability.

A 2-D sequence $x(n_1, n_2)$ is said to be *separable* if

$$x(n_1, n_2) = x_1(n_1)x_2(n_2). \tag{1.99}$$

The condition of separability plays an important role in 2-D signal processing. Separable signals and systems are easier to handle since they can be decomposed into 1-D functions, and then the rich variety of 1-D signal processing techniques become applicable.

The concept of periodicity of 2-D signals is quite interesting because there can be different types of periodicity. This is due to the presence of two independent variables or dimensions.

A 2-D sequence $x(n_1, n_2)$ is said to be *row-wise periodic* if

$$x(n_1, n_2) = x(n_1 + P, n_2)$$

for all (n_1, n_2), where P is an integer and is the period of the sequence. There are P signal values that repeat in each row. However, these values can be different from one row to another. Therefore, we can have an arbitrary number of signal values that repeat depending on the rows. In fact, if we have an infinite number of rows, then there can be an infinite number of values that repeat, but the periodicity is still 3. This is not possible in 1-D periodic signals. An example is given in Fig. 1.36(a) where the 2-D sequence has a period of 2. There are three rows with nonzero values. In each row the period is 2, but the values are different from one row to another. This is denoted by different symbols for the data.

A 2-D sequence $x(n_1, n_2)$ is said to be *column-wise periodic* if

$$x(n_1, n_2) = x(n_1, n_2 + Q)$$

for all n_1, n_2, where Q is an integer. An example of this type of periodicity is given in Fig. 1.36(b) which depicts two columns with a period of 3.

The third type of periodicity is the counterpart of 1-D periodicity, where there can be only one period in the entire sequence. A 2-D sequence $x(n_1, n_2)$ is said to be *periodic separable* if

$$x(n_1, n_2) = x(n_1 + P, n_2) = x(n_1, n_2 + Q) = x(n_1 + P, n_2 + Q)$$

for all (n_1, n_2), where P and Q are integers. The pair (P, Q) is called the period of the sequence. In this case, there are PQ signal values that repeat in the entire (n_1, n_2) plane. An example of a periodic separable sequence with period (2, 2) is shown in Fig. 1.36(c).

Now we present some examples on the manipulation of 2-D signals.

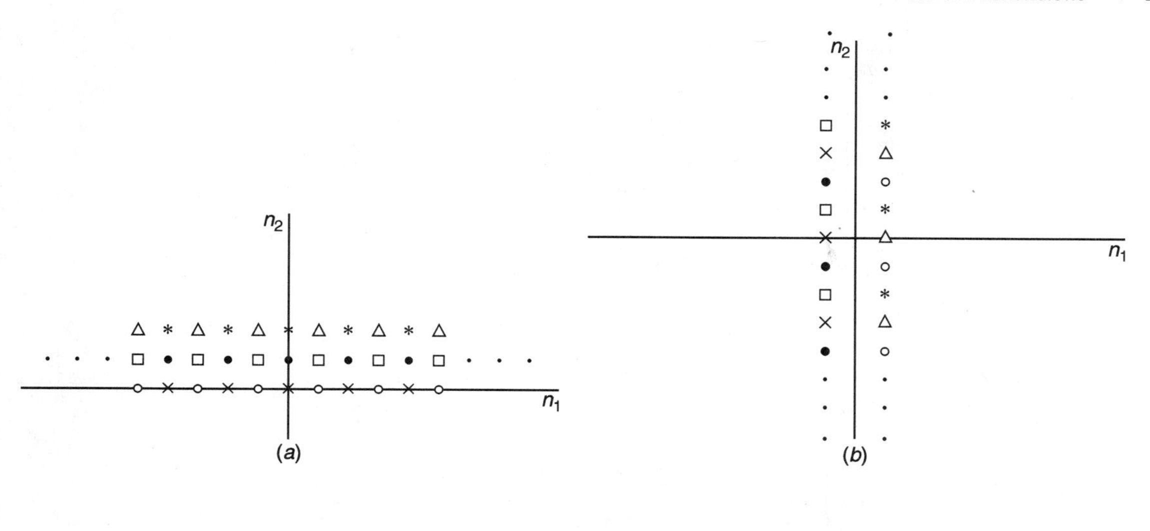

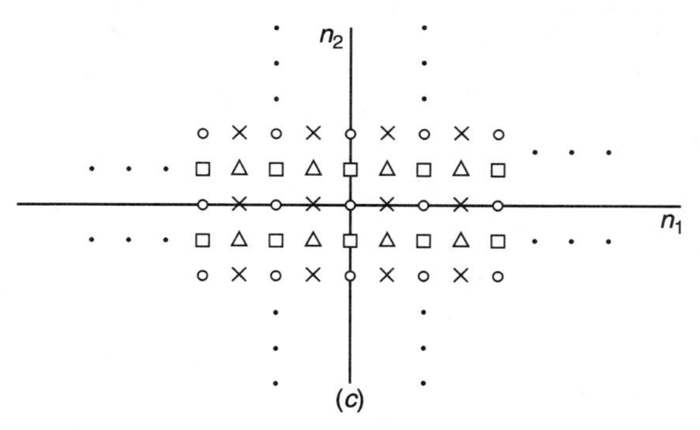

Figure 1.36 2-D periodicities: (a) Row-wise periodic with period 2; (b) column-wise with period 3; (c) periodic separable with period (2,2).

EXAMPLE 1.14 *Sketch the following sequences: (a)* $2\delta(n_1 - 1, n_2 - 1) + \delta_{2d}(n_1 - n_2)$; *(b)* $u_{2d}(n_1 - n_2)$; *(c)* $u(2 - n_1, 2 - n_2)\, u(n_1 + 1, n_2 + 1)$.

SOLUTION

(a) The first term represents the amplitude of 2 at (1, 1) and zero elsewhere. The second term can be analyzed by using (1.94) or (1.95). Accordingly, this function has an amplitude of unity for $n_1 - n_2 = 0$, that is, on the line through the origin with unit slope. When this function is added to the first, the resulting graph is shown in Fig. 1.37(*a*). Recall that if a pixel has no number in parentheses, its amplitude is unity.

(b) By the definitions in (1.97) and (1.98), this 2-D step function has a value of unity for $n_1 - n_2 \geq 0$, as shown in Fig. 1.37(*b*).

(c) This sequence is a product of two functions, both of which are shifted versions of the step function. The first function is unity for $2 - n_1 \geq 0$, $2 - n_2 \geq 0$; that is, for $n_1 \leq 2, n_2 \leq 2$, as shown in Fig. 1.37(*c*)(*i*). The second function is unity for $n_1 \geq -1, n_2 \geq -1$, as shown in Fig. 1.37(*c*)(*ii*). The product is the intersection of these two graphs as shown in Fig. 1.37(*c*)(*iii*).

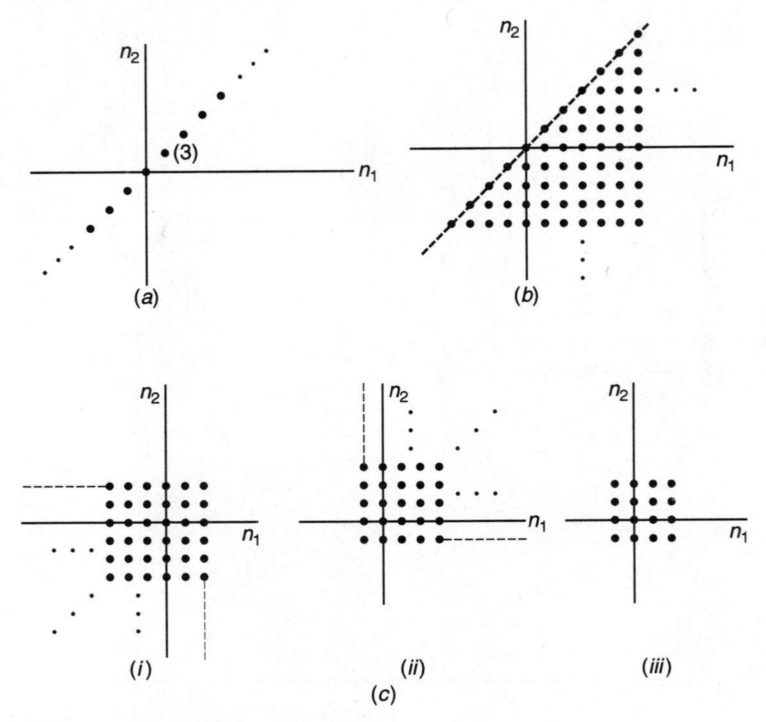

Figure 1.37 Plots for Example 1.14.

2-D Systems

For the most part, the fundamental theory of 2-D systems is a straightforward extension from 1-D. A 2-D system can be represented as

$$y(n_1, n_2) = \Re\{x(n_1, n_2)\}$$

where $x(n_1, n_2)$ and $y(n_1, n_2)$ are the input and output, respectively. The definitions of linearity, shift-invariance, causality, and BIBO stability are the same as in 1-D, except that now we have two independent variables. Let us illustrate using the following example.

▶ **EXAMPLE 1.15** *Consider a* 3×3 *mean filter described in Section 1.2. Determine whether this filter is linear, shift-invariant, causal, and BIBO stable.*

SOLUTION The mean filter is described by the difference equation

$$y(n_1, n_2) = \frac{1}{9} \sum_{i=-1}^{1} \sum_{j=-1}^{1} x(n_1 - i, n_2 - j). \tag{1.100}$$

This equation describes the 3×3 sliding window where the filtered version of the center pixel is the mean of the input window. To determine linearity, we first apply an input $x_1(n_1, n_2)$ to the system. The resulting output is

$$y_1(n_1,n_2) = \frac{1}{9}\sum_{i=-1}^{1}\sum_{j=-1}^{1} x_1(n_1 - i, n_2 - j).$$

If we apply $x_2(n_1,n_2)$ to the system, the output is

$$y_2(n_1,n_2) = \frac{1}{9}\sum_{i=-1}^{1}\sum_{j=-1}^{1} x_2(n_1 - i, n_2 - j).$$

Now we apply $x_1(n_1,n_2) + x_2(n_1,n_2)$ to the system, and the resulting output is

$$y_s(n_1,n_2) = \frac{1}{9}\sum_{i=-1}^{1}\sum_{j=-1}^{1} [x_1(n_1 - i, n_2 - j) + x_2(n_1 - i, n_2 - j)].$$

Since $y_s(n_1,n_2) = y_1(n_1,n_2) + y_2(n_1,n_2)$, the system is linear.

To check for shift-invariance, we apply a shifted version of the input to the system; that is, $x(n_1 - N_1, n_2 - N_2)$. Then the output is

$$\Re\{x(n_1 - N_1, n_2 - N_2)\} = \frac{1}{9}\sum_{i=-1}^{1}\sum_{j=-1}^{1} x(n_1 - N_1 - i, n_2 - N_2 - j)$$

where $\Re\{.\}$ is the system operator as before. The question we ask is the following. Is the above output equal to the original output shifted by (N_1, N_2)? If we replace (n_1, n_2) in (1.100) by $(n_1 - N_1, n_2 - N_2)$, we find the answer to be yes. That is, $y(n_1 - N, n_2 - N) = \Re\{x(n_1 - N_1, n_2 - N_2)\}$, and the filter is shift-invariant.

To determine causality, we ask if the system requires "future" inputs in order to calculate the output. To compute $y(n_1,n_2)$, the system requires $x(n_1 + 1, n_2), x(n_1, n_2 + 1)$ and $x(n_1 + 1, n_2 + 1)$, in addition to the present and past values of the input. Therefore, this system is noncausal.

In order to determine BIBO stability, we assume that the input is bounded; that is, $|x(n)| \leq \beta < \infty$. Then we take the absolute value of both sides of the given difference equation and proceed as follows:

$$\begin{aligned}
|y(n_1,n_2)| &= \left|\frac{1}{9}\sum_{i=-1}^{1}\sum_{j=-1}^{1} x(n_1 - i, n_2 - j)\right| \\
&\leq \frac{1}{9}\sum_{i=-1}^{1}\sum_{j=-1}^{1} |x(n_1 - i, n_2 - j)| \\
&\leq \frac{1}{9} 9\beta < \infty.
\end{aligned}$$

Since the output is bounded, the system is BIBO stable.

Now we discuss the concept of impulse response for 2-D systems. As in 1-D, the impulse response of a 2-D system is defined as the output of the system in response to an unit impulse. Symbolically, we write

$$h(n_1,n_2) = \Re\{\delta(n_1,n_2)\}.$$

Again, a 2-D FIR filter is defined as one whose impulse response is finite in length. The difference equation of a causal 2-D filter is given by

$$y(n_1, n_2) = \sum_{i=0}^{M_1} \sum_{j=0}^{M_2} b_{ij} x(n_1 - i, n_2 - j)$$

where M_1 and M_2 are positive integers. The filter coefficients are denoted by b_{ij}. The impulse response is obviously finite in length, and the amplitudes are given by the filter coefficients. Note a difference between this filter and its 1-D counterpart. For an M-order 1-D FIR filter, we have shown earlier that we need M initial conditions on the input. Now examine the 2-D equation carefully. If we calculate the outputs $y(n_1, 0)$ for $n_1 \geq 0$, we need $M_1 M_2$ initial conditions for each value of n_1. Since we have to calculate the output for an infinite number of values of n_1, we need an infinite number of initial conditions. The same is true for calculating the outputs $y(0, n_2)$. If the initial conditions are not specified, then we assume them to be zero as shown in Fig. 1.38.

The difference equation of a 2-D IIR filter is given by

$$\sum_{i=0}^{M_1} \sum_{j=0}^{M_2} a_{ij} y(n_1 - i, n_2 - j) = \sum_{i=0}^{M_1} \sum_{j=0}^{M_2} b_{ij} x(n_1 - i, n_2 - j), a_{00} = 1.$$

A simple example such as that used in the 1-D case can be used to show that the impulse response of this system is infinitely long. The reader is encouraged to perform this exercise. As for initial conditions, here we need an infinite number of initial conditions for both the input and the output. If the initial conditions are not as shown in Fig. 1.38, the system may not be linear or shift-invariant.

The concept of convolution and the output of a linear system are similar for 2-D systems. The following theorem is a straightforward extension from 1-D and is stated without proof.

Theorem 1.8 *If $x(n_1, n_2)$ and $h(n_1, n_2)$ are, respectively, the input and impulse response of a LSI system denoted by the operator $\Re\{.\}$, the output is obtained by the convolution*

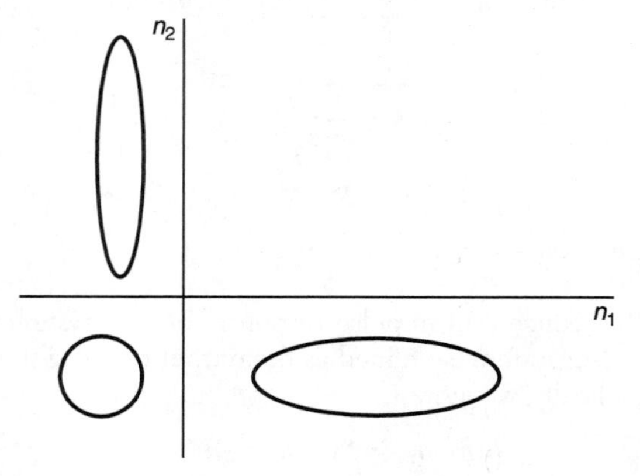

Figure 1.38 Typical initial conditions for a 2-D filter.

of these functions given by

$$y(n_1,n_2) = x(n_1,n_2) * h(n_1,n_2)$$

$$= \sum_{m_1=-\infty}^{\infty} \sum_{m_2=-\infty}^{\infty} x(m_1,m_2)h(n_1 - m_1, n_2 - m_2) \quad (1.101)$$

$$= \sum_{m_1=-\infty}^{\infty} \sum_{m_2=-\infty}^{\infty} h(m_1,m_2)x(n_1 - m_1, n_2 - m_2) \quad (1.102)$$

where $*$ denotes the convolution operation.

The 2-D convolution operation can be computed by the familiar "Switch-Shift-Sum" method, except that now we have to deal with two variables. The method is now illustrated with an example.

EXAMPLE 1.16 *Perform 2-D convolution of the functions $h(n_1,n_2)$ and $x(n_1,n_2)$ given in Fig. 1.39(a) and (b), respectively.*

SOLUTION In this example, (1.102) is used to compute the convolution. It is always useful to use the graphical method. As in 1-D convolution, we will use the Switch-Shift-Sum technique. However it is done in two dimensions as shown below. The independent variables in $h(n_1,n_2)$ and $x(n_1,n_2)$ are switched from n_1, n_2 to m_1, m_2 as given in Fig. 1.39(*c*) and (*d*). In this example, let us perform the "shift" operation on the input sequence $x(m_1,m_2)$. So, we first flip $x(m_1,m_2)$ over the vertical axis, which gives $x(-m_1,m_2)$. Then we flip $x(-m_1,m_2)$ over the horizontal axis to obtain $x(-m_1,-m_2)$. The sequences $x(-m_1,m_2)$ and $x(-m_1,-m_2)$ are plotted in Fig. 1.39(*e*) and (*f*), respectively. Replacing $(-m_1,-m_2)$ by (n_1-m_1,n_2-m_2) leads to a shift in the sequence $x(-m_1,-m_2)$ to $x(n_1-m_1,n_2-m_2)$. This sequence is plotted in Fig. 1.39(*g*)–(*i*) for $(n_1,n_2) = (0,0), (0,1)$, and (0,1), respectively, along with $h(m_1,m_2)$. There is no overlap of the sequences in any of these cases. It is easy to verify that a partial or full overlap between $x(n_1 - m_1, n_2 - m_2)$ and $h(m_1,m_2)$ occurs only when $1 \le n_1 \le 2$ and $-1 \le n_2 \le 1$. By sliding $x(n_1 - m_1, n_2 - m_2)$ appropriately, and multiplying and summing the overlapping points, we get the following results.

By sliding the sequence in Fig. 1.39(*g*) appropriately, we get

$$y(1,0) = (-1) \times 1 + 0.5 \times (-2) = -2;$$

$$y(2,0) = (-1) \times (-1) + 1 \times 1 + 0.5 \times 1 + 2 \times (-2) = -1.5;$$

$$y(3,0) = 1 \times (-1) + 2 \times 1 = 1.$$

Again, by sliding the sequences in Fig. 1.39(*h*) and (*i*) to the right, we get

$$y(1,-1) = (-1) \times (-2) = 2;$$

$$y(2,-1) = (-1) \times 1 + 1 \times (-2) = -3;$$

$$y(3,-1) = 1 \times 1 = 1.$$

$$y(1,1) = 0.5 \times 1 = 0.5;$$

$$y(2,1) = 0.5 \times (-1) + 2 \times 1 = 1.5;$$

$$y(3,1) = 2 \times (-1) = -2.$$

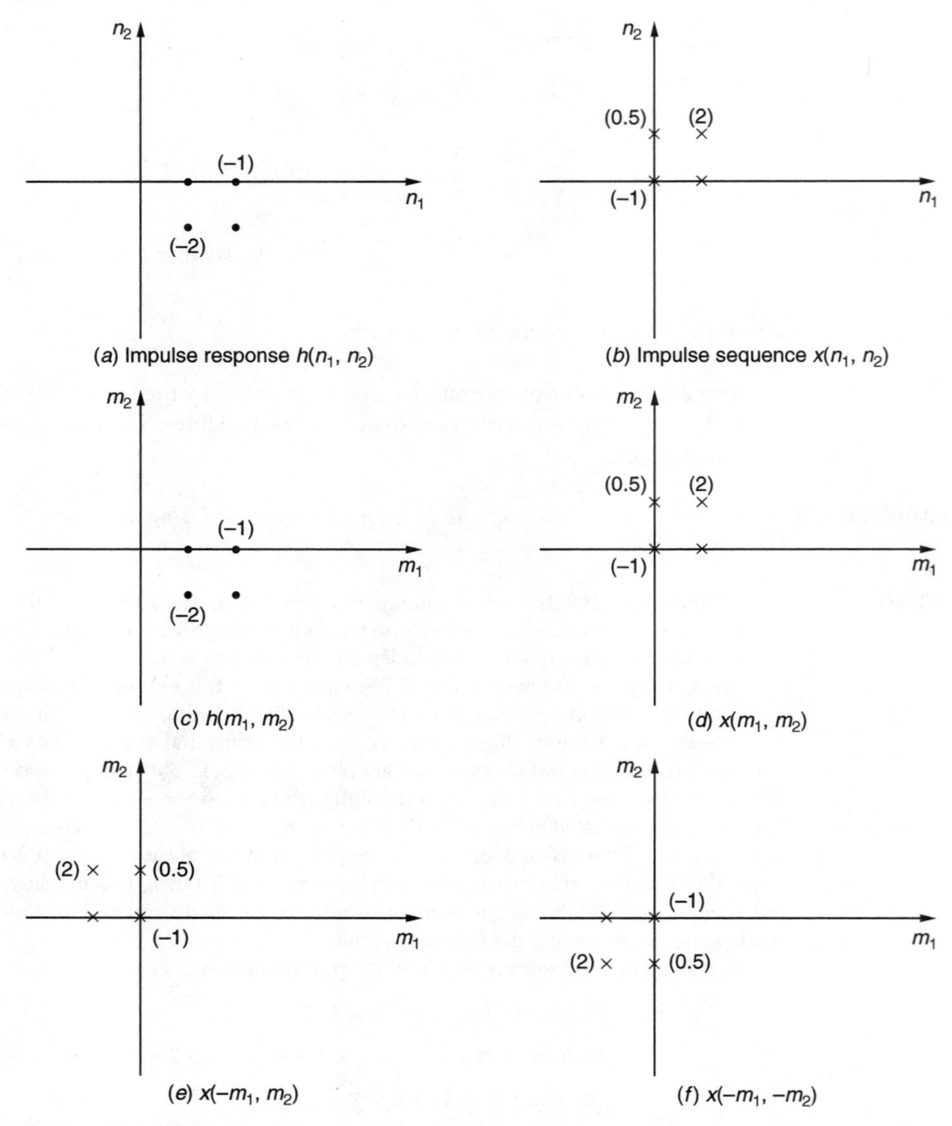

Figure 1.39 (*a*) Impulse response $h(n_1, n_2)$; (*b*) Input sequence $x(n_1, n_2)$; (*c*) $h(m_1, m_2)$; (*d*) $x(m_1, m_2)$; (*e*) $x(-m_1, m_2)$; (*f*) $x(-m_1, -m_2)$; (*g*) $n_1 = 0,\ n_2 = 1$; (*h*) $n_1 = 0,\ n_2 = -1$; (*i*) $n_1 = 0,\ n_2 = 1$; (*j*) Output $y(n_1, n_2)$

The final result of the convolution, $y(n_1, n_2)$ is plotted in Fig. 1.39(*j*). ◀

If one of the two functions is separable, then the 2-D convolution can be computed by a series of 1-D convolutions. This method saves a significant amount of computations over performing a straight 2-D convolution as in the above example. We will postpone a detailed analysis of this matter to Chapter 2.

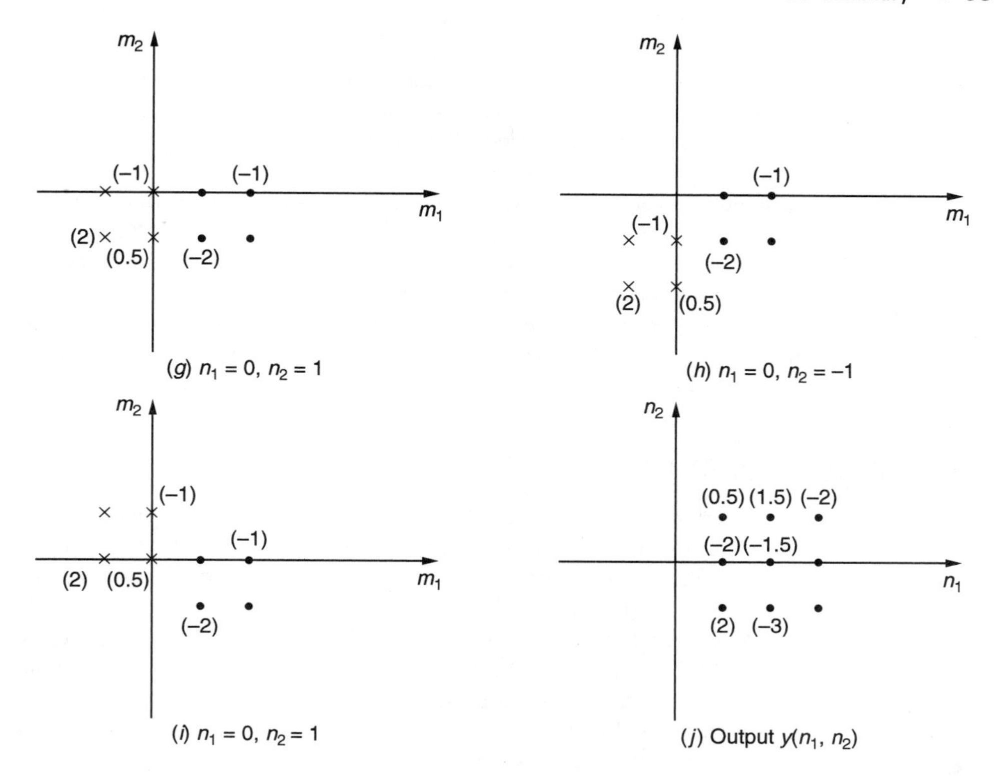

Figure 1.39 Continued

1.9 SUMMARY

This chapter began with an overview of signal processing and discussed the different types of signals. Then an overview of a typical DSP system was described in detail. The reader is encouraged to implement this system in MATLAB and to replicate the results. The subject of image processing was then introduced, and basic 2-D signal representations were presented. Simple 2-D mean and median filters were introduced for noise reduction in images. MATLAB functions are available for implementing mean and median filters. One of the computer projects at the end of this chapter deals with these functions.

The next three sections covered the fundamentals from 1-D signals and systems. Basic signals such as the impulse, step, and exponential were defined and illustrated with examples. The important relationship between the frequency of a continuous-time signal and that of its sampled counterpart was established. Linearity, shift-invariance, causality, and stability of systems were then defined and discussed with examples. LSI systems were then characterized in terms of difference equation and impulse response. It was then proved that the output of an LSI system is the convolution of its impulse response and the input. Two different ways of computing the convolution, namely, the graphical method and the vector method, were described with examples. We then proved that the necessary and sufficient condition for the stability of an LSI system is the absolute summability of its impulse response. All of the above concepts of linear system theory are very important to an in-depth understanding of DSP. The reader is therefore urged to get a solid grasp of this material. Several good books on linear systems can be used for further reading. These include texts by Chen [6], Ziemer/Tranter/Fannin [7], Lathi [8], Oppenheim/Willsky/Nawab [9], Cadzow [10], Soliman/Srinath [11], McGillem/Cooper [12], Sinha [13], and Gabel/Roberts [14].

Implementation of FIR and IIR filters was also discussed in this chapter. Only the basic Direct Form structures were presented. A computer project is given at the end of the chapter that deals with filter implementation. A comprehensive treatment of filter structures is given in Chapter 4. The basic concepts of random processes were also discussed briefly. For further reading on random processes, the reader is referred to some excellent books such as by Papoulis [15], Gardner [16], Leon-Garcia [17], Balakrishnan [18], Childers [19], and Gray [20]. Finally, the concepts of 1-D signals and systems were extended to 2-D in the last section. For further reading on 2-D system theory, the reader is referred to the texts by Lim [5], Mitra/Ekstrom [22], Dudgeon/Mersereau [23], and Huang [24].

1.10 MATLAB FUNCTIONS

This section lists some MATLAB functions relevant to this chapter. These functions may be useful for the computer projects given at the end of the chapter. A brief description is given for each function. The commands and variables are boldfaced and presented in the context of their usage. Further details about these functions can be obtained from the MATLAB help menu. At the end of some subsequent chapters, we will introduce more MATLAB functions and thereby build on this section. We assume that, in addition to basic MATLAB, the student has access to the following MATLAB toolboxes: (i) Signal Processing, (ii) Image Processing, and (iii) Control Systems. Students who are totally unfamiliar with MATLAB are referred to the texts by Etter [25], Math Works, Inc. [26], Buck/Daniel/Singer [27], and MacClellan/Schafer/Yoder [28].

MATLAB Help

help size; The **help** command describes the command following it. In this case, it will provide information on the **size** command. If the **help** command is not followed by another command, then it lists all the MATLAB commands.

Data Entry and Addressing

f = **input**('**enter the frequency**'); The phrase within quotes is displayed on the screen. The user enters a number that is assigned to **f**. The semicolon (;) denotes the end of the statement. If it is omitted, the program displays: "f = value entered."

x = **[a:del:b];** This is a way to generate a row vector. It samples the interval from **a** to **b** in steps of **del**, and generates the vector **[a a+del a+2del ... b]**. The value of **del** may be positive or negative, and integer or noninteger.

y = **x**'; The quote (') denotes the transpose operation. For example, if **x** is a row vector, then **y** is a column vector.

A = **[0 2 1; 3 7 2];** This generates a 2 × 3 matrix. The rows are separated by;

y = **A(:,i);** The colon represents a "wild card" operation. In this case, it means "all rows." Therefore, **y** becomes a column vector consisting of **i**th column of **A**. On the other hand, the command **x = A(i,:)** denotes "all columns" and assigns the **i**th row to **x**.

B = size(A); This command is used to determine the size (M × N) of a matrix. **B** becomes a 2 × 1 row vector as **B** = [M N].

L = length(x); This command provides the length of a row or column vector to the scalar **L**.

Matrix Operations

The following commands are described for matrices. However, they are valid for vectors or scalars since these are special cases of a matrix.

clear; Clear all data from the workspace. This is usually used as the first line of a MATLAB program.

clear A; Clear matrix A from the workspace. This command, used alone, clears all variables from the workspace.

who; This command displays all the matrices that are defined in the MATLAB workspace.

A ∗ B; This is the product of two matrices (or scalars). Dimensions must be compatible for matrix multiplication.

inv(A); computes the matrix inverse of **A**.

A. * B; This is an element-by-element multiplication of the matrices **A** and **B**. The command **./** would perform an elementwise division of **A** by **B**.

A + B; adds matrices of compatible dimensions. Same for subtraction.

abs(A); This command finds the absolute value of each element of a possibly complex matrix.

angle(A); This command finds the phase angle of each element of the complex matrix.

sum (X); For vectors, the result is the sum of the elements of **X**. For matrices, this command generates a row vector, with each element being the sum over each column. Therefore, for a matrix, **sum(sum(X))** adds all the elements of the matrix.

x = sin(2 ∗ pi ∗ f/f$_s$ ∗ n); This generates a sinusoid of frequency f and sampling frequency f_s where **n** = $[0\ 1\ 2 \ldots N]$.

Plotting Functions

plot (t,y); This command plots the vector **y** (row or column) versus the vector **t**. The vectors must agree in dimension. The command **plot(s1,t1,s2,t2,s3,t3)** plots the vectors **s1, s2**, and **s3** against **t1, t2**, and **t3**, respectively, on the same graph.

xlabel('samples'); inserts the phrase within quotes as the x-axis label of the plot.

ylabel('amplitude'); inserts the phrase within quotes as the y-axis label.

title('Sinusoidal signal'); inserts a title on the top of the plot.

Filter Commands

Some of the following commands will be with respect to the difference equation representation of a filter given by

$$\begin{aligned} y(n) + a_1 y(n-1) + a_2 y(n-2) + \cdots + a_N y(n-M) \\ = b_0 x(n) + b_1 x(n-1) + \cdots + b_M x(n-M). \end{aligned} \tag{1.103}$$

The nonrecursive coefficients are entered as the row $\mathbf{b} = [b_0\, b_1 \ldots b_M]$, and the recursive coefficients are entered as $\mathbf{a} = [1\, a_1 \ldots a_N]$.

y = filter(b,a,x); This command filters the input signal in vector **x** by the filter represented by vectors **b** and **a**.

h = impz(b,a,N); This command computes **N** samples of the impulse response of a filter described by **b** and **a**.

[H,f] = freqz(b,a,N,Fs); Given a sampling frequency **Fs** in Hz, this command computes the N-point frequency vector **f** in Hz and the N-point complex frequency response vector **H** of the filter described by (**b,a**). This command is therefore useful for plotting the magnitude and phase response of the filter. If the left-hand side [**H,f**] is omitted, then the magnitude response (dB) and the phase response are automatically plotted.

v = randn(size(x)); This command generates a normally distributed (Gaussian) random array of the same size as that of **x**. This is useful for generating white Gaussian noise.

y = conv(h,x); This command convolves the vectors **h** and **x**, and the result is stored in **y**.

xcorr; Cross-correlation function estimates.

C = xcorr(A,B); where A and B are length M vectors (M > 1), returns the length 2*M − 1 cross-correlation sequence C. If A and B are of different length, the shortest one is zero-padded. C will be a row vector if A is a row vector, and a column vector if A is a column vector.

xcorr(A); when A is a vector, is the auto-correlation sequence.

xcorr(...,SCALEOPT); normalizes the correlation according to SCALEOPT:

'biased'—scales the raw cross-correlation by 1/M.
'unbiased'—scales the raw correlation by 1/(M-abs(lags)).
'coeff'—normalizes the sequence so that the autocorrelations at zero lag are identically 1.0.
'none'—no scaling (this is the default).

xcov; Cross-covariance function estimates.

xcov(A,B); where A and B are length M vectors, returns the length 2*M − 1 cross-covariance sequence in a column vector. The rest of the usage is similar to the **xcorr** function.

Image Processing Commands

load clown; This command retrieves the image **clown.mat** from the hard disk to the MATLAB workspace. The matrix is stored in variable **X**. This image is one of many that are available in the MATLAB Image Processing toolbox.

colormap('gray'); sets the current image colormap to gray level; that is, it assumes a black and white image.

image(X); displays matrix **X** as an image.

imagesc(...); the same as **image**(...) except that the data are scaled to use the full colormap.

imread; Read image from graphics file.

A = imread(FILENAME, *FMT*); reads the image in FILENAME into A. If the file contains a grayscale intensity image, A is a two-dimensional array. If the file contains a truecolor (RGB) image, A is a three-dimensional (M-by-N-by-3) array. FILENAME is a string that specifies the name of the graphics file, and FMT is a string that specifies the format of the file. The file must be in the current directory or in a directory on the MATLAB path. If **imread** cannot find a file named FILENAME, it looks for a file named FILENAME.FMT. The possible values for FMT include:

'jpg' or 'jpeg' Joint Photographic Experts Group (JPEG)
'tif' or 'tiff' Tagged Image File Format (TIFF)
'gif' Graphics Interchange Format (GIF)
'bmp' Windows Bitmap (BMP)
'png' Portable Network Graphics
'hdf' Hierarchical Data Format (HDF)
'pcx' Windows Paintbrush (PCX)
'xwd' X Window Dump (XWD)
'cur' Windows Cursor resources (CUR)
'ico' Windows Icon resources (ICO)

Y = medfilt2(X,[3;3]; performs median filtering of the image **X** over a 3×3 window. The matrix **Y** is assigned the filtered image.

Y = conv2(H,X); performs the 2-D convolution of matrices **H** and **X**.

xcorr2(A,B); computes the cross-correlation of matrices A and B.

xcorr2(A); the autocorrelation function.

Loops and Conditions

Command structure for the **for** loop:

for i = p : Δ : q

statements

end;

Command structure for the **if** statement:

if condition

statements;
else
statements;
end

1.11 PROBLEMS

1. Classify the following signals as (1) continuous-time signal; (2) discrete-time signal or sequence; and (3) digital signal or sequence. If not classifiable, make the necessary specifications.

(a) Music information read by a computer from a CD-ROM.
(b) Temperature in the room measured by a thermometer.
(c) Signal recorded by a seismogram.
(d) Your exact weight every day at 8:00 A.M.

2. Plot the following signals or sequences, with sampling time $T_s = 0.25$ seconds.

(a) The continuous-time signal

$$x(t) = \sin\left(2\pi t + \frac{\pi}{4}\right), \quad -4 \le t \le 4.$$

(b) The discrete-time signal

$$x_s(nT_s) = x(t)|_{t=nT_s}, \quad -4 \le t \le 4.$$

(c) The discrete-time sequence

$$x(n) = x_s(nT_s), \quad -16 \le n \le 16.$$

(d) The digital sequence

$$Q\{x(n)\},$$

with $B = 3$ bits magnitude-truncation quantizer.

3. What is the difference between discrete-time signal and discrete-time sequence? What is the difference between analog signal and digital sequence?

4. Assume that the variances of a clean signal $s(n)$ and an independent noise source $v(n)$ are $\sigma_s^2 = 0.5$ and $\sigma_v^2 = 1$, respectively. The noisy signal $x(n)$ is generated by

$$x(n) = s(n) + Kv(n),$$

where K is a constant. Find the constant K, such that the signal to noise ratio (SNR) of $x(n)$ is 20dB. (*Hint*: $\sigma_x^2 = \sigma_s^2 + K^2\sigma_v^2$.)

5. Let $s(n) = \sin(0.25\pi n)$, and $v(n)$ be an independent Gaussian noise with zero mean and unit variance. The input and output of a system are given respectively by

$$x(n) = s(n) + 0.5v(n);$$

$$y(n) = \frac{1}{2}(x(n) + x(n-8)).$$

Find the signal to noise ratio (SNR) of $x(n)$ and $y(n)$ in dB. What is the signal to noise ratio improvement (SNRI) between $y(n)$ and $x(n)$? Explain your answer. (*Hint*: $\sigma_s^2 = 0.5$.

var$\{a+b\}$ = var$\{a\}$+var$\{b\}$ holds for deterministic signals. It also holds for random processes a and b when they are independent.)

6. Evaluate the following functions:

(a) $x_1(t) = \Pi(t-2)\delta(t)$.

(b) $x_2(t) = u(t-1)u(5-t)\delta(t-3)$.

(c) $x_3(t) = \int_{-\infty}^{\infty} e^{-3t}\delta(t-2)dt$.

(d) $x_3(t) = \int_{-\infty}^{\infty} e^{-2t}\delta'(t-2)dt$.

(e) $x_5(t) = \sin(2\pi t)\delta\left(2t - \frac{1}{4}\right)$

7. A discrete-time sequence $x(n)$ is defined as $x(n) = \left(\frac{1}{2}\right)^n u(n)$.

(a) What is the value of $x(n)\delta(n-5)$? Sketch it.

(b) What is the result of $\sum_{k=0}^{\infty} x(n)\delta(n-k)$? Sketch it.

8. Plot the following signals or sequences:

(a) $\Pi(2t) + \int_{-1}^{t} \Pi(4t-2)dt$.

(b) $\int_{-\infty}^{\infty} \delta(t-2)u(t)dt$.

(c) $a^n(u(n+2) - u(n-2))$.

(d) $3\sin(0.15\pi n + \frac{\pi}{6})u(-n)$.

9. Determine the even and odd parts of the following sequences:

(a) $x(n) = [\mathbf{1} \quad -6 \quad 3 \quad 4 \quad 9 \quad 2]$.

(b) $x(n) = [\mathbf{5} \quad -2 \quad -3 \quad 1 \quad 0 \quad 4 \quad 3]$.

10. Determine each of the following signals:

(a) $f(t) = e^{-3t-1}\delta(t)$.

(b) $\int_{-\infty}^{t} e^{-\tau}\delta'(\tau)d\tau$.

(c) $f(t) = \int_{-1}^{1} \delta(t^2-4)dt$.

(d) $f(t) = \int_{-\infty}^{\infty} \delta(t^2-4)dt$.

11. Determine the even and odd parts of the following real signal:

(a) $x(t) = A\cos(\omega_0 t) + B\sin(\omega_0 t)$.

(b) $x(t) = e^{-t} + t^2$.

12. Prove Properties 5 and 6 of the continuous-time impulse function.

13. A discrete-time sequence $x(n)$ is given in Fig. P1.1. Express this sequence in terms of: (a) impulse functions only; (b) step functions only; (c) a combination of impulse and step functions. Is the solution in each case unique? If not, give another solution.

14. If $x(n)$ and $y(n)$ are periodic sequences, is the sum $z(n) = x(n) + y(n)$ always periodic? If so, what is the period of this sequence?

15. Determine the frequency of the following discrete-time sequences:

(a) $\sin\left(\frac{\pi}{2}n + \frac{\pi}{4}\right)$.

(b) $\cos\left(\frac{3\pi}{7}n\right) + \sin\left(\frac{\pi}{4}n + \frac{\pi}{6}\right)$.

16. Is any discrete-time signal uniformly sampled from a periodic continuous-time signal periodic? If not, give an example. Otherwise, prove it.

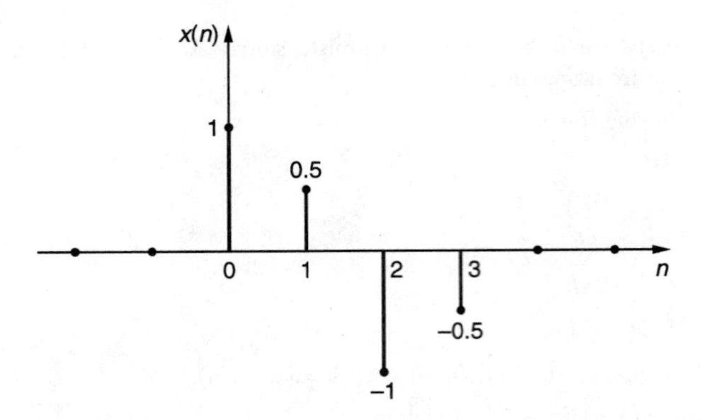

Figure P1.1

17. Determine which of the following discrete-time sinusoids is periodic. If periodic, find its fundamental period.

(a) $\cos\left(\frac{3}{7}\pi n + 0.45\right)$.

(b) $\sin(0.7n)$.

(c) $\sin(0.4\pi n) + \cos(0.4\pi n)$.

(d) $3\sin\left(0.5n + \frac{\pi}{2}\right)\cos\left(\frac{13}{15}\pi n\right)$.

18. If $x(n) = \cos\left(\frac{2\pi}{3}n\right) + 2\sin\left(\frac{16\pi}{3}n\right)$, determine whether $x(n)$ is periodic. If periodic, find its fundamental period.

19. Determine the fundamental period of the $x(n) = \cos(an)$, where a has the following values:

(a) 0.25π, (b) 0.34π (c) 0.75π.

20. The following sequences represent one period of cosine sequence of the form $x(n) = A\cos(\omega_0 n + \phi)$.

(a) $[0\ 2\ 0\ -2]$.

(b) $[1\ 1\ -1\ -1]$.

(c) $[2\ \sqrt{2}\ 0\ -\sqrt{2}\ -2\ -\sqrt{2}\ 0\ \sqrt{2}]$.

Determine the values of the parameters: A, ω_0, ϕ for each case.

21. A discrete-time periodic sequence has 20 samples per cycle, and the sampling frequency $F_s = 100$ Hz.

(a) What is the frequency f of the discrete-time sequence?

(b) What is the frequency F of the continuous-time signal from which the discrete-time sequence was obtained?

22. For a continuous-time sinusoid signal $x(t) = \cos\left(7.5\pi t + \frac{\pi}{3}\right)$, what value of sampling frequency F_s will result in a discrete-time sequence with a frequency $f = 0.3$ cycles/sample.

23. What is the Nyquist frequency of $x(t) = \sin(21\pi t) + \cos(16\pi t)$? Suppose that we sample this signal with $F_s = 50$ Hz. Can we uniquely reconstruct this signal? If so, give a reconstruction block diagram.

24. What is the Nyquist frequency of $x(t) = 0.5^t(\sin(21\pi t) + \cos(16\pi t))$? Can we uniquely reconstruct this signal?

25. For each of the following systems, determine whether or not the system is (1) linear; (2) shift-invariant; (3) causal; and (4) BIBO stable.

(a) $y(n) = \Re\{x(n)\} = 2x(n-1) - x(n) + 2x(n+1)$;

(b) $y(n) = \Re\{x(n)\} = 0.5^n x(n)$;

(c) $y(n) = \Re\{x(n)\} = x(n) + n$;

(d) $y(n) = \Re\{x(n)\} = y(n-1) + x(n)$; (*Hint*: Use recursion to find $y(n)$ when the input is $x(n) = \delta(n-m)$. If the output is a function of $(n-m)$, then the system is shift-invariant; otherwise not.)

(e) $y(n) = \Re\{x(n)\} = \sin(x(n) - x(n-1))$.

26. For each of the following systems, determine whether or not the system is (1) linear; (2) time-invariant; (3) causal; and (4) BIBO stable.

(a) $y(t) = \Re\{x(t)\} = \int_{-\infty}^{3t} x(\tau)d\tau$;

(b) $y(t) = \Re\{x(t)\} = \begin{cases} 0 & t < 0 \\ x(t) + x(t-1) & t \geq 0 \end{cases}$

(c) $y(t) = \Re\{x(t)\} = \begin{cases} 0 & x(t) < 0 \\ x(t) + x(t-1) & x(t) \geq 0 \end{cases}$

(d) $y(n) = \Re\{x(n)\} = x(n/2)$;

(e) $y(n) = \Re\{x(n)\} = x(n-1) - x(1-n)$;

27. Consider the system $y(n) = \min\{x(n-1), x(n), x(n+1)\}$; determine whether the system is (1) linear; (2) shift-invariant; (3) causal; and (4) BIBO stable.

28. For $w(0) = 0$ and $x(-1) = 0$, consider an adaptive filter

$$y(n) = w(n)x(n-1),$$

where $x(n)$ is the input, $y(n)$ is the output, and $w(n)$ is the filter weight. Let ε be a small positive number; the weight update equation is given by

$$e(n) = x(n) - w(n)x(n-1);$$
$$w(n+1) = w(n) + e(n)x(n)/(\varepsilon + x^2(n-1)).$$

Determine whether the system is (1) linear; (2) shift-invariant; and (3) causal.

29. An IIR system is characterized by a difference equation

$$y(n) = 0.5y(n-1) + x(n)$$

for $x(n) = u(n+1) - u(n-1)$ and $y(0) = 1$, compute $y(n)$ recursively.

30. An FIR system is characterized by a difference equation

$$y(n) = 0.5x(n-2) - x(n-1) + 0.5x(n).$$

(a) Find the impulse response $h(n)$.

(b) Using (a) and the linear and shift-invariant property of the system, find the step response of the system (without using the convolution formula).

31. An LSI system has impulse response

$$h(n) = \left(\frac{2}{3}\right)^n (u(n) - u(n-5)).$$

For the input $x(n) = \delta(n) - 2\delta(n-3)$, find the system response $y(n)$.

32. A sequence is defined by: $f(n) = 5f(n-1) - 6f(n-2), n \geq 2$ with $f(0) = 0, f(1) = -1$. Develop an exact formula to calculate $f(n)$ directly for any n. (*Hint*: Assume that the solution is $f(n) = a\lambda_1^n + b\lambda_2^n$. First find λ_1 and λ_2, which are the roots of the characteristic solution. Then substitute the initial conditions to find a and b.)

33. The impulse response of a continuous-time LTIV system is given by $h(t)$. Prove in general that the output of this system in response to any input $x(t)$ is given by $h(t) * x(t)$.

34. Without performing convolution, compute and plot the response of an LSI system whose impulse response $h(n)$ and input sequence $x(n)$ are given in Fig. P1.2.

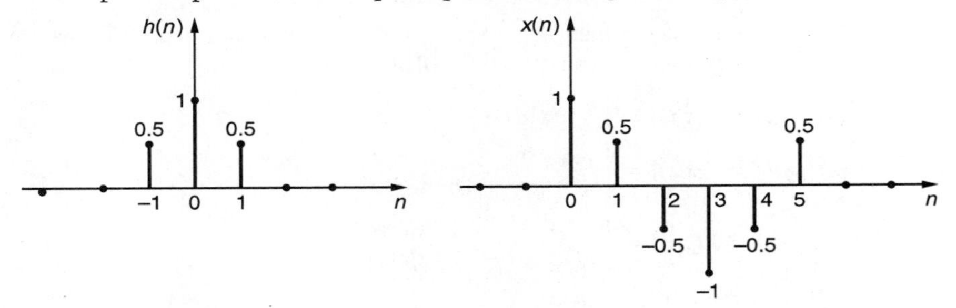

Figure P1.2

35. Compute the convolution of the following functions: $x(t) = u(t-2) - u(t-6), h(t) = e^{-2t}$.

36. Determine and sketch the convolution of the following two signals:

$$x(t) = \begin{cases} t+1, & 0 \leq t \leq 1 \\ 2-t, & 1 < t \leq 2 \\ 0, & \text{otherwise} \end{cases}$$

$$h(t) = \delta(t+1) + 3\delta(t+2).$$

37. An LSI system has impulse response

$$h(i) = \begin{cases} 2 & i = 0; \\ -1 & i = 1; \\ 2 & i = 2. \end{cases}$$

Determine the step response of the system by convolution.

38. Compute and plot the convolution for the sequences given in Fig. P1.3.

39. Compute $y(n) = h(n) * x(n)$ for $h(n) = a_1^n u(n)$ and $x(n) = a_2^n u(-n)$, where $0 < a_1 < 1$ and $-1 < a_2 < 0$.

40. Compute the convolution of the following pairs of signals:

(a) $x(n) = \left(-\frac{1}{2}\right)^n u(n-3), h(n) = 2^n u(3-n)$;

(b) $x(n) = h(n) = \alpha^n u(n)$.

41. Using the vector method, compute the convolution of the following finite-length sequences. Recall that the digit in bold represents the value at $n = 0$.

(a) $x(n) = [-1\ 2\ {-3}\ \mathbf{2}\ {-1}], \quad h(n) = [-0.5\ \mathbf{1}\ 1.5]$.

(b) $h(n) = [-5\ \mathbf{0.25}\ {-1}\ 0\ 2\ 7], \quad x(n) = [\mathbf{1}\ {-1.5}\ 1\ 2]$.

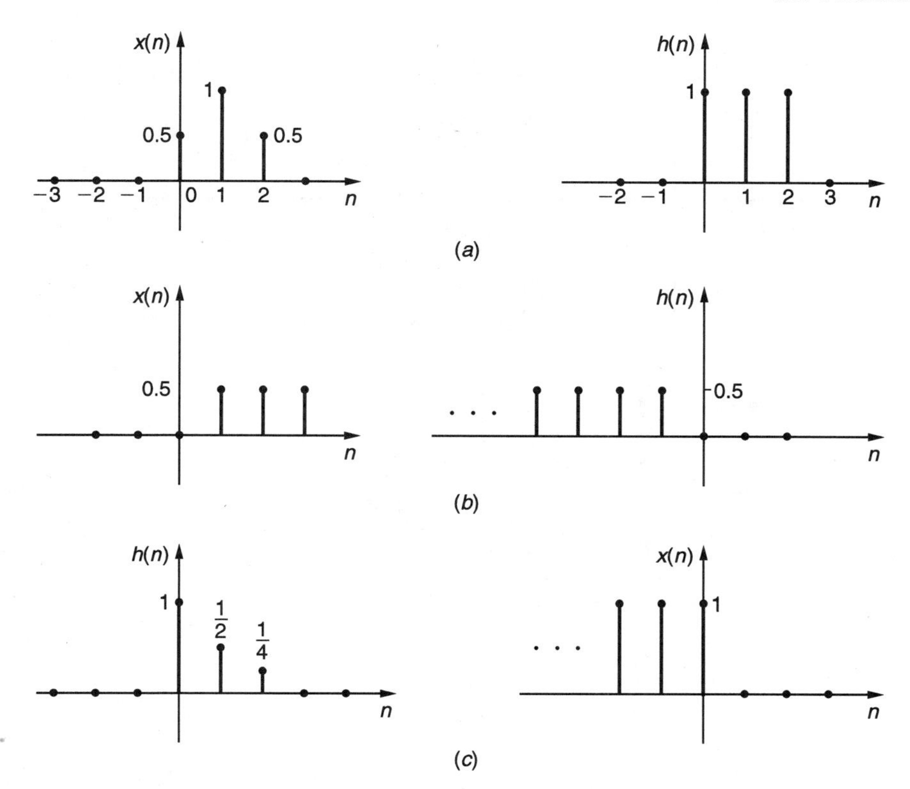

Figure P1.3

42. Prove that equation (1.40) and equation (1.41) are equivalent.

43. Prove Theorem 1.4.

44. Prove Theorem 1.5.

45. Consider the system given in Fig. P1.4. Determine the output $y(n)$ in terms of $h_i(n)$ and $x(n)$, where $i = 1, 2, \ldots 5$.

46. Determine whether or not the following sequences are bounded:

(a) $a(n) = A\beta^n$, where $|\beta| < 1$.

(b) $b(n) = A\beta^n u(n)$, where $|\beta| < 1$.

(c) $c(n) = A\beta^n u(-n)$, where $|\beta| > 1$.

(d) $d(n) = A\cos(\omega_0 n)$.

47. Show that the sequence $x(n) = \frac{1}{n}$ is not absolutely summable, for $n = 1, 2 \ldots$

48. Show that the sequence $x(n) = \frac{1}{n^2}$ is absolutely summable, for $n = 1, 2 \ldots$

49. A system is a cascade of two subsystems with impulse responses $h_1(n)$ and $h_2(n)$ as given in Fig. 1.4. Prove that if each impulse response is absolutely summable, then the overall impulse response $h(n)$ is absolutely summable.

50. Two FIR filters are shown in Fig. P1.5. Find their difference equation representations and corresponding impulse responses.

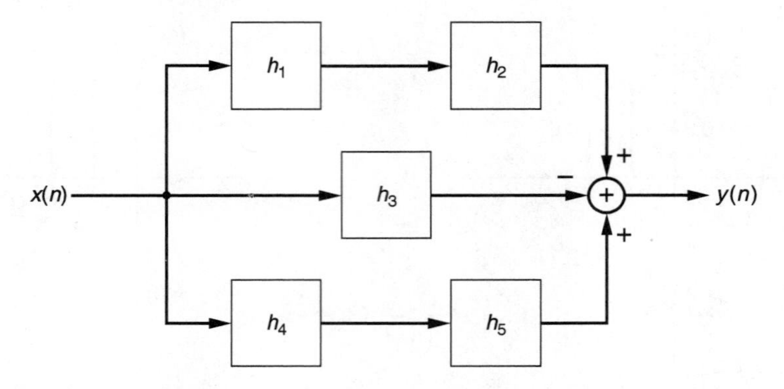

Figure P1.4

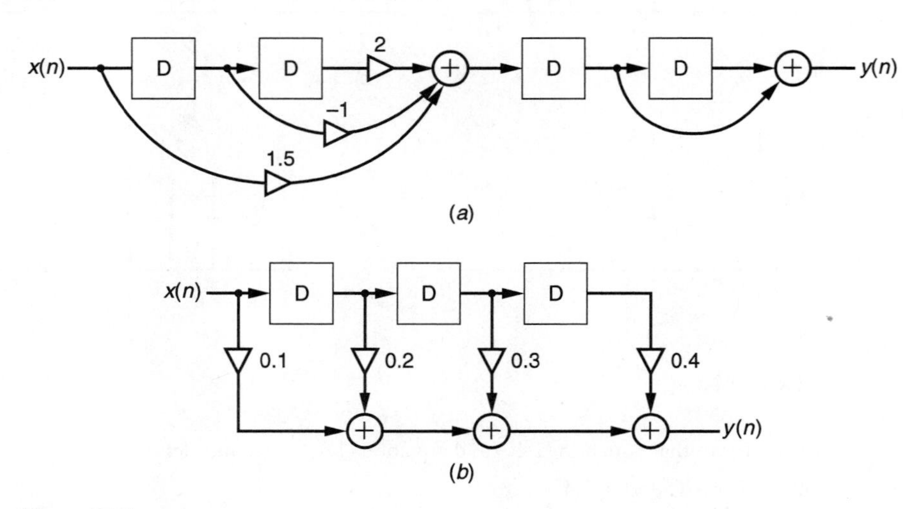

Figure P1.5

51. Consider a system given by

$$y(n) - 0.5y(n-1) + 0.3y(n-2) = x(n-1) - 0.8x(n)$$

where $x(n)$ and $y(n)$ are the input and output, respectively. Draw the flow diagram of the *Direct Form* I_T and *Direct Form* II_T structures.

52. Given the system

$$y(n) - 0.5y(n-1) + 0.3y(n-2) = x(n-1) - 0.8x(n)$$

where $x(n)$ and $y(n)$ are the input and output, respectively. Draw the flow diagram of the *Direct Form I* and *Direct Form II* structures.

53. Given the system

$$y(n) + 0.5y(n-1) + 0.3y(n-2) + 0.1y(n-3) = x(n) + 2x(n-1) + 3x(n-3)$$

where $x(n)$ and $y(n)$ are the input and output, respectively. Draw the flow diagram of the *Direct Form II* and *Direct Form* II_T structures.

54. Given the filter structures of Fig. P1.6, find the difference equations.

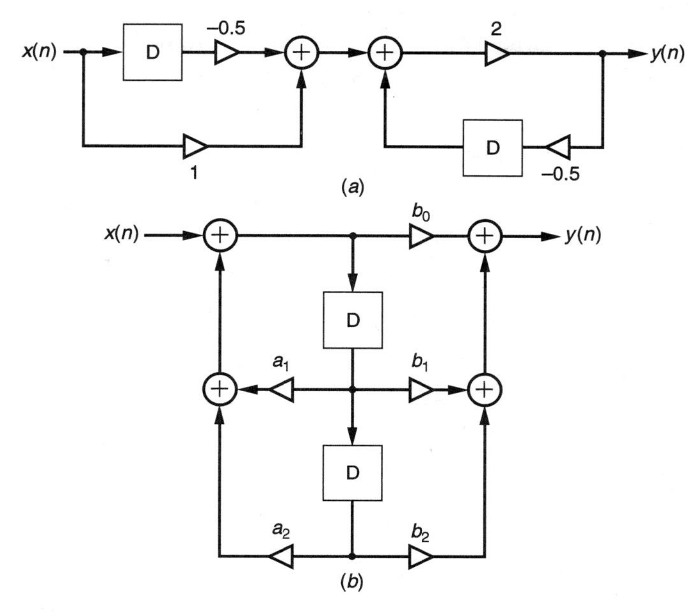

Figure P1.6

55. Prove Fact 2.

56. Find the unscaled time-autocorrelation of the sequence, $x(n) = [1 \ -2 \ 3 \ 5]$.

57. Show that the unscaled time-autocorrelation of the sequence $x(n) = \alpha^n u(n), 0 < \alpha < 1$ is

$$\tilde{r}_{xx}(k) = \frac{\alpha^{|k|}}{1 - \alpha^2}.$$

58. A received signal is obtained as the sum of the main signal and a reflected component that is delayed and possibly scaled. This phenomenon is called multipath. The received signal is given by

$$x(n) = s(n) + as(n - M).$$

(a) Find the unscaled time-autocorrelation $\tilde{r}_{xx}(k)$ of the received signal.

(b) Is it possible to obtain a and M from the $\tilde{r}_{xx}(k)$? Why?

59. Let $X(t)$ be a random process defined by $X(t) = At + B$, where A and B are independent random variables, each uniformly distributed on $[-1, 1]$. Find the mean and autocorrelation of $X(t)$.

60. Given the process $X(t) = X$, where X is a random variable uniformly distributed in $(-1, 1)$. Find the mean and autocorrelation of this random process.

61. Given the signal

$$Y(t) = \cos(2\pi f_0 t) \sin(X)$$

where X is a uniformly distributed random variable in the range $(-\pi, \pi)$.

(a) Find the mean of $Y(t)$.

(b) Find the autocorrelation of $Y(t)$.

(c) Is this process wide-sense stationary, cyclostationary, or neither?

62. For a wide-sense stationary random process $X(t)$, prove that $R_X(\tau) \leq R_X(0)$. *Hint*: Consider the inequality $E(\{X(t+\tau) \pm X(t)\}^2) \geq 0$.

63. A signal is modeled as a lowpass stationary process $X(t)$ whose probability density function at any time is given in Fig. P1.7. Find the mean and variance of this process.

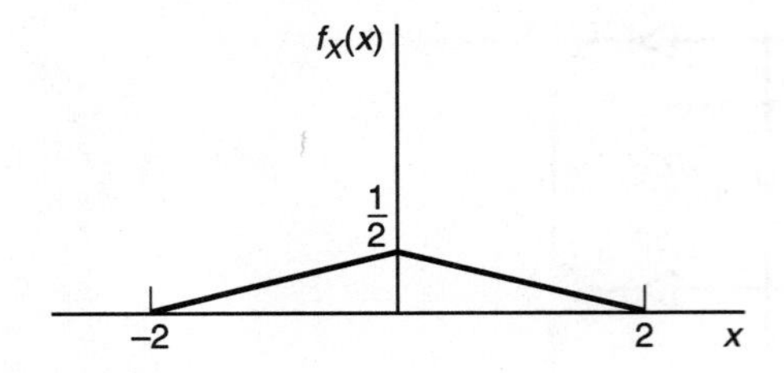

Figure P1.7

64. Sketch the following 2-D sequences:

(a) $u(n_1, n_2) - u(n_1 - 2, n_2 - 2)$.

(b) $\delta_{2d}(n_1 + n_2)u(n_1, -n_2)$.

(c) $a^{u(n_1,n_2)}u_{2d}(n_1 + n_2)$.

(d) $\left(\frac{1}{2}\right)^{n_1}\left(\frac{1}{3}\right)^{n_2} u(n_1, n_2)$.

(e) $u_{2d}(2n_1 + n_2 - 1)u(n_1, -n_2)$.

65. For each of the following 2-D systems, determine whether or not the system is (1) linear; (2) shift-invariant; (3) causal; and (4) BIBO stable.

(a) $y(n_1, n_2) = \Re\{x(n_1, n_2)\} = \frac{1}{5}\sum_{i=n_1-4}^{n_1} x(i, n_2)$;

(b) $y(n_1, n_2) = \Re\{x(n_1, n_2)\} = a^{n_1+n_2}x(n_1, n_2)$;

(c) $y(n_1, n_2) = \Re\{x(n_1, n_2)\} = x(n_1, n_2)u_{2d}(n_1 + n_2)$;

(d) $y(n_1, n_2) = \Re\{x(n_1, n_2)\} = 0.5y(n_1 - 1, n_2) + x(n_1, n_2)$. Assume zero initial conditions.

(e) $y(n_1, n_2) = \Re\{x(n_1, n_2)\} = x(n_1, n_2) + \delta(n_1, n_2)$.

66. Prove that a 2-D LSI system is BIBO stable if and only if the impulse response $h(n_1, n_2)$ is absolutely summable, that is,

$$\sum_{n_1=-\infty}^{\infty}\sum_{n_2=-\infty}^{\infty} |h(n_1, n_2)| < \infty.$$

67. A 2-D LSI system has impulse response

$$h(n_1, n_2) = \left(\frac{1}{3}\right)^{n_1}\left(\frac{1}{2}\right)^{n_2} u(n_1, n_2).$$

Find the system response $y(n_1,n_2)$ for the input $x(n_1,n_2) = \delta(n_1,n_2) - 2\delta(n_1-2,n_2) + 2\delta(n_1,n_2-1)$.

68. Consider an LSI system. When the input is the unit step sequence $u(n_1,n_2)$, the response of the system is $s(n_1,n_2)$.

Express $y(n_1,n_2)$, the output of the response of the system, in terms of the input $x(n_1,n_2)$ and the unit step response $s(n_1,n_2)$.

69. For each of the following cases, determine the region of support of $y(n_1,n_2) = x(n_1,n_2) * h(n_1,n_2)$

(a) $x(n_1,n_2)$ and $h(n_1,n_2)$ are both a fourth-quadrant support sequence.

(b) $x(n_1,n_2)$ is a first-quadrant support sequence, and $h(n_1,n_2)$ is a second-quadrant support sequence.

(c) $x(n_1,n_2)$ is a third-quadrant support sequence, and $h(n_1,n_2)$ is a first-quadrant support sequence.

70. Compute $x(n_1,n_2) * h(n_1,n_2)$ for each of the following sequences:

(a) $x(n_1,n_2) = \delta_{2d}(n_1+n_2)u(n_1,n_2)$,
$h(n_1,n_2) = 1$ for $0 \le n_1 \le 2, 0 \le n_2 \le 2$;

(b) $x(n_1,n_2) = \left(\frac{1}{2}\right)^{n_1}(u(n_1,n_2) - u(n_1,n_2-3))$,
$h(n_1,n_2) = u(n_1,n_2)$.

71. The vector method of convolution in 1-D can be easily extended to the 2-D case. Determine the 2-D convolution of the following finite-length sequences by such a method. As in 1-D convolution, the bold value represents the signal at $n_1 = n_2 = 0$.

(a)

$$x(n_1,n_2) = \begin{bmatrix} -1 & 2 & -1 \\ 2 & \mathbf{-1} & 2 \\ -1 & 2 & -1 \end{bmatrix}, \quad h(n_1,n_2) = \begin{bmatrix} \mathbf{1} & -1 \\ 1 & -1 \\ 1 & -1 \end{bmatrix};$$

(b)

$$h(n_1,n_2) = \begin{bmatrix} -0.5 & 0.25 & -0.125 \\ 1.25 & \mathbf{-1} & 1.25 \\ -0.5 & 0.25 & -0.125 \end{bmatrix}, \quad x(n_1,n_2) = \begin{bmatrix} -1 & 2 & 1 \\ 1 & -2 & \mathbf{-1} \end{bmatrix}.$$

72. Assume that $s(n_1,n_2)$ is a 2-D sequence, $v(n_1,n_2)$ is an independent Gaussian noise sequence, and

$$x(n_1,n_2) = s(n_1,n_2) + v(n_1,n_2)$$

represents the corrupted sequence. Let

$$s(n_1,n_2) = \begin{bmatrix} 10 & 10 & 10 & 10 & 10 \\ 10 & 0 & 0 & 0 & 10 \\ 10 & 0 & 10 & 0 & 10 \\ 10 & 0 & 0 & 0 & 10 \\ 10 & 10 & 10 & 10 & 10 \end{bmatrix},$$

$$v(n_1, n_2) = \begin{bmatrix} -0.4 & 1.2 & -0.2 & 0.1 & 0.3 \\ -1.8 & 1.2 & 0.7 & 1 & -1.3 \\ 0.1 & 0 & -0.5 & 0 & 0.7 \\ 0.3 & 0.3 & 2 & 0 & 1.6 \\ -1.1 & 0.2 & -0.1 & -0.8 & -0.7 \end{bmatrix},$$

(a) Compute the output of a (3×3) mean filter $y_1(n_1, n_2)$.
(b) Compute the output of a (3×3) median filter $y_2(n_1, n_2)$.
(c) Find the signal to noise ratio improvements (SNRI) in (a) and (b). Which filter is more efficient to eliminate Gaussian noise?

73. Suppose the same sequence $s(n_1, n_2)$ as in the last problem is corrupted by impulse noise so that the corrupted sequence is

$$X(n_1, n_2) = \begin{bmatrix} 10 & 10 & 10 & 10 & 10 \\ 10 & 0 & 32 & 0 & 10 \\ 10 & 32 & 10 & 32 & 10 \\ 10 & 32 & 0 & 0 & 10 \\ 10 & 10 & 10 & 10 & 10 \end{bmatrix}.$$

(a) Compute the output of a (3×3) mean filter $y_1(n_1, n_2)$.
(b) Compute the output of a (3×3) median filter $y_2(n_1, n_2)$.
(c) Find the signal to noise ratio improvement SNRI in (a) and (b).
(d) Which filter is more efficient to eliminate the impulse noise?

74. Show that if the input $x(n_1, n_2)$ is periodic with a period of (N_1, N_2), the output of an LSI system $y(n_1, n_2)$ is also periodic.

75. Prove Theorem 1.8.

1.12 COMPUTER EXPERIMENTS

Project 1

Data Generation. Generate the following data using MATLAB. You may store the data in files for later use. You can store them in ASCII (∗.dat), MATLAB (∗.mat) format, or any other format of your choice.

File	*Contents*
rectwin.mat	Unit amplitude for 10 samples.
square.mat	Square-wave with amplitude ± 1 and 20 samples/cycle; total of 3 cycles.
sinusoid.mat	Sinusoid with 40 samples/cycle; total of 2 cycles.
triangle.mat	Triangle pulse with 10 samples/cycle; total of 3 cycles.
hamming.mat	Hamming window; total of 30 points (see Table 4.1 in text).
blackman.mat	Blackman window; total of 40 points (see Table 4.1 in text).
hanning.mat	Hanning window; total of 50 points (see Table 4.1 in text).

Procedure

1. Write a program to convolve the following sequences. Do *not* use the MATLAB command **conv(..)**. Try to make the program as short and efficient as possible. Have your program determine the lengths of the inputs and outputs. Plot the inputs and output. You can plot both inputs and the output for each convolution on one graph.

   (a) rect $*$ triangle
   (b) square $*$ sinusoid
   (c) sinusoid $*$ square
   (d) sinusoid $*$ hamming
   (e) sinsusoid $*$ blackman
   (f) sinusoid $*$ hanning

2. We will now look at the algebraic properties of convolution.

   (a) Verify the *commutative* property of convolution by using the signals square and sinusoid.
   (b) Verify the *associative* property by showing

$$[h(n) * x_1(n)] * x_2(n) = h(n) * [x_1(n) * x_2(n)].$$

   Use the following signals and plot your results.

$x_1(n) =$ square
$x_2(n) =$ blackman
$h(n) =$ triangle

   (c) Write a short program to add two signals together (point by point). Use this program with the convolution program to show that

$$h(n) * [x_1(n) + x_2(n)] = h(n) * x_1(n) + h(n) * x_2(n),$$

   which is the *distributive* property. Use the same signals as above. Plot the signals $x_1(n) + x_2(n), h(n) * x_1(n), h(n) * x_2(n)$, and $h(n) * [x_1(n) + x_2(n)]$.

Questions

1. Find a general formula for determining the length of a sequence resulting from the convolution of an N-point sequence with an N-point sequence.
2. Suppose one of the signals in the convolution was of infinite length (or longer than you have memory in the computer to store it). Describe how you might modify your program to allow you to perform the convolution. (Do not implement it.)

3. An N-point real sequence is convolved with an M-point real sequence. Find the total number of multiplications and additions required.

Project 2

A digital filter is described by

$$y(n) - 2.56y(n-1) + 2.22y(n-2) - 0.65y(n-3) = x(n) + x(n-3)$$

where $x(n)$ is the input and $y(n)$ is the output. Assume all zero initial conditions.

1. Generate the input signal $x(n)$, which is a sinusoid of frequency 500 Hz sampled at 6 kHz.
2. Compute the first four cycles of the output by directly implementing the above difference equation. Plot the input and output on the same graph.
3. Implement the above filter by using the MATLAB "filter" function. Compare your result with (2). Comment on your result.
4. Plot the impulse response of the filter by using the MATLAB **impz** function.
5. The above filter is obviously an IIR filter; that is, the length of its impulse response is infinite. If we truncate this impulse response up to 32 points, then this impulse response can represent an FIR filter that approximates the given IIR filter. Let this FIR filter impulse response be represented by $\widehat{h}(n)$. Find the output signal of the FIR filter by using the MATLAB **conv** function. Compare your result with (2) or (3).
6. Generate a new input signal $x(n)$, which is a summation of two sinusoids with frequency 500 Hz and 1500 Hz sampled at 6 kHz. Repeat (3).
7. The frequency response of a filter can be obtained by using the MATLAB **freqz** function. Plot the magnitude and phase response of the IIR filter. Note that your horizontal axis must be scaled appropriately to represent frequency in Hz.
8. Comment on the results of (2) and (6), based on the frequency response of the filter.
9. Generate the input signal $x(n)$, which is a sinusoid with discrete-time frequency 0.1 cycles/sample, corrupted by a white Gaussian noise of approximately 5 dB.
10. Filter the input and compute the signal to noise ratio improvement (SNRI). Comment on your result.

Project 3

Using an image of your choice perform the following:

1. Corrupt the image with white Gaussian noise so that the SNR of the image is approximately 5 dB. Find the actual SNR of the corrupted image. Display the original and corrupted images.
2. Filter your clean image and your corrupted image with (3×3) and (5×5) mean and median filters. Display the results.

3. Find the SNRI for each case above.
4. Add 20% impulse noise to your original image randomly. Find the SNR of the corrupted image. Display the original and corrupted images. (*Explanation*: In order to add impulse noise, perform the following. Scan the image pixel by pixel. At every pixel, "roll a 5-sided dice, labeled 1 through 5." If it falls on one particular face, say 1, then the value of that pixel is changed to the 2^L, where L is the number of bits for the image. Otherwise, leave the pixel alone. This "rolling of the dice" can be performed by generating a random number in MATLAB. Use the $A = ([\text{M N})]$ command, where the image is $M \times N$. This will generate an $M \times N$ matrix, with each element randomly between 0 and 1. If $A_{ij} < 0.2$, then change the ijth pixel to 2^L.)
5. Repeat steps (2) and (3) above.
6. Discuss the performance of the mean and median filters in the above experiments.
7. Using an appropriate image, illustrate that a median filter tends to preserve the sharp edges of the image.
8. Determine whether a median filter is (a) linear; (b) shift-invariant; (c) causal; or (d) BIBO stable.

BIBLIOGRAPHY

[1] F.H. MITCHELL, Jr. and F.H. MITCHELL, Sr., *Introduction to Electronics Design*, Prentice Hall, Englewood Cliffs, NJ, 1992.
[2] H. NYQUIST, "Certain topics in telegraph transmission theory," *AIEE Trans.*, pp. 617–644, 1928.
[3] C.E. SHANNON, "Communication in the presence of noise," *Proc. IRE-37*, pp. 10–21, 1949.
[4] B. WIDROW and S.D. STEARNS, *Adaptive Signal Processing*, Prentice Hall, Englewood Cliffs, NJ, 1985.
[5] J.S. LIM, *Two-Dimensional Signal and Image Processing*, Prentice Hall, Englewood Cliffs, NJ, 1990.
[6] C.T. CHEN, *System and Signal Analysis*, Saunders College Publishing, 1994.
[7] R.E. ZIEMER, W.H. TRANTER, and D.R. Fannin, *Signals and Systems: Continuous and Discrete*, Prentice Hall, Englewood Cliffs, NJ, 1998.
[8] B.P. LATHI, *Linear Systems and Signals*, Berkley Cambridge Press, 1992.
[9] A.V. OPPENHEIM, A.S. WILLSKY, and S.H. NAWAB, *Signals and Systems*, Prentice Hall, Englewood Cliffs, NJ, 1997.
[10] J.A. CADZOW, *Discrete-Time Systems*, Prentice Hall, Englewood Cliffs, NJ, 1973.
[11] S.S. SOLIMAN and M.D. SRINATH, *Continuous and Discrete Signals and Systems*, Prentice Hall, Englewood Cliffs, NJ, 1998.
[12] C.D. MCGILLEM and G.R. COOPER, *Continuous & Discrete Signal & System Analysis*, Saunders College Publishing, 1991.
[13] N.K. SINHA, *Linear Systems*, Wiley, New York, 1991.
[14] R.A. GABEL and R.A. ROBERTS, *Signals and Linear Systems*, Wiley, New York, 1987.
[15] A. PAPOULIS, *Probability, Random Variables, and Stochastic Processes*, McGraw-Hill, New York, 1991.
[16] W.A. GARDNER, *Introduction to Random Processes*, McGraw-Hill, New York, 1990.
[17] A. LEON-GARCIA, *Probability and Random Processes for Electrical Engineering*, Addison-Wesley, Reading, MA, 1993.
[18] A.V. BALAKRISHNAN, *Introduction to Random Processes in Engineering*, Wiley, New York, 1995.
[19] D.G. CHILDERS, *Probability and Random Processes: Using MATLAB with Applications to Continuous and Discrete Time Systems*, Irwin, 1997.
[20] ROBERT M. GRAY, *Probability, Random Processes, and Ergodic Properties*, Springer-Verlag, New York, 1988.

[21] J.G. PROAKIS and D.G. MANOLAKIS, *Digital Signal Processing: Principles, Algorithms, and Applications*, Prentice-Hall, Upper Saddle River, NJ, 1996.
[22] S.K. MITRA and M.P. EKSTROM, eds., *Two-Dimensional Digital Signal Processing*, Dowden, Hutchinson and Ross, Stroudsburg, PA, 1978.
[23] D.E. DUDGEON and R.M. MERSEREAU, *Multidimensional Digital Signal Processing*, Prentice Hall, Englewood Cliffs, NJ, 1983.
[24] T.S. HUANG, ed., *Two-Dimensional Digital Signal Processing I*, in "Topics in Applied Physics," Vol. 42, Springer-Verlag, Berlin, 1981.
[25] D.M. ETTER, *Engineering Problem Solving with MATLAB*, Prentice Hall, Englewood Cliffs, NJ, 1996.
[26] Math Works, Inc., *Student Edition of MATLAB*, Prentice Hall, Englewood Cliffs, NJ, 1995.
[27] J.R. BUCK, M.M. DANIEL, and A.C. SINGER, *Computer Explorations in Signals and Systems Using MATLAB*, Prentice Hall, Englewood Cliffs, NJ, 1997.
[28] J.H. MCCLELLAN, R.W. SCHAFER, and M.A. YODER, *DSP First*, Prentice Hall, Englewood Cliffs, NJ, 1998.

CHAPTER 2

Fourier Analysis

The indispensability of Fourier analysis in signal processing will be made evident in this chapter. In the last chapter, the fundamental concepts of signal and image processing were presented. Most of the concepts were explained in the time or space (for images) domain. In this chapter, we transition to the frequency domain. Different types of Fourier techniques will be used for analyzing and designing signal processing systems in the frequency domain. We will begin with a brief review of the continuous-time Fourier transform (CTFT) in Section 2.1. The discrete-time Fourier transform (DTFT) and its properties are presented in Section 2.2. This will set the stage for discussions of the sampling theorem, frequency response, and decimation/interpolation, which are given in Sections 2.3–2.4. The frequency variable in the DTFT is continuous-time; therefore, it must be sampled for computation using a digital processor. The frequency sampled version of the DTFT is called the discrete Fourier transform (DFT). In Section 2.5, analysis and properties of the DFT are discussed in detail. The DFT is computationally not very efficient. Algorithms for efficiently computing the DFT are called fast Fourier transform (FFT) methods. Some FFT algorithms are presented in Section 2.6. In Section 2.7, these concepts are generalized to 2-D systems for applications in image processing.

Since this chapter is on Fourier analysis, it is appropriate to begin with some historical notes on the life of the man whose famous theorem made signal processing a reality.

In 1753, Bernoulli postulated that a vibrating string could be expressed as a weighted sum of sinusoids, but he was unable to find the coefficients. The solution of this problem and numerous others had to wait for Fourier. Jean-Baptiste-Joseph Fourier was born in Auxerre, France, on March 21, 1768 [1]. He was orphaned at the age of 8 and raised in a military school run by the Benedictines. This is where he discovered his love of mathematics and studied it day and night. At age 21, Fourier presented his famous work on the solution of numerical equations before the Academy in Paris. This work made him famous among the mathematical community in France. In 1794, Fourier was named the chair of mathematics at the newly formed Ecole Normale. After a short stay there, Fourier took up a position at the renowned Ecole Polytechnique. Fourier become known not only as a great mathematician but also as an outstanding teacher. He enlivened his lectures with historical allusions. In fact, he did a significant amount of research on the history of mathematics. In those days, Fourier was one of the first mathematicians to use applications to explain abstract mathematical concepts.

In 1798, Napoleon decided to invade Egypt in order to "civilize" the country, and he took along several well-known mathematicians and scientists. Among them was Fourier who was later placed in charge of building the first road from Grenoble to Turin. The journey to Egypt started with the famous Battle of the Pyramids, where the Egyptians proved no match for the French. Napoleon proceeded to invade Cairo. Having conquered Cairo, Napoleon offered cultural banquets for the Egyptians. These banquets included concerts and scientific lectures. Needless to say, the Egyptians were not the least bit interested in these banquets offered by their conquerors. In fact, they had a different plan. After one of the banquets, they attacked Napoleon's soldiers, killing 300 of his bravest men. Realizing that his own life was in danger also, Napoleon soon after took secret passage to France, deserting his loyal troops. Fourier was thus left behind in Egypt. Fourier was a bit of a maverick with rather unique ideas. For example, he felt that the desert heat was an ideal environment, and so he lived in overheated rooms while wearing excessively heavy clothing. In 1801, the French decided to bring back the troops and Fourier returned home.

On his return, Fourier was appointed prefect of the Department of Isere at Grenoble. It was here that he wrote his landmark *Theorie analytiquede la chaleur* (The Mathematical Theory of Heat). In this work, he proved that any periodic function $x(t)$ can be represented as a sum of sinusoids with frequency which are integer multiples of the frequency of $x(t)$. There is a set of conditions that ensure the existence of this representation, which are satisfied by most signals encountered in practice. These conditions are called the *Dirichlet*[1] *conditions* and are as follows:

1. $x(t)$ has a finite number of discontinuities within its time-period.
2. $x(t)$ has a finite number of maxima and minima within its time-period.
3. $x(t)$ is absolutely integrable within its period T_0; that is

$$\int_{t_0}^{t_0+T_0} |x(t)|dt < \infty.$$

Fourier's theorem is not only beautiful but also has applications in numerous areas of science and engineering. The famous physicist and engineer, Lord Kelvin, called it "a great mathematical poem." In 1807, Fourier presented this work at the Paris Academy of Science. This revolutionary and heretic idea was hotly debated by Lagrange, a well-known mathematician at the time.

The last few years of Fourier's life were spent in Paris where he was the permanent secretary of the Academy. He died of heart disease at age 59.

2.1 CONTINUOUS-TIME FOURIER ANALYSIS

In this section we briefly review the Fourier series and Fourier transform of continuous-time systems. Only the main equations and properties will be presented without proof.

[1] In honor of German mathematician Peter Gustaw Lejeune Dirichlet (1805–1857). In 1829, Dirichlet developed the series convergence properties.

For a detailed discussion of these topics, the reader is referred to any standard textbook on linear systems.

Fourier Series

From linear systems and our earlier discussion, we know that any periodic function can be represented as a sum of complex exponentials, that is, sines and cosines. Here we simply present the Fourier series formulas. If a continuous-time function $x(t)$ is periodic with a time-period of T_0, it can be represented by a Fourier series as

$$x(t) = \sum_{k=-\infty}^{\infty} c_k e^{j\Omega_0 kt} \tag{2.1}$$

where Ω_0 is the frequency of the function given by $\Omega_0 = \frac{2\pi}{T_0}$. The coefficients of the Fourier series are given by c_k and are possibly complex numbers with some absolute value and angle. Equation (2.1) is called the Fourier series *synthesis equation*, or simply a Fourier series representation of the periodic function. This equation states that any periodic function can be represented as a sum of complex exponentials of frequency that are integer multiples of Ω_0, that is, in terms of its harmonics. The absolute value and angle of coefficient c_k denote the magnitude and phase of the k-th harmonic of the signal. Graphs of the magnitude and phase of c_k against k are called magnitude spectrum and phase spectrum, respectively. The two graphs together are referred to as the *line spectrum* of the signal. In order to find a formula for the Fourier coefficients, the following result will be useful. This is called the orthogonal property of exponentials, and its proof is left as an exercise.

Lemma 2.1 (Orthogonality of continuous-time exponentials) *Let t_0 be a constant, k and m be integers, and $T_0 = \frac{2\pi}{\Omega_0}$. Then*

$$\int_{t_0}^{t_0+T_0} e^{j\Omega_0(k-m)t} dt = \begin{cases} 0, & k \neq m \\ T_0, & k = m. \end{cases} \tag{2.2}$$

The formula for the Fourier coefficients is derived as follows. Multiply both sides of (2.1) by $e^{-j\Omega_0 mt}$ and integrate from t_0 to $t_0 + T_0$ with respect to t. Interchange[2] the order of the $\int$ and the $\sum$ on the right-hand side and then use Lemma 2.1 to obtain the following formula for the coefficients:

$$c_k = \frac{1}{T_0} \int_{t_0}^{t_0+T_0} x(t) e^{-j\Omega_0 kt}\, dt. \tag{2.3}$$

This is called the Fourier series *analysis equation*. From this equation, it is easy to see that if the periodic function is real, then the coefficients satisfy $c_{-k} = c_k^*$; that is, $|c_{-k}| = |c_k|$.

[2]This technique of interchanging the summation and the integration, or for that matter two summations or two integrations, usually leads to wonderful results in signal processing.

This implies that the magnitude spectrum of a real signal is symmetric about the vertical axis.

▶ EXAMPLE 2.1 *Find the Fourier series representation of the following functions and sketch the spectrum in each case:*
(a) $x(t) = 2\cos 6t + 4\sin 9t$, *(b) the sawtooth function shown in Fig. 2.2(a).*

SOLUTION **(a)** The first question we ask is whether or not the sum of two (or more) periodic signals is periodic. This depends on the relationship between the time-periods T_1 and T_2. From the definition of periodicity, if a function is periodic with period T_1, it is also periodic with period n_1T_1, for any integer n_1. Moreover, if the sum of the two signals is periodic, they must have a common period, so that

$$n_1T_1 = n_2T_2.$$

That is, for the sum to be periodic, the ratio of the time-periods (or the ratio of the frequencies) must be a rational number (ratio of integers). Therefore, the fundamental time-period of the sum is the least common denominator (LCD) of T_1 and T_2. In other words, the fundamental frequency of the sum is the greatest common divisor (GCD) of Ω_0 and Ω_1.

Returning to our problem, we find that the two frequencies are 6 and 9 radians/second, respectively. The ratio of these numbers is rational, and so the sum of the sinusoids is periodic. The GCD of these two numbers is 3 radians/second, and this is the fundamental frequency. By using the familiar Euler's formula, we can represent the signal as

$$x(t) = e^{j6t} + e^{-j6t} + (2e^{j9t} + 2e^{-j9t})/j.$$

By rearranging these terms and writing the frequencies as multiples of the fundamental frequency $\Omega_0 = 3$, we get the Fourier series representation

$$x(t) = -\frac{2}{j}e^{-j3\Omega_0 t} + e^{-j2\Omega_0 t} + e^{j2\Omega_0 t} + \frac{2}{j}e^{j3\Omega_0 t}.$$

Comparing the above equation with the Fourier series synthesis equation (2.1), we have the following Fourier coefficients:

$$c_{-3} = 2j,\ \ c_{-2} = 1,\ \ c_{-1} = c_0 = c_1 = 0,\ \ c_2 = 1,\ \ c_3 = -2j.$$

The rest of the coefficients are all zero. The magnitude and phase spectra are shown in Fig. 2.1(a) and (b), respectively.

(b) From Fig. 2.2(a), it is easy to see that $T_0 = 1, \Omega_0 = 2\pi$, and

$$x(t) = t - mT_0 \quad \text{for} \quad mT_0 \le t < (m+1)T_0, m = 0, 1, \ldots$$

Letting $t_0 = 0$ in the analysis equation, we have

$$c_0 = \int_0^1 t\,dt = t^2/2|_0^1 = \frac{1}{2}$$

and

$$c_m = \int_0^1 (t-m)e^{-j2\pi mt}dt.$$

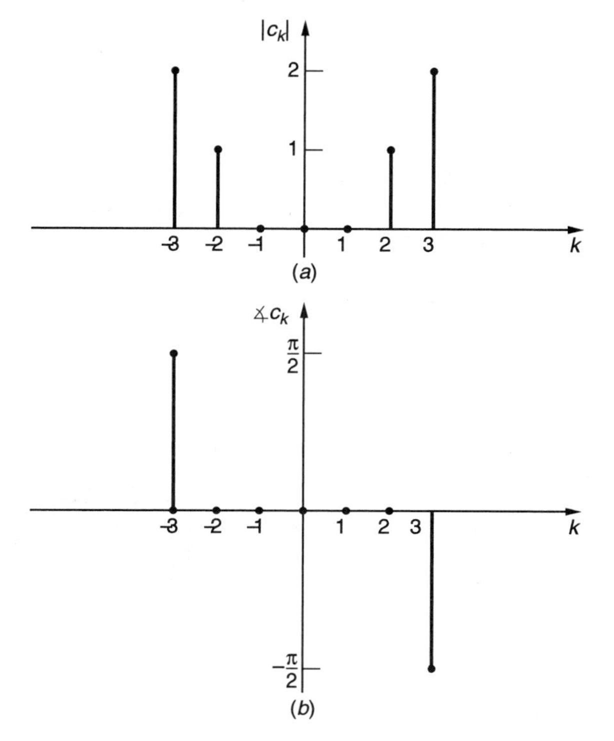

Figure 2.1 Example 2.1(a): (*a*) Magnitude spectrum, (*b*) phase spectrum.

Performing a substitution of variables with $(t - m)$ by τ gives

$$c_m = \int_0^1 \tau e^{-j2\pi m(\tau+m)} d\tau.$$

Since $e^{-j2\pi n} = 1$ for any integer n, the above becomes

$$c_m = \int_0^1 \tau e^{-j2\pi m\tau} d\tau.$$

Integrating by parts and using Lemma 2.1, we get

$$\begin{aligned} c_m &= \frac{1}{-j2\pi m}(\tau e^{-j2\pi m\tau}|_0^1 - \int_0^1 e^{-j2\pi m\tau} d\tau) \\ &= \frac{j}{2\pi m}, m \neq 0. \end{aligned}$$

Thus, the Fourier series representation of the function is

$$x(t) = \sum_{\substack{m=-\infty \\ m\neq 0}}^{\infty} \frac{j}{2\pi m} e^{j2\pi mt} + \frac{1}{2}.$$

The magnitude and phase spectrum are plotted in Fig. 2.2(b) and (c), respectively.

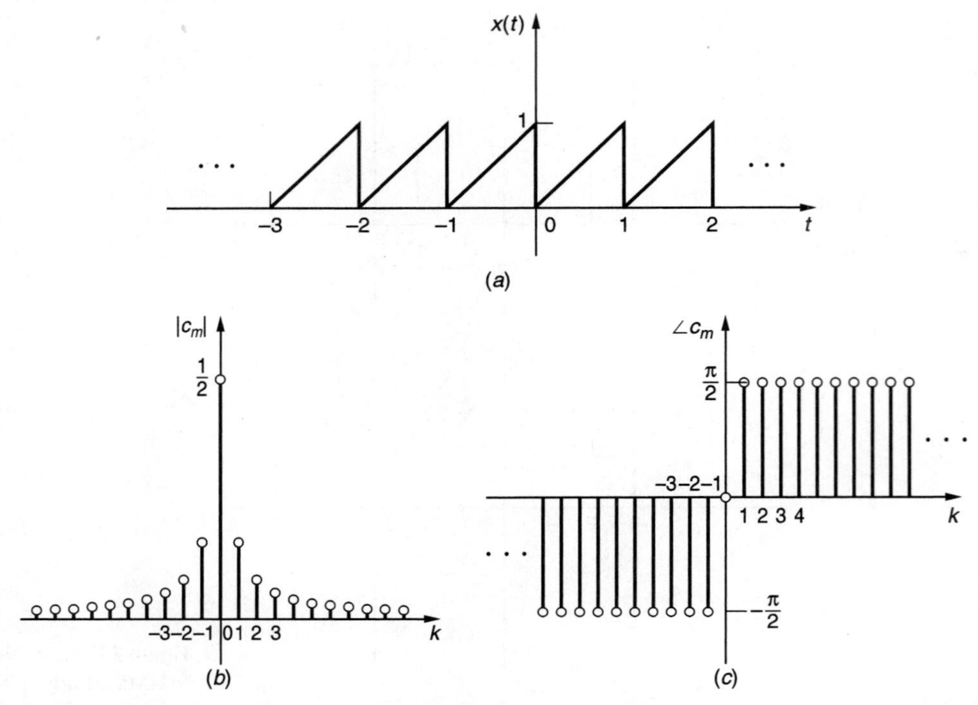

Figure 2.2 Example 2.1(b): (*a*) Given function; (*b*) magnitude spectrum; (*c*) phase spectrum.

Continuous-Time Fourier Transform (CTFT)

In the last subsection, we saw that any continuous-time periodic signal can be represented as a sum of complex exponentials with frequencies that are harmonics of the signal. Nonperiodic signals can also be represented in terms of complex exponentials. However, in this case the exponentials can assume any frequency and are not restricted to the harmonics. Therefore, in the formula for signal representation (synthesis equation), the summation is replaced by an integral. The continuous-time Fourier transform (CTFT) is as follows.

CTFT analysis equation

$$X(\Omega) = \int_{-\infty}^{\infty} x(t)e^{-j\Omega t}dt \tag{2.4}$$

CTFT synthesis equation

$$x(t) = \frac{1}{2\pi}\int_{-\infty}^{\infty} X(\Omega)e^{j\Omega t}d\Omega \tag{2.5}$$

The Fourier transform of a function $x(t)$ is also denoted by $\Im\{x(t)\}$, and a Fourier transform pair is usually written as $x(t) \longleftrightarrow X(\Omega)$. The CTFT is proved from the Fourier series equations by assuming that the time-period $T_0 \to \infty$. A rigorous proof is beyond the

TABLE 2.1 CTFT Properties

	Property	Function	CTFT
1.	Linearity	$\alpha_1 x_1(t) + \alpha_2 x_2(t)$	$\alpha_1 X_1(\Omega) + \alpha_2 X_2(\Omega)$
2.	Scaling	$x(at)$	$\frac{1}{\lvert a \rvert} X\left(\frac{\Omega}{a}\right)$
3.	Time-shifting	$x(t - t_0)$	$X(\Omega)e^{-j\Omega t_0}$
4.	Frequency-shifting	$x(t)e^{j\Omega_0 t}$	$X(\Omega - \Omega_0)$
5.	Differentiation	$\frac{dx(t)}{dt}$	$j\Omega X(\Omega)$
6.	Convolution	$x_1(t) * x_2(t)$	$X_1(\Omega)X_2(\Omega)$
7.	Multiplication	$x_1(t)x_2(t)$	$\frac{1}{2\pi} X_1(\Omega) * X_2(\Omega)$
8.	Duality	$X(t)$	$2\pi x(-\Omega)$

scope of this book and can be found in [2]. The important thing to realize at this point is that unlike the Fourier coefficients, the Fourier transform $X(\Omega)$ is defined for all Ω; that is, it is a continuous-time function of Ω. As pointed out in Chapter 1, we say continuous-time as opposed to continuous because the function does not have to be continuously differentiable. Since $X(\Omega)$ is possibly complex, it has a magnitude and phase, which are also continuous-time functions of Ω. By using the analysis and synthesis equations, it is straightforward to prove several properties of the CTFT. Some of the basic properties are given in Table 2.1, and the proofs are left as exercises. Now we find the Fourier transform of some simple functions.

EXAMPLE 2.2 *Find the Fourier transform of the following functions and sketch the spectrum in each case: (a) $\delta(t)$, (b) $e^{-\beta t}u(t)$, (c) 1, (d) $e^{j\Omega_0 t}$.*

SOLUTION **(a)** Using the analysis equation directly and the sifting property of an impulse, we get

$$\begin{aligned} \Im\{\delta(t)\} &= \int_{-\infty}^{\infty} \delta(t) e^{-j\Omega t} dt \\ &= 1. \end{aligned}$$

The Fourier transform is real, and the phase is therefore zero. The function and the transform are plotted in Fig. 2.3.

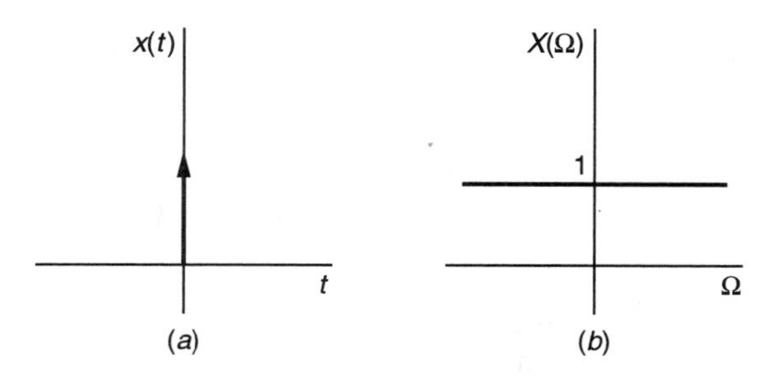

Figure 2.3 The continuous-time impulse and its CTFT.

(b)

$$
\begin{aligned}
\Im\{e^{-\beta t}u(t)\} &= \int_{-\infty}^{\infty} e^{-\beta t}u(t)e^{-j\Omega t}\,dt \\
&= \int_{0}^{\infty} e^{-(j\Omega+\beta)t}\,dt \\
&= \frac{e^{-j\Omega t}e^{-\beta t}}{-(j\Omega+\beta)}\Big|_0^{\infty} \\
&= \frac{1}{j\Omega+\beta}, \qquad \beta > 0.
\end{aligned}
$$

The condition $\beta > 0$ is required so that for $t = \infty$ the term $e^{-\beta t}$ becomes zero; otherwise it goes to infinity. The magnitude and phase of the transform are $\frac{1}{\sqrt{\Omega^2+\beta^2}}$ and $-\tan^{-1}\left(\frac{\Omega}{\beta}\right)$, respectively. The function and its spectrum are plotted in Fig. 2.4.

(c) The Fourier transform of 1 is found by invoking the *duality* property given in Table 2.1 and the Fourier transform pair found in part (a), that is, $\delta(t) \longleftrightarrow 1$. From duality we know that if $x(t) \longleftrightarrow X(\Omega)$ then $X(t) \longleftrightarrow 2\pi x(-\Omega)$. Let

$$\delta(t) = x(t) \longleftrightarrow X(\Omega) = 1.$$

Then

$$1 = X(t) \longleftrightarrow 2\pi x(-\Omega) = 2\pi\delta(-\Omega).$$

Since $\delta(-\Omega) = \delta(\Omega)$, we have $\Im\{1\} = 2\pi\delta(\Omega)$. Again, the transform is real and the phase is zero. The function and its transform are the same as in (a) except that the time-domain and frequency-domain plots are interchanged, and there is a scale factor of 2π.

(d) The Fourier transform of this complex exponential can be found by using the *frequency-shifting* property given by $x(t)e^{j\Omega_0 t} \longleftrightarrow X(\Omega - \Omega_0)$. If we use $x(t) = 1$ in this pair, then by using the result from part (c), we have $\Im\{e^{j\Omega_0 t}\} = 2\pi\delta(\Omega - \Omega_0)$. This transform is plotted as an impulse at $\Omega = \Omega_0$. Note that the Fourier transform in this case is zero everywhere except at the frequency Ω_0. This is expected because the complex exponential is a periodic function and its Fourier series representation is itself with just one Fourier coefficient. ◀

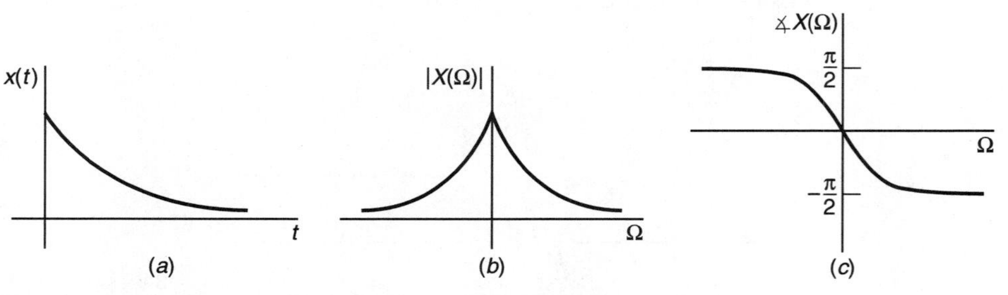

Figure 2.4 The function $e^{-\beta t}u(t)$, $\beta > 0$ and its magnitude and phase spectrum.

TABLE 2.2 CTFT Pairs

	$x(t)$	$X(\Omega)$
1.	$\delta(t)$	1
2.	1	$2\pi\delta(\Omega)$
3.	$e^{-\beta t}u(t)$	$\dfrac{1}{j\Omega+\beta}, \beta>0$
4.	$e^{j\Omega_0 t}$	$2\pi\delta(\Omega-\Omega_0)$
5.	$u(t)$	$\pi\delta(\Omega)+\dfrac{1}{j\Omega}$
6.	$\cos(\Omega_0 t)$	$\pi[\delta(\Omega-\Omega_0)+\delta(\Omega+\Omega_0)]$
7.	$e^{-\beta t}\cos(\Omega_0 t)$	$\dfrac{\beta+j\Omega}{(\beta+j\Omega)^2+\Omega_0^2}$

A list of common functions (including the ones above) and their CTFT are given in Table 2.2. The derivation of these pairs and the derivation of the properties in Table 2.1 should be familiar to the reader from a course on linear systems. If a refresher is required, the reader is encouraged to work on the relevant CTFT problems at the end of this chapter.

2.2 DISCRETE-TIME FOURIER TRANSFORM

In the last section, we reviewed the Fourier series and Fourier transform for periodic and nonperiodic continuous-time signals, respectively. These concepts can be extended to discrete-time signals as well. Discrete-time Fourier series (DTFS) is only useful for periodic sequences and can be obtained by sampling the frequency variable in the continuous-time Fourier series equations. The analysis is quite straightforward. Since most signals encountered in practice are not periodic, we routinely use the discrete-time Fourier transform (DTFT) which is applicable to nonperiodic sequences as well. Therefore, we will not present the DTFS. However, the DTFT will be covered in detail now.

The DTFT of a discrete sequence $x(n)$ will be denoted by $\Im\{x(n)\}$ or $X(\omega)$. Similar notation was used for the continuous-time Fourier transform. However, the meaning should be clear from the context because in a discrete-time sequence we use n instead of t, and the frequency is ω instead of Ω. Again, as in the continuous-time case, we will use the following notation to denote a DTFT pair:

$$x(n) \longleftrightarrow X(\omega).$$

The analysis and synthesis equations for DTFT are given by the following.

DTFT analysis equation

$$X(\omega) = \sum_{n=-\infty}^{\infty} x(n)e^{-j\omega n} \tag{2.6}$$

DTFT synthesis equation

$$x(n) = \frac{1}{2\pi}\int_{-\pi}^{\pi} X(\omega)e^{j\omega n}d\omega. \tag{2.7}$$

It remains to show that (2.6) and (2.7) are indeed a DTFT pair. In other words, given the analysis equation, we need to show that the synthesis equation holds, or vice versa. Let us now prove that the right-hand side (RHS) of the synthesis equation is indeed equal to $x(n)$. Consider the RHS and substitute for $X(\omega)$ from the analysis equation as

$$\frac{1}{2\pi}\int_{-\pi}^{\pi} X(\omega)e^{j\omega n}d\omega = \frac{1}{2\pi}\int_{-\pi}^{\pi}\sum_{m=-\infty}^{\infty} x(m)e^{-j\omega m}e^{j\omega n}d\omega \tag{2.8}$$

where we have used a new time index m for generality. Interchanging the order of the summation and the integral gives

$$\frac{1}{2\pi}\int_{-\pi}^{\pi} X(\omega)e^{j\omega n}d\omega = \frac{1}{2\pi}\sum_{m=-\infty}^{\infty} x(m)\int_{-\pi}^{\pi} e^{j\omega(n-m)}d\omega. \tag{2.9}$$

By Lemma 2.1 (orthogonality of exponentials), the integral on the RHS equals zero for $n \neq m$ and is equal to 2π for $n = m$. Therefore, for every value of m in the summation, the integral goes to zero except for $n = m$. Hence, the RHS reduces to $x(n)$.

Before we go into the mundane techniques of evaluating the DTFT of various sequences and establishing its various properties, we present some conceptual results that are very important in digital filter theory. These are presented as theorems. The proofs are very simple and use only the definition of the transform. First we show that the DTFT of the output of a linear system is the product of the DTFT of its input and that of its impulse response. In other words, convolution in the discrete-time domain is equivalent to multiplication in the DTFT domain.

Theorem 2.1 *Let $h(n)$ be the impulse response of an LSI system and $x(n)$ be the input. The DTFT of the output $y(n)$ can then be written as*

$$Y(\omega) = H(\omega)X(\omega). \tag{2.10}$$

Proof. The output of the system is given by the convolution

$$\begin{aligned} y(n) &= h(n) * x(n) \\ &= \sum_{m=-\infty}^{\infty} h(m)x(n-m). \end{aligned} \tag{2.11}$$

Taking the DTFT of the above gives

$$Y(\omega) = \sum_{n=-\infty}^{\infty}\sum_{m=-\infty}^{\infty} h(m)x(n-m)e^{-j\omega n}. \tag{2.12}$$

Now we perform a change of variables $p = n - m$ in (2.12) to get

$$\begin{aligned} Y(\omega) &= \sum_{p=-\infty}^{\infty} \sum_{m=-\infty}^{\infty} h(m)x(p)e^{-j\omega(m+p)} \\ &= \sum_{m=-\infty}^{\infty} h(m)e^{-j\omega m} \sum_{p=-\infty}^{\infty} x(p)e^{-j\omega p} \\ &= H(\omega)X(\omega) \end{aligned}$$

and the theorem is proved. ■

Theorem 2.2 *Let $h(n)$ represent the impulse response of an LSI system. If the input is a complex exponential $e^{j\omega_0 n}$, then its output is the same complex exponential, with magnitude and phase given by $|H(\omega)|$ and $\measuredangle H(\omega)$, respectively, at $\omega = \omega_0$.*

Proof. The output of the system is given by the convolution formula as

$$\begin{aligned} y(n) &= \sum_{m=-\infty}^{\infty} h(m)e^{j\omega_0(n-m)} \\ &= e^{j\omega_0 n} \sum_{m=-\infty}^{\infty} h(m)e^{-j\omega_0 m} \\ &= H(\omega_0)e^{j\omega_0 n}. \end{aligned} \tag{2.13}$$

Since $H(\omega_0)$ is possibly complex, it can be represented as $H(\omega_0) = Ae^{j\theta}$. So, the output becomes

$$y(n) = Ae^{j(\omega_0 n + \theta)}, \tag{2.14}$$

and the conclusion follows. ■

Theorem 2.2 is the basis for the concept of *frequency response* of a digital filter, which is defined as the DTFT of the impulse response. The frequency response $H(\omega)$ is a continuous-time function of the frequency variable ω. Plots of $|H(\omega)|$ and $\measuredangle H(\omega)$ are called the magnitude and phase spectrum, respectively. The above theorem states that if a sinusoidal sequence with frequency ω_0 is applied to a digital filter, the magnitude and phase of the output can be determined from the magnitude and phase spectra of the filter by reading off the respective values at $\omega = \omega_0$.

Theorem 2.3 *The DTFT of any discrete sequence is periodic with a period of 2π; that is,*

$$H(\omega) = H(\omega + 2\pi). \tag{2.15}$$

Proof. From the definition of DTFT, we have

$$\begin{aligned} H(\omega + 2\pi) &= \sum_{n=-\infty}^{\infty} h(n)e^{-j(\omega+2\pi)n} \\ &= \sum_{n=-\infty}^{\infty} h(n)e^{-j\omega n}e^{-j2\pi n}. \end{aligned} \tag{2.16}$$

Since $e^{-j2\pi n} = 1$, the RHS is equal to $H(\omega)$. ∎

Theorem 2.3 implies that the frequency response of a digital filter is periodic with a period of 2π, which in turn implies that the magnitude and phase spectra are also periodic. An example of the magnitude response of a digital filter is shown in Fig. 2.5. This is clearly an ideal lowpass filter with a cutoff frequency of $\frac{\pi}{2}$. Such a filter is called a *half-band* filter and is very useful in many signal processing applications. Note that unlike continuous-time filters, the frequency response repeats indefinitely. Therefore, the spectrum of a digital filter or that of a discrete-time sequence is usually plotted between $-\pi$ and π only. Recall from Chapter 1 that the frequency of the discrete-time sequence is related to that of its continuous-time counterpart by

$$\omega = \Omega T_s, \tag{2.17}$$

where T_s is the sampling time-period. The frequency Ω is in radians/seconds, and the sampling period T_s is in seconds. Therefore, ω is in radians as depicted in Fig. 2.5. If we substitute $\Omega = \Omega_s$ in the above formula, we get $\omega_s = \Omega_s T_s = 2\pi$. This is an important result and deserves a formal statement.

Fact: *The frequency of 2π radians of a discrete-time sequence corresponds to the sampling frequency Ω_s rads/sec.*

The frequency response of the half-band digital filter can also be plotted against the continuous-time frequency, as shown by the second axis in Fig. 2.5. These representations of the frequency response in different frequency domains will be useful later on in this chapter.

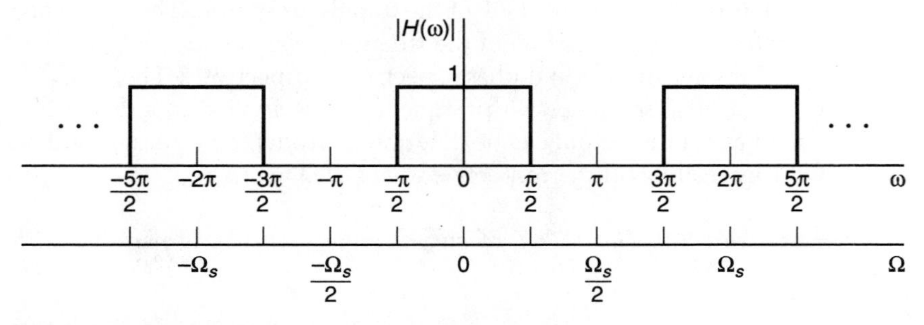

Figure 2.5 Magnitude spectrum of a half-band digital filter.

DTFT Properties

In this subsection, we present several properties of the DTFT. Some of these properties will be proved, while the others will be left as exercises. Some examples of evaluating the DTFT for some common sequences will also be given.

First, we discuss the convergence condition of the DTFT. This transform is an infinite sum, and therefore it must converge in order for the transform to exist. The following theorem gives a sufficient condition for the existence of the DTFT.

Theorem 2.4 *The DTFT of a sequence $x(n)$ exists if the sequence is absolutely summable, that is, if*

$$\sum_{n=-\infty}^{\infty} |x(n)| < \infty. \tag{2.18}$$

Proof. Taking the absolute value of the DTFT formula and using the triangular inequality, we get

$$\begin{aligned} |X(\omega)| &= \left| \sum_{n=-\infty}^{\infty} x(n)e^{-j\omega n} \right| \\ &\leq \sum_{n=-\infty}^{\infty} |x(n)||e^{-j\omega n}| \\ &= \sum_{n=-\infty}^{\infty} |x(n)| < \infty. \end{aligned}$$

Thus, if the given condition is satisfied, the DTFT is bounded. ■

Now we present a series of properties of the DTFT by using the notation $x(n) \longleftrightarrow X(\omega)$. These properties are listed in Table 2.3 for convenient reference.

TABLE 2.3 DTFT Properties

	Property	Function	DTFT
1.	Linearity	$ax_1(n) + bx_2(n)$	$aX_1(\omega) + bX_2(\omega)$
2.	Time reversal	$x(-n)$	$X(-\omega)$
3.	Time shift	$x(n-n_0)$	$e^{-j\omega n_0}X(\omega)$
4.	Frequency shift	$x(n)e^{j\omega_0 n}$	$X(\omega-\omega_0)$
5.	Convolution in time	$h(n) * x(n)$	$H(\omega)X(\omega)$
6.	Convoluion in frequency	$x(n)h(n)$	$\frac{1}{2\pi}\int_{-\pi}^{\pi} X(\gamma)H(\omega-\gamma)d\gamma$
7.	Differentiation in frequency	$nx(n)$	$j\frac{dX(\omega)}{d\omega}$
8.	Parseval's theorem	$\sum_{n=-\infty}^{\infty} x(n)y^*(n)$	$= \frac{1}{2\pi}\int_{-\pi}^{\pi} X(\omega)Y^*(\omega)d\omega$

Property 1 (Linearity): The DTFT satisfies the principle of superposition; that is, for sequences $x_1(n)$ and $x_2(n)$ and constants a and b, we have

$$ax_1(n) + bx_2(n) \longleftrightarrow aX_1(\omega) + bX_2(\omega).$$

Proof. This property can be proved by a straightforward application of the DTFT definition. ■

Property 2 (Time reversal): A reversal of the index in the time domain causes a reversal of the frequency in Fourier domain, that is,

$$x(-n) \longleftrightarrow X(-\omega).$$

Proof. Taking the DTFT of $x(-n)$ and performing a change of variables gives

$$\begin{aligned}
\Im\{x(-n)\} &= \sum_{n=-\infty}^{\infty} x(-n)e^{-j\omega n} \\
&= \sum_{m=\infty}^{-\infty} x(m)e^{j\omega m} \\
&= \sum_{m=-\infty}^{\infty} x(m)e^{-jm(-\omega)} \\
&= X(-\omega).
\end{aligned}$$

■

Property 3 (Time shift): A shift in the time domain by n_0 samples causes a multiplication in the frequency domain by $e^{-j\omega n_0}$; that is,

$$x(n - n_0) \longleftrightarrow e^{-j\omega n_0}X(\omega).$$

Proof. Again, by taking the DTFT of $x(n - n_0)$ and performing a change of variables, we get

$$\begin{aligned}
\Im\{x(n - n_0)\} &= \sum_{n=-\infty}^{\infty} x(n - n_0)e^{-j\omega n} \\
&= \sum_{m=-\infty}^{\infty} x(m)e^{-j\omega(m+n_0)} \\
&= e^{-j\omega n_0} \sum_{m=-\infty}^{\infty} x(m)e^{-j\omega m} \\
&= e^{-j\omega n_0}X(\omega).
\end{aligned}$$

■

Property 4 (Frequency shift): A shift in the frequency domain by ω_0 causes a multiplication in the time domain by $e^{j\omega_0 n}$; that is,

$$x(n)e^{j\omega_0 n} \longleftrightarrow X(\omega - \omega_0).$$

The proof is similar to that of Property 3 and is omitted.

Property 5 (Convolution in time): Convolution in discrete-time is equivalent to multiplication in the DTFT domain; that is,

$$h(n) * x(n) \longleftrightarrow H(\omega)X(\omega).$$

The proof was given in Theorem 2.1.

The above property is very important in linear filter theory. We know from Chapter 1 that when two filters are cascaded, the impulse response of the overall filter is the convolution of the individual impulse responses. From the above property, we now know that the frequency response of the overall filter is the product of the individual frequency responses. It also says that the frequency response of the output of a linear system is the product of the frequency response of the filter and that of the input.

The following three properties can be proved directly from the definition of the DTFT. The proofs are assigned as problems at the end of the chapter.

Property 6 (Convolution in frequency): Multiplication in the discrete-time domain is equivalent to convolution in the frequency domain; that is,

$$x(n)h(n) \longleftrightarrow \frac{1}{2\pi}\int_{-\pi}^{\pi} X(\gamma)H(\omega-\gamma)d\gamma.$$

Property 7 (Differentiation in frequency):

$$nx(n) \longleftrightarrow j\frac{dX(\omega)}{d\omega}$$

Property 8 (Parseval's theorem): For two sequences $x(n)$ and $y(n)$,

$$\sum_{n=-\infty}^{\infty} x(n)y^*(n) = \frac{1}{2\pi}\int_{-\pi}^{\pi} X(\omega)Y^*(\omega)d\omega.$$

The energy of a signal (defined by $\sum_{n=-\infty}^{\infty} x(n)x^*(n)$) can be determined in the Fourier domain by using this property. By substituting $y(n) = x(n)$ in Parseval's theorem, the energy of a signal is

$$\sum_{n=-\infty}^{\infty} |x(n)|^2 = \frac{1}{2\pi}\int_{-\pi}^{\pi} |X(\omega)|^2 d\omega. \tag{2.19}$$

Now we give some examples for evaluating the DTFT of some simple sequences. These evaluations are done using the DTFT definition and/or some of the above properties.

EXAMPLE 2.3 *Evaluate the DTFT of the following discrete-time sequences and sketch the frequency spectrum:*
(a) $\delta(n)$; (b) 1; (c) $\beta^n u(n), |\beta| < 1$; (d) $\cos(\omega_0 n)$; (e) $rect_M(n) \triangleq \begin{cases} 1, & -M \le n \le M \\ 0, & \text{otherwise} \end{cases}$

SOLUTION In these examples, we will first find the DTFT, $X(\omega)$. If $X(\omega)$ is real, then we will plot its amplitude only. If it is complex, then its magnitude and phase spectra will be plotted separately.

(a) By using the definition of DTFT and applying the sifting property of the discrete-time impulse, we get

$$\begin{aligned}\Im\{\delta(n)\} &= \sum_{n=-\infty}^{\infty} \delta(n)e^{-j\omega n} \\ &= 1.\end{aligned}$$

The sequence and its DTFT are shown in Fig. 2.6. Note that since the DTFT is periodic with a period of 2π, it is only plotted between $-\pi$ and π. Since the DTFT is real and positive, the phase response is zero.

(b) The sequence $x(n) = 1$ does not satisfy the sufficient condition for the existence of DTFT given by Theorem 2.4. However, if we can show that there is a function $X(\omega)$ whose inverse DTFT is equal to 1, then we will say that this function is the DTFT of 1. We now show that $1 \longleftrightarrow 2\pi \sum_{m=-\infty}^{\infty} \delta(\omega - 2\pi m)$ form a DTFT pair. From the inverse DTFT formula, we have

$$\begin{aligned}x(n) &= \frac{1}{2\pi}\int_{-\pi}^{\pi} X(\omega)e^{j\omega n}\, d\omega \\ &= \frac{1}{2\pi}\int_{-\pi}^{\pi} 2\pi \sum_{m=-\infty}^{\infty} \delta(\omega - 2\pi m)e^{j\omega n}\, d\omega.\end{aligned} \tag{2.20}$$

Interchanging the order of the integral and summation, we get

$$x(n) = \sum_{m=-\infty}^{\infty} \int_{-\pi}^{\pi} \delta(\omega - 2\pi m)e^{j\omega n}\, dw. \tag{2.21}$$

The integral can be evaluated by using the sifting property of the continuous-time impulse function. Except $m = 0$, for every other value of m the impulse appears outside the limits of integration yielding a result of zero. For $m = 0$ the integral is unity. The sequence and its DTFT are plotted in Fig. 2.7. The phase response is again zero.

(c) Using the definition again,

$$\begin{aligned}X(\omega) &\triangleq \Im\{\beta^n u(n)\} = \sum_{n=0}^{\infty} \beta^n e^{-j\omega n} \\ &= \sum_{n=0}^{\infty} (\beta e^{-j\omega})^n.\end{aligned} \tag{2.22}$$

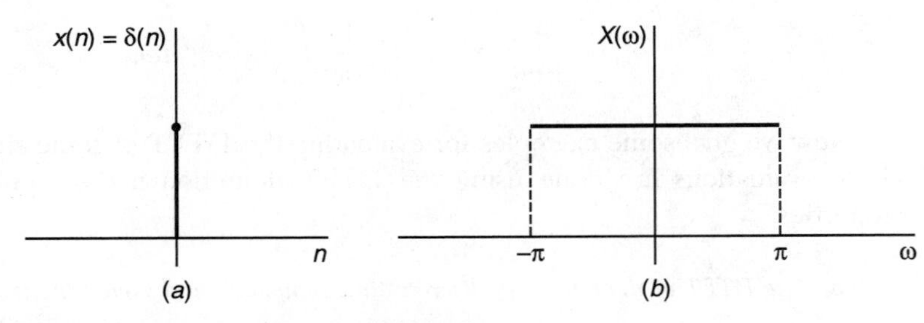

Figure 2.6 Discrete-time impulse and its DTFT.

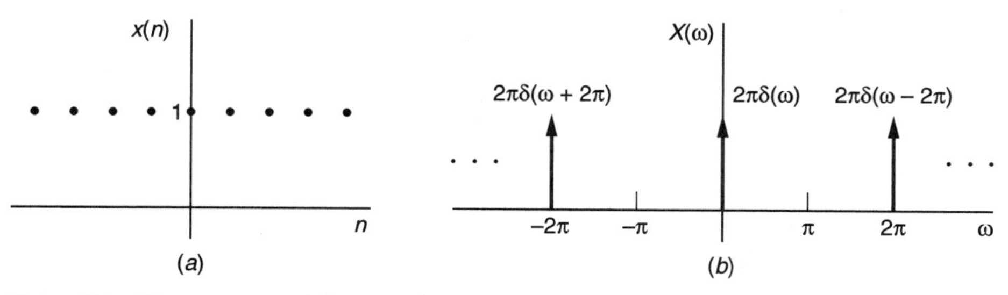

Figure 2.7 The sequence $x(n) = 1$ and its DTFT.

Now, we use the formula for the geometric sum given by

$$\sum_{n=0}^{\infty} r^n = \frac{1}{1-r}, \quad |r| < 1. \tag{2.23}$$

The DTFT thus becomes

$$\begin{aligned} X(\omega) &= \sum_{n=0}^{\infty} (\beta e^{-j\omega})^n \\ &= \frac{1}{1-\beta e^{-j\omega}}, \quad |\beta| < 1. \end{aligned} \tag{2.24}$$

$X(\omega)$ was obtained using the DTFT formula, which was shown to be periodic. Therefore, the above function is valid in the interval $(-\pi, \pi)$ and repeats thereafter with a period of 2π. The magnitude and phase response are given by

$$|X(\omega)| = \sqrt{\frac{1}{(1-\beta\cos\omega)^2 + (\beta\sin\omega)^2}} \tag{2.25}$$

and

$$\measuredangle X(\omega) = -\tan^{-1}\frac{\beta\sin\omega}{1-\beta\cos\omega}. \tag{2.26}$$

The magnitude and phase spectra for $\beta = 0.9$ are plotted in Fig. 2.8. Again, the plots are only shown in the interval $(-\pi, \pi)$ and are assumed to repeat at every integer multiple of 2π.

(d) This function can be written by Euler's formula as $x(n) = \frac{1}{2}e^{j\omega_0 n} + \frac{1}{2}e^{-j\omega_0 n}$. Let $g(n) = 1$. Then from part (b), $G(\omega) = 2\pi\sum_{m=-\infty}^{\infty}\delta(\omega - 2\pi m)$. Now $x(n) = \frac{1}{2}g(n)e^{j\omega_0 n} + \frac{1}{2}g(n)e^{-j\omega_0 n}$. The frequency-shifting property gives

$$\begin{aligned} X(\omega) &= \frac{1}{2}G(\omega-\omega_0) + \frac{1}{2}G(\omega+\omega_0) \\ &= \pi\sum_{m=-\infty}^{\infty}\delta(\omega - 2\pi m - \omega_0) + \pi\sum_{m=-\infty}^{\infty}\delta(\omega - 2\pi m + \omega_0). \end{aligned} \tag{2.27}$$

The sequence and its DTFT are shown in Fig. 2.9. Notice that in the interval $(-\pi, \pi)$ the impulses appear at ω_0 and $-\omega_0$.

(e) The DTFT of this function is

$$X(\omega) \triangleq \Im\{rect_M(n)\} = \sum_{n=-M}^{M} e^{-j\omega n}. \tag{2.28}$$

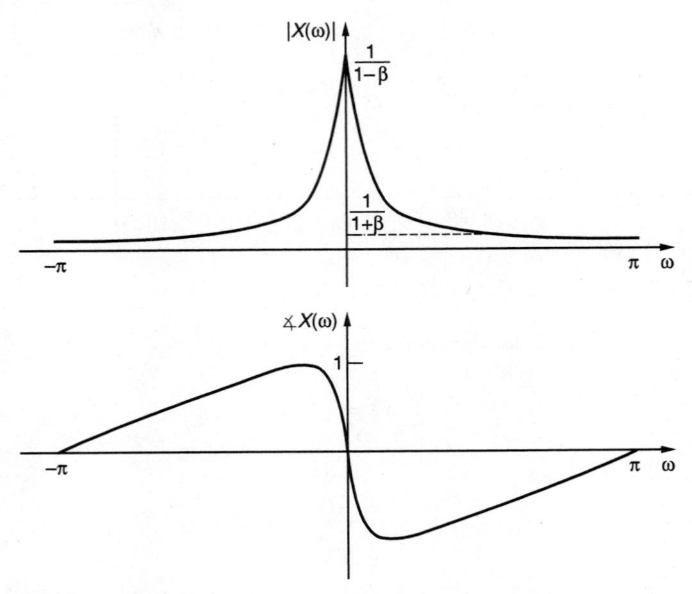

Figure 2.8 Magnitude and phase spectra for $x(n) = \beta^n u(n)$ with $\beta = 0.9$.

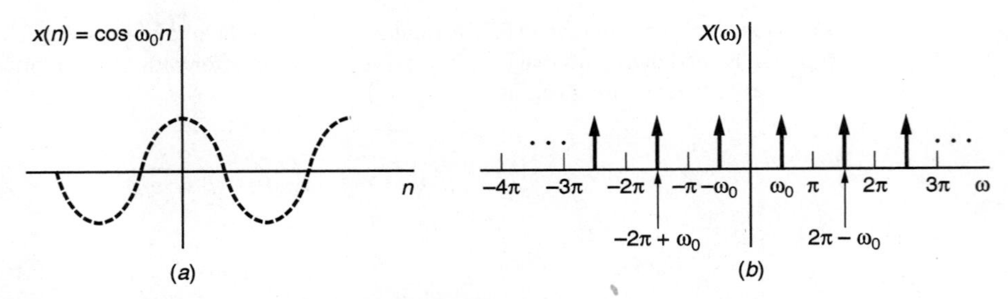

Figure 2.9 The cosine sequence and its DTFT.

The above can be evaluated by using the formula for the finite sum given in Chapter 1 and is repeated here for convenience:

$$\sum_{n=K_1}^{K_2} \gamma^n = \frac{\gamma^{K_1} - \gamma^{K_2+1}}{1-\gamma}, \quad K_2 \geq K_1. \tag{2.29}$$

Using this formula, we find that the desired DTFT becomes

$$\begin{aligned} X(\omega) &\triangleq \Im\{rect_M(n)\} \\ &= \frac{e^{j\omega M} - e^{-j\omega(M+1)}}{1-e^{-j\omega}} \\ &= \frac{e^{-j\omega/2}(e^{j\omega(M+1/2)} - e^{-j\omega(M+1/2)})}{e^{-j\omega/2}(e^{j\omega/2} - e^{-j\omega/2})} \\ &= \frac{\sin{(M+\frac{1}{2})\omega}}{\sin\frac{\omega}{2}}, \quad \omega \neq 0. \end{aligned} \tag{2.30}$$

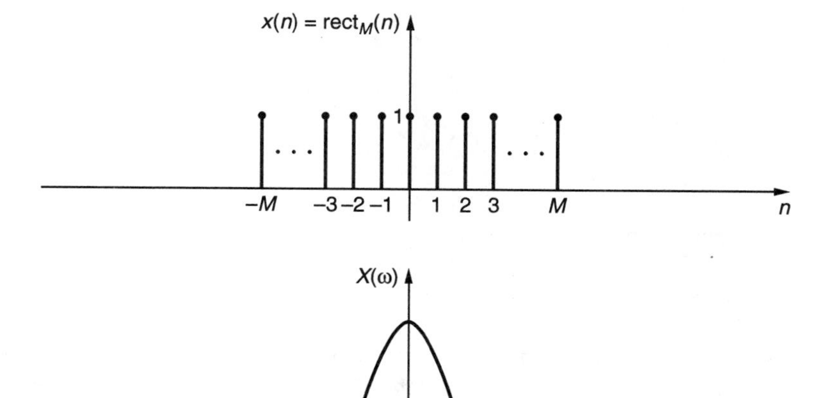

Figure 2.10 The rectangular function and its DTFT.

For $\omega = 0, X(\omega) = 2M + 1$. The DTFT in this case is real. So, the phase spectrum is 0 if $X(\omega)$ is positive and π if $X(\omega)$ is negative. The sequence and its DTFT are plotted in Fig. 2.10.

EXAMPLE 2.4 *Find the impulse response of an ideal lowpass filter (LPF) with a cutoff frequency at ω_c radians per second.*

SOLUTION The frequency response of an ideal LPF is described by

$$H(\omega) = \begin{cases} 1, & |\omega| \le \omega_c \\ 0, & \omega_c < \omega \le \pi. \end{cases} \tag{2.31}$$

This is depicted in Fig. 2.11(a). Recall that the frequency response of a digital filter repeats indefinitely with a period of 2π. Therefore, it is customary to display it only in the interval $(-\pi, \pi)$ or $(0, 2\pi)$. The inverse DTFT of the above is found easily from the synthesis equation as follows.

$$\begin{aligned} h(n) &= \frac{1}{2\pi}\int_{-\omega_c}^{\omega_c} e^{j\omega n} d\omega \\ &= \frac{1}{2\pi}\frac{e^{j\omega n}}{jn}\bigg|_{-\omega_c}^{\omega_c}, \quad n \neq 0 \\ &= \frac{1}{\pi n}\frac{e^{j\omega_c n} - e^{-j\omega_c n}}{2j} \\ &= \frac{\sin \omega_c n}{\pi n}. \end{aligned} \tag{2.32}$$

Equation (2.32) is not valid for $n = 0$ since for this value the second equation does not exist. To evaluate the integral for this value, we substitute $n = 0$ in the original integrand to get $h(0) = \frac{\omega_c}{\pi}$.

So, the impulse response of the ideal LPF is

$$h(n) = \begin{cases} \dfrac{\sin \omega_c n}{\pi n}, & n \neq 0 \\ \dfrac{\omega_c}{\pi}, & n = 0. \end{cases} \tag{2.33}$$

The impulse response is plotted in Fig. 2.11(b). ◀

From Example 2.4, observe that the impulse response of an ideal LPF is infinite in length. To implement this filter using an FIR filter, we must approximate the impulse response by a finite number of coefficients. One way to do this is by simply truncating the impulse response to $2M+1$ coefficients, M in each of the positive and negative directions. This is equivalent to multiplying $h(n)$ by the $rect_M(n)$ function defined in the previous example. We have just designed a lowpass FIR filter! The designed filter is

$$y(n) = \sum_{i=-M}^{M} h(i)x(n-i) \tag{2.34}$$

where $y(n)$ and $x(n)$ are the output and input of the filter, respectively, and the filter coefficients are given by (2.33). This method of FIR filter design is called the method of *windowing*. In this instance, we have used the rectangular window. There are several

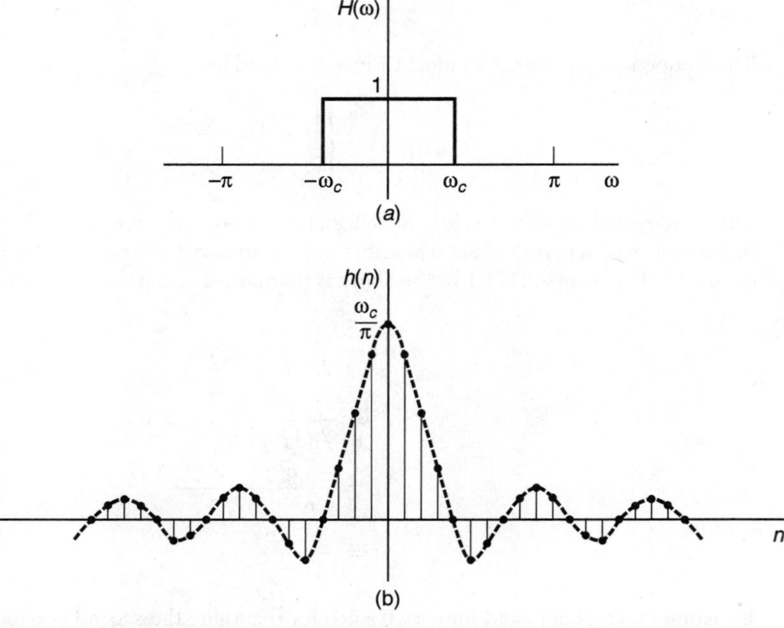

Figure 2.11 Ideal lowpass filter: (a) Frequency response; (b) impulse response.

different types of windows with relative pros and cons. More on this topic will appear in Chapter 4.

EXAMPLE 2.5 *Consider a FIR filter with the impulse response*

$$x(n) = [\mathbf{1}\ 2\ 3\ 2\ 1].$$

Evaluate the following: (a) $X(0)$; *(b)* $X(\pi)$; *(c)* $\angle X(\omega)$; *(d)* $\int_{-\pi}^{\pi} |X(\omega)|^2\, dw$.

SOLUTION As in Chapter 1, the element in boldface represents x(0). All the parts of this problem can be solved without computing the DTFT as follows.

(a) From the DTFT formula, $X(0) = \sum_{n=-\infty}^{\infty} x(n) = 9$. This is the dc gain of this sequence.

(b) Again, from the DTFT formula, we have $X(\pi) = \sum_{n=-\infty}^{\infty} x(n)e^{-j\pi n} = \sum_{n=-\infty}^{\infty} x(n)(-1)^n = 1$.

(c) Note that the above filter coefficients are symmetric about $n = 2$. Therefore, the DTFT can be written as

$$\begin{aligned}
X(\omega) &= 1 + 2e^{-j\omega} + 3e^{-j2\omega} + 2e^{-j3\omega} + e^{-j4\omega} \\
&= e^{-j2\omega}\left[e^{j2\omega} + 2e^{j\omega} + 3 + 2e^{-j\omega} + e^{-j2\omega}\right] \\
&= e^{-j2\omega}\left[3 + (2e^{j\omega} + 2e^{-j\omega}) + (e^{j2\omega} + e^{-j2\omega})\right] \\
&= e^{-j2\omega}\left[3 + 4\cos\omega + 2\cos 2\omega\right] \\
&= e^{-j2\omega} Y(\omega).
\end{aligned}$$

In this equation, the quantity within square brackets (defined by $Y(\omega)$) is real and therefore has a phase of zero or π. The overall phase of this filter is $\angle X(\omega) = -2\omega + \beta$, where $\beta = 0$ if $Y(\omega) \geq 0$, and $\beta = \pi$ if $Y(\omega) < 0$. We could have come to this conclusion without computing the DTFT, simply from the fact that the point of symmetry of the impulse response is $n = 2$. Such a filter is called a *linear phase* filter since the phase is a linear function of ω as long as $Y(\omega) < 0$ or $Y(\omega) \geq 0$. Further discussion of this topic will be postponed until Chapter 3.

(d) This can be evaluated from Parseval's theorem as

$$\begin{aligned}
\int_{-\pi}^{\pi} |X(\omega)|^2 d\omega &= 2\pi \sum_{n=-\infty}^{\infty} |x(n)|^2 \\
&= 38\pi.
\end{aligned}$$

◀

Power Spectral Density

Now we briefly discuss the concept of power spectral density and some of its properties with respect to stationary random processes and linear systems.

Definition 2.1 *The Power Spectral Density (PSD) of a stationary random process is defined as the discrete-time Fourier transform (DTFT) of its autocorrelation function; that is,*

$$S(\omega) = \sum_{l=-\infty}^{\infty} r(l)e^{-j\omega l} \tag{2.35}$$

where $r(l)$ *is the autocorrelation of the process.*

The following two facts are given without proof. The proofs are obtained directly from the definition given earlier and are assigned as problems.

Fact 1: *The PSD of a stationary random process is real.*

Fact 2: *If the stationary random process is real, the PSD is an even function.*

The following result gives a formula for the PSD of the output of a linear system in terms of that of its input.

Fact 3: *Let $S_x(\omega)$ denote the PSD of a stationary process, which is the input to a real linear filter with frequency response $H(\omega)$. Then the PSD of the output of the filter is*

$$S_y(\omega) = |H(\omega)|^2 S_x(\omega). \tag{2.36}$$

Proof. The autocorrelation of the output of the filter was obtained in equation (1.92) as

$$r(k) = E\{y(n)y^*(n-k)\} = \sum_{m=-\infty}^{\infty} \sum_{l=-\infty}^{\infty} h(m)h(l)r_x(l+k-m) \tag{2.37}$$

where $r_x(k)$ is the autocorrelation of the input. Taking the DTFT of both sides of the above gives

$$\begin{aligned} S_y(\omega) &= \sum_{k=-\infty}^{\infty} \left(\sum_{m=-\infty}^{\infty} \sum_{l=-\infty}^{\infty} h(m)h(l)r_x(l+k-m) \right) e^{-j\omega k} \\ &= \sum_{m=-\infty}^{\infty} h(m) \sum_{l=-\infty}^{\infty} h(l) \sum_{k=-\infty}^{\infty} r_x(l+k-m)e^{-j\omega k}. \end{aligned} \tag{2.38}$$

Now we perform the substitution of variables, $p = l + k - m$ in (2.38) to get

$$\begin{aligned} S_y(\omega) &= \sum_{m=-\infty}^{\infty} h(m) \sum_{l=-\infty}^{\infty} h(l) \sum_{p=-\infty}^{\infty} r_x(p)e^{-j\omega(p+m-l)} \\ &= \sum_{m=-\infty}^{\infty} h(m)e^{-j\omega m} \sum_{l=-\infty}^{\infty} h(l)e^{j\omega l} \sum_{p=-\infty}^{\infty} r_x(p)e^{-j\omega p} \\ &= H(\omega)H^*(\omega)S_x(\omega). \\ &= |H(\omega)|^2 S_x(\omega). \end{aligned} \tag{2.39}$$

■

Definition 2.2 *A stationary random process is said to be white if its PSD is constant; that is,*

$$S(\omega) = \sigma^2, \quad -\pi \le \omega \le \pi. \tag{2.40}$$

Recall from (2.35) that the PSD and the autocorrelation function form a DTFT pair. Since $S(\omega)$ is a constant, the autocorrelation is therefore an impulse represented by

$$r(k) = \sigma^2 \delta(k). \tag{2.41}$$

This implies that $r(0) = \sigma^2$, which is the variance of a zero-mean process. This definition makes sense because for a white noise sequence, the elements are uncorrelated with each other. So, the autocorrelation function is nonzero only for zero shift.

2.3 SAMPLING THEORY

In this section, we use Fourier analysis to examine what happens to the spectrum of a continuous-time signal when it is sampled. Then we present the topics of decimation and interpolation and the effect of these operations on the spectrum.

The following discussion will lead to the famous *sampling theorem*. Consider the block diagram of Fig. 2.12. This is similar to the one given in Section 1.1.2, except that we omitted the A/D and D/A converters because they do not have any effect on the spectrum. In addition, we omitted the digital filter block, which means that in this case the filter does nothing. Let the continuous-time signal and its CTFT be denoted by $x_c(t)$ and $X_c(\Omega)$, respectively. Let the spectrum be as shown in Fig. 2.13(a), where Ω_h denotes the highest frequency component of the signal. So, the spectrum of the time-domain signal can be represented using the CTFT synthesis equation as

$$x_c(t) = \frac{1}{2\pi} \int_{-\infty}^{\infty} X_c(\Omega) e^{j\Omega t} d\Omega. \tag{2.42}$$

This signal is sampled to obtain the discrete-time signal $x(nT_s)$, where T_s is the sampling time-period. As noted in Chapter 1, this can be represented as the sampled sequence $x_s(n) \triangleq x(nT_s)$. Thus, we can substitute $t = nT_s$ in (2.42) to get

$$x_s(n) = x_c(nT_s) = \frac{1}{2\pi} \int_{-\infty}^{\infty} X_c(\Omega) e^{j\Omega nT_s} d\Omega. \tag{2.43}$$

Since integration is a linear operation, we can subdivide the integral into an infinite number of small integrals and add them up as follows:

$$x_s(n) = x_c(nT_s) = \frac{1}{2\pi} \sum_{k=-\infty}^{\infty} \int_{(2k-1)\pi/T_s}^{(2k+1)\pi/T_s} X_c(\Omega) e^{j\Omega nT_s} d\Omega. \tag{2.44}$$

Note that the length of each integral interval is $\frac{2\pi}{T_s}$. If we replace Ω by $\Omega - \frac{2\pi k}{T_s}$ and adjust the limits accordingly, we get

$$x_s(n) = x_c(nT_s) = \frac{1}{2\pi} \sum_{k=-\infty}^{\infty} \int_{-\pi/T_s}^{\pi/T_s} X_c\left(\Omega - \frac{2\pi k}{T_s}\right) e^{j(\Omega - \frac{2\pi k}{T_s})nT_s} d\Omega. \tag{2.45}$$

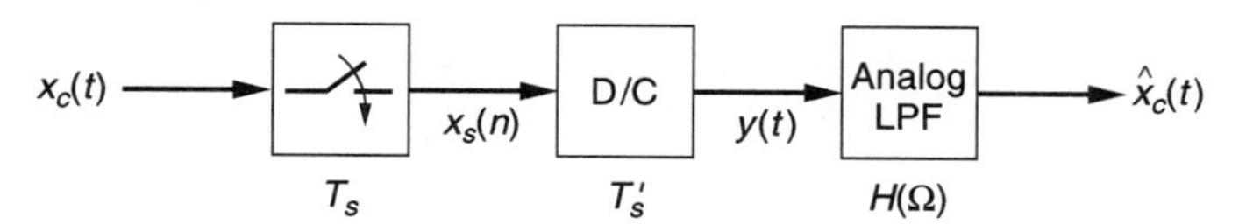

Figure 2.12 Sampling and recovery of a continuous-time signal.

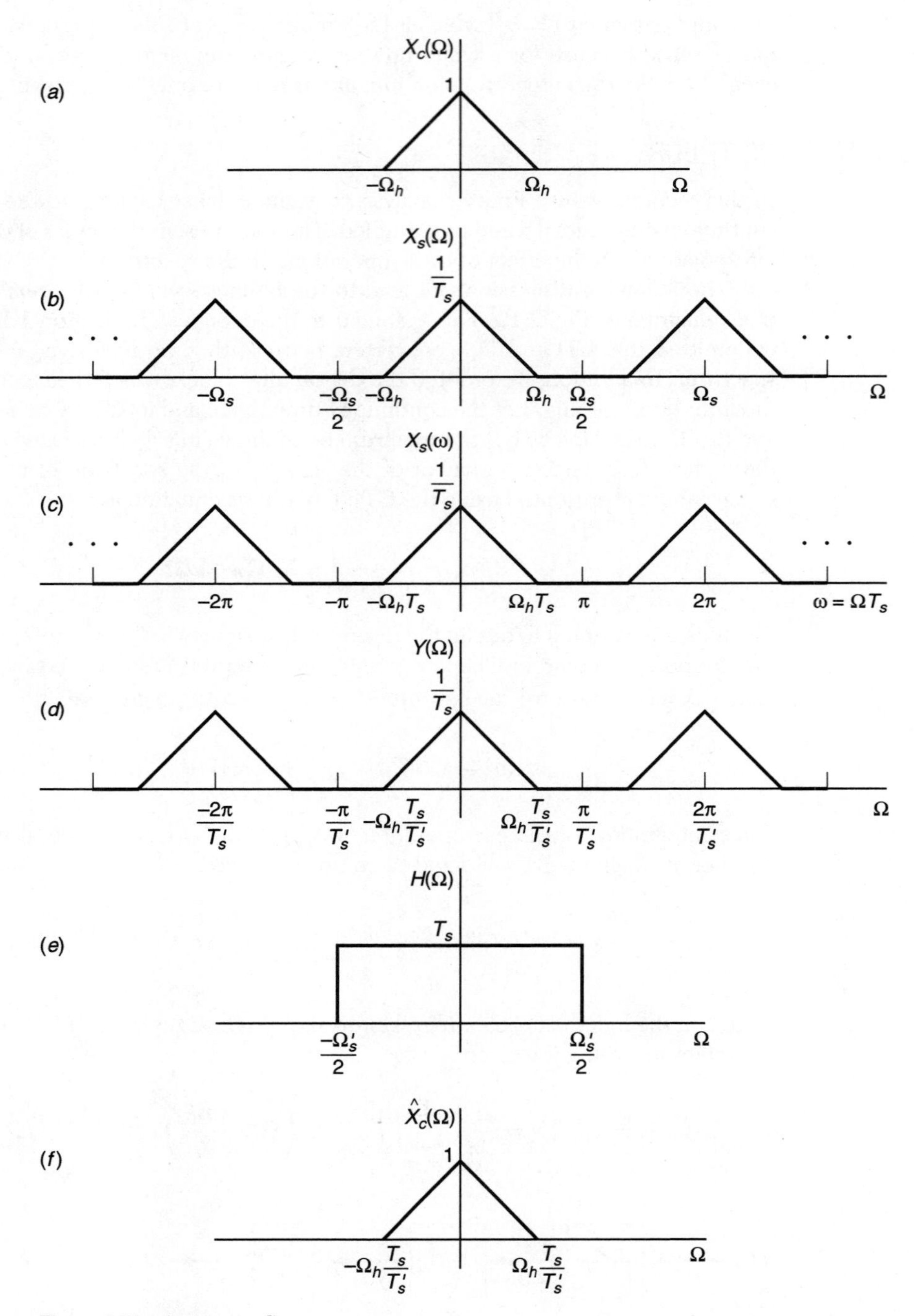

Figure 2.13 *(Continued)*

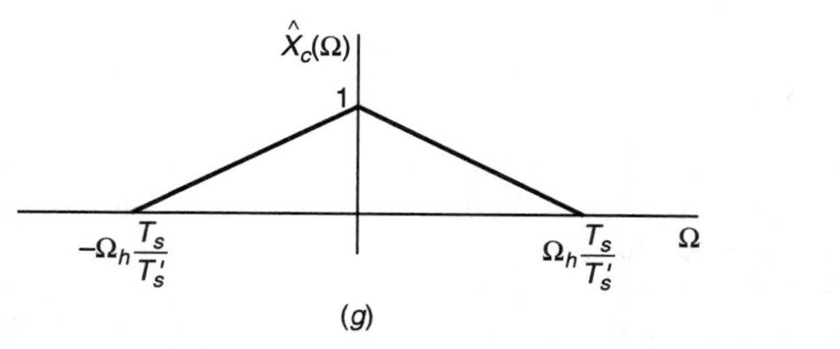

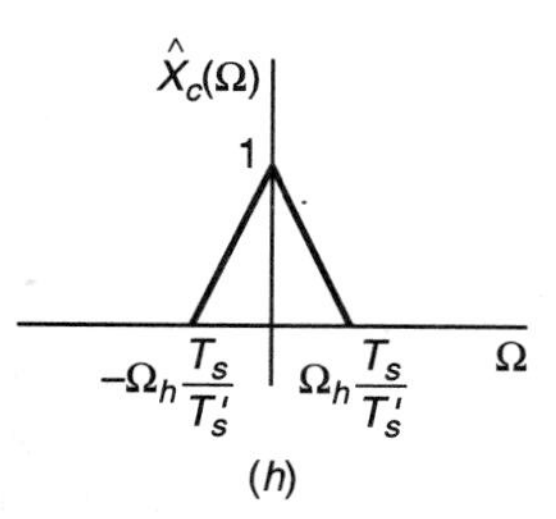

Figure 2.13 Frequency spectra of (*a*) continuous-time signal; (*b*) sampled signal against Ω; (*c*) sampled signal against ω; (*d*) after D/C conversion; (*e*) analog LPF (*f*) reconstructed signal; (*g*) $T_s > T_s'$ (*h*) $T_s < T_s'$.

Now we use the fact that $e^{j2\pi kn} = 1$ in the exponent above to get

$$x_s(n) = x_c(nT_s) = \frac{1}{2\pi} \sum_{k=-\infty}^{\infty} \int_{-\pi/T_s}^{\pi/T_s} X_c\left(\Omega - \frac{2\pi k}{T_s}\right) e^{j\Omega n T_s} d\Omega. \tag{2.46}$$

We know that the frequency of the continuous-time signal and that of its discrete-time counterpart are related by $\Omega = \frac{\omega}{T_s}$. Performing this change of variables in (2.46), we get

$$x_s(n) = x_c(nT_s) = \frac{1}{2\pi} \sum_{k=-\infty}^{\infty} \int_{-\pi}^{\pi} X_c\left(\frac{\omega - 2\pi k}{T_s}\right) e^{j\omega n} \frac{d\omega}{T_s}. \tag{2.47}$$

Again, we perform the all-useful swapping of the summation and integral to get

$$x_s(n) = x_c(nT_s) = \frac{1}{2\pi} \int_{-\pi}^{\pi} \left[\frac{1}{T_s} \sum_{k=-\infty}^{\infty} X_c\left(\frac{\omega - 2\pi k}{T_s}\right)\right] e^{j\omega n} d\omega. \tag{2.48}$$

Leaving the above for a moment, let us revisit the DTFT synthesis equation:

$$x_s(n) = \frac{1}{2\pi} \int_{-\pi}^{\pi} X_s(\omega) e^{j\omega n} d\omega. \tag{2.49}$$

Comparing (2.48) and (2.49), we get

$$X_s(\omega) = \frac{1}{T_s} \sum_{k=-\infty}^{\infty} X_c\left(\frac{\omega - 2\pi k}{T_s}\right). \tag{2.50}$$

This is the spectrum of the discrete-time signal in terms of that of the continuous-time signal. The frequency variable used is for the sampled signal. If we substitute $\Omega = \frac{\omega}{T_s}$ in the RHS, then we have

$$X_s(\Omega) = \frac{1}{T_s} \sum_{k=-\infty}^{\infty} X_c\left(\Omega - \frac{2\pi k}{T_s}\right). \tag{2.51}$$

Note that the LHS is written as $X_s(\Omega)$ instead of $X_s(\Omega T_s)$ for simplicity and to avoid any confusion in terms of the frequency variable. By substituting $\Omega_s = \frac{2\pi}{T_s}$, equation (2.51) becomes

$$X_s(\Omega) = \frac{1}{T_s} \sum_{k=-\infty}^{\infty} X_c(\Omega - k\Omega_s). \tag{2.52}$$

Both equations (2.50) and (2.52) represent the sampling theorem. The only difference between these equations is in the frequency variable. Now let us examine (2.52) carefully. This equation states that the spectrum of the sampled signal is composed of shifted versions of the spectrum of the continuous-time signal and is then scaled by $\frac{1}{T_s}$. As noted earlier, let the spectrum of the continuous-time be as shown in Fig. 2.13(a). The spectrum of the sampled signal as represented by (2.52) is given in Fig. 2.13(b). To change the frequency axis to ω, we simply scale the axis according to $\omega = \Omega T_s$. The resulting graph is shown in Fig. 2.13(c). Note that the sampling frequency in the discrete domain is $\omega_s = \Omega_s T_s = 2\pi$, as was discussed in Section 2.2. Now consider the process of converting back to the continuous-time domain using a discrete-to-continuous (D/C) converter. This element puts the time variable back in the discrete sequence by outputting each sample every T_s' seconds. Of course, T_s' may be equal to T_s. The spectrum of the output of the D/C converter is shown in Fig. 2.13(d) where the frequency axis is converted back to Ω by using $\Omega = \frac{\omega}{T_s'}$. This spectrum has many high-frequency components and is not the same as that of the original signal. To recover the original signal, we must filter the signal with an analog lowpass filter (LPF) shown in Fig. 2.13(e). The cutoff frequency of this LPF should be $\frac{\pi}{T_s'}$, which is the same as $\frac{\Omega_s'}{2}$. The resulting spectrum is shown in Fig. 2.13(f). If $T_s = T_s'$, the final spectrum is exactly the same as that of the original and the signal is said to have been recovered perfectly. If $T_s < T_s'$, the final spectrum is a compressed version of the original. On the other hand, if $T_s > T_s'$, the final spectrum is a stretched version of the original. These cases are shown in Fig. 2.13(g) and (h), respectively.

Now let us look at Fig. 2.13(b) and see what happens if the sampling frequency Ω_s is low enough so that Ω_h is greater than $\frac{\Omega_s}{2}$. The triangles overlap, and the result is shown in Fig. 2.14(b). In the other parts of this figure, we repeat the steps as before. In this case, we realize that we cannot recover the original spectrum by lowpass filtering. This is the well-known phenomenon called *aliasing*. That is, instead of the original signal, its alias[3] has appeared at the output. This phenomenon does not occur if the triangles do not overlap, that is, if $\frac{\Omega_s}{2} > \Omega_h$. This is the celebrated sampling theorem and is formally stated as follows.

Theorem 2.5 *(Sampling theorem): A continuous-time signal must be sampled at a rate that is greater than twice its highest frequency component, so that it can be uniquely determined from its samples.*

To illustrate the above concept a little more, consider the example of a single cosine function. We borrow this example from [3]. Let the continuous-time signal be $x_c(t) = \cos \Omega_0 t$. From Table 2.2, the CTFT of this signal is $X_c(\Omega) = \pi[\delta(\Omega - \Omega_0) + \delta(\Omega + \Omega_0)]$, which is plotted in Fig. 2.15(a). The spectrum is thus an impulse at Ω_0 and its "ghost" at $-\Omega_0$, which is denoted by a broken arrow. We call it a ghost because it appears

[3] **a·li·as** n. **1**. An assumed name. **2**. *Electronics*. A false signal in telecommunication links from beats between signal frequency and sampling frequency. (From the American Heritage Dictionary.)

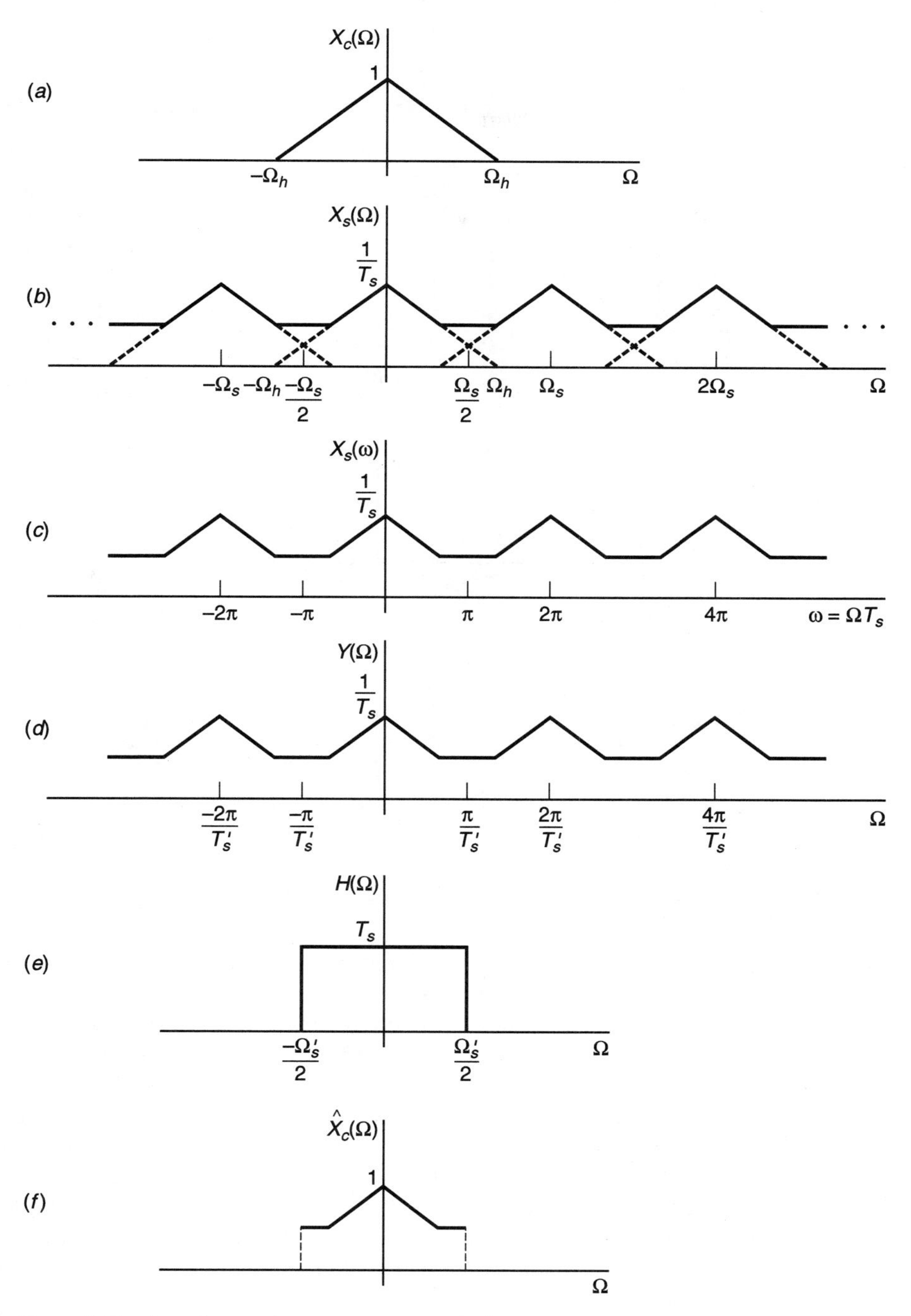

Figure 2.14 Sampling and reconstruction with aliasing.

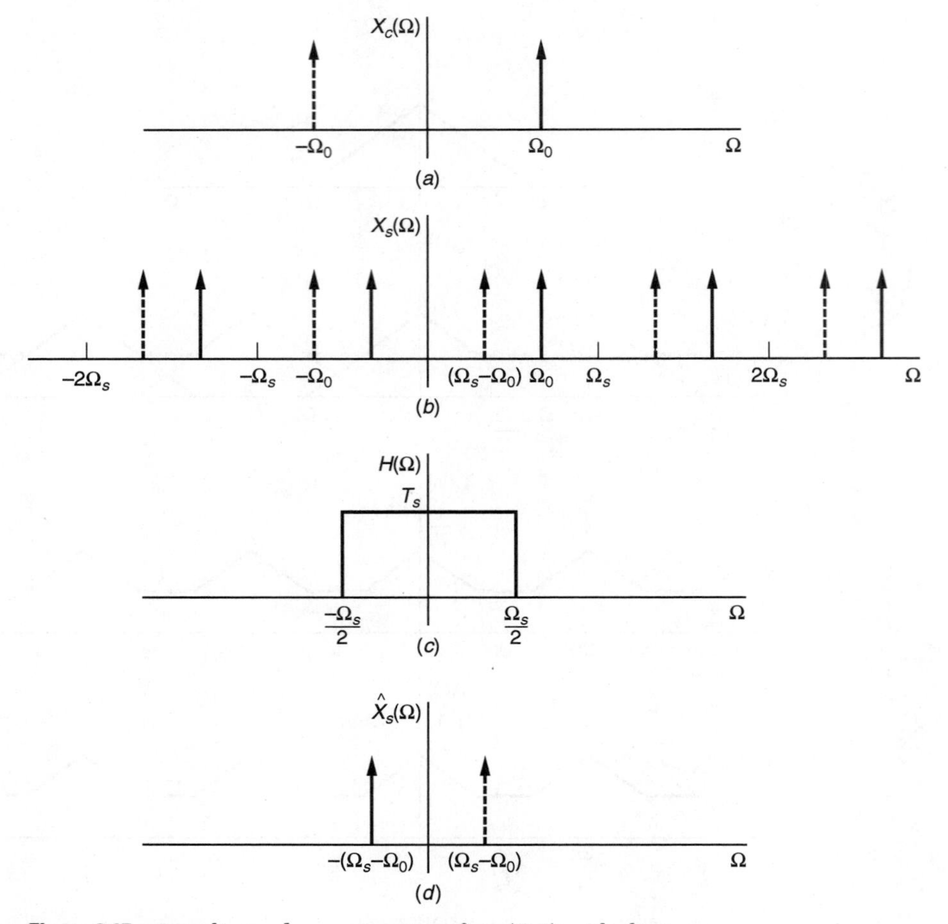

Figure 2.15 Sampling and reconstruction of $\cos(\Omega_0 t)$, with aliasing.

at the negative frequency.[4] Notice that we do not have π in the figure because it is not the amplitude of the impulse. It is simply a constant multiplier of the impulse function. The amplitude of the continuous-time impulse function is undefined; only its area is defined as unity. In any case, let us sample this signal with a sampling frequency that is less than $2\Omega_0$. In the spectrum of the sampled signal shown in Fig. 2.15(b), we get an impulse on either side of every integer multiple of Ω_s. Now we convert back to continuous-time using a D/C converter with the same sampling frequency, and then we use an analog LPF as before. The spectra of the resulting signals are shown in Fig. 2.15(c) and (d). In the output signal spectrum, we see that not only is the ghost on the right, but it also appears at a different position, which is at $\Omega_s - \Omega_0$. This spectrum tells us that the reconstructed signal is $\widehat{x}_c(t) = \cos(\Omega_s - \Omega_0)$. Thus, the reconstructed signal has a different frequency

[4]Negative frequency actually does have physical significance. If we consider the clockwise rotation of a wheel at Ω_0 rads/sec, then a counterclockwise rotation at the same rate would be $-\Omega_0$ rads/sec.

than that of the original. Instead of the original signal, its alias has appeared! If we sample a 60 Hz sinusoidal signal at 100 Hz and perform the above reconstruction procedure, the resulting signal is a sinusoid at 40 Hz. What happens if we sample the signal at a rate exactly equal to the highest frequency component of the signal? In this case, if we sample the sinusoid at 60 samples/second, then we essentially have one sample per cycle. That would give us a constant output sequence. If the sample happens to be at the zero crossing of the sinusoid, then we simply get a sequence of zeros. This is why the sampling theorem explicitly requires the sampling rate to be higher than twice the highest frequency component.

2.4 DECIMATION AND INTERPOLATION

In many situations in digital signal processing, it becomes necessary to reduce or increase the data rate or the sampling rate. This can be done in two ways. One way is to convert the discrete-time signal to continuous-time (by the method in the last section) and then resample at the desired frequency. The other way is to change the sampling rate in the discrete domain. This is the preferred method because it avoids going back to the continuous-time domain and in the process adding more hardware in terms of A/D and D/A converters and associated filters. Reducing the sampling rate by an integer factor is called *downsampling*. As we will see shortly, prefiltering must be done before downsampling. This combination of prefiltering and downsampling is called *decimation*. On the other hand, the process of increasing the sampling rate by an integer factor is called *upsampling*. Upsampling is followed by a postfilter in order to preserve the spectrum of the original signal. This combination of upsampling and postfiltering is called *interpolation*. In this section, we will examine the effects of downsampling and upsampling in the frequency domain.

Downsampling

Downsampling a sequence $x(n)$ by a factor of M is simply a matter of picking every Mth sample in the sequence and discarding the rest. The downsampling process is denoted by $(\downarrow M)x(n)$. If $x(n) = [\ldots, x(-1), x(0), x(1), x(2), \ldots]$, the downsampled sequence is given by $[\ldots, x(-2M), x(-M), x(0), x(M), x(2M), \ldots]$. In another representation, we have

$$x_{(M)}(n) \triangleq (\downarrow M)x(n) = x(Mn), \quad n = 0, \pm 1, \pm 2, \ldots. \tag{2.53}$$

Let us consider the simple case of downsampling by a factor of 2, and derive the relationship between the spectra of the original and downsampled sequences. This derivation is a little tricky and uses the same idea as in [4]. The following definitions will be useful in the subsequent discussion:

$$x_{(2)}(n) \triangleq [\ldots, x(-2), \mathbf{x(0)}, x(2), x(4), \ldots]$$

$$v(n) \triangleq [\ldots, x(-2), 0, \mathbf{x(0)}, 0, x(2), 0, x(4), \ldots].$$

First note that $v(n)$ can be written in terms of $x(n)$ as

$$v(n) = \frac{1}{2}[x(n) + (-1)^n x(n)]. \tag{2.54}$$

Since $-1 = e^{-j\pi}$, (2.54) can be written as

$$v(n) = \frac{1}{2}[x(n) + e^{-j\pi n}x(n)]. \tag{2.55}$$

Using the frequency shifting property of DTFT (Property 4), we find that the above equation in the frequency domain is

$$V(\omega) = \frac{1}{2}[X(\omega) + X(\omega + \pi)]. \tag{2.56}$$

Now let us find the relationship between the spectra of $v(n)$ and $x_{(2)}(n)$. The DTFT of $x_{(2)}(n)$ is

$$\begin{aligned} X_{(2)}(\omega) &= \sum_{n=-\infty}^{\infty} x_{(2)}(n)e^{-j\omega n} \\ &= \sum_{n=-\infty}^{\infty} v(2n)e^{-j\omega n} \\ &= \sum_{m=-\infty}^{\infty} v(m)e^{-j\frac{\omega}{2}m} && (2.57) \\ &= V\left(\frac{\omega}{2}\right). && (2.58) \end{aligned}$$

Note that in (2.57) the index is $m = 2n$, which is even, and the validity of the subsequent equation may be questioned. But this is okay because all the odd components of $v(m)$ are zero anyway. Combining (2.56) and (2.58), we get the desired result

$$X_{(2)}(\omega) = \frac{1}{2}\left[X\left(\frac{\omega}{2}\right) + X\left(\frac{\omega}{2} + \pi\right)\right]. \tag{2.59}$$

Equation (2.59) states that the spectrum of the downsampled signal is the sum of the frequency scaled and shifted versions of the original spectrum. An example is given in Fig. 2.16. Part (*a*) shows the spectrum of the original discrete sequence. Note that $\Omega_h T_s$ is chosen to be approximately $\frac{\pi}{2}$. In terms of the continuous-time frequency, Ω_h is approximately equal to $\frac{\Omega_s}{4}$. This is done on purpose, and the reason will be clear shortly. The first component of (2.59), $X(\frac{\omega}{2})$, is shown in (*b*). Note that the frequency is halved; that is, the horizontal scale is compressed. In other words, each label on the horizontal axis is multiplied by 2. The second component of (2.59), $X(\frac{\omega+2\pi}{2})$, is shown in (*c*). This is simply a result of shifting the figure in (*b*) by 2π. The spectrum of the downsampled sequence shown in (*d*), is the sum of the figures in (*b*) and (*c*), amplitude scaled by $\frac{1}{2}$. The spectra of the original signal and the downsampled signal are repeated in Fig. 2.17(*a*) and (*b*), with the horizontal axes in the same scale. It turns out that if we stretch each triangle in (*a*) to double its width, then we get (*b*). Now we make an important observation from the final spectrum. The triangles are at the edge of overlapping; that is, we are operating at the borderline of aliasing. This is because Ω_h is approximately equal to $\frac{\Omega_s}{4}$, and downsampling by a factor of 2 halves the sampling rate. So, Ω_h is now

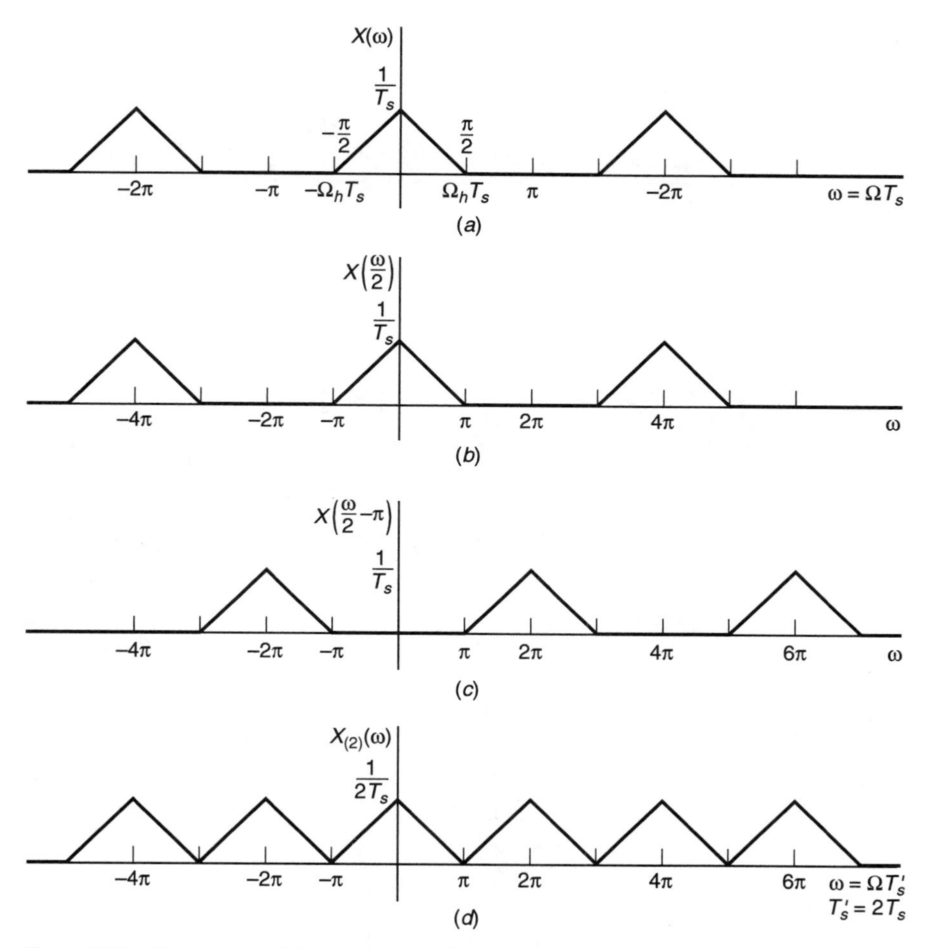

Figure 2.16 Derivation of the sepectrum of a decimated signal.

at $\frac{\Omega'_s}{2}$, where Ω'_s is the sampling rate after downsampling. This is essentially at the edge of aliasing as we have seen before. If $\Omega_h > \frac{\Omega_s}{4}$, downsampling by a factor of 2 will cause aliasing. In general, we can state the following fact.

Fact: *If a discrete sequence is downsampled by a factor of M, we must satisfy $\Omega_h \leq \frac{\Omega_s}{2M}$ in order to avoid aliasing. Since $\Omega_s = 2\pi/T_s$, this condition can also be written as $\Omega_h T_s \leq \frac{\pi}{M}$ or $\omega_h \leq \frac{\pi}{M}$.*

The method of deriving (2.59) can be easily extended to downsampling by a factor of M. This will give the general spectral relationship between a sequence and its downsampled counterpart. The following theorem summarizes the result. The proof is left as an exercise.

Theorem 2.6 *Let $x(n)$ and $X(\omega)$ denote a sequence and its DTFT. Let $x_{(M)}(n)$ and $X_{(M)}(\omega)$ denote the downsampled signal and its DTFT, where M is the decimation factor. Then*

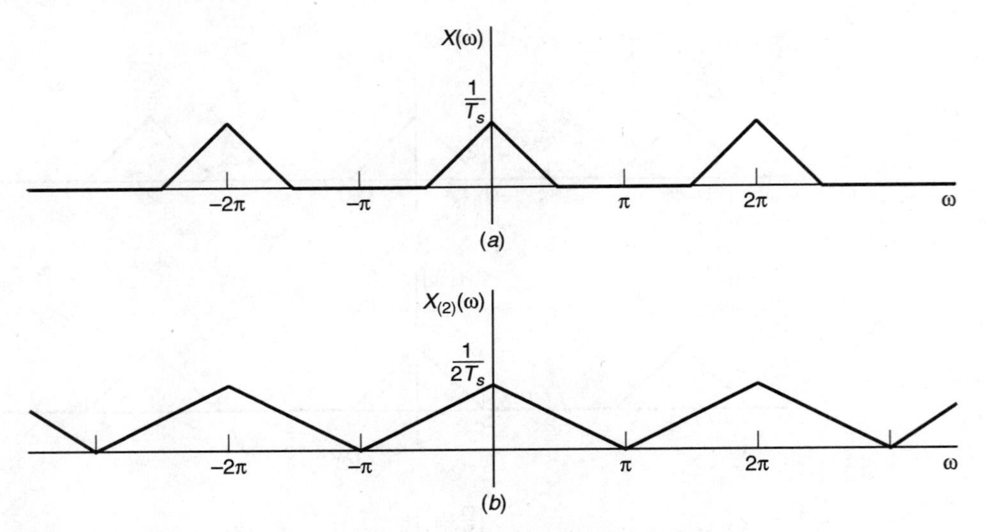

Figure 2.17 Spectrum of $(a)\ x(n)$; $(b)\ (\downarrow 2)x(n)$. No aliasing condition.

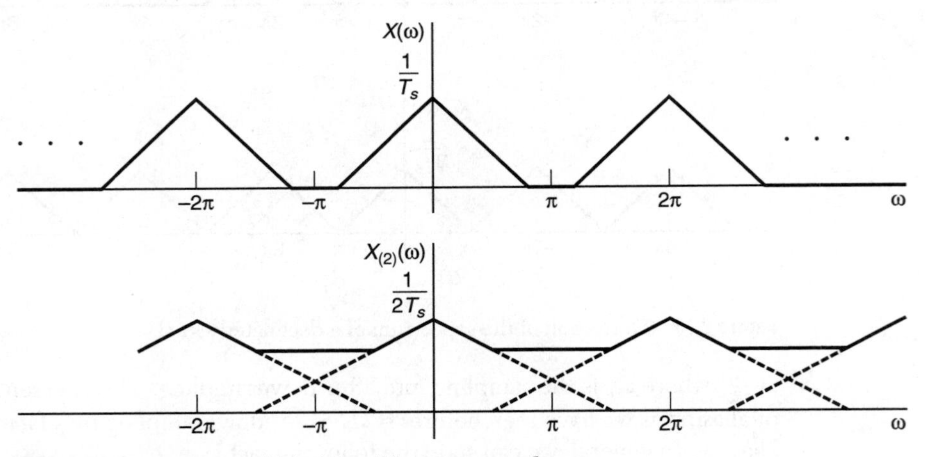

Figure 2.18 Spectrum of $(a)\ x(n)$: (b) aliasing condition.

$$X_{(M)}(\omega) = \frac{1}{M}\sum_{k=0}^{M-1} X\left(\frac{\omega}{M} + \frac{2\pi k}{M}\right). \tag{2.60}$$

As explained before, in order to obtain $X_{(M)}(\omega)$ all we need to do is stretch each spectral component of $X(\omega)$ to M times its width. An example of downsampling (by a factor of 2) with aliasing is shown in Fig. 2.18. To avoid aliasing due to downsampling, the spectrum of the signal must be appropriately bandlimited by filtering. This can be done by prefiltering the signal by an appropriate lowpass filter. For downsampling by a factor of M, the signal may be filtered by a lowpass digital filter with a cutoff frequency of $\frac{\pi}{M}$. This is shown in Fig. 2.19. This process is called decimation.

Upsampling

Upsampling is the process of increasing the sampling frequency of a signal. In order to upsample by a factor M, all we need to do is insert $M - 1$ zeroes between every two successive samples. As an example, if we upsample a sequence $x(n)$ by a factor of 4, we get the sequence $\{\ldots, x(-1), 0, 0, 0, \mathbf{x(0)}, 0, 0, 0, x(1), 0, 0, 0, x(2), \ldots\}$.
The output of an upsampler is denoted by

$$y(n) = (\uparrow M)x(n)$$

where M is the upsampling factor. This is shown as a block diagram in Fig. 2.20. An example of upsampling by a factor of 4 is shown in Fig. 2.21. But as we shall see shortly, this introduces high-frequency components in the spectrum of the upsampled signal and

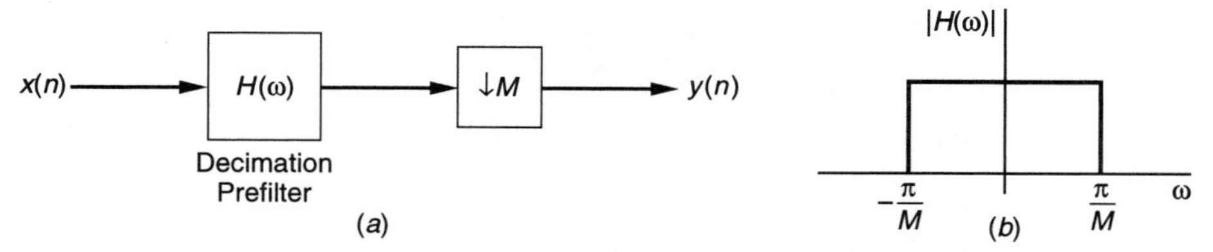

Figure 2.19 (a) Prefiltering scheme for decimation; (b) ideal prefilter magnitude response.

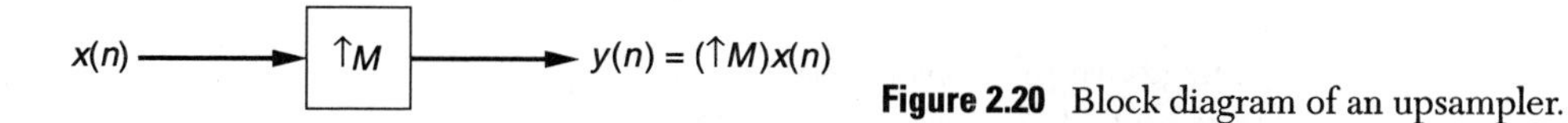

Figure 2.20 Block diagram of an upsampler.

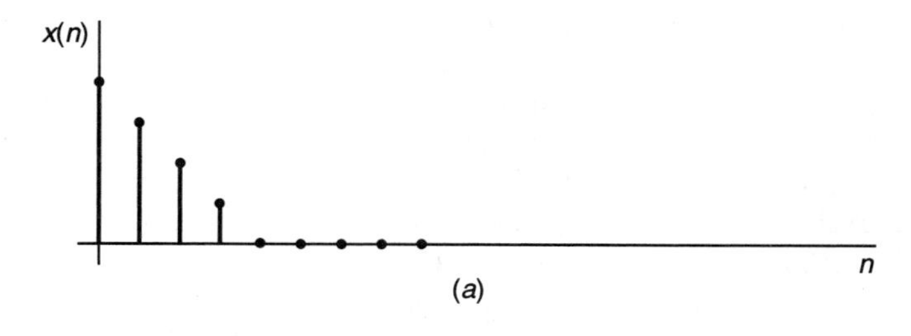

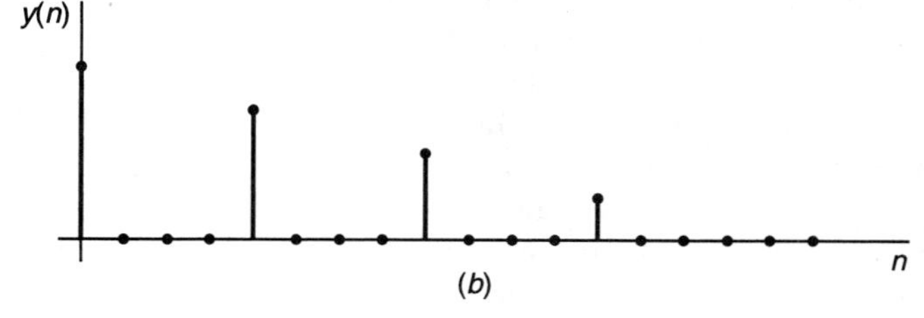

Figure 2.21 (a) Given sequence; (b) sequence after upsampling by a factor of 4.

must be filtered out. Let $(\uparrow M)x(n)$ or $x^{(M)}(n)$ denote upsampling of the sequence $x(n)$ by a factor of M. Then upsampling can be mathematically described as

$$(\uparrow M)x(n) = x^{(M)}(n) \triangleq \begin{cases} x(n/M), n = 0, \pm M, \pm 2M, \ldots \\ 0, \text{ otherwise.} \end{cases} \tag{2.61}$$

Now we derive the relationship between the DTFT of an interpolated sequence to that of the original sequence. Let $x(n)$ and $X(\omega)$ denote a discrete-time sequence and its DTFT. The DTFT of the interpolated sequence is

$$\begin{aligned} X^{(M)}(\omega) &= \sum_{n=-\infty}^{\infty} x^{(M)}(n)e^{-j\omega n} \\ &= \ldots + x(-1)e^{j\omega M} + x(0) + x(1)e^{-j\omega M} + x(2)e^{j\omega 2M} + \ldots \\ &= \sum_{n=-\infty}^{\infty} x(n)e^{-j\omega Mn} \\ &= X(\omega M). \end{aligned} \tag{2.62}$$

The above relationship is formally stated in Theorem 2.7.

Theorem 2.7 *Let $x(n)$ and $X(\omega)$ denote a discrete-time sequence and its DTFT. When this sequence is upsampled by a factor of M, the DTFT of the upsampled sequence is related to that of the original by*

$$X^{(M)}(\omega) = X(\omega M).$$

The above relationship states that the spectrum of an upsampled sequence is obtained from that of the original sequence by stretching the frequency axis. Therefore, this is just the opposite of decimation. As an example, let the spectrum of a signal, $x(n)$, be as shown in Fig. 2.22(a). Let the sequence be upsampled by a factor of 3. The spectrum of the upsampled sequence is obtained by stretching the frequency axis to three times in length; that is, we divide each label on the frequency axis by 3. The resulting spectrum is shown in Fig. 2.22(b). It is clear that upsampling compresses each spectral component; therefore, there is no danger of aliasing. However, we observe that in the interval $(-\pi, \pi)$, we have three spectral components. But the original spectrum had only one! These extra components are called *images* and must be filtered out. The required filter is a lowpass digital filter that cuts off at $\frac{\pi}{3}$, as shown in Fig. 1.22(c). The spectrum of the filter output is shown in part (d) of the same figure. Now we ask, can we use an analog filter? The answer is yes, of course. However, we must perform a discrete-to-continuous (D/C) conversion before we can do that. If further processing after interpolation is intended to be in the discrete-domain, then a digital filter should be used. Otherwise, we convert to continuous-time and then use an analog filter. Recall that an analog filter is necessary anyway after D/C conversion, in order to remove the high-frequency components introduced by sampling. In Fig. 2.22(e), we show the result of D/C conversion. Note that the sampling time-period used for the D/C conversion is

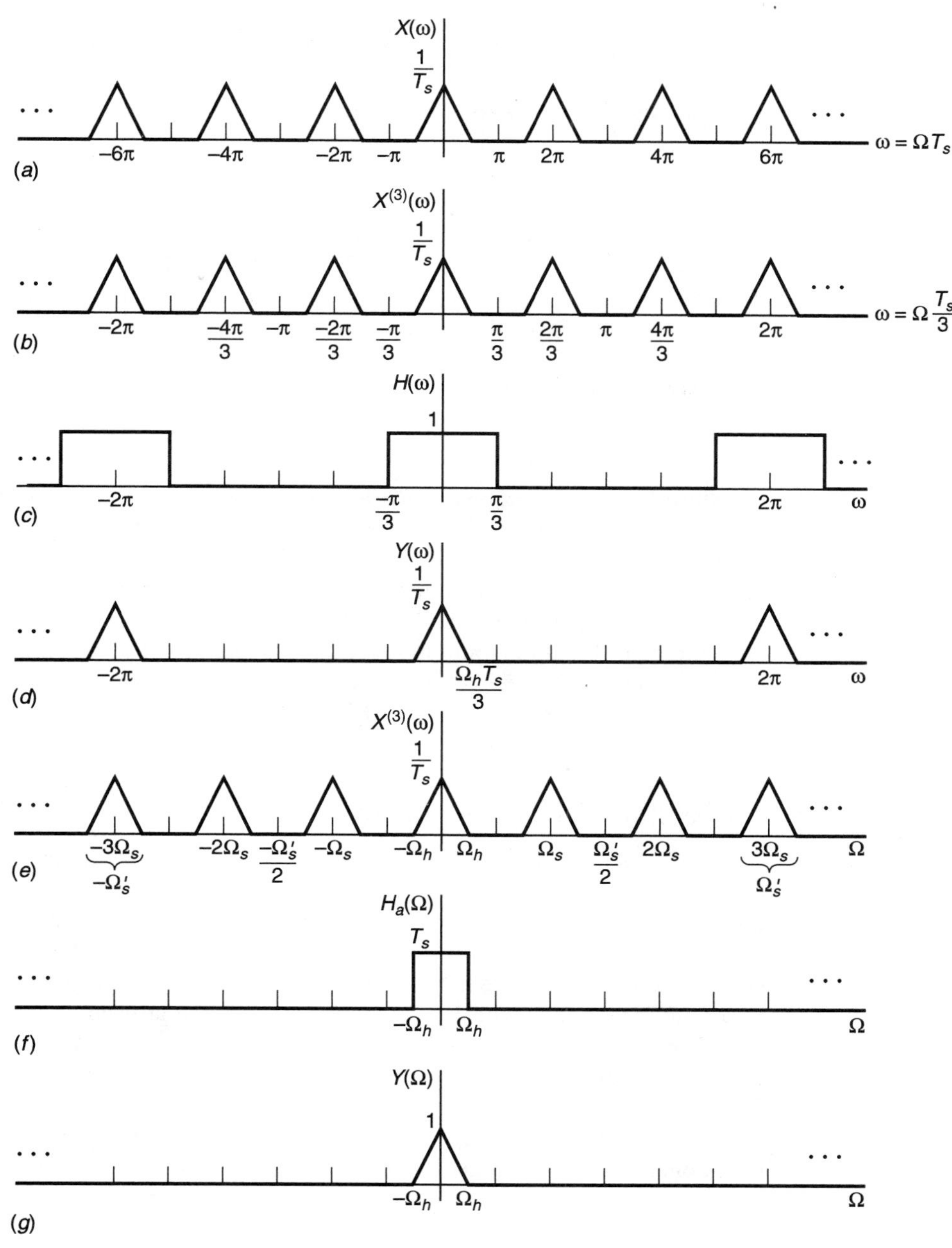

Figure 2.22 Spectrum of (*a*) original sequence, (*b*) upsampled sequence; (*c*) post-processing digital filter; (*d*) output of digital filter; (*e*) upsampled signal in the analog frequency domain; (*f*) post-processing analog filter; (*g*) output of analog filter.

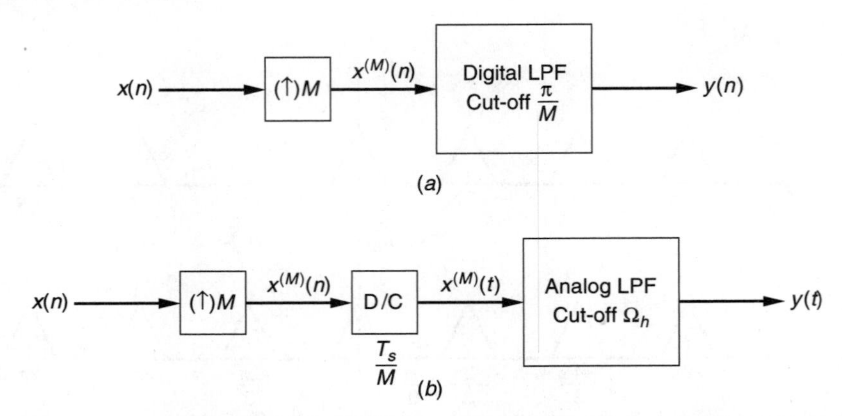

Figure 2.23 Upsampling followed by post-processing in (a) digital domain (b) analog domain.

one-third that of the original. That is, the new sampling frequency is $\Omega_s' = 3\Omega_s$. The spectrum of the desired analog LPF is shown in (f), and the final output spectrum is given in (g).

We just described two different techniques for post-processing after upsampling. The application at hand would dictate the choice between the two. These two techniques are denoted by block diagrams in Fig. 2.23. The process of postfiltering after upsampling is called *interpolation*. A comprehensive discussion of decimation and interpolation will appear in Chapter 5.

▶ 2.5 DISCRETE FOURIER TRANSFORM (DFT)

As we have learned in the preceding sections, the DTFT is an important tool for analyzing discrete-time sequences and systems. As noted earlier, the DTFT is the transform of a discrete sequence $x(n)$, but the transform $X(\omega)$ itself is a continuous-time function of the frequency ω. Since all signals in a digital computer (including the ones that are computed) are discrete in nature, the frequency variable ω in the DTFT is discretized in the interval $(0, 2\pi)$ and the transform is calculated at each of these discrete points. The result is a discrete sequence $X(k)$, and this is called the discrete Fourier transform (DFT). The DFT has its own set of properties and algorithms for efficient computation. These will be discussed in this section.

The DFT is obtained by simply discretizing the frequency variable ω in the DTFT equation. Reconsider the DTFT analysis equation:

$$X(\omega) = \sum_{n=-\infty}^{\infty} x(n)e^{-j\omega n}.$$

In order to discretize ω in the interval $(0, 2\pi)$, we must first decide on the number of points we want in this interval. If N represents this number, then the discrete values of

ω are 0, $\frac{2\pi}{N}, \frac{2\cdot 2\pi}{N}, \frac{3\cdot 2\pi}{N}, \ldots, \frac{(N-1)\cdot 2\pi}{N}$. Substituting these discrete values of ω in the DTFT equation, we get the following.

DFT analysis equation

$$X(k) = \sum_{n=0}^{N-1} x(n) e^{-j\frac{2\pi}{N}kn}, \quad 0 \le k \le N-1. \tag{2.63}$$

Equation (2.63) is called the N-point DFT of the sequence $x(n)$. For each value of k, the value of $X(k)$ represents the Fourier transform at the frequency $\omega = \frac{2\pi}{N}k$. As for the synthesis equation, we know that the synthesis equation for the DTFT has an integral and that the power of the exponential is positive. Since ω is discrete in the DFT, we expect the integral to be replaced by a sum. The DFT synthesis equation is as follows.

DFT synthesis equation

$$x(n) = \frac{1}{N} \sum_{k=0}^{N-1} X(k) e^{j\frac{2\pi}{N}kn}, \quad 0 \le n \le N-1. \tag{2.64}$$

The proof of the above (see Problem 40) follows by simply substituting the expression for $X(k)$ from (2.63) into the RHS of (2.64) and showing that it simplifies to $x(n)$. Note that in this simplification process, we will need to swap the summations and then use the orthogonality property of discrete-time exponentials, which is given below.

Lemma 2.2 *(Orthogonality of discrete exponentials) For integers N, n, r and l,*

$$\sum_{n=0}^{N-1} e^{j\frac{2\pi}{N}nl} = \begin{cases} N, & l = rN \\ 0, & \text{otherwise.} \end{cases} \tag{2.65}$$

Proof. By using the formula for the finite sum given in (2.29), we have

$$\sum_{n=0}^{N-1} (e^{j\frac{2\pi}{N}l})^n = \frac{1 - (e^{j\frac{2\pi}{N}l})^N}{1 - e^{j\frac{2\pi}{N}l}}. \tag{2.66}$$

Formula (2.66) is valid as long as we don't have a $\frac{0}{0}$ condition, which is undefined. If $l = rN$, we have this undefined condition. For this case, we use the original formula directly to get $\sum_{n=0}^{N-1} e^{j\frac{2\pi}{N}nl} = \sum_{n=0}^{N-1} e^{j\frac{2\pi}{N}nrN} = N$, which is one part of the desired result. If $l \ne rN$, we use (2.66) and find that the denominator is nonzero, but the numerator is zero, and we get the other part of the result. ■

Now let us revisit the DFT analysis equation and examine its computation. To compute $X(k)$ for some k, we multiply each $x(n)$ by a unit vector (complex exponential with unit magnitude) and sum up the products. For example, if $N = 8$, in order to find $X(0)$, we multiply $x(0), x(1), \ldots, x(7)$, each by $e^{j0} = 1$, and sum them up. To find $X(1)$, each $x(n)$ is multiplied, respectively, by each of the vectors $e^{j0}, e^{-j\frac{2\pi}{8}\cdot 1}, e^{-j\frac{2\pi}{8}\cdot 2}, \ldots, e^{-j\frac{2\pi}{8}\cdot 7}$ shown in Fig. 2.24. The products are then summed up to yield $X(1)$. To find $X(2)$, we multiply $x(n)$

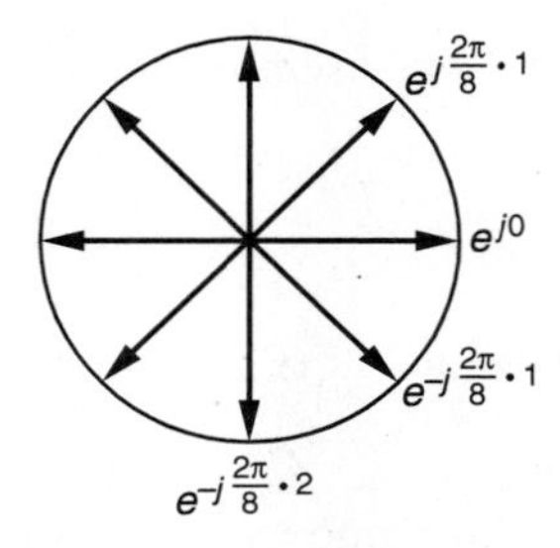

Figure 2.24 Unit vectors for an 8-point DFT.

successively by the vectors in the same figure, but this time we take every other vector as we go clockwise around the circle. Of course, we have to go around the circle twice so as to pick up the eight vectors. For $X(3)$, we pick every third vector and go around the circle three times; for $X(4)$ four times and so on. From this discussion it is clear that to compute $X(k)$ for each k, it requires N complex multiplications[5] and $N-1$ complex additions.[6] So, for N values of k, that is, for the entire DFT, it requires N^2 complex multiplications and $N(N-1) \simeq N^2$ complex additions. Thus, the DFT is computationally very intensive. As mentioned earlier, some algorithms compute the DFT at a much lower computational cost. These are called fast Fourier transforms (FFTs) and will be discussed later in this chapter.

Properties of the DFT

The properties of the DFT are similar to those of the DTFT, with one basic difference. This involves the concept of circular shift of a sequence. A circularly shifted sequence is denoted by $x((n-L))_N$. As an example, consider the sequence shown in Fig. 2.25(a). Let $N=5$ in this case. To find the circularly shifted sequence $x((n-1))_5$, we first perform a linear shift operation as shown in part (b) of the figure. A circularly shifted sequence is only valid for the same domain as the original sequence, which is $0 \le n \le 4$. So, we take $x(5)$ and rotate it to the left to form $x(0)$. A similar procedure is used to find $x((n-2))_5$. In this case, $x(6)$ is rotated to position $x(1)$, and then $x(5)$ is rotated to position $x(0)$. These linear and circularly shifted sequences are shown in Fig. 2.25(d) and (e), respectively. Now consider finding $x((-n))_N$. The linearly shifted sequence is shown in Fig. 2.25(f). Since four of the samples are to the left of the desired domain of n, they have to be rotated to the right. $x(-4)$ rotates to position $x(1)$, $x(-3)$ to position $x(2)$, and so on. The final sequence is shown in Fig. 2.25(g).

Now, we can mathematically represent a circularly shifted sequence as follows. If $x(n)$ is in the domain $N_1 \le n \le N_2$, and r is an integer, then

$$\begin{aligned} x_c(n) &\triangleq x((n-L))_N \\ &= x([n-L] \bmod N) \\ &= x(n-L+rN), \quad N_1 \le n-L+rN \le N_2. \end{aligned} \tag{2.67}$$

[5]Consider the multiplication of two complex numbers $(a+jb)(c+jd) = ac - bd + j(bc+ad)$. This requires four real multiplications and two real additions.
[6]A complex addition $(a+jb)+(c+jd) = (a+c)+j(b+d)$ requires two real additions.

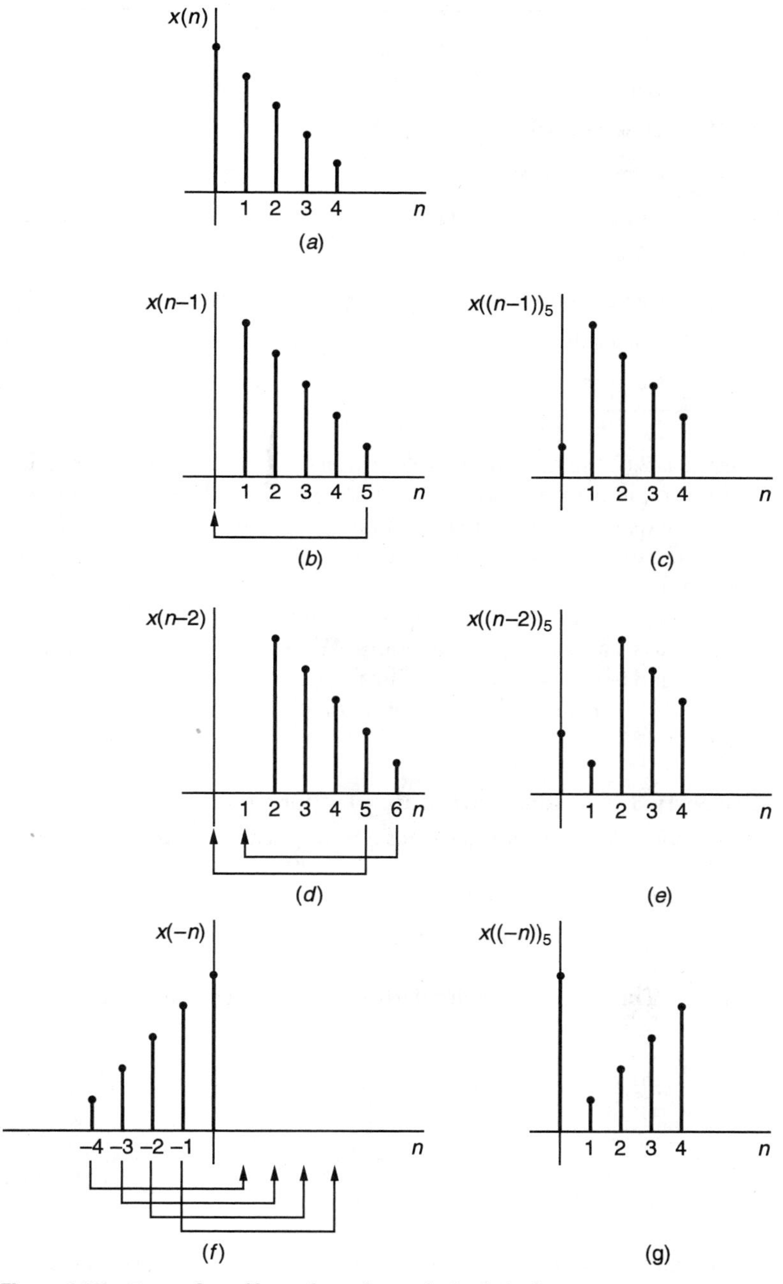

Figure 2.25 Examples of linearly and circularly shifted sequences.

TABLE 2.4 DFT Properties

	Property	Function	DFT
1.	Linearity	$ax_1(n) + bx_2(n)$	$aX_1(k) + bX_2(k)$
2.	Time reversal	$x(-n)$	$X(-k)$
3.	Circular time shift	$x((n-n_0))_N$	$e^{-j\frac{2\pi}{N}kn_0}X(k)$
4.	Circular frequency shift	$e^{j\frac{2\pi}{N}k_0 n}x(n)$	$X((k-k_0))_N$
5.	Circular convolution in time	$(x(n) * h(n))_N$	$X(k)H(k)$
6.	Cicular convolution in frequency	$x(n)h(n)$	$\frac{1}{N}(X(k) * H(k))_N$
7.	Conjugation in time	$x^*(n)$	$X^*((-k))_N$
8.	Conjugation in frequency	$x^*((-n))_N$	$X^*(k)$

Formula (2.67) suggests that, as long as $n-L$ is within the original domain, the circularly shifted sequence is the same as the linearly shifted sequence. Otherwise, the circularly shifted sequence is obtained by finding an integer r such that $n-L+rN$ is in the same domain as the original sequence. The reader is encouraged to use this formula to determine the circularly shifted sequences in Fig. 2.25.

A list of the simple DFT properties is given in Table 2.4; these can be proved using the DFT analysis and synthesis equations. We will prove a couple of them, and the reader is encouraged to verify the others. The notation for a sequence and its DFT will be denoted by $x(n) \leftrightarrow X(k)$, and unless otherwise mentioned the sequence will be assumed to be of length N.

Property 3 (Circular shift in time): For an N-point sequence, a circular shift in the time domain by n_0 samples causes a multiplication in the frequency domain by $e^{-j\frac{2\pi}{N}kn_0}$; that is,

$$x((n-n_0))_N \leftrightarrow e^{-j\frac{2\pi}{N}kn_0}X(k). \tag{2.68}$$

Proof. The DFT of the circularly shifted sequence is given by

$$\begin{aligned} DFT\{x((n-n_0))_N\} &= \sum_{n=0}^{N-1} x((n-n_0))_N e^{-j\frac{2\pi}{N}kn} \\ &= \sum_{n=n_0}^{N-1} x(n-n_0)e^{-j\frac{2\pi}{N}kn} + \sum_{n=0}^{n_0-1} x((n-n_0))_N e^{-j\frac{2\pi}{N}kn}. \end{aligned}$$

Note that if $n_0 > 0$, the sequence shifts to the right by n_0 points. So the shifted sequence in the interval $(n_0, N-1)$ is still in the domain. This allows us to replace the circular shift by a linear shift in the first term of the above equation. For $n_0 < 0$, the limits in the equation need to be modified, but the same technique of proof works as well. Now we replace the circular shift in the second term by the formula in (2.67), to get

$$DFT\{x((n-n_0))_N\} = \sum_{n=n_0}^{N-1} x(n-n_0)e^{-j\frac{2\pi}{N}kn} + \sum_{n=0}^{n_0-1} x(N+n-n_0)e^{-j\frac{2\pi}{N}kn}. \tag{2.69}$$

Now we perform a change of variables with $m = n - n_0$ in the first summation and $m = N + n - n_0$ in the second. Some further simplifications yield the desired result

$$\begin{aligned} DFT\{x((n-n_0))_N\} &= \sum_{m=0}^{N-1-n_0} x(m)e^{-j\frac{2\pi}{N}k(m+n_0)} + \sum_{m=N-n_0}^{N-1} x(m)e^{-j\frac{2\pi}{N}k(m+n_0-N)} \\ &= e^{-j\frac{2\pi}{N}kn_0} \sum_{m=0}^{N-1-n_0} x(m)e^{-j\frac{2\pi}{N}km} + e^{-j\frac{2\pi}{N}kn_0}e^{j2\pi k} \sum_{m=N-n_0}^{N-1} x(m)e^{-j\frac{2\pi}{N}km} \\ &= e^{-j\frac{2\pi}{N}kn_0} \sum_{m=0}^{N-1} x(m)e^{-j\frac{2\pi}{N}km} \\ &= e^{-j\frac{2\pi}{N}kn_0} X(k). \end{aligned}$$

■

Property 5 (Circular convolution): The product of two sequences in the frequency domain corresponds to circular convolution in the discrete-time domain; that is,

$$(x(n) * h(n))_N \leftrightarrow X(k)H(k)$$

where the left-hand side denotes circular convolution.

Proof. The circular convolution operation is written as

$$(x(n) * h(n))_N = \sum_{m=0}^{N-1} x(m)h((n-m))_N. \tag{2.70}$$

Taking the DFT gives

$$\begin{aligned} DFT\{(x(n) * h(n))_N\} &= \sum_{n=0}^{N-1}\sum_{m=0}^{N-1} x(m)h((n-m))_N e^{-j\frac{2\pi}{N}kn} \\ &= \sum_{m=0}^{N-1} x(m) \sum_{n=0}^{N-1} h((n-m))_N e^{-j\frac{2\pi}{N}kn}. \end{aligned}$$

Now we use the same technique as in the last example in order to decompose the summation involving $h((n-m))_N$ into two summations that involve linear shifts and then perform a change of variables, as follows:

$$
\begin{aligned}
DFT\{(x(n) * h(n))_N\} &= \sum_{m=0}^{N-1} x(m) \left[\sum_{n=m}^{N-1} h(n-m) e^{-j\frac{2\pi}{N}kn} + \sum_{n=0}^{m-1} h(n-m+N) e^{-j\frac{2\pi}{N}kn} \right] \\
&= \sum_{m=0}^{N-1} x(m) \left[\sum_{l=0}^{N-m-1} h(l) e^{-j\frac{2\pi}{N}(m+l)k} + \sum_{l=N-m}^{N-1} h(l) e^{-j\frac{2\pi}{N}(l+m-N)k} \right] \\
&= \sum_{m=0}^{N-1} x(m) \left[\sum_{l=0}^{N-1} h(l) e^{-j\frac{2\pi}{N}mk} e^{-j\frac{2\pi}{N}lk} \right] \\
&= \sum_{m=0}^{N-1} x(m) e^{-j\frac{2\pi}{N}mk} \sum_{l=0}^{N-1} h(l) e^{-j\frac{2\pi}{N}lk} \\
&= X(k)H(k).
\end{aligned}
$$

■

This above property is very important because of its significance in linear systems. As we know, the output of a linear system is the linear convolution of the input and the impulse response. In the DFT domain, the product $X(k)H(k)$ corresponds to circular convolution and not linear convolution. That is, if we perform an inverse DFT of $X(k)H(k)$, the result is the circular convolution $(x(n) * h(n))_N$. However, there is a way to zero-pad the input and the impulse response so that the inverse DFT yields linear convolution. We will return to this issue after a few examples on computing the DFT and circular convolution.

▶ **EXAMPLE 2.6** *Determine the DFT of the following: (a)* $\delta(n)$; *(b)* $u(n)u(N-1-n)$; *(c)* $(0.9)^n, 0 \leq n \leq 31$.

SOLUTION

(a) $X(k) = DFT\{\delta(n)\} = \sum_{n=0}^{N-1} \delta(n) e^{-j\frac{2\pi}{N}kn} = 1$. The sequence and its DFT are shown in Fig. 2.26.

(b) This sequence has unit amplitude for $0 \leq n \leq N-1$. The DFT is therefore

$$
X(k) = \sum_{n=0}^{N-1} e^{-j\frac{2\pi}{N}kn}.
$$

The solution of the above equation is given by Lemma 2.2 as $X(k) = N$ for $k = 0$, and zero otherwise. The sequence and its DFT are plotted in Fig. 2.27.

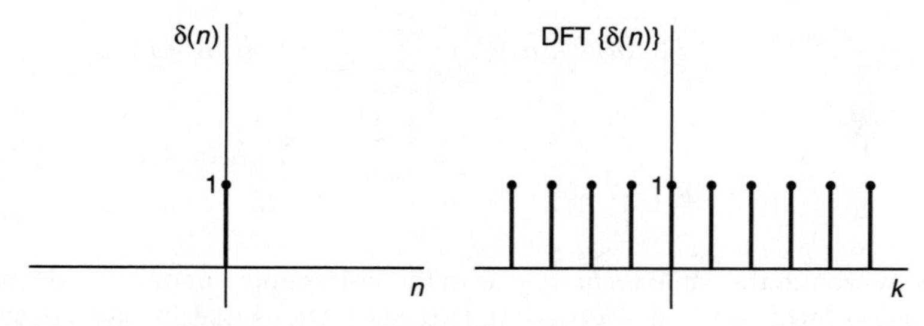

Figure 2.26 The discrete impulse and its DFT.

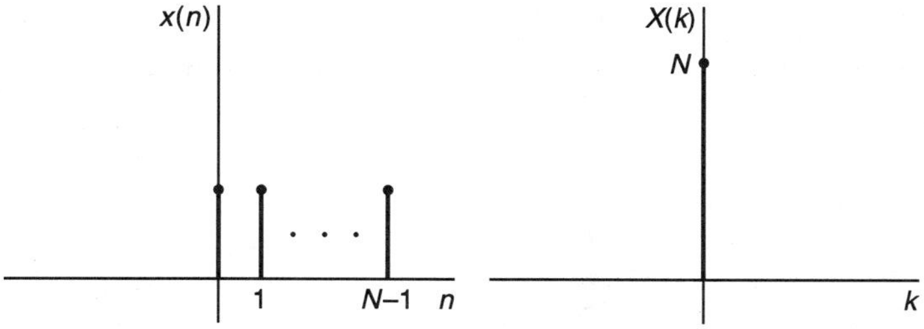

Figure 2.27 A constant sequence and its DFT.

(c) The DFT for this sequence is obtained by using the DFT formula and then using the finite geometric sum formula, as

$$X(k) = \sum_{n=0}^{31} (0.9)^n e^{-j\frac{2\pi}{32}kn}$$

$$= \sum_{n=0}^{31} (0.9e^{-j\frac{2\pi}{32}k})^n$$

$$= \frac{1-(0.9e^{-j\frac{2\pi}{32}k})^{32}}{1-0.9e^{-j\frac{2\pi}{32}k}}, \quad 0 \leq k \leq 31.$$

The magnitude and phase of the above are shown in Fig. 2.28. This sequence $x(n)$ is the same as that in Example 2.3(c), and its DTFT was plotted in Fig. 2.8. Note that the DFT of the magnitude and phase are the sampled versions of the DTFT. In this example, the DFT consists of 32 points and spans the frequencies of $\omega = 0$ to $\omega = \frac{2\pi}{32} \cdot 31$.

EXAMPLE 2.7 *Perform the circular convolution of the following sequences:*

$$x_1(n) = [\mathbf{-1}\ 2\ {-3}\ 2], \quad x_2(n) = [\mathbf{-0.5}\ 1\ 1.5],$$

where the digits in bold represent the value at $n = 0$.

SOLUTION The circular convolution can be easily calculated by using the vector method. The basic steps of "Switch-Shift-Sum" described in Chapter 1 for linear convolution are also applicable here, except that the "Shift" operation is done in a rotating fashion. First, let's flip the sequence $x_1(n)$ and switch the independent variable from n to m; that is,

$$x_1(-m) = [2\ {-3}\ 2\ \mathbf{-1}];$$

Rotating $x_1(-m)$ around to the same domain as the original sequence gives the circularly shifted sequence

$$x_1((-m))_4 = [\mathbf{-1}\ 2\ {-3}\ 2];$$

Similarly, we can find the following circularly shifted sequences which are as follows.

$$x_1((1-m)_4) = [\mathbf{2}\ {-1}\ 2\ {-3}],$$

$$x_1((2-m)_4) = [\mathbf{-3}\ 2\ {-1}\ 2],$$

$$x_1((3-m)_4) = [\mathbf{2}\ {-3}\ 2\ {-1}].$$

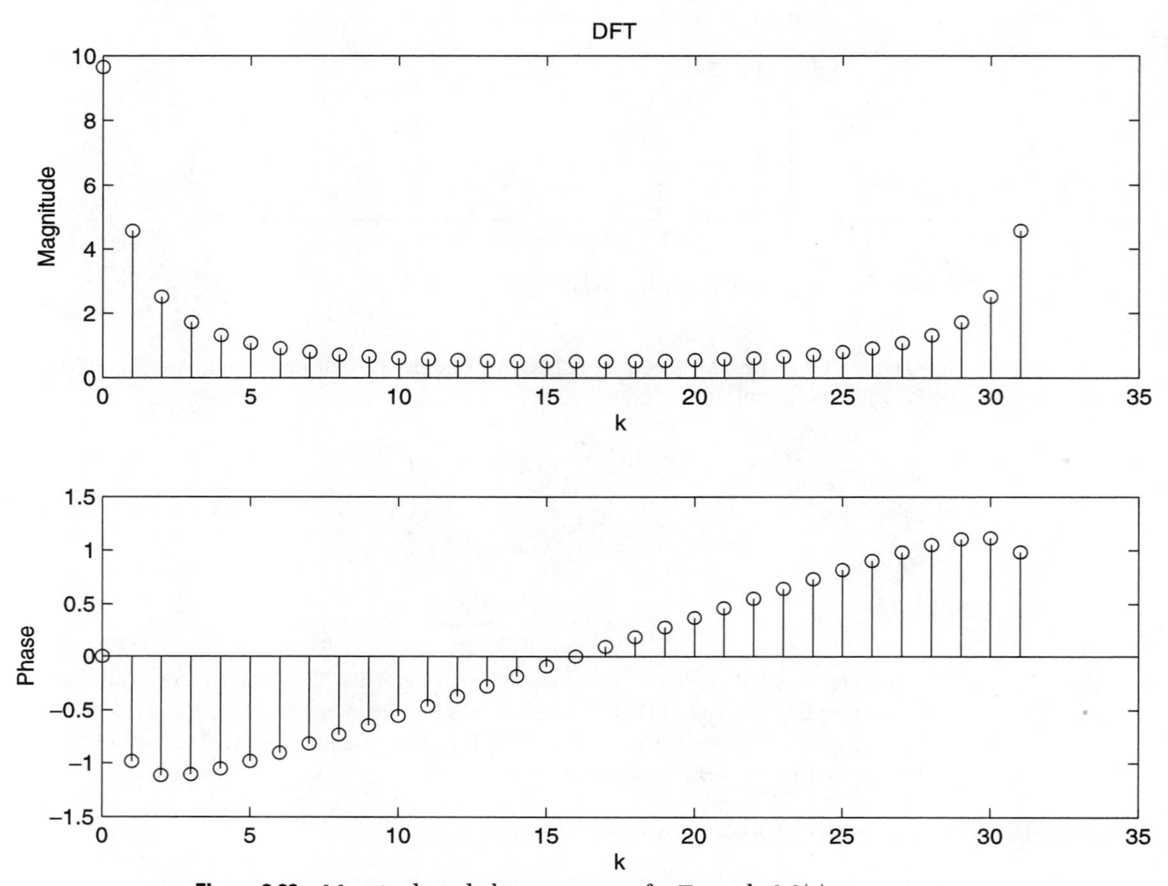

Figure 2.28 Magnitude and phase responses for Example 2.6(c).

The result of the circular convolution is obtained by multiplying each of these sequences by the unshifted sequence $x_2(m) = [\,-\mathbf{0.5}\;\;1\;\;1.5]$ and summing the products as follows:

$$
\begin{aligned}
y(0) &= (-1) \times (-0.5) + 2 \times 1 + (-3) \times 1.5 = -2,\\
y(1) &= 2 \times (-0.5) + (-1) \times 1 + 2 \times 1.5 = 1,\\
y(2) &= (-3) \times (-0.5) + 2 \times 1 + (-1) \times 1.5 = 2,\\
y(3) &= 2 \times (-0.5) + (-3) \times 1 + 2 \times 1.5 = -1,
\end{aligned}
$$

That is, $y(n) = [\mathbf{-2}\;\;1\;\;2\;\;-1]$.
This example is graphically illustrated in Fig. 2.29. ◀

From Example 2.7 and the earlier discussion on circular shift, we know that circular convolution yields a different result than linear convolution. This is because a circularly shifted sequence is different from a linearly shifted sequence. Let us consider two sequences of length N and M, with $N > M$. The result of their linear convolution is

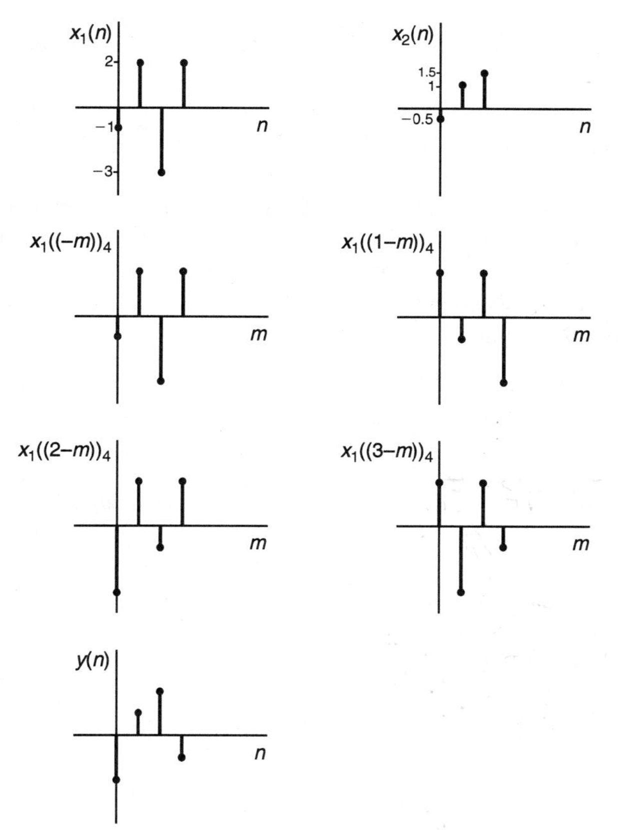

Figure 2.29 Example of circular convolution.

of length $M + N - 1$, whereas that of circular convolution is of length N. If we can somehow make the result of circular shift the same as that of linear shift, then the two convolutions will yield the same answer. In the circular shift of a N-point sequence, we first linearly shift the sequence and then rotate those elements that leave the original N-point domain of $n(0 \leq n \leq N - 1)$ into the original domain. These rotated elements are the ones that are in different positions in a linearly shifted sequence. This tells us that if we insert an appropriate number of zeros at the end of the original sequence and then perform the circular convolution, then we would only rotate the zeros around and the result would be the same as the linearly shifted sequence. The number of zeros to be padded must be such that the length of the sequence is increased to $N + M - 1$. For example, consider the sequence $x(n) = [\mathbf{1}\ \ 2\ \ 3\ \ 4]$, where the boldfaced element is the value for $x(0)$. The linearly shifted sequence $x(n-3) = [\mathbf{0}\ \ 0\ \ 0\ \ 1\ \ 2\ \ 3\ \ 4]$. The circularly shifted sequence is $x((n-3))_4 = [\mathbf{2}\ \ 3\ \ 4\ \ 1]$. Since we have a shift of 3, zero-pad the original sequence as $x(n) = [\mathbf{1}\ \ 2\ \ 3\ \ 4\ \ 0\ \ 0\ \ 0]$. Now the circularly shifted sequence in $x((n-3))_7 = [\mathbf{0}\ \ 0\ \ 0\ \ 1\ \ 2\ \ 3\ \ 4]$, which is the same as the linearly shifted sequence. We can now formally state the following result.

Fact: *Let the sequences $x(n)$ and $h(n)$ be of length N and M, respectively. Zero-pad each sequence so that they are both of length $N + M - 1$. Let $X(k)$ and $H(k)$ represent the DFT* *of the sequences. Then the inverse DFT of $X(k)H(k)$ is equal to the linear convolution $x(n) * h(n)$.*

2.6 FAST FOURIER TRANSFORM

As mentioned earlier, the fast Fourier transform (FFT) refers to algorithms that compute the discrete Fourier transform (DFT) in a numerically efficient manner. Although there are many such FFT algorithms, we will present only two such algorithms: *decimation in time* and *decimation in frequency*. For notational convenience, we define $W_N = e^{-j\frac{2\pi}{N}}$ and rewrite the DFT equation as

$$X(k) = \sum_{n=0}^{N-1} x(n)W_N^{kn}, \quad 0 \le k \le N-1. \tag{2.71}$$

We will use the following properties in the derivation of the algorithm.

Lemma 2.3 *1.* $W_N^2 = W_{\frac{N}{2}}$; *2.* $W_N^{k+\frac{N}{2}} = -W_N^k$.

Proof. The proofs are obtained directly from the definition as follows:

1. $W_N^2 = e^{-j\frac{2\pi}{N}\cdot 2} = e^{-j\frac{2\pi}{N/2}} = W_{\frac{N}{2}}$.
2. $W_N^{k+\frac{N}{2}} = e^{-j\frac{2\pi}{N}\frac{N}{2}}e^{-j\frac{2\pi}{N}k} = -e^{-j\frac{2\pi}{N}k} = -W_N^k$. ■

Decimation in Time FFT

The decimation in time algorithm uses the "divide and conquer" approach. It is assumed that N is a power of 2, that is, $N = 2^p$. Then the N-point DFT is performed as several 2-point DFTs. Let us first consider a 2-point DFT. The equations are

$$X(0) = x(0) + x(1)$$
$$X(1) = x(0) + x(1)e^{-j\pi} = x(0) - x(1).$$

Note that the 2-point DFT requires two additions and no multiplications. The flow diagram is shown in Fig. 2.30. Now consider decimating the sequence by a factor of 2; that is, consider the even and odd components separately. Equation (2.71) can be written as

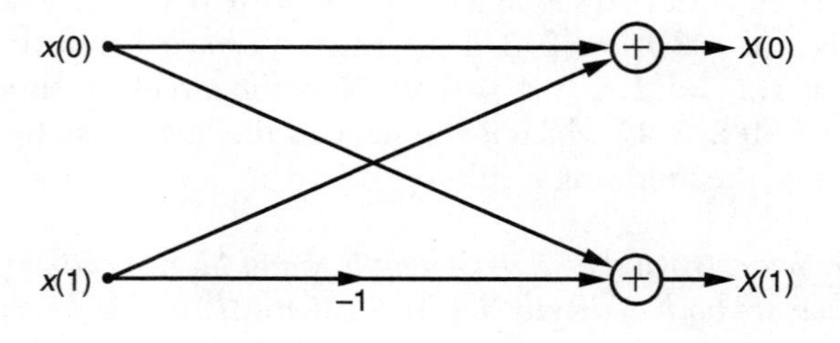

Figure 2.30 A multiplierless 2-point DFT.

$$X(k) = \sum_{n=0}^{\frac{N}{2}-1} x(2n)W_N^{k2n} + \sum_{n=0}^{\frac{N}{2}-1} x(2n+1)W_N^{k(2n+1)}$$

$$= \sum_{n=0}^{\frac{N}{2}-1} x(2n)W_{\frac{N}{2}}^{kn} + W_N^k \sum_{n=0}^{\frac{N}{2}-1} x(2n+1)W_{\frac{N}{2}}^{kn}$$

$$= P(k) + W_N^k S(k), \quad 0 \leq k \leq N-1 \tag{2.72}$$

where $P(k)$ and $S(k)$ are $\frac{N}{2}$-point DFTs of the even-numbered samples and odd-numbered samples, respectively. These DFTs yield $P(0), P(1), \ldots, P(\frac{N}{2}-1)$ and $S(0), S(1), \ldots, S(\frac{N}{2}-1)$. For the overall N-point DFT, we need $P(k)$ and $S(k)$ for $k = 0, 1, \ldots, \frac{N}{2}-1, \ldots, N-1$. However, $P(k)$ and $S(k)$ are periodic with period $\frac{N}{2}$, and so for $k = \frac{N}{2}, \ldots, N-1$, the values are the same as those for $N = 0, 1, \ldots, \frac{N}{2}-1$. The flow diagram for the decomposition of (2.72) is shown in Fig. 2.31. This is referred to as stage 1 of the decomposition. Now look at each of the computations needed to go from the signal pair $(P(k), S(k))$ to form the pair $X(k), X(k+\frac{N}{2}))$. From (2.72), the equations are

$$X(k) = P(k) + W_N^k S(k), \quad 0 \leq k \leq \frac{N}{2} - 1 \tag{2.73}$$

$$X(k + \frac{N}{2}) = P(k) + W_N^{k+\frac{N}{2}} S(k), \quad 0 \leq k \leq \frac{N}{2} - 1$$

$$= P(k) - W_N^k S(k). \tag{2.74}$$

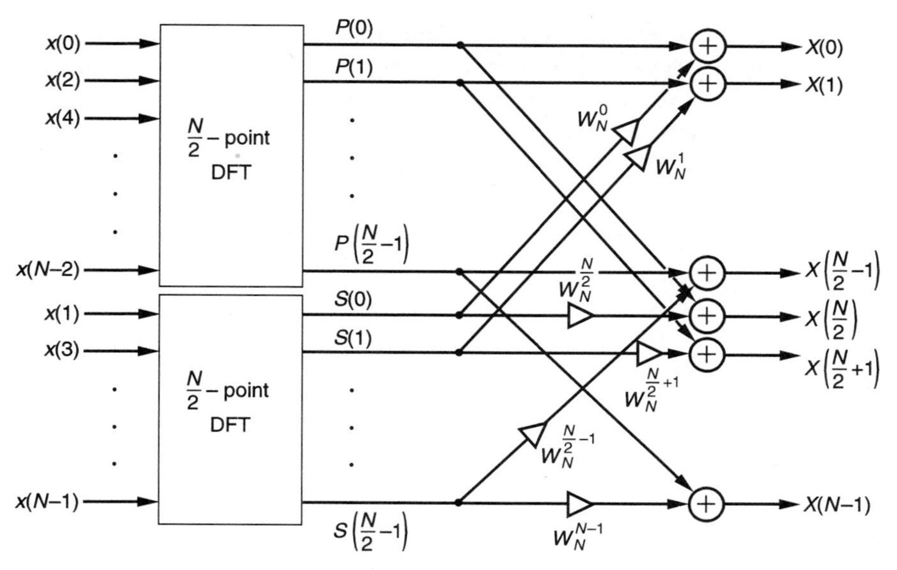

Figure 2.31 First stage of decomposition of the decimation in time FFT.

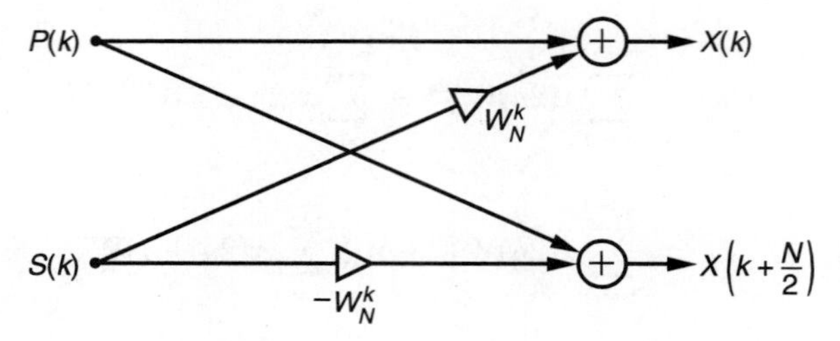

Figure 2.32 Butterfly.

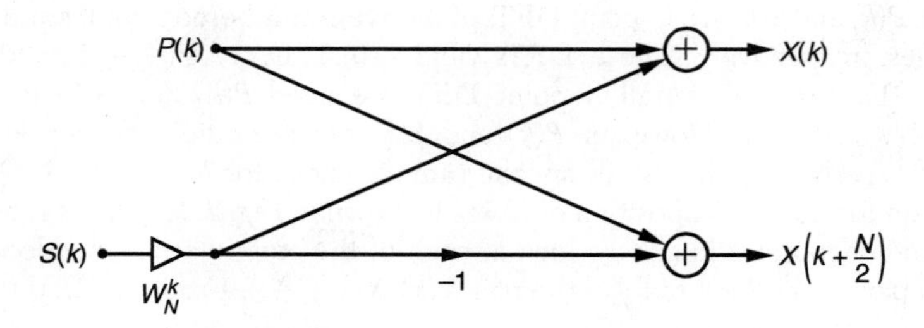

Figure 2.33 Reduced butterfly.

The flow diagram for (2.73) and (2.74) is given in Fig. 2.32. This basic computation is called the *butterfly* and is used in every stage of this algorithm. The butterfly has two multiplications and two additions. If we move the multiplier before the node $S(k)$ as shown in Fig. 2.33, the final result is the same, but we save one multiplication. This is called the *reduced butterfly*. A total of $\frac{N}{2}$ reduced butterflies are needed for this stage.

Now let us look at the computational count for this decomposition. Recall from the previous section, that a $\frac{N}{2}$-point DFT requires $(\frac{N}{2})^2$ complex multiplications and approximately the same number of complex additions.[7] Calculation of the $\frac{N}{2}$ reduced butterflies requires $\frac{N}{2}$ complex multiplications and N complex additions. The complex computations required for the above decomposition can be derived as follows.

$$\begin{aligned}\text{No. of computations for stage 1} &= \frac{N}{2} \text{ mults} + N \text{ adds} + 2 \times \left(\frac{N}{2}\text{-points DFTs}\right)\\ &= \frac{N}{2} \text{ mults } + N \text{ adds } +\\ &\quad 2 \times \left(\left(\frac{N}{2}\right)^2 \text{ mults } + \left(\frac{N}{2}\right)^2 \text{ adds}\right)\\ &= \left(\frac{N}{2} + \frac{N^2}{2}\right) \text{ mults and } \left(N + \frac{N^2}{2}\right) \text{ adds.} \end{aligned} \quad (2.75)$$

[7] Not all the multipliers are complex, for example, W_N^0. For simplicity, we will approximate by assuming all complex multipliers.

This should be compared to N^2 multiplications and N^2 additions for a direct DFT calculation. For $N \leq 3$, the computation of (2.75) requires fewer computations than the direct DFT. However, this is only part of the story. Instead of computing the $\frac{N}{2}$-point DFTs $P(k)$ and $S(k)$, we can decompose them as $\frac{N}{4}$-point DFTs and reduced butterflies in the same way as above. This is stage 2. In the next stage, the $\frac{N}{4}$-point DFTs are decomposed into $\frac{N}{8}$-point DFTs. We continue this process until we end up with 2-point DFTs. This would require $(p-1)$ stages where $N = 2^p$ or $p = \log_2 N$. The computational count at each of the stages is as follows.

Stage 1

$$\text{No. of computations} = \frac{N}{2}\text{ mults } + N\text{ adds } + 2 \times \left(\frac{N}{2}\text{-point DFTs}\right).$$

Stage 2

$$\begin{aligned}\text{No. of computations} &= \frac{N}{2}\text{ mults } + N\text{ adds } + 2 \times \left(\frac{N}{4}\text{ mults } + \frac{N}{2}\text{ adds } + \right.\\ &\qquad \left. 2 \times \left(\frac{N}{4}\text{-point DFTs}\right)\right)\\ &= 2\left(\frac{N}{2}\right)\text{ mults } + 2N\text{ adds } + 4 \times \left(\frac{N}{4}\text{-point DFTs}\right).\end{aligned}$$

Stage 3

$$\begin{aligned}\text{No. of computations} &= 2\left(\frac{N}{2}\right)\text{ mults } + 2N\text{ adds } + 4 \times \left(\frac{N}{8}\text{ mults } + \frac{N}{4}\text{ adds } + \right.\\ &\qquad \left. 2 \times \left(\frac{N}{8}\text{-point DFTs}\right)\right)\\ &= 3\left(\frac{N}{2}\right)\text{ mults } + 3N\text{ adds } + 8 \times \left(\frac{N}{8}\text{-point DFTs}\right).\end{aligned}$$

Stage $(p-1)$, where $p = \log_2 N$:

$$\begin{aligned}\text{No. of computations} &= (p-1)\frac{N}{2}\text{ mults} + (p-1)N\text{ adds} + 2^{(p-1)} \times (\text{2-point DFTs})\\ &= (p-1)\frac{N}{2}\text{ mults} + (p-1)N\text{ adds} + \frac{N}{2} \times (\text{0 mults and 2 adds})\\ &= (p-1)\frac{N}{2}\text{ mults and } pN\text{ adds}\\ &= \frac{N}{2}(\log_2 N - 1)\text{ multiplications and } N\log_2 N\text{ additions.}\end{aligned} \quad (2.76)$$

The decimation in time algorithm is now illustrated with an example of an 8-point FFT of a sequence $x(n)$. The first stage consists of separating the even and odd components of the signal and performing the 4-point DFTs and multiplications given by (2.72). This is shown in Fig. 2.34. In the second stage, the 4-point DFTs are decomposed into 2-point DFTs as shown in Fig. 2.35. The final flow diagram for the FFT is given in Fig. 2.36.

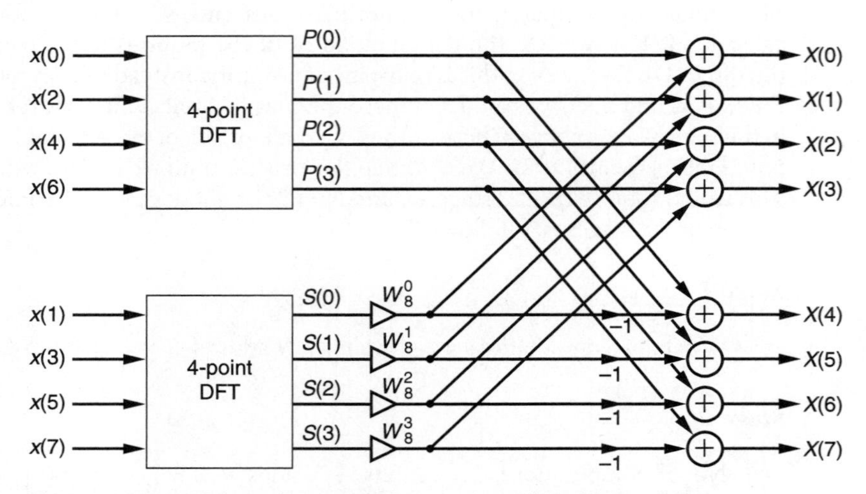

Figure 2.34 First stage of an 8-point FFT using reduced butterflies.

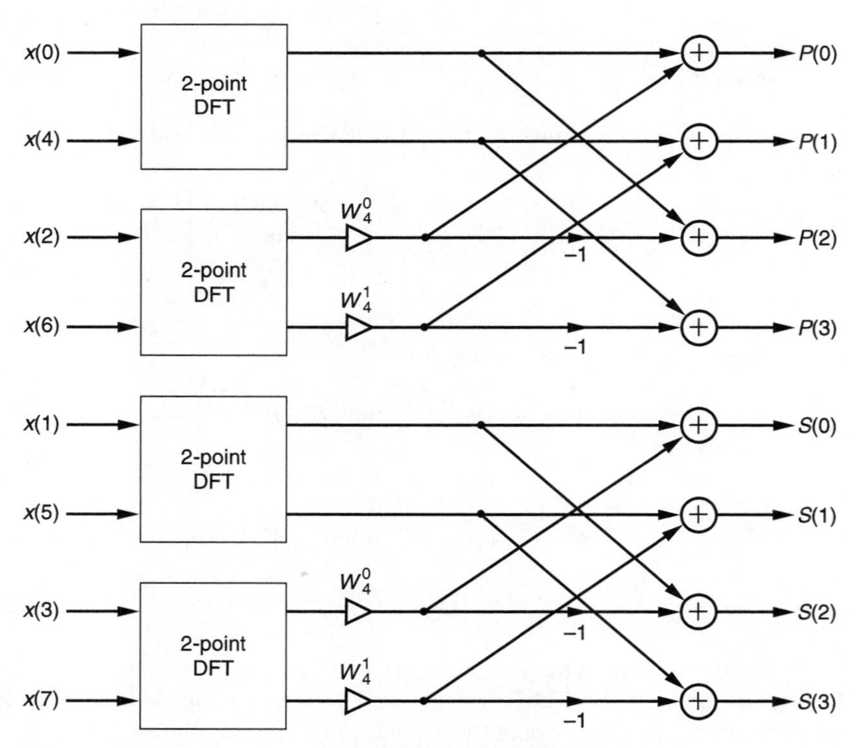

Figure 2.35 Second stage of an 8-point FFT using reduced butterflies.

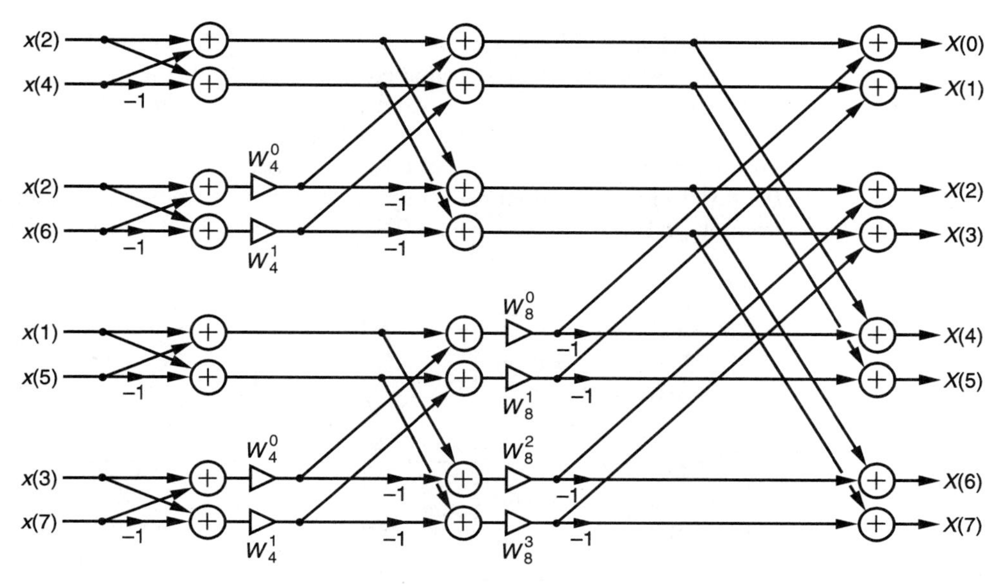

Figure 2.36 Complete 8-point decimation in the FFT.

Besides computations, another advantage comes naturally from this algorithm, namely, a reduction in storage requirement. Looking at the flow graph of Fig. 2.36, we see that the algorithm is performed in stages. Each stage is nothing but a set of butterfly computations shown in Fig. 2.33. The butterfly operates on a pair of complex numbers and produces another pair of complex numbers. Once the output pair is computed, we don't need the input pair anymore. Therefore, the output pair can be stored in the same locations as the input pair. So, for an N-point DFT, we need N complex registers or $2N$ real registers for the first stage. The same registers will be used for the subsequent stages. This is called *in-place computations* in the literature. In the case of the 8-point FFT, the butterfly computations are performed on the following pairs: $x(0)$ and $x(4)$, $x(2)$ and $x(6)$, $x(1)$ and $x(5)$, and $x(3)$ and $x(7)$. First label the input signal samples as follows:

$$
\begin{aligned}
X_0(0) &= x(0)\\
X_0(1) &= x(4)\\
X_0(2) &= x(2)\\
X_0(3) &= x(6)\\
X_0(4) &= x(1)\\
X_0(5) &= x(5)\\
X_0(6) &= x(3)\\
X_0(7) &= x(7).
\end{aligned}
$$

So, $X_i(n)$ refers to the data at the ith stage. The first set of butterflies will yield the data $X_1(0), X_1(1), \ldots, X_1(7)$. Now the butterflies will be performed on the (once) decimated data pairs given by $X_1(0)$ and $X_1(2), X_1(1)$ and $X_1(3), X_1(4)$ and $X_1(6), X_1(5)$ and $X_1(7)$.

This will yield the data set $X_2(0), X_2(1), \ldots, X_2(7)$. The butterflies in the final stage will be computed on the (twice decimated) data pairs $X_2(0)$ and $X_2(4), X_2(2)$ and $X_2(6), X_2(1)$ and $X_2(5), X_2(3)$, and $X_2(7)$, to yield the final DFT values. This scheme is pictorially illustrated in Fig. 2.37. Now let us look at the original data samples with the sample number represented in binary (3 bits in this case). As shown in Fig. 2.38, if we reverse the order of the bits, then we get the location where we want the data to be stored. For example, $x(1)$ is represented as $x(001)$. If we reverse the bits, we get $x(100)$, which means

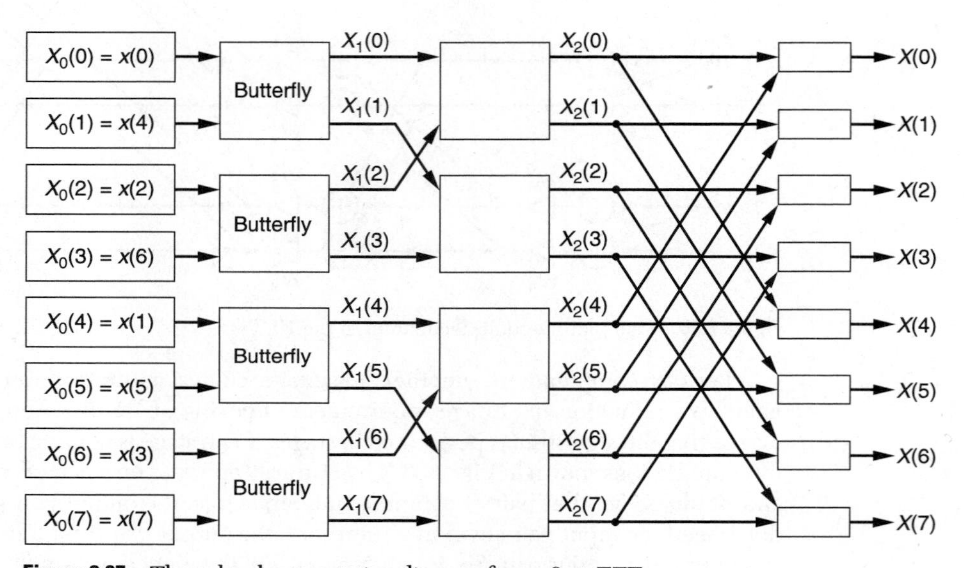

Figure 2.37 Three-level computation diagram for an 8-pt FFT.

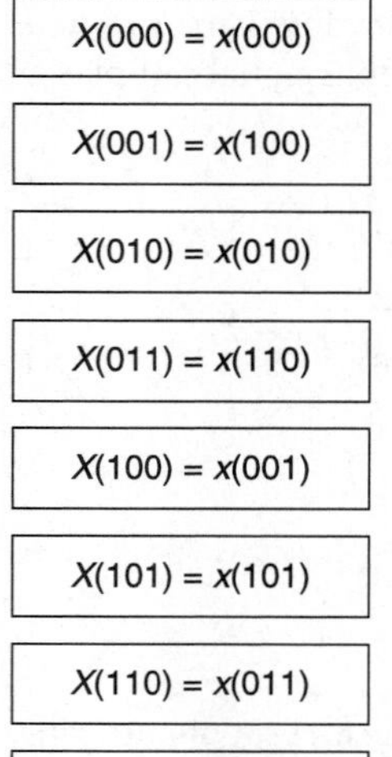

Figure 2.38 A set of N complex registers for bit-reversed storage of an 8-pt FFT.

that $x(1)$ will be stored in $X_0(4)$. Similarly, the bit-reversed version of $x(3) = x(011)$ is $x(110)$, which means that $x(3)$ will be stored in $X_0(6)$, and so on. Thus, if the input data samples $x(n)$ are stored in bit-reversed order, then the final DFT is obtained in natural order; that is, $X(k)$, $k = 0, 1, \ldots, 7$. It is easy to show that if the input data samples are stored in natural order, then the final DFT is obtained in bit-reversed order.

Decimation in Frequency FFT

In this algorithm, the output sequence $X(k)$ is divided into smaller and smaller subsequences in the same manner as the decimation in time algorithm. Again, we consider N to be a power of two, so that $N = 2^p$. The goal is to separately compute the even-numbered frequency samples and the odd-numbered frequency samples. We start with the DFT formula and decompose it as follows.

$$\begin{aligned} X(k) &= \sum_{n=0}^{N-1} x(n)W_N^{kn}, \quad k = 0, 1, \ldots, N-1 \\ &= \sum_{n=0}^{N/2-1} x(n)W_N^{kn} + \sum_{n=N/2}^{N-1} x(n)W_N^{kn} \\ &= \sum_{n=0}^{N/2-1} x(n)W_N^{kn} + \sum_{n=0}^{N/2-1} x\left(n+\frac{N}{2}\right) W_N^{k(n+N/2)} \\ &= \sum_{n=0}^{N/2-1} x(n)W_N^{kn} + W_N^{kN/2} \sum_{n=0}^{N/2-1} x\left(n+\frac{N}{2}\right) W_N^{kn}. \end{aligned} \tag{2.77}$$

Notice that $W_N^{kN/2} = (-1)^k$. Therefore,

$$X(k) = \sum_{n=0}^{N/2-1} \left[x(n) + (-1)^k x\left(n+\frac{N}{2}\right)\right] W_N^{kn}, \quad k = 0, 1, \ldots, N-1 \tag{2.78}$$

Now, we separate the above into even- and odd-numbered samples. From Lemma 2.3, we know that $W_N^2 = W_{N/2}$. Using this, the even-numbered samples can be written as

$$\begin{aligned} X(2k) &= \sum_{n=0}^{N/2-1} \left[x(n) + (-1)^{2k} x\left(n+\frac{N}{2}\right)\right] W_N^{2kn} \\ &= \sum_{n=0}^{N/2-1} \left[x(n) + x\left(n+\frac{N}{2}\right)\right] W_{N/2}^{kn}, \quad k = 0, 1, \ldots, \frac{N}{2}-1 \end{aligned} \tag{2.79}$$

The odd-numbered samples can be written as

$$X(2k+1) = \sum_{n=0}^{N/2-1} \left\{ \left[x(n) - x\left(n+\frac{N}{2}\right)\right] W_N^{n} \right\} W_{N/2}^{kn} \tag{2.80}$$

Define

$$e(n) = x(n) + x\left(n + \frac{N}{2}\right) \tag{2.81}$$

and

$$o(n) = \left[x(n) - x\left(n + \frac{N}{2}\right)\right] W_N^n. \tag{2.82}$$

Then the even- and odd-numbered samples are the $\frac{N}{2}$ point DFTs of $e(n)$ and $o(n)$, given, respectively, by

$$X(2k) = \sum_{n=0}^{N/2-1} e(n) W_{N/2}^{kn} \tag{2.83}$$

$$X(2k+1) = \sum_{n=0}^{N/2-1} o(n) W_{N/2}^{kn}, \quad k = 0, 1, \ldots, \frac{N}{2} - 1. \tag{2.84}$$

The flow graph for the above computations is shown in Fig. 2.39 for an 8-point DFT. The basic computation is a butterfly shown in Fig. 2.40. For the 8-point DFT example, each 4-point DFT is further decomposed into butterflies and 2-point DFTs as shown in Fig. 2.41. The final 2-point DFTs are implemented by simply adding and subtracting the inputs. The computational complexity can be found in the same way as for the decimation in time algorithm. It requires $\frac{N}{2}\log_2 N$ complex multiplications and

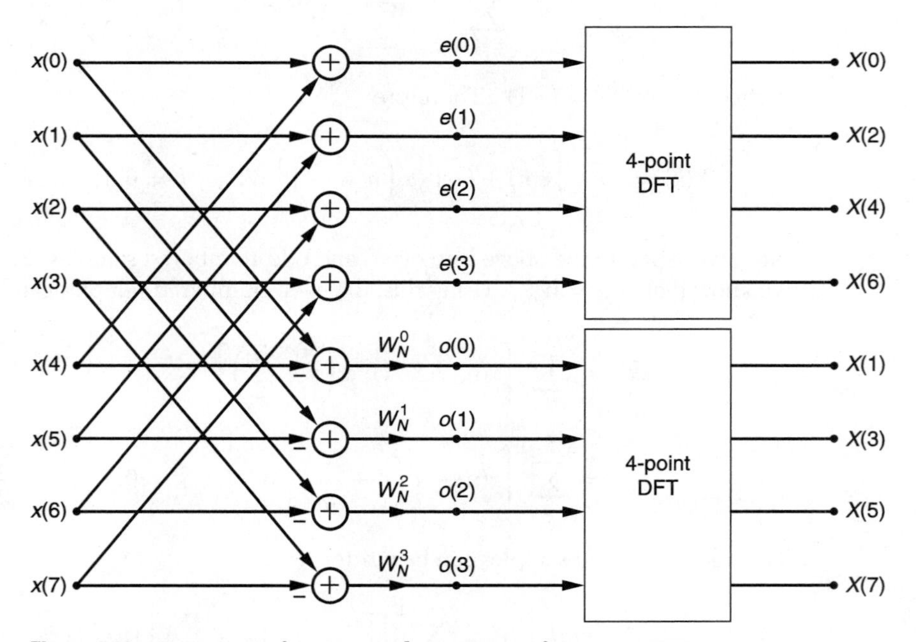

Figure 2.39 First stage of an 8-point decimation in frequency FFT.

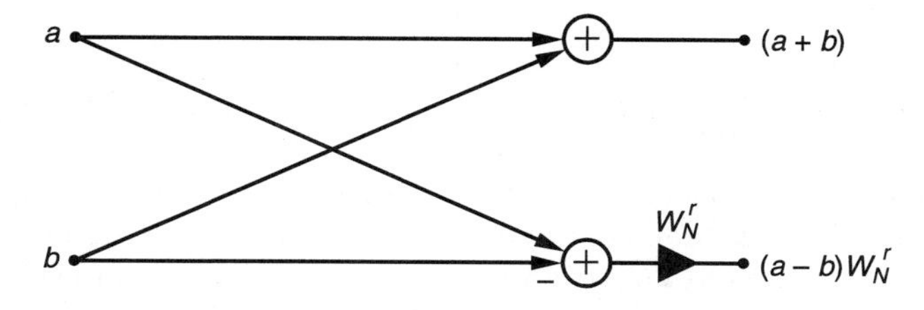

Figure 2.40 Basic butterfly computation in decimation in frequency FFT.

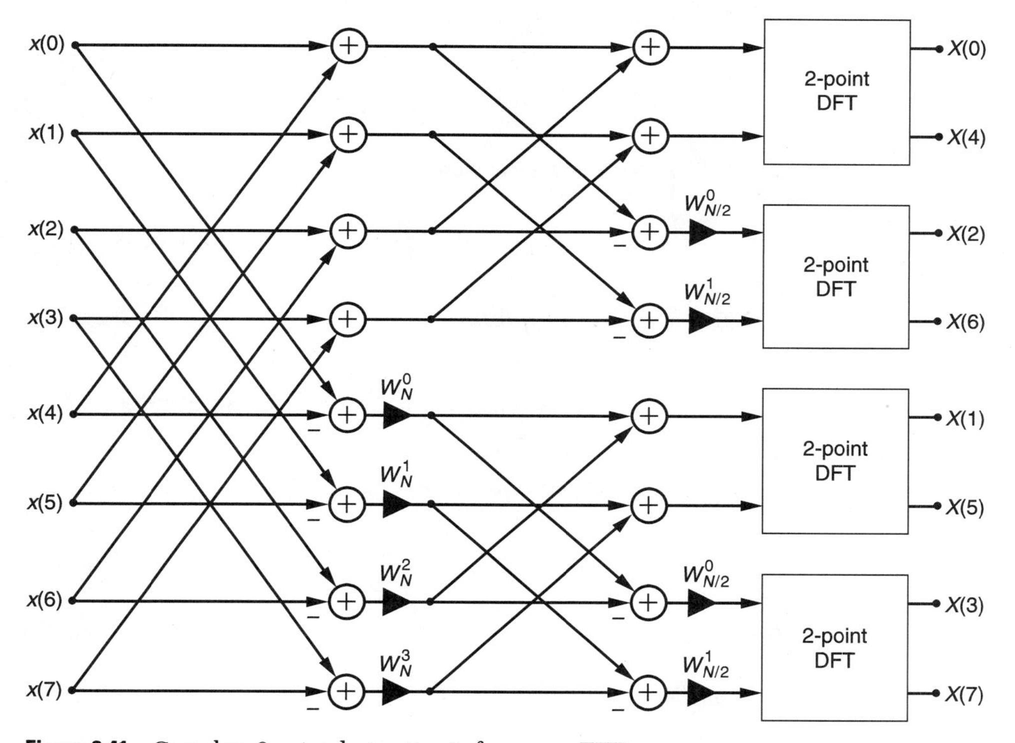

Figure 2.41 Complete 8-point decimation in frequency FFT.

$N \log_2 N$ complex additions. In-place computations can also be done in this method for minimal storage requirements.

2.7 2-D EXTENSIONS

Some of the results of the foregoing sections will now be generalized for two-dimensional signals and systems. Most of these results are stated without proof since the proofs are straightforward extensions of the 1-D counterparts. The material is organized into two subsections. In the first subsection, 2-D Fourier transforms are presented along with a discussion of sampling and reconstruction of 2-D signals. The second subsection presents some computational aspects of 2-D convolution and Fourier transform algorithms.

2-D Fourier Analysis

Let $x(t_1, t_2)$ and $X(\Omega_1, \Omega_2)$ denote a 2-D continuous-time signal and its CTFT, respectively. Also let $x(n_1, n_2)$ and $X(\omega_1, \omega_2)$ denote a 2-D sequence and its DTFT, respectively. Since a 2-D signal has two spatial (or time) dimensions, there are also two frequency variables. Again as before, (Ω_1, Ω_2) denotes the frequency pair for the continuous-time signal and (ω_1, ω_2) that of its discrete-time counterpart. The 2-D CTFT and DTFT equations are simple extensions of 1-D and are given in the following.

2-D CTFT analysis equation

$$X(\Omega_1, \Omega_2) = \int_{-\infty}^{\infty} \int_{-\infty}^{\infty} x(t_1, t_2) e^{-j\Omega_1 t_1} e^{-j\Omega_2 t_2} dt_1 dt_2 \tag{2.85}$$

2-D CTFT synthesis equation

$$x(t_1, t_2) = \frac{1}{(2\pi)^2} \int_{-\infty}^{\infty} \int_{-\infty}^{\infty} X(\Omega_1, \Omega_2) e^{j\Omega_1 t_1} e^{j\Omega_2 t_2} d\Omega_1 d\Omega_2 \tag{2.86}$$

2-D DTFT analysis equation

$$X(\omega_1, \omega_2) = \sum_{n=-\infty}^{\infty} \sum_{n=-\infty}^{\infty} x(n_1, n_2) e^{-j\omega_1 n_1} e^{-j\omega_2 n_2} \tag{2.87a}$$

2-D DTFT synthesis equation

$$x(n_2, n_2) = \frac{1}{(2\pi)^2} \int_{-\pi}^{\pi} \int_{-\pi}^{\pi} X(\omega_1, \omega_2) e^{j\omega_1 n_1} e^{j\omega_2 n_2} d\omega_1 d\omega_2 \tag{2.88}$$

Now, we present some examples of computing the 2-D DTFT and the inverse DTFT.

▶ **EXAMPLE 2.8** *Compute the 2-D DTFT for the following sequences:*

(a) $x(n_1, n_2) = \delta(n_1, n_2)$;

(b) $x(n_1, n_2) = u(n_1, n_2) - u(n_1 - 2, n_2 - 2)$;

(c) $x(n_1, n_2) = \begin{bmatrix} -1 & 2 & -1 \\ -2 & \mathbf{4} & -2 \\ -1 & 2 & -1 \end{bmatrix}$, *where the digit in bold represents the value at* $n_1 = 0$ *and* $n_2 = 0$. *Note that the spectrum of this sequence is lowpass in one direction and highpass in the other.*

(d) $x(n_1, n_2) = a^{n_1} b^{n_2} u(n_1, n_2)$, *with* $|a| < 1$ *and* $|b| < 1$.

SOLUTION

(a) $X(\omega_1, \omega_2) = 1, \quad$ for all (ω_1, ω_2).

(b) Since

$$x(n_1, n_2) = \begin{cases} 1 & \text{for } 0 \le n_1 \le 1 \text{ and } 0 \le n_2 \le 1 \\ 0 & \text{otherwise} \end{cases},$$

we have

$$X(\omega_1, \omega_2) = \sum_{n_1=0}^{1}\sum_{n_2=0}^{1} e^{-j\omega_1 n_1} e^{-j\omega_2 n_2}$$

$$= (1 + e^{-j\omega_1})(1 + e^{-j\omega_2}).$$

(c)

$$X(\omega_1, \omega_2) = \sum_{n_1=-1}^{1}\sum_{n_2=-1}^{1} x(n_1, n_2) e^{-j\omega_1 n_1} e^{-j\omega_2 n_2}.$$

In this case, the given $x(n_1, n_2)$ is separable. It can be written as $x(n_1, n_2) = x_1(n_1)x_2(n_2)$, where $x_1(n_1) = [-1 \ \ 2 \ \ -1]$, and $x_2(n_2) = [1 \ \ 2 \ \ 1]$. Therefore

$$X(\omega_1, \omega_2) = \sum_{n_1=-1}^{1} x_1(n_1) e^{-j\omega_1 n_1} \sum_{n_2=-1}^{1} x_2(n_2) e^{-j\omega_2 n_2}.$$

$$= (-e^{j\omega_1} + 2 - e^{-j\omega_1})(e^{j\omega_2} + 2 + e^{-j\omega_2}).$$

(d)

$$X(\omega_1, \omega_2) = \sum_{n_1=0}^{\infty}\sum_{n_2=0}^{\infty} a^{n_1} b^{n_2} e^{-j\omega_1 n_1} e^{-j\omega_2 n_2}$$

$$= \sum_{n_1=0}^{\infty} a^{n_1} e^{-j\omega_1 n_1} \sum_{n_2=0}^{\infty} b^{n_2} e^{-j\omega_2 n_2}$$

$$= \frac{1}{1 - ae^{-j\omega_1}} \frac{1}{1 - be^{-j\omega_2}}.$$

The 2-D DTFT magnitude responses are plotted in Figs. 2.42 to 2.45. The phase responses are omitted. ◀

▶ **EXAMPLE 2.9** *Find the impulse response of the following: (a) 2-D separable ideal LPF; and (b) 2-D circularly symmetric ideal LPF.*

SOLUTION There are actually several types of 2-D lowpass filters, depending on the way we connect the cutoff frequencies on the two axes. The two common types are called separable LPF and circularly symmetric LPF. The frequency response of these are shown in Fig. 2.46 and Fig. 2.47, respectively. Now we find the impulse responses of these filters.

(a) The frequency response of the 2-D separable ideal LPF is described by

$$H(\omega_1, \omega_2) = \begin{cases} 1, & |\omega_1| \le \omega_{c_1} \ \& \ |\omega_2| \le \omega_{c_2} \\ 0, & \omega_{c_1} < |\omega_1| \le \pi \ \& \ \omega_{c_2} < |\omega_2| \le \pi. \end{cases} \tag{2.89}$$

As in the 1-D case, the impulse response is easily obtained by using the DTFT synthesis formula as

$$h(n_1, n_2) = \frac{\sin \omega_{c_1} n_1}{\pi n_1} \frac{\sin \omega_{c_2} n_2}{\pi n_2}. \tag{2.90}$$

The surface plot for this impulse response for $\omega_{c_1} = \omega_{c_2} = \pi/3$ is shown in Fig. 2.48(a). The ripples are clearly aligned along the axes which exhibit separability (or rectangularly symmetric).

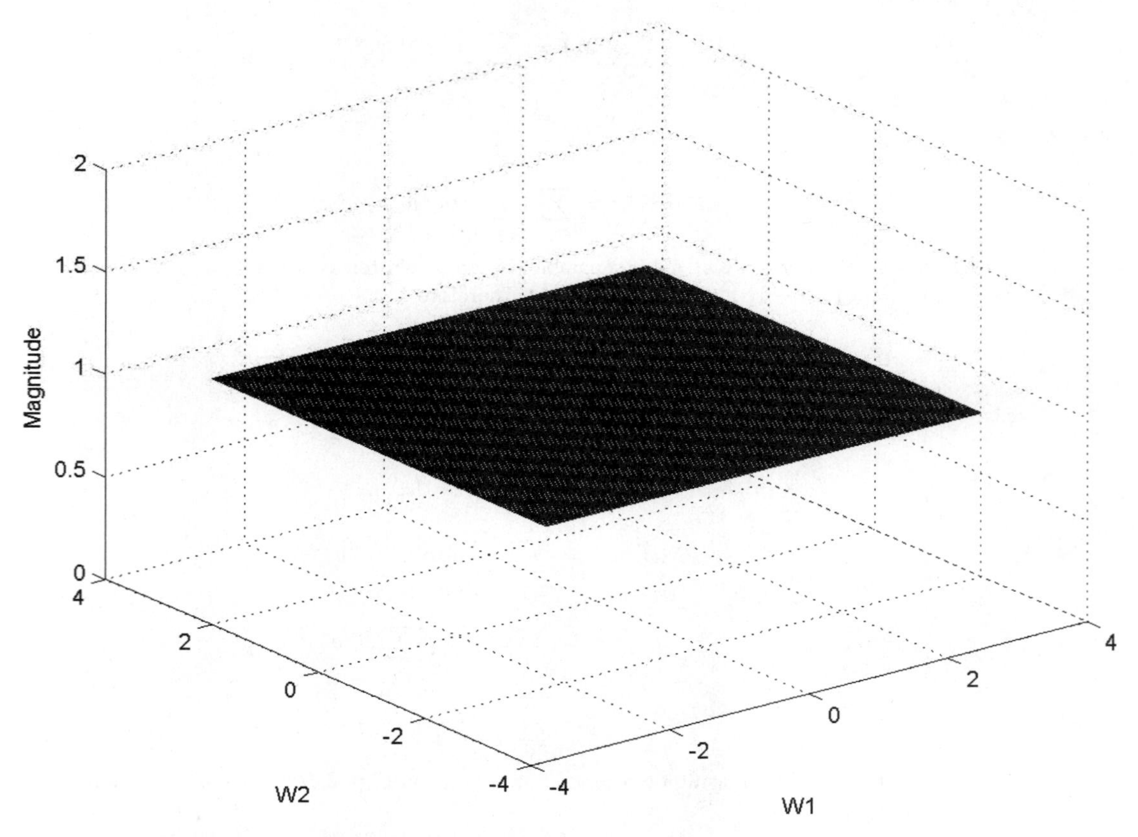

Figure 2.42 2-D DTFT magnitude response.

(b) The frequency response of the 2-D circularly symmetric ideal LPF is given by

$$H(\omega_1, \omega_2) = \begin{cases} 1, & \sqrt{\omega_1^2 + \omega_2^2} < \omega_c \\ 0, & \sqrt{\omega_1^2 + \omega_2^2} \geq \omega_c, \text{ and } |\omega_1| \leq \pi \text{ \& } |\omega_2| \leq \pi. \end{cases} \tag{2.91}$$

The calculation of the inverse 2-D DTFT of the above requires quite a bit of manipulation, which does not provide much insight and is therefore omitted. Interested readers are referred to [5] for a derivation. The impulse response is given by

$$h(n_1, n_2) = \frac{\omega_c}{2\pi\sqrt{n_1^2 + n_2^2}} J_1\left(\omega_c\sqrt{n_1^2 + n_2^2}\right) \tag{2.92}$$

where $J_1(.)$ represents the Bessel function of the first kind and first order [6]. The surface plot of this function is shown in Fig. 2.48(b). The ripples are seen to be circularly symmetric. ◀

The above impulse responses can be truncated (as in 1-D) for implementation as FIR filters. Other windows and design methods are discussed in Chapter 4. We now state

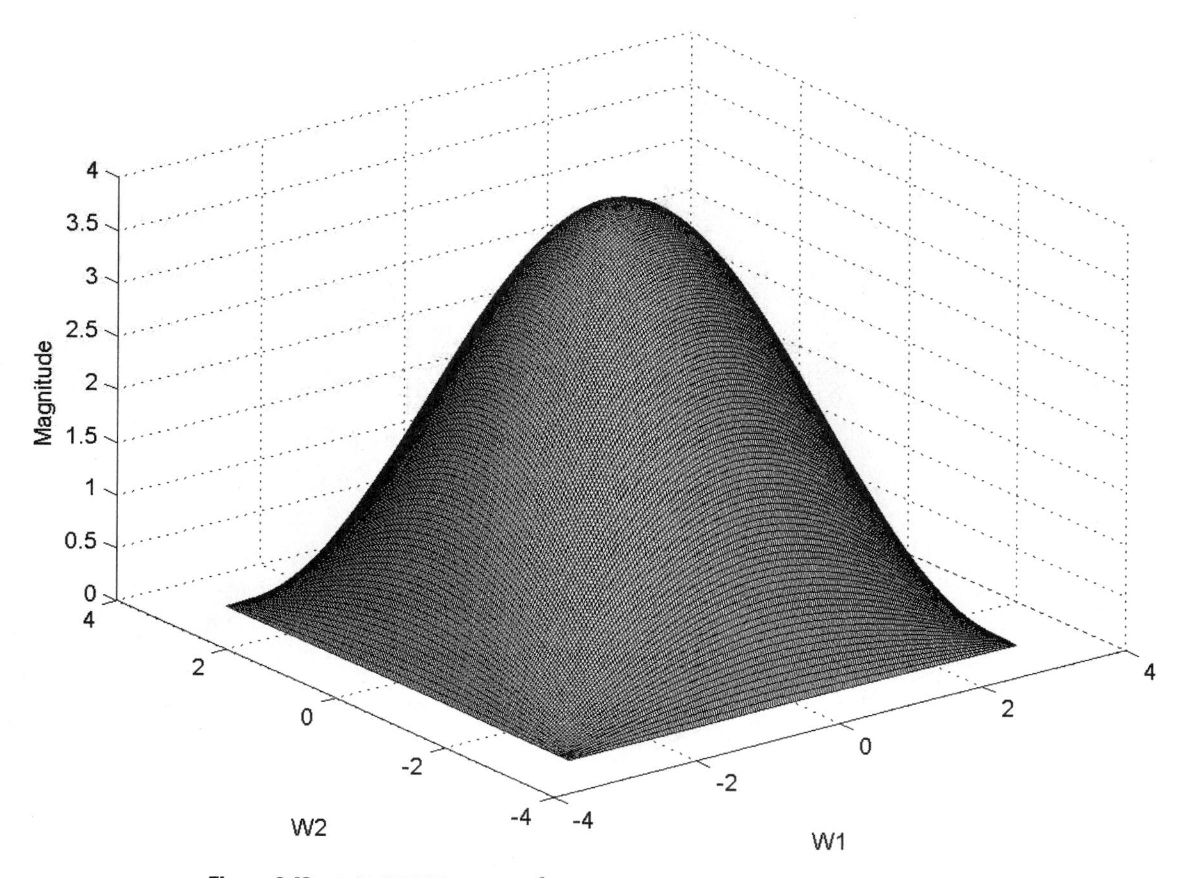

Figure 2.43 2-D DTFT magnitude response.

three important results on the 2-D DTFT. The proofs are similar to the 1-D cases and are therefore omitted.

Theorem 2.8 *The 2-D DTFT of a sequence $x(n_1, n_2)$ exists if the sequence is absolutely summable, that is, if*

$$\sum_{n_1=-\infty}^{\infty} \sum_{n_2=-\infty}^{\infty} |x(n_1, n_2)| < \infty. \tag{2.93}$$

Theorem 2.9 *Let $h(n_1, n_2)$ represent the impulse response of a 2-D LSI system. If the input is a 2-D complex exponential $e^{j\omega_x n_1} e^{j\omega_y n_2}$, then its output is the same complex exponential with magnitude and phase given by $|H(\omega_1, \omega_2)|$ and $\measuredangle H(\omega_1, \omega_2)$, respectively, at $(\omega_1, \omega_2) = (\omega_x, \omega_y)$.*

Theorem 2.9 is the basis for the concept of *2-D frequency response* of a digital filter, which is defined as the DTFT of the impulse response.

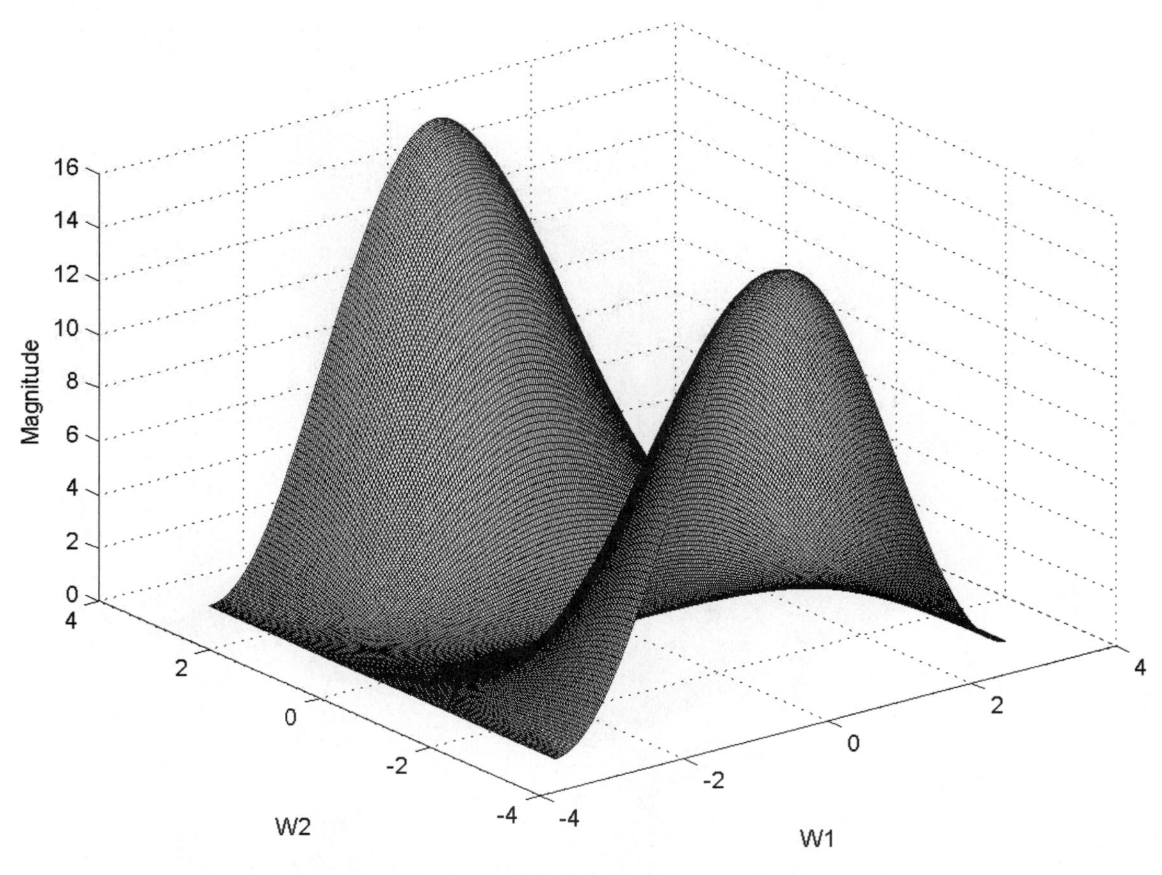

Figure 2.44 2-D DTFT magnitude response.

Theorem 2.10 *The 2-D DTFT of any discrete-time sequence is periodic with a period of* $(2\pi, 2\pi)$*; that is,*

$$H(\omega_1, \omega_2) = H(\omega_1 + 2\pi, \omega_2) = H(\omega_1, \omega_2 + 2\pi) = H(\omega_1 + 2\pi, \omega_2 + 2\pi). \tag{2.94}$$

The last theorem suggests that the 2-D DTFT is periodic separable (recall definition in Chapter 1) with a period of $(2\pi, 2\pi)$. So, the magnitude and phase response of any 2-D digital filter are periodic separable. For example, the frequency response of an ideal 2-D circularly symmetric LPF is periodic as shown in Fig. 2.49. The frequency response $H(\omega_1, \omega_2)$ is unity within the circles and zero elsewhere. Because of the periodicity, it is sufficient to depict the frequency response in the square with corners at $(-\pi, -\pi), (\pi, -\pi), (-\pi, \pi)$ and (π, π) as shown in the figure.

The properties of the 2-D DTFT are direct extensions of the 1-D properties and are therefore omitted. We state one property, however, that does not have a 1-D counterpart.

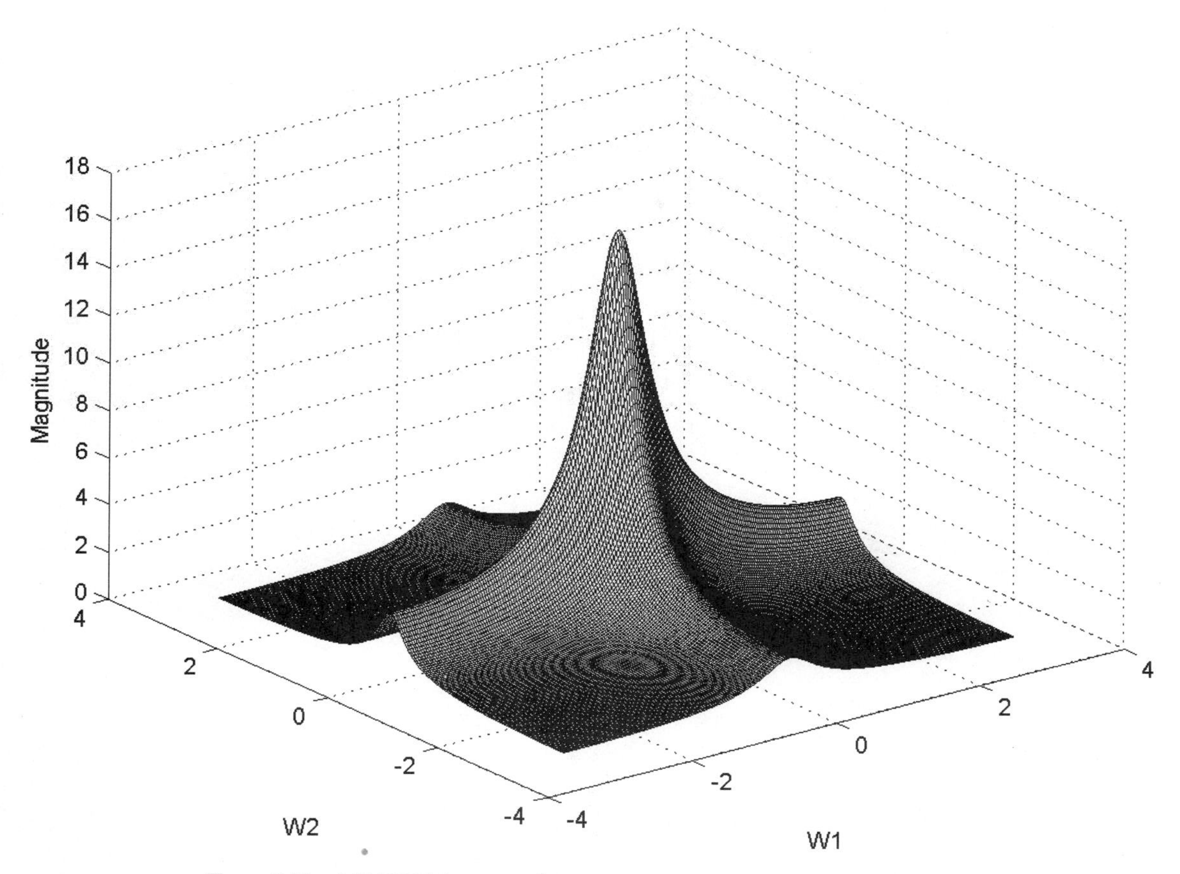

Figure 2.45 2-D DTFT magnitude response.

This property can be proved by a direct application of the DTFT definition and is assigned as a problem at the end of the chapter.

Property (Separable DTFT): The 2-D DTFT of a separable sequence is the product of the 1-D DTFT of the individual sequences; that is,

$$x(n_1, n_2) = x_1(n_1)x_2(n_2) \leftrightarrow X_1(\omega_1)X_2(\omega_2). \tag{2.95}$$

Recall from the 1-D properties that a product in the time domain corresponds to a convolution is the frequency domain. However, for 2-D separable sequences this is not true, and this is what makes the above property unique and interesting. It is unique because there is a different index in each term of the product. Therefore, the sequences act like vectors instead of matrices.

Now we present the 2-D discrete Fourier transforms (DFT). As in the 1-D case, the analysis equation is obtained by discretizing the frequencies ω_1 and ω_2 in the interval

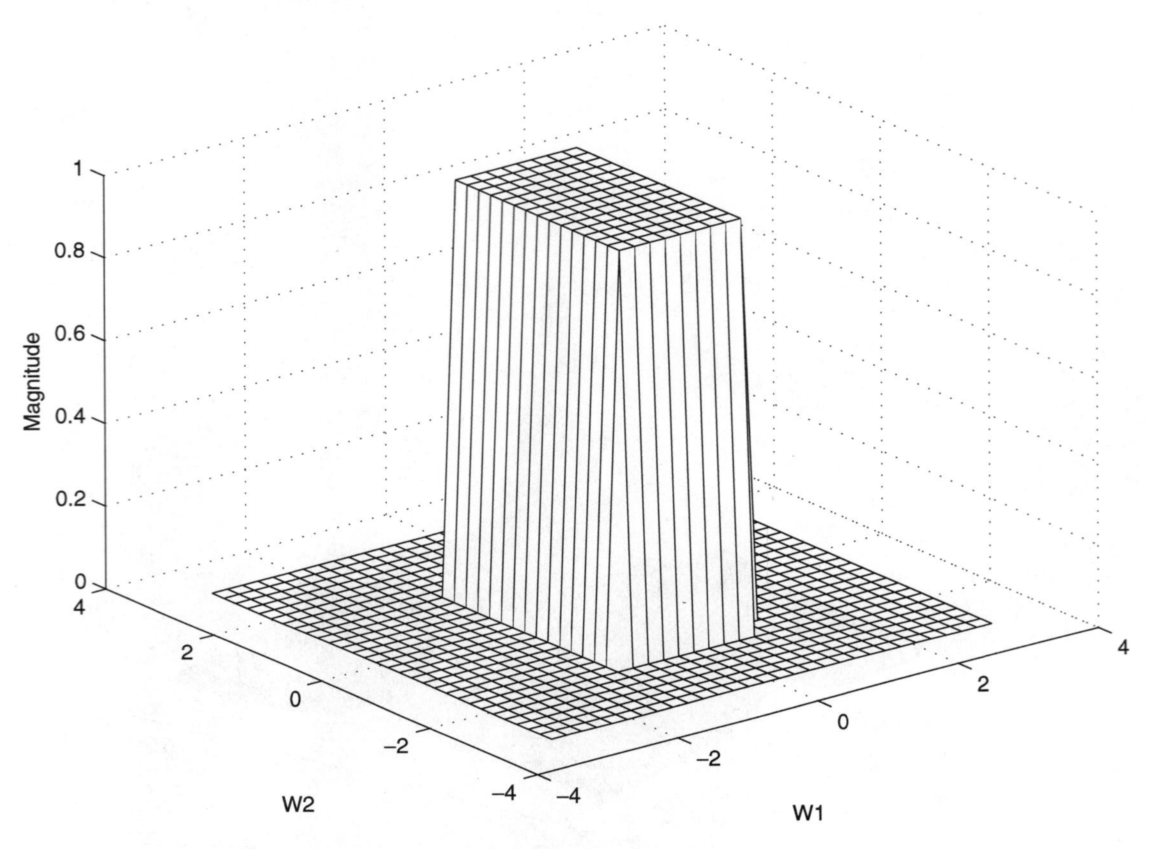

Figure 2.46 2-D ideal separable filter.

$(0, 2\pi)$. Because there are two frequency axes, the number of samples can be different for each frequency. The analysis and synthesis equations are given in the following.

2-D DFT analysis equation

$$X(k_1, k_2) = \sum_{n_1=0}^{N_1-1} \sum_{n_2=0}^{N_2-1} x(n_1, n_2) e^{-j\frac{2\pi}{N_1}k_1 n_1} e^{-j\frac{2\pi}{N_2}k_2 n_2}, \tag{2.96}$$

$0 \leq k_1 \leq N_1 - 1, 0 \leq k_2 \leq N_2 - 1$. The above is called the (N_1, N_2)-point DFT of the sequence $x(n_1, n_2)$. For each value of the pair (k_1, k_2), the value of $X(k_1, k_2)$ represents the Fourier transform at the frequency $(\omega_1, \omega_2) = (\frac{2\pi}{N_1}k_1, \frac{2\pi}{N_2}k_2)$. The synthesis equation is as follows.

2-D DFT synthesis equation

$$x(n_1, n_2) = \frac{1}{N_1 N_2} \sum_{k_1=0}^{N_1-1} \sum_{k_2=0}^{N_2-1} X(k_1, k_2) e^{j\frac{2\pi}{N_1}k_1 n_1} e^{j\frac{2\pi}{N_2}k_2 n_2}, \tag{2.97}$$

$0 \leq n_1 \leq N_1 - 1, 0 \leq n_2 \leq N_2 - 1$.

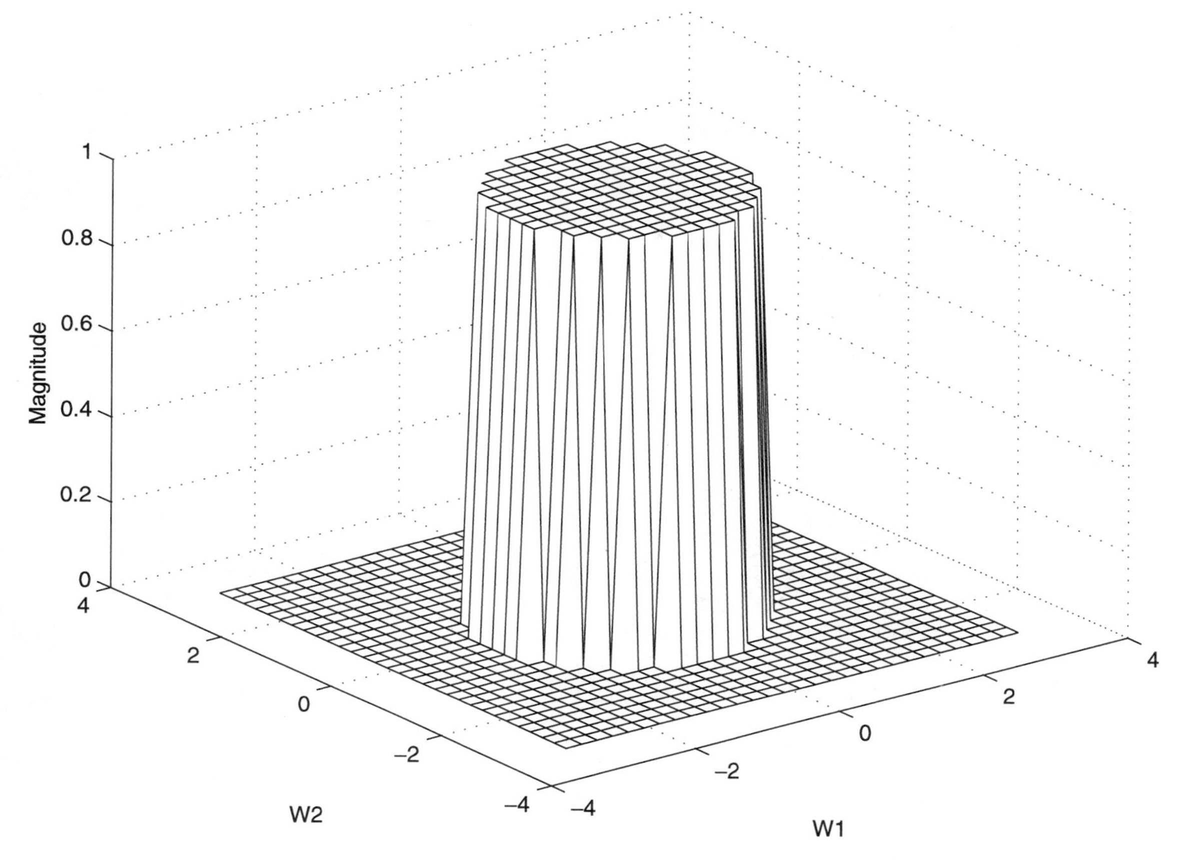

Figure 2.47 2-D ideal circularly symmetric filter.

Again, the properties of the 2-D DFT are straightforward extensions to the 1-D DFT properties, including those on circular shift and circular convolution. The DFT property of separable sequences is the same as that for the 2-D DTFT in (2.95) property, except that ω_1, ω_2 are replaced by k_1, k_2.

Now we consider the application of the 2-D DFT on a lowpass filter. A causal 3×3 mean filter is given by

$$y(n_1, n_2) = \frac{1}{9} \sum_{i=0}^{2} \sum_{j=0}^{2} x(n_1 - i, n_2 - j).$$

The impulse response of this filter, $h(n_1, n_2)$, is a 3×3 matrix with all the elements equal to $\frac{1}{9}$. In order to find the frequency response of this filter, a 2-D DFT is required. For sufficient resolution, let us perform a (200, 200)-point 2-D DFT. This is the same as evaluating the 2-D DTFT and then sampling the two frequencies in the interval $(0, 2\pi)$ with 200 points in the interval. So, the point $H(0, 0)$ is the value of the spectrum at zero

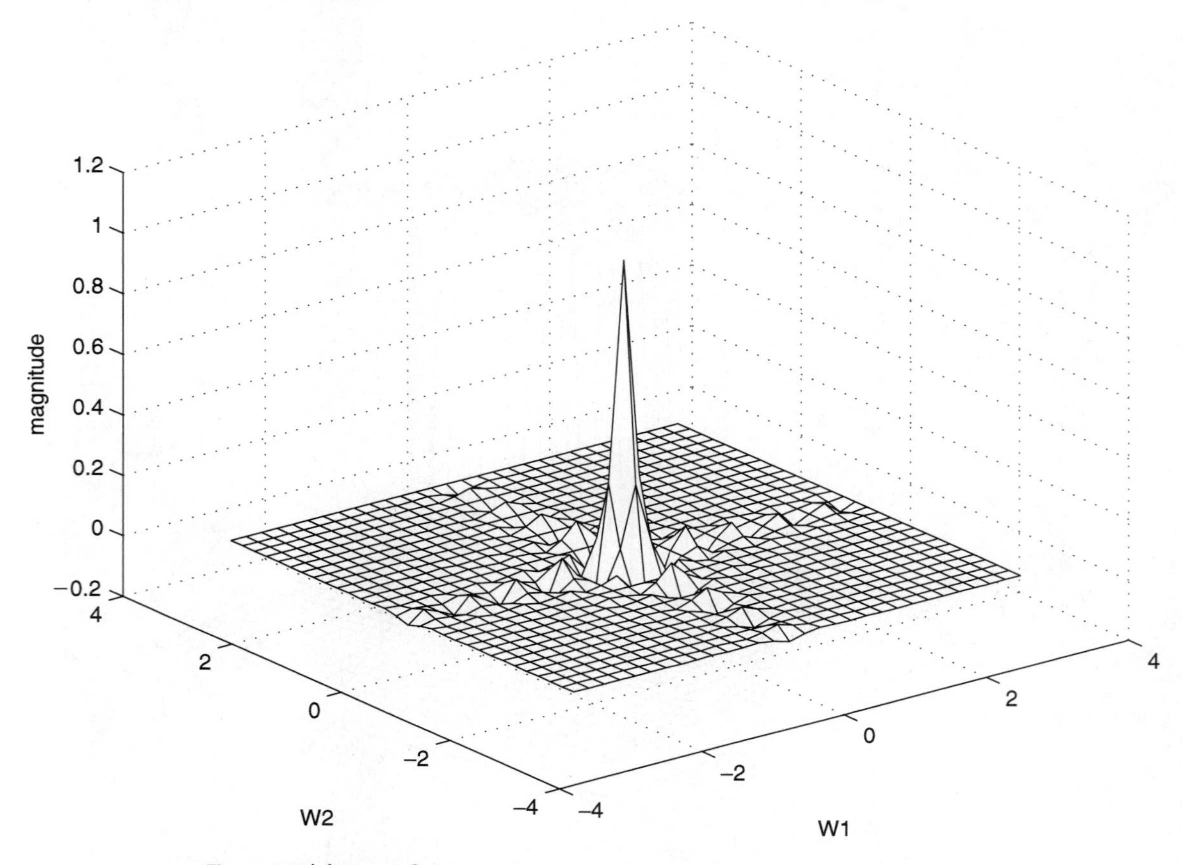

Figure 2.48(a) Impulse response.

frequency, and the point $H(199, 199)$ is that at $(\omega_1, \omega_2) = (199 \cdot \frac{2\pi}{200}, 199 \cdot \frac{2\pi}{200})$. The DFT is a 200×200 matrix represented in Fig. 2.50(a) with the element indices and the radial frequencies indicated. The letters L and H refer to low-frequency (ω near zero) and high-frequency (ω near π) regions for each of the two frequencies. As we know, the frequency response is periodic with a period of $(2\pi, 2\pi)$, and it is customary to represent it in the interval $(-\pi, \pi)$ for each frequency. To do this, all we need to do is swap the first and third quadrants and the second and fourth quadrants to get the frequency response matrix shown in Fig. 2.50(b). The absolute value of the DFT matrix (magnitude response of the mean filter) is shown in Fig. 2.51(a) as a surface plot. After the quadrant swaps, the magnitude response is shown in Fig. 2.51(b). This figure shows that the filter is lowpass. It is not a very good lowpass filter because of its large transition band and large ripples in the stopband.

Now we present the sampling theorem for 2-D systems. It is proved in just the same way as for 1-D systems except that now we have two time indices and two frequencies. Let $x_c(t_1, t_2)$ denote the continuous-time 2-D signal and $x_s(n_1, n_2)$ the sampled sequence.

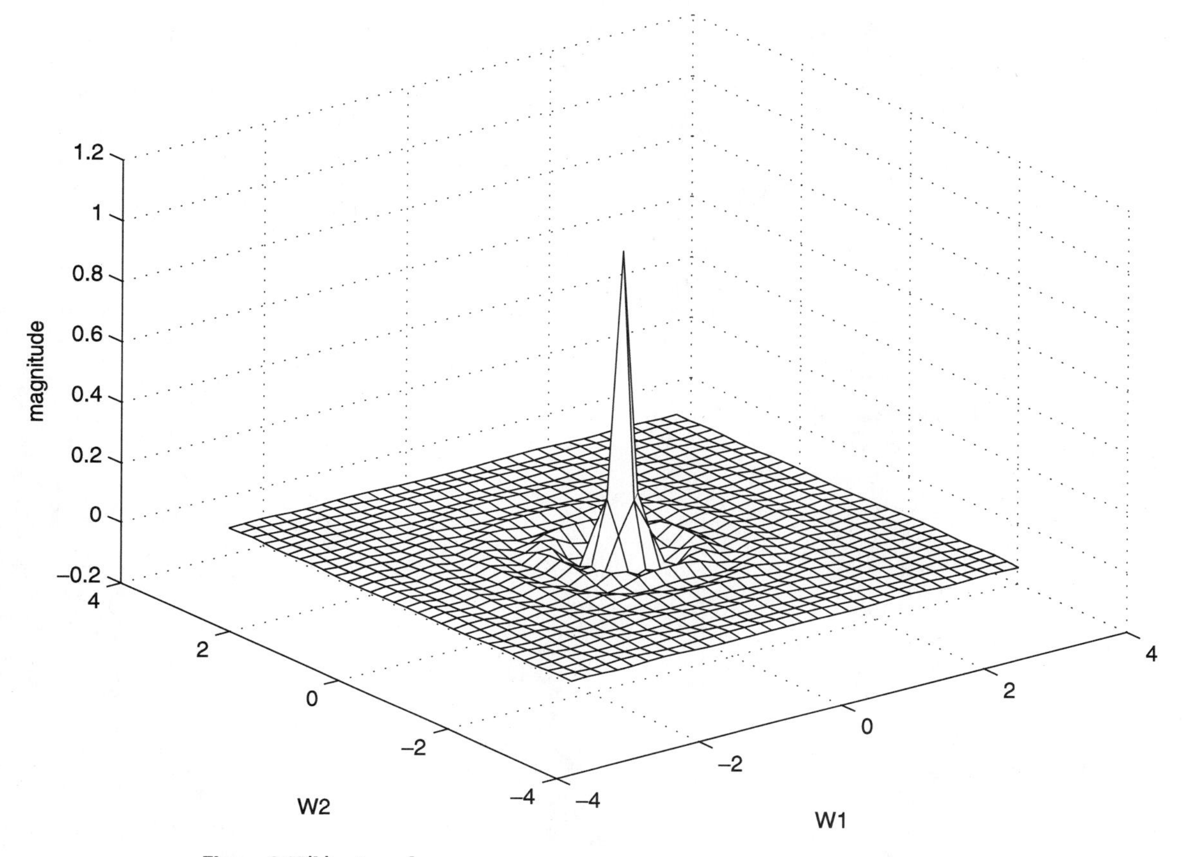

Figure 2.48(b) Impulse response.

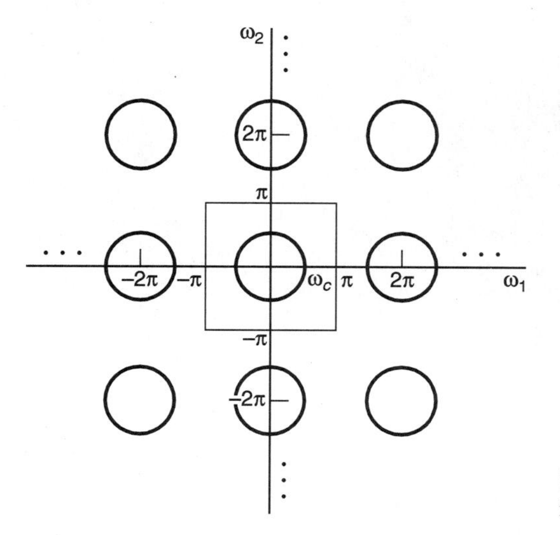

Figure 2.49 Frequency response of 2-D ideal circularly symmetric lowpass filter.

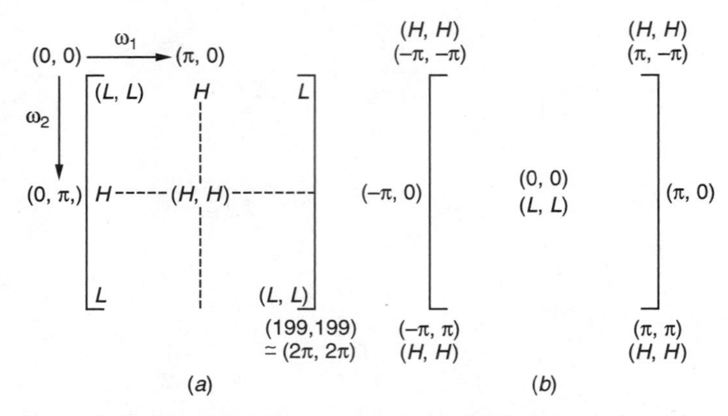

Figure 2.50 (a) 2-D DFT matrix; (b) 2-D DFT matrix with quadrant swaps.

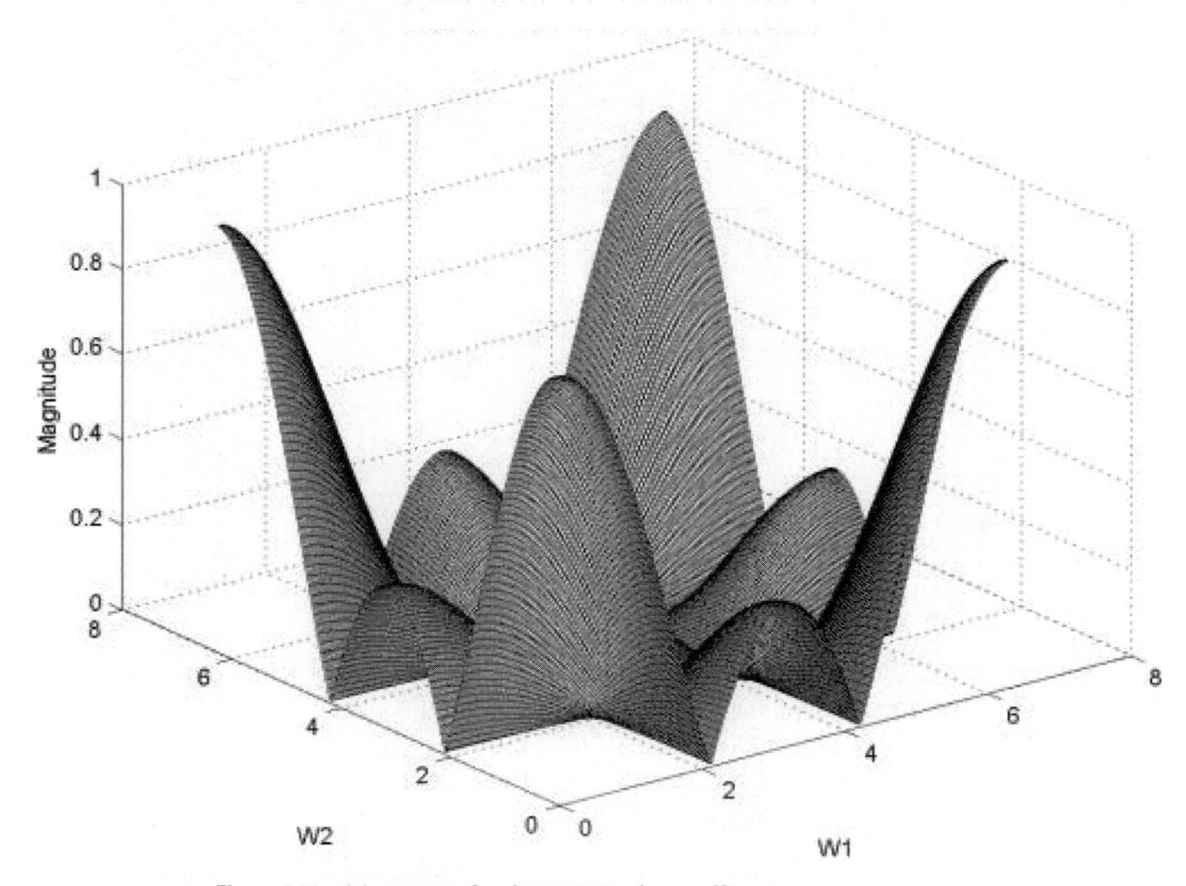

Figure 2.51 (*a*) Magnitude of 2-D DFT of mean filter;

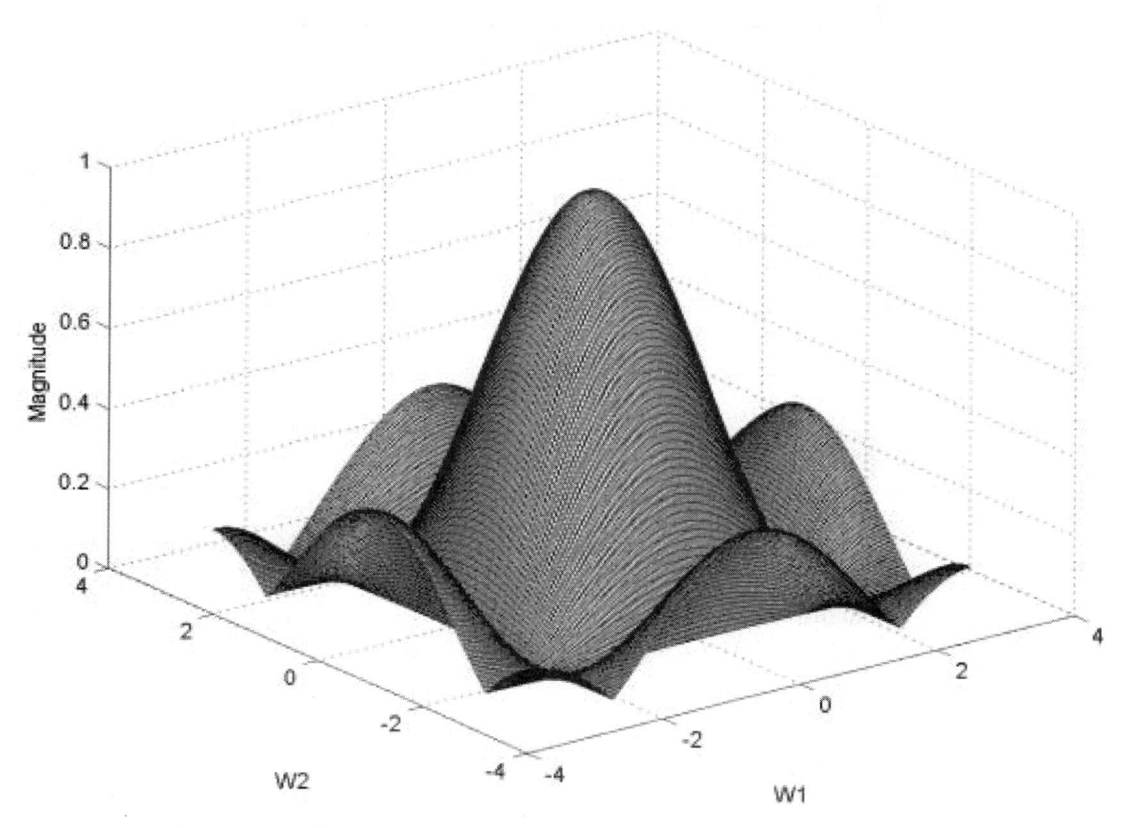

Figure 2.51 (*b*) Magnitude response of mean filter after quadrant swaps.

Let $\Omega_{s_1}, \Omega_{s_2}$ and T_{s_1}, T_{s_2} denote the sampling frequencies and time-periods, respectively for the two time axes. The main equation that leads to the 2-D sampling theorem is the 2-D counterpart of equation (2.52) given by

$$X_s(\Omega_1, \Omega_2) = \frac{1}{T_{s_1} T_{s_2}} \sum_{k_1=-\infty}^{\infty} \sum_{k_2=-\infty}^{\infty} X_c(\Omega_1 - k_1\Omega_{s_1}, \Omega_2 - k_2\Omega_{s_2}). \tag{2.98}$$

From this equation we see that the spectrum of the sampled signal is composed of shifted versions of the spectrum of the continuous-time signal and then scaled by $\frac{1}{T_{s_1} T_{s_2}}$. Let the spectrum of the continuous-time signal be as shown in Fig. 2.52(*a*). The spectrum of the sampled signal is given in Fig. 2.52(*b*). Aliasing occurs if the spectral components overlap; that is, if $\frac{\Omega_{s_1}}{2} < \Omega_{h_1}$ or $\frac{\Omega_{s_2}}{2} < \Omega_{h_2}$, the original signal cannot be recovered from its samples. This is the 2-D sampling theorem.

Let us now examine the effect of aliasing on an image. Because we do not have a continuous-time image, we use a digital image and decimate it. For example, consider

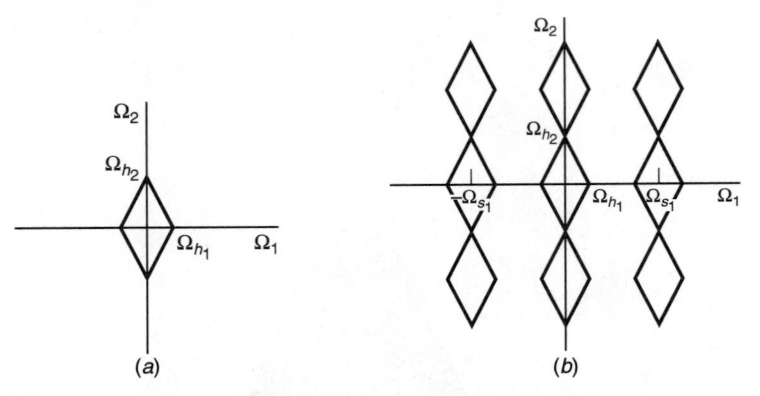

Figure 2.52 Frequency response of (*a*) continuous-time signal; (*b*) sampled signal.

the 200 × 320 "mandrill" image shown in Fig. 2.53(*a*). The image is decimated by a factor of (6, 6) and is shown in Fig. 2.53(*b*), where the effect of aliasing is clear. Aliasing causes blocking effects, which results in the loss of detail. The blocking effect can also be considered as the introduction of high-frequency noise. Now filter the original image by a 2-D 3 × 3 noncausal mean filter. The filtered image is then decimated by the same factor and shown in Fig. 2.53(*c*). The effect of aliasing is significantly reduced by the prefiltering process. If we use a filter with a sharper frequency response, then the effect of aliasing can be further reduced.

Computational Aspects

This section considers the computational aspects regarding 2-D convolution and 2-D DFT and presents techniques for performing these operations that significantly reduce computations.

2-D Separable Convolution

First consider the 2-D convolution of a $N \times N$ sequence $x(n_1, n_2)$ with the $M \times M$ sequence $h(n_1, n_2)$. These matrices are shown in Fig. 2.54(*a*) and (*b*), respectively. Let $y(n_1, n_2)$ denote the result of convolution. Recall the method of 2-D convolution described earlier. We flip one of the sequences twice (over the vertical axis and then over the horizontal axis) and then slide it over the other sequence. The overlapping samples are multiplied and added. This suggests that the size of $y(n_1, n_2)$ is $(N+M-1) \times (N+M-1)$. This is depicted in Fig. 2.54(*c*). The 2-D convolution equations from Chapter 1 are

$$\begin{aligned} y(n_1, n_2) &= x(n_1, n_2) * h(n_1, n_2) \\ &= \sum_{m_1=-\infty}^{\infty} \sum_{m_2=-\infty}^{\infty} x(m_1, m_2) h(n_1 - m_1, n_2 - m_2) \qquad (2.99) \\ &= \sum_{m_1=-\infty}^{\infty} \sum_{m_2=-\infty}^{\infty} h(m_1, m_2) x(n_1 - m_1, n_2 - m_2). \qquad (2.100) \end{aligned}$$

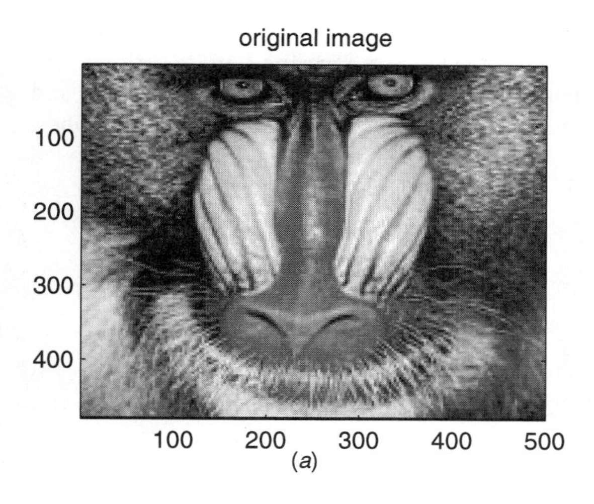

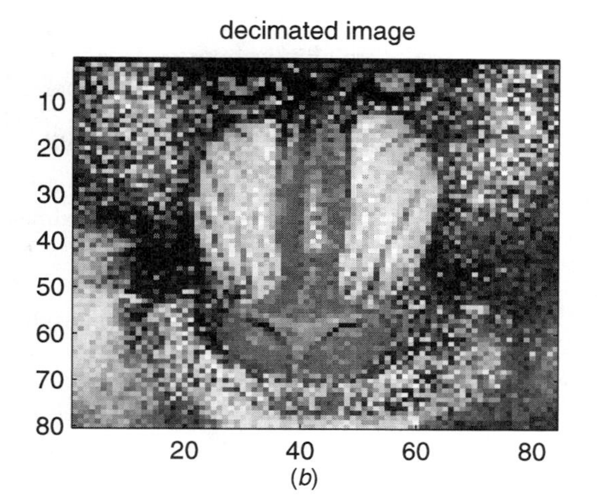

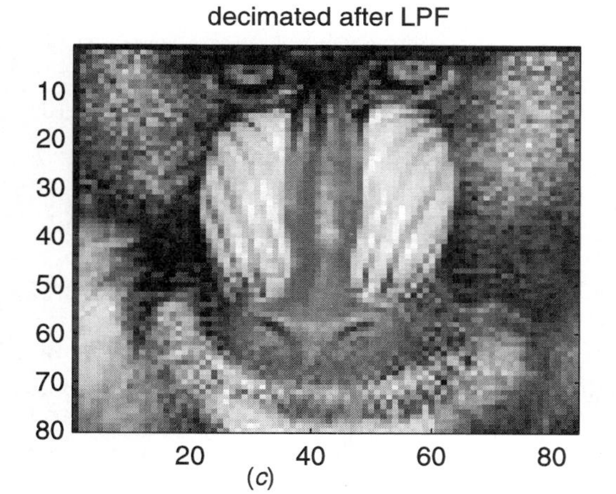

Figure 2.53 Illustration of decimation of an image.

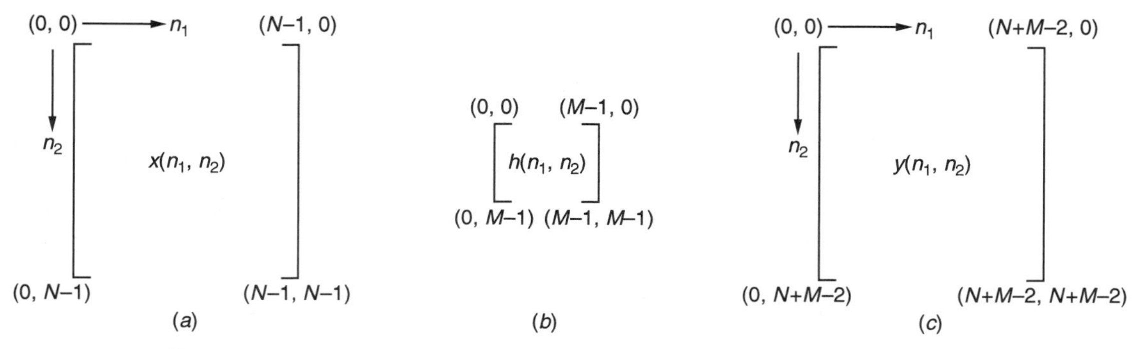

Figure 2.54 Sizes of matrices in a 2-D convolution where $x(n_1, n_2)*h(n_1, n_2) = y(n_1, n_2)$.

From the above, we see that the computation of each element of $y(n_1, n_2)$ requires M^2 multiplications and $(M-1)^2$ additions. For the elements near the border of the matrix we need a smaller number of computations. However, we will assume, $N \gg M$, and then we can ignore these few computations in comparison to the overall computations. To compute all the $(N+M-1) \times (N+M-1)$ elements in the $y(n_1, n_2)$ matrix, we need $M^2(N+M-1)^2$ multiplications and $(M-1)^2(N+M-1)^2$ additions. Since $N \gg M$, we have the following approximations.

2-D direct convolution

$$\text{No. of multiplications} \simeq M^2N^2; \quad \text{no. of Additions} \simeq M^2N^2.$$

Now, we present a technique to reduce the computations if one of the sequences is separable. Let $h(n_1, n_2)$ be separable; that is, $h(n_1, n_2) = h_1(n_1)h_2(n_2)$. Substituting this in the convolution equation, we get

$$\begin{aligned} y(n_1, n_2) &= \sum_{m_1=-\infty}^{\infty} \sum_{m_2=-\infty}^{\infty} h_1(n_1 - m_1)h_2(n_2 - m_2)x(m_1, m_2) \\ &= \sum_{m_1=-\infty}^{\infty} h_1(n_1 - m_1)\left[\sum_{m_2=-\infty}^{\infty} h_2(n_2 - m_2)x(m_1, m_2)\right] \end{aligned} \tag{2.101}$$

Observe that the term inside the square brackets is evaluated for a fixed m_1. With this fixed index, the term inside the brackets is the 1-D convolution

$$f(m_1, n_2)|_{\textit{fixed } m_1} = h_2(n_2) * [x(m_1, n_2)]_{\textit{fixed } m_1}. \tag{2.102}$$

The above represents a 1-D convolution of a column of x with $h_2(n_2)$. Since each column of x has length N and $h_2(n_2)$ has length M, this convolution yields a vector of length $N+M-1$ and requires approximately $(N+M-1)M$ multiplications and approximately the same number of additions. For N values of n_1 (columns of x), the computation of f requires $N(N+M-1)M$ computations. The procedure for finding f is illustrated in Fig. 2.55. The size of f is thus $N \times (N+M-1)$. Now we combine equations (2.101) and (2.102) to get

$$y(n_1, n_2) = \sum_{m_1=-\infty}^{\infty} h_1(n_1 - m_1)f(m_1, n_2). \tag{2.103}$$

For each fixed n_2, the above is a 1-D convolution of f and h_1. Since n_2 is now fixed, we convolve each row of f by $h_1(n_1)$ (Fig. 2.55). This again yields a length $M+N-1$ vector and requires $M(M+N-1)$ multiplications and the same number of additions, for each value of n_2. For $(M+N-1)$ values of n_2, equation (2.103) requires $M(N+M-1)^2$ computations. Now, we summarize the procedure and simplify the computations using $N \gg M$.

1. Convolve each column of x with $h_2(n_2)$ and form the columns of f.

$$\text{no. of computations} = MN(N+M-1) \simeq MN^2.$$

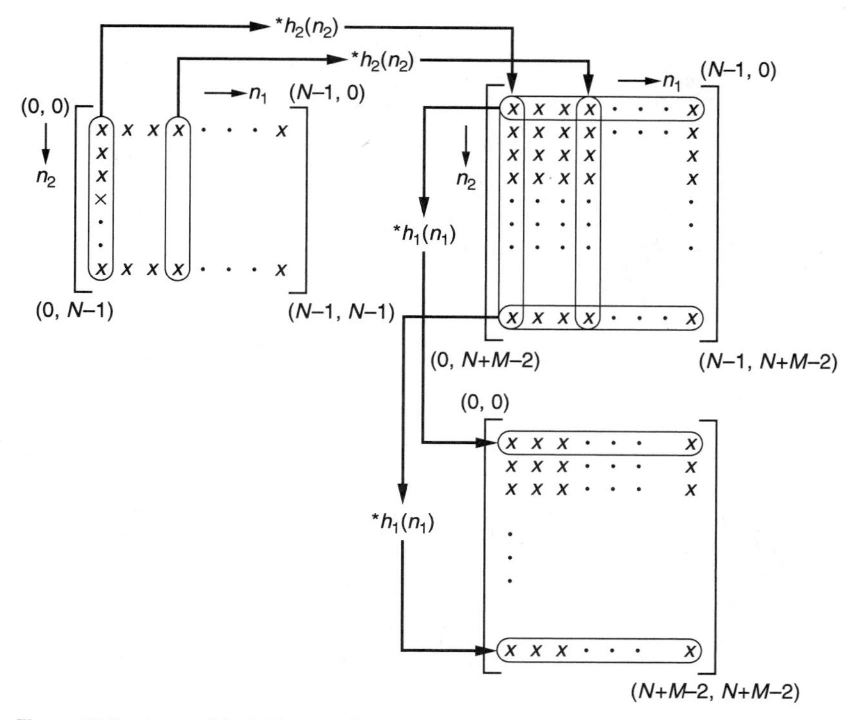

Figure 2.55 Separable 2-D convolution.

2. Convolve each row of f with $h_1(n_1)$ and form the rows of y.

$$\text{no. of computations} = M(N+M-1)^2 \simeq MN^2.$$

2-D separable convolution computations:

$$\text{no. of multiplications} \simeq 2MN^2; \text{ no. of Additions} \simeq 2MN^2.$$

When compared to direct convolution, separable convolution is more efficient by a factor of $\frac{M^2N^2}{2MN^2} = \frac{M}{2}$. As an example, if we use a simple 4×4 separable filter h, separable convolution will require only $\frac{1}{2}$ the number of computations as that of direct convolution. Since 2-D filtering requires a tremendous amount of computations, we must be sensitive to the computational complexity of the algorithms that we implement. By designing 2-D separable FIR filters, we can use separable convolution and thereby achieve significant computational savings.

2-D DFT

In this subsection, we consider the computational complexity of the 2-D DFT and repeat the 2-D DFT formula here for convenience. For a $N_1 \times N_2$ sequence $x(n_1, n_2)$, the DFT is given by

$$X(k_1, k_2) = \sum_{n_1=0}^{N_1-1} \sum_{n_2=0}^{N_2-1} x(n_1, n_2) e^{-j\frac{2\pi}{N_1}k_1 n_1} e^{-j\frac{2\pi}{N_2}k_2 n_2}, \tag{2.104}$$

$0 \leq k_1 \leq N_1-1$, $0 \leq k_2 \leq N_2-1$. For a known pair (N_1, N_2), the product $e^{-j\frac{2\pi}{N_1}k_1n_1}e^{-j\frac{2\pi}{N_2}k_2n_2}$ can be computed and stored for all the necessary combinations for (k_1, k_2) and (n_1, n_2). Then the total number of complex multiplications is $N_1^2 N_2^2$, and the number of complex additions is $N_1N_2(N_1N_2 - 1) \simeq N_1^2 N_2^2$. As before, the number of multiplications and additions is approximately the same, and so we will consider the multiplications only. The computational count can be significantly reduced by performing a so-called *row-column decomposition*. This is done by decomposing the DFT equation as follows:

$$X(k_1, k_2) = \sum_{n_1=0}^{N_1-1} \left[\sum_{n_2=0}^{N_2-1} x(n_1, n_2) e^{-j\frac{2\pi}{N_2}k_2n_2} \right] e^{-j\frac{2\pi}{N_1}k_1n_1} \tag{2.105}$$

$$= \sum_{n_1=0}^{N_1-1} [f(n_1, k_2)] e^{-j\frac{2\pi}{N_1}k_1n_1}. \tag{2.106}$$

In equation (2.105), for some fixed n_1, the term inside the brackets is the 1-D DFT of the n_1th column of x. It is therefore defined as $f(n_1, k_2)$. So, the 1-D DFT of each column of x forms the matrix f. For some fixed k_2, equation (2.106) represents the 1-D DFT of the k_2th row of $f(n_1, k_2)$. The 1-D DFT of each row of f forms the final 2-D DFT matrix X. This process is similar to the 2-D separable convolution discussed before. Now we proceed with the computational count. Recall that an N-point 1-D DFT requires N^2 complex multiplications and $N(N-1) \simeq N^2$ complex additions. So, the formation of f requires $N_1N_2^2$ computations. The formation of X from f requires $N_1^2N_2$ computations. The total number of computations is therefore $N_1N_2^2 + N_1^2N_2 = N_1N_2(N_1 + N_2)$. Compared to a direct computation of the 2-D DFT, this represents a computational saving by a factor of $\frac{N_1N_2}{N_1+N_2}$. Since N_1 and N_2 are typically very large numbers for images, this leads to a significant amount of computational saving. Here is a summary of the *row-column decomposition* method.

1. Perform 1-D DFT of each column of x to form matrix f. no of Computations $= N_1N_2^2$.
2. Perform 1-D DFT of each row of f to form matrix X. no. of Computations $= N_1^2N_2$.

Total no. of Multiplications $= N_1N_2(N_1+N_2)$; Total no. of Additions $= N_1N_2(N_1+N_2)$.

Instead of performing 1-D DFT on the columns of x and the rows of f, if we perform 1-D FFT, the computational count for the row-column decomposition method goes down even further. Recall that an N-point 1-D FFT requires $\frac{N}{2}(\log_2 N - 1)$ multiplications and $N\log_2 N$ additions. To compute f from x, we require N_1 1-D DFTs with N_2 points each. This requires $N_1\frac{N_2}{2}(\log_2 N_2 - 1)$ multiplications and $N_1N_2 \log_2 N_2$ additions. Computing X from f requires N_2 N_1-point 1-D DFTs, which requires $N_2\frac{N_1}{2} \log_2 N_1 - 1)$ multiplications and $N_2N_1 \log_2 N_1$ additions. This gives a total of $\frac{N_1N_2}{2}(\log_2 (N_1N_2) - 2)$ multiplications and $N_1N_2\log_2(N_1N_2)$ additions.

The computational costs derived in this section for different algorithms are summarized in Table 2.5.

TABLE 2.5 Computations for 2-D algorithms

Algorithms	no. of Complex Multiplications	no. of Complex Additions
Direct convolution	M^2N^2	M^2N^2
Separable convolution	$2MN^2$	$2MN^2$
Direct DFT	$N_1^2N_2^2$	$N_1^2N_2^2$
Row-column decomposition with Direct DFT	$N_1N_2(N_1+N_2)$	$N_1N_2(N_1+N_2)$
Row-column decomposition with FFT	$\frac{N_1N_2}{2}(\log_2(N_1N_2)-2)$	$N_1N_2\log_2(N_1N_2)$

2.8 SUMMARY

This chapter began with a presentation of the continuous-time Fourier series (CTFS) and the continuous-time Fourier transform (CTFT). A link was established between the two, and the concept of the spectrum of a signal was explained. The discrete-time Fourier transform (DTFT) and its properties were then presented in detail. Having acquired a background in the concepts of CTFT and DTFT, we went on to discuss the sampling theorem and the important topics of decimation and interpolation. Since a typical DSP system has different data rates within the system, it is important to understand how the signal spectrum is modified as a result of downsampling and upsampling. The theory that deals with various sampling rates is called multirate signal processing. We presented a very rudimentary discussion of this topic; a detailed discussion is given in Chapter 5. Some fine books on this topic include those by Crochiere and Rabiner [7], Vaidyanathan [8], Suter [9], and Fliege [10].

The next sections of this chapter dealt with the DFT and its fast computation, namely, the fast Fourier transform (FFT). There are many algorithms for performing the FFT. We described only two of these algorithms: the decimation in time and the decimation in frequency. The computational count for a direct DFT is N^2, whereas by using the decimation in time algorithm, we can reduce the count to $N\log_2 N$, if N is a power of 2. The impact of the FFT algorithms in signal processing was first recognized following publication of the paper by Cooley and Tukey [11]. Since then, Singleton [12], Duhamel and Hollman [13], Goertzel [14], Rader and Brenner [15], Good [16], and Winograd [17], to name a few, have advanced a number of algorithms.

The last section of this chapter was on 2-D extensions of some of the earlier topics. We presented the 2-D CTFT, DTFT, DFT and some associated examples. Computational aspects of 2-D algorithms were subsequently discussed. It was shown that the computational cost of 2-D separable convolution is significantly lower than that of general convolution. This is a good reason to design 2-D separable FIR filters where possible. The row-column decomposition algorithm was also presented for the fast computation of a 2-D DFT.

2.9 MATLAB COMMANDS

The following MATLAB commands are relevant to this chapter and may be useful for computer projects. Some of the commands will touch on the difference equation representation of a filter given by

$$\begin{aligned} y(n) + a_1y(n-1) + a_2y(n-2) + \cdots + a_Ny(n-N) \\ = b_0x(n) + b_1x(n-1) + \cdots + a_Mx(n-M). \end{aligned} \tag{2.107}$$

The nonrecursive coefficients are entered as row $\mathbf{b} = [b_0\, b_1 \ldots b_M]$, and the recursive coefficients are entered as $\mathbf{a} = [1\, a_1 \ldots a_N]$.

[H,f] = freqz(b,a,N,Fs,'whole'); Given a sampling frequency Fs in Hz, this command computes the N-point frequency vector **f** in Hz and the N-point complex frequency response vector **H** of the filter (**b,a**). The entire unit circle is used. Therefore, the resulting vector **f** will have frequencies from 0 to **Fs**. If the option **'whole'** is left out, then only half the unit circle is used; that is, vector **f** will have frequencies from 0 to **Fs/2**. This command is useful for plotting the magnitude and phase response of the filter.

[H,w] = freqz(b,a,N,'whole'); uses N points around the whole unit circle. The N-point frequency response is stored in vector **H**. Vector **w** contains the angular frequencies from 0 to 2π. Again, if the option **'whole'** is left out, then only half the unit circle is used, and vector **w** contains frequencies from 0 to π.

freqz(b,a,...); with no output arguments plots the magnitude and unwrapped phase of the filter. A normalized frequency axis is used; that is, it goes from 0 to 1, where 1 represents the frequency of π radians or equivalently **Fs/2** Hertz.

Y = fft(X); computes the discrete Fourier transform (DFT) of vector **X** and saves it in vector **Y**, which is possibly complex. The magnitude and phase of this vector can be plotted if desired. If the length of **X** is a power of two, a fast radix-2 FFT algorithm is used. If the length of **X** is not a power of two, a slower non-power-of-two algorithm is employed. For matrices, the FFT operation is applied to each column.

fft(X,N); computes the N-point FFT, padded with zeros if **X** has less than N points and truncated if it has more.

Important note on **fft:** As we know from the discussion in this chapter, the DFT samples the unit circle from 0 to 2π. However, it is customary to plot the frequency response from $-\pi$ to π; that is, the DC component is in the center of the spectrum. MATLAB has a command for this purpose, which is given below.

fftshift (Y); For vectors, this command swaps the left and right halves of **Y**. For matrices (2-D signals), this command swaps the first and third quadrants and the second and fourth quadrants. Thus, this command is useful for visualizing the Fourier transform with the DC component in the middle of the spectrum.

ifft(X); the inverse DFT of **X**.

ifft(X,N); the N-point inverse DFT.

Y = decimate(X,M); resamples the sequence in vector **X** at $1/M$ times the original sample rate. The resulting resampled vector **Y** is M times shorter. This function also filters the data with an eighth-order Chebyshev type I lowpass filter with cutoff frequency of $\frac{0.8\pi}{M}$ or $0.8^*(\mathrm{Fs}/2)/M$, before resampling. Other types of filters can also be chosen with options. See the help command for details.

Y = interp(X,M); resamples the sequence in vector **X** at M times the original sample rate. The resulting resampled vector **Y** is M times longer. An appropriate lowpass interpolation filter is used.

Y = fft2(X); computes the 2-D DFT of a matrix **X**.

Y = fft2(X,M,N); computes the $(M \times N)$-point 2-D DFT of matrix **X** and pads with zeros in necessary.

ifft2(F); returns the two-dimensional inverse Fourier transform of matrix **F**.

ifft2(F,M,N); pads matrix **F** with zeros to size M × N before transforming.

2.10 PROBLEMS

1. Prove Lemma 1, the orthogonality of continuous-time exponentials.

2. Obtain the Fourier series representation and sketch the magnitude and phase spectra of the functions given in Fig. P2.1.

3. Find the Fourier series representation of $f(t) = 3\sin(4t) + 2\cos(7t) + \sin(14t)$. Sketch the line spectrum (magnitude and phase) for this function.

4. Determine the Fourier series representation of $f(t)$

(a) $f(t) = \begin{cases} \frac{E}{2} & kT \le t < kT + \frac{T}{2} \\ -\frac{E}{2} & kT + \frac{T}{2} \le t < (k+1)T \end{cases}$

(b) $f(t) = \begin{cases} \frac{2}{T}(t - kT) & kT \le t < kt + \frac{T}{2} \\ 0 & kT + \frac{T}{2} \le t < (k+1)T \end{cases}$

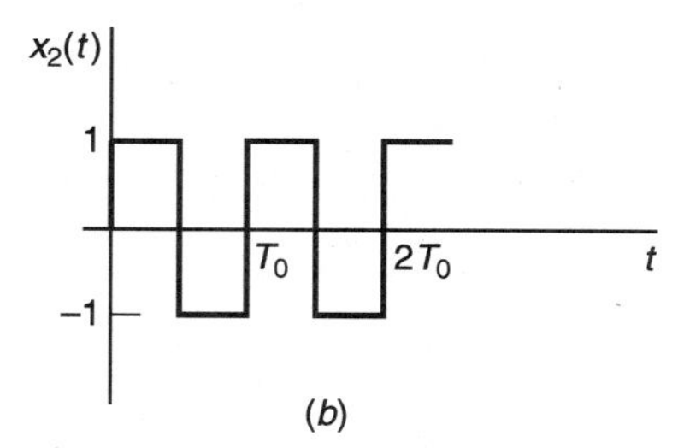

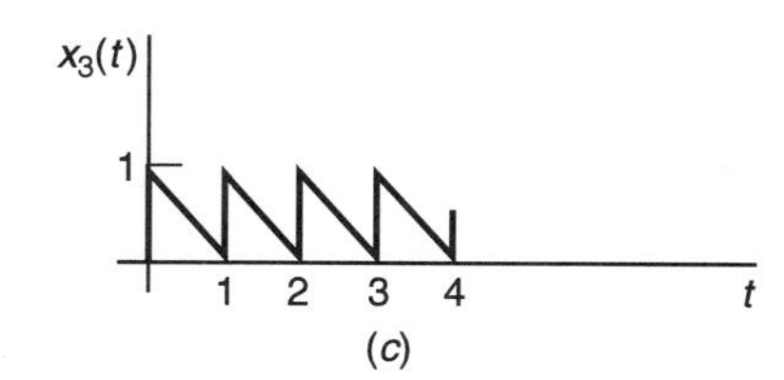

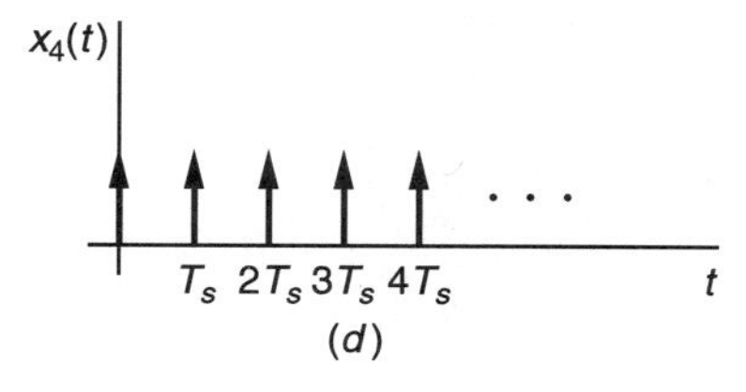

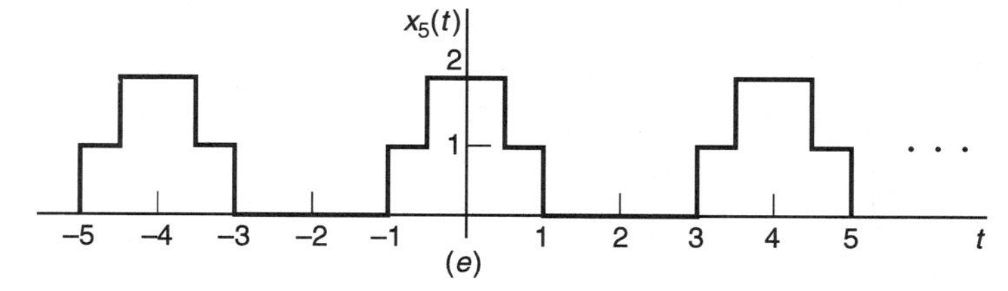

Figure P2.1

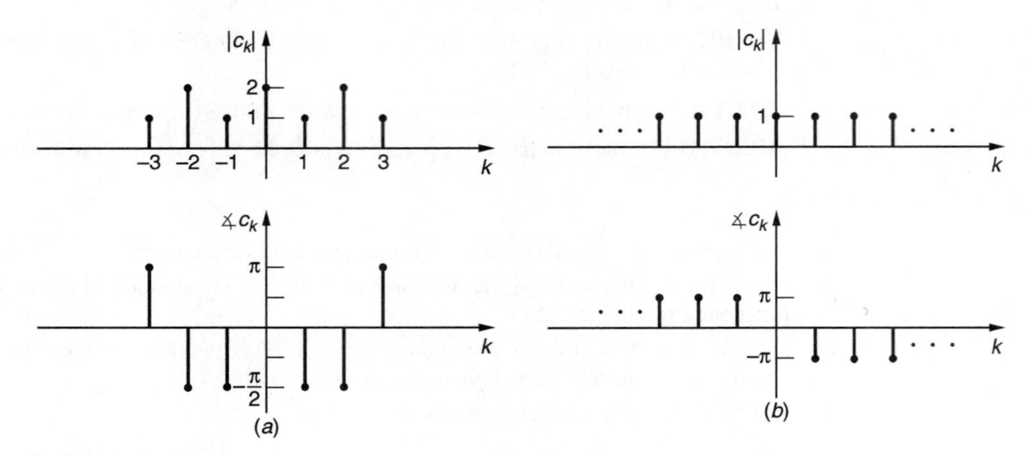

Figure P2.2

5. Give an example to show that a periodic function may not have a Fourier series representation.
6. The line spectra for some functions are given in Fig. P2.2. Find the time-domain functions.
7. Prove the Linearity, Scaling, Time-shifting, and Frequency shifting properties of the CTFT.
8. Prove the Differentiation and Duality properties of the CTFT.
9. Prove the Convolution and Multiplication properties of the CTFT.
10. Find the CTFT of the following functions: (a) $\cos(\Omega_0 t)$, (b) the impulse train represented by $\sum_{m=-\infty}^{\infty} \delta(t - mT_s)$, (c) $sgn(t) \triangleq \begin{cases} 1, & t > 0 \\ -1, & t < 0. \end{cases}$, (d) $u(t)$, (e) $\pi(\frac{t}{\beta})$.
11. Find the CTFT for the following functions. Sketch and label the magnitude response of each transform.

(a) $f(t) = e^{2+t}u(-t+1)$.
(b) $f(t) = e^{-3t}[u(t+2) - u(t-3)]$.
(c) $f(t) = e^{-3|t|} \sin(2t)$.

12. Find the CTFT for the following functions. Sketch and label the magnitude response of each transform.

(a) $\delta(t+\alpha) + \delta(t-\alpha)$, where $\alpha > 0$ is a constant.
(b) $\sin\left(2\pi t + \dfrac{\pi}{3}\right)$.

13. Find the CTFT of $e^{-|t|}$.
14. Find the CTFT for the following functions. Sketch and label the magnitude response of each transform.

(a) $f(t) = \begin{cases} t - 0.5 & for \quad 0 \le t < 1 \\ 0 & \text{otherwise} \end{cases}$.

(b) $f(t) = (\sin(2t) + \cos(0.5t))u(t)$.

15. Find the inverse CTFT of the following functions:

(a) $F(\Omega) = \dfrac{1}{(j\Omega + 2)^2}$.

(b) $F(\Omega) = 1$ for $\Omega > 0$.

16. Prove that if $f(t) = \dfrac{t^{n-1}}{(n-1)!}e^{-at}u(t), a > 0$, then $F(\Omega) = \dfrac{1}{(a + j\Omega)^n}$.

17. Prove the Convolution and Differentiation properties of the DTFT.

18. Prove Parseval's theorem for the DTFT.

19. If $X(\omega)$ is the DTFT of $x(n)$, Derive the DTFT of the following in terms of $X(\omega)$. Assume that $x(n)$ is possibly complex.

(a) $\mathrm{Re}\{x(n)\}$.

(b) $x^*(-n)$.

(c) $x(Mn)$, where M is a positive integer.

20. Find the DTFT of the following sequences: (a) $x_1(n) = u(n) - u(n-M)$ (*Hint*: The DTFT of $u(n)$ is $\frac{1}{1-e^{-j\omega}} + \sum_{k=-\infty}^{\infty} \pi\delta(\omega + 2\pi k)$.), (b) $x_2(n) = u(M-n)$, (c) $x_3(n) = \alpha^n \cos(\omega_0 n + \theta)u(n)$, (d) $x_4(n) = [1\ 2\ 3\ \mathbf{0}\ 3\ 2\ 1]$, (e) $x_5(n) = \delta(n) - \delta(n-2)$. Plot the magnitude and phase spectrum in each case. Choose appropriate values for α, θ and M, where necessary.

21. Find the DTFT of each of the following sequences. Sketch the magnitude and phase spectrum in each case: (a) $x_1(n) = 2^n u(-n)$, (b) $x_2(n) = (\frac{1}{2})^n u(n+5)$, (c) $x_3(n) = (-1)^n$, (d) $x_4(n) = 2 - \frac{1}{4}n$, (e) $x_3(n) = \cos(\frac{\pi n}{20})$.

22. Use properties of the DTFT to show that

$$(n+1)\alpha^n u(n) \longleftrightarrow \frac{1}{(1-\alpha e^{-j\omega})^2}.$$

23. Given the sequence $x(n) = [4\ \ -1\ 3\ \mathbf{-2}\ 3\ \ -1\ 4\ ]$. Find the following without explicitly evaluating the DTFT: (a) $X(0)$, (b) $X(-\pi)$, (c) $\int_{-\pi}^{\pi} X(\omega)d\omega$, (d) $\int_{-\pi}^{\pi} |X(\omega)|^2 d\omega$, (e) $\angle X(\omega)$.

24. $x(n)$ and $h(n)$ are real causal sequences; prove that

$$\frac{1}{2\pi}\int_{-\pi}^{\pi} X(\omega)H(\omega)d\omega = \left\{\frac{1}{2\pi}\int_{-\pi}^{\pi} X(\omega)d\omega\right\}\left\{\frac{1}{2\pi}\int_{-\pi}^{\pi} H(\omega)d\omega\right\}.$$

25. Find the inverse DTFT of the following transforms:

(a) $X(\omega) = \dfrac{1}{(1-\alpha e^{-j\omega})^N}$. (*Hint*: Use the differentiation property).

(b) $X(\omega) = \dfrac{e^{j\omega} - e^{-j\omega(M+1)}}{1 - e^{j\omega}}$.

(c) $A\sin^2\omega$.

(d) $\Pi\left(\dfrac{\omega - \omega_c}{\beta}\right) + \Pi\left(\dfrac{\omega + \omega_c}{\beta}\right)$.

26. Evaluate the inverse DTFT of the following:

(a) $X(\omega) = \begin{cases} 1 & \text{for } \pi \geq |\omega| \geq \omega_0 \\ 0 & \text{for } |\omega| < \omega_0 \end{cases}$.

(b) $X(\omega) = \sum_{n=-2}^{2} e^{-2n\omega}$.

27. Find the impulse response of the filters with frequency response given in Fig. P2.3.

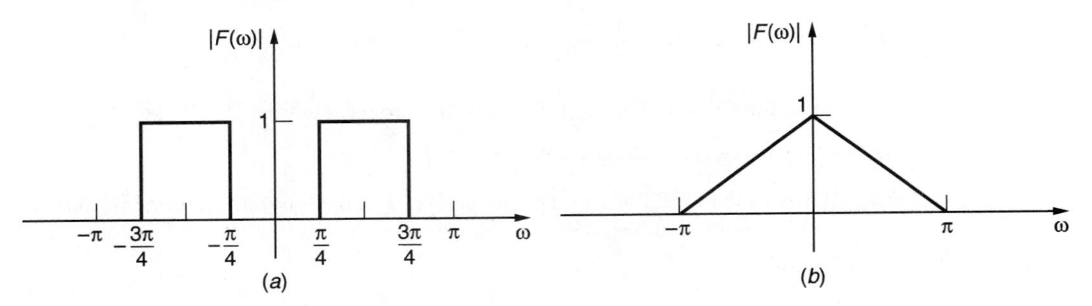

Figure P2.3

28. Prove Fact 1 and Fact 2.

29. The impulse response of a LSI system is $h(n) = (\frac{1}{4})^n u(n)$.

(a) Find and sketch the magnitude and phase spectrum of this system.

(b) If the input to this system is $\cos(\frac{7\pi n}{20})$, find the output sequence. Also, sketch the magnitude and phase spectra of the input and the output.

(c) Repeat (b) for

$$x(n) = \begin{cases} n, & 0 \leq n \leq 5 \\ 10 - n, & 6 \leq n \leq 10 \\ 0, & \text{otherwise.} \end{cases}$$

30. The difference equation of a comb filter is given by

$$y(n) = x(n) - x(n-6).$$

(a) Compute and sketch the magnitude and phase spectrum of this filter.

(b) What frequencies are notched out by this filter?

(c) Find the output of this filter if the input is $x(n) = 20 + 10\cos(\frac{2\pi n}{6} + \frac{\pi}{6})$.

31. Given the IIR filter

$$y(n) - 0.5y(n-1) = 0.5x(n)$$

(a) Find the frequency response of this filter and sketch its magnitude and phase spectrum.

(b) Find the 3-dB frequency.

(c) Find the output sequence if the input is $x(n) = (-1)^n u(n)$. Sketch the magnitude and phase response of the input and output.

32. A continuous-time signal $x(t)$ has the Fourier transform shown in Fig. P2.4. This signal is sampled as $x_s(n) \triangleq x(nT_s)$, with $T_s = 2\pi/\Omega_0$ seconds.

(a) Sketch the spectrum of the sampled signal, $X_s(\omega)$ as a function of the discrete-time frequency ω.

(b) Now we perform a D/C conversion at the same sampling frequency to get a continuous-time signal, $y(t)$. Sketch $|Y(\Omega)|$ against Ω.
(c) Sketch the frequency response of the ideal analog filter that can be used to recover the original signal.

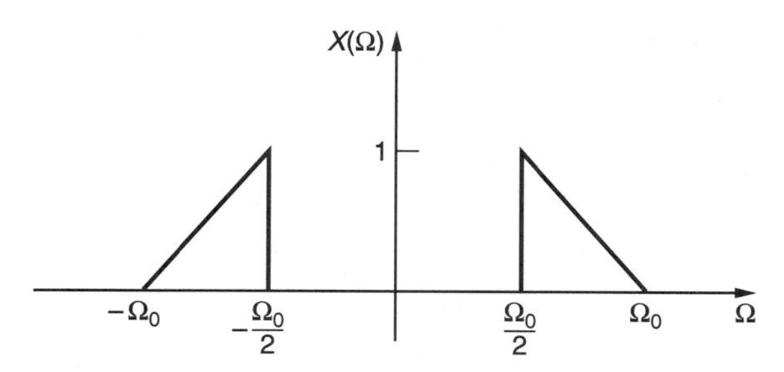

Figure P2.4

33. The continuous-time signal $x(t) = \cos(100t) + \sin(200t)$ is sampled to yield the sequence $x_s(n)$. Sketch the transforms $X(\Omega)$ and $X_s(\omega)$ for the following sampling frequencies: (a) $\Omega_s = 500$, (b) $\Omega_s = 300$. For case (b), is it possible to reconstruct the original signal after a D/C conversion? If so, sketch the frequency response of the analog filter required.

34. A complex continuous-time signal $x(t)$ has the Fourier transform shown in Fig. P2.5. This signal is sampled to produce the sequence $x_s(n)$.
(a) What is the lowest sampling frequency Ω_s required to avoid aliasing?
(b) Sketch $X_s(\Omega)$ and $X_s(\omega)$ for a no-aliasing condition.
(c) Sketch $X_s(\Omega)$ and $X_s(\omega)$ for an aliasing condition.
(d) Draw the block diagram of a system that can recover $x(t)$ from $x_s(n)$ if there is no aliasing. Sketch the spectrum of any filter used in your diagram.

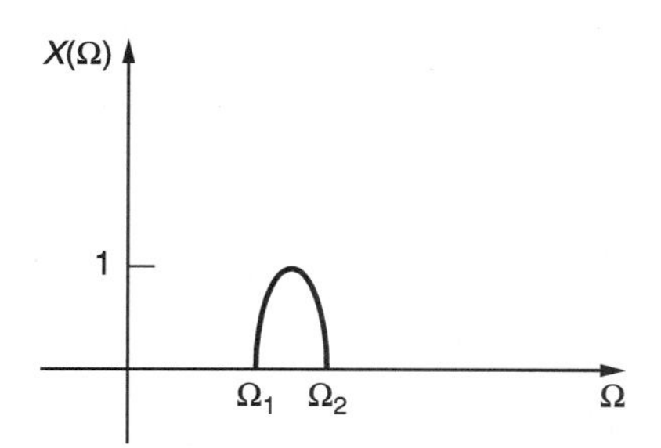

Figure P2.5

35. Let $x(n) = (1/2)^n u(n)$.
(a) Sketch the magnitude spectrum of this sequence.
(b) The sequence is passed through an ideal lowpass half-band digital filter to produce $y(n)$. Sketch the magnitude spectrum of $y(n)$.
(c) The sequence $y(n)$ is downsampled by a factor of 3 to produce $y_{(3)}(n)$. Sketch the magnitude spectrum of this decimated signal.
(d) If $y(n)$ is upsampled by a factor of 2, sketch the magnitude spectrum of $y^{(2)}(n)$.

36. Consider the block diagram of Fig. P2.6(a), where $x(t)$ is bandlimited as shown in Fig. P2.6(b). The digital filter $H(\omega)$ is an ideal HPF with a cutoff frequency of 0.8π. Sketch the spectra $X_s(\omega), Y(\omega), V(\omega)$, and $G(\omega)$.

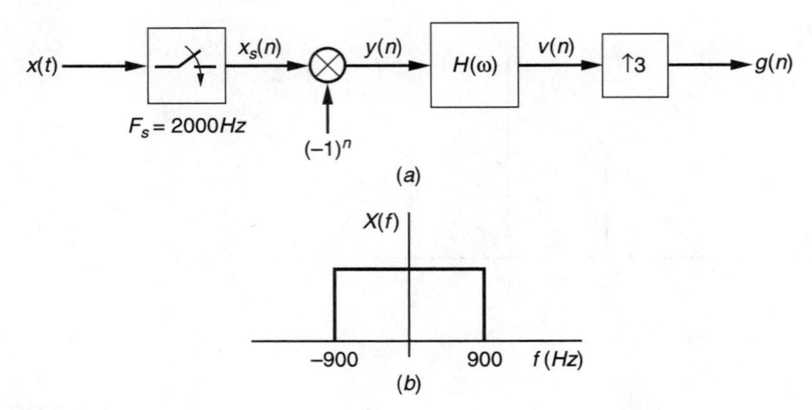

Figure P2.6

37. The spectrum of a sequence $x(n)$ is shown in Fig. P2.7. This is downsampled by a factor of 2 to produce $v(n)$. (a) Sketch the spectrum of $v(n)$. (b) What can you say about the form of the sequence $x(n)$?

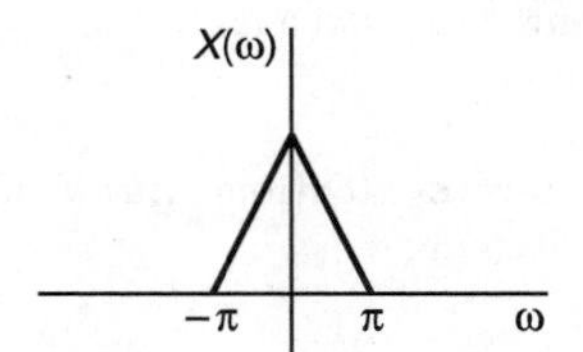

Figure P2.7

38. Prove Theorem 2.6 in general.

39. In Fig. P2.8, find the output $y(n)$ in terms of the input $x(n)$.

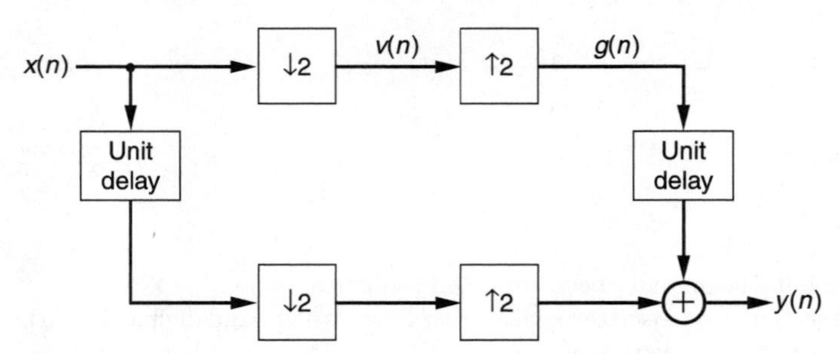

Figure P2.1

40. Prove the DFT synthesis equation given by (2.64).

41. Prove the Time-reversal and the Circular Frequency Shift properties of the DFT (Properties 2 and 4 in Table 2.4).

42. Prove the Circular Convolution in Frequency property of the DFT (Property 6 in Table 2.4).

43. Prove the Conjugation in Time and Conjugation in Frequency properties of the DFT (Properties 7 and 8 in Table 2.4).

44. Prove the Parseval's relation for the DFT given by

$$\sum_{n=0}^{N-1} |x(n)|^2 = \frac{1}{N}\sum_{k=0}^{N-1} |X(k)|^2.$$

45. Find the circular convolution and linear convolution of the sequences

$$x(n) = [\mathbf{1} \quad 3 \quad 2 \quad -1 \quad 4]$$

$$h(n) = [\mathbf{2} \quad 0 \quad 1 \quad 7 \quad -3]$$

using the time-domain approach.

46. Find the circular convolution and linear convolution of the sequences

$$x(n) = [\mathbf{1} \quad 2 \quad 7 \quad -2 \quad 3 \quad -1 \quad 5]$$

$$h(n) = [\mathbf{-1} \quad 3 \quad 5 \quad -3 \quad 1]$$

using the time-domain approach.

47. Use the DFT approach to find the circular convolution and the linear convolution of the sequences given in the preceding two problems.

48. Prove the following properties of circular convolution:

(a) *Commutativity*: $(x(n) * h(n))_N = (h(n) * x(n))_N$.

(b) *Associativity*: $((x(n) * h(n))_N * g(n))_N = (x(n) * (h(n) * g(n))_N)_N$.

49. Find the energy of the N-point sequence

$$x(n) = \sin\frac{2\pi kn}{N}, \quad 0 \le n \le N-1.$$

50. Compute the N-point DFT of the following sequences:

(a) $\delta(n - n_0)$.

(b) $\cos\frac{2\pi}{N}mn, \ 0 \le n \le N-1$.

(c) $$x(n) = \begin{cases} 1, & n \text{ even} \\ 0, & n \text{ odd} \end{cases} \quad 0 \le n \le N-1.$$

(d) $$x(n) = \begin{cases} 1, & 0 \le n \le N/2 - 1 \\ -1, & N/2 \le n \le N-1. \end{cases}$$

(e) $e^{j\frac{2\pi}{N}k_0 n}, \ 0 \le n \le N-1$.

51. Let $x(n) = [1 \quad 2 \quad 0 \quad 3 \quad -2 \quad 4 \quad 7 \quad 5]$. Evaluate the following: (a) $X(0)$, (b) $X(4)$, (c) $\sum_{k=0}^{7} X(k)$, (d) $\sum_{k=0}^{7} |X(k)|^2$.

52. Determine the 5-point DFT of the following sequence: $x(n) = [-1 \quad 2 \quad 5 \quad -1 \quad 1]$.

53. If we perform the DFT of an N-length sequence four times, find the resulting sequence.

54. The first four samples of the 7-point DFT of a real length-7 sequence are given as follows: $9.0000; -7.0000 + j14.54; -7.0000 + j5.58; -7.0000 + j1.6$. Find the remaining samples.

55. A continuous-time signal has a bandwidth of 5 MHz. An $N = 2^m$-point DFT must be performed to compute the spectrum of this signal with a resolution of at most 600 Hz. Find (a) the minimum sampling rate, (b) the minimum number of samples required, (c) the minimum length of the continuous-time signal in seconds, and (d) the exact frequency resolution of the spectrum.

56. A sequence $x(n)$ is upsampled by a factor of M to produce $y(n)$. Find the MN-point DFT of $y(n)$ in terms of the N-point DFT of $x(n)$.

57. A sequence $x(n)$ is decimated by a factor of M to produce $y(n)$. Find the N/M-point DFT of $y(n)$ in terms of the N-point DFT of $x(n)$, assuming that M divides N.

58. Construct a 16-point radix-2 decimation in time FFT algorithm. Label all multipliers appropriately.

59. Construct a 16-point radix-2 decimation in frequency FFT algorithm. Label all multipliers appropriately.

60. Assume that we have some 8-point FFT chips. Show how to interconnect four of these chips in order to obtain the 32-point FFT.

61. Determine the 2-D DTFT of each of the following sequences by using the basic definition:

(a) $\delta(n_1 - 1, n_2)$.

(b) $a^{n_1+n_2} u_{2d}(n_2)\delta(n_1 - n_2), \quad |a| < 1$.

62. Determine the 2-D DTFT of each of the following sequences by using the basic definition:

(a) $n_2 a^{n_1+n_2} u(n_1, n_2), \quad |a| < 1$.

(b) $a^{n_1} b^{n_2} u(n_1, n_2)\delta_{2d}(n_1 - 4n_2), \quad |a| < 1, |b| < 1$.

63. Plot the impulse responses of circularly symmetric lowpass filters with cut frequencies of $\frac{\pi}{4}$ and $\frac{\pi}{2}$.

64. Plot the impulse responses of the following separable lowpass filters:

(a) $\omega_{c1} = \dfrac{\pi}{4}, \quad \omega_{c2} = \dfrac{\pi}{4}$.

(b) $\omega_{c1} = \dfrac{\pi}{4}, \quad \omega_{c1} = \dfrac{\pi}{2}$.

65. Prove the 2-D separable DTFT property given by

$$x(n_1, n_2) = x_1(n_1)x(n_2) \leftrightarrow X_1(\omega_1)X_2(\omega_2).$$

66. Prove Theorem 2.8.

67. Prove Theorem 2.9.

68. Prove Theorem 2.10.

69. Perform the 2-D separable convolution of the following. Note that $h(n_1, n_2)$ is separable:

(a) $x(n_1, n_2) = \begin{bmatrix} \mathbf{-1} & 2 & -1 \\ 2 & -1 & 2 \\ -1 & 2 & -1 \end{bmatrix}$, $h(n_1, n_2) = \begin{bmatrix} \mathbf{-1} & 1 \\ 1 & -1 \end{bmatrix}$;

(b) $x(n_1, n_2) = \begin{bmatrix} \mathbf{-0.5} & 0.25 & -0.125 \\ 1.25 & -1 & 1.25 \\ -0.5 & 0.25 & -0.125 \end{bmatrix}$, $h(n_1, n_2) = \begin{bmatrix} -1 & 1 & -2 \\ 1 & -1 & \mathbf{2} \end{bmatrix}$.

70. Solve the preceding problem using 2-D DFT. Zero-pad appropriately for linear convolution.

71. Repeat Problem 2.69 for 2-D circular convolution.

72. Repeat Problem 2.69 for 2-D circular convolution using 2-D separable DFT.

73. Give an example of an image (any size matrix) for which most of the energy is not concentrated in the low-frequency regions.

74. Show that

$$\sum_{n_1=-\infty}^{\infty} \sum_{n_2=-\infty}^{\infty} |x(n_1, n_2)|^2 = \frac{1}{(2\pi)^2} \int_{\omega_1=-\pi}^{\pi} \int_{\omega_2=-\pi}^{\pi} |X(\omega_1, \omega_2)|^2 d\omega_1 d\omega_2.$$

2.11 COMPUTER PROJECTS

Project 1

1. Generate a 10-kHz sinusoid sampled at 100 kHz. Plot four cycles of this signal. Also plot the spectrum of this signal in the interval $(-\pi, \pi)$. Use the **fft** and **fftshift** commands for this purpose. The horizontal axis must be scaled appropriately to represent the interval $(-\pi, \pi)$.
2. Decimate this signal by a factor of 2. Plot the time-domain signal. Also plot the spectrum of the decimated signal in the interval $(-\pi, \pi)$. What frequency in Hz does π represent? Comment on your result.
3. Upsample the original sinusoidal sequence by a factor of 3. Plot the time-domain signal. Also plot the spectrum of the upsampled signal in the interval $(-\pi, \pi)$. What frequency does π represent now?
4. Use the **fir1** command to design an appropriate LPF for filtering the upsampled signal in order to recover the original spectrum. Plot the magnitude response and the phase response of the designed filter. Filter the upsampled signal with this filter and plot the output. Also plot the spectrum of the filtered signal in the interval $(-\pi, \pi)$. Comment on your result.

Project 2

1. Write a program to compute the N-point DFT of a sequence. Do not use the MATLAB **fft** command. Write your own routine for the DFT. Use your program to find the 200-point DFT of the following sequences: (a) $x(n) = 2\cos(\frac{2\pi}{10}n) + \cos(\frac{2\pi}{5}n)$, (b) $x(n) = n$. Plot the magnitude response of these sequences. The horizontal axis must be scaled appropriately to represent the interval $(0, \pi)$. Comment on the validity of your results.

2. Write a program to compute the N-point inverse-DFT (IDFT) of a sequence. Do not use the MATLAB **ifft** command. Test your program on the DFTs of the sequences in the previous part.
3. Write a program to compute the circular convolution of two length-N sequences via the DFT approach. Use your DFT and IDFT programs for this purpose. Use your programs to find the circular convolution of the following pairs of sequences: (a) $x(n) = [\mathbf{1} \quad 3 \quad -2 \quad 4 \quad 7], h(n) = [\mathbf{3} \quad 11 \quad 21 \quad -33 \quad 9]$; (b) $x(n) = n,\ h(n) = (1/2)^n,\ 0 \leq n \leq 10$; (c) $x(n) = \sin(\pi/20), h(n) = n(1/4)^n,\ 0 \leq n \leq 20$.
4. Write a program to perform circular convolution in the time domain. Test your program on the sequences in (3).
5. Using MATLAB, plot the magnitude (in dB) response of the FFT of a square window, triangle window, Hamming window, Blackman window, and Hanning window. Make the windows 64 points long, and use a 512-point FFT. Use the MATLAB command **fft** in this part. Again, the horizontal axis must be scaled appropriately to represent the interval $(0, \pi)$.
6. Generate 4096 samples of the sequence in equation (2.33) in text with $\omega_c = \pi/2$. Window this sequence with each of the five windows. Plot the magnitude (in dB) response of the windowed signals in each case.

Questions

1. Comment on the main-lobe width and side-lobe (ripple) level of each of the windows. Which windows have a constant side lobe level?
2. Is the amplitude of the FFT of the sinusoids what you expected?
3. Explain the use of windows, and discuss the variables of window design. Refer to the results (6) above.

Project 3

Using an image of your choice, perform the following:

1. Corrupt the image with white Gaussian noise so that the SNR of the image is approximately 5 dB. Find the actual SNR of the corrupted image. Display the original and corrupted images.
2. Construct a 2-D separable FIR filter (h) from the following 1-D sequences: $h_1 = \begin{bmatrix}1 & 0 & -1\end{bmatrix}, \quad h_2 = \begin{bmatrix}1 & 0 & 1\end{bmatrix}$.
3. Filter your clean image and your corrupted image with the above filter in the time domain. In other words, convolve the image with the filter. You must exploit the separability of the filter and write your own routine to perform the 2-D convolution efficiently as a sequence of 1-D convolutions.
4. Display the filtered images. What kind of filter do you think it is. Find the SNRI of the filtered images.

Project 4

Using an image of your choice, perform the following:

1. Corrupt the image with white Gaussian noise so that the SNR of the image is approximately 5 dB. Find the actual SNR of the corrupted image. Display the original and corrupted images.
2. Design a 8×8 Gaussian lowpass filter with a standard deviation of 0.6, using the MATLAB command **fspecial**.
3. Filter the corrupted image with the above filter in the DFT domain. Make sure you zero-pad in order to avoid circular convolution. The DFT of each 2-D sequence must be performed efficiently using a sequence of 1-D FFTs. So, you must write your own program to do so.
4. Display the original image, the corrupted image, and the filtered image. Find the SNRI of the filtered image.

BIBLIOGRAPHY

[1] E.T. BELL, *Men of Mathematics*, Simon and Schuster, New York, 1937.

[2] R.V. CHURCHILL and C.W. BROWN, *Fourier Series and Boundary Value Problems*, McGraw-Hill, New York, 1978.

[3] A.V. OPPENHEIM, R.W. SCHAFER, and J.R. BUCK, *Discrete Time Signal Processing*, Prentice Hall, New York, 1999.

[4] G. STRANG and T. NGUYEN, *Wavelets and Filter Banks*, Wellesley-Cambridge Press, Wellesley, MA, 1996.

[5] J.S. LIM, *Two-dimensional Signal and Image Processing*, Prentice Hall, Englewood Cliffs, NJ, 1990.

[6] M. ABRAMOWITZ and I.A. STEGUN, *Handbook of Mathematical Functions*, Dover Publications, New York, 1964.

[7] R.E. CROCHIERE and L.R. RABINER, *Multirate Digital Signal Processing*, Prentice Hall, Englewood Cliffs, NJ, 1983.

[8] P.P. VAIDYANATHAN, *Multirate Systems and Filter Banks*, Prentice Hall, Englewood Cliffs, NJ, 1993.

[9] B. SUTER, *Multirate and Wavelet Signal Processing*, Academic Press, New York, 1998.

[10] N. FLIEGE, *Multirate Digital Signal Processing: Multirate Systems, Filter Banks, Wavelets*, Wiley, New York, 1994.

[11] J.W. COOLEY and J.W. TUKEY, "An algorithm for the machine computation complex Fourier series," *Math. Comp*. 19, pp. 297–301, April 1965.

[12] R.C. SINGLETON, "An algorithm for computing the mixed radix fast Fourier transform," *IEEE Trans. on Audio and Electroacoustics*, AU-17, pp. 93–103, June 1969.

[13] P. DUHAMEL and H. HOLLMAN, "Split-radix FFT algorithm," *Electronic Letters* 20, pp. 14–16, January 1984.

[14] G. GOERTZEL, "An algorithm for the evaluation of finite trigonometric series," *Am. Math. Monthly* 65, pp. 34–35, January 1968.

[15] C.M. RADER and N.M. BRENNER, "A new principle for fast Fourier transformation," *IEEE Trans. on Acous., Speech, and Signal Processing*, ASSP-24, pp. 264–266, June 1976.

[16] I.J. GOOD, "The relationship between two fast Fourier transforms," *IEEE Trans. on Computers*, C-20, pp. 310–317, 1971.

[17] S. WINOGRAD, "On computing the discrete Fourier transform," *Math. Comp*. 32, pp. 177–199, 1978.

CHAPTER 3

Z-Transform and Digital Filters

In the last chapter, several transforms in the Fourier domain were presented for continuous-time and discrete-time signals and systems. The Z-transform is an important tool for analysis of discrete-time systems. It is actually more general than the DTFT discussed in the last chapter. It turns out that the Z-transform evaluated on the unit circle ($z = e^{j\omega}$) is the DTFT. Therefore, the frequency-domain characteristics of a system can also be evaluated from the Z-transform. Digital filters are usually represented in terms of Z-transforms. This leads to simplicity and convenience in the analysis and design of filters.

In this chapter, we first present the definition of the Z-transform, its properties, its inverse, and several examples. The different types of filters are then discussed. The concept of linear phase filters is explained in detail. The basic theory of state-space digital filters is then discussed. Linear filtering can be done using FFT for computational complexity. The overlap-add and overlap-save methods are presented for this purpose. On the other hand, the DFT can be computed using linear filtering. The Goertzel algorithm is useful for this purpose in certain applications. A generalized Z-transform called the Chirp-Z transform is also described. Some of these topics are then extended to 2-D systems.

3.1 Z-TRANSFORM

The Z-transform of a sequence $x(n)$ is defined as

$$X(z) = Z\{x(n)\} = \sum_{n=-\infty}^{\infty} x(n)z^{-n} \tag{3.1}$$

where $z = a + jb = re^{j\omega}$. The sequence and its Z-transform will be denoted by the pair $x(n) \longleftrightarrow X(z)$. For the special case of $z = e^{j\omega}$, the right-hand side of equation (3.1) reduces to the definition of the DTFT. Since the Z-transform is an infinite sum, it must converge in some domain in the complex plane of z for the transform to exist. This domain is referred to as the region of convergence (ROC) and must be specified for every transform. The ROC for the transform is specified in the complex plane of z. This will be illustrated with an example following the definition.

Definition 3.1 *A sequence is said to be right-sided if it has nonzero sample values for all $n_0 \leq n \to \infty$, where n_0 is some integer. A sequence is said to be left-sided if it has nonzero sample values for all $-\infty \longleftarrow n \leq n_0$.*

These two types of sequences have significance in the convergence of the Z-transform and its ROC, as will be clear shortly.

▶ EXAMPLE 3.1 *The Z-transform of the right-sided sequence*[1] *$h_1(n) = \beta^n u(n)$ can be found directly from Definition 3.1. In the evaluation, we use the formula for the infinite geometric sum and its convergence condition given in Chapter 2. The result is*

$$
\begin{aligned}
Z\{\beta^n u(n)\} &= \sum_{n=0}^{\infty} \beta^n z^{-n} \\
&= \sum_{n=0}^{\infty} (\beta z^{-1})^n \\
&= \frac{1}{1-\beta z^{-1}}, \quad |\beta z^{-1}| < 1 \text{ or } |z| > |\beta|. \\
&= \frac{z}{z-\beta}, \quad |z| > |\beta|.
\end{aligned} \tag{3.2}
$$

The region of convergence is given by the condition on z and is plotted in Fig. 3.1(a). If the sequence is left-sided, that is, $h_2(n) = \beta^n u(-n-1)$, the Z-transform is

$$
\begin{aligned}
Z\left\{\beta^n u(-n-1)\right\} &= \sum_{n=-1}^{-\infty} \beta^n z^{-n} \\
&= \sum_{m=1}^{\infty} (\beta^{-1} z)^m \\
&= \sum_{m=0}^{\infty} (\beta^{-1} z)^n - 1 \\
&= \frac{1}{1-\beta^{-1} z} - 1, \quad |\beta^{-1} z| < 1 \text{ or } |z| < \beta \\
&= \frac{-z}{z-\beta}, \quad |z| < |\beta|.
\end{aligned} \tag{3.3}
$$

The ROC is shown in Fig. 3.1(b).

If $h_1(n)$ and $h_2(n)$ given above denote the impulse responses of digital filters, then the Z-transforms given by (3.2) and (3.3) denote the transfer functions of these filters,

[1] Note that right-sided or left-sided sequences only exist for real systems.

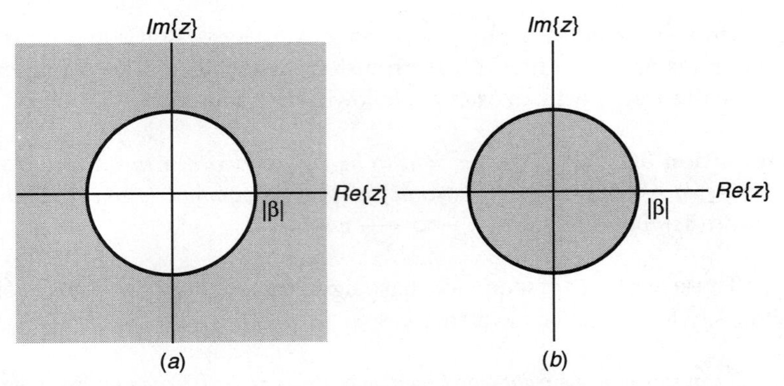

Figure 3.1 ROC for the Z-transform of (a) $\beta^n u(n)$; (b) $\beta^n u(-n-1)$.

respectively. Both of these transfer functions have a pole at $z = \beta$. The circle with radius $|\beta|$ is therefore called the *pole circle*. From this example, we observe that for a right-sided impulse response, the ROC extends outside the pole circle, and for a left-sided sequence, the ROC is inside the pole circle. Now consider a transfer function $H(z)$ with several poles that all correspond to right-sided sequences. The ROC for each pole is the region that extends outside that pole circle. The ROC of $H(z)$ is the intersection of the ROCs for each of the poles and therefore lies outside the outermost pole circle. Similarly, if a transfer function has several poles that correspond to left-sided sequences, then the ROC lies inside the innermost pole circle. This is an important fact and deserves a formal statement.

Fact: Consider a transfer function with poles that correspond to both right-sided and left-sided sequences. The ROC for the right-sided sequences is the region outside the outermost pole circle, say R_r. The ROC for the left-sided sequences is the region inside the innermost pole circle, say R_l. The ROC of the overall transfer function is then the intersection of these two regions given by $R_r \frown R_l$.

The Z-transforms of some simple sequences are derived in the following example.

▶ **EXAMPLE 3.2** *Find the Z-transform of the following sequences:* (a) $\delta(n)$, (b) $u(n)$, (c) $u(-n-1)$, (d) $\cos(\omega_0 n)u(n)$, (e) $\sin(\omega_0 n)u(n)$.

SOLUTION (**a**) The Z-transform can be easily obtained by using the definition of Z-transform and then applying the sifting property of the discrete impulse. The result is

$$
\begin{aligned}
Z\{\delta(n)\} &= \sum_{n=-\infty}^{\infty} \delta(n) z^{-n} \\
&= 1, \qquad \forall z. \qquad (3.4)
\end{aligned}
$$

Since we did not have to use the infinite sum formula here, the ROC is essentially everywhere in the z-plane.

(**b**) The transforms of (b) and (c) can be found by substituting $\beta = 1$ in Example 3.1.

$$
Z\{u(n)\} = \frac{z}{z-1}, \qquad |z| > 1. \qquad (3.5)
$$

(c)

$$Z\{u(-n)\} = \frac{-z}{z-1}, \quad |z| < 1. \tag{3.6}$$

(**d**) The Z-transform of cos can be found by using Euler's formula and using the definition as follows.

$$\begin{aligned} Z\{cos(\omega_0 n)u(n)\} &= \sum_{n=0}^{\infty} \frac{1}{2}\left(e^{j\omega_0 n} + e^{-j\omega_0 n}\right) z^{-n} \\ &= \frac{1}{2}\left(\frac{z}{z - e^{j\omega_0}} + \frac{z}{z - e^{-j\omega_0}}\right), \quad |z| > |e^{j\omega_0}| = 1 \\ &= \frac{z}{2}\left(\frac{2z - e^{-j\omega_0 n} - e^{j\omega_0 n}}{z^2 - z(e^{j\omega_0} + e^{-j\omega_0}) + 1}\right) \\ &= \frac{z(z - \cos\omega_0)}{z^2 - 2z\cos\omega_0 + 1}, \quad |z| > 1. \end{aligned} \tag{3.7}$$

(e) The Z-transform of sin can be found in a similar way as for cos. The result is

$$\{\sin(\omega_0 n)u(n)\} = \frac{z\sin\omega_0}{z^2 - 2z\cos\omega_0 + 1}, \quad |z| > 1. \tag{3.8}$$

3.2 Z-TRANSFORM PROPERTIES

As with the various Fourier transforms, we can derive several useful properties of the Z-transform. Most of these properties are easily obtained from the definition of the transform. However, we must take care in deriving the ROC in these properties. From the examples in the last section, it is clear that the ROC is denoted by an inequality condition on $|z|$. Since z is a complex number, this inequality represents a region bounded inward or outward by a circle. In the following properties, we will assume that the Z-transform pair is $x(n) \longleftrightarrow X(z)$. The ROC is denoted by R_x and is assumed to be $r_i < |z| < r_o$, where r_i and r_o are the radii of the inner and outer circle, respectively, and can be possibly zero or infinite.

Property 1 (**Linearity**): For constants a and b,

$$ax_1(n) + bx_2(n) \longleftrightarrow aX_1(z) + bX_2(z), \quad R_{x_1} \cap R_{x_2}.$$

Proof. The proof comes from a direct application of the definition. For the existence of the Z-transform of the overall function, the transforms of both $x_1(n)$ and $x_2(n)$ must converge. Therefore, the ROC is the intersection (common region) of the ROC of the two individual sequences. ■

As an example, consider the sequence $x(n) = 2^n u(n) + (-3)^n u(-n-1)$. From the last section, the Z-transform is $X(z) = \frac{z}{z-2} - \frac{z}{z+3}$. The ROC for the first term is $|z| > 2$, and that for the second term is $|z| < 3$. Hence, the ROC for $X(z)$ is $2 < |z| < 3$. This is shown in Fig. 3.2.

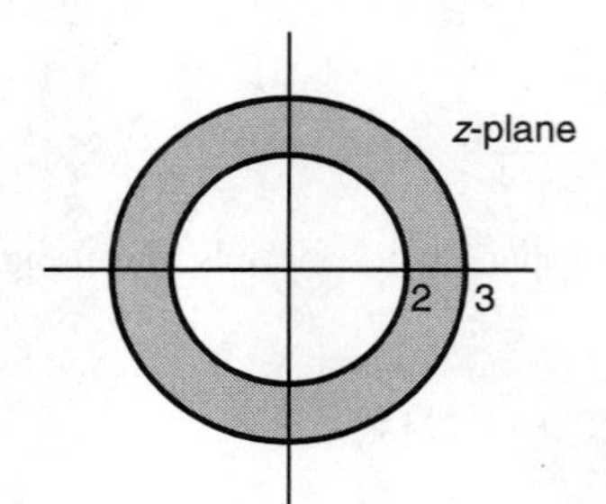

Figure 3.2 ROC for the Z-transform of the sequence $2^n u(n) + (-3)^n u(-n-1)$.

Property 2 (Time shifting): A shift in the time domain by n_0 corresponds to a multiplication in the Z domain by z^{-n_0}; that is,

$$x(n-n_0) \longleftrightarrow z^{-n_0}X(z), \quad ROC = R_x \text{ (possible changes for } z=0 \text{ or } z=\infty).$$

Proof. The property is proved by applying the Z-transform definition and then performing a change of variables $m = n - n_0$, as follows:

$$\begin{aligned} Z\{x(n-n_0)\} &= \sum_{n=-\infty}^{\infty} x(n-n_0)z^{-n} \\ &= \sum_{m=-\infty}^{\infty} x(m)z^{-m-n_0} \\ &= z^{-n_0}X(z). \end{aligned}$$

The ROC depends on how the factor z^{-n_0} changes the ROC of R_x. First assume that $X(z)$ does not have any poles or zeros at $z = 0$; that is, there is no factor z^p in the numerator or denominator. For $n_0 > 0$, we then have a negative power in z^{-n_0}, and so the ROC would not include $z = 0$. For $n_0 < 0$, $z = \infty$ would not be part of the ROC. If $X(z)$ has poles or zeros at $z = 0$, the factor z^{-n_0} may cancel these poles or zeros, thereby altering the ROC only for $z = 0$ or $z = \infty$. ■

▶ **EXAMPLE 3.3** *Consider the FIR digital filter given by the difference equation*

$$y(n) = b_0x(n) + b_1x(n-1) + \cdots + b_Mx(n-M).$$

The transfer function is defined as $H(z) = Y(z)/X(z)$. *Using Property 2, we obtain it as*

$$\begin{aligned} Y(z) &= b_0X(z) + b_1z^{-1}X(z) + \cdots + b_Mz^{-M}X(z), \\ H(z) &= \frac{Y(z)}{X(z)} = b_0 + b_1z^{-1} + \cdots + b_Nz^{-M}. \end{aligned} \tag{3.9}$$

Now, consider the IIR filter described by the difference equation

$$y(n) + a_1y(n-1) + \cdots + a_Ny(n-N) = b_0x(n) + b_1x(n-1) + \cdots + b_Mx(n-M).$$

The transfer function can be found by applying Property 2 to both sides of the above as

$$H(z) = \frac{b_0 + b_1z^{-1} + \cdots + b_Mz^{-M}}{1 + a_1z^{-1} + \cdots + a_Nz^{-N}}. \tag{3.10}$$

Property 3 (Multiplication by discrete exponential):
Let the ROC of $X(z)$ be $r_i < |z| < r_o$. Then

$$a^n x(n) \longleftrightarrow X\left(\frac{z}{a}\right), \quad ROC: |a|r_i < |z| < |a|r_o.$$

Proof. By using the definition of Z-transform, we have

$$\begin{aligned} Z\{a^n x(n)\} &= \sum_{n=-\infty}^{\infty} a^n x(n) z^{-n} \\ &= \sum_{n=-\infty}^{\infty} x(n) \left(\frac{z}{a}\right)^{-n} \\ &= X\left(\frac{z}{a}\right). \end{aligned}$$

From the definition of the ROC of $X(z)$, the ROC of the above is $r_i < |\frac{z}{a}| < r_o$ which leads to the conclusion. ■

EXAMPLE 3.4 *The Z-transforms of $r^n \cos(\omega_0 n)$ and $r^n \sin(\omega_0 n)$ are important because these are the typical responses of a second-order discrete system with complex conjugate poles. The transforms are obtained by using Property 3 and the transforms of cos and sin given in the last section, that is, by substituting $\frac{z}{r}$ in (3.7) and (3.8), respectively. The results are given in Table 3.1 along with the transforms of some other common sequences.*

Property 4 (Differentiation in the Z domain):

$$nx(n) \longleftrightarrow -z\frac{dX(z)}{dz}, \quad ROC = R_x \ \text{(possible changes for } z = 0 \text{ or } z = \infty).$$

TABLE 3.1 Common Z-Transform Pairs

	Sequence	Transform	ROC
1.	$\delta(n)$	1	$\forall z$
2.(a)	$a^n u(n)$	$\frac{z}{z-a}$	$\|z\| > \|a\|$
(b)	$a^n u(-n-1)$	$\frac{-z}{z-a}$	$\|z\| < \|a\|$
3.(a)	$na^n u(n)$	$\frac{az}{(z-a)^2}$	$\|z\| > \|a\|$
(b)	$na^n u(-n-1)$	$\frac{-az}{(z-a)^2}$	$\|z\| < \|a\|$
4.	$r^n \cos(\omega_0 n) u(n)$	$\frac{z(z - r\cos\omega_0)}{z^2 - (2r\cos\omega_0)z + r^2}$	$\|z\| > r$
5.	$r^n \sin(\omega_0 n) u(n)$	$\frac{zr\sin\omega_0}{z^2(2r\cos\omega_0)z + r^2}$	$\|z\| > r$

Proof. Consider the definition of the Z-transform and differentiate with respect to z, to get

$$\begin{aligned} X(z) &= \sum_{n=-\infty}^{\infty} x(n)z^{-n} \\ \frac{dX(z)}{dz} &= -\sum_{n=-\infty}^{\infty} nx(n)z^{-n-1} \\ &= -\frac{1}{z}\sum_{n=-\infty}^{\infty} nx(n)z^{-n} \\ &= -\frac{1}{z}Z\{nx(n)\}. \end{aligned}$$

The ROC is essentially the same as that for R_x. However, by differentiating $X(z)$ we introduce the term z^{-1}, which will alter the number of poles and zeros of $X(z)$ at $z = 0$. ■

▶ **EXAMPLE 3.5** *Here we find the Z-transform of $na^n u(n)$ by using the above property. Let $x(n) = a^n u(n)$. Then $X(z) = \frac{z}{z-a}$, $|z| > |a|$. Using Property 4, we have*

$$\begin{aligned} Z\{na^n u(n)\} &= -z\frac{d}{dz}\left\{\frac{z}{z-a}\right\} \\ &= \frac{az}{(z-a)^2}, \quad |z| > a. \end{aligned} \tag{3.11}$$

By setting $x(n) = na^n u(n)$ in the above property, we can find the Z-transform of $n^2 a^n u(n)$. Then by setting $x(n) = n^2 a^n u(n)$, we can find the Z-transform of $n^3 a^n u(n)$, and so on.

The following properties are stated without proof. They can be proved by applying the Z-transform definition and then performing some simple manipulations as above.

Property 5 (Time reversal):

$$x(-n) \longleftrightarrow X(z^{-1}), \quad \frac{1}{r_0} < |z| < \frac{1}{r_i}$$

Property 6 (Convolution in time): A convolution in the time domain corresponds to multiplication in the z domain, that is,

$$x(n) * h(n) \longleftrightarrow X(z)H(z), \quad ROC = R_x \cap R_h.$$

Property 7 (Complex conjugation):

$$x^*(n) \longleftrightarrow [X(z^*)]^*, \quad ROC = R_x$$

The proof of the following two properties requires the inverse Z-transform formula for $x(n)$ in terms of a contour integration. This formula is derived in the next section, and the result is in equation (3.15). The formula for evaluating the contour integration will follow shortly.

Property 8 (Multiplication in time):

$$x(n)h(n) \longleftrightarrow \frac{1}{2\pi j}\oint_{C_1} X(v)H\left(\frac{z}{v}\right)v^{-1}\,dv$$

$$\text{or } \frac{1}{2\pi j}\oint_{C_2} H(v)X\left(\frac{z}{v}\right)v^{-1}\,dv$$

where the contour of integration C_1 is in the common region of the ROC of $X(v)$ and $H(\frac{z}{v})$, and C_2 is in the common region of the ROC of $H(v)$ and $X\left(\frac{z}{v}\right)$.

Property 9 (Parseval's theorem):

$$\sum_{n=-\infty}^{\infty} x(n)h^*(n) = \frac{1}{2\pi j}\oint_{C_1} X(v)H^*\left(\frac{1}{v^*}\right)v^{-1}\,dv$$

$$\text{or } \frac{1}{2\pi j}\oint_{C_2} H^*(v)X\left(\frac{1}{v}\right)v^{-1}\,dv$$

where the contour of integration C_1 is in the common region of the ROC of $X(v)$ and $H^*\left(\frac{1}{v^*}\right)$, and C_2 is in the common region of the ROC of $H^*(v)$ and $X\left(\frac{1}{v}\right)$. Note that from Property 7, the ROC of $X^*\left(\frac{1}{v^*}\right)$ is the same as that for $X\left(\frac{1}{v}\right)$.

Parseval's theorem can be used to find the energy of a sequence $x(n)$, which is defined by Energy $\{x(n)\} = \sum_{n=-\infty}^{\infty} x(n)x^*(n)$. The above theorem then gives

$$\text{Energy }\{x(n)\} = \frac{1}{2\pi j}\oint_{C} X(v)X^*\left(\frac{1}{v^*}\right)v^{-1}\,dv. \tag{3.12}$$

The contour of integration must lie in the intersection of the ROC of $X(v)$ and $X^*\left(\frac{1}{v^*}\right)$. From Property 7, the ROC of $X(v)$ is the same as that of $X\left(\frac{1}{v}\right)$. If the ROC of $X(v)$ is $r_i < |v| < r_o$, the ROC of $X\left(\frac{1}{v}\right)$ is $\frac{1}{r_o} < |v| < \frac{1}{r_i}$. Observe that if the ROC of $X(v)$ does not include the unit circle, then the above two ROCs do not have any overlapping region, and the energy of the signal is zero. If there is an overlapping region, then it always includes the unit circle. In Fig. 3.3, we illustrate these two cases with examples. If there is an overlapping region, then we pick the unit circle as the contour of integration.

The application of Properties 8 and 9 requires the evaluation of a contour integration. This is done by using Cauchy's Residue theorem as follows.

Theorem 3.1 *(Cauchy's Residue theorem)*

$$\frac{1}{2\pi j}\oint_{C} G(z)dz = \sum [\textit{residues of } G(z) \textit{ at the poles inside } C].$$

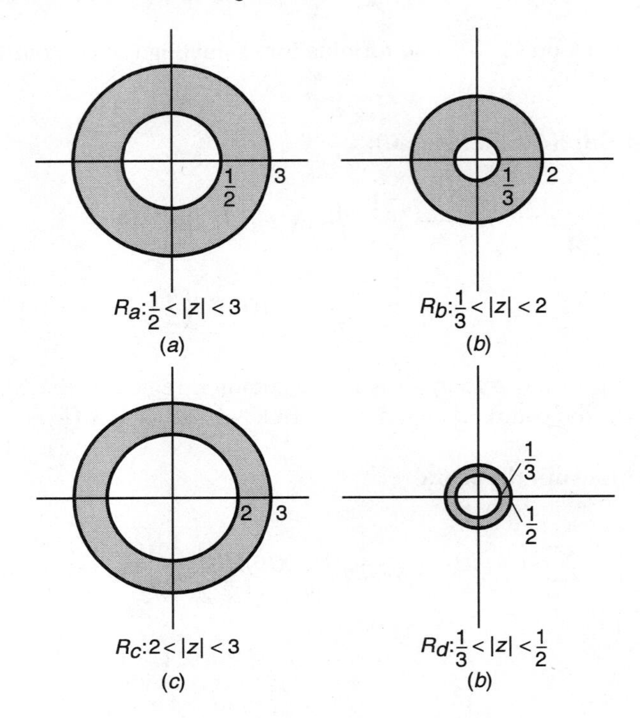

Figure 3.3 $R_a \cap R_b$ is the region $\frac{1}{3} < |z| < 2$. $R_c \cap R_d$ does not exist.

Now we explain how to find the residues. Let $G(z)$ have a simple pole (nonrepeated) at $z = p_0$; that is, it has a factor $(z - p_0)$ in the denominator. The residue is found by removing this factor and evaluating the rest of $G(z)$ at $z = p_0$. Mathematically, we can write

$$\text{Res}\,[G(z) \text{ at simple pole } z = p_0] = [(z - p_0)G(z)]_{z=p_0}. \tag{3.13}$$

If the pole is repeated, then we use a different formula. If we have a factor $(z - p_0)^2$ in the denominator (double pole), we remove it from $G(z)$, differentiate the rest, and then evaluate it at $z = p_0$. For a triple pole we differentiate twice, for a fourth-order pole we differentiate three times, and so on. There is also a scaling factor involved. The general formula for residues is

$$\text{Res}\,[G(z) \text{ at the } Nth \text{ order pole } z = p_0] = \frac{1}{(N-1)!}\left[\frac{d^{N-1}}{dz^{N-1}}\{(z - p_0)^N G(z)\}\right]_{z=p_0}. \tag{3.14}$$

▶ **EXAMPLE 3.6** *(a) Given the sequences,* $x(n) = 2^n u(n)$ *and* $h(n) = \left(\frac{1}{2}\right)^n u(n)$*, find the Z-transform of the product.*
(b) Find the energy of the sequence $h(n)$*.*

SOLUTION (a) $X(z) = \frac{z}{z-2}$, $|z| > 2$, and $H(z) = \frac{z}{z-1/2}$, $|z| > 1/2$. From Property 8, we have

$$\begin{aligned} Z\{x(n)h(n)\} &= \frac{1}{2\pi j}\oint_{C_1}\left(\frac{v}{v-2}\right)\left(\frac{z/v}{z/v-1/2}\right)v^{-1}\,dv \\ &= \frac{1}{2\pi j}\oint_{C_1}\frac{-2z}{(v-2)(v-2z)}\,dv \\ &= \sum \text{Res}\left[\frac{-2z}{(v-2)(v-2z)} \text{ at the poles inside } C_1\right]. \end{aligned}$$

The contour C_1 lies in the intersection region of the ROC of $X(v)$ and $H(z/v)$, which are $|v| > 2$ and $|v| < 2|z|$, respectively. Therefore, C_1 encompasses only the pole at $v = 2$. Therefore, we have

$$\begin{aligned} Z\{x(n)h(n)\} &= \text{Res}\left\{\frac{-2z}{(v-2)(v-2z)} \text{ at } v = 2\right\} \\ &= \left.\frac{-2z}{v-2z}\right|_{v=2} \\ &= \frac{z}{z-1}. \end{aligned}$$

The solution is exactly what we expect for the Z-transform of $u(n)$.

(b) From (2.19), we have

$$\begin{aligned} \text{Energy }\{h(n)\} &= \frac{1}{2\pi j}\oint_C H(v)H^*\left(\frac{1}{v^*}\right)v^{-1}\,dv \\ &= \frac{1}{2\pi j}\oint_C\left(\frac{v}{v-1/2}\right)\left(\frac{1/v^*}{1/v^*-1/2}\right)^* v^{-1}\,dv \\ &= \frac{1}{2\pi j}\oint_C\frac{-2}{(v-1/2)(v-2)}\,dv \\ &= \sum \text{Res}\left[\frac{-2}{(v-1/2)(v-2)} \text{ at poles inside } C\right]. \end{aligned}$$

The ROC of $H(v)$ is $|v| > 1/2$ and that of $H^*\left(\frac{1}{v^*}\right)$ is $|v| < 2$. The contour of integration must lie in the overlapping region $\frac{1}{2} < |v| < 2$. Therefore, the only pole inside C is at $v = 1/2$. The above then reduces to

$$\begin{aligned} \text{Energy }\{h(n)\} &= \text{Res}\left[\frac{-2}{(v-1/2)(v-2)} \text{ at } v = 1/2\right] \\ &= \left.\frac{-2}{v-2}\right|_{v=1/2} \\ &= 4/3. \quad \clubsuit \end{aligned}$$

These properties of the Z-transform are tabulated in Table 3.2 for convenient reference.

3.3 INVERSE Z-TRANSFORM

Several methods may be used to find the inverse Z-transform of a sequence. These include partial fraction expansion, the method of residues, and power series. The power

TABLE 3.2 Z-Transform Properties

Property	Time Domain	Z-domain
1. Linearity	$ax_1(n) + bx_2(n)$	$aX_1(z) + bX_2(z)$
2. Time shifting	$x(n - n_0)$	$z^{-n_0}X(z)$
3. Multiplication by discrete exponential	$a^n x(n)$	$X\left(\frac{z}{a}\right)$
4. Differentiation in the Z-domain	$nx(n)$	$-z\frac{dX(z)}{dz}$
5. Time reversal	$x(-n)$	$X(z^{-1})$
6. Convolution in time	$x(n) * h(n)$	$X(z)H(z)$
7. Complex conjugation	$x^*(n)$	$(X(z^*))^*$
8. Multiplication in time	$x(n)h(n)$	$\frac{1}{2\pi j}\oint_{C_1} X(v)H\left(\frac{z}{v}\right)v^{-1}\,dv$
9. Parseval's theorem	$\sum_{n=-\infty}^{\infty} x(n)h^*(n)$	$= \frac{1}{2\pi j}\oint_{C_1} X(v)H^*\left(\frac{1}{v^*}\right)v^{-1}dv$

series method is simply a matter of performing a long division of the numerator by the denominator and then finding the inverse Z-transform of each term in the quotient. The answer is not always found in closed form. This method is tedious and does not always yield a closed-form solution. Therefore, it will not be presented. The method of residues deals with a direct evaluation of the inverse Z-transform formula by using Cauchy's Residue theorem. This, too, can be rather tedious and cumbersome except for very simple cases. Therefore, this method is rarely used. However, it is sometimes useful in special situations such as in computing the quantization noise power, as will be shown in Chapter 6. The method of partial fractions is the most commonly used technique and will be discussed in detail. It is most useful for analysis of practical FIR and IIR filters. We begin with the derivation of the formula for the inverse Z-transform.

Inverse Z-transform formula

$$x(n) = \frac{1}{2\pi j}\oint_C X(z)z^{n-1}\,dz \tag{3.15}$$

where the contour C lies in the ROC of $X(z)$. This formula can be proved by substituting the definition of $X(z)$ in the right-hand side and then manipulating it so that it equals $x(n)$. To this end we write

$$\begin{aligned}\frac{1}{2\pi j}\oint_C X(z)z^{n-1}dz &= \frac{1}{2\pi j}\oint_C \sum_{k=-\infty}^{\infty} x(k)z^{-k}z^{n-1}\,dz \\ &= \frac{1}{2\pi j}\sum_{k=-\infty}^{\infty} x(k)\oint_C z^{n-k-1}\,dz.\end{aligned} \tag{3.16}$$

In order to evaluate the above, we need the following theorem, which can be found in any standard book on complex analysis such as [1].

Theorem 3.2 *(Cauchy's integral theorem) For a complex variable* z,

$$\oint_C z^{-l} dz = \begin{cases} 2\pi j, & l = 1 \\ 0, & l \neq 1 \end{cases}$$

where the counterclockwise contour circles the origin.

Now we apply this theorem to equation (3.16). For every value of k in the summation, the contour integral is zero except when $-n + k + 1 = 1$ for which it equals $2\pi j$. Equation (3.16) therefore reduces to $x(n)$, and the inverse Z-transform formula of (3.15) is established. As stated earlier, in order to evaluate the inverse Z-transform using this formula, a contour integral must be evaluated. This is done by using Cauchy's Residue theorem as described in the following section.

Inverse Z-Transform Using Residues

Using the formula for the inverse Z-transform (3.15) and Cauchy's residue theorem, we have

$$\begin{aligned} x(n) &= \frac{1}{2\pi j}\oint_c X(z)z^{n-1}\, dz \\ &= \sum \{ \text{ residue of } X(z)z^{n-1} \text{ at the poles inside } c \}. \end{aligned} \tag{3.17}$$

where the contour c lies in the ROC of $X(z)$. Equations (3.13) and (3.14) can be used appropriately to find the residues.

Let us now examine (3.17) for two cases. First, if $n \geq 1$, the term z^{n-1} does not contribute a pole at $z = 0$. Second, if $n \leq 0$, the term z^{n-1} contributes a pole at $z = 0$ of multiplicity $n - 1$. For example, if $n = 0$, we have a simple pole at $z = 0$; for $n = -1$, we have a double pole; for $n = -2$, we have a triple pole; and so on. The residues have to be calculated for all n, and therefore the process is quite cumbersome. To overcome this problem, we modify (3.17) in the manner given in [3]. We use the substitution of variables $z = s^{-1}$ so that (3.15) becomes

$$x(n) = \frac{1}{2\pi j}\oint_{\bar{c}} X(s^{-1})s^{-n+1}(-s^{-2}\, ds).$$

Note that the direction of the contour of $\bar{c}$ is opposite to that of c since the residue of $\bar{c}$ is the reciprocal of that of c. If the negative sign in the above is taken out, the direction of $\bar{c}$ is reversed again to give

$$\begin{aligned} x(n) &= \frac{1}{2\pi j}\oint_{\bar{c}} X(s^{-1})s^{-n-1} ds \\ &= \sum \{ \text{ residue of } X(s^{-1})s^{-n-1} \text{ at the poles inside } \bar{c} \}. \end{aligned} \tag{3.18}$$

For $n \leq -1, s^{-n-1}$ does not contribute a pole at $s = 0$. For $n = 0$, there is a simple pole at $s = 0$. Now let us use (3.17) and (3.18) in an example.

▶ **EXAMPLE 3.7** *First, we have the inverse Z-transform of*

$$X(z) = \frac{z}{z-a}, \qquad |z| > |a|.$$

Using (3.15), we have

$$x(n) = \frac{1}{2\pi j} \oint_C \frac{z^n}{z-a} dz$$

where the contour c has radius greater than $|a|$. For $n \geq 0, z^n$ does not provide any poles at $z = 0$, and so we can use this formula. Therefore,

$$\begin{aligned} x(n) &= \left\{ \text{Res of } \frac{z^n}{z-a} \text{ at } z = a \right\} \\ &= a^n, \quad n \geq 0. \end{aligned}$$

For $n \leq -1$, we use (3.18) to get

$$x(n) = \frac{1}{2\pi j} \oint_{\bar{C}} \frac{s^{-n-1}}{1-as} ds.$$

The contour $\bar{C}$ has radius less than $1/|a|$. So, the pole at $s = 1/a$ is outside the contour and does not contribute to the residue. For $n \leq -1$, there are no poles at $s = 0$ and so $x(n) = 0$. The inverse Z-transform is therefore $x(n) = a^n, \quad n \geq 0$.

Partial Fraction Expansion

The method of partial fractions involves simplification of the transfer function into individual parts such that the inverse can be read off from Table 3.1. The method is different if the transfer function has simple (nonrepeated) poles versus multiple (repeated) poles, as discussed below.

Simple Poles. Consider a transfer function $H(z)$ of the form

$$\begin{aligned} H(z) &= \frac{b_0 z^M + b_1 z^{M-1} + \cdots + b_M}{z^N + a_1 z^{N-1} + \cdots + a_N} \\ &= \frac{b_0 z^M + b_1 z^{M-1} + \cdots + b_M}{(z-p_1)(z-p_2)\ldots(z-p_N)} \end{aligned} \tag{3.19}$$

where p_i represents a pole that may be real or complex. A partial fraction expansion of the above requires the rational polynomial to be *proper*, that is, $N > M$. If this is not satisfied, then we must first perform a long division to satisfy this requirement. The partial fraction of the above is

$$H(z) = \frac{A_1}{z-p_1} + \frac{A_2}{z-p_2} + \cdots + \frac{A_N}{z-p_N}.$$

From Table 3.1, we have $(p_i)^n u(n) \leftrightarrow \frac{z}{z-p}$, $|z| > |p_i|$ and $(p_i)^n u(-n-1) \leftrightarrow \frac{-z}{z-p}$, $|z| < |p_i|$. In order to be able to readily evaluate the impulse response $h(n)$ for each term, we need a z in the numerator. Therefore, we should perform the partial fraction expansion on $\frac{H(z)}{z}$ instead. This yields

$$\begin{aligned}\frac{H(z)}{z} &= \frac{b_0 z^M + b_1 z^{M-1} + \cdots + b_M}{z(z-p_1)(z-p_2)\ldots(z-p_N)} \\ &= \frac{A_0}{z-p_0} + \frac{A_1}{z-p_1} + \frac{A_2}{z-p_2} + \cdots + \frac{A_N}{z-p_N}, \quad p_0 = 0. \end{aligned} \tag{3.20}$$

The constants A_i are found as follows:

$$A_i = (z-p_i)\frac{H(z)}{z}\bigg|_{z=p_i}, \qquad i = 0, 1, 2, \ldots \tag{3.21}$$

So, the constants are nothing but the residues at the poles of the function $\frac{H(z)}{z}$.

EXAMPLE 3.8 *Find the inverse Z-transform of*

$$H(z) = \frac{z^2 + z + 1}{(z^2 + 3z + 2)}, \qquad 1 < |z| < 2.$$

SOLUTION We perform a partial fraction expansion of $\frac{H(z)}{z}$, which is

$$\begin{aligned}\frac{H(z)}{z} &= \frac{z^2 + z + 1}{z(z+1)(z+2)} \\ &= \frac{A_0}{z} + \frac{A_1}{z+1} + \frac{A_2}{z+2}.\end{aligned}$$

Note that the degree of the numerator is less than that of the denominator, and so $\frac{H(z)}{z}$ is proper. The constants are found by using equation (3.21),

$$\begin{aligned}A_0 &= \frac{z^2 + z + 1}{(z+1)(z+2)}\bigg|_{z=0} = \frac{1}{2}, \\ A_1 &= \frac{z^2 + z + 1}{z(z+2)}\bigg|_{z=-1} = -1, \\ A_2 &= \frac{z^2 + z + 1}{z(z+1)}\bigg|_{z=-2} = \frac{3}{2}.\end{aligned}$$

The transfer function now becomes

$$H(z) = \frac{1}{2} - \frac{z}{z+1} + \frac{\frac{3}{2}z}{z+2}.$$

The ROC is outside the pole circle $|z| = 1$, and inside the pole circle $|z| = 2$. Therefore, the pole $z = -1$ corresponds to a right-sided sequence, and the pole $z = -2$ corresponds to a left sided sequence. The inverse Z-transform is

$$h(n) = \frac{1}{2}\delta(n) - (-1)^n u(n) - \frac{3}{2}(-2)^n u(-n-1).$$

The above impulse response is clearly noncausal. ◀

▶ **EXAMPLE 3.9** *Find the inverse Z-transform of the following:*

$$H(z) = \frac{z^2}{z^2 + 2z + 9}, \qquad |z| > 2.$$

SOLUTION The poles of $\frac{H(z)}{z}$ are $p_1 = -1 + j\sqrt{3}, p_2 = -1 - j\sqrt{3}$. In phasor form, the poles are $p_1 = 2e^{j2\pi/3}, p_2 = 2e^{-j2\pi/3}$. The partial fraction expansion is

$$\frac{H(z)}{z} = \frac{A}{z - p_1} + \frac{B}{z - p_2}$$

where

$$A = \frac{z}{z + 1 + j\sqrt{3}}\bigg|_{z=-1+j\sqrt{3}} = \frac{1}{2} + j\frac{1}{2\sqrt{3}},$$

$$B = \frac{z}{z + 1 + j\sqrt{3}}\bigg|_{z=-1-j\sqrt{3}} = \frac{1}{2} - j\frac{1}{2\sqrt{3}}.$$

Here we have $A = B^*$, but this is not an accident. It can be easily shown to be true for all complex conjugate pole pairs. In phasor form, these constants are $A = \frac{1}{\sqrt{3}}e^{j\pi/6}, B = \frac{1}{\sqrt{3}}e^{-j\pi/6}$. The inverse Z-transform is

$$\begin{aligned}
h(n) &= A(p_1)^n + B(p_2)^n \\
&= \left(\frac{1}{\sqrt{3}}e^{j\pi/6}(2)^n e^{j\frac{2\pi}{3}n} + \frac{1}{\sqrt{3}}e^{-j\pi/6}(2)^n e^{-j\frac{2\pi}{3}n}\right)u(n) \\
&= \frac{1}{\sqrt{3}}2^n\left(e^{j(\frac{2\pi}{3}n + \pi/6)} + e^{-j(\frac{2\pi}{3}n + \pi/6)}\right)u(n) \\
&= \frac{2}{\sqrt{3}}2^n\cos\left(\frac{2\pi}{3}n + \pi/6\right)u(n).
\end{aligned}$$

The above impulse response is clearly unbounded, and the given filter is therefore unstable. ◀

▶ **EXAMPLE 3.10** *Find the inverse Z-transform of the following:*

$$H(z) = \frac{z}{(z-1)(z^2 + 2z + 4)}, \qquad |z| > 2.$$

SOLUTION The poles are at 1 and $-1 \pm j\sqrt{3} = 2e^{\pm j\pi/3}$. This problem can be solved by performing a partial fraction of $\frac{H(z)}{z}$ as a sum of three terms and then finding the inverse transform. But let us take a different approach this time. We will perform the partial fraction expansion, keeping the second-order factor intact as follows:

$$\frac{H(z)}{z} = \frac{A}{z-1} + \frac{Bz + C}{z^2 + 2z + 4}. \tag{3.22}$$

The constant A can be obtained as before, that is,

$$A = \frac{1}{z^2 + 2z + 4}\bigg|_{z=1} = \frac{1}{7}.$$

In order to find the constants B and C, we first write

$$\frac{1}{(z-1)(z^2+2z+4)} = \frac{1/7}{z-1} + \frac{Bz+C}{z^2+2z+4} = \frac{\frac{1}{7}(z^2+2z+4)+(Bz+C)(z-1)}{(z-1)(z^2+2z+4)}.$$

Now equate the like powers of z in the numerator of both sides of the above to get

$$\frac{1}{7} + B = 0$$

$$\frac{2}{7} + C - B = 0.$$

Solving the above gives $B = -\frac{1}{7}$ and $C = -\frac{3}{7}$. We now have

$$H(z) = \frac{1}{7}\frac{z}{z-1} - \frac{1}{7}\frac{z(z+3)}{z^2+2z+4}. \tag{3.23}$$

The inverse Z-transform of the first term is $\frac{1}{7}u(n)$. We now manipulate the second term so that it matches items 4 and 5 in Table 3.1. Let us rewrite them for convenient reference.

$$\frac{z(z+3)}{z^2+2z+4} \quad \text{compare to:} \quad \frac{z(z-r\cos\omega_0)}{z^2-(2r\cos\omega_0)z+r^2} \quad \text{or} \quad \frac{zr\sin\omega_0}{z^2-(2r\cos\omega_0)z+r^2}$$

Comparing the denominators, we get

$$r^2 = 4, \quad 2r\cos\omega_0 = 2.$$

The above are solved easily to get $r = 2$ and $\omega_0 = \pi/3$. These are the same as the magnitude and angle of the pole. Let us now find the following: $r\cos\omega_0 = 1$, and $r\sin\omega_0 = \sqrt{3}$. The numerator of (3.23) can be written as

$$z(z+3) = z(z - 1 + \frac{4}{\sqrt{3}}\sqrt{3}).$$

Now we have

$$\begin{aligned}\frac{z(z+3)}{z^2+2z+4} &= \frac{z(z-1)}{z^2+2z+4} + \frac{4}{\sqrt{3}}\frac{z\sqrt{3}}{z^2+2z+4} \\ &\leftrightarrow 2^n\cos(\sqrt{3}n)u(n) + \frac{4}{\sqrt{3}}2^n\sin(\sqrt{3}n)u(n) \\ &= 2^n\sqrt{\frac{19}{3}}\cos\left(\sqrt{3}n - \tan^{-1}4/\sqrt{3}\right)u(n).\end{aligned}$$

The final answer is therefore[2]

$$h(n) = \frac{1}{7}\left[1 - 2^n\sqrt{\frac{19}{3}}\cos\left(\sqrt{3}n - \tan^{-1}4/\sqrt{3}\right)\right]u(n).$$

[2]We use the following formula:

$$A\cos\theta + B\sin\theta = \sqrt{A^2+B^2}\cos\left(\theta - \tan^{-1}\frac{B}{A}\right)$$

Multiple Poles. The partial fraction expansion of transfer functions with multiple poles is performed a little differently. Consider a proper rational function given by

$$\frac{H(z)}{z} = \frac{P(z)}{(z-p)^m}$$

where m is the multiplicity of the pole $z = p$. The pole may be real or complex. The partial fraction expansion is then given by

$$\frac{H(z)}{z} = \frac{A_m}{z-p} + \frac{A_{m-1}}{(z-p)^2} + \cdots + \frac{A_2}{(z-p)^{m-1}} + \frac{A_1}{(z-p)^m}. \tag{3.24}$$

To find the coefficient A_1, simply evaluate $P(z)$ at $z = p$. To find A_2, differentiate $P(z)$ with respect to z, evaluate it at $z = p$, and divide the result by 1! For A_3, differentiate $P(z)$ twice, evaluate the result at $z = p$, divide by 2!, and so on for the rest of the constants. The general formula is

$$A_i = \frac{1}{(i-1)!} \left. \frac{d^{i-1}P(z)}{dz^{i-1}} \right|_{z=p}, \quad i = 1, 2, \ldots m. \tag{3.25}$$

▶ **EXAMPLE 3.11** *Find the Z-transform of*

$$H(z) = \frac{1}{(z-1)^2}, \quad |z| < 1.$$

SOLUTION We perform a partial fraction expansion of $\frac{H(z)}{z}$ as follows:

$$\begin{aligned} \frac{H(z)}{z} &= \frac{1}{z(z-1)^2} \\ &= \frac{1}{z} + \frac{A_2}{z-1} + \frac{A_1}{(z-1)^2} \end{aligned}$$

The constants are obtained by using the formula in (3.25) as

$$\begin{aligned} A_1 &= \left. \frac{1}{z} \right|_{z=1} = 1, \\ A_2 &= \left. \frac{d}{dz} \left\{ \frac{1}{z} \right\} \right|_{z=1} \\ &= \left. -\frac{1}{z^2} \right|_{z=1} = -1. \end{aligned}$$

The ROC is such that the sequence is left-sided. We now rewrite the transfer function in order to match items 2(b) and 3(b) in Table 3.1, as follows:

$$H(z) = 1 + \frac{-z}{z-1} - \frac{-z}{(z-1)^2}.$$

The inverse Z-transform then becomes

$$h(n) = \delta(n) + u(-n-1) - nu(-n-1).$$ ◀

3.4 CAUSALITY AND STABILITY

In this section, we present the concepts of causality and stability of linear systems in relation to the Z-transform and its region of convergence.

Recall from Chapter 1 that a system is said to be causal if the output at $n = n_0$ does not depend on the input for any $n > n_0$. That is, the output does not depend on "future" inputs. Let $h(n)$ and $x(n)$ represent the impulse response and input of a linear system. The impulse response may be finite or infinite. The output of the system, $y(n)$, is given by the convolution of the input and the impulse response as

$$\begin{aligned} y(n) = & \cdots + h(-2)x(n+2) + h(-1)x(n+1) + h(0)x(n) \\ & + h(1)x(n-1) + h(2)x(n-2) + \cdots . \end{aligned}$$

If the impulse response is such that $h(n) = 0$ for all $n < 0$, then the output does not depend on future inputs and the system is causal. In other words, a causal system has a right-sided impulse response. If $h(n) \neq 0$ for any $n < 0$, then the system is noncausal. If the impulse response is left-sided, that is, $h(n) = 0$ for all $n \geq 0$, the system is then said to be *anticausal*. Note that anticausality is a special case of noncausality. These observations have direct significance for the ROC of causal, anticausal, and noncausal systems. By recalling the ROC for left-sided and right-sided sequences, we have the following results.

Fact 1: The ROC of a causal sequence extends outward from the outermost pole circle.

Fact 2: The ROC of an anticausal sequence lies inside the innermost pole circle.

Fact 3: The ROC of a noncausal (but not anticausal) sequence is bounded by two different pole circles.

Now we move on to stability. Recall from Chapter 1 that a system is said to be stable if and only if its impulse response is absolutely summable. This will be used along with the definition of the Z-transform in order to find the condition of stability in terms of the ROC of the transfer function. The transfer function $H(z)$ of a system can be written as

$$H(z) = \sum_{n=-\infty}^{\infty} h(n) r^{-n} e^{-j\omega n} \tag{3.26}$$

where we have used $z = re^{j\omega}$. The above Z-transform is said to exist if in some region of the z-plane $|H(z)| < \infty$, that is, if

$$|H(z)| = \left| \sum_{n=-\infty}^{\infty} h(n) r^{-n} e^{-j\omega n} \right| \leq \sum_{n=-\infty}^{\infty} |h(n)| r^{-n} < \infty.$$

If $r = 1$, the above is the absolute summability condition of the impulse response. Therefore, the system is stable if and only if the ROC includes the circle with radius $r = 1$, that is, the unit circle. This is an important result and is formally stated below.

Theorem 3.3 *An LSI system with transfer function $H(z)$ is stable if and only if the ROC of the transform includes the unit circle.*

As a special case, consider a causal system. From the preceding analysis, we know that a causal system has a right-sided impulse response and the ROC of the transfer function lies outside the outermost pole circle. In order for the unit circle to lie in this ROC, the unit circle must be outside the outermost pole circle. In other words, we have the following all-familiar result.

Fact 4: A causal system is stable if and only if all the poles are inside the unit circle.

By similar reasoning, it can be said that an anticausal system is stable if and only if all the poles are outside the unit circle. Now consider a system with ROC $\frac{1}{2} < |z| < 2$, which is clearly noncausal. The ROC includes the unit circle, and the system is therefore stable. For FIR filters, the ROC is everywhere in the z-plane except possibly at $z = 0$ or $z = \infty$. Also, FIR filters have no poles except possibly at $z = 0$ (Why?). Therefore, the ROC always includes the unit circle, and FIR filters are always stable.

Digital filters are often implemented as parallel or cascade connections of second-order systems as discussed in future chapters. Therefore, it is important to study the stability of the second-order filter as a special case.

The Stability Triangle

Now we derive the condition for stability of a causal second-order filter in terms of its coefficients. A second-order digital filter is given by the transfer function

$$H(z) = \frac{P(z^{-1})}{1 + a_1 z^{-1} + a_2 z^{-2}} = \frac{z^2 P(z^{-1})}{z^2 + a_1 z + a_2} \tag{3.27}$$

where $P(z^{-1})$ is a polynomial in z^{-1}. If the poles are denoted by p_1 and p_2, then (3.27) can be written as

$$H(z) = \frac{z^2 P(z^{-1})}{(z - p_1)(z - p_2)} = \frac{z^2 P(z^{-1})}{z^2 - (p_1 + p_2)z + p_1 p_2}. \tag{3.28}$$

Comparing (3.27) and (3.28), we have

$$a_1 = -(p_1 + p_2) \tag{3.29}$$

$$a_2 = p_1 p_2. \tag{3.30}$$

The causal filter is stable if and only if the poles are inside the unit circle, that is, $|p_1| < 1$ and $|p_2| < 1$. By using (3.30), we get

$$|a_2| = |p_1 p_2| = |p_1||p_2| < 1. \tag{3.31}$$

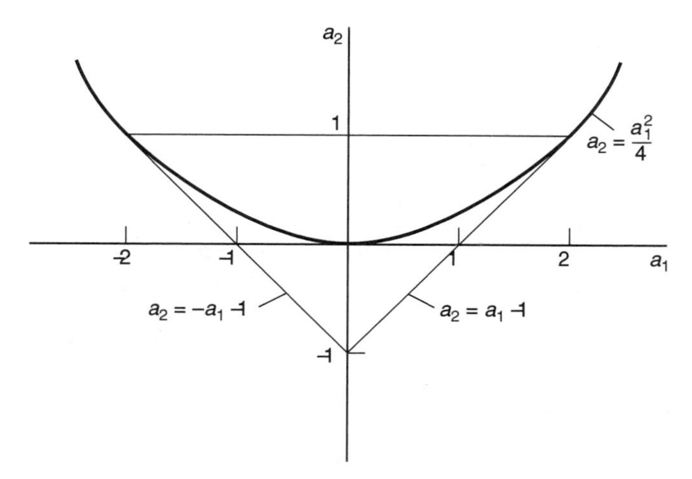

Figure 3.4 The stability triangle.

By using the quadratic formula, the poles are

$$p_{1,2} = -\frac{a_1}{2} \pm \sqrt{\frac{a_1^2}{4} - a_2}. \tag{3.32}$$

In order to find a condition involving the other filter coefficient, we consider the following cases.

Case 1: Complex conjugate poles. From the quadratic formula, we have for this case

$$a_2 > \frac{a_1^2}{4}$$

which is the region above the parabola shown in Fig. 3.4. The region must be bounded by the line $a_2 = 1$ due to the condition in (3.31).

Case 2: Real Poles. In this case, we have $a_2 \leq \frac{a_1^2}{4}$, which is the region under the parabola. Using the condition that the poles must be less than unity in absolute value, we get

$$-1 < \frac{-a_1}{2} \pm \sqrt{\frac{a_1^2}{4} - a_2} < 1.$$

Since the quantity under the radical is nonnegative, we have the two following conditions:

$$\frac{-a_1}{2} + \sqrt{\frac{a_1^2}{4} - a_2} < 1$$

and

$$\frac{-a_1}{2} - \sqrt{\frac{a_1^2}{4} - a_2} > -1.$$

Squaring both sides of the above inequalities and simplifying, we get the desired conditions

$$a_2 > a_1 - 1 \tag{3.33}$$

$$a_2 > -a_1 - 1. \tag{3.34}$$

The stability region for this case is therefore below the parabola and bounded above by the above two conditions as shown in Fig. 3.4. The entire stability region is the union of the regions for cases 1 and 2, which is the stability triangle. In the design of second-order filters, the parameters must be restricted to within the stability triangle.

3.5 ONE-SIDED Z-TRANSFORM

The significance of the one-sided (or unilateral) Z-transform is in the solution of difference equations with initial conditions. In the definition of the two-sided (or bilateral) Z-transform, the limits of the summation go from $n = -\infty$ to ∞. This means that the digital filter is operating for all n. This is not always the case. In many cases, the digital filter starts at $n = n_0$ with the required initial conditions before that. The two-sided Z-transform will not be able to solve this system. The limits of the one-sided Z-transform go from $n = 0$ to ∞, allowing the use of initial conditions as we will show shortly.

The one-sided Z-transform is defined as

$$X_u(z) \triangleq Z_u\{x(n)\} = \sum_{n=0}^{\infty} x(n)z^{-n} \tag{3.35}$$

where we have used the subscript u with Z to denote the unilateral Z-transform. The properties of this transform are derived in the same way as for the bilateral transform. In fact, all of the properties of the bilateral transform are also valid for the unilateral case except those involving a time shift.

Property A: Let $x(n) \leftrightarrow X(z)$ represent a sequence and its unilateral Z-transform. The unilateral Z-transform of the right-shifted sequence is given by

$$Z_u\{x(n-M)\} = z^{-M}X_u(z) + z^{-M} \sum_{m=-M}^{-1} x(m)z^{-m}. \tag{3.36}$$

Proof. By using the definition of the unilateral Z-transform and then performing the change of variables $m = n - M$, we have

$$\begin{aligned} Z_u\{x(n-M)\} &= \sum_{n=0}^{\infty} x(n-M)z^{-n} \\ &= \sum_{m=-M}^{\infty} x(m)z^{-(m+M)} \end{aligned}$$

$$= z^{-M}\sum_{m=0}^{\infty} x(m)z^{-m} + z^{-M}\sum_{m=-M}^{-1} x(m)z^{-m}$$

$$= z^{-M}X_u(z) + z^{-M}\sum_{m=-M}^{-1} x(m)z^{-m}.$$

By using a similar analysis as above, it is straightforward to derive the following property for a left-shifted sequence.

Property B: Let $x(n) \leftrightarrow X_u(z)$ represent a sequence and its unilateral Z-transform. The unilateral Z-transform of the left-shifted sequence is given by

$$Z_u\{x(n+M)\} = z^M X_u(z) - z^M \sum_{m=0}^{M-1} x(m)z^{-m}. \tag{3.37}$$

Now we explicitly write down some special cases of the above properties for convenient reference.

$$\begin{aligned}
&(\text{A}.1) \quad Z_u\{x(n-1)\} = z^{-1}X_u(z) + x(-1)\\
&(\text{A}.2) \quad Z_u\{x(n-2)\} = z^{-2}X_u(z) + z^{-1}x(-1) + x(-2)\\
&(\text{B}.1) \quad Z_u\{x(n+1)\} = zX_u(z) - zx(0)\\
&(\text{B}.2) \quad Z_u\{x(n+2)\} = z^2X_u(z) - z^2x(0) - zx(1).
\end{aligned}$$

EXAMPLE 3.12 *Given a causal system represented by the following difference equation:*

$$y(n) + \frac{3}{4}y(n-1) + \frac{1}{8}y(n-2) = x(n) + x(n-1), \quad y(-1) = 1,\ y(-2) = 0, \quad x(-1) = -1.$$

Find the output of the system if $x(n) = \delta(n)$.

SOLUTION The unilateral Z-transform of both sides of the system equation gives

$$\begin{aligned}
Y_u(z) &+ \frac{3}{4}\left(z^{-1}Y_u(z) + y(-1)\right) + \frac{1}{8}\left(z^{-2}Y_u(z) + z^{-1}y(-1) + y(-2)\right)\\
&= X_u(z) + z^{-1}X_u(z) + x(-1).
\end{aligned}$$

The system output is

$$\begin{aligned}
y(n) &= Z^{-1}\left\{\frac{x(-1) - \frac{3}{4}y(-1) - \frac{1}{8}y(-2) - \frac{1}{8}y(-1)z^{-1}}{1 + \frac{3}{4}z^{-1} + \frac{1}{8}z^{-2}}\right\}\\
&+ Z^{-1}\left\{\frac{(1+z^{-1})X_u(z)}{1 + \frac{3}{4}z^{-1} + \frac{1}{8}z^{-2}}\right\}.
\end{aligned} \tag{3.38}$$

The first term on the right-hand side is called the *zero-input response*, while the second term is called the *zero-state response*.[3] The appropriate initial conditions can now be applied. Since the

[3] The term *state* refers to initial condition in this instance.

input is an impulse, we have $X_u(z) = 1$. The inverse Z-transforms can then be easily obtained by the method of partial fractions discussed earlier. The final result is left as an exercise.

In the above example, the required initial conditions $\{y(-1), y(-2), x(-1)\}$ were given. What if some initial conditions other than the ones required were given? Then the required initial conditions must be first obtained by using the system difference equation before the problem can be solved. For example, assume that we are given $y(-5), y(-4)$, and $x(-4)$. Then we have to use the difference equation repeatedly to find $y(-3), y(-2)$, and $y(-1)$ and then use (3.38) to find the solution. A similar method can be used to solve difference equations in advance form, that is, where the difference equation is specified in terms of $x(n+i)$ and $y(n+i)$. ◀

▶ 3.6 THE LINEAR PHASE CONCEPT

In this section, we will discuss the significance of linear phase in digital filters. As we know, a digital filter frequency response has a magnitude and a phase given by $|H(\omega)|$ and $\measuredangle H(\omega)$, respectively. A digital filter is said to have *linear phase* if its phase response satisfies

$$\measuredangle H(\omega) = -\alpha\omega \tag{3.39}$$

where α is a constant. If α is positive, then the system delays the signal; otherwise it is an advance. The phase is thus a linear function of frequency, similar to $y = mx + b$, where $-\alpha$ is the slope and $b = 0$. Consider a sinusoid with a frequency of ω_0 radians/second or a time period of T_0 seconds, where $\omega_0 = \frac{2\pi}{T_0}$. The full time-period corresponds to 2π radians. A phase shift of ϕ_0 radians therefore corresponds to a time delay of $\tau_0 = \frac{\phi_0}{2\pi}T_0$ seconds. Now if $\phi_0 = -\alpha\omega_0$ as above, then the time delay is $\tau_0 = \frac{\alpha\omega_0}{2\pi}T_0 = \alpha$. The time delay is independent of the frequency. So if a digital filter satisfies the linear phase condition of (3.39), then all frequencies are delayed by the same amount. How nice! This means that the output of the digital filter is simply a delayed version of the filtered input signal. On the other hand, if the filter does not have linear phase, then different frequency components of the input signal are delayed by different amounts. This leads to a distortion of the shape of the signal at the output. This is not desirable in most applications.

Now consider a frequency response given by $H(\omega) = e^{-j\alpha\omega}$. The magnitude response is simply unity, and the phase response is $-\alpha\omega$ radians, which is linear phase. As an example, let $H(z) = z^{-2}$, which is simply a delay of 2. The phase response is shown in Fig. 3.5. Note that by convention (as discussed in Chapter 2) the phase is plotted in the interval $(-\pi, \pi)$. This is what causes the "wrapping" of the phase plot, which is actually linear. Now consider the frequency response given by

$$H(\omega) = e^{-j\alpha\omega}G(\omega) \tag{3.40}$$

where $G(\omega)$ is real. Since $G(\omega)$ is real, it only causes a change in the amplitude of the signal. The term $e^{-j\alpha\omega}$ simply causes a phase shift in the input signal. If $G(\omega) > 0$, the phase is $\phi(\omega) = -\alpha\omega$ and the system has linear phase. If $G(\omega) < 0$, the phase is $\phi(\omega) = -\alpha\omega + \pi$. In this case, the time delay is not independent of frequency, and the system is not strictly linear phase in the sense of the definition given above. However, we can write the function as $H(\omega) = -[e^{-j\alpha\omega}|G(\omega)|]$. The function within brackets is

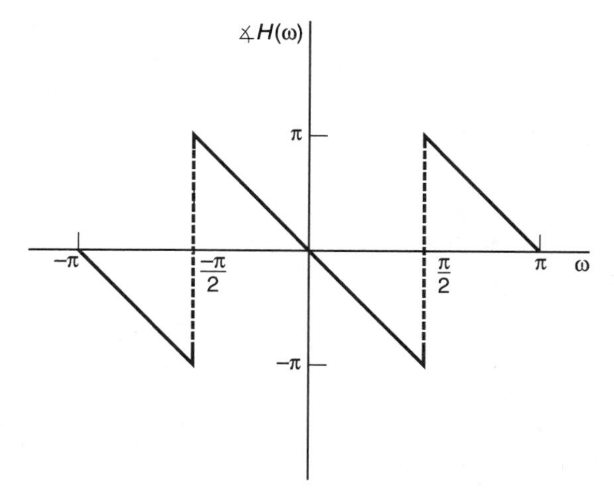

Figure 3.5 Phase response of a double delay.

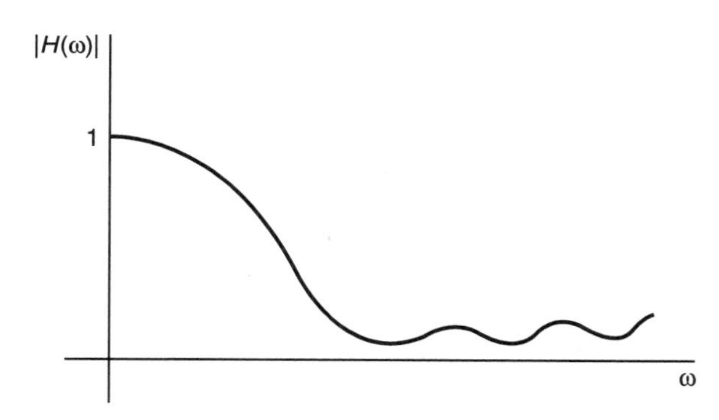

Figure 3.6 Typical LPF magnitude response.

now linear phase and does not distort the waveform. The negative sign simply flips the waveform. But if $G(\omega)$ changes sign, the waveform can get distorted. In view of this, consider a typical lowpass filter magnitude response shown in Fig. 3.6. $G(\omega)$ changes sign only in the vicinity of the horizontal axis, that is, in the stopband. So, the system distorts the waveform only in the stopband where the signal is very much attenuated anyway. Therefore, a system with frequency response given by (3.40) does not distort the signal anywhere in the passband. Such a system is also referred to as a linear phase system. It should be mentioned in passing that an analog filter cannot possibly admit linear phase characteristics, except an approximately linear phase over a small frequency band.

Linear Phase FIR Filters

FIR filters can be designed easily to achieve the linear phase characteristic. This feature makes FIR filters very attractive. As discussed earlier, FIR filters are also always stable.

On the other hand, the design of linear phase IIR filters is not as straightforward. Quite often we have to live with linearity of phase only in some range of frequencies within the passband. The advantage of IIR filters is that we can achieve the same magnitude response specification with much fewer coefficients than with FIR.

Consider a causal FIR filter described by

$$H(z) = h(0) + h(1)z^{-1} + h(2)z^{-2} + \cdots + h(M-1)z^{-(M-1)} + h(M)z^{-M} \tag{3.41}$$

where the length of the filter is $M + 1$ and the filter coefficients are represented by $h(i)$. We will show that linear phase is achieved by making the filter coefficients symmetric or antisymmetric. Symmetry of the coefficients implies $h(n) = h(M - n)$. Antisymmetry of the coefficients implies $h(n) = -h(M - n)$. Let us first consider an example before we write the equations for the general cases.

Let $M = 5$; that is, there are six coefficients. The coefficients are symmetric as shown in Fig. 3.7(a). So we have $h(0) = h(5), h(1) = h(4)$, and $h(2) = h(3)$. The impulse response is therefore symmetric around the point $n = \frac{M}{2} = \frac{5}{2}$. The frequency response of this filter is

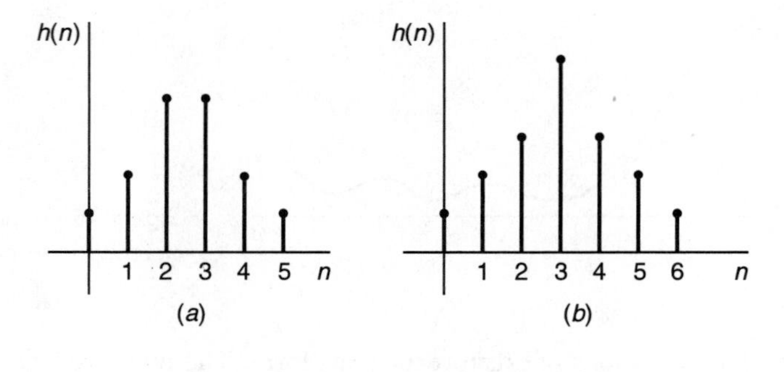

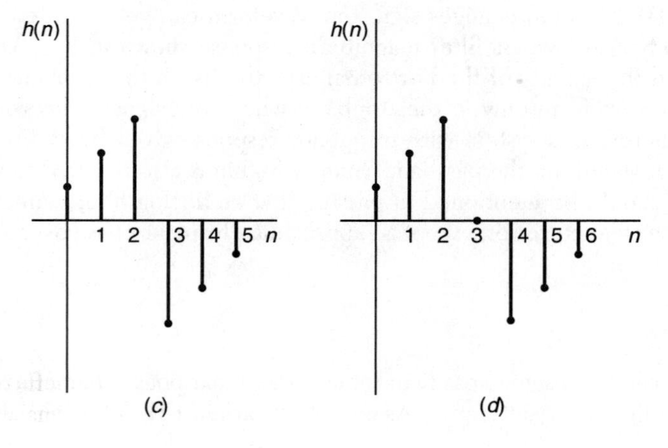

Figure 3.7 Examples of FIR filter impulse responses: (a) FIR-I; (b) FIR-II; (c) FIR-III; (d) FIR IV.

$$\begin{aligned}
H(\omega) &= h(0) + h(1)e^{-j\omega} + h(2)e^{-2j\omega} + h(3)e^{-3j\omega} + h(4)e^{-4j\omega} + h(5)e^{-5j\omega} \\
&= h(0) + h(1)e^{-j\omega} + h(2)e^{-2j\omega} + h(2)e^{-3j\omega} + h(1)e^{-4j\omega} + h(0)e^{-5j\omega} \\
&= e^{-j\frac{5}{2}\omega}\left(h(0)\left[e^{j\frac{5}{2}\omega} + e^{-j\frac{5}{2}\omega}\right] + h(1)\left[e^{j\frac{3}{2}\omega} + e^{-j\frac{3}{2}\omega}\right] + h(2)\left[e^{j\frac{1}{2}\omega} + e^{-j\frac{1}{2}\omega}\right]\right) \\
&= e^{-j\frac{5}{2}\omega}\left(2h(0)\cos(\frac{5}{2}\omega) + 2h(1)\cos(\frac{3}{2}\omega) + 2h(2)\cos(\frac{1}{2}\omega)\right) \\
&= 2e^{-j\frac{5}{2}\omega}\sum_{i=0}^{2} h(i)\cos\left(\left(\frac{5}{2} - i\right)\omega\right).
\end{aligned} \tag{3.42}$$

For the general case of odd M, we have the following.

FIR-I (*M odd, symmetric coefficients*)

$$H(\omega) = 2e^{-j\frac{M}{2}\omega}\sum_{i=0}^{\frac{M-1}{2}} h(i)\cos\left(\frac{M}{2} - i\right)\omega. \tag{3.43}$$

Note that the term inside the summation is always real; therefore, equation (3.43) is in the same form as (3.40), and the FIR-I filter is linear phase.

Now let $M = 6$; that is, there are seven coefficients that are symmetric, as shown in Fig. 3.7(b). The equation for the frequency response now becomes

$$\begin{aligned}
H(\omega) &= h(0) + h(1)e^{-j\omega} + h(2)e^{-2j\omega} + h(3)e^{-3j\omega} + h(4)e^{-4j\omega} + h(5)e^{-5j\omega} + h(6)e^{-6j\omega} \\
&= h(0) + h(1)e^{-j\omega} + h(2)e^{-2j\omega} + h(3)e^{-3j\omega} + h(2)e^{-4j\omega} + h(1)e^{-5j\omega} + h(0)e^{-6j\omega} \\
&= e^{-j3\omega}\left(h(3) + h(0)\left[e^{j3\omega} + e^{-j3\omega}\right] + h(1)\left[e^{j2\omega} + e^{-2j\omega}\right] + h(2)\left[e^{j\omega} + e^{-j\omega}\right]\right) \\
&= e^{-j3\omega}\left(h(3) + 2h(0)\cos(3\omega) + 2h(1)\cos(2\omega) + 2h(2)\cos(\omega)\right) \\
&= e^{-j3\omega}\left(h(3) + 2\sum_{i=0}^{2} h(i)\cos(3 - i)\omega\right).
\end{aligned} \tag{3.44}$$

As a general case of the above, we have the following.

FIR-II (*M even, symmetric coefficients*)

$$H(\omega) = e^{-j\frac{M}{2}\omega}\left(h(\frac{M}{2}) + 2\sum_{i=0}^{\frac{M}{2}} h(i)\cos\left(\frac{M}{2} - i\right)\omega\right). \tag{3.45}$$

Again, equation (3.45) is in the same form as (3.40), and the FIR-II filter is linear phase.

Now let us consider what happens if the coefficients are antisymmetric, that is, if $h(n) = -h(M - n)$. For the case of $M = 5$, an example of the coefficients is shown in Fig. 3.7(c). The difference in this case is that the signs become alternating, which in turn leads to sin terms instead of cos along with a j in the numerator. The general formula is given by the following.

FIR-III (*M odd, antisymmetric coefficients*)

$$H(\omega) = 2e^{j[-\frac{M}{2}\omega+\frac{\pi}{2}]}\sum_{i=0}^{\frac{M-1}{2}} h(i)\sin\left(\frac{M}{2}-i\right)\omega. \tag{3.46}$$

For M even, note that the central coefficient $h\left(\frac{M}{2}\right)$ must be the negative of itself and is therefore zero. For $M = 6$ a set of coefficients is given in Fig. 3.7 (d). The general formula for this case is the following.

FIR-IV (*M even, antisymmetric coefficients*)

$$H(\omega) = 2e^{j[-\frac{M}{2}\omega+\frac{\pi}{2}]}\sum_{i=0}^{\frac{M}{2}-1} h(i)\sin\left(\frac{M}{2}-i\right)\omega. \tag{3.47}$$

The FIR-III and FIR-IV filters are not in the same form as equation (3.40) and are therefore not linear phase. The phase response of these filters is $\phi(\omega) = -\frac{M}{2}\omega + \frac{\pi}{2}$. Filters with phase response in the form of

$$\phi(\omega) = -\frac{M}{2}\omega + \beta \tag{3.48}$$

are said to have *generalized linear phase*. The cases with $\beta = 0$ or π are special cases and are linear phase, as discussed before. Generalized linear phase filters are also useful in many applications, including narrowband filters, and in the demodulation of communication signals [2]. These filters are said to have constant *group delay* or time delay, which is defined as follows:

$$\text{Group Delay: } \tau(\omega) = -\frac{d}{d\omega}\{\phi(\omega)\}. \tag{3.49}$$

Examples are now given for the four types of filters given above.

▶ **EXAMPLE 3.13** *Consider the impulse response of an ideal lowpass filter derived earlier as*

$$h(n) = \begin{cases} \dfrac{\sin\omega_c n}{\pi n}, & n \neq 0 \\ \omega_c/\pi, & n = 0. \end{cases}$$

where ω_c is the cutoff frequency. To make this realizable and causal, we shift it to the right and then multiply it by a $M+1$ point window. The amount of shift must be equal to M/2 so that the resulting sequence is symmetric. The finite impulse response then becomes

$$h(n) = \begin{cases} \dfrac{\sin\omega_c\left(n-\frac{M}{2}\right)}{\pi\left(n-\frac{M}{2}\right)}, & 0 \leq n \leq M, \quad n \neq M/2 \\ \omega_c/\pi, & n = M/2, \quad \textit{if } M/2 \textit{ is an integer}. \end{cases} \tag{3.50}$$

If $M = 3$, the point of symmetry is about $n = \frac{3}{2}$ and we have an FIR-I filter. For $\omega_c = \frac{\pi}{2}$, the impulse response, magnitude response, phase response, and group delay are shown in Fig. 3.8(a),

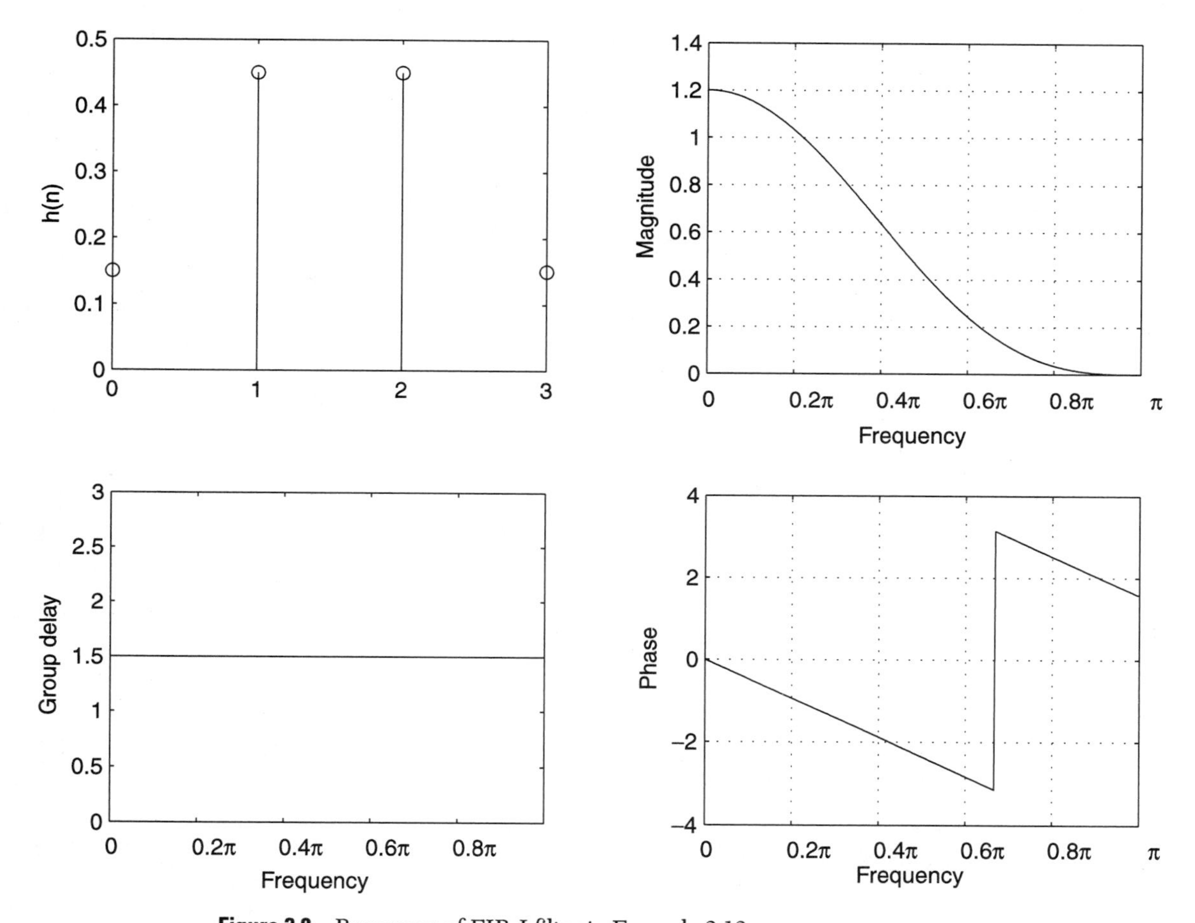

Figure 3.8 Responses of FIR-I filter in Example 3.13.

(b), (c), and (d), respectively. For $M = 4$, we have an FIR-II filter. The corresponding responses are shown in Fig. 3.9. Now let the impulse response be

$$h(n) = \delta(n) - \delta(n-1)$$

which is a very simple highpass filter. In this case $M = 1$, and this is a FIR-III filter. The impulse response, magnitude response, phase response, and group delay are shown in Fig. 3.10. To get a FIR-IV filter, we insert a zero between the two spikes of the impulse response. The responses are shown in Fig. 3.11.

Zero Locations of Linear Phase Filters

In this section, we will establish some general rules about the zero locations of linear phase filters. For linear phase filters, we know that $h(n) = \pm h(M - n)$. The transfer function is therefore

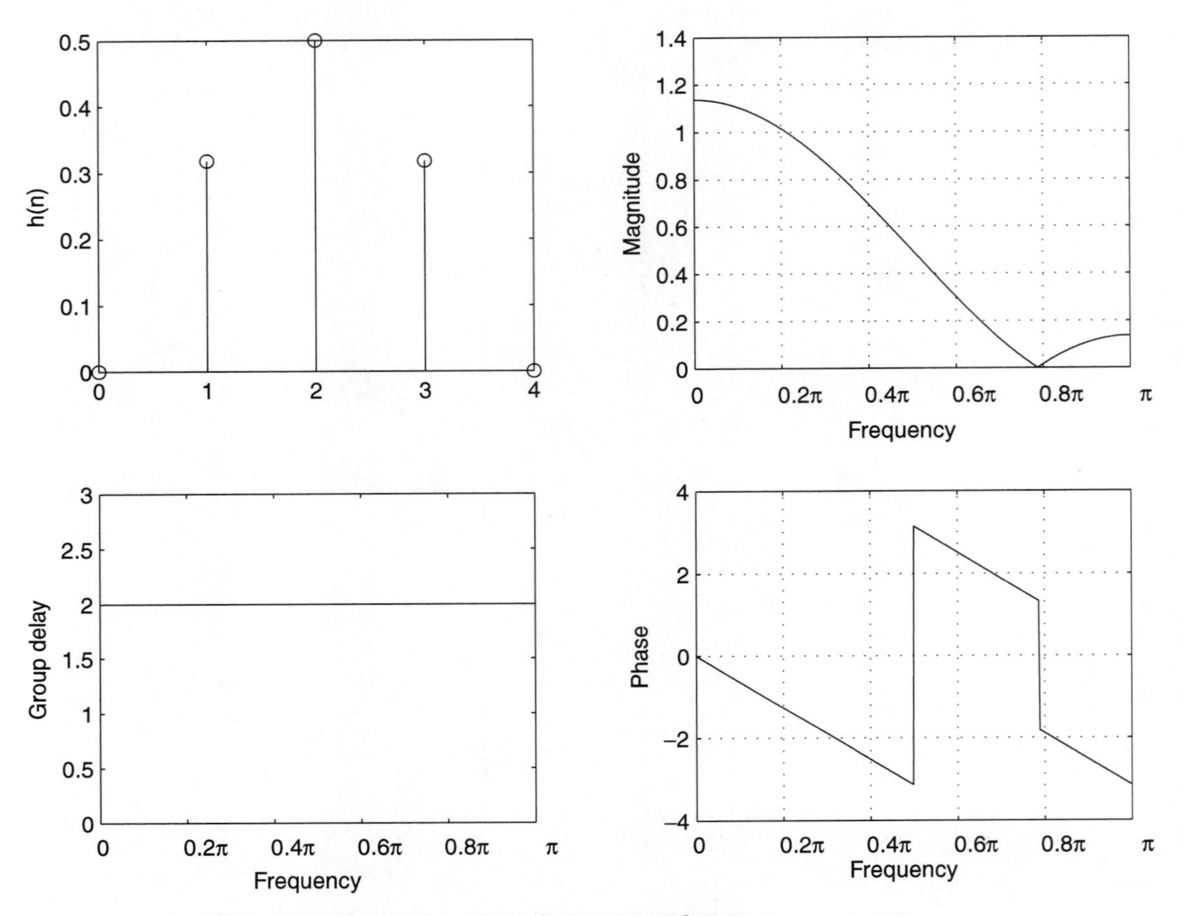

Figure 3.9 Responses of FIR-II filter in Example 3.13.

$$H(z) = \sum_{n=0}^{M} h(n)z^{-n} = \pm \sum_{n=0}^{M} h(M-n)z^{-n}. \tag{3.51}$$

A simple substitution of variables gives

$$H(z) = \pm z^{-M} H(z^{-1}). \tag{3.52}$$

From this identity, we make the following observation.

Fact 1: If z_0 is a zero of $H(z)$, then we have

$$H(z_0) = \pm z_0^{-M} H(z_0^{-1}) = 0,$$

which implies that $H(z_0^{-1}) = 0$. That is, the reciprocal z_0^{-1} is also a zero of the filter.

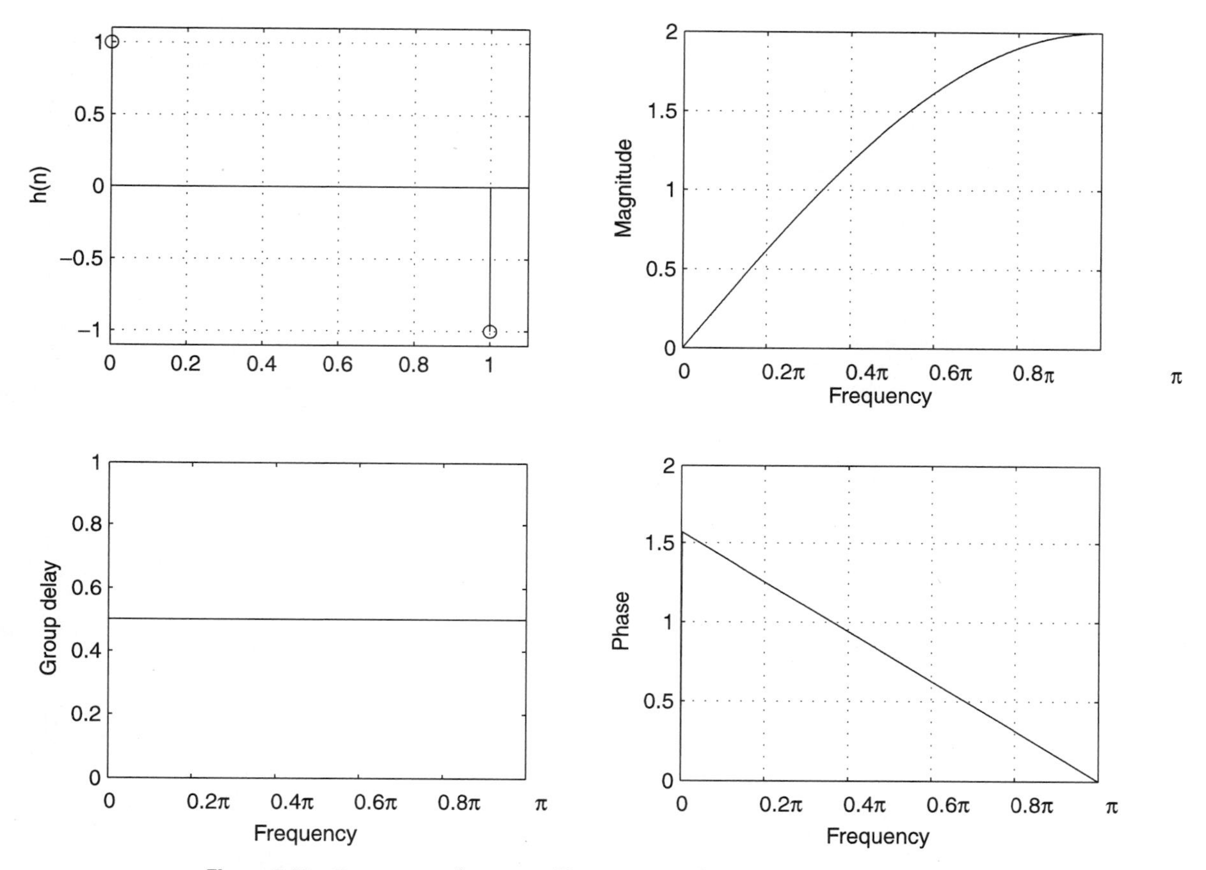

Figure 3.10 Responses of FIR-III filter in Example 3.13.

Fact 2: Real zeros not on the unit circle will always occur in pairs such as $(1 - rz^{-1})(1 - r^{-1}z^{-1})$. This is a direct consequence of Fact 1.

Fact 3: If the zero is complex, $z_0 = re^{j\theta}$, then its conjugate $re^{-j\theta}$ is also a zero, assuming that $h(n)$ is real. By Fact 1, the reciprocals will also be zeros, $r^{-1}e^{-j\theta}$ and $r^{-1}e^{j\theta}$. Thus, complex zeros not on the unit circle always occur in quadruples.

Fact 4: If M is odd and $h(n) = h(M - n)$, then $H(z)$ must have a zero at $z = -1$.

Fact 5: If $h(n) = -h(M - n)$, then $H(z)$ must have a zero at $z = 1$, for both even and odd M.

Fact 6: If $h(n) = -h(M - n)$, and M is even, then $H(z)$ must have a zero at $z = -1$.

The proofs of Facts 4, 5, and 6 are left as exercises. The possible zero locations of a linear phase filter are illustrated in Fig. 3.12.

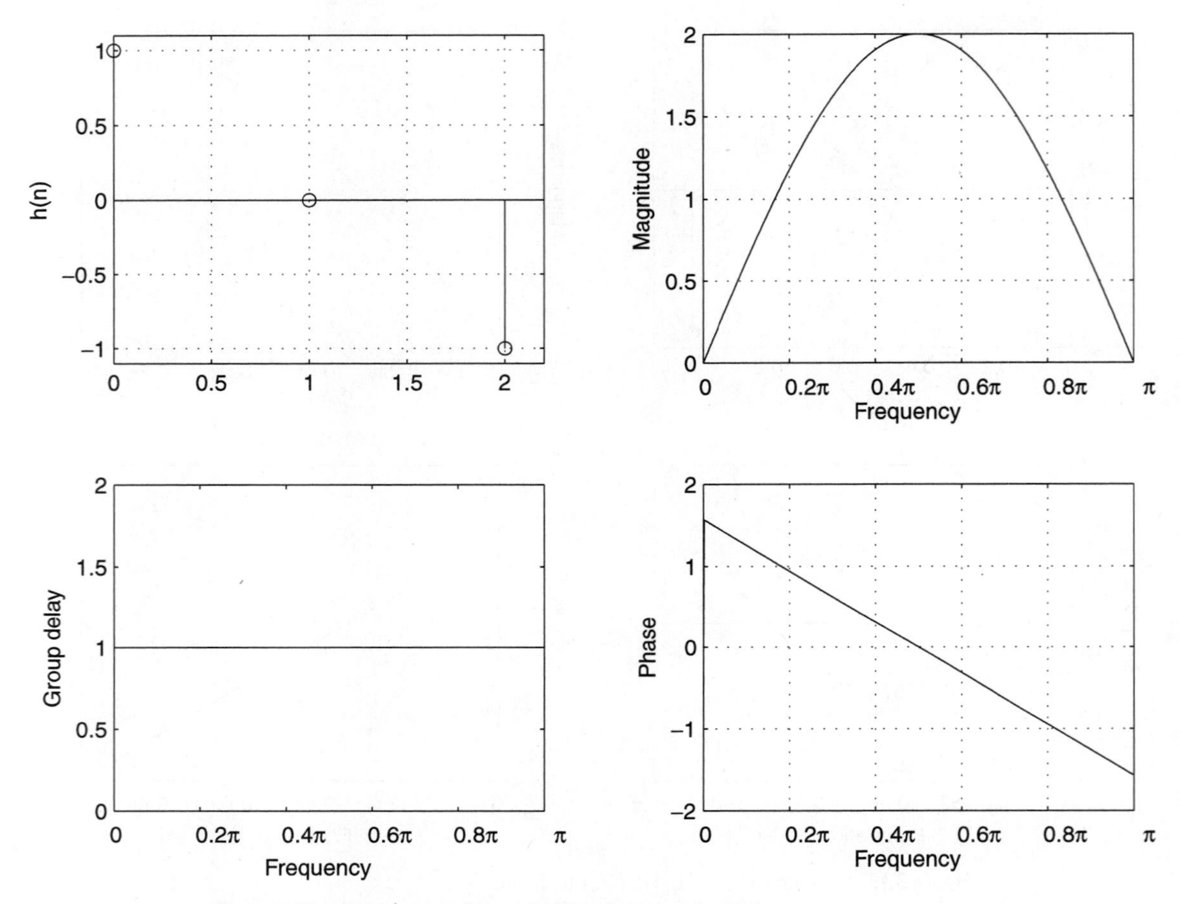

Figure 3.11 Responses of FIR-IV filter in example 3.13.

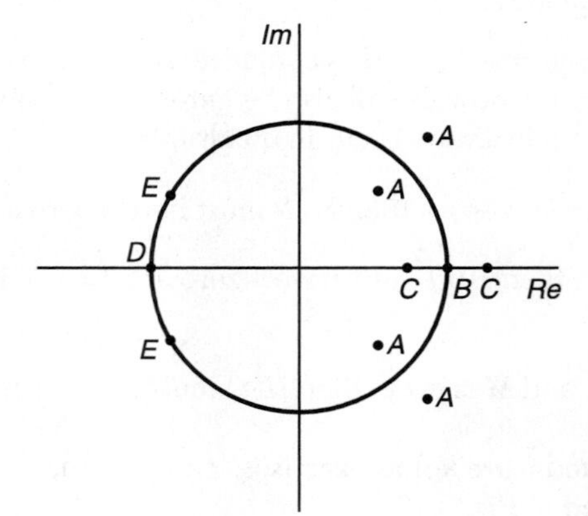

Figure 3.12 Possible zero positions for a linear phase FIR filter. The alphabets refer to the zeros that appear together.

3.7 STATE-SPACE ANALYSIS

The *state* of a system at any discrete-time index, n_0, is the minimum amount of information, which together with any input specified in the interval $n_0 \leq n \leq n_1$ is adequate to determine uniquely the system output for $n_0 \leq n \leq n_1$. The state is comprised of *state variables*. The state variables not only allow a calculation of the system output but also describe the internal states of a system that may be of interest. In continuous-time circuits, the state variables are defined as the current or voltage of energy storage elements that do not change instantaneously, that is, capacitor voltages or inductor currents. In discrete-time systems, the state variables are typically defined as the outputs of delay elements. Solving for the state variables therefore gives a description of some internal signals that would not otherwise be available from the transfer function. In addition, it will be seen shortly that a given digital filter (transfer function) can be represented by different state-space models. They all implement the same filter but have different flow diagrams. These different state-space models tend to have different properties when implemented in finite wordlength digital processors. This topic will be discussed in detail in Chapter 6.

State-Space Description

The state-space model of an LSIV single input single output (SISO) system is written as

$$\mathbf{v}(n+1) = \mathbf{A}\mathbf{v}(n) + \mathbf{B}x(n) \tag{3.53}$$

$$y(n) = \mathbf{C}\mathbf{v}(n) + Dx(n). \tag{3.54}$$

Let N denote the order of the system. The vector $\mathbf{v}(n) = [v_1(n)\, v_2(n) \ldots v_N(n)]$ is called the state vector, and its elements are called state variables. $x(n)$ and $y(n)$ are the input and output, respectively. $\mathbf{A}$ is a $N \times N$ matrix, $\mathbf{B}$ is a $N \times 1$ vector, $\mathbf{C}$ is $1 \times N$, and D is a scalar. Equation (3.53) is called the *state equation*, and (3.54) is called the *output equation*. The state-space model of a digital filter can be derived from its flow diagram as we now show with an example. Consider the Direct Form II implementation of a second-order digital filter given in Chapter 1. The difference equation is given by

$$y(n) + a_1 y(n-1) + a_2 y(n-2) = b_0 x(n) + b_1 x(n-1) + b_2 x(n-2).$$

The flow diagram is repeated in Fig. 3.13 for convenience. As discussed earlier, we define the state variables as the outputs of the delays as shown in the figure. This is a second-order system and consequently has two state variables $v_1(n)$ and $v_2(n)$. The problem at hand is to write $v_1(n+1), v_2(n+1)$, and $y(n)$ in terms of the state variables and the input. The simplest way to do as is to write the equations by inspection of the flow diagram. First, it is easy to see that $v_1(n+1) = v_2(n)$. Then we consider the equation for $v_2(n+1)$, which is the output of the adder. There are three inputs to the adder, and they are summed to form $v_2(n+1)$. The required state equations are therefore

$$\begin{aligned} v_1(n+1) &= v_2(n) \\ v_2(n+1) &= -a_2 v_1(n) - a_1 v_2(n) + x(n). \end{aligned}$$

The output is formed by three signals added together as $y(n) = b_2 v_1(n) + b_1 v_2(n) + b_0 v_2(n+1)$. By substituting for $v_2(n+1)$ from the above equation, we get the output equation

$$y(n) = (b_2 - b_0a_2)v_1(n) + (b_1 - b_0a_1)v_2(n) + b_0x(n).$$

The state and output equations are usually written in the matrix format of (3.53) and (3.54) as follows:

$$\begin{bmatrix} v_1(n+1) \\ v_2(n+2) \end{bmatrix} = \begin{bmatrix} 0 & 1 \\ -a_2 & -a_1 \end{bmatrix} \begin{bmatrix} v_1(n) \\ v_2(n) \end{bmatrix} + \begin{bmatrix} 0 \\ 1 \end{bmatrix} x(n) \tag{3.55}$$

$$y(n) = \begin{bmatrix} b_2 - b_0a_2 & b_1 - b_0a_1 \end{bmatrix} \begin{bmatrix} v_1(n) \\ v_2(n) \end{bmatrix} + [b_0]x(n). \tag{3.56}$$

If we consider another implementation of the same filter such as the Direct Form II$_T$, then the elements of the matrices in the state and output equations will be different. Therefore, the state-space model denoted by $\{\mathbf{A}, \mathbf{B}, \mathbf{C}, D\}$ describes the structural implementation of a digital filter. A comprehensive treatment of filter structures and their state-space models will be given in Chapter 4. Different implementations of the same filter do not change the input-output characteristic. Therefore, all the state-space models for the different structures are equivalent. It is possible to convert one model $\{\mathbf{A}, \mathbf{B}, \mathbf{C}, D\}$ to another $\{\widehat{\mathbf{A}}, \widehat{\mathbf{B}}, \widehat{\mathbf{C}}, \widehat{D}\}$, as given by the following theorem.

Theorem 3.4 *(Similarity Transformation) Let $\{\mathbf{A}, \mathbf{B}, \mathbf{C}, D\}$ represent a state-space model of a digital filter. If there exists a nonsingular matrix $\mathbf{P}$, then this model can be transformed to an equivalent model $\{\widehat{\mathbf{A}}, \widehat{\mathbf{B}}, \widehat{\mathbf{C}}, \widehat{D}\}$ by the following equations:*

$$\begin{aligned} \widehat{\mathbf{A}} &= \mathbf{P}^{-1}\mathbf{A}\mathbf{P} \\ \widehat{\mathbf{B}} &= \mathbf{P}^{-1}\mathbf{B} \\ \widehat{\mathbf{C}} &= \mathbf{C}\mathbf{P} \\ \widehat{D} &= D. \end{aligned} \tag{3.57}$$

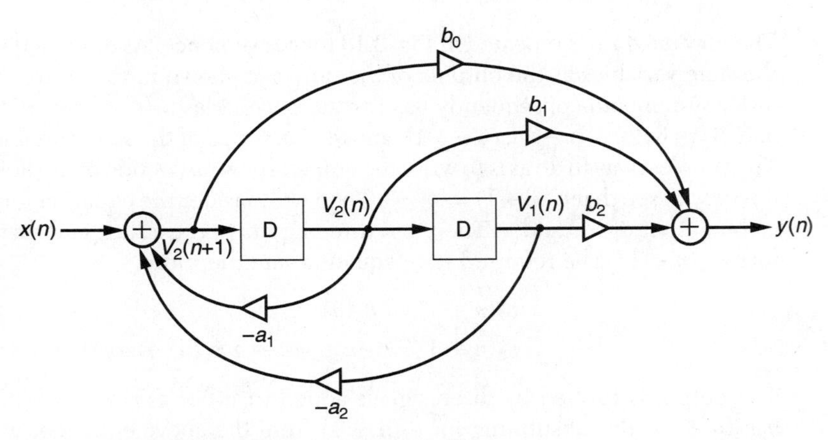

Figure 3.13 Second-order Direct Form II filters.

Proof. The state and output equations are

$$\mathbf{v}(n+1) = \mathbf{A}\mathbf{v}(n) + \mathbf{B}x(n)$$
$$y(n) = \mathbf{C}\mathbf{v}(n) + Dx(n).$$

Define the state vector transformation $\mathbf{v}(n) = \mathbf{P}\widehat{\mathbf{v}}(n)$. Substituting this in the above and simplifying gives

$$\widehat{\mathbf{v}}(n+1) = (\mathbf{P}^{-1}\mathbf{A}\mathbf{P})\widehat{\mathbf{v}}(n) + (\mathbf{P}^{-1}\mathbf{B})x(n)$$
$$y(n) = \mathbf{C}\mathbf{P}\widehat{\mathbf{v}}(n) + Dx(n).$$

In comparing these sets of equations, the theorem is proved. ■

Solution of the State-Space Equation

The state variables are internal variables and provide further information about the system. Therefore, it is important to find a solution of the state equation. The easiest way to do so is to use the Z-transform. Taking the unilateral Z-transform of the state equation (3.53), we get

$$z\mathbf{V}(z) - z\mathbf{v}(0) = \mathbf{A}\mathbf{V}(z) + \mathbf{B}X(z).$$

Simple manipulation of the above gives

$$\mathbf{V}(z) = (z\mathbf{I} - \mathbf{A})^{-1}\mathbf{B}X(z) + (z\mathbf{I} - \mathbf{A})^{-1}z\mathbf{v}(0) \tag{3.58}$$

where $\mathbf{I}$ is the identity matrix and $\mathbf{v}(0)$ is the initial condition of the state variables. In the time domain, the above is

$$\mathbf{v}(n) = Z^{-1}\left\{(z\mathbf{I} - \mathbf{A})^{-1}\mathbf{B}X(z)\right\} + Z^{-1}\left\{(z\mathbf{I} - \mathbf{A})^{-1}z\mathbf{v}(0)\right\} \tag{3.59}$$

where Z^{-1} denotes the inverse Z-transform. The above is the solution of the state equation. The output of the system in the Z domain is found by substituting (3.58) in the Z-transform of the output equation (3.54). The result is

$$Y(z) = \mathbf{C}(z\mathbf{I} - \mathbf{A})^{-1}\mathbf{B}X(z) + \mathrm{C}(z\mathbf{I} - \mathbf{A})^{-1}z\mathbf{v}(0) + DX(z). \tag{3.60}$$

The time-domain output is simply the inverse transform of the above. The transfer function of the system is the ratio of the output to the input in the Z domain with zero initial conditions. The transfer function formula is therefore given by

$$H(z) = \frac{Y(z)}{X(z)} = \mathbf{C}(z\mathbf{I} - \mathbf{A})^{-1}\mathbf{B} + D. \tag{3.61}$$

The term $(z\mathbf{I} - \mathbf{A})^{-1}$ is a matrix inverse that is obtained by finding its adjoint matrix and then dividing it by its determinant. Therefore, we can write this as

$$(z\mathbf{I} - \mathbf{A})^{-1} = \frac{\mathrm{Adj}\{z\mathbf{I} - \mathbf{A}\}}{\det\{z\mathbf{I} - \mathbf{A}\}}$$

where Adj and det denote the adjoint and determinant, respectively. The transfer function then becomes

$$H(z) = \frac{Y(z)}{X(z)} = \mathbf{C}\frac{\text{Adj}\{\mathbf{zI}-\mathbf{A}\}}{\det\{\mathbf{zI}-\mathbf{A}\}}\mathbf{B} + D.$$

This equation is a ratio of polynomials with the denominator given by det $\{z\mathbf{I}-\mathbf{A}\}$. As we know, the roots of the denominator are the poles of the transfer function, which are given by the determinant. This is formally stated below.

Fact: Let $\mathbf{A}$ be the system matrix of a digital filter. The polynomial given by det $(z\mathbf{I}-\mathbf{A})$ is called the *characteristic polynomial*, and its roots are the poles of the filter transfer function. The poles are therefore also referred to as the *characteristic roots* of the filter.

Under similarity transformation, the transformed filter has the same transfer function as the original. Therefore, the characteristic roots of det $(z\mathbf{I}-\mathbf{A})$ are the same as those of det $(z\mathbf{I}-\mathbf{P}^{-1}\mathbf{A}\mathbf{P})$.

▶ **EXAMPLE 3.14** *The state variable model* $\{\mathbf{A}, \mathbf{B}, \mathbf{C}, D\}$ *of a causal digital filter is given as*

$$\mathbf{A} = \begin{bmatrix} 0 & 1 \\ -0.02 & -0.3 \end{bmatrix}, \quad \mathbf{B} = \begin{bmatrix} 0 \\ 1 \end{bmatrix}, \quad \mathbf{C} = \begin{bmatrix} 1 & 2 \end{bmatrix}, \quad D = 0.$$

Find the poles of the filter, the transfer function, the difference equation, and the solution of the state equation for a step input and zero initial condition.

SOLUTION By comparing the given matrices with equations (3.55) and (3.56), it is clear that this model represents a Direct Form II implementation. The poles of the filter are given by the roots of the characteristic equation:

$$\det\left\{z\begin{bmatrix} 1 & 0 \\ 0 & 1 \end{bmatrix} - \begin{bmatrix} 0 & 1 \\ -0.02 & -0.3 \end{bmatrix}\right\} = 0.$$

The above simplifies to $z^2 + 0.3z + 0.02 = 0$. The poles are $z = -0.1$ and $z = -0.2$. The transfer function is obtained as follows[4]:

$$\begin{aligned} H(z) &= \mathbf{C}(z\mathbf{I}-\mathbf{A})^{-1}\mathbf{B} + D \\ &= \begin{bmatrix} 1 & 2 \end{bmatrix}\begin{bmatrix} z & -1 \\ 0.02 & z+0.3 \end{bmatrix}^{-1}\begin{bmatrix} 0 \\ 1 \end{bmatrix} \\ &= \begin{bmatrix} 1 & 2 \end{bmatrix}\frac{1}{z^2+0.3z+0.02}\begin{bmatrix} z+0.3 & 1 \\ -0.02 & z \end{bmatrix}\begin{bmatrix} 0 \\ 1 \end{bmatrix} \\ &= \frac{2z+1}{z^2+0.3z+0.02}. \end{aligned}$$

[4]The adjoint of a 2×2 matrix is found by swapping the diagonal elements and changing signs of the off-diagonal elements. The inverse is then the adjoint divided by the determinant or the characteristic polynomial.

The solution of the state equation is given by

$$\mathbf{v}(n) = Z^{-1}\{(z\mathbf{I} - \mathbf{A})^{-1}\mathbf{B}X(z)\} + Z^{-1}\{(z\mathbf{I} - \mathbf{A})^{-1}z\mathbf{v}(0)\}.$$

The input is a unit step, and so $X(z) = \frac{z}{z-1}$. By substituting this value of $X(z)$ and $\mathbf{v}(0) = 0$ in the above and after simplification, we get

$$\begin{aligned}\mathbf{v}(n) &= Z^{-1}\left\{\frac{1}{z^2 + 0.3z + 0.02}\begin{bmatrix}1\\ z\end{bmatrix}\frac{z}{z-1}\right\}\\ &= \begin{bmatrix}\{0.91(-0.1)^n - 1.66(-0.2)^n + 0.76\}u(n)\\ \{-0.091(-0.1)^n + 0.33(-0.2)^n + 0.76\}u(n)\end{bmatrix}.\end{aligned}$$

3.8 LINEAR FILTERING USING DFT/FFT

A linear filtering operation is a convolution of an input sequence $x(n)$ with the filter impulse response. Let us assume that the filter has a finite impulse response, $h(n)$. If the input sequence is also finite, then the filter output, $y(n)$ can be obtained by a simple linear convolution operation, $y(n) = h(n) * x(n)$. Alternatively, this linear convolution can be efficiently computed by using the DFT after appropriately zero-padding $x(n)$ and $h(n)$. This procedure was described in Chapter 2.

In most real-time applications, the input samples of $x(n)$ arrive at the filter continuously. Therefore, the input can be thought of as an infinite-length sequence. In this section, we address the problem of efficient real-time filtering of this infinitely long input sequence. There are two well-known solutions: the overlap-add method and the overlap-save method.

Overlap-Add Method

Let the filter impulse response $h(n)$ have length M. Since the input sequence is infinitely long, we need to segment it. Let N be the length of each of these segments, which can be labeled as $x_0(n), x_1(n), \ldots$ and so on. This segmentation of a given sequence is depicted in Fig. 3.14. Each of the segments is assumed to start at the zeroth sample. Therefore, the original input sequence can be represented as

$$x(n) = \sum_{i=0}^{\infty} x_i(n - iN). \tag{3.62}$$

Since the filter is assumed to be linear and time-invariant, the output $y(n)$ is therefore composed of a sum of the segments $y_i(n)$. Each of these segments is the linear convolution

$$y_i(n) = h(n) * x_i(n). \tag{3.63}$$

The final output is then

$$y(n) = \sum_{i=0}^{\infty} y_i(n - iN). \tag{3.64}$$

There remains one issue to resolve here. When the length-M sequence $h(n)$ is convolved with the length-N segment $x_0(n)$, the resulting sequence $y_0(n)$ is of length $(N+M-1)$. Again, the convolution of $h(n)$ with $x_1(n)$ yields another length $(N+M-1)$ sequence $y_1(n)$. As shown in Fig. 3.14, there is a $(M-1)$ point overlap between $y_0(n)$ and $y_1(n)$, which must be added together. This process is repeated for all the segments. The reason for the name overlap-add is now obvious.

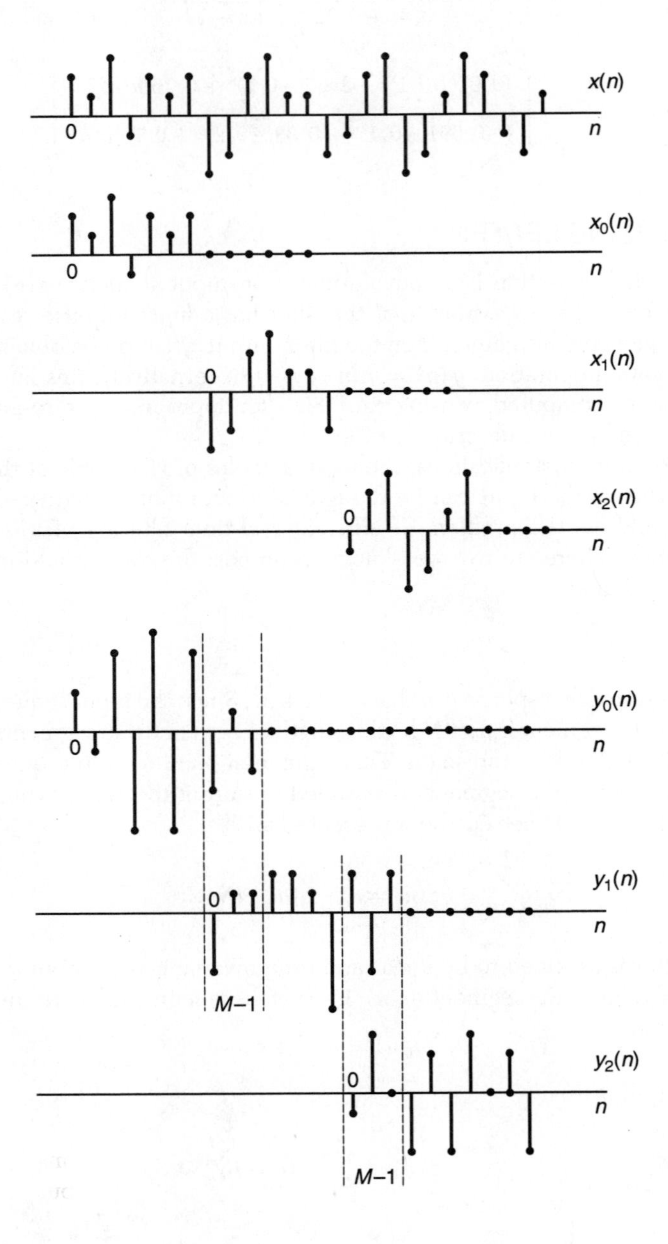

Figure 3.14 Illustration of the overlap-add method.

The convolution operations of (3.63) can be performed as direct linear convolution or more efficiently by using the zero-padded FFT method. The relative computational costs is straightforward to derive and is left as an exercise.

Overlap-Save Method

The overlap-save method is designed to further reduce the computational complexity of the overlap-add method. The overlap-add method requires the computation of two $(N+M-1)$-point DFTs/FFTs and one $(N+M-1)$-point IDFT/IFFT. The overlap-save method is designed to reduce these operations to N-point transforms. However, without zero-padding, the operation $F^{-1}\{F\{x(n)\} \cdot F\{h(n)\}\}$ results in circular convolution. The trick then is to extract the linear convolution part from the resulting circular convolution. The following result will be used for this purpose. The proof of this fact is left as an exercise.

Fact : *If a length-M sequence is circularly convolved with a length-N sequence, where $N > M$, the last $N-M+1$ samples of the result correspond to linear convolution, whereas the first $M-1$ samples do not.*

Now consider the segmentation of the infinitely long sequence $x(n)$ as shown in Fig. 3.15. The segmentation is done with $(M-1)$-point overlaps, the reason for which will be clear shortly. Circular convolution is now performed as

$$v_i(n) = x_i(n) \circledast h(n). \tag{3.65}$$

According to the foregoing Fact, the first $(M-1)$ points of $v_0(n)$ are discarded and the last $(N-M+1)$ points are saved. Therefore, $x_0(n)$ must be shifted by $(M-1)$ points so that this rejection of data has no impact. In the next circular convolution $v_1(n)$, again the first $(M-1)$ points are discarded and the last $(N-M+1)$ points are saved. This process is repeated as shown in Fig. 3.15. The reason for the overlapped segmentation of $x(n)$ is now clear.

The circular convolution of (3.65) can be performed either directly or via the FFT. In the FFT method, it will require the computation of two N-point FFTs and one N-point IFFT.

3.9 COMPUTING THE DFT USING LINEAR FILTERING

In the last chapter, we discussed methods to compute the DFT and their relative computational costs. In particular, we presented the Direct DFT computation, the decimation in time, and the decimation in frequency FFT algorithms. The DFT can also be computed using some linear filtering approaches. The two linear filtering techniques discussed here are the Goertzel algorithm and the Chirp Z-transform. The Goertzel algorithm is computationally more efficient than the FFT when a relatively small number of values of the DFT are desired. It will be shown shortly that these values can be obtained by implementing appropriate IIR filters.

We know that the DFT is a special case of the Z-transform computed on the unit circle. In some applications, the Z-transform is required to be computed at points other

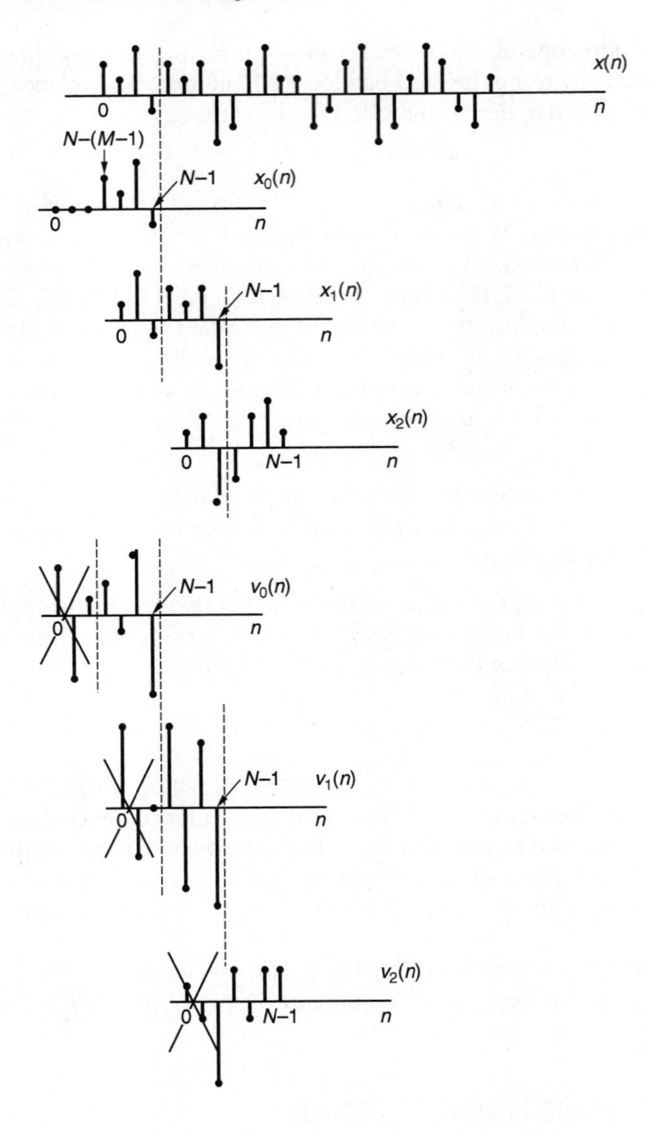

Figure 3.15 Illustration of the overlap-save method.

than the unit circle. The Chirp Z-transform is also a linear filtering approach that allows us to compute the Z-transform of a sequence on a variety of contours in the z-plane.

The Goertzel Algorithm

The Goertzel algorithm [3] is actually a linear filtering technique that is computationally even more efficient than the FFT when the DFT must be computed for a relatively small number of frequency points. Before we derive the Goertzel algorithm, let us review the computational complexities of the DFT and the FFT. Recall that an N-point Direct DFT computation uses the formula

$$X(k) = \sum_{n=0}^{N-1} x(n) W_N^{kn}, \quad k = 0, 1, \ldots, N-1 \tag{3.66}$$

where $W_N = e^{\frac{-j2\pi}{N}}$.

Assuming that $x(n)$ is a complex signal, the above requires N complex multiplications and approximately N complex additions for each k (frequency bin) and N^2 complex multiplications for the entire DFT. Recall from Chapter 2 that the FFT requires $\frac{N}{2}\log_2 N$ complex multiplications and $N \log_2 N$ complex additions for the entire DFT. In the FFT, it is not possible to compute the transform for individual frequency bins. Using the fact that each complex multiplication requires four real multiplications and two real adds, and each complex addition requires two real additions, we can find the total computational counts in Table 3.3.

The Goertzel algorithm exploits the periodicity of the sequence W_N^k in order to reduce computations. The goal is to express the DFT formula (3.66) as a convolution of two sequences so that a linear filter can be realized. Since $W_N^{-kN} = 1$, we multiply the DFT by this factor to get

$$\begin{aligned} X(k) &= W_N^{-kN} \sum_{m=0}^{N-1} x(m) W_N^{km} \\ &= \sum_{m=0}^{N-1} x(m) W_N^{-k(N-m)}. \end{aligned} \tag{3.67}$$

Because $x(m)$ is a finite-length sequence in $0 \leq m \leq N-1$, we can change the limits in this summation as

$$X(k) = \sum_{m=-\infty}^{\infty} x(m) W_N^{-k(N-m)}. \tag{3.68}$$

Now define the sequence at the kth bin as

$$y_k(n) = \sum_{m=-\infty}^{\infty} x(m) W_N^{-k(n-m)}. \tag{3.69}$$

so that we have

$$X(k) = y_k(n)\,|_{n=N}. \tag{3.70}$$

TABLE 3.3 Real Multiplications and Additions (Approx.)

	For 1 freq. bin		For entire DFT	
ALGORITHM	No. of Mults	No. of Adds	No. of Mults	No. of Adds
Direct DFT	$4N$	$4N$	$4N^2$	$4N^2$
FFT	–	–	$2N \log_2 N$	$3N \log_2 N$
Goertzel	$2N$	$4N$	N^2	$2N^2$

Notice that (3.69) is a convolution of the sequences $x(n)$ and W_N^{-kn}. That is,

$$y_k(n) = W_N^{-kn} * x(n). \tag{3.71}$$

Therefore, (3.71) can be realized as a linear filter with impulse response

$$h_k(n) = (W_N^{-k})^n \tag{3.72}$$

and input $x(n)$. The DFT at the kth bin, $X(k)$, can be found by computing the output of this filter at the Nth sample, which is $y_k(N)$. The transfer function of this filter is found by simply taking the Z-transform of $h_k(n)$, which gives

$$H_k(z) = \frac{1}{1 - W_N^{-k} z^{-1}}. \tag{3.73}$$

The flow diagram of this first-order filter is shown in Fig. 3.16. This filter requires N complex multiplications to compute the output at the Nth sample. To reduce computations, we multiply the numerator and the denominator by $1 - W_N^k z^{-1}$ to get

$$\begin{aligned} H_k(z) &= \frac{1 - W_N^k z^{-1}}{(1 - W_N^{-k} z^{-1})(1 - W_N^k z^{-1})} \\ &= \frac{1 - W_N^k z^{-1}}{1 - 2\cos\left(\frac{2\pi k}{N}\right) z^{-1} + z^{-2}}. \end{aligned} \tag{3.74}$$

The Direct Form II realization for this second-order filter is shown in Fig. 3.17, and its difference equations are given by

$$v_k(n) = 2\cos\left(\frac{2\pi k}{N}\right) v_k(n-1) - v_k(n-2) + x(n) \tag{3.75}$$

$$y_k(n) = v_k(n) - W_N^k v_k(n-1). \tag{3.76}$$

The goal is to find $y_k(n)$ with the least amount of computations. So we compute (3.75) for $n = 0, 1, \ldots, N$, but we compute (3.76) only for $n = N$. If $x(n)$ is complex, then $v_k(n)$ is also complex. This requires $2(N+1)$ real multiplications and $4(N+1)$ real additions for (3.75). Computing (3.76) once requires four real multiplications and four real additions. Therefore, we have a total of $2N+6$ real multiplications and $4N+8$ real additions. For large N, these can be approximated to $2N$ and $4N$, respectively, as shown in Table 3.3. Thus, for each frequency bin, the Goertzel algorithm requires half the real multiplications and

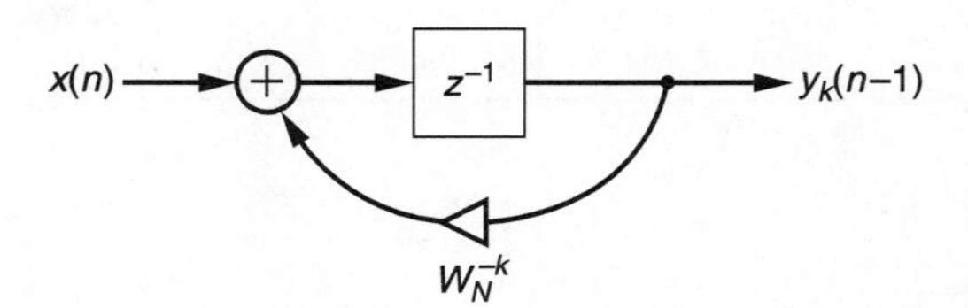

Figure 3.16 First-order Goertzel filter.

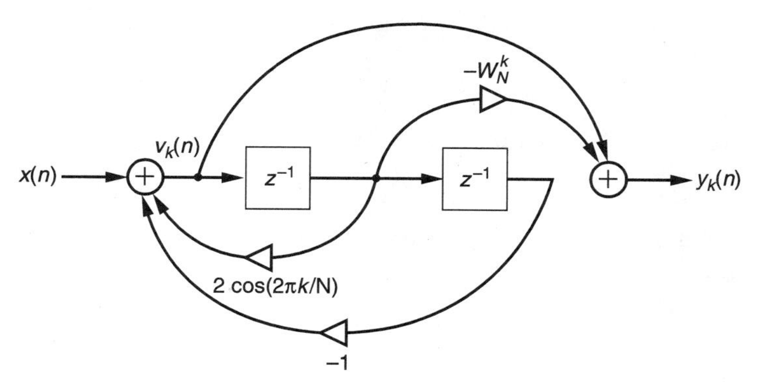

Figure 3.17 Second-order Goertzel filter to compute $X(k)$.

the same number of real additions as the Direct DFT. Multiplications are more complex operations in hardware than additions, and so the Goertzel algorithm has an advantage. The other advantage is that even for a single bin, the Direct DFT requires all the N values W_N^{kn} to be stored. However, the Goertzel algorithm requires only W_N^k to be stored for the kth bin, which is only one complex storage register as opposed to N registers for the Direct DFT.

Now if we want to compute $X(N-k)$, then from equation (3.74) we see that the structure for $H_{N-k}(n)$ has the same pole as that for $H_k(n)$. Therefore, computations for (3.75) are used for both $X(k)$ and $X(N-k)$. However, (3.76) will have to be computed separately, and will cost four real multiplications and four real additions. However, for large N, these are ignored anyway. So, for an entire N-point DFT, the Goertzel algorithm requires approximately N^2 real multiplications and $2N^2$ real additions. This reflects a savings over the Direct DFT method but is inefficient when compared with FFT. If we want to compute only M frequency bins, then the Goertzel algorithm requires $2MN$ real multiplications and $4MN$ real additions, without taking advantage of the possible savings in computations if $X(k)$ and $X(N-k)$ are present in the given M bins. Compared with the FFT, the Goertzel algorithm is more efficient if $2MN \leq 2N\log_2 N$ for multiplications and $4MN \leq 4N\log_2 N$ for additions, which implies $M \leq \log_2 N$.

The Chirp Z-transform (CZT)

The Chirp Z-transform algorithm was invented by Rabiner, Schafer, and Rader [4] in 1969 and finds applications even today. The CZT algorithm is used for computing the Z-transform of a sequence at M points in the z-plane which lie on circular or spiral contours beginning at any arbitrary point in the z-plane. To derive this algorithm, let us begin with the computation of the Z-transform at a finite number of points z_k by using the formula

$$X(k) = X(z_k) = \sum_{n=0}^{N-1} x(n) z_k^{-n}. \tag{3.77}$$

Equation (3.77) reduces to the DFT if

$$z_k = e^{j\frac{2\pi k}{N}} = W_N^{-k}, \quad k = 0, 1, \ldots, N-1 \tag{3.78}$$

where $W_N = e^{-j\frac{2\pi}{N}}$ as before. The CZT uses the more general contour given by

$$z_k = AB^k, \quad k = 0, 1, \ldots, M-1 \tag{3.79}$$

where M is the number of points to be computed and A and B are complex numbers of the form

$$A = A_o e^{j\theta_o} \tag{3.80}$$

and

$$B = B_o e^{j\phi_o} \tag{3.81}$$

The special case of $A = 1, B = e^{j\frac{2\pi}{N}}$, and $M = N$ corresponds to the DFT, and the contour is the unit circle. It is easy to see that if $A_o = B_o = 1$, the contour is an arc of the unit circle. The general z-plane contour begins at the point $z = A$. If $B_o < 1$, the contour spirals in toward the origin. If $B_o > 1$, the contour spirals out. Figure 3.18 shows some examples of contours on which the Z-transform may be evaluated. Along these general contours, one transform is

$$X(k) = \sum_{n=0}^{N-1} x(n) A^{-n} B^{-kn}, \quad k = 0, 1, \ldots, M-1. \tag{3.82}$$

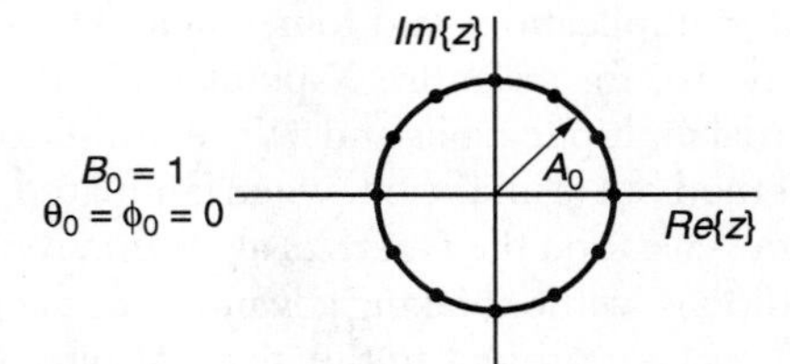

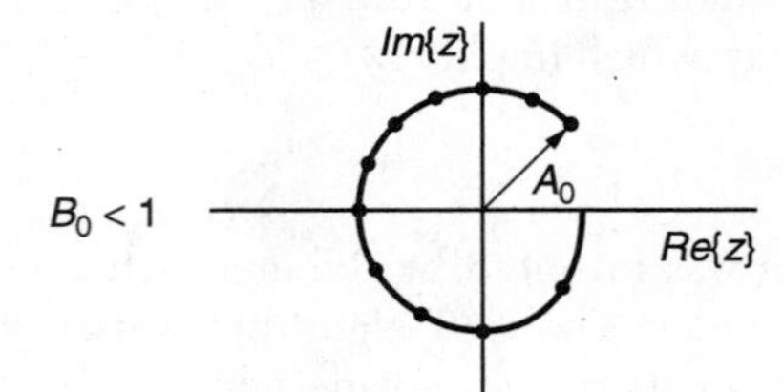

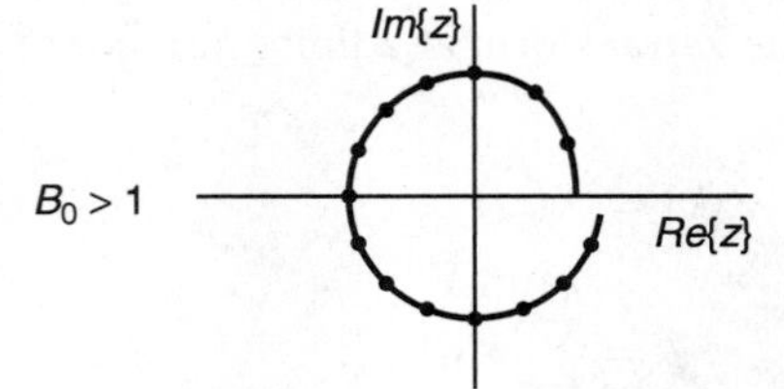

Figure 3.18 Sample contours for the CZT.

Formula (3.82) requires NM complex multiplications and additions. In order to reduce computations, the CZT algorithm reformulates (3.82) in the form of a convolution (linear filtering). To achieve this, in (3.82) we use the substitution

$$nk = \frac{n^2 + k^2 - (k-n)^2}{2}. \tag{3.83}$$

This gives

$$\begin{aligned} X(k) &= \sum_{n=0}^{N-1} x(n) A^{-n} B^{-n^2/2} B^{-k^2/2} B^{(k-n)^2/2} \\ &= B^{-k^2/2} \sum_{n=0}^{N-1} \left[x(n) A^{-n} B^{-n^2/2} \right] B^{(k-n)^2/2} \qquad (3.84) \\ &= B^{-k^2/2} \sum_{n=0}^{N-1} g(n) h(k-n) \\ &= B^{-k^2/2} (g(k) * h(k)), \qquad (3.85) \end{aligned}$$

where $*$ denotes linear convolution. The newly defined sequences are

$$g(n) = x(n) A^{-n} B^{-n^2/2} \tag{3.86}$$

$$h(n) = B^{n^2/2}. \tag{3.87}$$

The computation of (3.85) is the CZT algorithm, which is illustrated by the flow graph of Fig. 3.19. The main concept of the CZT algorithm is now complete. The sequence $h(n) = B^{n^2/2}$ for $B_o = 1$ is a complex sinusoidal of linearly increasing frequency. Such waveforms are used in radar systems where they are called "chirp" signals—hence the origin of the name.

Now we find the computational complexity of the CZT algorithm. For simplicity, we will count the complex multiplications only. The input signal is assumed to be complex. The algorithm is divided into the following three parts.

Step 1. Formation of $g(n)$: It is assumed that the multipliers $A^{-n}B^{-n^2/2}$ are precomputed and stored. Then the formation of $g(n)$ requires N complex multiplications.

Step 2. Linear Convolution: This is the computationally intensive step. Therefore, FFT will be used to perform the filtering. However, we know from Chapter 2 that a product of

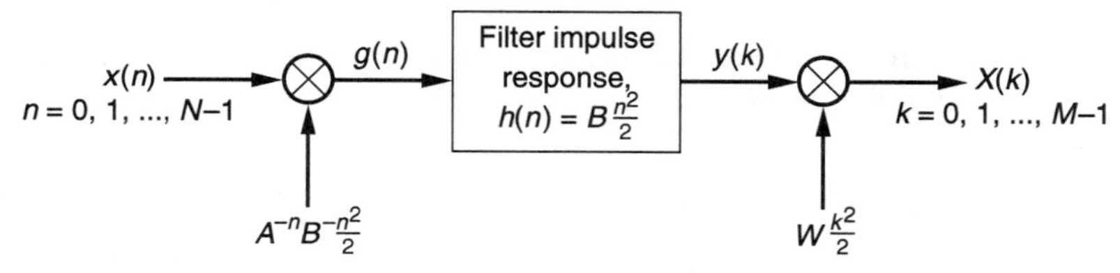

Figure 3.19 The CZT algorithm.

FFTs corresponds to circular convolution. So, we must zero-pad appropriately to obtain linear convolution. Let L be the smallest integer greater than $N+M-1$ so that it is a power of two. This is done in order to use radix-2 FFT.

(2a) Zero-pad $g(n)$ so that it has length L. Compute its FFT. This costs $\frac{L}{2}\log_2 L$ complex multiplications.

(2b) The filter impulse response, $h(n)$, is infinite length and therefore cannot be zero-padded. It must be used to construct a signal $\widehat{h}(n)$ of length L in such a manner that

$$IFFT\{FFT\{\widehat{h}(n)\}FFT\{g(n)\}\} = h(n) * g(n). \tag{3.88}$$

In other words,

$$\widehat{h}(n) \circledast g(n) = h(n) * g(n). \tag{3.89}$$

It can be shown (exercise) that we need

$$\widehat{h}(n) = \begin{cases} h(n), & 0 \le n \le M-1 \\ h(n-L), & L-N+1 \le n \le L \\ \text{arbitrary}, & \text{otherwise}. \end{cases} \tag{3.90}$$

This step requires two FFTs and one IFFT, each of which requires $\frac{L}{2}\log_2 L$ complex multiplications, for a total of $\frac{3}{2}L\log_2 L$. A direct convolution would require LN complex multiplications.

Step 3. CZT coefficients: The final CZT coefficients are obtained by multiplying $y(k)$ by $W^{k^2/2}$ for $k = 0, 1, \ldots, M-1$. This costs M complex multiplications.

The total computational cost for the entire algorithm is

$$\text{no. of Complex Multiplications } = N + \frac{3}{2}L\log_2 L + M.$$

In closing, we summarize the salient features of the CZT.

1. The CZT allows the efficient evaluation of the Z-transform at M points in the z-plane which lie on circular or spiral contours.
2. The number of time samples (N) does not have to be equal to the number of Z-transform samples (M).
3. The angular spacing of z_k is arbitrary.
4. The CZT algorithm can be used for the enhancement of poles in spectral analysis, high-resolution analysis, interpolation of data in the time domain for a change in the sampling rate, and many other applications.

3.10 2-D EXTENSIONS

In this section, we first extend the concepts of Z-transforms and ROC to two dimensions. Then we discuss the fundamentals of the stability of 2-D systems.

2-D Z-transform

The 2-D Z-transform is defined as

$$X(z_1, z_2) = Z\{x(n_1, n_2)\} = \sum_{n_1=-\infty}^{\infty} \sum_{n_2=-\infty}^{\infty} x(n_1, n_2) z_1^{-n_1} z_2^{-n_2} \qquad (3.91)$$

where $z_1 = a_1 + jb_1 = r_1 e^{j\omega_1}$ and $z_2 = a_2 + jb_2 = r_2 e^{j\omega_2}$. The sequence and its Z-transform will be denoted by the pair $x(n_1, n_2) \longleftrightarrow X(z_1, z_2)$. The ROC is the region in the z_1 plane and the z_2 plane for which the above sum converges. In contrast to the 1-D case, here we have two different planes. The ROC is the intersection of these two regions. We now present some examples. Some of the 2-D signal definitions that were given in Chapter 1 will be used.

EXAMPLE 3.15 *Find the Z-transform of the following sequences:* $(a)\,\delta(n_1, n_2)$, $(b)\ \delta_{2d}(n_2)u(n_1, n_2)$, $(c)\ u(n_1, n_2)$, $(d)\ u_{2d}(n_1)\delta_{2d}(n_1 - n_2)$, $(e)\ a^{n_1} b^{n_2} u(-n_1, -n_2)$.

SOLUTION All of the above examples can be easily found by using the definition.

(a) In this case, we simply use the definitions of the Z-transform and the 2-D impulse to get

$$Z\{\delta(n_1, n_2)\} = \sum_{n_1=-\infty}^{\infty} \sum_{n_2=-\infty}^{\infty} \delta(n_1, n_2) z_1^{-n_1} z_2^{-n_2} = 1, \qquad \text{ROC: } \forall z_1, z_2.$$

(b) The sequence is sketched in Fig. 3.20(a), and the transform is

$$\begin{aligned} Z\{\delta_{2d}(n_2)u(n_1, n_2)\} &= \sum_{n_1=-\infty}^{\infty} \sum_{n_2=-\infty}^{\infty} \delta_{2d}(n_2)u(n_1, n_2) z_1^{-n_1} z_2^{-n_2} \\ &= \sum_{n_1=0}^{\infty} z_1^{-n_1} \\ &= \frac{1}{1 - z_1^{-1}}, \qquad \text{ROC: } |z_1| > 1, \forall z_2. \end{aligned}$$

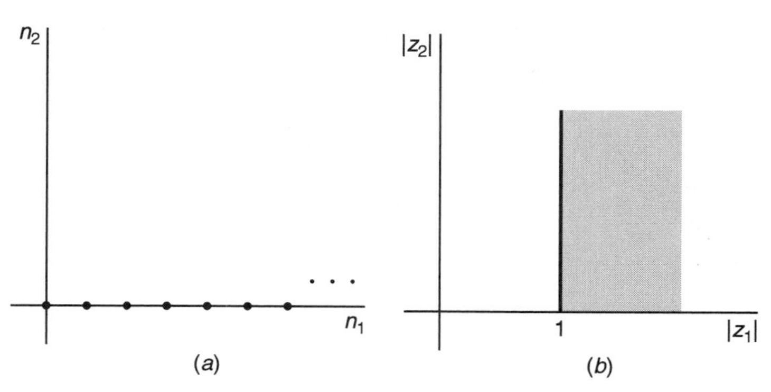

Figure 3.20 (a) $\delta_{2d}(n_2)u(n_1, n_2)$; (b) ROC.

The ROC has four dimensions, namely, the magnitudes and angles of z_1 and z_2. So, it is customary to denote the ROC in the $(|z_1|, |z_2|)$ quarter plane as shown in Fig. 3.20(b).

(c)

$$\begin{aligned} Z\{u(n_1, n_2)\} &= \sum_{n_1=0}^{\infty}\sum_{n_2=0}^{\infty} z_1^{-n_1} z_2^{-n_2} \\ &= \frac{1}{1-z_1^{-1}} \frac{1}{1-z_2^{-1}}, \qquad \text{ROC: } |z_1| > 1, |z_2| > 1. \end{aligned}$$

The sequence and its ROC are shown in Fig. 3.21(a) and (b), respectively.

(d) The sequence is sketched in Fig. 3.22(a). The Z-transform is

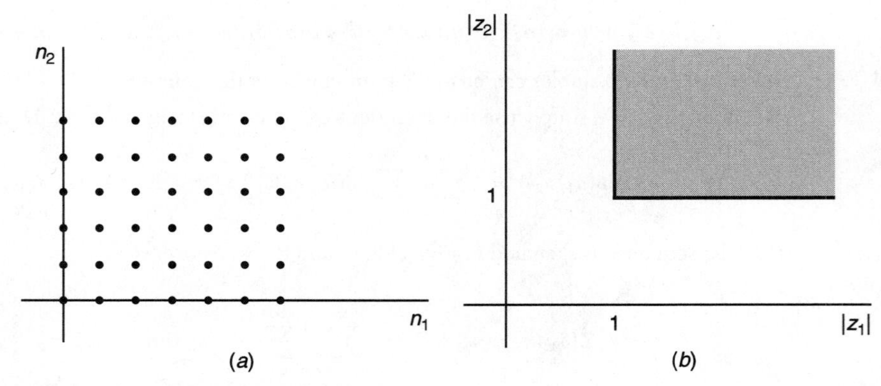

Figure 3.21 (a) $u(n_1, n_2)$; (b) ROC.

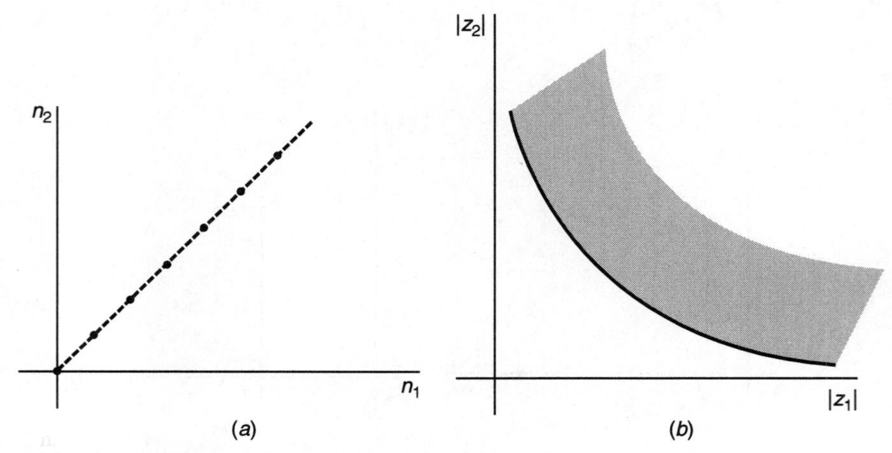

Figure 3.22 (a) $u_{2d}(n_1)\delta_{2d}(n_1 - n_2)$; ($b$) ROC.

$$
\begin{aligned}
Z\{u_{2d}(n_1)\delta_{2d}(n_1 - n_2)\} &= \sum_{n_1=-\infty}^{\infty} \sum_{n_2=-\infty}^{\infty} u_{2d}(n_1)\delta_{2d}(n_1 - n_2) z_1^{-n_1} z_2^{-n_2} \\
&= \sum_{n_1=0}^{\infty} z_1^{-n_1} z_2^{-n_2} \Big|_{n_2=n_1} \\
&= \frac{1}{1 - z_1^{-1} z_2^{-1}}, \qquad \text{ROC} : |z_1||z_2| > 1.
\end{aligned}
$$

The ROC is sketched in Fig. 3.22(b).

(e)

$$
\begin{aligned}
Z\{a^{n_1} b^{n_2} u(-n_1, -n_2)\} &= \sum_{n_1=0}^{-\infty} \sum_{n_2=0}^{-\infty} a^{n_1} b^{n_2} z_1^{-n_1} z_2^{-n_2} \\
&= \sum_{n_1=0}^{\infty} a^{-n_1} z_1^{n_1} \sum_{n_2=0}^{\infty} b^{-n_2} z_2^{n_2} \\
&= \frac{1}{1 - a^{-1} z_1} \frac{1}{1 - b^{-1} z_2}, \qquad \text{ROC} : |z_1| < |a|, \quad |z_2| < |b|.
\end{aligned}
$$

The ROC is sketched in Fig. 3.23.

The properties of the 2-D Z-transform are straightforward extensions of 1-D and can be derived from the definition. Table 3.4 summarizes the basic properties. The property of separable sequences is unique to 2-D and is given below. It is similar to the property on the DTFT of 2-D separable sequences.

Property : Let $x(n_1, n_2) = x_1(n_1)x_2(n_2)$. Then

$$X(z_1, z_2) \leftrightarrow X_1(z_1)X_2(z_2).$$

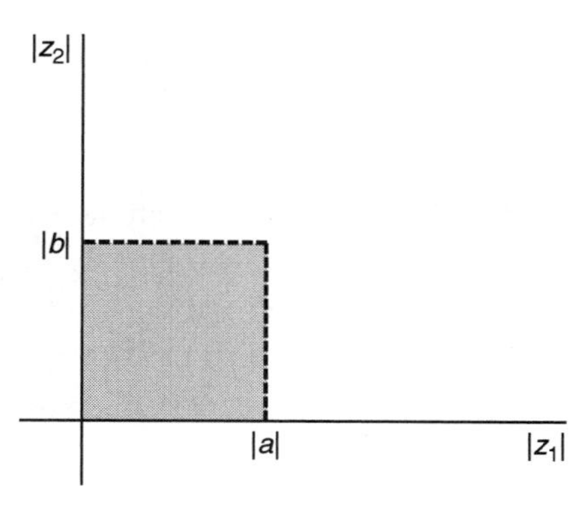

Figure 3.23 ROC of the Z-transform of $a^{n_1} b^{n_2} u(-n_1, -n_2)$.

TABLE 3.4 Properties of the 2-D Z-Transform

Property	Time Domain	Z-domain
1. Linearity	$ax_1(n_1,n_2)+bx_2(n_1,n_2)$	$aX_1(z_1,z_2)+bX_2(z_1,z_2)$
2. Time shifting	$x(n_1-N_1,n_2-N_2)$	$z_1^{-N_1}z_2^{-N_2}X(z_1,z_2)$
3. Differentiation	$-n_1x(n_1,n_2)$	$\frac{\partial X(z_1,z_2)}{\partial z_1}\cdot z_1$
	$-n_2x(n_1,n_2)$	$\frac{\partial X(z_1,z_2)}{\partial z_2}\cdot z_2$
4. Time reversal	$x(-n_1,-n_2)$	$X(z_1^{-1},z_2^{-1})$
5. Convolution	$x(n_1,n_2)*h(n_1,n_2)$	$X(z_1,z_2)H(z_1,z_2)$
6. Separability	$x(n_1,n_2)=x_1(n_1)x_2(n_2)$	$X_1(z_1)X_2(z_2)$
7. Complex conjugation	$x^*(n_1,n_2)$	$[X(z_1^*,z_2^*)]^*$

Proof. From the definition of the 2-D Z-transform,

$$
\begin{aligned}
Z\{x(n_1,n_2)\} &= \sum_{n_1=-\infty}^{\infty}\sum_{n_2=-\infty}^{\infty} x(n_1,n_2)z_1^{-n_1}z_2^{-n_2} \\
&= \sum_{n_1=-\infty}^{\infty}\sum_{n_2=-\infty}^{\infty} x_1(n_1)x_2(n_2)z_1^{-n_1}z_2^{-n_2} \\
&= \sum_{n_1=-\infty}^{\infty} x_1(n_1)z_1^{-n_1}\sum_{n_2=-\infty}^{\infty} x_2(n_2)z_2^{-n_2} \\
&= X_1(z_1)X_2(z_2).
\end{aligned}
$$

■

Stability

In this section, we will consider the stability of 2-D IIR digital filters with the difference equation expressed as

$$
\sum_{i=0}^{N_1}\sum_{j=0}^{N_2} a_{ij}y(n_1-i,n_2-j)=\sum_{i=0}^{M_1}\sum_{j=0}^{M_2} b_{ij}x(n_1-i,n_2-j), \quad a_{00}=1. \tag{3.92}
$$

By taking the Z-transform of both sides with zero initial conditions, we have the transfer function

$$
H(z_1,z_2)=\frac{\sum_{i=0}^{M_1}\sum_{j=0}^{M_2} b_{ij}z_1^{-i}z_2^{-j}}{\sum_{i=0}^{N_1}\sum_{j=0}^{N_2} a_{ij}z_1^{-i}z_2^{-j}}, \quad a_{00}=1 \tag{3.93}
$$

$$
\triangleq \frac{B(z_1,z_2)}{A(z_1,z_2)}. \tag{3.94}
$$

The impulse response of this filter will be denoted by $h(n_1, n_2)$. Thus, we are considering the stability of 2-D rational polynomials. The theory of stability of such 2-D systems is considerably more difficult in general than its 1-D counterpart. In order to explain why, the following definitions are needed for the rational polynomial $H(z_1, z_2) = \frac{B(z_1,z_2)}{A(z_1,z_2)}$.

Definition 3.2 *A root of $B(z_1, z_2)$, which is not simultaneously a root of $A(z_1, z_2)$, is called a nonessential singularity of the first kind (NSFK).*

Definition 3.3 *When the sets of roots for $A(z_1, z_2)$ and $B(z_1, z_2)$ intersect, each common root is called a nonessential singularity of the second kind (NSSK).*

In 1-D, the numerator and denominator polynomials can always be factored, and the common factors can be removed so that we do not have any NSSK. However, 2-D polynomials cannot always be factored, and we may have NSSK. This means that the root of numerator polynomial $B(z_1, z_2)$ may be the same as that in the denominator. This may affect the stability of the overall system. The presence of NSSK complicates the stability analysis of 2-D systems. A comprehensive analysis is beyond the scope of this book. In order to simplify the analysis, we will consider systems with first-quadrant support; that is, the impulse response has all its nonzero values in the first quadrant $n_1 \geq 0$, $n_2 \geq 0$. This means that we are dealing with causal systems.

In Chapter 1 we proved that a 1-D system is stable if and only if its impulse response is absolutely summable. A straightforward extension of this gives us the following result. The proof follows along the same lines as given in Chapter 1 and is left as an exercise.

Theorem 3.5 *A 2-D LSI system is BIBO stable if and only if its impulse response is absolutely summable; that is,*

$$\sum_{n_1=-\infty}^{\infty} \sum_{n_2=-\infty}^{\infty} |h(n_1, n_2)| < \infty. \tag{3.95}$$

This theorem deals with only the impulse response and is therefore similar to 1-D. However, when we consider rational polynomials, we may have NSSK to complicate matters. Now consider the following example.

EXAMPLE 3.16 *Find the zeros and poles of the following: (a) $H(z_1, z_2) = \frac{z_1 z_2}{z_1 z_2 - 1}$, (b) $H(z_1, z_2) = \frac{(1-z_1^{-1})(1-z_2^{-1})}{2-z_1^{-1}-z_2^{-1}}$.*

SOLUTION (a) The zeros are the points in the sets

$$B_1 = \{(z_1, z_2) : z_1 = 0, \quad z_2 \text{ arbitrary}\}$$

$$B_2 = \{(z_1, z_2) : z_2 = 0, \quad z_1 \text{ arbitrary}\}.$$

The poles are the points in the set:

$$A = \{(z_1, z_2) : z_1 z_2 = 1\}.$$

Clearly, this function has an infinite number of poles and zeros.

(b) The zeros are the points in the sets:

$$B_1 = \{(z_1, z_2) : z_1 = 1, \quad z_2 \text{ arbitrary}\}$$

$$B_2 = \{(z_1, z_2) : z_2 = 1, \quad z_1 \text{ arbitrary}\}.$$

The poles are the points in the set

$$A = \{(z_1, z_2) : z_1^{-1} + z_2^{-1} = 2\}.$$

The poles and zeros are again infinitely many. In this case, note that the point $(z_1, z_2) = (1, 1)$ is both a zero and a pole. This point is therefore a NSSK.

Now we will discuss the stability of 2-D filters in the Z domain. A few theorems will be presented for this purpose. Since some of the proofs are quite involved and the proofs do not provide further insight into our intended subject matter, the proofs are omitted. Interested readers may look up the references cited. We know that a causal 1-D filter described by $H(z) = B(z)/A(z)$ is BIBO stable if and only if all the poles are inside the unit circle. In other words, the filter is stable if and only if there are no poles on or outside the unit circle; that is, $A(z) \neq 0$, $|z| \geq 1$. Shanks has presented a 2-D version of this theorem [5]. ◀

Theorem 3.6 *(Shanks) A 2-D digital filter described by $H(z_1, z_2) = 1/A(z_1, z_2)$ is BIBO stable if and only if*

$$A(z_1, z_2) \neq 0, \quad |z_1| \geq 1, \quad |z_2| \geq 1.$$

This theorem states that the necessary and sufficient condition for the filter $H(z_1, z_2) = 1/A(z_1, z_2)$ to be stable is that there are no poles on or outside the *unit bidisk* defined by $|z_1| = |z_2| = 1$. Note the difference between this theorem and its 1-D counterpart. In the 2-D case, we have used $B(z_1, z_2) = 1$. This was needed in order to avoid NSSK. A more general theorem is given by Goodman [6]. The following definitions will be useful.

$$S_c = \{(z_1, z_2) : |z_1| \geq 1, \quad |z_2| \geq 1\} \tag{3.96}$$

$$S_o = \{(z_1, z_2) : |z_1| > 1, \quad |z_2| > 1\} \tag{3.97}$$

$$T = \{(z_1, z_2) : |z_1| = 1, \quad |z_2| = 1\}. \tag{3.98}$$

Theorem 3.7 *(Goodman) Consider a 2-D digital filter described by $H(z_1, z_2) = B(z_1, z_2)/A(z_1, z_2)$. Then*

(1) $\quad A(z_1, z_2) \neq 0$ *in* $S_c \Longrightarrow$ *Stable*

(2) $\quad$ *Stable* $\Longrightarrow A(z_1, z_2) \neq 0$ *in* S_o.

In other words, the conditions can be written as follows:

1. *Sufficient condition:* If there are no poles on or outside the unit bidisk, then the system is stable.
2. *Necessary condition:* If the system is stable, then there are no poles outside the unit bidisk.

From the above we observe that unlike in 1-D, a 2-D system with pole(s) on the unit bidisk may be stable. In this case, the poles on the unit bidisk are really NSSK. So, the presence of the numerator in the transfer function has the effect of relaxing the necessary condition. For practical purposes it is not so crucial. So, we will assume that the transfer function does not have any NSSK on the unit bidisk. The above theorems are theoretically very elegant. However, they are difficult to apply in practice to test the stability of a given filter. This is because we have two variables, z_1 and z_2, each of which has a real and an imaginary part. This leads to performing a 4-D search to find whether or not all the roots are inside the unit bidisk. We will now present some theorems useful in the development of an algorithm for testing the stability of a given digital filter. The first one is credited to Huang [7].

Theorem 3.8 *(Huang) A 2-D causal digital filter $H(z_1, z_2) = B(z_1, z_2)/A(z_1, z_2)$ with no NSSK on the unit bidisk is stable if and only if either of the following sets of conditions is satisfied:*

Set I

$$\text{I. 1.} \quad A(z_1, \infty) \neq 0 \; for \; |z_1| \geq 1$$

$$\text{I. 2.} \quad A(z_1, z_2) \neq 0 \; for \; |z_1| = 1 \; and \; |z_2| \geq 1.$$

Set II

$$\text{II. 1.} \quad A(\infty, z_2) \neq 0 \; for \; |z_2| \geq 1$$

$$\text{II. 2.} \quad A(z_1, z_2) \neq 0 \; for \; |z_2| = 1 \; and \; |z_1| \geq 1.$$

Note that $A(z_1, z_2)$ is a polynomial involving the powers of z_1^{-1} and z_2^{-1}. So, in the above theorem when z_1 or z_2 is set to ∞, the inverse power simply goes to zero. Consider an example now to see how Theorem 3.8 can be applied.

EXAMPLE 3.17 *Let*

$$A(z_1, z_2) = 1 - 0.5z_1^{-1} + 2z_2^{-1} + z_1^{-1}z_2^{-1}.$$

To test for condition I.1, we substitute $z_2 = \infty$, or $z_2^{-1} = 0$, to get

$$A(z_1, \infty) = 1 - 0.5z_1^{-1}.$$

The root of the above polynomial is $z_1 = 0.5$, and condition I.1 is satisfied. Now to test for condition I.2, we write $z_1^{-1} = e^{-j\omega_1}$, to get

$$A(e^{-j\omega_1}, z_2) = 1 - 0.5e^{-j\omega_1} + 2z_2^{-1} + e^{-j\omega_1}z_2^{-1}.$$

For some chosen value of ω_1, the above becomes a polynomial in z_2 only. Condition I.2 states that for all ω_1 in the above, there should be no roots for z_2 outside the unit circle. How do we check for all ω_1? Since $e^{-j\omega_1}$ is periodic with period 2π, the simplest way to do this is to sample ω_1 finely between 0 and 2π and for each value of ω_1 solve for the roots of the resulting polynomial to check if they are all within the unit circle. This condition (or condition II.2) is obviously a computationally intensive test. If any root for any chosen value of ω_1 is on or outside the unit circle, then the system is unstable. Otherwise it is stable. In this example, notice that condition I.1 was satisfied, and therefore we must perform test condition I.2. If condition I.1 had failed, then we could declare the system to be unstable and stop. In order to potentially avoid the computationally intensive condition I.2 (or II.2), it is worthwhile to test condition II.1 in addition to condition I.1 first. If either of these fails, then the system is unstable and we can stop. If not, then we can test for condition I.2 or II.2. If we apply condition II.1 to the above example, we get

$$A(\infty, z_2) = 1 + 2z_2^{-1}.$$

Solving the above gives $z_2 = -2$, and condition II.1 fails. Therefore, the system is unstable, and there is no need to perform the computationally intensive test of condition I.2 or II.2. Based on Huang's theorem and the above discussion, we have the following algorithm for stability testing.

Algorithm (Based on Huang's theorem)

1. Test condition I.1. If it fails, system is unstable. Otherwise go to next step.
2. Test condition II.1. If it fails, system is unstable. Otherwise go to next step.
3. Test condition I.2 or II.2. If it fails, system is unstable. Otherwise the system is stable.

Efficient Computation of Condition I.2 or II.2

From the foregoing discussion we know that the testing of these conditions requires finding the roots of several I-D polynomials, one for each value of ω_1 or ω_2. Let us consider condition I.2 for discussion. From (3.93), we have

$$A(z_1, z_2) = \sum_{i=0}^{N_1}\sum_{j=0}^{N_2} a_{ij} z_1^{-i} z_2^{-j}, \quad a_{00} = 1$$

where a_{ij} are the coefficients of the recursive part of the filter. The above can be rewritten as

$$A(z_1, z_2) = \sum_{n_1=0}^{N_1}\sum_{n_2=0}^{N_2} a(n_1, n_2) z_1^{-n_1} z_2^{-n_2}, \quad a(0,0) = 1 \tag{3.99}$$

where $a(n_1 n_2)$ are now the coefficients of the recursive part of the filter. These coefficients can be represented as a matrix as shown in Fig. 3.24. For condition I.2, we have

$$A(e^{j\omega_1}, z_2) = \sum_{n_1=0}^{N_1}\sum_{n_2=0}^{N_2} a(n_1, n_2) e^{-j\omega_1 n_1} z_2^{-n_2}. \tag{3.100}$$

Now we sample ω_1 from 0 to 2π. That is, we substitute $\omega_1 = \frac{2\pi}{N}k$, where $N+1$ is the number of points around the circle and $k = 0, 1, \ldots, N$. Equation (3.100) can then be simplified as

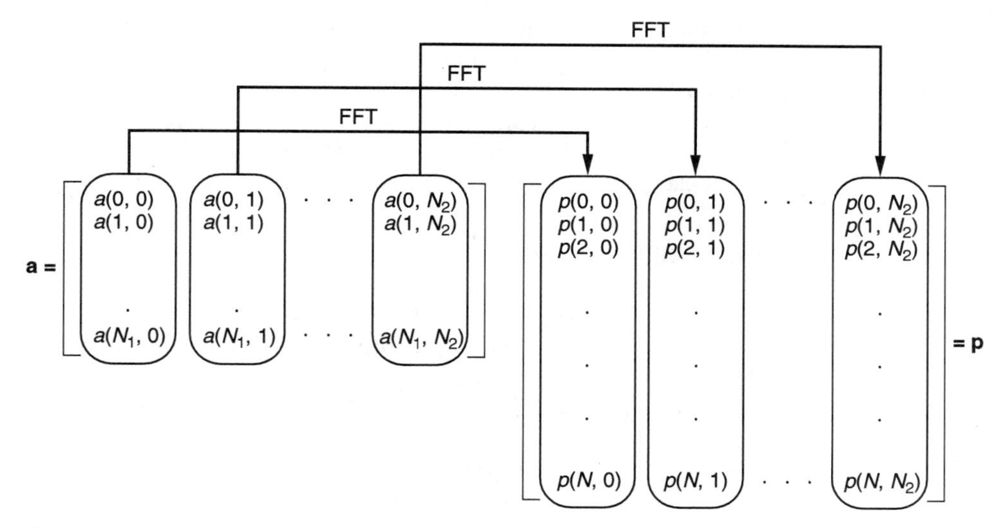

Figure 3.24 Efficient computation of condition I.2 in Huang's theorem.

$$
\begin{aligned}
A(e^{-j\omega_1}, z_2) &= \sum_{n_1=0}^{N_1}\sum_{n_2=0}^{N_2} a(n_1,n_2)e^{-j\frac{2\pi}{N}kn_1}z_2^{-n_2} \\
&= \sum_{n_2=0}^{N_2}\left(\sum_{n_1=0}^{N_1} a(n_1,n_2)e^{-j\frac{2\pi}{N}kn_1}\right)z_2^{-n_2} \\
&= \sum_{n_2=0}^{N_2} p(k,n_2)z_2^{-n_2}, \quad 0 \le k \le N. \qquad (3.101)
\end{aligned}
$$

In (3.101) we have used the definition

$$
p(k,n_2) = \sum_{n_1=0}^{N_1} a(n_1,n_2)e^{-j\frac{2\pi}{N}kn_1}. \qquad (3.102)
$$

The reason for the arguments of $p(k, n_2)$ is that there is a summation on n_2 before that and for every fixed value of n_2, we perform the summation of (3.102) for $k = 0, 1, \ldots, N$. Note that for some fixed n_2 equation (3.102) is nothing but an N-point 1-D DFT of a vector of coefficients $[a(0,n_2), a(1,n_2), \ldots, a(N_1,n_2)]^T$. That is, we perform an N-point DFT of the n_2th column of the coefficient matrix. The length of this column is N_1. Typically, we want a fine grid along the circle, and so $N \gg N_1$. Therefore, in order to perform the DFT, we need to pad the coefficient vector with zeros so that its length is equal to $N+1$. For each value of n_2 in the range $0 \le n_2 \le N_2$, we perform an N-point DFT and get a vector of length N. That is, we perform an N-point DFT of each column of the coefficient matrix. So, $p(k, n_2)$ is a matrix of size $(N+1) \times N_2$ as shown in Fig. 3.24. This matrix is now substituted in equation (3.101). From this equation we see that the elements in each row of the matrix $p(k, n_2)$ form the coefficients of a polynomial in z_2^{-1}. Each of these polynomials is then solved for its roots to check whether they are all inside the unit circle. Therefore, we have to find N_2 roots each for a total of N polynomials. The computational complexity of this algorithm can be reduced significantly by performing FFT instead of DFT of each row of the coefficient matrix.

More Stability Results

The stability testing algorithm we have described is based on reducing the 2-D polynomial to 1-D and then finding the roots of 1-D polynomials. The results described in this section are not useful in simplifying the algorithm. However, they are theoretically interesting and are useful in quickly testing the stability of simple 2-D polynomials. The first result is credited to Strintzis [8].

Theorem 3.9 *A 2-D causal digital filter $H(z_1, z_2) = B(z_1, z_2)/A(z_1, z_2)$ with no NSSK on the unit bidisk is stable if and only if*

1. $A(z_1, 1) \neq 0$ *for* $|z_1| \geq 1$, and
2. $A(1, z_2) \neq 0$ *for* $|z_2| \geq 1$, and
3. $A(z_1, z_2) \neq 0$ *for* $|z_1| = |z_2| = 1$.

At this time it is useful to compare the last condition of this theorem with that of condition I.2 or II.2 of Huang's theorem. In I.2 of Huang's theorem, we have $|z_1| = 1$ and $|z_2| \geq 1$. The magnitude of z_1 is restricted, but the magnitude and phase of z_2 are both unrestricted. This corresponds to a 3-D search. In Strintzis' theorem, the magnitude of both z_1 and z_2 is restricted to unity. This corresponds to a 2-D search and is therefore theoretically simpler than Huang's theorem. But in practice it is easier to perform 1-D stability tests or root solving of 1-D polynomials. In this respect, Huang's theorem turns out to be simpler. However, the Strintzis theorem proves to be useful in quickly checking the stability of simple 2-D polynomials without resorting to root solvers. This is illustrated by the following examples.

▶ **EXAMPLE 3.18** *Let $A(z_1, z_2) = 1 - 0.1z_1^{-1} - 0.2z_1^{-1}z_2^{-1}$. Now we apply each condition of Theorem 3.9.*

Condition 1: $A(z_1, 1) = 1 - 0.3z_1^{-1}$. The root is $z_1 = 0.3$, which is less than unity in magnitude. Test passes!

Condition 2: $A(1, z_2) = 0.9 - 0.2z_2^{-1}$. The root is $z_2 = \frac{2}{9}$, which is less than unity in magnitude. Test passes!

Condition 3: $A(z_1, z_2)\big|_{|z_1|=|z_2|=1} = A(e^{j\omega_1}, e^{j\omega_2}) = 1 - 0.1e^{-j\omega_1} - 0.2e^{-j\omega_1}e^{-j\omega_2}$. Taking the absolute value of both sides gives

$$
\begin{aligned}
|A(e^{j\omega_1}, e^{j\omega_2})| &= |1 - 0.1e^{-j\omega_1} - 0.2e^{-j\omega_1}e^{-j\omega_2}| \\
&\geq 1 - 0.1 - 0.2 \\
&= 0.7.
\end{aligned}
$$

Thus, the absolute value is greater than or equal to 0.7, which means that $A(e^{j\omega_1}, e^{j\omega_2})$ cannot be equal to zero. Therefore, this test also passes, and the system is stable.

EXAMPLE 3.19 *Let* $A(z_1, z_2) = 1 - 0.3z_1^{-1} - 0.5z_2^{-1} + 0.22z_1^{-1}z_2^{-1}$. *Now we apply each condition of Theorem 3.9.*

Condition 1: $A(z_1, 1) = 0.5 - 0.08z_1^{-1}$. The root is $z_1 = 0.16$, which is less than unity in magnitude. Test passes!

Condition 2: $A(1, z_2) = 0.7 - 0.28z_2^{-1}$. The root is $z_2 = 0.4$, which is less than unity in magnitude. Test passes!

Condition 3:

$$\begin{aligned} |A(e^{j\omega_1}, e^{j\omega_2})| &= |1 - 0.3e^{-j\omega_1} - 0.5e^{-j\omega_2} + 0.22e^{-j\omega_1}e^{-j\omega_2}| \\ &\geq 1 - 0.3 - 0.5 - 0.22 \\ &= -0.02. \end{aligned}$$

The above does not provide any conclusion to whether or not $A(e^{j\omega_1}, e^{-j\omega_2})$ can be equal to zero. Therefore, in this case we have to resort to the algorithm discussed earlier to come to a conclusion about its stability. ■

A theorem similar to theorem 3.9 is given next.

Theorem 3.10 *(DeCarlo, Murray, and Saeks [9]) A 2-D causal digital filter* $H(z_1, z_2) = B(z_1, z_2)/A(z_1, z_2)$ *with no NSSK on the unit bidisk is stable if and only if*

1. $A(z_1, z_1) \neq 0$ *for* $|z_1| \geq 1$, and
2. $A(z_1, z_2) \neq 0$ *for* $|z_1| = |z_2| = 1$.

The first condition of this theorem differs slightly from that of the Strintzis theorem. This condition may therefore be used to find unstable systems.

3.11 SUMMARY

In this chapter, we first presented the bilateral Z-transform and its properties. The concept of region of convergence (ROC) was also discussed. It is important to specify the ROC for the Z-transform of any discrete-time sequence. The concepts of causality and stability were discussed in relation to the ROC. The formula for the inverse Z-transform was then derived, and it involves a contour integration. The partial fraction expansion method is the simplest way to find the inverse Z-transform of given functions. This method has been presented in detail. The unilateral or one-sided Z-transform was then presented. It is useful for the analysis of linear systems with a finite number of initial conditions. Jury [10] has written a classic text on Z-transform and its applications on linear systems.

The concept of linear phase and group delay for digital filters was subsequently presented. Linear phase filters are important because they cause the same time delay for every frequency and hence do not distort the signal. Four different types of FIR filters with symmetrical/antisymmetrical coefficients were defined and shown to have linear phase and/or constant group delay. These filter definitions are essentially the same as those given in the text by Oppenheim and Schafer [11]. The subsequent section dealt with the state-space analysis. The state-space model of an LSIV system was given. The similarity transformation was derived. The matrices in the state-space model of a digital filter actually define the structure of the filter. That is, different structures (Direct I, II, cascade, parallel, etc.) for the same filter yield different matrices in the state-space model, but they all have the same transfer function. The state equation can be solved to find an equation for each state variable of the system.

Linear filtering techniques using the DFT/FFT were discussed. The overlap-add and overlap-save methods were described in detail. These methods are used for linear filtering at a low computational cost. In addition, the DFT can be computed using linear filtering methods. The Goertzel algorithm was presented and shown to have a computational advantage over the FFT when only a few bins of DFT are needed. The Chirp Z-transform is a more general method than the Z-transform. It was shown that the Chirp Z-transform can be computed using a linear filtering method in order to save computations.

Some of the 1-D concepts were generalized to 2-D in the last section of the chapter. First, the 2-D Z-transform and its properties were discussed. The theory of stability of 2-D rational polynomials was then introduced. It was shown that NSSK can affect the stability of a given IIR filter. That is, a filter with poles on the unit bidisk can be stabilized by appropriate NSSK on the unit bidisk. This is an advanced topic, and a detailed analysis is beyond the scope of this book. We have assumed that the filters do not have NSSK on the unit bidisk and thereby developed an algorithm based on Huang's theorem. This is a simple but practical algorithm, and its implementation is assigned as a computer project. Considerable work on the subject of stability theory of 2-D rational polynomials can be credited to Shanks, et al. [5], Goodman [6], Huang [7], [15], Strintzis [8], DeCarlo et al. [9], Jury [12], Bose [13], Siljak [14], and many others. The topic of 2-D state-space analysis has been omitted here. The most popular state-space models are called the Roesser model [16] and the Fornasini-Marchesini model [17]. The theory of stability of 2-D systems can be addressed in terms of these state-space models. References on this topic include the work of Lu and Lee [18], [19], Anderson et al. [21], El-Agizi and Fahmy [20], and Bose et al. [22], [23], among many others.

▶ 3.12 MATLAB COMMANDS

The following MATLAB commands are relevant to this chapter and may be useful for the computer projects that follow. The transfer function of a 1-D digital filter is typically written as

$$H(z) = \frac{b_0 + b_1 z^{-1} + b_2 z^{-2} + \cdots + b_M z^{-M}}{1 + a_1 z^{-1} + a_2 z^{-2} + \cdots + a_N z^{-N}}. \tag{3.103}$$

The numerator and denominator are represented as the following row vectors:

$$\text{num} = \begin{bmatrix} b_0 & b_1 & \ldots & b_M \end{bmatrix}$$

$$\text{den} = \begin{bmatrix} 1 & a_1 & \ldots & b_N \end{bmatrix}$$

roots(den); finds the roots of the polynomial described by the row **den**.

poly (v); The argument **v** is a row containing the roots of a polynomial (possibly complex). This command returns a row that describes the polynomial, which has the roots in **v; roots** and **poly** are inverse functions of each other.

zplane(Z,P); plots the zeros and poles supplied in column vectors **Z** and **P** with the unit circle for reference.

zplane(num,den); plots the zeros and poles of the transfer function represented by the rows in **num** and **den**.

[R, P, K]= residue(B,A); finds the residues, poles, and direct term of a partial fraction expansion of the ratio of two polynomials $B(z)/A(z)$, where rows B and A specify the coefficients of the numerator and denominator in descending powers of z.

If there are no multiple roots,

$$\frac{B(z)}{A(z)} = \frac{R(1)}{z - P(1)} + \frac{R(2)}{z - P(2)} + \cdots + \frac{R(n)}{z - P(n)} + K(z).$$

The residues are returned in column vector **R**, the pole locations in column vector **P**, and the direct terms in row vector **K**. The direct term coefficient vector **K** is empty if length(**B**) < length(**A**); otherwise length(**K**) = length(**B**) − length(**A**)+1.

If P (j) = ... = P(j + m − 1) is a pole of multiplicity m, then the expansion includes terms of the form

$$\frac{R(j)}{z - P(j)} + \frac{R(j+1)}{(z - P(j))^2} + \cdots + \frac{R(j+m-1)}{(z - P(j))^m}.$$

As an example, consider finding the impulse response of a causal filter with transfer function

$$H(z) = \frac{z(z+5)}{z^2 + 3z + 2}.$$

Recall from the section on partial fractions that we perform a partial fraction of $\frac{H(z)}{z}$ so that the resulting $H(z)$ is in a form so that the table of Z-transforms can be used. Thus, we have

$$\frac{H(z)}{z} = \frac{z+5}{z^2 + 3z + 2}$$

where we have **B** = [1 5] and **A** = [1 3 2]. The result of running the command **[R,P,K]** = **residue (B,A)** returns the following: $\mathbf{R} = [-3\ 4]^T$, $\mathbf{P} = [-2\ -1]^T$, and **K** = []. That is, we have the poles at $z = -2$ and $z = -1$, and the residues are -3 and 4, respectively. The partial fraction expansion is

$$\frac{H(z)}{z} = \frac{-3}{z+2} + \frac{4}{z+1}$$

which implies

$$H(z) = \frac{-3z}{z+2} + \frac{4z}{z+1}.$$

The impulse response is then easily found as $h(n) = [-3(-2)^n + 4(-1)^n]u(n)$.

h = impz(num,den,N); calculates the impulse response of the digital filter transfer function represented by **num** and **den**. N points of the impulse response are stored in vector **h**.

[A,B,C,D]=tf2ss(num,den); converts from transfer function to state-space model for the Direct Form II structure. The **num** and **den** are as given in (3.103), but **num** must be padded with zeros to make it the same length as **den**.

[num,den]=ss2tf(A,B,C,D); converts from state-space to transfer function. The resulting **num** and **den** represent the transfer function as given by (3.103).

3.13 PROBLEMS

1. Using the definition, find the Z-transform for the following sequences. Find the ROC in each case and sketch it.

(a) $x_1(n) = \delta(n + N)$;
(b) $x_2(n) = (\frac{1}{3})^n u(n)$;
(c) $x_3(n) = u(N - n)u(n - M)$, where M and N are positive integers, and $N > M$;
(d) $x_4(n) = 2^n u(-n - 1)$;
(e) $x_5(n) = e^n u(2 - n) + (1/e)^n u(n - 5)$;
(f) $x_6(n) = n[u(n) - u(n - N)],\ N > 0$;
(g) $x_7(n) = n^2 u(n - 1)$.

2. Find the Z-transform and the ROC for the following sequences:

(a) $x_1(n) = n\alpha^n \cos(\omega_0 n)u(n)$;
(b) $x_2(n) = n^2\alpha^n u(n)$;
(c) $x_3(n) = \alpha^n \cos(\omega_0 n + \beta)u(n)$;
(d) $x_4(n) = n(-1)^n u)(n)$.

3. Find the Z-transform and ROC for the sequence $x(n) = \beta^{|n|}$.

4. Determine the Z-transform and ROC of the following sequences. Also determine whether or not the Fourier transform of the sequence exists.

(a) $x_1(n) = \left(\frac{1}{4}\right)^n \{u(n+3) - u(n-4)\}$;
(b) $x_2(n) = n\left(\frac{1}{4}\right)^{|n|}$;
(c) $x_3(n) = \left(\frac{1}{2}\right)^n \cos(\frac{\pi}{3}n + \frac{\pi}{6})u(n)$.

5. Find the Z-transform and the ROC for the following sequences:

(a) $x_1(n) = [\mathbf{1},\ 0,\ -1,\ 1,\ 0,\ -1,\ \ldots]$;
(b) $x_2(n) = n(-1)^n \sin(\omega_0 n)u(n)$;
(c) $x_3(n) = \begin{cases} n, & 0 \le n \le 10 \\ 20 - n, & 11 \le n \le 20 \\ 0, & \text{otherwise.} \end{cases}$

6. Sketch the ROC of the Z-transform of left-sided, right-sided, and two-sided sequences.

7. Determine the Z-transform of the following:

(a) $x_1(n) = (-1)^{n+1}\frac{a^n}{n}u(n - 1)$;
(b) $x_2(n) = u(-n - 1) + (\frac{1}{2})^n u(n)$;
(c) $x_3(n) = a^n u(n) + b^n u(n) + c^n u(-n - 1),\ |a| < |b| < |c|$.

8. Prove Property 5 (Time Reversal).

9. A sequence $x(n)$ is even with Z-transform $X(z)$.

(a) Show that $X(z) = X(z^{-1})$.
(b) What can you say about the poles and zeros of $X(z)$?

10. Prove Property 6 (Convolution in time).

11. Prove Property 7 (Complex conjugation).

12. Prove Property 8 (Multiplication in time).

13. Prove Property 9 (Parseval's theorem).

14. Prove the following properties of the Z-transform of a causal sequence.

(a) Initial Value Theorem:

$$\lim_{z \to \infty} X(z) = x(0).$$

(b) Final Value Theorem:

$$\lim_{z\to 1}[(z-1)X(z)] = \lim_{n\to\infty} x(n).$$

15. Formulate and prove the initial and final value theorems for anticausal signals.

16. If $X(z)$ is the Z-transform of $x(n)$, prove the following:

(a) $\mathrm{Re}\,\{x(n)\} \longleftrightarrow \frac{1}{2}\{X(z) + X^*(z^*)\}$.

(b) $\mathrm{Im}\,\{x(n)\} \longleftrightarrow \frac{1}{2j}\{X(z) - X^*(z^*)\}$.

(c) If $x(n)$ is real and z_0 is a zero of $X(z)$, then z_0^* is also a zero of $X(z)$.

17. The autocorrelation function of a sequence $x(n)$ is defined by

$$r(n) = \sum_{k=-\infty}^{\infty} x(k)x(n+k).$$

Show that the Z-transform of this function is

$$R(z) = X(z)X(z^{-1}).$$

Find the region of convergence for $R(z)$.

18. Compute the convolution of the following sequences using the Z-transform approach: $x(n) = (\frac{1}{2})^n u(n)$, $\quad h(n) = u(n) - u(n-N), N > 0$.

19. Use the residue method to find the inverse Z-transform of

$$H(z) = \frac{z}{(z-a)^2},\ |z| > |a|.$$

20. Use the residue method to find the inverse Z-transform of

(a) $H_1(z) = \dfrac{1}{1-z^{-3}}, \quad |z| > 1.$

(b) $H_2(z) = \dfrac{1}{1-z^{-2}}, \quad |z| > 1.$

21. Find the inverse Z-transform of the following using partial fractions:

(a) $H_1(z) = \dfrac{1}{(1-\frac{1}{3}z^{-1})(1+\frac{1}{4}z^{-1})}, \quad |z| > 1.$

(b) $H_2(z) = \dfrac{1}{1-z^{-2}}, \quad |z| > 1.$

22. Find the inverse Z-transform of the following.

(a) $X_1(z) = \dfrac{z}{z-2}, \quad |z| > 2.$

(b) $X_2(z) = \dfrac{z+1}{z+2}, \quad |z| < 2.$

(c) $X_3(z) = \dfrac{z(z+\frac{1}{2})}{z^2+\frac{3}{4}z+\frac{1}{8}}, \quad \frac{1}{4} < |z| < \frac{1}{2}.$

(d) $X_4(z) = \dfrac{z+4}{z^2+z+1}, \quad$ ROC appropriate for a causal sequence.

(e) $X_5(z) = \dfrac{z+\frac{1}{4}}{z^2+\frac{1}{2}}, \quad |z| > \dfrac{1}{\sqrt{2}}.$

23. Determine the inverse Z-transform of the following:

$$X(z) = (z+1)(1+z^{-1}+z^{-2})(1+3z^2).$$

24. Find all possible sequences associated with the Z-transform:

$$X(z) = \frac{7z}{(z+\frac{1}{4})(5-z^{-2})}.$$

25. Find the causal sequence whose Z-transform is given by the following:

(a) $X_1(z) = \dfrac{z+1}{(z+\frac{1}{2})^2(z+3)(z+4)}.$

(b) $X_2(z) = \dfrac{z-a}{1-az^{-1}}.$

(c) $X_3(z) = \dfrac{z+1}{(z-2)^3}.$

(d) $X_4(z) = \cos(z^{-1}).$

(e) $X_5(z) = \ln(1-3z^{-1}).$

26. Determine the inverse Z-transform of the following using the method of residues and also partial fractions. Assume that the sequence is right-sided.

$$H(z) = \frac{z-1}{(z-0.1)(z-0.2)}.$$

27. Determine the sequence whose Z-transform is given by

$$X(z) = e^{z}.$$

28. Given a Linear Shift Invariant (LSIV) causal system with the transfer function

$$H(z) = \frac{z+2}{z+\frac{3}{4}}.$$

(a) Find the impulse response of the system.

(b) If the input to the system is $x(n) = u(n+2) + (\frac{1}{2})^n u(n)$, find the output of the system for all n. Use the Z-transform method and check your answer using time-domain convolution.

(c) Is the system stable? Why?

29. Given a Linear Shift Invariant (LSIV) system with transfer function

$$H(z) = \frac{z(z+1)}{(z^2+\frac{9}{4}z+\frac{1}{2})}, \quad \frac{1}{4} < |z| < 2.$$

(a) Find the impulse response of the system.

(b) Is the impulse response absolutely summable? Why?

(c) Find the response of the system if the input is $x(n) = u(10-n)u(n+5)$. Use the Z-transform method and then check your answer using time-domain convolution.

30. Prove Property B for the one-sided Z-transform.

31. The step response of a filter is given by

$$y(n) = \left(\frac{1}{2}\right)^n u(n) + 4\left(\frac{1}{3}\right)^n u(n-1).$$

(a) Find the impulse response of the filter.
(b) Is the filter stable? Is it causal? Give reasons.

32. Use the one-sided Z-transform to solve the following difference equations:

(a) $y(n) + \frac{3}{4}y(n-1) + \frac{1}{8}y(n-2) = 0; \quad y(-1) = 1, \quad y(-2) = 0.$
(b) $y(n+2) + \frac{9}{4}y(n+1) + \frac{1}{2}y(n) = \delta(n); \quad y(-1) = -1, \quad y(0) = 1.$
(c) $y(n) + \frac{1}{5}y(n-1) = u(n); \quad y(-1) = 1.$

33. Use the one-sided Z-transform to find $y(n), n \geq 0$ for the following:

(a) $y(n) + \frac{1}{2}y(n-1) - \frac{1}{4}y(n-2) = 0; \quad y(-2) = y(-3) = 1.$
(b) $y(n) + \frac{1}{2}y(n-1) = x(n), \quad x(n) = (\frac{1}{4})^n u(n), \quad y(-1) = 2.$
(c) $y(n) - \frac{3}{2}y(n-1) + \frac{1}{2}y(n-2) = u(n), \quad y(-2) = 1, \quad y(-1) = 0.$

34. Determine the step response of the following system:

$$y(n) = -\frac{1}{2}y(n-1) + x(n), \quad y(-1) = 1.$$

35. The impulse response of a digital filter is given by

$$h(n) = 2^n u(-n-1) + (\frac{1}{3})^n u(n).$$

(a) Find the step response of the filter.
(b) Is the filter stable? Is it causal?

36. The pole-zero pattern of a filter is given in Fig. P3.1.

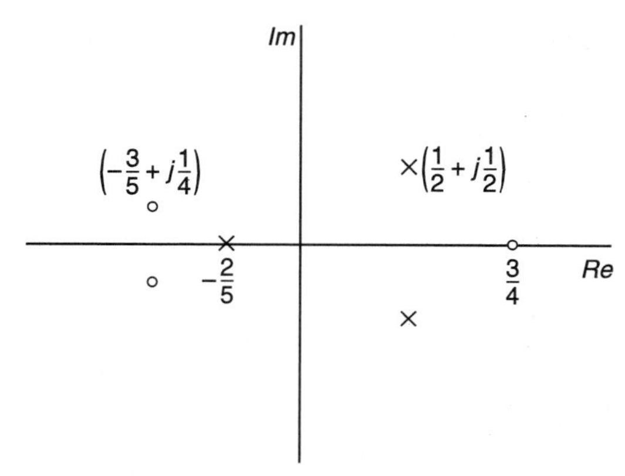

Figure P3.1

(a) If possible, determine the ROC for the system to be stable and causal.
(b) Find ROC for the system to be stable.
(c) Determine all possible ROCs (not necessarily stable or causal) for this filter.

37. Find the one-sided Z-transform for the following sequences:

(a) $\delta(n+2)$.
(b) $\delta(n-1)$.
(c) $\alpha^{n-N}u(n-N)$.
(d) $\alpha^n u(-n)$.
(e) $(\frac{1}{2})^n u(N-n)$.

38. The pole-zero plot and ROC for an LSIV filter is given in Fig. P3.2.

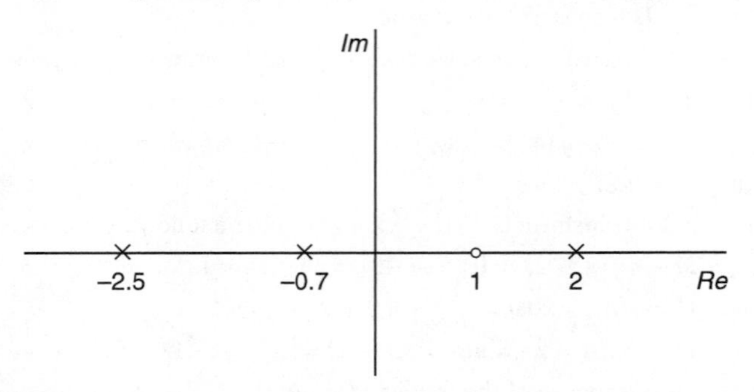

Figure P3.2

(a) Determine the ROC if the system is known to be stable.
(b) Is it possible for the system to be causal and stable? If so, find the corresponding ROC.
(c) What are all the different possible systems associated with this pole-zero plot?

39. Let $h_{LP}(n)$ denote the impulse response of a lowpass filter. Consider the following transformation:

$$h(n) = (-1)^n h_{LP}(n).$$

(a) What type of filter is represented by $h(n)$?
(b) Let the lowpass filter have generalized linear phase. If the transformed filter is required to be symmetric, which of the four types (I, II, III, or IV) must the lowpass filter be designed as? Give reasons.

40. An allpass filter is given by the transfer function

$$H(z) = \frac{z^{-1} - \alpha}{1 - \alpha z^{-1}}, \qquad \alpha \text{ real.}$$

(a) Prove that this system is indeed an allpass filter, that is, that it has a frequency response with unit magnitude.
(b) Find the difference equation and impulse response of the filter.
(c) Find the range of values for α so that the filter is stable. Assume causality.
(d) If a phase of ψ is needed at frequency ω_0, find an expression for α.

41. The state-space model for a causal LSIV filter is described by the following matrices:

$$\mathbf{A} = \begin{bmatrix} 0 & 1 \\ -0.1 & -0.2 \end{bmatrix}, \quad \mathbf{B} = \begin{bmatrix} 0 \\ 1 \end{bmatrix}, \quad \mathbf{C} = [1 \;\; 2], \quad D = 0.2.$$

(a) Find the transfer function of the filter.
(b) Plot the pole-zero diagram of the filter.
(c) Find the zero-state solution of the state equation when the input is a unit step.
(d) Find the zero-input solution of the state equation for initial condition $\mathbf{x}(\mathbf{0}) = [1 \quad -1]^T$.
(e) Sketch the flow diagram for the filter.

42. Find the step response of the state equation of a filter described by

$$\mathbf{v}(n+1) = \begin{bmatrix} 0.5 & -0.1 \\ 0.1 & 0.5 \end{bmatrix} \mathbf{v}(n) + \begin{bmatrix} 0 \\ 1 \end{bmatrix} x(n).$$

43. Find the solution of the following zero-input state equation:

$$\mathbf{v}(n+1) = \begin{bmatrix} 0 & 1 \\ -0.1 & -0.2 \end{bmatrix} \mathbf{v}(n), \quad \mathbf{v}(0) = \begin{bmatrix} 1 \\ -1 \end{bmatrix}.$$

44. The state-space model for a causal LSIV filter is described by the following matrices:

$$\mathbf{A} = \begin{bmatrix} 0.5 & -0.1 \\ 0.1 & 0.5 \end{bmatrix}, \quad \mathbf{B} = \begin{bmatrix} 0 \\ 1 \end{bmatrix}, \quad \mathbf{C} = [1 \ \ 2 \ ], \quad D = 0.2.$$

(a) Find the transfer function of the filter.
(b) Plot the pole-zero diagram of the filter.
(c) Find the zero-state solution of the state equation when the input is a unit step.
(d) Find the zero-input solution of the state equation for initial condition $\mathbf{x}(\mathbf{0}) = [1 \quad -1]^T$.
(e) Sketch the flow diagram for the filter. Comment on the structure of this filter in relation to Example 3.14. Comment on the number of multipliers and adders for each structure.

45. Determine the Direct Form II state-space representation $\{\mathbf{A}, \mathbf{B}, \mathbf{C}, D\}$ of the filter given the transfer function

$$H(z) = \frac{z(z+1)}{(z^2 + \frac{9}{4}z + \frac{1}{2})}.$$

46. Determine the Direct Form II state-space representation $\{\mathbf{A}, \mathbf{B}, \mathbf{C}, D\}$ of the filter given the transfer function

$$H(z) = \frac{z(z+1)}{(z^3 + \frac{1}{4}z^2 + \frac{9}{4}z + \frac{1}{2})}.$$

47. Prove that if a polynomial has real coefficients, then its complex roots occur in conjugate pairs. Show with an example that a polynomial with complex coefficients does not necessarily have its complex roots in conjugate pairs.

48. Find the transfer function $H(z)$ and the difference equation for the filter that uses the Goertzel algorithm to compute the DFT value $X(N-k)$.

49. We would like to implement linear convolution of a 15,000-point sequence with a 150-tap FIR filter. The convolution is to be implemented using DFTs and inverse DFTs of length 256.

(a) For the overlap-add method, find the number of DFTs and IDFTs required.
(b) For the overlap-save method, find the number of DFTs and IDFTs required.
(c) If FFTs and IFFTs are used, find the number of multiplications and additions required for (i) overlap-add, (ii) overlap-save, (iii) direct convolution.

50. Prove Theorem 3.5, which is a necessary and sufficient condition for the stability of a 2-D LSIV filter.

51. Determine the Z-transform and its ROC for each of the following sequences:

(a) $\delta(n_1 - 3, 1 - n_2)$.
(b) $(\frac{1}{4})^{n_1}(2)^{n_2}u(n_1 - 1, -n_2 + 3)$.

(c) $u_{2d}(2n_1 - n_2)u(n_1, n_2)$.

(d) $\alpha^{n_2}u(-n_1, n_2)u_{2d}(n_1 + 2n_2)$.

(e) $\cos(0.5\pi n)\delta(2n_1, n_2 - 1)$.

52. Prove the following properties of the 2-D Z-transform:

(a) Time shifting.
(b) Multiplication by a discrete exponential.
(c) Differentiation in the Z domain.

53. Prove the following properties of the 2-D Z-transform:

(a) Time reversal.
(b) Convolution in time.
(c) Parseval's theorem.

54. A separable 2-D system is given by $H(z_1, z_2) = H(z_1)H(z_2)$. Show how the stability of the 2-D system is related to the stability of the two 1-D systems $H(z_1)$ and $H(z_2)$.

55. The impulse response of a system is given by

$$h(n_1, n_2) = \alpha^{n_1}\beta^{n_2}u(n_1, n_2).$$

(a) Determine the condition for BIBO stability by imposing the absolute summability condition.
(b) Determine the ROC of $H(z_1, z_2)$, and using a property of the ROC confirm the condition of (a).

56. Determine the inverse Z-transform of each of the following sequences. Assume each of the systems to be stable.

(a) $H(z_1, z_2) = \dfrac{z_2}{1 + az_1^{-1}z_2^{-1}}$.

(b) $H(z_1, z_2) = \dfrac{1}{(1 - \alpha z_1^{-1})(1 + \beta z_2^{-1})}$.

(c) $H(z_1, z_2) = \dfrac{z_1^{-3}}{(1 - 4z_1^{-1}z_2^{-1})}$.

57. The step response of a 2-D LSI system is given by

$$y(n_1, n_2) = \left(\frac{1}{4}\right)^{n_2} u(n_1, n_2 - 2).$$

(a) Find the transfer function of the system, $H(z_1, z_2)$.
(b) Find the impulse response of the system.
(c) Comment on the stability of the system.

58. Find whether or not the following filters are stable.

(a) $H(z_1, z_2) = \dfrac{1}{1 - 0.9z_1^{-1} + 0.81z_1^{-1}z_2^{-1}}$.

(b) $H(z_1, z_2) = \dfrac{1}{1 + 0.9z_1^{-1} - 0.9z_2^{-1} - 0.81z_1^{-1}z_2^{-1}}$.

59. Show that the first-order filter given by

$$H(z_1, z_2) = \frac{1}{1 + az_1^{-1} + bz_2^{-1} + cz_1^{-1}z_2^{-1}}$$

is stable if and only if

$$|a + b| - 1 < c < 1 - |a - b|.$$

3.14 COMPUTER PROJECTS

Project 1

Filter implementation. Consider the digital filter with state-space description given by

$$\mathbf{A} = \begin{bmatrix} 0.5 & -0.1 \\ 0.1 & 0.5 \end{bmatrix}, \quad \mathbf{B} = \begin{bmatrix} 0 \\ 1 \end{bmatrix}, \quad \mathbf{C} = \begin{bmatrix} 1 & 2 \end{bmatrix}, \quad D = 0.2.$$

(a) Implement the state equation and the output equation in MATLAB. Find the state vector $\mathbf{x}(n)$ for about 100 samples of n with zero input and the initial condition $\mathbf{x}(\mathbf{0}) = [1 \quad -1]^T$. Plot $x_1(n)$ versus $x_2(n)$. This is called the phasor diagram. Also plot the output $y(n)$.

(b) Find the transfer function of the filter using an appropriate MATLAB command. Obtain a pole-zero plot of this filter.

(c) Find the Direct Form II state-space model for the filter from its transfer function. You may again use a MATLAB command to do so.

(d) Implement this model in MATLAB and repeat (a). Comment on the plots of the phasor diagram and the output for this model in relationship to those obtained in (a).

Project 2

Fast Convolution

Procedure

1. Design a 1024-tap FIR filter with a cutoff frequency of $\pi/2$ using the command: $\mathbf{h} = \mathbf{fir1}(\mathbf{1024}, \mathbf{0.5})$. Use the **freqz** command to plot its frequency response.
2. Generate a 500,000-point square-wave (**x**) with 20 samples per cycle. Plot its frequency response.
3. Use the **fftfilt** command to perform fast convolution of **x** and **h**. This command uses the overlap-add algorithm. Use a segment size of 2000 points. Using the MATLAB functions **tic** and **toc**, calculate the execution time for the overlap-add algorithm to process the entire input data. Plot a representative segment of the output. Plot the frequency response of the output.
4. Write a program (in C or MATLAB) to filter the square-wave data generated in (2) using the FFT-based overlap-save method. Plot a representative segment of the output. Using the MATLAB functions **tic** and **toc**, calculate the execution time for the overlap-save algorithm to process the entire input data.
5. Perform the filtering using the MATLAB **conv**( ) command which performs direct convolution. Compute the execution time for the direct convolution operation to process the entire input data.

Questions

1. Do the spectra plotted in (1), (2), and (3) above make sense? Why?
2. How do the execution times compare for the three different algorithms?

3. Let M denote the filter length and N the segment size, where $N > M$. Find the number of multiplications required for direct convolution, and FFT-based overlap-add and overlap-save methods. For the fast convolution methods, assume that Radix-2 FFT is used.
4. For the values of M and N used in this lab, find the actual number of multiplications for the three algorithms.

Project 3

Goertzel Algorithm

Procedure

1. Generate a 1024-point sequence of (a) a square-wave with 10 samples per cycle and (b) a triangular wave with five samples per cycle. (c) Plot the magnitude response of these sequences.
2. Write a program (in C or MATLAB) to find the N-point DFT of a signal using the Goertzel algorithm. The inputs of the program must be the input signal, the value of N, and a vector of desired frequency bins. The output of the program must consist of the vectors: (a) the values of the DFT at the specified frequency bins; and (b) the values of the DFT at the frequency bins $N - k$, where k is the desired frequency bin.
3. Using the program in (2), find the 1024-point DFT of the signals generated in (1) for the frequency bins at $k = 50$, 100, 150, 200, and 250. Verify your results using the **fft** command. Also verify that the DFT values for the $(1024 - k)$ frequency bins are correct. Plot the magnitude and phase of your results obtained from the Goertzel algorithm and the FFT.

Questions

1. What is the number of real multiplications and additions required for the Goertzel algorithm to compute the N-point DFT value at a single frequency bin? How can this be compared to the Radix-2 FFT algorithm?
2. Find the actual number of real multiplications and additions involved in (3) above.
3. After the Goertzel algorithm has computed the N-point DFT value at a single frequency bin k, how many real multiplications and additions are necessary to compute the N-point DFT value at the frequency bin $N - k$? Show your work.

Project 4

2-D Filter Stability Testing Algorithm. Write a program for testing the stability of IIR filters with no nonessential singularities of the second kind (NSSK). The algorithm should be based on Huang's stability theorem as given in Section 3.10. Test the algorithm on several different filters of your choice. Make sure you choose filters so that some are stable and the others unstable.

BIBLIOGRAPHY

[1] R.W. CHURCHILL, *Complex Variables and Applications*, McGraw-Hill, New York, 1960.

[2] J.G. PROAKIS and D.G. MANOLAKIS, *Digital Signal Processing: Principles, Algorithms, and Applications*, Prentice Hall, Upper Saddle River, NJ, 1996.

[3] G. GOERTZEL, "An algorithm for the evaluation of finite trigonometry series," *Am. Math. Monthly* 65, pp. 34–35, January 1958.

[4] L.R. RABINER, R.W. SCHAFER, and C.M. RADER, "The Chirp z-Transform," *IEEE Transactions on Audio and Electroacoustics*, AU-17, pp. 86–92, 1969.

[5] J.L. SHANKS, S. TRIETEL, and J.H. JUSTICE, "Stability and synthesis of two-dimensional recursive filters," *IEEE Trans. on Audio Electroacoustics*, AU-20, pp. 115–128, June 1972.

[6] D. GOODMAN, "Some stability properties of two-dimensional linear shift-invariant digital filters," *IEEE Trans. on Circuits and Systems*, CAS-24, pp. 210–208, April 1977.

[7] T.S. HUANG, "Stability of two-dimensional recursive filters," *IEEE Trans. on Audio Electroacoustics*, AU-20, pp. 158–163, June 1972.

[8] M.G. STRINTZIS, "Tests of stability of multidimensional filters," *IEEE Trans. on Circuits and Systems*, CAS-24, pp. 432–437, August 1977.

[9] R.A. DECARLO, J. MURRAY, and R. SAEKS, "Multivariable Nyquist theory," *Intl. J. Control* 25, pp. 657–675, 1977.

[10] E.I. JURY, *Theory and Applications of the Z-transform Method*, Wiley, New York, 1964.

[11] A.V. OPPENHEIM and R.W. SCHAFER, *Discrete-time Signal Processing*, Prentice Hall, New York, 1989.

[12] E.I. JURY, "Stability of multidimensional scalar and matrix polynomials," *Proc. IEEE* 66, pp. 1018–1047, September 1978.

[13] N.K. BOSE, *Applied Multidimensional Systems Theory*, Van Nostrand Rheinhold, New York, 1982.

[14] D.D. SILJAK, "Stability criteria for two-variable polynomials," *IEEE Trans. on Circuits and Systems*, CAS-22, pp. 185–189, March 1975.

[15] B.T. O'CONNOR and T.S. HUANG, "Stability of general two-dimensional recursive filters," *Topics in Applied Physics* 42, pp. 85–154, Springer Verlag, New York, 1981.

[16] R.P. ROESSER, "A discrete state space model for linear image processing," *IEEE Trans. on Automatic Control*, AC-20, pp. 1–10, February 1975.

[17] E. FORNASINI and G. MARCHESINI, "State-space realization theory of two-dimensional filters," *IEEE Trans. on Automatic Control*, AC-21, pp. 484–492, August 1976.

[18] W.-S. LU and E.B. LEE, "Stability analysis for two-dimensional systems," *IEEE Trans. on Circuits and Systems*, CAS-30, pp. 455–461, July 1983.

[19] W.-S. LU and E.B. LEE, "Stability analysis for two-dimensional systems via a Lyapunov approach," *IEEE Trans. on Circuits and Systems*, CAS-32, pp. 61–68, January 1985.

[20] N.G. EL-AGIZI and M.M. FAHMY, "Two-dimensional digital filters with no overflow oscillations," *IEEE Trans. on Acous., Speech, and Signal Processing*, ASSP-27, pp. 465–469, June 1979.

[21] B.D.O. ANDERSON, P. AGATHOKLIS, E.I. JURY, and M. MANSOUR, "Stability and the matrix Lyapunov equations for discrete two-dimensional systems," *IEEE Trans. on Circuits and Systems*, CAS-33, pp. 261–267, March 1986.

[22] T. BOSE and D.A. TRAUTMAN, "Two's complement quantization in two-dimensional state-space digital filters," *IEEE Transactions on Signal Processing* 41, pp. 2589–2592, October 1992.

[23] T. BOSE, G.F. XU, and J. SCHROEDER, "A New Stability Testing Algorithm for 2-D Discrete Systems," *Proc. IEEE Intl. Symp. on Circuits and Systems*, Orlando, FL, May 1999.

CHAPTER 4

Filter Design and Implementation

This chapter focuses on the design and implementation of both 1-D and 2-D filters. The design of a filter starts with a set of specifications that can be either in the discrete-time or frequency domain. In the discrete-time domain, the desired impulse response of the filter is usually specified. In the frequency domain, the specifications include certain parameters of the magnitude response and/or the phase response. The goal is then to design the digital filter transfer function, $H(z)$. Of course, the design is not unique. Many different design techniques yield filters with respective advantages and disadvantages. Once the filter is designed, it must be implemented in a signal processor. Again, there are many different ways of implementing a transfer function; these are referred to as filter structures. The choice of a filter structure has many consequences, such as the computational complexity of the filter, its noise characteristics in finite wordlength implementations, and numerical stability. Therefore, the filter structure must be carefully chosen depending on the application.

In this chapter, we begin with the topic of FIR filter design. We present the windowing method and the frequency sampling method in detail and briefly discuss other design techniques. In Section 4.2, the design of IIR filters is presented. We describe the popular method, whereby the digital filter is designed via an analog lowpass (LP) prototype. Transformations are then used to convert from lowpass to highpass (HP), bandpass (BP), or bandstop (BS). Transformations from the analog-to-digital domain are also discussed. In Section 4.3, various filter structures are derived for implementation, and their respective advantages and disadvantages are discussed. In particular, we present the following filter structures: Direct Form (DF), Coupled Form (or normal form), Cascade, Parallel, and Lattice. Section 4.4 deals with the design of 2-D FIR filters. The 2-D FIR filter design is achieved by straightforward extensions of the 1-D methods. The design of 2-D IIR filters is then discussed in Section 4.5. The section on MATLAB presents many useful commands for designing 1-D and 2-D filters.

4.1 FIR FILTER DESIGN

Numerous methods are available for designing digital filters. Most of these methods trade off between computational complexity and how closely the designed filter meets the specifications. The FIR filter design techniques can be classified into the following categories: (1) design using windows; (2) frequency sampling method; (3) optimal equiripple design; and (4) least-squares design. In this chapter, the first two methods will be covered thoroughly, whereas the others will be discussed briefly.

Causality and linear phase are important considerations in the design. Most filters are required to be causal and linear phase. As we know, a causal FIR filter is described by the equation

$$H(z) = h(0) + h(1)z^{-1} + h(2)z^{-2} + \cdots + h(M-1)z^{-(M-1)} + h(M)z^{-M} \tag{4.1}$$

where the length of the filter is $M+1$ and the filter coefficients are represented by $h(i)$. In the last chapter, it was shown that linear phase can be achieved by making the filter coefficients symmetric or antisymmetric. Symmetry of the coefficients implies $h(n) = h(M-n)$. Antisymmetry of the coefficients implies $h(n) = -h(M-n)$. Recall from Chapter 3 that symmetry and antisymmetry do not guarantee the strictly linear phase definition given by

$$\phi(\omega) \triangleq \measuredangle H(\omega) = -\alpha\omega. \tag{4.2}$$

However, the symmetry and antisymmetry conditions satisfy the generalized linear phase definition given by

$$\phi(\omega) = -\frac{M}{2}\omega + \beta. \tag{4.3}$$

Before we describe the filter design methods, it is useful to derive the ideal impulse response of a lowpass filter (LPF). An ideal causal LP frequency response is given by

$$H_{LP}(\omega) = \begin{cases} e^{-j\omega M/2}, & -\omega_c \le \omega \le \omega_c \\ 0, & \text{otherwise} \end{cases} \tag{4.4}$$

where ω_c is the cutoff frequency. The magnitude response is shown in Fig. 4.1(a). A delay of $M/2$ samples has been incorporated so that the filter length is $M+1$. The impulse response of this ideal filter can be found by taking the inverse DTFT of the above. The result is

$$\begin{aligned} h_{LP}(n) &= \frac{1}{2\pi}\int_{-\omega_c}^{\omega_c} e^{-j\omega M/2} e^{j\omega n}\, d\omega \\ &= \begin{cases} \dfrac{\sin\left(\omega_c\left(n - \frac{M}{2}\right)\right)}{\pi\left(n - \frac{M}{2}\right)}, & n \ne \dfrac{M}{2} \\ \dfrac{\omega_c}{\pi}, \quad n = \dfrac{M}{2}. \end{cases} \end{aligned} \tag{4.5}$$

The ideal impulse response given in (4.5) is of infinite length. It is also noncausal because of the nonzero coefficients for $n < 0$. The concept of the windowing is to multiply $h_{LP}(n)$ by a function that is of finite length.

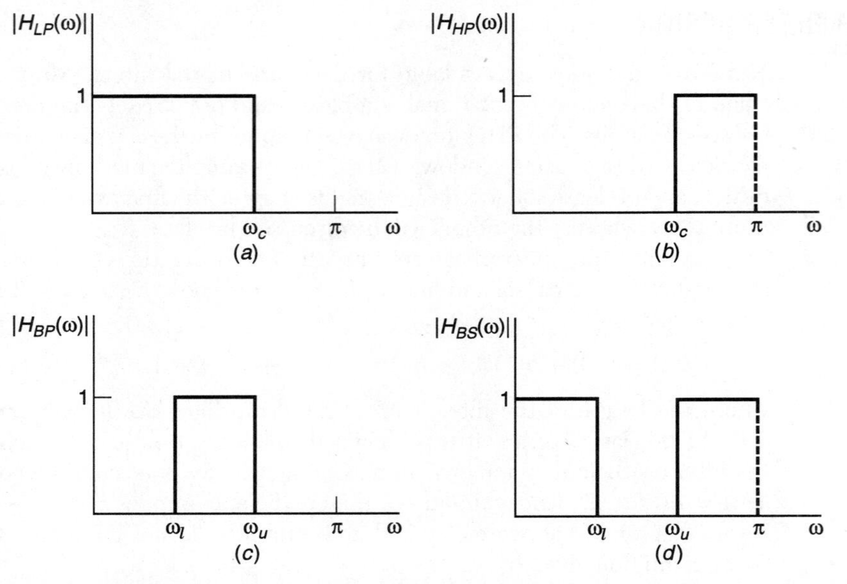

Figure 4.1 Ideal frequency responses of lowpass, highpass, bandpass, and band stop filters. Positive frequency scale shown only.

The frequency response of an ideal causal highpass filter (HPF) is simply

$$H_{HP}(\omega) = 1 - H_{LP}(\omega). \tag{4.6}$$

The magnitude response of this filter is shown in Fig. 4.1(b). A bandpass filter (BPF) has two cutoff frequencies, namely, a lower cutoff frequency ω_l and an upper cutoff frequency, ω_u. The frequency response of an ideal bandpass (BP) filter can be written as

$$H_{BP}(\omega) = H_{LP}(\omega)|_{\text{cutoff}=\omega_u} - H_{LP}(\omega)|_{\text{cutoff}=\omega_l}. \tag{4.7}$$

Similarly, an ideal bandstop filter (BSF) can be written as

$$H_{BS}(\omega) = 1 - H_{BP}(\omega). \tag{4.8}$$

The magnitude responses of the ideal BPF and BSF are given in Fig. 4.1(c) and (d), respectively. By taking the inverse DTFT of the above equations, we get the time-domain relationships:

$$h_{HP}(n) = \delta(n) - h_{LP}(n), \tag{4.9}$$

$$h_{BP}(n) = h_{LP}(n)|_{\text{cutoff}=\omega_u} - h_{LP}(n)|_{\text{cutoff}=\omega_l}, \tag{4.10}$$

$$h_{BS}(n) = \delta(n) - h_{BP}(n). \tag{4.11}$$

Equations (4.9)–(4.11) will be useful in FIR filter design by windowing.

Design Using Windows

The concept of the window method is to multiply an ideal impulse response, $h_I(n)$, by a finite-length window, $\omega(n)$. The question now is, what characteristic should this window have? The impulse response of the designed filter is

$$h(n) = h_I(n)w(n). \tag{4.12}$$

Taking the DTFT of the above gives the convolution

$$H(\omega) = H_I(\omega) * W(\omega). \tag{4.13}$$

Now, we would, of course, like $H_I(\omega)$ to be equal to $H(\omega)$. This means that $W(\omega)$ must be equal to $\delta(\omega)$, which in turn implies that $w(n) = 1$, for all n. This gives us the infinite-length ideal impulse response that cannot be realized in practice. The point here is that we seek a finite-length window $w(n)$ whose DTFT $W(\omega)$ is close to an impulse. The meaning of this will be explained shortly.

The simplest method of windowing is to truncate the infinite-length ideal impulse response, $h_I(n)$. This is called a *rectangular window* and can be written as

$$w(n) = \begin{cases} 1, & n = 0, 1, \ldots, M \\ 0, & \text{otherwise.} \end{cases} \tag{4.14}$$

Now we need to find out what its frequency response looks like. The DTFT of this window can be easily found as follows.

$$\begin{aligned} W(\omega) &= \sum_{n=-\infty}^{\infty} w(n)e^{-j\omega n} \\ &= \sum_{n=0}^{M} e^{-j\omega n} \\ &= \frac{1 - e^{-j\omega(M+1)}}{1 - e^{-j\omega}} \\ &= e^{-j\omega M/2} \frac{\sin(\omega \frac{M+1}{2})}{\sin(\frac{\omega}{2})}. \end{aligned} \tag{4.15}$$

The magnitude and phase are

$$|W(\omega)| = \left| \frac{\sin\left(\omega \frac{M+1}{2}\right)}{\sin(\frac{\omega}{2})} \right| \tag{4.16}$$

and

$$\measuredangle W(\omega) = -\frac{\omega M}{2} + \beta \tag{4.17}$$

where $\beta = 0$ or π, depending on the sign of $\frac{\sin(\omega \frac{M+1}{2})}{\sin(\frac{\omega}{2})}$. The frequency response of the rectangular window therefore has (generalized) linear phase. This is important because

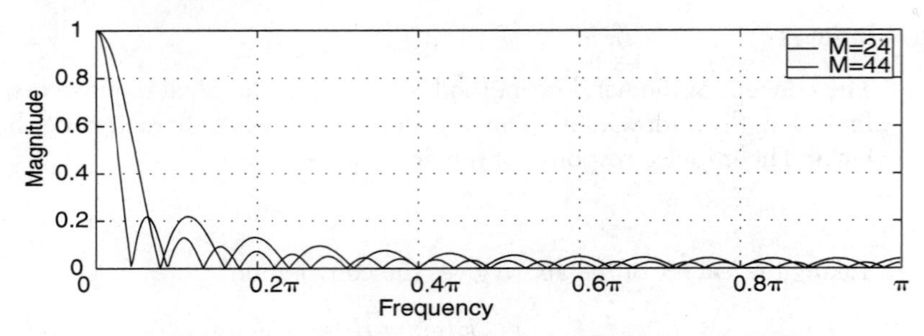

Figure 4.2 Magnitude response of the rectangular window. Windows' lengths are 25 and 45.

we want the designed filter to have linear phase. It is now instructive to plot the magnitude response of this window. The plots for $M = 24$ and 44 are shown in Fig. 4.2 in absolute scale. First, we observe that the response is "impulse like"; that is, $W(\omega)$ is a narrow pulse, and it falls off sharply. It has a main lobe and side lobes[1] that have ripples. When this function is convolved with one of the ideal frequency responses of Fig. 4.1, the ripples in the side lobes cause ripples in the resulting filter. Therefore, it is desirable to use windows with low side-lobe ripples. In addition, the narrower the main lobe, the narrower the transition band of the filter, which is also desirable. Second, we observe in Fig. 4.2 that for the longer window, we have a narrower main-lobe width. For the longer window, we have more ripples with lower height. However, the maximum ripple height in both cases is the same. The main-lobe width of the rectangular window can be easily calculated from (4.15) by finding the zero crossings on either side of the origin. This turns out to be $4\pi/(M+1)$. This explains the fact that the main-lobe width decreases with increasing filter length.

▶ EXAMPLE 4.1 *Consider the design of an LPF with a cutoff frequency of* 0.5π *using rectangular windows of length* 25 $(M = 24)$ *and length* 45 $(M = 44)$. *The impulse response (filter coefficients) for the longer filter is plotted in Fig. 4.3. The impulse response of the shorter filter is essentially the same plot centered at* $n = 12$ *and has only 12 coefficients on either side. The magnitude responses are plotted in absolute and log scales. These are given in Fig. 4.4* $(M = 24)$ *and Fig. 4.5* $(M = 44)$. *Clearly, the transition band of the longer filter is narrower, as expected. The stopband attenuation in each case is about the same and only about 30 dB. Such a low attenuation in the stopband is not acceptable for most applications. Another drawback of the rectangular window is that the designed filter exhibits oscillations in the passband and stopband. This is called the Gibbs phenomenon. The amplitude of the oscillations decreases for longer filters, but their frequency increases. This is analogous to the Fourier series representation, where we get a better approximation with more coefficients. However, the height of the largest ripple is independent of the filter length. This is consistent with the magnitude response of rectangular windows, as discussed earlier. For most applications, these oscillations are usually acceptable in the stopband but not in the passband. This makes the rectangular window unattractive, in spite of its simplicity.*

In order to increase the stopband attenuation and reduce the Gibbs phenomenon, we need windows that taper smoothly to zero unlike the "brickwall" in the rectangular

[1] Only the side lobes for the positive frequency are shown. The other side lobes are symmetric.

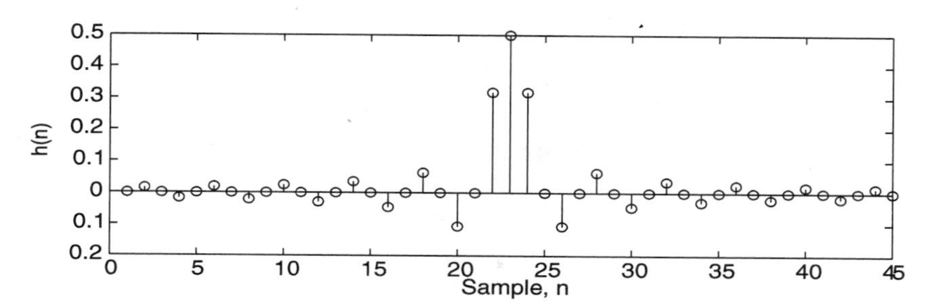

Figure 4.3 Impulse response of LPF designed using a rectangular window of length-45.

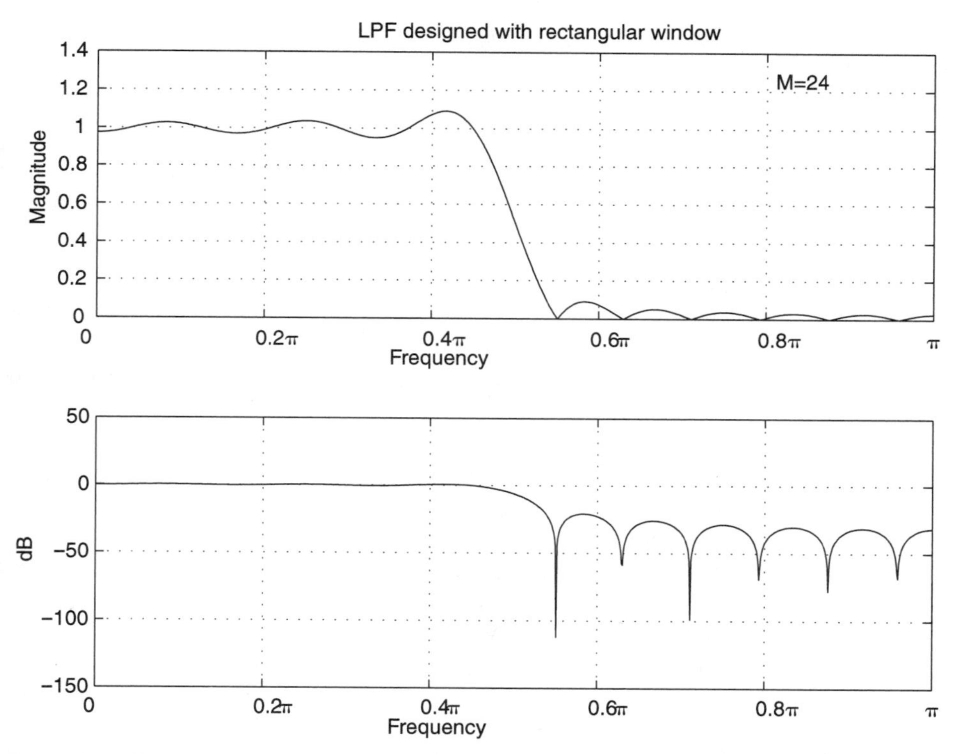

Figure 4.4 Magnitude response of a LPF designed using a rectangular window of length-25.

window. Such smooth transition will reduce the height of the ripples in the frequency response of the window. This in turn will reduce the ripples in the designed filter and also increase the stopband attenuation.

Researchers have proposed many different types of windows. All of these windows trade off between main-lobe width and stopband attenuation. The choice of window is often dictated by the application at hand. The formulas for some popular windows are given in Table 4.1. The magnitude responses of some of these windows are given in Fig. 4.6 for $M = 44$. The rectangular window has the narrowest main-lobe width. On the other

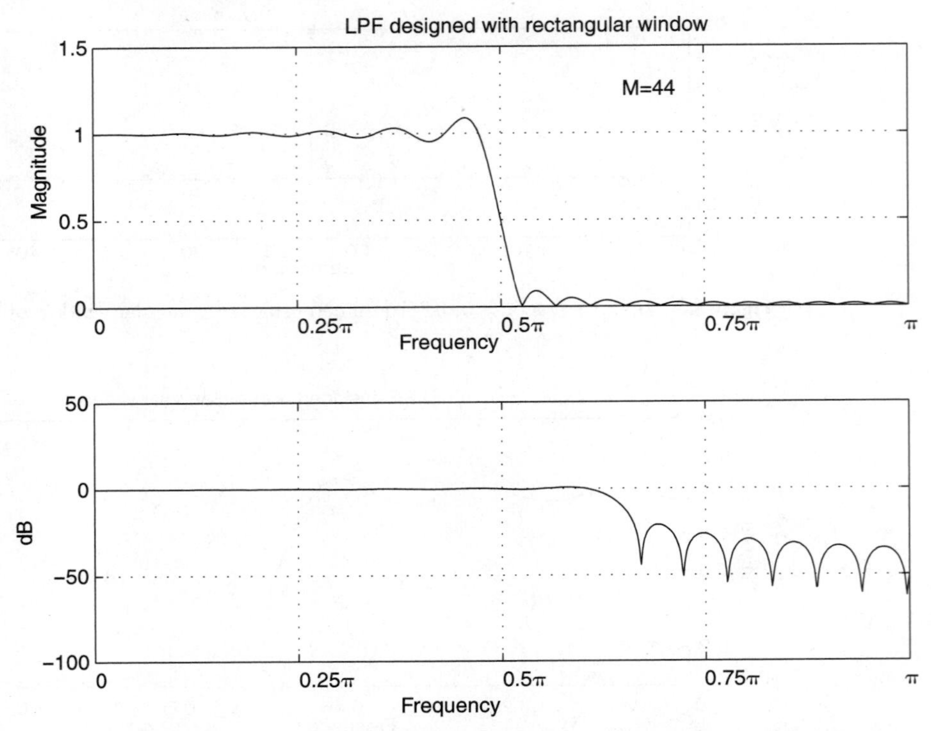

Figure 4.5 Magnitude response of a LPF designed using a rectangular window of length-45.

hand, the other windows have better side-lobe attenuation. The side-lobe attenuation tapers down with frequency for the Bartlett and the Blackman windows, whereas it is quite constant for the Hamming window. The attenuation for the peak side lobe is best for the Blackman window. Unfortunately, the Blackman window has the widest main lobe. These results are tabulated in Table 4.2 [10]. The peak side lobe of a window is independent of the window length, whereas the main lobe width is inversely proportional to it. Therefore, we can always reduce the main-lobe width by increasing the filter length. However, this increases the number of computations[2] per sample.

▶ EXAMPLE 4.2 *Now we design lowpass filters with $M = 44$ using the Hamming and Blackman windows. The results are shown in Figs. 4.7 and 4.8, respectively. At this time it is important to compare these designs with the design obtained using the rectangular window. The filter designed by the rectangular window had significant ringing in the passband and stopband as we saw in Fig. 4.5(a). Figures 4.7(a) and 4.8(a) illustrate that these ringing effects are eliminated by the use of the Hamming and Blackman windows. From Figs. 4.5(b), 4.7(b), and 4.8(b), we can compare the peak side-lobe attenuation. The rectangular window yielded a filter with only about 25-dB attenuation. On the other hand, the peak side-lobe attenuation for the Hamming and Blackman designs are 60 dB and 75 dB, respectively. Thus, these windows produced filters with no ringing and much better side-lobe attenuation.*

[2] Recall that a FIR filter is nothing but a sum of products. Therefore, the longer the filter, the higher is the computational complexity.

TABLE 4.1 Commonly Used Windows (Domain $0 \leq n \leq M$)

1. Rectangular

$$w(n) = \begin{cases} 1, & 0 \leq n \leq M \\ 0, & \text{otherwise} \end{cases}$$

2. Bartlett

$$w(n) = 1 - \frac{2\left|n - \frac{M}{2}\right|}{M}$$

3. Blackman

$$w(n) = 0.42 - 0.5 \cos \frac{2\pi n}{M} + 0.08 \cos \frac{4\pi n}{M}$$

4. Hamming

$$w(n) = 0.54 - 0.46 \cos \frac{2\pi n}{M}$$

5. Hanning

$$w(n) = 0.5 - 0.5 \cos \frac{2\pi n}{M}$$

6. Kaiser

$$w(n) = \frac{I_0\left\{\beta\sqrt{1 - \left(\frac{n-M/2}{M/2}\right)^2}\right\}}{I_0\left\{\beta\right\}}$$

However, all this comes at a price. These filters have transition bands that are significantly wider than those designed by the rectangular window. The transition band of the filter produced by the Blackman window is the widest. This is consistent with our earlier discussion about the main-lobe width and Table 4.2.

The windows discussed so far are called fixed windows; that is, there is no parameter (except the length) that can be adjusted. For a selected window, the transition bandwidth is a function of the specified cutoff frequency. No formula has been derived that can be used to meet specifications on transition bandwidth or the ripple; this has to be done by trial and error.

Some adjustable windows have a parameter that can be varied in order to trade off between main-lobe width and side-lobe attenuation. Lanczos, Tukey, Dolph-Chebyshev [2] and Kaiser windows are examples. The Kaiser window is by far the most popular window design.

Kaiser Window Design

The Kaiser window is given by

$$w(n) = \frac{I_0\left\{\beta\sqrt{1 - \left(\frac{n-M/2}{M/2}\right)^2}\right\}}{I_0\{\beta\}}, \quad 0 \leq n \leq M \tag{4.18}$$

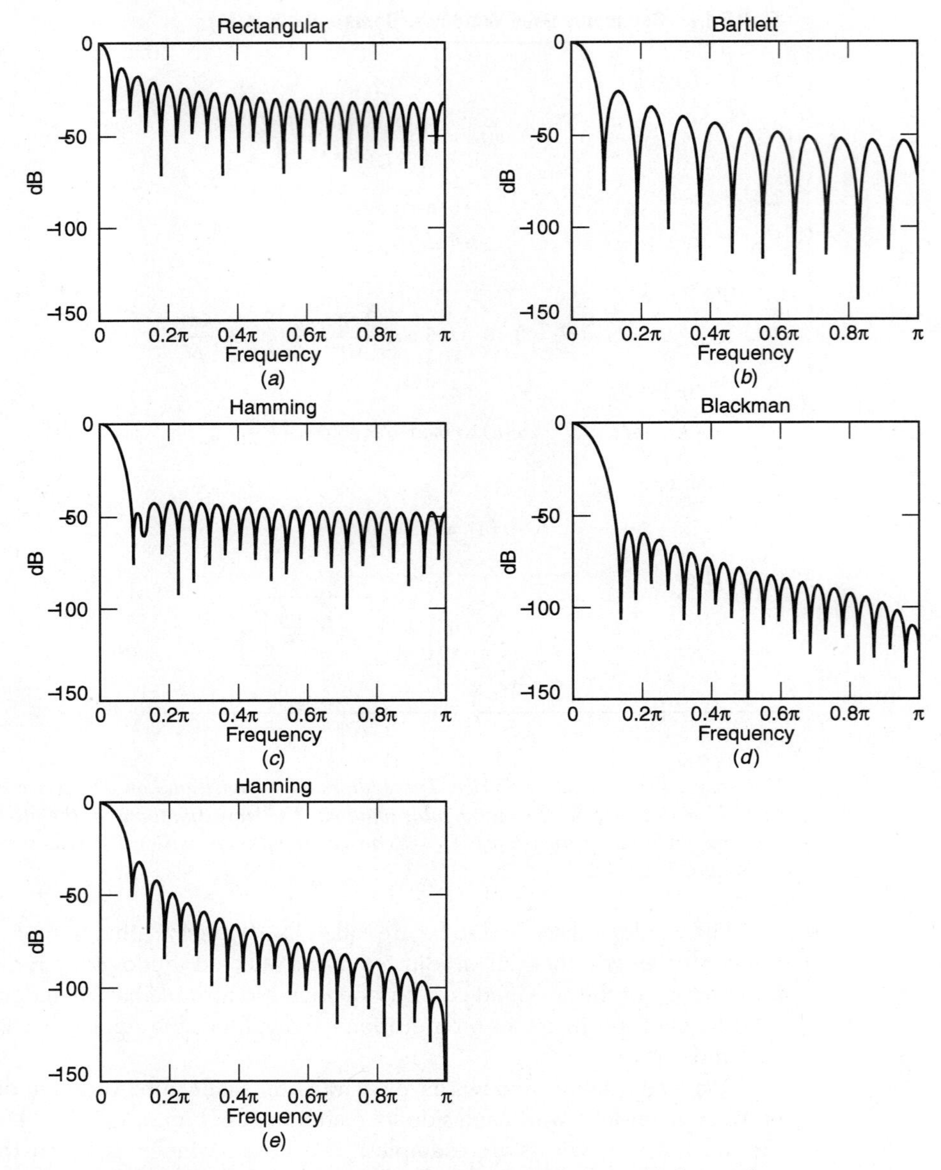

Figure 4.6 Comparison of the magnitude responses of some popular windows.

where $I_0\{x\}$ is the modified zeroth-order Bessel function, given by

$$I_0\{x\} = 1 + \sum_{k=1}^{\infty} \left[\frac{(x/2)^k}{k!} \right]^2 . \tag{4.19}$$

In practice, the above series can be truncated for desired accuracy. The parameter β can be adjusted to trade off between the side-lobe attenuation and main-lobe width. Kaiser

TABLE 4.2 Frequency Domain Characteristics of Commonly Used Windows

Windows	Mainlobe width	Sidelobe attenuation (dB)
Rectangular	$4\pi/M$	−13
Bartlett	$8\pi/M$	−27
Harning	$8\pi/M$	−32
Hamming	$8\pi/M$	−43
Blackman	$12\pi/M$	−58

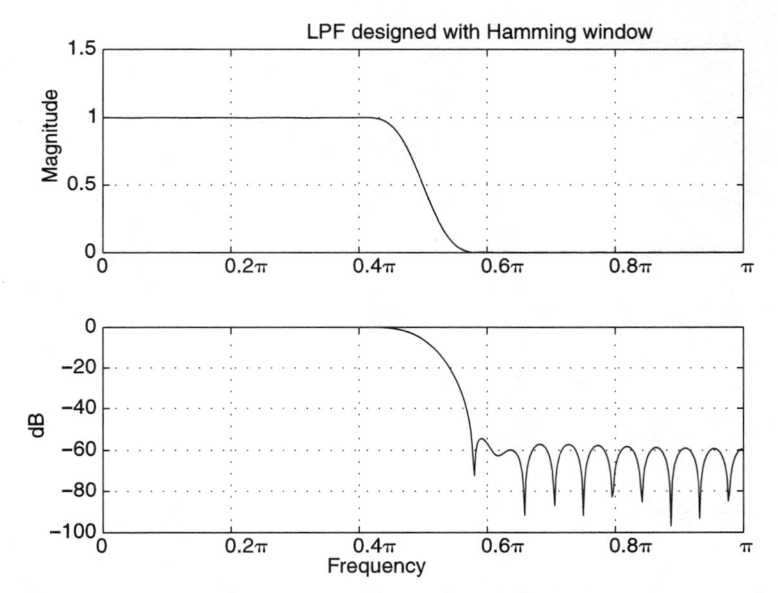

Figure 4.7 Magnitude response of a LPF designed using a Hamming window.

has developed some empirical design formulas [1] to calculate the filter length M and β, such that some given specifications are satisfied. The specifications for a LPF filter design using a Kaiser window are the following.

1. Passband edge frequency, ω_p.
2. Stopband edge frequency, ω_s.
3. Passband ripple, δ_p.
4. Stopband ripple, δ_s.

These specifications are depicted in Fig. 4.9(a). The specifications are similar for the HPF, except that $\omega_s < \omega_p$. For a BSF the specifications are shown in Fig. 4.9(b), which indicates that it has two extra edge frequencies. The design formulas for the Kaiser window are as follows.

(a) Minimum ripple in decibels, $A = -20 \log_{10}\{\min\{\delta_s, \delta_p\}\}$ dB.

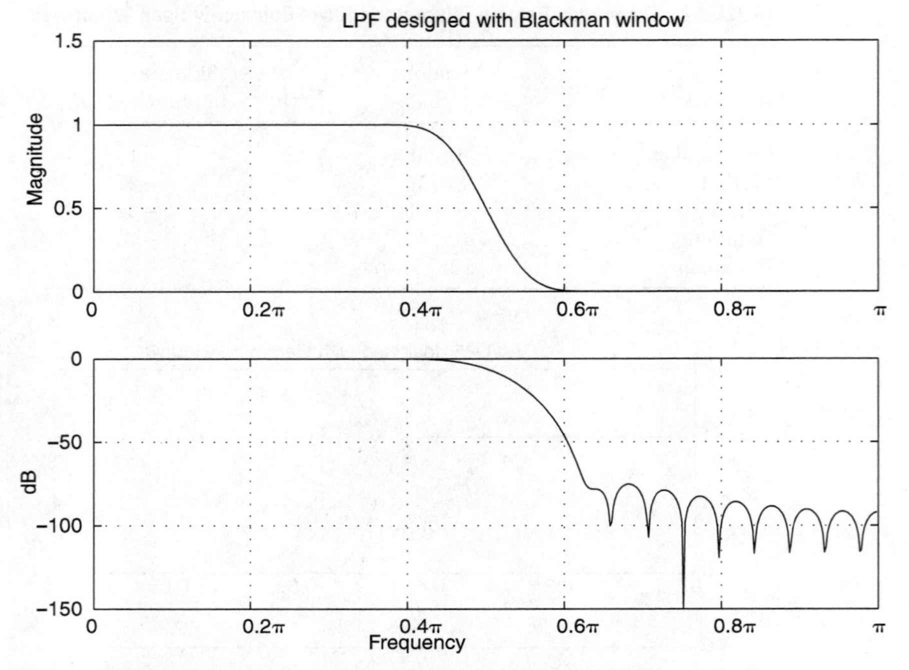

Figure 4.8 Magnitude response of a LPF designed using a Blackman window.

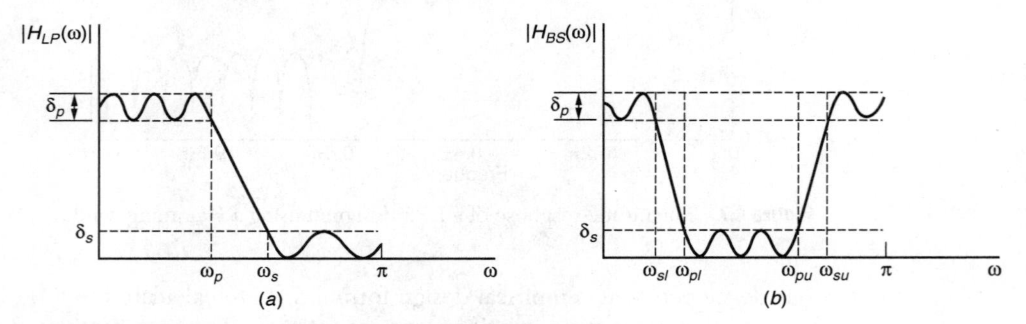

Figure 4.9 Filter design specifications for (*a*) LPF, (*b*) BSF.

(b) Normalized transition bandwidth, $\Delta\omega = \omega_s - \omega_p$.

(c) Window parameter, $\beta = \begin{cases} 0.1102(A - 8.7), & A > 50 \\ 0.5842(A - 21), & 21 \le A \le 50 \\ 0 & A < 21 \end{cases}$

(d) Length of filter, $M + 1 = \begin{cases} \dfrac{A - 7.95}{2.285\Delta\omega} + 1, & A > 21 \\ \dfrac{5.79}{\Delta\omega} + 1, & A \le 21 \end{cases}$ rounded to the nearest integer.

(e) Design Kaiser window using formula (4.18).

These formulas can be used to meet given specifications on the transition bandwidth and the ripples, without any need for trial and error. Once the Kaiser window is designed by these formulas, the final filter is obtained by multiplying the window by the ideal impulse response, $h_I(n)$, which is obtained by using one of (4.5), (4.9), (4.10), or (4.11) for LPF, HPF, BPF, or BSF, respectively. These formulas require the cutoff frequency ω_c. From the specifications given above for this design, the cutoff frequency is approximated by $\omega_c = \frac{\omega_p+\omega_s}{2}$.

EXAMPLE 4.3 *In this example we consider the design of a bandpass filter using the Kaiser window. The specifications are as follows. (1) lower stopband edge frequency, $\omega_{sl} = 0.2\pi$, (2) lower passband edge frequency, $\omega_{pl} = 0.4\pi$, (3) upper passband edge frequency, $\omega_{pu} = 0.6\pi$, (4) upper stopband edge frequency, $\omega_{su} = 0.8\pi$, (5) passband ripple, $\delta_p = 0.001$, and (6) stopband ripple, $\delta_s = 0.001$. Now we use formulas $(a - e)$ to design the Kaiser window.*

(a) $A = 60\,dB$.

(b) *Since this is a BPF, we have two transition bands. Notice that the bandwidth is the same for both transition bands. If they were not, then we would choose the narrower of the two bands for our design. In this case, we have $\Delta\omega = 0.2\pi$.*

(c) $\beta = 0.1102(A - 8.7) = 5.65$.

(d) *Length of filter, $M + 1 = 38$.*

(e) *Design the Kaiser[3] window, $w(n)$ by using (4.18) for $0 \leq n \leq M$.*

The ideal BPF impulse response, $h_I(n)$, is computed using equations (4.10) and (4.5). The final filter is then obtained as $h(n) = w(n)h_I(n)$. The magnitude and phase response are plotted in Fig. 4.10. Note that we have used the normalized frequency in these plots, where 1.0 is equivalent to π radians, or one-half the sampling frequency (Nyquist frequency). The phase response is linear in the passband region, as desired. A computer project involving the Kaiser window is assigned at the end of the chapter.

Frequency Sampling Design

In the frequency sampling method of FIR filter design, we specify the magnitude response of the desired filter at N equally spaced points in the interval $(0, 2\pi)$. This will yield a length-N filter. That is, if $H_d(\omega)$ represents the frequency response of the desired filter, then the specification is as follows.

$$\text{Specification: } |H_d(\omega_k)|_{\omega_k = \frac{2\pi}{N}k} = \left|H_d\left(\frac{2\pi}{N}k\right)\right|, \quad k = 0, 1, 2, \ldots, N-1. \tag{4.20}$$

This is clearly a sampling of the discrete time Fourier transform (DTFT), which is actually the discrete Fourier transform (DFT). Since we want to design a linear phase filter of length N, we assume a phase of $\phi(\omega) = -j\omega(N-1)/2$. Therefore, the DFT coefficients are

$$H_d(k) = \left|H_d\left(\frac{2\pi}{N}k\right)\right| e^{-j\frac{2\pi}{N}k\frac{(N-1)}{2}}, \quad k = 0, 1, 2, \ldots, N-1. \tag{4.21}$$

[3] *The Kaiser window can be computed using the MATLAB function **kaiser(..)**. See the MATLAB section at the end of the chapter for details.*

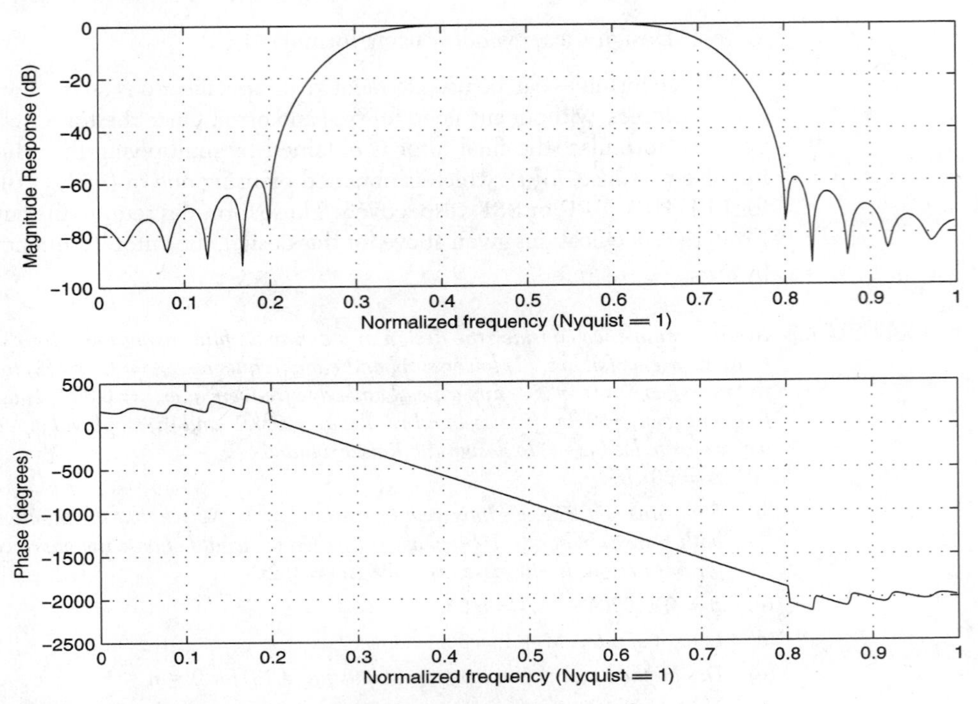

Figure 4.10 Magnitude response of BPF using the Kaiser window.

Note that in equation (4.21), $H_d(k)$ does not have conjugate symmetry about the folding index $N/2$, for $k = 0, 1, 2, \ldots, N-1$. This symmetry is necessary for the inverse DFT to yield a real filter. Therefore, we force conjugate symmetry as follows. For N odd, we have

$$H_d(k) = \left| H_d\left(\frac{2\pi}{N}k\right) \right| e^{-j\frac{2\pi}{N}k\frac{(N-1)}{2}}, \quad k = 0, 1, 2, \ldots, \left\lfloor \frac{N-1}{2} \right\rfloor .$$

$$H_d(N-k) = H_d^*(k), \; k = 0, 1, 2, \ldots, \left\lfloor \frac{N-1}{2} \right\rfloor . \tag{4.22}$$

For N even, we have the same condition and in addition the requirement that $H_d(\frac{N}{2})$ must be real.

The filter coefficients, $h(n)$, are obtained simply by taking the inverse DFT of $H_d(k)$. Let $H(\omega)$ be the DTFT of this filter. The problem at hand now is to show that

$$H\left(\frac{2\pi}{N}k\right) = H_d\left(\frac{2\pi}{N}k\right), \; k = 0, 1, 2, \ldots, N-1. \tag{4.23}$$

That is, we need to prove that the magnitude response of the desired and the designed filters match exactly at the sampled (specified) points.

Since $h(n)$ is obtained by an inverse DFT, we have

$$h(n) = \frac{1}{N} \sum_{k=0}^{N-1} H_d(k) e^{j\frac{2\pi}{N}nk}, \quad n = 0, 1, \ldots, N-1. \tag{4.24}$$

Taking the Z-transform of equation (4.24) gives

$$
\begin{aligned}
H(z) &= \sum_{n=0}^{N-1} h(n) z^{-n} \\
&= \sum_{n=0}^{N-1} \left(\frac{1}{N} \sum_{k=0}^{N-1} H_d(k) e^{j\frac{2\pi}{N}nk} \right) z^{-n}.
\end{aligned} \tag{4.25}
$$

Interchanging the order of the summations and performing some simple manipulations, we get

$$
\begin{aligned}
H(z) &= \frac{1}{N} \sum_{k=0}^{N-1} H_d(k) \sum_{n=0}^{N-1} e^{j\frac{2\pi}{N}nk} z^{-n} \\
&= \frac{1}{N} \sum_{k=0}^{N-1} H_d(k) \sum_{n=0}^{N-1} \left(e^{j\frac{2\pi}{N}k} z^{-1} \right)^n \qquad (4.26) \\
&= \frac{1}{N} \sum_{k=0}^{N-1} H_d(k) \frac{1 - \left(e^{j\frac{2\pi}{N}k} z^{-1} \right)^N}{1 - e^{j\frac{2\pi}{N}k} z^{-1}} \\
&= \frac{1}{N} \sum_{k=0}^{N-1} H_d(k) \frac{1 - z^{-N}}{1 - e^{j\frac{2\pi}{N}k} z^{-1}}. \qquad (4.27)
\end{aligned}
$$

The above is the Z-transform of the designed filter. It is an important equation, and we will return to it shortly. At this point, we find the frequency response of the designed filter by substituting $z = e^{j\omega}$ in (4.26), to get

$$
H(\omega) = \frac{1}{N} \sum_{k=0}^{N-1} H_d(k) \sum_{n=0}^{N-1} e^{j\frac{2\pi}{N}kn} e^{-j\omega n}. \tag{4.28}
$$

Now, let us examine what happens at the sampled frequencies, $\omega_m = \frac{2\pi}{N}m$, $m = 0, 1, \ldots, N-1$. The result is

$$
H\left(\frac{2\pi}{N} m \right) = \frac{1}{N} \sum_{k=0}^{N-1} H_d(k) \sum_{n=0}^{N-1} e^{j\frac{2\pi}{N}(k-m)n}. \tag{4.29}
$$

The second summation should be familiar from Chapter 2. Its evaluation gives

$$
\sum_{n=0}^{N-1} e^{j\frac{2\pi}{N}(k-m)n} = \begin{cases} N, & k = m \\ 0 & k \neq m. \end{cases} \tag{4.30}
$$

Substituting this in (4.29), we get

$$
H\left(\frac{2\pi}{N} m \right) = H_d(m), \quad m = 0, 1, 2, \ldots, N-1, \tag{4.31}
$$

which is the desired result. An example is now in order.

▶ **EXAMPLE 4.4** *The specifications are the sampled magnitude response points shown in Fig. 4.11(a) for the full range of $(0, 2\pi)$. These are the magnitudes of the DFT points.*

$$
\begin{aligned}
|H_d(0)| &= |H_d(1)| = |H_d(2)| = |H_d(8)| = |H_d(9)| = 1; \\
|H_d(3)| &= |H_d(7)| = 0.5; \\
|H_d(4)| &= |H_d(5)| = |H_d(6)| = 0.
\end{aligned}
$$

The filter is of length-10. The actual DFT points are then obtained from equation (4.22). The final FIR filter is obtained by taking the inverse DFT of $H_d(k), 0 \leq k \leq 9$. The coefficients are found to be

$$h(n) = [0.0128, 0.0157, -0.1, 0.0606, 0.5108, 0.5108, 0.0606, -0.1, 0.0157, 0.0128].$$

The magnitude response of this filter is shown in Fig. 4.11(b). The magnitude response exactly matches the specified points. The phase response is not shown and is of course linear by design.

Now we return to equation (4.27), which is the Z-transform of the designed FIR filter in terms of the frequency response samples of the desired filter. This equation is rewritten as

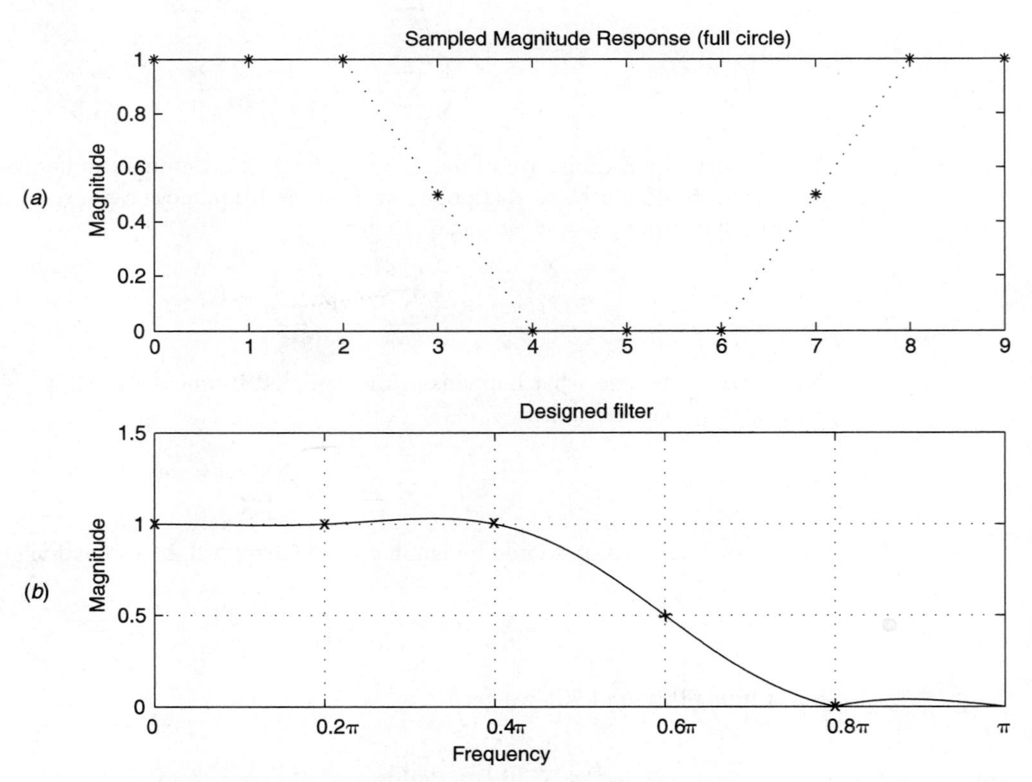

Figure 4.11 Filter design using the frequency sampling method: (*a*) specifications; (*b*) magnitude response of designed filter.

$$H(z) = \frac{1 - z^{-N}}{N} \sum_{k=0}^{N-1} \frac{H_d(k)}{1 - e^{j\frac{2\pi}{N}k} z^{-1}}. \tag{4.32}$$

Equation (4.32) is a cascade of two transfer functions as follows.

$$H(z) = H_{\text{fir}}(z) H_{\text{iir}}(z) \tag{4.33}$$

where

$$H_{\text{fir}}(z) = \frac{1 - z^{-N}}{N} \tag{4.34}$$

and

$$H_{\text{iir}}(z) = \sum_{k=0}^{N-1} \frac{H_d(k)}{1 - e^{j\frac{2\pi}{N}k} z^{-1}}. \tag{4.35}$$

The overall implementation is shown in Fig. 4.12. $H_{\text{iir}}(z)$ is a parallel realization of N first-order complex IIR sections. These sections have complex poles on the unit circle and are therefore referred to as resonators. These complex poles are located at $e^{j\frac{2\pi}{N}k}$, $k = 0, 1, \ldots, N-1$. The transfer function $H_{\text{fir}}(z)$ has N zeros, which can be found as follows.

$$z^{-N} = 1 = e^{-j2\pi k}$$

$$\Longrightarrow z_k = e^{j\frac{2\pi}{N}k}, \quad k = 0, 1, \ldots, N-1.$$

The poles of $H_{\text{iir}}(z)$ (on the unit circle) are therefore canceled by the zeros of $H_{\text{fir}}(z)$, thereby making the overall transfer function stable.

The natural question now is, what is the purpose of the implementation of Fig. 4.12? For one thing, this is an alternate implementation of the FIR filter. This implementation has some pros and cons.

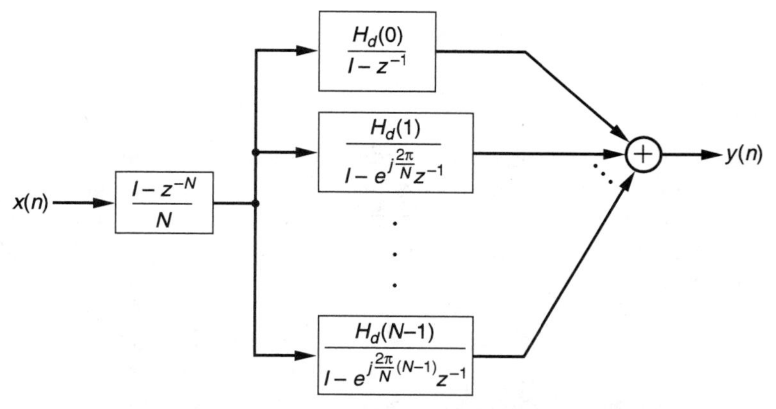

Figure 4.12 Frequency sampling design of FIR filter.

Advantages

1. Figure 4.12 is suitable for implementation in parallel processors, where each processor performs a maximum of two complex multiplications per sample.
2. The design does not require an inverse DFT, which can be computationally intensive for long filters.

Disadvantages

1. For single-processor implementation, Fig. 4.12 requires $2N-1$ complex multiplications and 1 real multiplication per sample, whereas a Direct Form FIR filter requires only N real multiplications.
2. The pole-zero cancellation may not actually happen when the filter in implemented in a finite wordlength processor. This may lead to instability. However, this disadvantage may be overcome by slightly reducing the magnitude of the poles and zeros. This is done by multiplying them with a real number β slightly less than unity. The resulting transfer function then becomes

$$H(z) = \frac{1-(\beta z)^{-N}}{N} \sum_{k=0}^{N-1} \frac{H_d(k)}{1-e^{j\frac{2\pi}{N}k}(\beta z)^{-1}}. \tag{4.36}$$

With β very close to unity, there is little variation in the frequency response.

At this time, it is natural to bring up a new type of filter called the *comb filter*, which is described by

$$H_{\text{comb}}(z) = 1 - z^{-N}. \tag{4.37}$$

The zeros of this filter have already been discussed. The difference equation is simply

$$y(n) = x(n) - x(n-N). \tag{4.38}$$

The magnitude responses are given in Fig. 4.13 for $N=8$ and 9. The reason for the name comb filter is now obvious. Such a filter is useful to notch out a set of unwanted frequencies from a spectrum.

Other FIR Design Methods

The frequency sampling method and the window method are simple but powerful techniques for FIR filter design. However, these methods do not provide precise control of the critical frequencies, nor do they yield equiripple in the passband and stopband. The optimal equiripple method designs the filter so that the following cost function is minimized:

$$E = \frac{1}{2\pi} \int_{-\pi}^{\pi} \beta(\omega) |H_d(\omega) - H(\omega)|^2 d\omega, \tag{4.39}$$

where $H_d(\omega)$ and $H(\omega)$ are the desired and designed frequency response, respectively, and $\beta(\omega)$ is a nonnegative weighting function. Many algorithms can be designed to

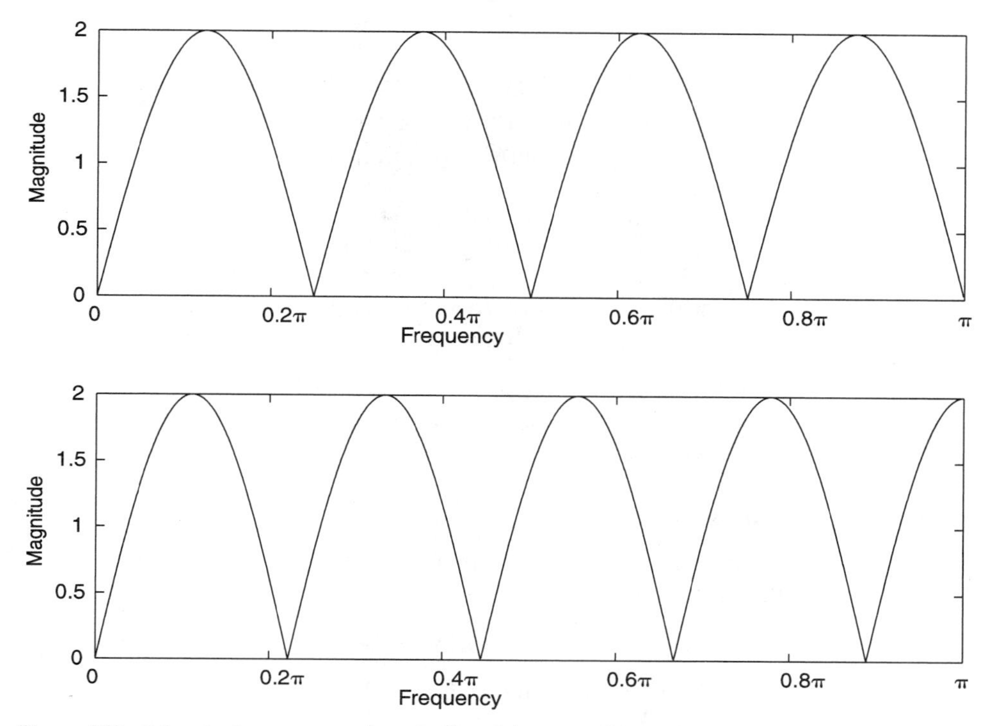

Figure 4.13 Magnitude response of comb filter (*a*) $N = 8$; (*b*) $N = 9$.

solve this problem. The Parks-McClellan algorithm is the most popular, where the weighted approximation error is spread evenly across the passband and the stopband. This algorithm is derived based on the alternation theorem in the theory of Chebyshev approximation. Detailed derivations of the Parks-McClellan algorithm can be found in [2], [26], and [27].

The least-squares method is based on minimizing a cost function that is a weighted squared error between the desired and designed impulse responses. This method will be described in the context of 2-D filters in a later section.

4.2 IIR FILTER DESIGN

The specifications for an IIR digital filter are given in terms of the passband and stopband edge frequencies, and the ripples in these bands, just as in the Kaiser window method. The design methods can be classified into two main categories as follows: (a) *design via an analog prototype*; (b) *least-squares design* in the digital domain. In this section, we will present the first method only. The least-squares design method will be described in the context of 2-D filters in a later section. The method can be easily adapted to 1-D filters.

When designing digital filters via an analog prototype, the following steps are performed.

Design Steps

1. Transform the given digital filter frequencies to analog domain.
2. Design an appropriate prototype LP analog filter, $H_{\text{LP}}(s)$.
3. Perform either (a) or (b), given below.

   (a) (i) Perform *analog-to-analog transformation* to convert $H_{\text{LP}}(s)$ to HP, BP, or BS, if needed; the final analog filter is $H(s)$; (ii) perform *analog-to-digital transformation* to convert $H(s)$ to $H(z)$.

   (b) (i) Transform $H_{\text{LP}}(s)$ to $H_{\text{LP}}(z)$; (ii) perform *digital-to-digital* transformation to convert $H_{\text{LP}}(z)$ to HP, BP, or BS, if needed.

This design algorithm requires several transformations that must be discussed. First, we present analog-to-digital transformation, or vice versa. There are two major techniques: (1) *impulse invariant* and (2) *bilinear transformation*.

Impulse Invariant Transformation

Let $h_a(t)$ be the impulse response of the analog filter. The transformation must be such that the impulse response of the digital filter, $h(n)$, is a sampled version of $h(t)$, that is,

$$h(n) = h_a(t)|_{t=nTs}, \quad n = 0, 1, 2, \ldots \tag{4.40}$$

where T_s is the sampling time period. Consider an analog filter $H_a(s)$ with N distinct poles, given by

$$H_a(s) = \sum_{i=1}^{N} \frac{b_i}{s - p_i} \tag{4.41}$$

where p_i represents a real or complex pole. The impulse response is obtained by an inverse Laplace transform as

$$h_a(t) = \sum_{i=1}^{N} b_i e^{p_i t} u(t). \tag{4.42}$$

The impulse response of the digital filter is therefore

$$h(n) = \sum_{i=1}^{N} b_i e^{p_i n} u(n). \tag{4.43}$$

Taking the Z-transform of the above gives the digital filter transfer function

$$H(z) = \sum_{i=1}^{N} \frac{b_i}{1 - e^{p_i T_s} z^{-1}}. \tag{4.44}$$

Comparing (4.44) and (4.41), we get the transformation

$$\frac{1}{s - p_i} \longleftrightarrow \frac{1}{1 - e^{p_i T_s} z^{-1}}. \tag{4.45}$$

By using a similar method, it is straightforward to derive the transformation for poles with multiplicity m. The result is

$$\frac{1}{(s-p_i)^m} \longleftrightarrow \frac{(-1)^{m-1}}{(m-1)!} \frac{d^{m-1}}{ds_i^{m-1}} \left\{ \frac{1}{1-e^{p_i T_s} z^{-1}} \right\}. \tag{4.46}$$

Formulas (4.45) and (4.46) can be used to transform any analog transfer function into the Z-domain. From these formulas, we have the mapping

$$\begin{aligned} z &= e^{sT_s} \\ &= e^{(\sigma + j\Omega)T_s}. \end{aligned} \tag{4.47}$$

Therefore

$$|z| = e^{\sigma T_s} \begin{cases} < 1, & \sigma < 0 \\ = 1, & \sigma = 0 \\ > 1, & \sigma > 0. \end{cases} \tag{4.48}$$

The above implies that the left half of the s-plane maps to the inside of the unit circle of the z-plane. Also, the right half of the s-plane maps to the outside of the unit circle, and the imaginary axis maps to the unit circle. Therefore, a stable analog filter yields a stable digital filter by this transformation. In this sense, this is a desirable transformation.

Because the impulse response is sampled, we know from Chapter 2 that the frequency response consists of replicas of the original spectrum appearing at multiples of the sampling frequency, Ω_s. That is, the original spectrum $H_a(\Omega)$ appears at $0, \pm\Omega_s, \pm 2\Omega_s, \pm 3\Omega_s, \ldots$. There is no aliasing (overlap of spectra) if $\Omega_s > 2\Omega_h$, where Ω_h is the highest frequency component of the analog filter spectrum. However, no practical analog filter is bandlimited. Therefore, the impulse invariant method of transformation will always cause aliasing. This is not signal aliasing as described in Chapter 2. Here, the impulse response has aliasing, which implies that the frequency response of the designed filter will be distorted. This is a serious drawback of this transformation. Therefore, this transformation can be used only when the effect of aliasing is small. Aliasing is small when the sampling frequency is high and we are designing lowpass or bandpass filters, where the stopbands overlap. However, in highpass and bandstop filters, the passbands overlap, thereby magnifying the effect of aliasing. The impulse invariant method is therefore unsuitable for designing highpass and bandstop filters.

Bilinear Transformation

The bilinear transformation is used for transforming an analog filter to a digital filter. It was first applied to discrete systems by Tustin [3]. Therefore, it is often referred to as *Tustin's rule*. It uses the trapezoidal rule for integrating a continuous-time function. To derive this transformation, consider the derivative

$$\frac{dy(t)}{dt} = x(t). \tag{4.49}$$

The Laplace transform of the above with zero initial conditions gives

$$sY(s) = X(s). \tag{4.50}$$

To find $y(t)$ within the limits $(n-1)T_s$ and nT_s, we integrate both sides of (4.49) with respect to t. This gives

$$\int_{(n-1)T_s}^{nT} \frac{dy(t)}{dt} dt = \int_{(n-1)T_s}^{nTs} x(t)\, dt \tag{4.51}$$

or

$$y(nT_s) - y((n-1)T_s) = \int_{(n-1)T_s}^{nTs} x(t) dt. \tag{4.52}$$

Now, we use the trapezoidal rule to approximate the RHS. This rule states that if T_s is small, the integral (area) can be approximated by the average height of $x(t)$ between the limits and then multiplying by the width. Equation (4.52) then becomes

$$y(nTs) - y((n-1)Ts) = \frac{x(nT_s) + x((n-1)T_s)}{2} T_s. \tag{4.53}$$

Define $x(n) \triangleq x(nT_s)$ and $y(n) \triangleq y(nT_s)$, which is standard practice in signal processing. Equation (4.53) now becomes

$$y(n) - y(n-1) = \frac{T_s}{2}[x(n) + x(n-1)]. \tag{4.54}$$

Taking the Z-transform of the above gives

$$X(z) = \frac{2}{T_s} \frac{1 - z^{-1}}{1 + z^{-1}} Y(z). \tag{4.55}$$

Comparing (4.55) with (4.50), we get the transformation

$$s = \frac{2}{T_s} \frac{1 - z^{-1}}{1 + z^{-1}}. \tag{4.56}$$

The transformation from s to z is

$$z = \frac{1 + \frac{T_s}{2} s}{1 - \frac{T_s}{2} s}. \tag{4.57}$$

By substituting $s = \sigma + j\Omega$, it is easy to show that the s-plane maps to the z-plane in just the same way as in the impulse invariant transformation. Therefore, a stable analog filter yields a stable digital filter.

Now let us examine how the analog frequency maps to digital. Let $z = e^{j\omega}$; that is, ω is the frequency of the digital system. By substituting $z = e^{j\omega}$ and $s = j\Omega$ in (4.56), we get

$$\begin{aligned} j\Omega &= \frac{2}{T_s} \frac{1 - e^{-j\omega}}{1 + e^{-j\omega}} \\ &= \frac{2}{T_s} \frac{e^{-j\frac{\omega}{2}}(e^{j\frac{\omega}{2}} - e^{-j\frac{\omega}{2}})}{e^{-j\frac{\omega}{2}}(e^{j\frac{\omega}{2}} + e^{-j\frac{\omega}{2}})}. \end{aligned} \tag{4.58}$$

With some simple manipulations, (4.58) then becomes

$$\Omega = \frac{2}{T_s} \tan \frac{\omega}{2} \tag{4.59}$$

or

$$\omega = 2 \tan^{-1} \frac{\Omega T_s}{2}. \tag{4.60}$$

This mapping from the analog imaginary s-axis to the digital Z unit circle and is shown in Fig. 4.14 for two different sampling frequencies. The mapping is obviously not linear and is called *frequency warping*. In a digital filter design problem, the specifications are given in terms of the digital frequency. If we design the filter via an analog prototype, then the corresponding analog frequencies must be first determined from (4.59). This process is called *prewarping*. Prewarping will preserve the edge frequencies but not the exact shape of the magnitude response. This is illustrated in Fig. 4.15, where we transform

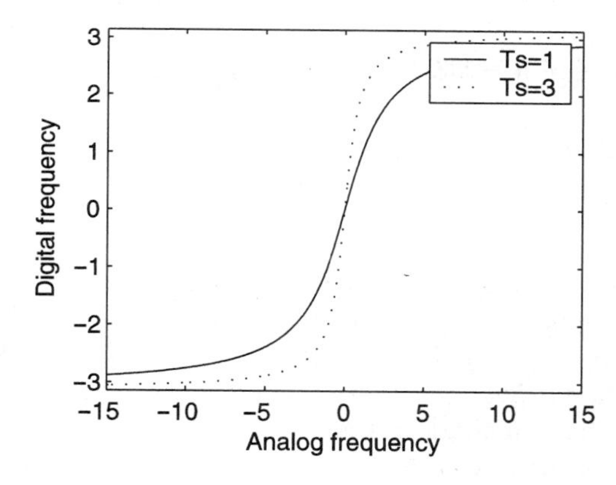

Figure 4.14 Frequency mapping in the bilinear transformation.

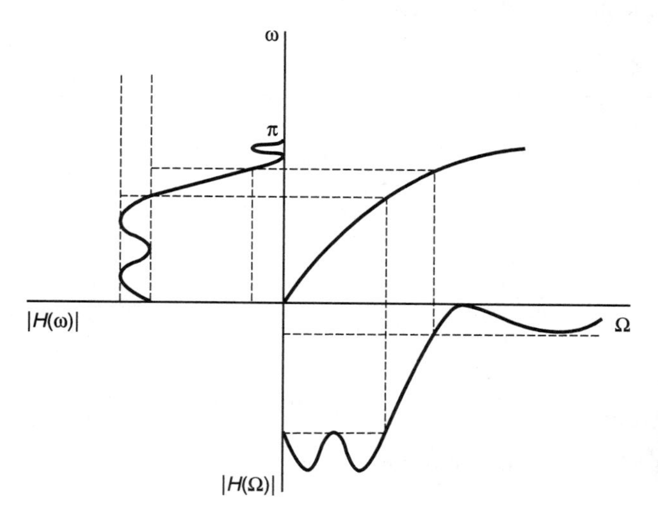

Figure 4.15 Mapping of an analog LPF to digital via the bilinear transformation.

an analog LPF to digital. For lower frequencies, the mapping is quite linear, and so the passband is hardly distorted. Therefore, this distortion of frequency response is not a problem for most lowpass filters. However, for higher frequencies, the mapping is quite nonlinear, and hence the stopband ripples are distorted. Therefore, this distortion of frequency response is not a problem for most lowpass filters. For other types of filters, this distortion may be too severe to be acceptable in many cases. Compared to the impulse invariant transformation, the advantage of the bilinear transformation is the absence of aliasing. The disadvantage is, of course, the distortion of the frequency response.

Analog Filter Design

The field of analog filter design is well established, and many papers and books have been written on this subject. There are several popular types of analog filters, some of which will be discussed briefly. Only lowpass analog filters will be discussed, followed by a presentation of the transformation methods to other types of filters.

Butterworth Filter

The magnitude response of an LP Butterworth filter is given by

$$|H(\Omega)| = \frac{1}{\sqrt{1 + \left(\frac{\Omega}{\Omega_c}\right)^{2N}}} \tag{4.61}$$

where Ω_c is the 3-dB or cutoff frequency and N is a positive integer. The term $(\frac{\Omega}{\Omega_c})^{2N}$ in the denominator has an important implication. It ensures that $H(0) = 1$. In addition, all derivatives $H'(0) = 1, H''(0) = 1, \ldots$. This means that the function is as flat as possible at $\Omega = 0$, without the function being equal to unity everywhere, and that the Butterworth filter is *maximally flat*. The magnitude response is plotted in Fig. 4.16(a) for several values of N. The transition band clearly narrows with increasing N. In addition, the magnitude response is monotonic in both the passband and stopband. As we know, the specifications include the passband and stopband edge frequencies (Ω_p, Ω_s in analog) and

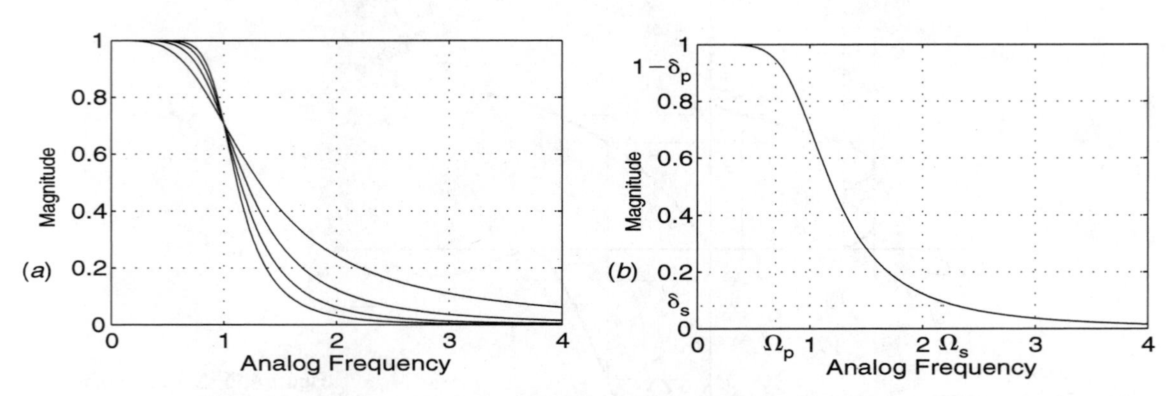

Figure 4.16 (a) Magnitude response of Butterworth filters. (b) Filter specifications in the analog domain.

the corresponding ripples (δ_p, δ_s). These specifications are illustrated in Fig. 4.16(b) with $N = 3$ as an example. Using (4.61), we can find a formula for N so that these specifications are satisfied. For the passband edge, we have

$$\frac{1}{1+\left(\frac{\Omega_p}{\Omega_c}\right)^{2N}} = (1-\delta_p)^2 \tag{4.62}$$

or

$$\left(\frac{\Omega_p}{\Omega_c}\right)^{2N} = \frac{1}{(1-\delta_p)^2} - 1. \tag{4.63}$$

Similarly, the stopband edge gives

$$\left(\frac{\Omega_s}{\Omega_c}\right)^{2N} = \frac{1}{\delta_s^2} - 1. \tag{4.64}$$

Dividing (4.63) by (4.64) and a few simple manipulations gives

$$N = \frac{1}{2}\frac{\log\left\{\frac{1}{(1-\delta_p)^2} - 1\right\} - \log\left\{\frac{1}{\delta_s^2} - 1\right\}}{\log\left\{\frac{\Omega_p}{\Omega_s}\right\}}. \tag{4.65}$$

Of course, N must be rounded to the nearest integer. Once N is found, the 3-dB frequency can be obtained from (4.63). The final formula is

$$\Omega_c = \frac{\Omega_p}{\left(\frac{1}{(1-\delta_p)^2} - 1\right)^{\frac{1}{2N}}}. \tag{4.66}$$

The task now is to find the transfer function $H(s)$ of the analog Butterworth filter. We first note that

$$\begin{aligned} H(s)H(-s)|_{s=j\Omega} &= |H(\Omega)|^2 \\ &= \frac{1}{1+\left(\frac{\Omega}{\Omega_c}\right)^{2N}} \\ &= \frac{1}{1+\left(\frac{j\Omega}{j\Omega_c}\right)^{2N}} \\ &= \frac{1}{1+\left(\frac{s}{j\Omega_c}\right)^{2N}}. \end{aligned} \tag{4.67}$$

The roots of the denominator are

$$s = (-1)^{\frac{1}{2N}} \left(j\Omega_c\right). \tag{4.68}$$

Since $-1 = e^{j\pi(2k+1)}, k = 0, 1, \ldots$ and $j = e^{j\pi/2}$, the pole positions are given by

$$s_k = e^{j\frac{\pi(2k+1)}{2N}} e^{j\pi/2}\Omega_c, k = 0, 1, \ldots, 2N-1. \tag{4.69}$$

The poles are therefore on a circle with radius Ω_c and are placed at angles

$$\theta_k = \frac{\pi}{N}k + \frac{\pi}{2N} + \frac{\pi}{2}, \quad k = 0, 1, \ldots, 2N-1. \tag{4.70}$$

For $N = 3$ and $N = 4$, the poles are shown in Fig. 4.17. These are the poles of $H(s)H(-s)$. To synthesize the final transfer function $H(s)$, we choose the left-half plane poles (for stability).

▶ EXAMPLE 4.5 *Design a lowpass Butterworth digital filter with the following specifications: (a) passband edge frequency, $\omega_p = 0.4\pi$; (b) stopband edge frequency, $\omega_s = 0.6\pi$; (c) passband ripple, $\delta_p = 0.15$; and (d) stopband ripple, $\delta_s = 0.25$. Use the method of bilinear transformation.*

SOLUTION The first step is to prewarp the digital edge frequencies to the analog domain by using (4.59). This gives

$$\Omega_p = \frac{2}{T_s} \tan\frac{0.4\pi}{2} = \frac{1.45}{T_s}$$

and

$$\Omega_s = \frac{2}{T_s} \tan\frac{0.6\pi}{2} = \frac{2.75}{T_s}.$$

Since we are given normalized digital frequencies, it is convenient to use $T_s = 1$. The filter order is found by using (4.65),

$$\begin{aligned} N &= \frac{1}{2}\frac{\log\left\{\frac{1}{0.25^2} - 1\right\} - \log\left\{\frac{1}{0.85^2} - 1\right\}}{\log\left\{\frac{2.75}{1.45}\right\}} \\ &= 2.86 \simeq 3. \end{aligned}$$

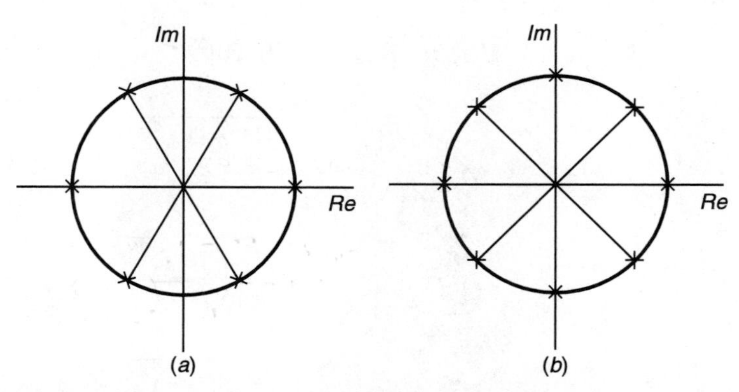

Figure 4.17 Pole locations for an analog Butterworth filter of order (*a*) $N = 3$ (*b*) $N = 4$.

Using (4.66), we find that the 3-dB frequency is

$$\Omega_c = \frac{1.45}{\left(\frac{1}{(1-\delta_p)^2} - 1\right)^{\frac{1}{2N}}}$$
$$= 1.7.$$

Therefore, the analog lowpass prototype filter, $H(s)$, has three poles with a magnitude of 1.7 and are located on the left-half plane as shown in Fig. 4.17(a). These poles are

$$s_1 = -1.7$$
$$s_{2,3} = 1.7e^{\pm j\frac{2\pi}{3}}.$$

The filter transfer function then becomes

$$H(s) = \frac{1}{(s-s_1)(s-s_2)(s-s_3)}$$
$$= \frac{1}{s^3 + 3.4s^2 + 5.78s + 4.91}.$$

Now we adjust the numerator so that the gain at zero-frequency is unity. The final LP analog prototype is

$$H(s) = \frac{4.91}{s^3 + 3.4s^2 + 5.78s + 4.91}.$$

Bilinear transformation is now performed to convert the analog filter to digital. The final digital filter is

$$H(z) = H(s)|_{s=2\frac{1-z^{-1}}{1+z^{-1}}}$$
$$= \frac{0.129 + 0.3867z^{-1} + 0.3869z^{-2} + 0.129z^{-3}}{1 - 0.2971z^{-1} + 0.3564z^{-2} - 0.0276z^{-3}}.$$

The magnitude and phase responses of this filter are shown in Fig. 4.18. The specifications are clearly satisfied. The phase response is nonlinear; however, in the passband it is quite linear. The above calculations are somewhat tedious. Fortunately, the MATLAB signal processing toolbox has commands that make the design of Butterworth filters almost a trivial task. The commands **butterord(...)** and **butter(...)** used in conjunction can achieve any given filter design task. These commands are described at the end of the chapter.

The monotonic response in the passband and stopband are desirable characteristics of the Butterworth filter. The Butterworth response is maximally flat, which is also desirable. However, the transition band is rather wide. A sharper transition band requires a higher filter order. This is the disadvantage of the Butterworth filter.

Chebyshev Filter

The Chebyshev polynomial yields *equiripple* behavior in the passband or the stopband. Let us explain this now. Consider the problem of approximating a desired magnitude

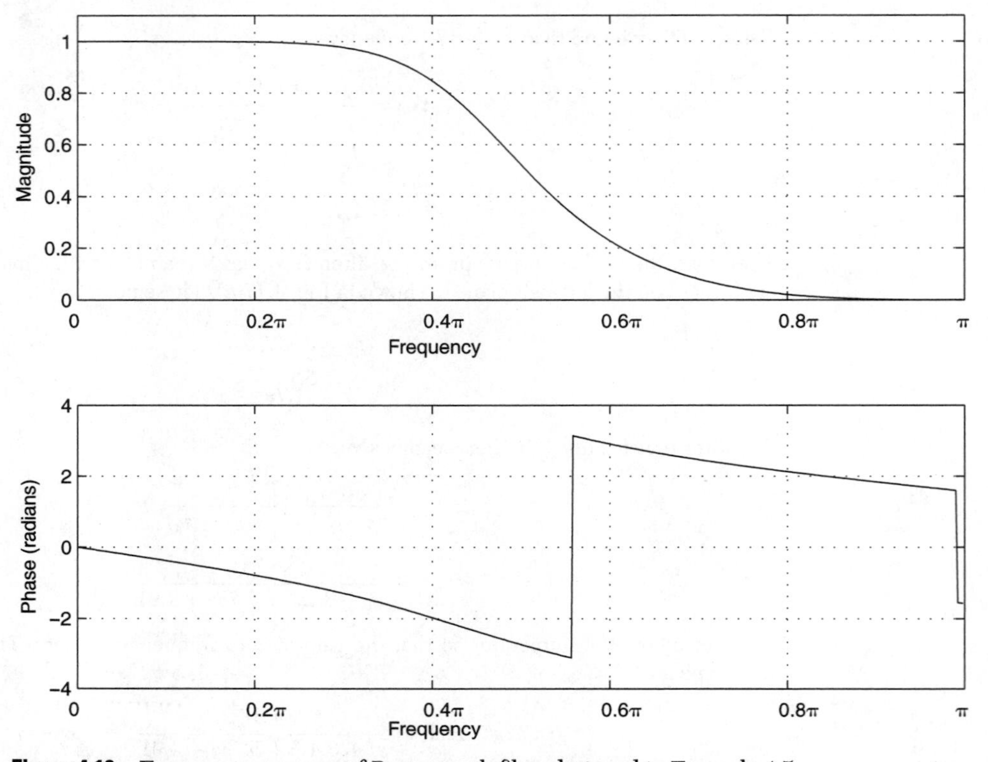

Figure 4.18 Frequency response of Butterworth filter designed in Example 4.5.

response. The designed filter should closely match the desired response but is expected to have some error. Equiripple behavior means that the maxima and minima of this error have equal magnitudes. The Chebyshev filter has ripples that are uniform in the passband or the stopband. There are two types, Chebyshev-I and Chebyshev-II filters. The Chebyshev-I filter is equiripple in the passband and monotonic in the stopband, whereas the Chebyshev-II is just the opposite.

The magnitude response of the Chebyshev-I filter is given by

$$|H(\Omega)| = \frac{1}{\sqrt{1+\epsilon^2 C_N^2(\Omega/\Omega_c)}}, \tag{4.71}$$

where ϵ is a parameter that controls the ripple amplitude and $C_N(x)$ is the Chebyshev polynomial of the first kind and degree N given by [4]:

$$C_N(x) = \begin{cases} \cos(N\cos^{-1}x), & 0 \le x \le 1 \\ \cosh(N\cosh^{-1}x), & \text{otherwise.} \end{cases} \tag{4.72}$$

It can be shown (see Problem 16) that the Chebyshev polynomial may be generated by using the recursive equations:

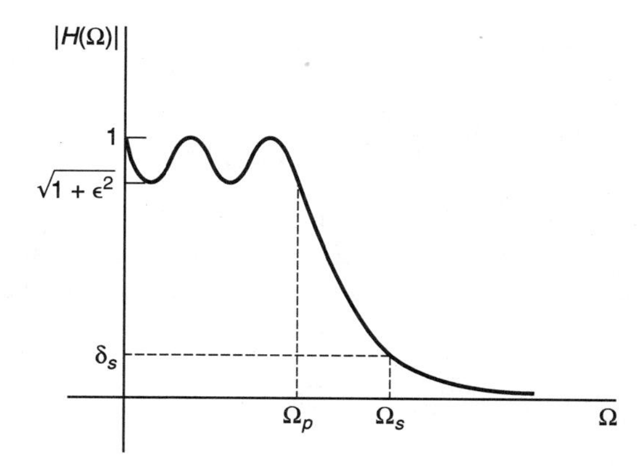

Figure 4.19 Typical magnitude response of a Chebyshev-I filter.

$$C_N(x) = 2xC_{N-1}(x) - C_{N-2}(x),$$
$$\text{where } C_0(x) = 1, \quad C_1(x) = x. \tag{4.73}$$

The magnitude response of the Chebyshev-I filter is shown in Fig. 4.19 for N odd. We note that $|H(0)| = 1$. The filter has uniform ripple (equiripple) in the passband and is monotonic in the stopband. For N even, the only difference is that the ripple begins at the minimum, that is, $|H(0)| = 1/\sqrt{1+\epsilon^2}$. The poles of this filter lie on an ellipse in the s-plane. There are formulas for the pole positions. Here we will simply present these formulas without derivations. The formulas are given with respect to the angular pole positions of the Butterworth filter given by

$$\theta_k = \frac{\pi}{N}k + \frac{\pi}{2N} + \frac{\pi}{2}, \quad k = 0, 1, \ldots, N-1. \tag{4.74}$$

The Chebyshev-I poles are

$$s = a\Omega_c \cos\theta_k \pm jb\Omega_c \sin\theta_k, \quad k = 0, 1, \ldots, N-1. \tag{4.75}$$

The constants a and b are given by

$$a = 0.5(\gamma^{1/N} - \gamma^{-1/N}),$$
$$b = 0.5(\gamma^{1/N} + \gamma^{-1/N}),$$
$$\text{where } \gamma = \frac{1}{\epsilon} + \sqrt{1 + \frac{1}{\epsilon^2}}. \tag{4.76}$$

These formulas require that we first calculate the filter order N and the ripple factor ϵ. These parameters can be derived as follows. For the analog lowpass prototype filter, we have the passband ripple δ_p, the stopband ripple δ_s, the passband edge frequency Ω_p, and the stopband edge frequency Ω_s. From the magnitude response (4.71), at the passband edge, we have

$$1 - \delta_p^2 = \frac{1}{1 + \epsilon^2 C_N^2\left(\Omega_p/\Omega_c\right)}. \tag{4.77}$$

At the stopband edge, we have

$$\delta_s^2 = \frac{1}{1+\epsilon^2 C_N^2\left(\Omega_s/\Omega_c\right)}. \tag{4.78}$$

In the Butterworth case, we divided the formulas that correspond to the two above, and Ω_c canceled out. In this case, such a cancellation does not happen. Therefore, for simplicity, we choose $\Omega_c = \Omega_p$. With this substitution, from (4.77), we get

$$\epsilon = \sqrt{\frac{1}{\left(1-\delta_p\right)^2} - 1}. \tag{4.79}$$

By substituting $\Omega_c = \Omega_p$ in (4.78) and a few simple manipulations, we get

$$N = \frac{\cosh^{-1}\left\{\left(\frac{1}{\epsilon}\right)\sqrt{\frac{1}{\delta_s^2} - 1}\right\}}{\cosh^{-1}\left(\Omega_s/\Omega_p\right)}. \tag{4.80}$$

This completes the set of formulas needed for the design. An example is now in order.

▶ **EXAMPLE 4.6** *Design a bandpass Chebyshev digital filter with the following specifications: (a) lower stopband edge frequency, $\omega_{sl} = 0.2\pi$; (b) lower passband edge frequency, $\omega_{pl} = 0.3\pi$; (c) upper passband edge frequency, $\omega_{pu} = 0.6\pi$; (d) upper stopband edge frequency, $\omega_{su} = 0.7\pi$; (e) passband ripple, $\delta_p = 0.1$; and (f) stopband ripple, $\delta_s = 0.1$. Use the method of bilinear transformation.*

The first step is to prewarp the digital edge frequencies to analog through the bilinear transformation formula (4.59). The analog frequencies are $\Omega_{sl}, \Omega_{pl}, \Omega_{pu}, \Omega_{su}$. Then we design the lowpass prototype with passband edge frequency: $\Omega_p = \Omega_{sl}$, stopband edge frequency: $\Omega_s = \Omega_{pl}$, and passband ripple and stopband ripple of 0.1. The lowpass prototype filter, $H(s)$, is designed using the formulas given in (4.80), (4.79), (4.75), and (4.76). Then we convert this analog lowpass to analog bandpass using a transformation given in the next subsection. Next, the analog bandpass is transformed to digital using the bilinear transformation. All of these steps can be performed painlessly by the MATLAB commands **cheb1ord(...)** and **cheby1(...)**. The details of these commands are given at the end of the chapter. The first command yields the filter order $N=4$. This is passed on to the second command, which designs the bandpass filter. The magnitude and phase responses are shown in Fig. 4.20. We clearly have equiripple in the passband and monotonic stopbands. The specifications on the frequency edges and ripples are satisfied. The phase response in the passband is quite linear but not as linear as in the case of the Butterworth filter. The Chebyshev filter has sharper transition bands than the Butterworth. This comes at the expense of ripples in the passband.

The Chebyshev-II filter is equiripple in the stopband and monotonic in the passband. The magnitude response of this filter is given by

$$|H(\Omega)| = \frac{1}{\sqrt{1+\epsilon^2 C_N^2(\Omega_s/\Omega_p)/C_N^2(\Omega_s/\Omega)}}. \tag{4.81}$$

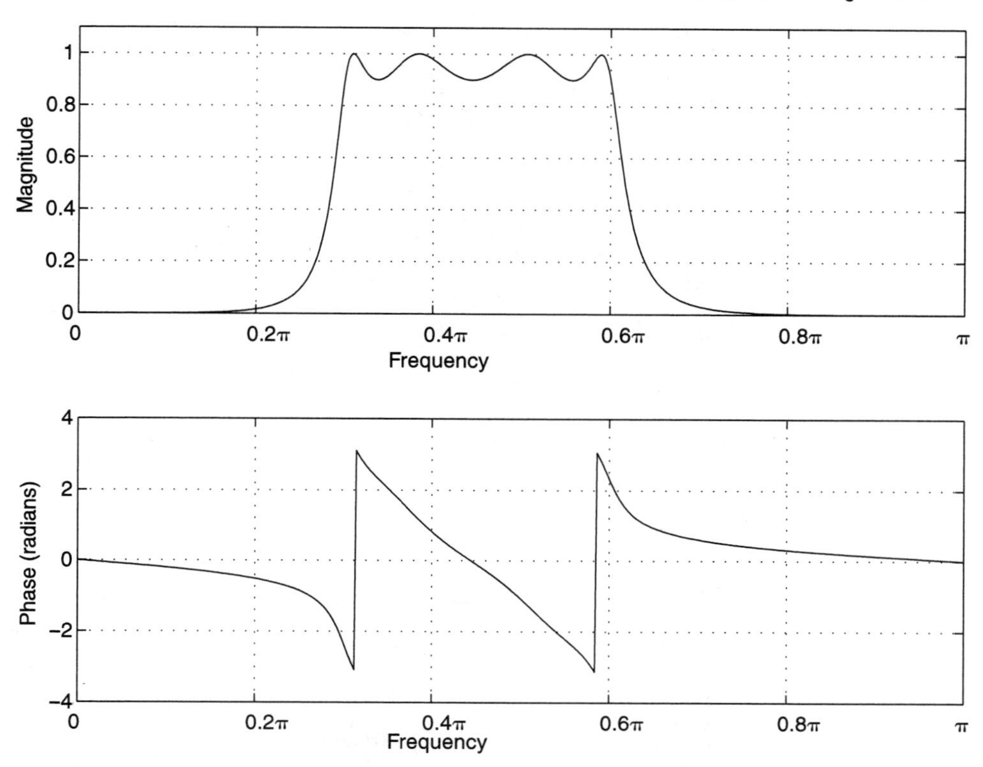

Figure 4.20 Frequency response of Chebyshev-I filter designed in Example 4.6.

The design formulas for calculating N, ϵ, and the pole positions can be obtained in a similar fashion as for the Chebyshev-I filter. These derivations and formulas are omitted. The MATLAB commands **cheb2ord(...)** and **cheby2(...)** can be used to design this type of filter. The above example is now used for a Chebyshev-II filter design. The lowpass prototype has an order of $N = 4$. The frequency response of the final filter is given in Fig. 4.21. The specifications are again satisfied, and we have equiripple in the stopband.

Elliptic Filter

The elliptic filter is also known as the *Cauer* filter. This filter exhibits a sharper transition band (for the same filter order) than Butterworth and Chebyshev filters. The sharp transition band comes at the expense of equiripple behavior in both the passband and stopband and a more nonlinear phase response in the passband. The magnitude response of the elliptic filter is given by

$$|H(\Omega)| = \frac{1}{\sqrt{1 + \epsilon^2 U_N^2(\Omega/\Omega_c)}} \tag{4.82}$$

where $U_N^2(.)$ is a Jacobian elliptic function of order N. The theory of elliptic functions and integrals is quite involved and beyond the scope of this book. For further information

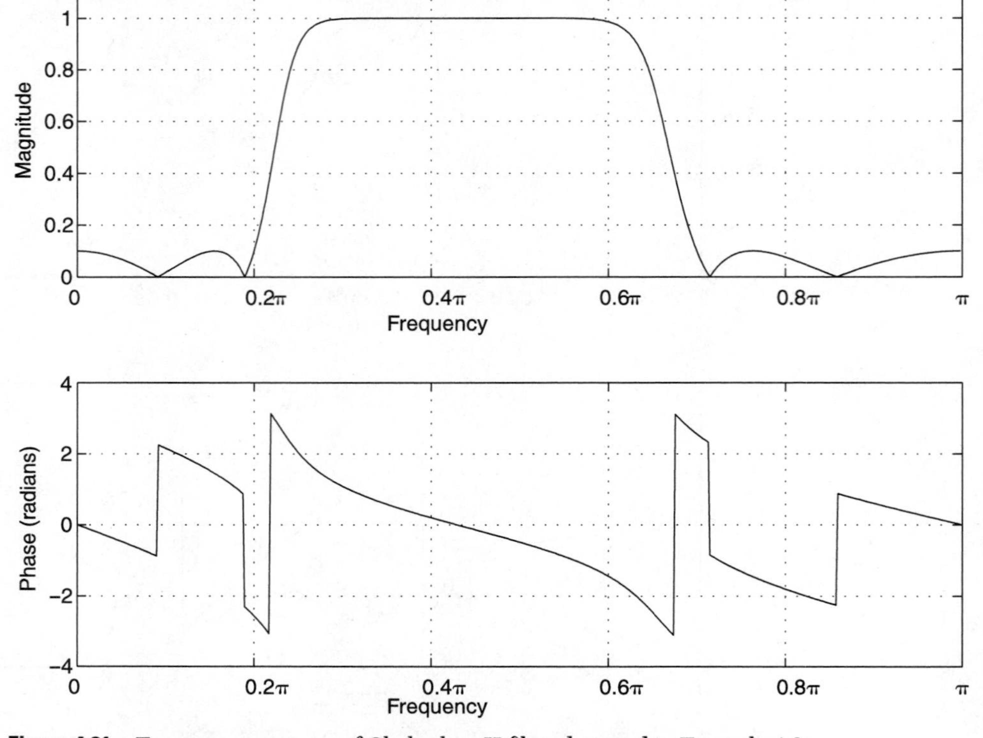

Figure 4.21 Frequency response of Chebyshev-II filter designed in Example 4.6.

on this topic, the reader is referred to texts by Antoniou [5], [6], and Parks/Burrus [7]. Given a set of specifications, the formula for calculating the filter order N involves an elliptic integral of the first kind. Several approximate formulas for finding N have been proposed. Without getting into the details, we will simply say that the MATLAB commands **ellipord**(...) and **ellip**(...) can be conveniently used for designing elliptic filters. These commands are now used to design the bandpass filter of the last example. The order of the analog lowpass prototype is found to be $N = 3$. Recall that the order was $N = 4$ in the case of the Chebyshev filters. The frequency response of the final filter is shown in Fig. 4.22.

Bessel Filter

The transfer function of the analog Bessel filter is given by

$$H(s) = \frac{1}{B_N(s)} \tag{4.83}$$

where $B_N(s)$ is the Bessel polynomial of Nth order. These polynomials can be generated in following ways.

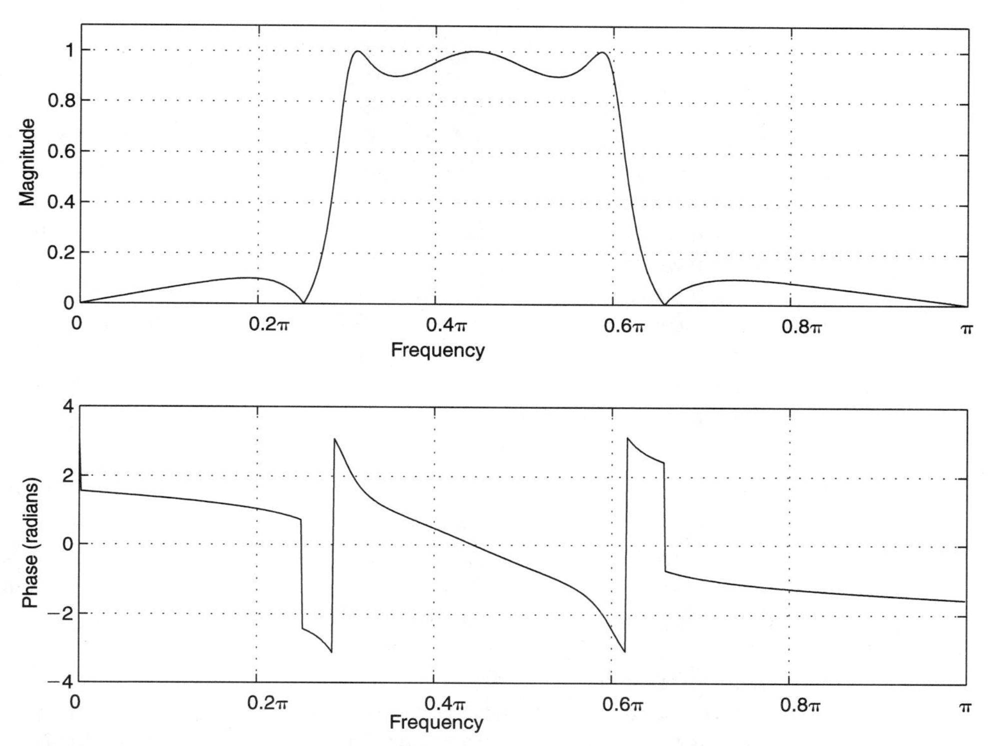

Figure 4.22 Frequency response of sixth order elliptic BPF.

1. The Bessel polynomial is

$$B_N(s) = b_0 + b_1 s + b_2 s^2 + \cdots + b_N s^N \tag{4.84}$$

where the coefficients are given by

$$b_k = \frac{(2N-k)!}{2^{N-k} k!(N-k)!}. \tag{4.85}$$

2. From the above it can be shown that the Bessel polynomials can be generated recursively as

$$\begin{aligned} B_N(s) &= (2N-1)B_{N-1}(s) + s^2 B_{N-2}(s), \\ \text{with } B_0(s) &= 1, \quad B_1(s) = s + 1. \end{aligned} \tag{4.86}$$

The magnitude response of the Bessel filter can now be written as

$$|H(\Omega)| = \frac{1}{\sqrt{(b_0 - b_2\Omega^2 + b_4\Omega^4 - \cdots)^2 + (b_1\Omega - b_3\Omega^3 + b_5\Omega^5 - \cdots)^2}}. \tag{4.87}$$

The order of the filter is chosen so that the magnitude response satisfies the specifications on the passband and stopband edges and the corresponding ripples. Then $H(s)$ can be transformed to HP, BP, or BS as required (next subsection). The analog filter is then transformed to digital by impulse invariant or bilinear transformation. The design method is therefore quite simple. However, there is no convenient MATLAB command that can be used to directly design a Bessel digital filter. The command **besself**(...) designs an analog prototype. This command can be used in developing a MATLAB program to design a Bessel digital filter.

The most important characteristic of the Bessel filter is its linear phase response in the passband. This comes at the cost of a terrible transition band. For a given filter order, the Bessel filter has the widest transition band of all the filters discussed. If we use bilinear transformation to transform the analog Bessel filter to digital, the linear phase response in the passband may be destroyed. This is due to the warping of the frequency axis in bilinear transformation (Fig. 4.15). If the passband of the filter is in the linear part of the frequency mapping, then the linear phase characteristic is somewhat preserved.

Spectral Transformations

Spectral transformation is the process of converting a lowpass filter to highpass, bandpass, or bandstop. In fact, it can also transform a lowpass filter to another lowpass filter with a different cutoff frequency. These transformations can be performed in the analog or digital domain. Of course, the formulas are different for each case.

Analog Transformation

Let $H(s)$ be the analog lowpass prototype filter with a passband edge frequency of Ω_p. The lowpass-to-lowpass transformation is simply a scaling of the analog frequency axis. If Ω_p^{LP} is the desired passband edge frequency of the new LP filter, then the transformation takes the form

$$s \longrightarrow \frac{\Omega_p}{\Omega_p^{\mathrm{LP}}} s \tag{4.88}$$

where $\longrightarrow$ is read as "is replaced by."

The lowpass-to-highpass transformation is simply achieved by replacing s by $1/s$. If the desired HP filter has a passband edge of Ω_p^{HP}, the transformation is

$$s \longrightarrow \frac{\Omega_p \Omega_p^{\mathrm{HP}}}{s}. \tag{4.89}$$

For a lowpass-to-bandpass transformation, it should be recognized that a BP filter is essentially a combination of an LPF and a HPF. For the BPF, let Ω_l^{BP} and Ω_u^{BP} denote the lower and upper passband edge frequencies. The transformation is given by

$$s \longrightarrow \Omega_p \frac{s^2 + \Omega_l^{\mathrm{BP}} \Omega_u^{\mathrm{BP}}}{s(\Omega_u^{\mathrm{BP}} - \Omega_l^{\mathrm{BP}})}. \tag{4.90}$$

Because of the quadratic nature of the right-hand side, for any value of the old frequency, we have two values for the new frequency. That is, the mapping is performed via two different curves as shown in Fig. 4.23 [8]. Note that the dc of the LPF is mapped to one point, which is the geometrical center of the BPF. This center is given by $\Omega_m = \sqrt{\Omega_l^{\mathrm{BP}} \Omega_u^{\mathrm{BP}}}$.

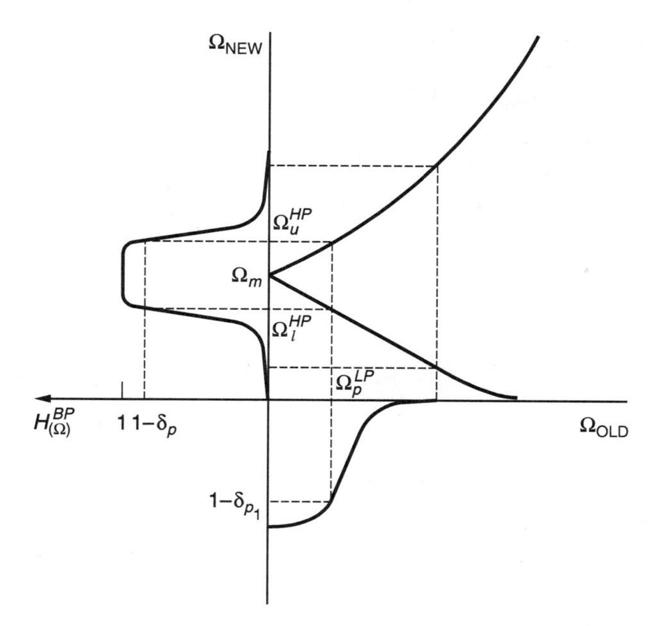

Figure 4.23 Mapping for LP to BP transformation.

Also, note that Ω_p^{LP} maps to two different points, which are Ω_l^{BP} and Ω_u^{BP}. It is also easy to see that the stopband of the LPF maps to the two stopbands of the BPF. The passband of the LPF maps to the passband of the BPF. Although this mapping is nonlinear, it causes very little distortion of the shape of the magnitude as seen from the figure.

The lowpass to bandstop transformation now follows from the above as

$$s \longrightarrow \Omega_p \frac{s(\Omega_u^{\text{BS}} - \Omega_l^{\text{BS}})}{s^2 + \Omega_l^{\text{BS}} \Omega_u^{\text{BS}}}. \tag{4.91}$$

These analog-to-analog transformations are summarized in Table 4.3. The MATLAB commands to implement these functions are given later in the chapter.

Digital Transformation and the Allpass Filter

A set of transformations can be derived for converting an LP digital prototype filter $H(z)$ with a passband edge frequency of ω_p to another digital filter, which may be LP, HP, BP, or BS. This transformation is achieved by replacing z^{-1} by a function $F(\hat{z}^{-1})$, that is,

$$z^{-1} \rightarrow F(\hat{z}^{-1}). \tag{4.92}$$

The transformation must satisfy the following set of conditions.

1. The mapping must transform the LP prototype filter to the desired type of filter.
2. The inside of the unit circle is mapped into itself.
3. The unit circle is mapped into itself.

On the unit circle, we have $z = e^{j\omega}$ and $\hat{z} = e^{j\hat{\omega}}$. The third condition implies that

$$e^{-j\omega} = F(e^{-j\hat{\omega}}) \tag{4.93}$$

TABLE 4.3 Analog-to-Analog Spectral Transformations

Ω_p: Passband edge frequency of prototype analog LPF.

LP–LP

$$s \longrightarrow \frac{\Omega_p}{\Omega_p^{\mathrm{LP}}} s$$

LP–HP

$$s \longrightarrow \frac{\Omega_p \Omega_p^{\mathrm{HP}}}{s}$$

LP–BP

$$s \longrightarrow \Omega_p \frac{s^2 + \Omega_l^{\mathrm{BP}} \Omega_u^{\mathrm{BP}}}{s(\Omega_u^{\mathrm{BP}} - \Omega_l^{\mathrm{BP}})}$$

LP–BS

$$s \longrightarrow \Omega_p \frac{s(\Omega_u^{\mathrm{BS}} - \Omega_l^{\mathrm{BS}})}{s^2 + \Omega_l^{\mathrm{BS}} - \Omega_u^{\mathrm{BS}}}$$

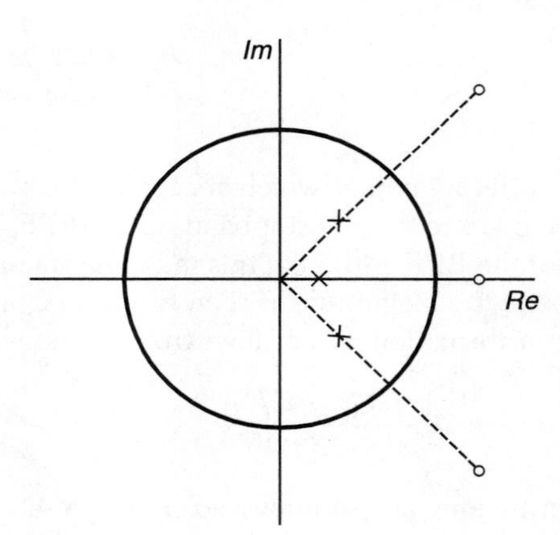

Figure 4.24 Pole-zero locations for an all-pass filter.

which in turn implies that $|F(e^{j\hat{\omega}})| = 1$. Therefore, $F(\hat{z}^{-1})$ is said to be an *allpass* system. It passes all frequencies without attenuation. Allpass filters are useful for modifying the phase response of a system. Now, we will digress a little from the digital transformation and briefly discuss allpass filters.

A simple first-order allpass filter has the transfer function

$$H_{ap}(z) = \frac{z^{-1} - \alpha^*}{1 - \alpha z^{-1}} \tag{4.94}$$

where the pole of the system is at α and the zero is at $1/\alpha^*$; that is, the pole and the zero are conjugate reciprocals of each other (see Fig. 4.24). For a causal system to be stable, the pole must be inside the unit circle. It is easy to show (Problem 17) that the magnitude response of this filter is unity. The phase response can be changed by varying α. The phase responses of this filter for some selected real values of α are shown in Fig. 4.25. The general form of an allpass filter can be written as

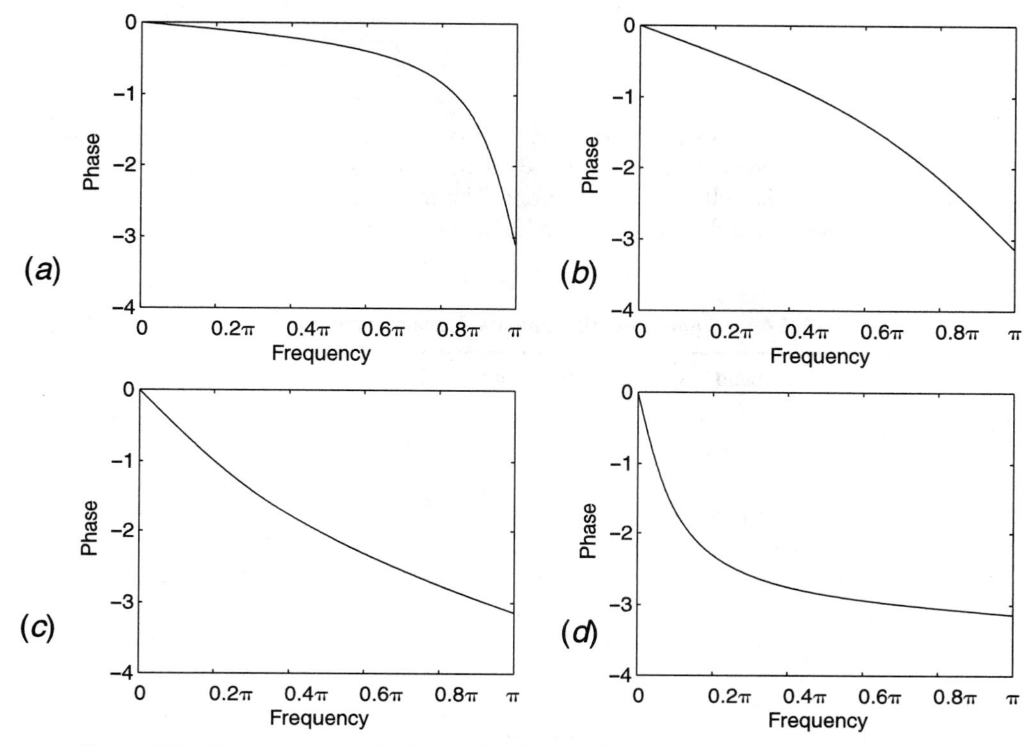

Figure 4.25 Phase response of a first-order all-pass filter with (*a*) $\alpha = -0.75$, (*b*) $\alpha = -0.25$, (*c*) $\alpha = 0.25$, and (d) $\alpha = 0.75$.

$$H_{ap}(z) = \prod_{k=1}^{N} \frac{z^{-1} - \alpha_k^*}{1 - \alpha_k z^{-1}} \tag{4.95}$$

where N is the number of poles and zeros of the filter. If the filter must have a real impulse response, all complex poles must occur as conjugate pairs. In other words, if β is a complex pole of the filter, then β^* must also be a pole. This general form can be used to design allpass filters with a variety of different phase responses. Another useful representation of allpass filters is

$$H_{ap}(z) = \frac{z^{-N}A(z^{-1})}{A(z)} \tag{4.96}$$

where $A(z) = 1 + a_1 z^{-1} + a_2 z^{-2} + \cdots + a_N z^{-N}$. It is easy to see that the numerator is

$$z^{-N}A(z^{-1}) = a_N + a_{N-1}z^{-1} + \cdots + a_1 z^{-(N-1)} + z^{-N}. \tag{4.97}$$

The numerator and denominator polynomials have the same coefficients, except that they appear in reverse order.

Now we return to the digital-to-digital transformation. It can be shown [9] that the requirements (1–3) given earlier for this transformation are satisfied by the mapping

$$z^{-1} = \prod_{k=1}^{N} \frac{\hat{z}^{-1} - \alpha_k}{1 - \alpha_k \hat{z}^{-1}} \tag{4.98}$$

where α_k are real numbers. It is not difficult to show that this mapping satisfies the required conditions (2) and (3). This general mapping function can be used to generate a variety of filter transformations. The transformations of a lowpass prototype filter to LP, HP, BP, and BS are given in Table 4.4.

TABLE 4.4 Digital-to-Digital Spectral Transformations

ω_p: Passband edge frequency of digital LPF.

LP–LP

$$z^{-1} = \frac{\hat{z}^{-1} - \alpha}{1 - \alpha \hat{z}^{-1}}, \qquad \alpha = \frac{\sin\left(\frac{\omega_p^{LP} - \omega_p}{2}\right)}{\sin\left(\frac{\omega_p^{LP} + \omega_p}{2}\right)}$$

LP–HP

$$z^{-1} = -\frac{\hat{z}^{-1} + \alpha}{1 + \alpha \hat{z}^{-1}}, \qquad \alpha = \frac{\cos\left(\frac{\omega_p^{HP} + \omega_p}{2}\right)}{\cos\left(\frac{\omega_p^{HP} - \omega_p}{2}\right)}$$

LP–BP

$$z^{-1} = -\frac{\hat{z}^{-2} - a_1 \hat{z}^{-1} + a_2}{a_2 \hat{z}^{-2} - a_1 \hat{z}^{-1} + 1}, \qquad K = \cot\left(\frac{\omega_u^{BP} - \omega_l^{BP}}{2}\right) \tan\left(\frac{\omega_p}{2}\right)$$

$$\alpha = \frac{\cos\left(\frac{\omega_u^{BP} + \omega_l^{BP}}{2}\right)}{\cos\left(\frac{\omega_u^{BP} - \omega_l^{BP}}{2}\right)}$$

$$a_1 = \frac{2\alpha K}{K+1}$$

$$a_2 = \frac{K-1}{K+1}$$

LP–BS

$$z^{-1} = \frac{\hat{z}^{-2} - a_1 \hat{z}^{-1} + a_2}{a_2 \hat{z}^{-2} - a_1 \hat{z}^{-1} + 1}, \qquad K = \tan\left(\frac{\omega_u^{BS} - \omega_l^{BS}}{2}\right) \tan\left(\frac{\omega_p}{2}\right)$$

$$\alpha = \frac{\cos\left(\frac{\omega_u^{BS} + \omega_l^{BS}}{2}\right)}{\cos\left(\frac{\omega_u^{BS} - \omega_l^{BS}}{2}\right)}$$

$$a_1 = \frac{2\alpha}{K+1}$$

$$a_2 = \frac{1-K}{K+1}$$

4.3 FILTER STRUCTURES

The first and second subsections of this section cover FIR and IIR filter structures, respectively. In the third subsection we present lattice structures, which are applicable to both types of filters.

FIR Structures

The transfer function and difference equation for an Mth order FIR filter are given, respectively, by

$$H(z) = \frac{Y(z)}{X(z)} = h(0) + h(1)z^{-1} + h(2)z^{-2} + \cdots + h(M)z^{-M} \tag{4.99}$$

and

$$y(n) = h(0)x(n) + h(1)x(n-1) + h(2)x(n-2) + \cdots + h(M)x(n-M) \tag{4.100}$$

where $x(n)$ and $y(n)$ are the input and output, respectively. The flow diagram for this filter was given in Chapter 1 and is repeated for convenience in Fig. 4.26. This is the simplest and most direct implementation of this filter and is called the *Direct Form* FIR structure. It requires a total of $(M+1)$ multiplications. The most important feature of an FIR filter is that it can be easily designed to exhibit a linear phase (or generalized linear phase) characteristic. Recall from Chapter 3 that this can be achieved in four different ways, defined as FIR-I through FIR-IV.

In FIR-I, we have M odd. This implies that the total number of coefficients $(M+1)$ is even. In addition, the coefficients are symmetric; that is, $h(n) = h(M-n)$. The point of symmetry is $M/2$, which equals an integer plus one-half. Thus, $h(0) = h(M)$, $h(1) = h(M-1)$, $h(2) = h(M-2), \ldots, h(\frac{M-1}{2}) = h(\frac{M+1}{2})$. (see Fig. 3.7($a$) for an example.) This symmetry can be used to simplify the filter equation as follows:

$$\begin{aligned}
y(n) &= h(0)x(n) + h(1)x(n-1) + h(2)x(n-2) + \cdots \\
&\quad + h(M-2)x(n-M+2) + h(M-1)x(n-M+1) + h(M)x(n-M) \\
&= h(0)x(n) + h(1)x(n-1) + h(2)x(n-2) + \cdots \\
&\quad + h(2)x(n-M+2) + h(1)x(n-M+1) + h(0)x(n-M) \\
&= h(0)[x(n) + x(n-M)] + h(1)[x(n-1) + x(n-M+1)] \\
&\quad + h(2)[x(n-2) + x(n-M+2)] + \cdots \\
&\quad + h\left(\frac{M-1}{2}\right)\left[x\left(n - \frac{M-1}{2}\right) + x\left(n - \frac{M+1}{2}\right)\right].
\end{aligned} \tag{4.101}$$

This filter structure is shown in Fig. 4.27. It requires only one-half the number of multiplications as that of the Direct Form structure.

For the FIR-II filter, M is even. That is, we have an odd number of filter coefficients that are symmetric about the integer point $M/2$. The coefficient $h(\frac{M}{2})$ is unpaired, whereas

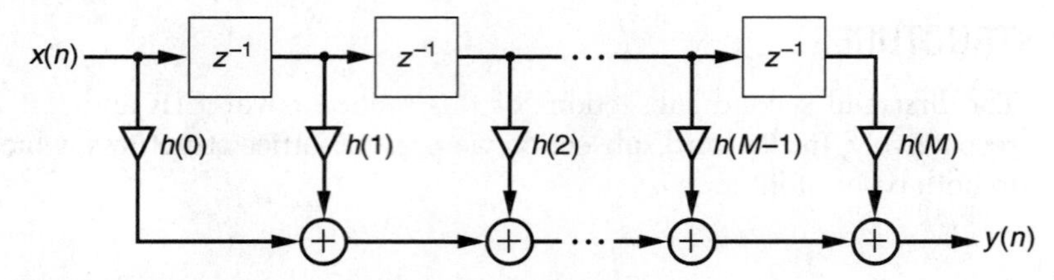

Figure 4.26 Direct Form structure of FIR filter.

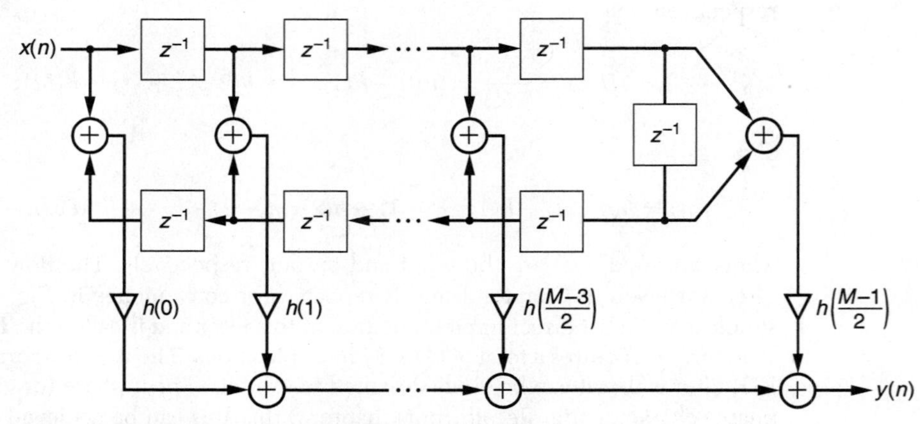

Figure 4.27 Implementation of FIR-I filter (M odd).

all the other coefficients are paired (see Fig. 3.7(b)). An analysis similar to the above yields the simplified equation

$$\begin{aligned} y(n) &= h\left(\frac{M}{2}\right)x\left(n-\frac{M}{2}\right) \\ &\quad + h(0)[x(n)+x(n-M)] + h(1)[x(n-1)+x(n-M+1)] \\ &\quad + h(2)[x(n-2)+x(n-M+2)] + \cdots \\ &\quad + h\left(\frac{M}{2}-1\right)\left[x\left(n-\frac{M}{2}+1\right)+x\left(n-\frac{M}{2}-1\right)\right]. \end{aligned} \tag{4.102}$$

This structure for the FIR-II filter is shown in Fig. 4.28.

FIR-III (M odd) and FIR-IV (M even) filters have antisymmetric coefficients, that is, $h(n) = -h(M-n)$. The filter structures are therefore the same as those for FIR-I and FIR-II, respectively, with the adders replaced by subtractors.

IIR Structures

The Direct Form IIR filter structures were derived in Chapter 1 and are called Direct Form I, II, I_T, and II_T, where the subscript T denotes the transposed form. It has been shown that the Direct Form I and I_T structures have redundant delay elements and are in

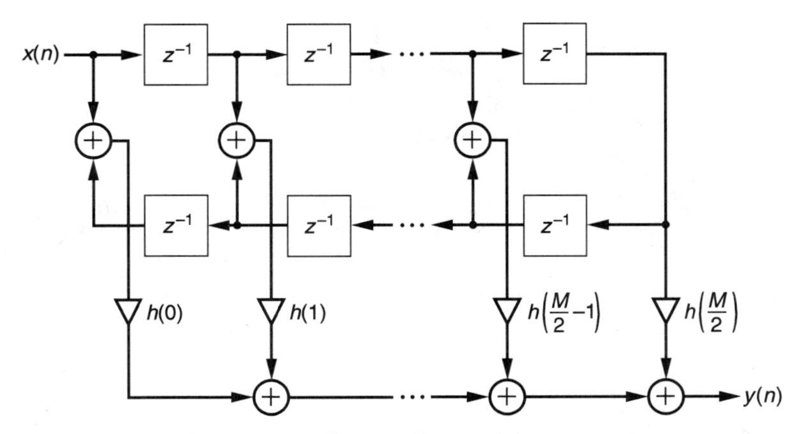

Figure 4.28 Implementation of FIR-II (M even).

general not used in practice. For convenience, we will repeat the derivation of the Direct Form II$_T$ and II structures for a second-order filter. Then we will present the Coupled Form structure and the parallel and cascade structures. For each of these structures, we will derive the state-space model.

Direct Form II$_T$

Consider the second-order IIR filter difference equation:

$$y(n) + a_1 y(n-1) + a_2 y(n-2) = b_0 x(n) + b_1 x(n-1) + b_2 x(n-2) \tag{4.103}$$

where $x(n)$ and $y(n)$ are the input and output, respectively. If z^{-1} denotes the delay operator, this equation can be rewritten as

$$\begin{aligned} y(n) &= b_0 x(n) + \{b_1 x(n-1) - a_1 y(n-1)\} + \{b_2 x(n-2) - a_2 y(n-2)\} \\ &= b_0 x(n) + z^{-1}\{b_1 x(n) - a_1 y(n)\} + z^{-2}\{b_2 x(n) - a_2 y(n)\} \\ &= b_0 x(n) + z^{-1}\{[b_1 x(n) - a_1 y(n)] + z^{-1}\{b_2 x(n) - a_2 y(n)\}\}. \end{aligned} \tag{4.104}$$

The flow diagram of this structure is shown in Fig. 4.29. This structure has five multipliers and four (two-input) adders.[4]

Now we derive the state-space model for this structure. The first step is to define the state variables. Since we have a second-order system, we have two state variables. These state variables are usually defined at the outputs of the delay elements as shown in the figure.[5] The state equation is found by writing equations for $v_1(n+1)$ and $v_2(n+1)$ in terms of $v_1(n), v_2(n)$, and $x(n)$. The output equation is obtained by writing an equation

[4] When filters are implemented in DSP, addition is performed with two signals at a time via the accumulator. Therefore, in the adder count, we always compute the number of 2-input additions.

[5] It is conventional to define $v_1(n)$ at the output of the rightmost delay, but it is not imperative.

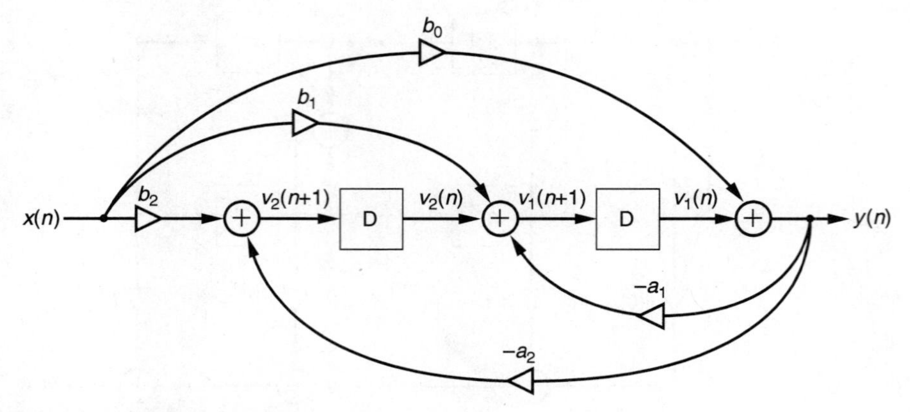

Figure 4.29 Direct Form II_T structure.

for $y(n)$ in terms of $v_1(n), v_2(n)$, and $x(n)$. These equations are easily obtained from the flow diagram as follows.

$$\begin{aligned} v_1(n+1) &= v_2(n) - a_1 y(n) + b_1 x(n) \\ &= v_2(n) - a_1[v_1(n) + b_0 x(n)] + b_1 x(n) \\ &= -a_1 v_1(n) + v_2(n) + [b_1 - a_1 b_0] x(n) \end{aligned} \tag{4.105}$$

$$\begin{aligned} v_2(n+1) &= -a_2 y(n) + b_2 x(n) \\ &= -a_2[v_1(n) + b_0 x(n)] + b_2 x(n) \\ &= -a_2 v_1(n) + [b_2 - a_2 b_0] x(n) \end{aligned} \tag{4.106}$$

$$y(n) = v_1(n) + b_0 x(n). \tag{4.107}$$

These equations can be written in matrix vector format to obtain the state-space model as follows.

State equation

$$\overbrace{\begin{bmatrix} v_1(n+1) \\ v_2(n+1) \end{bmatrix}}^{\mathbf{v}(n+1)} = \overbrace{\begin{bmatrix} -a_1 & 1 \\ -a_2 & 0 \end{bmatrix}}^{\mathbf{A}} \overbrace{\begin{bmatrix} v_1(n) \\ v_2(n) \end{bmatrix}}^{\mathbf{v}(n)} + \overbrace{\begin{bmatrix} b_1 - a_1 b_0 \\ b_2 - a_2 b_0 \end{bmatrix}}^{\mathbf{B}} x(n) \tag{4.108}$$

Output equation

$$y(n) = \underbrace{\begin{bmatrix} 1 & 0 \end{bmatrix}}_{\mathbf{C}} \underbrace{\begin{bmatrix} v_1(n) \\ v_2(n) \end{bmatrix}}_{\mathbf{v}(n)} + \underbrace{[b_0]}_{\mathbf{D}} x(n). \tag{4.109}$$

The importance of the state-space model is that it uniquely describes the filter structure. In addition, it gives valuable information about the internal states of the filter. These state variables can be easily solved for from the state equation and the initial conditions.

Direct Form II

The Direct Form II filter structure is obtained by using the algorithm derived in Chapter 1, Section 1.6. Accordingly, we introduce a new variable $g(n)$. The output variables $y(.)$ in (4.103) are replaced by $g(.)$, and the result is set equal to $x(n)$, to obtain the first equation. The second equation is obtained by replacing the input variables $x(.)$ in (4.103) by $g(.)$ and setting the result equal to $y(n)$. The resulting equations are

$$g(n) + a_1 g(n-1) + \cdots + a_N g(n-N) = x(n) \tag{4.110}$$

$$y(n) = b_0 g(n) + b_1 g(n-1) + \cdots + b_M g(n-M). \tag{4.111}$$

The flow diagram is easily obtained from these equations and is shown in Fig. 4.30. Again, this structure has five multipliers and four adders.

By defining the state variables as shown in the figure, we can obtain the state variable model as

$$\mathbf{v}(n+1) = \begin{bmatrix} 0 & 1 \\ -a_2 & -a_1 \end{bmatrix} \mathbf{v}(n) + \begin{bmatrix} 0 \\ 1 \end{bmatrix} x(n) \tag{4.112}$$

$$y(n) = \begin{bmatrix} b_2 - a_2 b_0 & b_1 - a_1 b_0 \end{bmatrix} \mathbf{v}(n) + [b_0] x(n). \tag{4.113}$$

Coupled Form

The Coupled Form structure is also called a *normal form* because the system matrix ($\mathbf{A}$) of its state-space model turns out to be normal. A real matrix is said to be normal if it satisfies the condition $\mathbf{A}^T\mathbf{A} = \mathbf{A}\mathbf{A}^T$. Because of this property, the Coupled Form filter structure has some advantages when implemented in a finite wordlength processor in

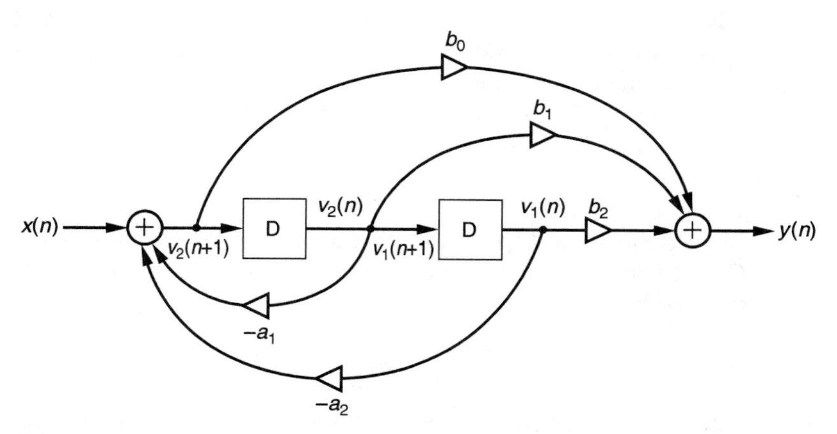

Figure 4.30 Direct Form II structure.

fixed point arithmetic. In practical implementation of filters, noise is introduced in the adders and multipliers due to the finite register lengths. This is called *roundoff noise*. In addition, with zero input, the filter can exhibit perpetual oscillations even though the filter has been designed to be stable. These are a result of accumulator overflow during addition and are called *overflow oscillations*. It can be shown that Coupled Form filters are completely free of overflow oscillations and have much lower roundoff noise than the Direct Form structures. These advantages come at the cost of increased multiplications as we shall see shortly. A comprehensive analysis of finite wordlength effects is given in Chapter 6.

We know from Chapter 3 that if $\{\mathbf{A}, \mathbf{B}, \mathbf{C}, D\}$ is a state-space model of a filter, then we can convert this to another state-space model $\{\widehat{\mathbf{A}}, \widehat{\mathbf{B}}, \widehat{\mathbf{C}}, \widehat{D}\}$ by a similarity transformation given by

$$\begin{aligned} \widehat{\mathbf{A}} &= \mathbf{P}^{-1}\mathbf{A}\mathbf{P} \\ \widehat{\mathbf{B}} &= \mathbf{P}^{-1}\mathbf{B} \\ \widehat{\mathbf{C}} &= \mathbf{C}\mathbf{P} \\ \widehat{D} &= D, \end{aligned} \tag{4.114}$$

where $\mathbf{P}$ is a nonsingular matrix. Reconsider the second-order digital filter of (4.103) and let the poles of the filter be $\sigma \pm j\omega$. Let the transformation matrix be

$$\mathbf{P} = \frac{1}{\omega}\begin{bmatrix} 0 & 1 \\ -\omega & \sigma \end{bmatrix}. \tag{4.115}$$

Using this matrix, we now perform a similarity transformation on the state-space model of the Direct Form II structure given by (4.112) and (4.113).The new system matrix then becomes

$$\widehat{\mathbf{A}} = \frac{1}{\omega}\begin{bmatrix} -\omega(\sigma + a_1) & a_2 + \sigma(\sigma + a_1) \\ -\omega^2 & \sigma\omega \end{bmatrix}. \tag{4.116}$$

At this point, it is necessary to derive formulas for (a_1, a_2) in terms of σ and ω. This is easily done using the quadratic formula to obtain $a_1 = -2\sigma$ and $a_2 = \sigma^2 + \omega^2$. Substituting these values in (4.116), we get

$$\widehat{\mathbf{A}} = \begin{bmatrix} \sigma & \omega \\ -\omega & \sigma \end{bmatrix}. \tag{4.117}$$

It is easy to check that $\widehat{\mathbf{A}}^T\widehat{\mathbf{A}} = \widehat{\mathbf{A}}\widehat{\mathbf{A}}^T$, that is, that the system matrix is normal. The matrices $\widehat{\mathbf{B}}$, $\widehat{\mathbf{C}}$, and $\widehat{D}$ can also be found using the similarity transformation equations. Let these be denoted by

$$\widehat{\mathbf{B}} \triangleq \begin{bmatrix} \hat{b}_1 & \hat{b}_2 \end{bmatrix}^T, \widehat{\mathbf{C}} \triangleq \begin{bmatrix} \hat{c}_1 & \hat{c}_2 \end{bmatrix}, \widehat{D} \triangleq \hat{d}. \tag{4.118}$$

The structure described by $\{\widehat{\mathbf{A}}, \widehat{\mathbf{B}}, \widehat{\mathbf{C}}, \widehat{D}\}$ given above has nine multipliers. In fact, this is more than is really needed. It can be shown that the same transfer function can be realized

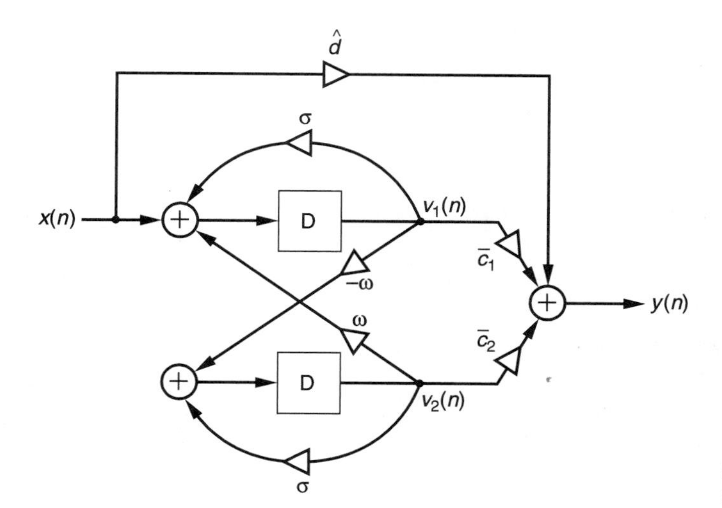

Figure 4.31 Coupled Form filter structure.

with $\overline{B} = [1\ 0]^T$ and $\overline{\mathbf{C}} \triangleq [\bar{c}_1\ \bar{c}_2]$, where $\bar{c}_1$ and $\bar{c}_2$ can be found in terms of $\hat{b}_1, \hat{b}_2, \hat{c}_1$ and $\hat{c}_2$ (see Problem 31). The structure described by $\{\widehat{\mathbf{A}}, \widehat{\mathbf{B}}, \widehat{\mathbf{C}}, \widehat{D}\}$ is the final Coupled Form filter shown in Fig. 4.31. This filter structure has seven multipliers and five adders. Coupled Form filters are very popular for implementation with short wordlengths and in fixed point arithmetic.

Cascade and Parallel Structures

In many applications, it is useful to implement digital filters as cascade or parallel connections of first- and second-order filter sections. These structures have different noise characteristics and are useful for modular implementation. The parallel structure is suitable for implementation in parallel processors. The cascade and parallel structures are obtained by simple manipulations of the transfer function.

By factorization of the numerator and the denominator, the transfer function can be expressed in the form

$$H(z) = \frac{\prod_{k=1}^{M_1}(1 + d_k z^{-1}) \prod_{k=1}^{M_2}(b_{0k} + b_{1k}z^{-1} + b_{2k}z^{-2})}{\prod_{k=1}^{N_1}(1 + c_k z^{-1}) \prod_{k=1}^{N_2}(a_{0k} + a_{1k}z^{-1} + a_{2k}z^{-2})}. \tag{4.119}$$

Now we can implement the above as a cascade of many subsystems $H_1(z)H_2(z)H_3(z)\ldots\ldots$ In fact, we have complete freedom in choosing these subsystems, and there are many possible combinations.

To obtain a parallel structure, we perform a partial fraction expansion of the transfer function to get the form

$$H(z) = \sum_{k=0}^{P} A_k z^{-k} + \sum_{k=1}^{F} \frac{B_k}{1 + c_k z^{-1}} + \sum_{k=1}^{S} \frac{b_{0k} + b_{1k}z^{-1}}{1 + a_{1k}z^{-1} + a_{2k}z^{-2}}. \tag{4.120}$$

The final structure is then simply a parallel connection, with the output being the sum of the outputs of each individual section.

▶ **EXAMPLE 4.7** *In this example, we obtain the cascade and parallel realizations of the Butterworth filter designed in Example 4.5. The transfer function of the filter was found to be*

$$H(z) = \frac{0.129 + 0.3867z^{-1} + 0.3869z^{-2} + 0.129z^{-3}}{1 - 0.2971z^{-1} + 0.3564z^{-2} - 0.0276z^{-3}}.$$

Simple factorization of the numerator and the denominator yields

$$H(z) = \frac{0.129(1 + 0.8947z^{-1})(1 + 2.103z^{-1} + 1.117z^{-2})}{(1 - 0.0815z^{-1})(1 - 0.2156z^{-1} + 0.3388z^{-2})}.$$

For the cascade implementation, let us choose the decomposition

$$\begin{aligned} H(z) &= \left[\frac{0.129 + 0.1154z^{-1}}{1 - 0.0815z^{-1}}\right]\left[\frac{1 + 2.103z^{-1} + 1.117z^{-2}}{1 - 0.2156z^{-1} + 0.3388z^{-2}}\right] \\ &= H_1(z)H_2(z). \end{aligned}$$

For the second-order filter $H_2(z)$, we have a choice of the filter structure. The cascade implementation is shown in Fig. 4.32, where we have used the Direct Form II for the second-order section. In order to obtain the parallel structure, we need to perform partial fraction expansion. Because the numerator and the denominator polynomials in $H(z)$ have the same order, we divide them to get

$$H(z) = 0.129 + \frac{0.425z^{-1} + 0.3409z^{-2} + 0.1325z^{-3}}{1 - 0.2971z^{-1} + 0.3564z^{-2} - 0.0276z^{-3}}.$$

Now multiply the numerator and the denominator of the second term by z^3. This will give us a form that we are familiar with for performing a partial fraction expansion. The result is

$$\begin{aligned} H(z) &= 0.129 + \frac{0.425z^2 + 0.3409z + 0.1325}{z^3 - 0.2971z^2 + 0.3564z - 0.0276} \\ &= 0.129 + \frac{0.4974}{z - 0.0815} + \frac{-0.0724z + 0.4423}{z^2 - 0.2156z + 0.3388} \\ &= 0.129 + \frac{0.4974z^{-1}}{1 - 0.0815z^{-1}} + \frac{-0.0724z^{-1} + 0.4423z^{-2}}{1 - 0.2156z^{-1} + 0.3388z^{-2}}. \end{aligned}$$

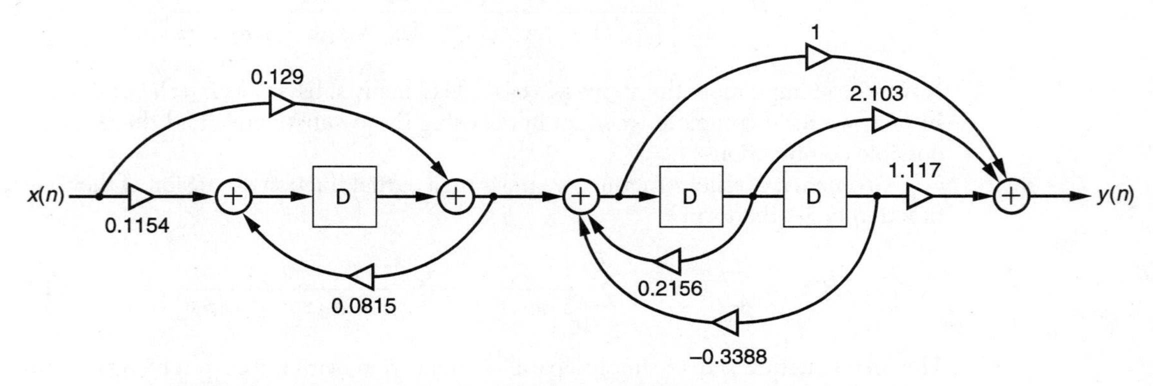

Figure 4.32 Cascade implementation of a third-order Butterworth filter.

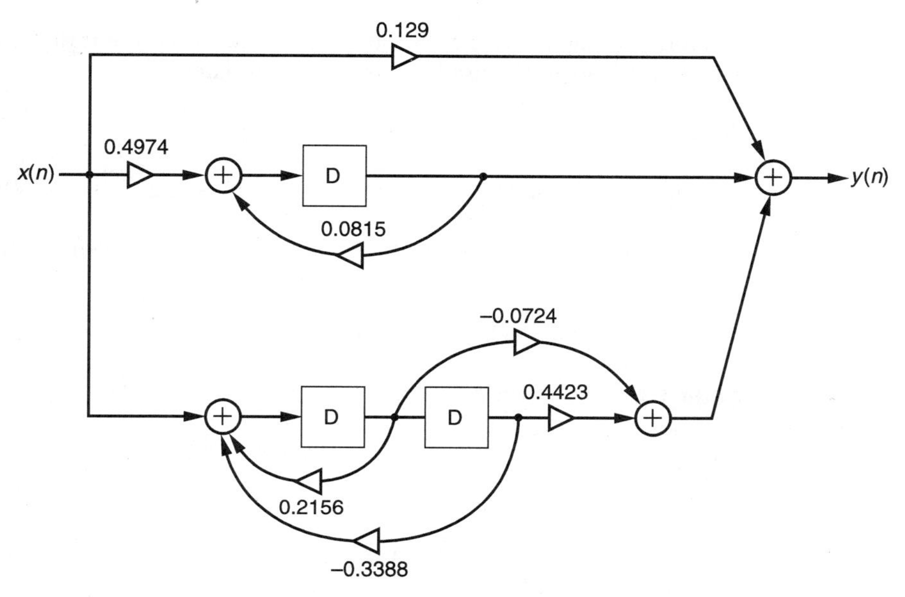

Figure 4.33 Parallel implementation of a third-Butterworth filter.

The parallel structure consists of three sections: a constant gain, a first-order section, and a second-order section. The overall implementation is shown in Fig. 4.33, where we have used the Direct Form II structure again for the second-order section.

Lattice Structures

The lattice structure has many advantages, making it very attractive in signal processing. These advantages are as follows.

1. In adaptive filtering applications [11], we usually try to model a system or a signal using a nonrecursive or recursive filter. The filter coefficients are calculated using a minimization algorithm. Let us say that we have modeled a system using an FIR filter and some given algorithm. Now, if we want to increase the filter order in an effort to get better modeling, we must re-run the algorithm on the higher order model. This will yield new coefficients of the filter. But if we have a lattice structure as the model, then its order can be simply increased by adding extra stage(s) and calculating the coefficients of the new stage(s) only. The coefficients of the older stages remain the same. This makes the upgrading of filter orders simple and computationally inexpensive.
2. Lattice filters are less sensitive to coefficient quantization (finite wordlength) than Direct Form filter structures.
3. Lattice filters are computationally more efficient than other filter structures for the implementation of wavelet transforms using filter banks [12].
4. In linear prediction (Chapter 9), it will be shown that lattice structures yield the forward and backward prediction errors simultaneously.

Before we begin deriving the lattice structures, we present the following definition and lemma. These will be used in our derivations later.

Definition 4.1 *Let $A_N(z)$ be a polynomial of degree-N given by*

$$A_N(z) = a(0) + a(1)z^{-1} + a(2)z^{-2} + \cdots + a(N-1)z^{-(N-1)} + a(N)z^{-N}. \quad (4.121)$$

The reverse polynomial A_N^R is the above polynomial with the coefficients in reverse order, that is,

$$A_N^R(z) = a(N) + a(N-1)z^{-1} + \cdots + a(1)z^{-(N-1)} + a(0)z^{-N}. \quad (4.122)$$

Lemma 4.1 *(i) For some constant K, let*

$$P(z) = A_N(z) + Kz^{-1}A_N^R(z). \quad (4.123)$$

Then

$$P^R(z) = KA_N(z) + z^{-1}A_N^R(z). \quad (4.124)$$

(ii)

$$A_N^R(z) = z^{-N}A_N(z^{-1}). \quad (4.125)$$

This lemma can be easily proved using Definition 4.1 and some straightforward algebraic manipulations. This is assigned as Problem 47.

Now, we begin the derivation of the lattice structure of an FIR filter. The FIR filter transfer function can be written as

$$A(z) = 1 + \sum_{i=1}^{m} a_m(i)z^{-i} \quad (4.126)$$

where m is the filter order and $a_m(0) = 1$. For $m = 1$, the difference equation for the filter is

$$y(n) = x(n) + a_1(1)x(n-1). \quad (4.127)$$

The lattice structure for this first-order filter is given in Fig. 4.34. This structure has two outputs, which are called the upper and lower channels and are denoted by $f_1(n)$ and

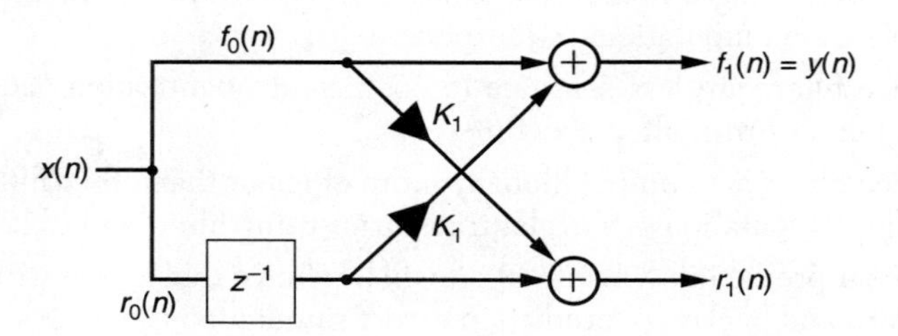

Figure 4.34 First-order lattice.

$r_1(n)$.[6] The subscripts denote the filter order. This is also called a single-stage lattice. The inputs to this lattice are

$$f_0(n) = r_0(n) = x(n). \tag{4.128}$$

The outputs are

$$f_1(n) = f_0(n) + K_1 r_0(n-1) \tag{4.129}$$

$$r_1(n) = K_1 f_0(n) + r_0(n-1). \tag{4.130}$$

From these equations, we see that the upper channel yields the output of the filter if $K_1 = a_1(1)$. Now consider the second-order FIR filter, which has the difference equation

$$y(n) = x(n) + a_2(1)x(n-1) + a_2(2)x(n-2). \tag{4.131}$$

The lattice structure for this filter is simply a cascade of two single-stage lattices (see Fig. 4.35). Again, the upper channel yields the filter output. By writing the equation for $f_2(n)$ in terms of $x(n)$ and comparing with the above difference equation, it is easy to show that

$$K_2 = a_2(2), \quad K_1 = \frac{a_2(1)}{1 + a_2(2)}. \tag{4.132}$$

In a similar fashion, an Mth order FIR filter can be implemented by a lattice structure with M stages as shown in Fig. 4.36. The mth stage (shown in Fig. 4.37) will now be used to derive a method for obtaining the lattice coefficients. These are often referred to as the *reflection coefficients*. The input-output relationships of the mth stage in the Z domain are

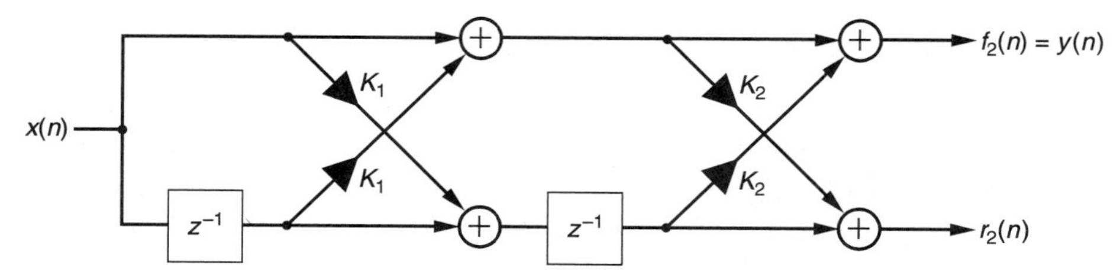

Figure 4.35 Second-order lattice.

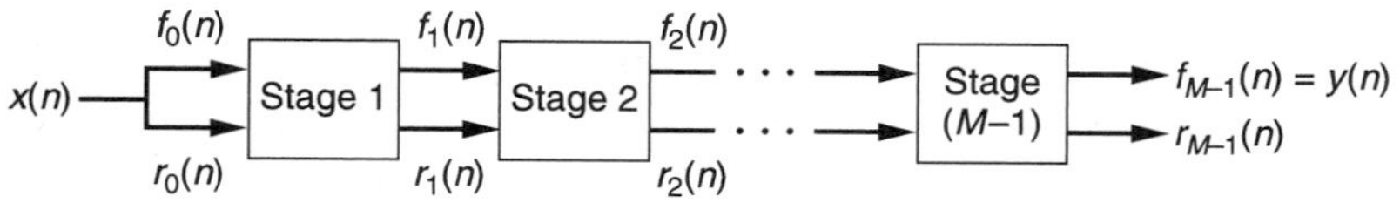

x(n)
$f_0(n)$ $r_0(n)$ Stage 1 $f_1(n)$ $r_1(n)$ Stage 2 $f_2(n)$ $r_2(n)$ . . . Stage (M−1) $f_{M-1}(n) = y(n)$ $r_{M-1}(n)$

Figure 4.36 M-Stage lattice.

[6] We will see in Chapter 9 that these are the forward and backward prediction errors of a linear predictor—hence the reason for these notations.

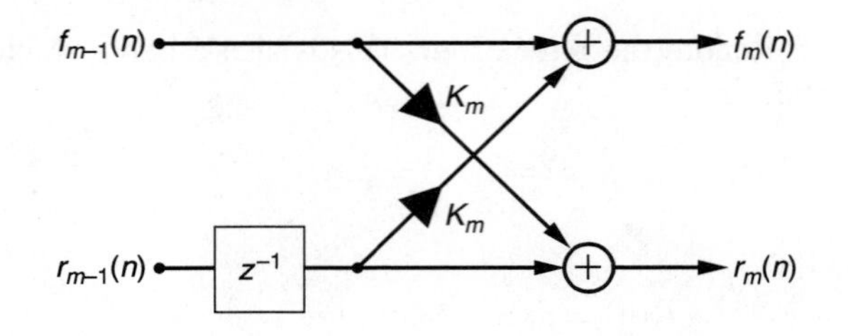

Figure 4.37 The m-th stage of a cascaded lattice structure.

$$F_m(z) = F_{m-1}(z) + K_m z^{-1} R_{m-1}(z) \tag{4.133}$$

$$R_m(z) = K_m F_{m-1}(z) + z^{-1} R_{m-1}(z). \tag{4.134}$$

The upper channel gives the filter output, and so the transfer function of the mth order filter is $A_m(z) = F_m(z)/X(z)$. Let us define the transfer function of the lower channel as $C_m(z) \triangleq R_m(z)/X(z)$. Now, dividing equations (4.133) and (4.134) by $X(z)$ and using these definitions, we get

$$A_m(z) = A_{m-1}(z) + K_m z^{-1} C_{m-1}(z) \tag{4.135}$$

$$C_m(z) = K_m A_{m-1}(z) + z^{-1} C_{m-1}(z). \tag{4.136}$$

Since $F_0(z) = R_0(z) = X(z)$, we have $A_0(z) = 1$ and $C_0(z) = 1$. We are now ready to present the following important theorem.

Theorem 4.1 *The transfer functions of the upper and lower channels of an mth order lattice can be written as*

$$A_m(z) = A_{m-1}(z) + K_m z^{-1} A^R_{m-1}(z) \tag{4.137}$$

$$C_m(z) = A^R_m(z). \tag{4.138}$$

Proof. The theorem is proved by the method of mathematical induction. That is, we first show that (4.137) and (4.138) hold for $m = 1$. Then we *assume* that the results are valid for $m - 1$. Next we show that with this assumption, the results are valid for m.

Let $m = 1$. Then we have

$$\begin{aligned} A_1(z) &= A_0(z) + K_1 z^{-1} C_0(z) \\ &= 1 + K_1 z^{-1}, \end{aligned} \tag{4.139}$$

$$\begin{aligned} C_1(z) &= K_1 A_0(z) + z^{-1} C_0(z) \\ &= K_1 + z^{-1}. \end{aligned} \tag{4.140}$$

Comparing (4.139) and (4.140) and using the definition in (4.122), it is easy to see that

$$C_1(z) = A^R_1(z). \tag{4.141}$$

Also, since $A_0(z) = 1$, we have $A_0^R(z) = 1$. Thus, (4.139) can be written as

$$A_1(z) = A_0(z) + K_1 z^{-1} A_0^R(z). \tag{4.142}$$

The results (4.137) and (4.138) hold for $m = 1$. Now suppose that the results hold for $m - 1$. That is,

$$A_{m-1}(z) = A_{m-2}(z) + K_{m-1} z^{-1} A_{m-2}^R(z), \tag{4.143}$$

$$C_{m-1}(z) = A_{m-1}^R(z). \tag{4.144}$$

Then using (4.135), we get

$$\begin{aligned} A_m(z) &= A_{m-1}(z) + K_m z^{-1} C_{m-1}(z) \\ &= A_{m-1}(z) + K_m z^{-1} A_{m-1}^R(z) \end{aligned} \tag{4.145}$$

which is one part of the desired result. Next, using (4.136), we get

$$C_m(z) = K_m A_{m-1}(z) + z^{-1} C_{m-1}(z) \tag{4.146}$$

$$= K_m A_{m-1}(z) + z^{-1} A_{m-1}^R(z). \tag{4.147}$$

Using Lemma 4.1(i), we see that the above becomes

$$C_m(z) = A_m^R(z) \tag{4.148}$$

which is the second part of the desired result and the proof is complete. ■

Theorem 4.1 forms the basis for deriving algorithms to convert from a lattice structure to Direct Form or vice versa. Substituting the polynomial $A_m(z) = 1 + \sum_{i=1}^{m} a_m(i) z^{-i}$ in equation (4.137), we find it is straightforward to derive the coefficients of the Direct Form filter (Problem 49). The resulting algorithm is as follows.

Conversion from Lattice to Direct Form

For $m = 1, 2, \ldots, M$ do

$$a_m(0) = 1$$

$$a_m(m) = K_m$$

$$a_m(i) = a_{m-1}(i) + a_m(m) a_{m-1}(m - i), \quad 1 \le i \le m - 1.$$

Conversion from Direct Form to Lattice

In this case, we are given the coefficients $a_M(i)$, $i = 0, 1, \ldots, M$ of the polynomial $A_M(z)$. Since we know that $a_M(M) = K_M$, we immediately have the lattice coefficient for the last stage. Now, we have to successively find the coefficients for the previous stages with $m = M - 1, M - 2, \ldots, 1$. To find the necessary formula, consider the lattice formula (4.135) and substitute for $C_{m-1}(z)$ from equation (4.146). Then convert to time domain to get the following.

$$\text{For } m = M, M-1, \ldots 2, 1 \text{ do}$$

$$K_m = a_m(m)$$

$$a_{m-1}(i) = \frac{a_m(i) - a_m(m)a_m(m-i)}{1 - K_m^2}, \quad 1 \le i \le m-1.$$

Observe that this recursion fails if $K_m = 1$ at any stage. This is an indication that there is a zero on the unit circle. If this happens, we can simply factor out this root from $A_M(z)$ and the above recursion can be performed for the reduced order system [2]. Now, we turn to the computational complexity. An M-order FIR filter with $a(0) = 1$, implemented in Direct Form, requires M multipliers. On the other hand, its lattice structure requires M stages, that is, $2M$ multipliers. Fortunately, a one-multiplier lattice is shown in Fig. 4.38. If $\gamma = K/(1-K)$, with simple algebraic manipulations, it can be shown (see Problem 50) that the one multiplier lattice yields the same outputs as the two-multiplier lattice, up to a constant. That is, if f_i and r_i are the outputs of the upper and lower channels of the two multiplier lattice, respectively, then the one-multiplier lattice yields $\alpha_i f_i$ and $\alpha_i r_i$ as the outputs. Therefore, every stage needs to be scaled by $1/\alpha_i$. Instead of using a multiplier per stage for this scaling, the obvious thing to do is to use one single multiplier at the output of the final stage. For an N-stage filter, the multiplier is $\frac{1}{\alpha_1\alpha_2\ldots\alpha_N}$. The one-multiplier lattice has one extra adder though.

▶ EXAMPLE 4.8 *Let the lattice coefficients of an FIR lattice be* $K_1 = 0.1,\ K_2 = 0.2,\ K_3 = 0.3$. *Find the coefficients of the Direct Form FIR filter.*

SOLUTION The solution is obtained by using the formulas for conversion from lattice to Direct Form. In this case, $M = 3$ and we have the following.

$$\text{For } m = 1:$$

$$a_1(0) = 1$$

$$a_1(1) = K_1 = 0.1.$$

$$\text{For } m = 2:$$

$$a_2(0) = 1$$

$$a_2(2) = K_2 = 0.2$$

$$a_2(i) = a_1(i) + a_2(2)a_1(2-i), i = 1$$

$$\Rightarrow \quad a_2(1) = 0.1 + (0.2)(0.1) = 0.12.$$

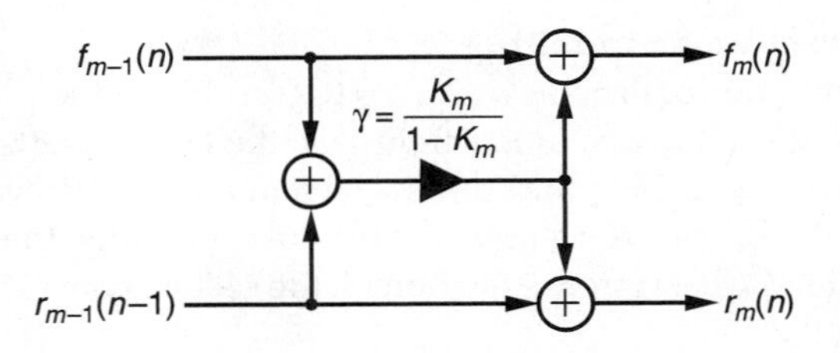

Figure 4.38 One-multiplier lattice.

$$\begin{aligned}
\text{For } m &= 3: \\
a_3(0) &= 1 \\
a_3(3) &= K_3 = 0.3 \\
a_3(i) &= a_2(i) + a_3(3)a_2(3-i), i = 1,2 \\
\Rightarrow \quad & a_3(1) = 0.12 + (0.3)(0.2) = 0.18 \\
& a_3(2) = 0.2 + (0.3)(0.12) = 0.56.
\end{aligned}$$

The Direct Form digital filter is therefore $A(z) = 1 + 0.18z^{-1} + 0.56z^{-2} + 0.3z^{-3}$.

The formulas for converting from lattice to Direct Form and vice versa can be easily programmed. In fact, the MATLAB commands **latc2tf** and **tf2latc** perform these functions.

IIR Lattice Structures

Lattice structures for IIR filters are obtained by first deriving the lattice for an all-pole filter and inserting appropriate multipliers and adders to realize the zeros. Consider the all-pole transfer function

$$H(z) = \frac{1}{A(z)} = \frac{1}{1 + \sum_{i=1}^{N} a(i)z^{-i}}. \tag{4.149}$$

In the last subsection, we have derived the lattice structure for the polynomial $A(z)$. The function $H(z) = 1/A(z)$ essentially has the same structure except that the input and the output are interchanged. That is, $f_N(n)$ is the input and $f_0(n)$ is the output. So, given the input $f_N(n) = x(n)$, we must successively find $f_{N-1}(n), f_{N-2}(n), \ldots, f_0(n)$. The lattice equations for the mth stage as given earlier are

$$f_m(n) = f_{m-1}(n) + K_m r_{m-1}(n-1) \tag{4.150}$$

$$r_m(n) = K_m f_{m-1}(n) + r_{m-1}(n-1). \tag{4.151}$$

For the all-pole filter, all we need to do is compute (4.150) in reverse order as

$$f_{m-1}(n) = f_m(n) - K_m r_{m-1}(n-1). \tag{4.152}$$

The lattice represented by equations (4.152) and (4.151) is shown in Fig. 4.39. When N of these stages are cascaded, we get the complete structure of the Nth order all-pole filter as given in Fig. 4.40.

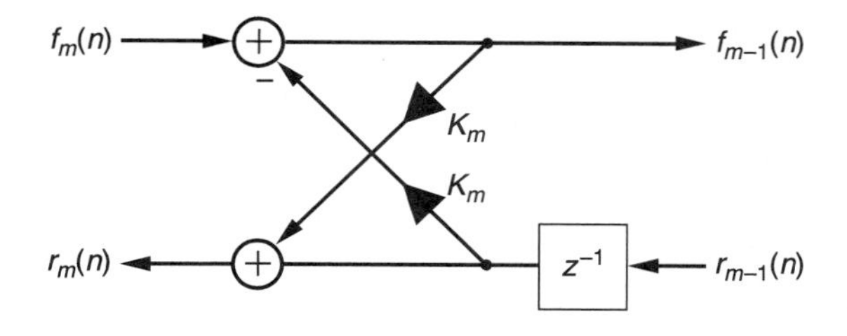

Figure 4.39 m-th stage of an all-pole lattice.

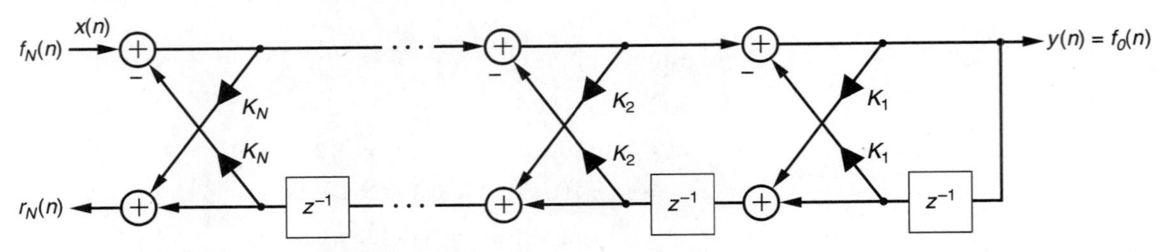

Figure 4.40 N-th order all-pole lattice.

Now we consider a pole-zero system with input $x(n)$ and output $v(n)$ given by

$$H(z) = \frac{B(z)}{A(z)} = \frac{\sum_{i=0}^{M} b_i z^{-i}}{1 + \sum_{i=1}^{M} a(i) z^{-i}} \tag{4.153}$$

$$= \frac{1}{A(z)} \cdot B(z). \tag{4.154}$$

Equation (4.154) can be decomposed as a cascade of an all-pole filter, $1/A(z)$ followed by an FIR filter, $B(z)$ as shown in Fig. 4.41. The all-pole filter can be implemented as a lattice as shown before. The FIR filter $B(z)$ is nothing but a linear combination of the present and past outputs $(y(n), y(n-1), \dots)$ of the all-pole filter. Since the all-pole filter is a lattice, where do we obtain these delayed outputs? Actually, these outputs are available in $r_m(n)$ in some linear combination. In order to implement the FIR part, we simply form a linear combination of the $r_m(n)$ signals as shown in Fig. 4.42. This structure is called the *lattice-ladder realization*. It remains to find the new coefficients $\beta_i, i = 0, 1, \dots, N$. Toward this end, we use equation (4.138) in Theorem 4.1 to write

$$R_m(z) = A_m^R(z)Y(z)$$

$$= z^{-m} A_m(z^{-1})Y(z). \tag{4.155}$$

Note that on the right-hand side we have $Y(z)$ instead of $X(z)$ because the lattice implements $1/A(z)$, thereby reversing the input and the output. From Fig. 4.42, the output of the pole-zero filter is

$$v(n) = \sum_{i=0}^{M} \beta_i r_m(n). \tag{4.156}$$

Taking the Z-transform of the above and using (4.155), we get

$$V(z) = \sum_{i=0}^{M} \beta_i z^{-i} A_i(z^{-1}) Y(z) \tag{4.157}$$

which in turn gives the transfer function of the FIR part as

$$\frac{V(z)}{Y(z)} = \sum_{i=0}^{M} \beta_i z^{-i} A_i(z^{-1})$$

$$= B(z). \tag{4.158}$$

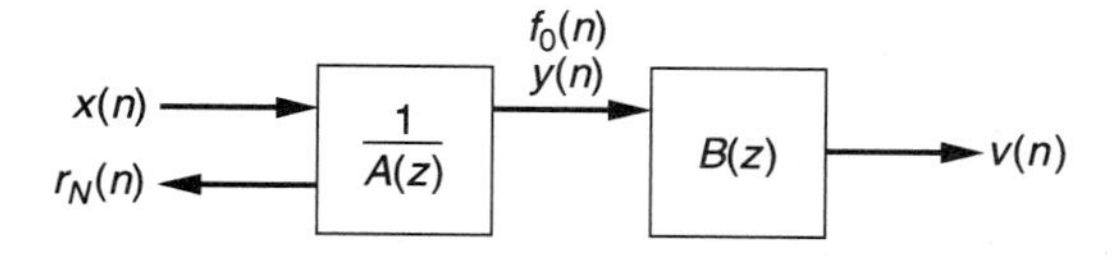

Figure 4.41 Cascade of an all-pole lattice with an FIR section $B(z)$.

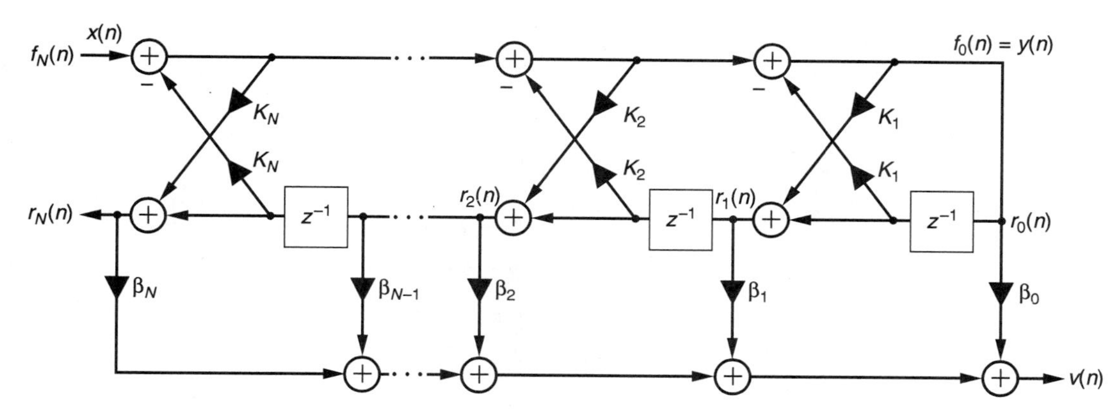

Figure 4.42 Lattice-ladder structure of an IIR filter.

Equating the formulas for $B(z)$ from (4.158) and (4.153), we get

$$\sum_{i=0}^{M} \beta_i z^{-i} A_i(z^{-1}) = \sum_{i=0}^{M} b_i z^{-i}. \tag{4.159}$$

Now substituting $A_m(z) = 1 + \sum_{i=1}^{m} a_m(i) z^{-i}$ in equation (4.159), it is not difficult (see Problem 51) to derive formulas for β_m in terms of b_m and $a_m(i)$. The results are

$$\beta_i = b_i - \sum_{m=i+1}^{M} \beta_m a_m(m-i), i = M, M-1, \ldots, 0. \tag{4.160}$$

An example is now in order.

EXAMPLE 4.9 *Consider the Butterworth filter designed in Example 4.5. The transfer function of the filter is*

$$H(z) = \frac{0.129 + 0.3867 z^{-1} + 0.3869 z^{-2} + 0.129 z^{-3}}{1 - 0.2971 z^{-1} + 0.3564 z^{-2} - 0.0276 z^{-3}},$$

where, according to our notations,

$$A(z) = 1 - 0.2971 z^{-1} + 0.3564 z^{-2} - 0.0276 z^{-3},$$

$$B(z) = 0.129 + 0.3867 z^{-1} + 0.3869 z^{-2} + 0.129 z^{-3}.$$

The lattice coefficients for A(z) can be found using the formulas given earlier as

$$\begin{aligned} a_3(3) &= K_3 = -0.0276, \\ a_3(2) &= 0.3564, \\ a_3(1) &= -0.2971, \\ a_2(2) &= K_2 = 0.3485, \\ a_2(1) &= -0.2875, \\ a_1(1) &= K_1 = -0.2132. \end{aligned}$$

Now we use formula (4.160) to find the forward tap coefficients as follows.

$$\beta_3 = b_3 = 0.129,$$

$$\beta_2 = b_2 - \beta_3 a_3(1) = 0.4252,$$

$$\beta_1 = b_1 - \beta_2 a_2(1) - \beta_3 a_3(2) = 0.4630,$$

$$\beta_0 = b_0 - \beta_1 a_1(1) - \beta_2 a_2(2) - \beta_3 a_3(3) = 0.0831.$$

4.4 2-D FIR FILTER DESIGN

In this section, we extend the window-based design method for 2-D filters. The extension is quite straightforward. The ideal impulse response is multiplied by a suitable window, which is 2-D in this case and can be either rectangular or circularly symmetric. The ideal impulse responses for lowpass rectangular and circularly symmetric filters were given in Chapter 2 and are repeated here for convenience.

(a) The frequency response of the 2-D separable ideal LPF is described by

$$H_{\mathrm{LPF}}(\omega_1, \omega_2) = \begin{cases} 1, & |\omega_1| \le \omega_{c_1} \ \& \ |\omega_2| \le \omega_{c_2} \\ 0, & \omega_{c_1} < |\omega_1| \le \pi \ \& \ \omega_{c_2} < |\omega_2| \le \pi. \end{cases} \tag{4.161}$$

The impulse response is

$$h_{\mathrm{LPF}}(n_1, n_2) = \frac{\sin \omega_{c_1} n_1}{\pi n_1} \frac{\sin \omega_{c_2} n_2}{\pi n_2}. \tag{4.162}$$

(b) The frequency response of the 2-D circularly symmetric ideal LPF is given by

$$H_{\mathrm{LPF}}(\omega_1, \omega_2) = \begin{cases} 1, & \sqrt{\omega_1^2 + \omega_2^2} < \omega_c \\ 0, & \sqrt{\omega_1^2 + \omega_2^2} \ge \omega_c, \text{ and } |\omega_1| \le \pi \ \& \ |\omega_2| \le \pi. \end{cases} \tag{4.163}$$

The impulse response is given by

$$h_{\mathrm{LPF}}(n_1, n_2) = \frac{\omega_c}{2\pi \sqrt{n_1^2 + n_2^2}} J_1\left(\omega_c \sqrt{n_1^2 + n_2^2}\right) \tag{4.164}$$

where $J_1(.)$ represents the Bessel function of the first kind and first order [4]. The ideal responses for the HPF, BPF, and BSF can be found as given earlier in (4.6)–(4.8) for the 1-D case.

As in the 1-D case, the 2-D FIR filter is obtained by multiplying the ideal impulse response, $h_I(n_1, n_2)$ by a window function, $w(n_1, n_2)$. That is,

$$h(n_1, n_2) = h_I(n_1, n_2)w(n_1, n_2) \tag{4.165}$$

and

$$H(\omega_1, \omega_2) = W(\omega_1, \omega_2) * H_I(\omega_1, \omega_2). \tag{4.166}$$

We would like to choose windows with frequency response $W(\omega_1, \omega_2)$ that have narrow main lobes and low side lobes. This will lead to filters with sharp transition bands and low ripples. The 2-D windows can be obtained directly from 1-D windows as follows.

Separable window

$$w(n_1, n_2) = w_1(n_1)w_2(n_2) \tag{4.167}$$

where $w_1(n_1)$ and $w_2(n_2)$ are 1-D windows. The size of the window will be denoted by $(N_1 \times N_2)$. For noncausal filters, the support region for the window is $-N_1 \le n_1 \le N_1, -N_2 \le n_2 \le N_2$. This is shown in Fig. 4.43(a).

Circularly symmetric window

In this case, we need the analog 1-D window function $w(t)$. The 2-D window is then obtained by rotating the 1-D window and then sampling it. This is called Huang's [28] method of 2-D window design and is as follows:

$$w(t_1, t_2) = w(t)\big|_{t=\sqrt{t_1^2+t_2^2}} \tag{4.168}$$

$$w(n_1, n_2) = w(t_1, t_2)|_{n_1=t_1, n_2=t_2}. \tag{4.169}$$

The size of this window is denoted by its radius N such that $\sqrt{n_1^2 + n_2^2} \le N$. The region of support of this window is shown in Fig. 4.43(b).

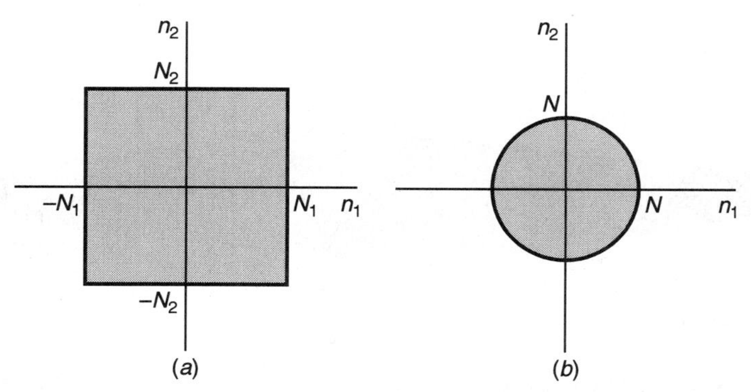

Figure 4.43 Region of support for the two basic windows: (a) Separable; (b) circularly symmetric.

The common 1-D window functions were presented earlier in the chapter, and their characteristics were described. That discussion is easily extended to 2-D. Rectangular windows have high side lobes and narrow main lobes, which lead to sharp transition bands but large ripples. The frequency responses of the two basic 2-D rectangular windows are given in Fig. 4.44, where $N_1 = N_2 = 10$, and $N = 10$. Notice the orientation of the ripples in the respective windows. In the circularly symmetric case, the ripples are somewhat circular but not perfectly symmetric. The 2-D Fourier transform destroys the symmetry to some extent because the transform itself is not circularly symmetric. This is not the case for the separable window, where the 2-D Fourier transform itself is actually separable.

Now, we give some examples of simple filter designs. We have two types of ideal impulse response, namely, the separable impulse response (4.162) and the circularly symmetric impulse response (4.164). The type of ideal impulse response used gives the

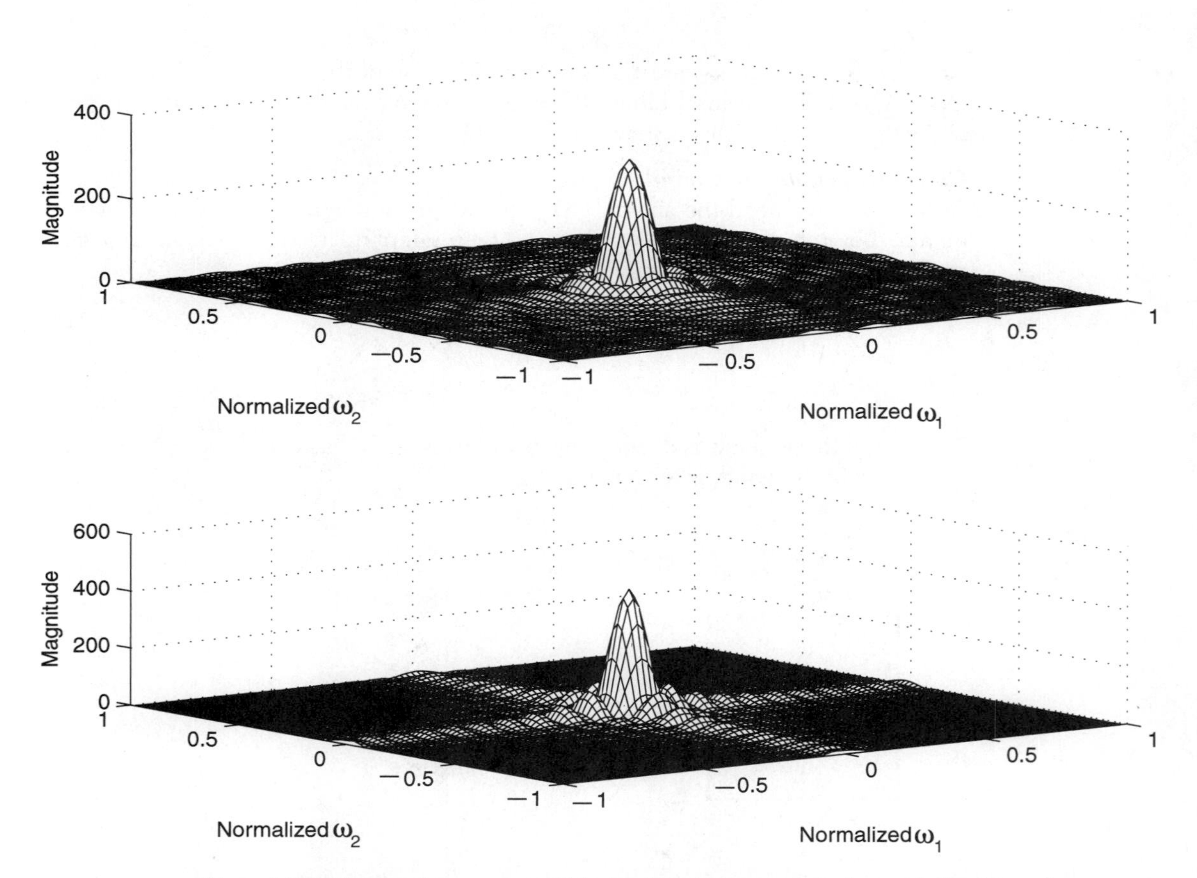

Figure 4.44 Magnitude response of (*a*) circularly symmetric window with $N = 10$, and (*b*) separable window of size 10×10.

name to the filter. That is, if we use an ideal separable impulse response, the designed filter will be called a separable filter. We also have two different types of windows, the separable and the circularly symmetric. Either of these windows can be multiplied with either type of ideal impulse response to obtain a filter design. Let us begin with a low-pass circularly symmetric filter designed with a circularly symmetric rectangular window. A circularly symmetric rectangular window is really a misnomer, since there is no rectangle in it. The name is derived from its 1-D counterpart. This window is simply a circle with all ones inside and zeros outside. For a cutoff frequency of $\omega_c = 0.5\pi$ and a window radius of $N = 10$, the impulse response and the magnitude response of the filter are shown in Fig. 4.45. In the magnitude response, the large ripples in the passband are quite noticeable, as is the sharp transition band. In Fig. 4.46, we design the same filter with a Hamming window. The ripples in the passband and stopband are much lower in this case. However, the transition band is not as sharp as with the rectangular window. Next, we design a separable bandpass filter using the Hamming window. The ideal impulse response is obtained using equations (4.162) and (4.6)–(4.8), with the lower and upper cutoff frequencies as 0.3π and 0.6π. This impulse response is then windowed by

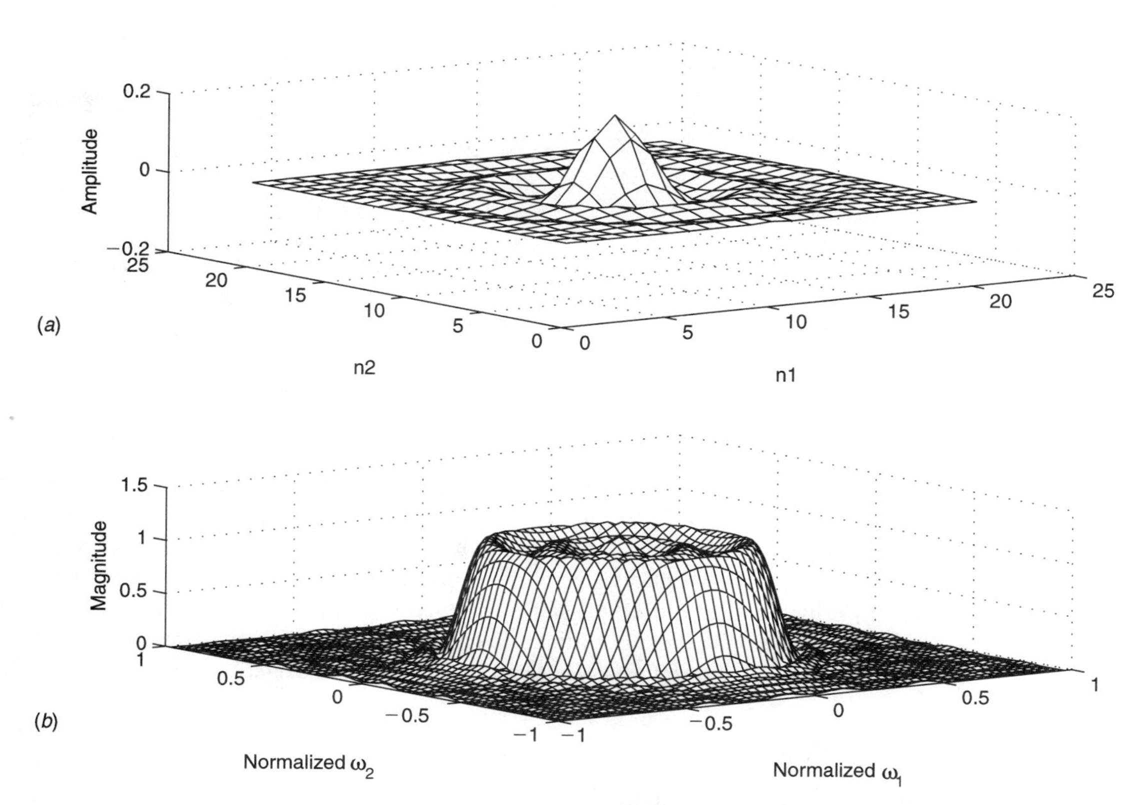

Figure 4.45 Circularly symmetric lowpass filter with $N = 10$ and $w_c = 0.5\pi$, designed using a circularly symmetric rectangular window: (*a*) Impulse response; (*b*) Magnitude response.

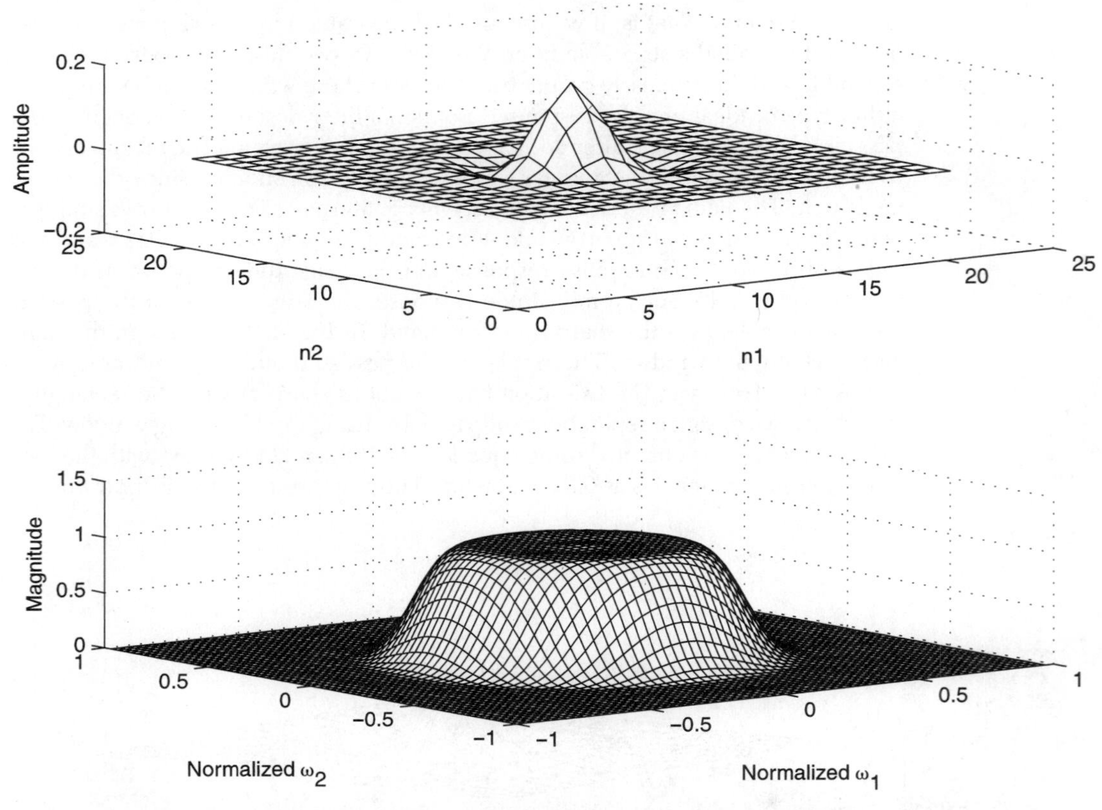

Figure 4.46 Circularly symmetric lowpass filter with $N = 10$, designed using a circularly symmetric Hamming window: (a) Impulse response; (b) magnitude response.

a Hamming window of size 10×10. The filter responses are given in Fig. 4.47. For the case of the separable filter design, the lower and upper cutoff frequencies can be chosen to be different in each frequency axis, if desired.

We have used the window method for designing zero-phase filters. However, this can easily be generalized to linear phase filters by simply shifting the impulse response to the first quadrant of support. The window method is simple but not optimal. Other methods for designing 2-D FIR filters give better approximations to the desired specifications as we have discussed in the 1-D case. These methods may also be extended to 2-D filter design. For example, it is not difficult to extend the 1-D frequency sampling method of design to 2-D. This is left as an exercise for interested readers.

4.5 2-D IIR FILTER DESIGN

As in 1-D, numerous methods can be used for designing 2-D IIR filters. Some are extensions of 1-D methods, whereas others are unique to 2-D filters. In this section we will discuss several methods of design. These methods have not been discussed in the

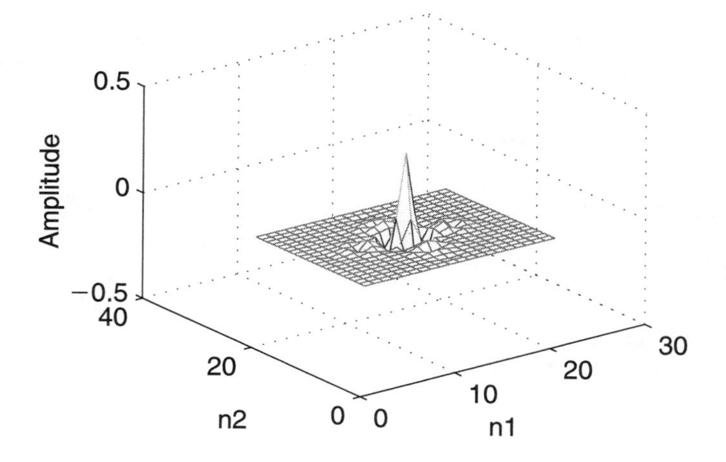

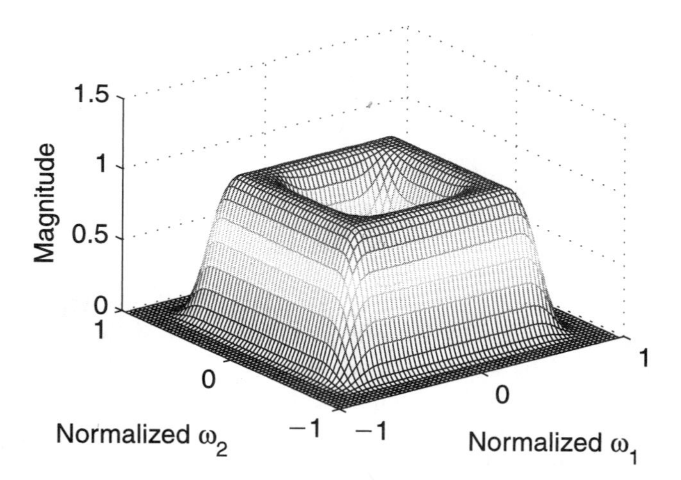

Figure 4.47 Separable BPF design with a separable Hamming window of size 10×10 The lower and upper cut-off frequencies in each direction are 0.3π and 0.7π, respectively: (*a*) Impulse response; (*b*) magnitude response.

context of 1-D filter design. However, these methods can actually be specialized for 1-D. Both of these methods are designed in the spatial domain; in other words, the specification is the desired impulse response of the filter. Of course, since we are designing IIR filters, the desired impulse response is a truncated version of the infinite 2-D sequence. A major advantage of the spatial domain filter designed is that the desired magnitude and phase functions are approximated simultaneously. This is especially important in systems where preservation of the phase or some spatial domain characteristic is desirable. What if the specification is given in the frequency domain in terms of passband and stopband edge frequencies and ripples? Then these specifications must be interpolated to obtain a complete magnitude response. Assuming zero phase or linear phase, the inverse DFT of this magnitude response can then be computed to obtain the 2-D desired impulse response. This impulse response will then be used for the spatial domain design methods as described next.

Shanks's Method

Shanks's method [15] is perhaps the most popular method for 2-D IIR filter design in the spatial domain. This method yields the filter design equations in closed form. However, it yields a suboptimal filter, which is often used as an initial condition for iterative design methods that give better filters. This will be shown in the next subsection where we present an iterative method of design.

Let $h_d(n_1, n_2)$ be the desired (specified) impulse response. The goal is to design a 2-D filter with transfer function

$$H(z_1, z_2) = \frac{B(z_1, z_2)}{A(z_1, z_2)} \tag{4.170}$$

where

$$B(z_1, z_2) = \sum_{(k_1,k_2)\epsilon R_b}\sum b(k_1, k_2) z_1^{-k_1} z_2^{-k_2} \tag{4.171}$$

and

$$A(z_1, z_2) = 1 + \sum_{(k_1,k_2)\epsilon R_a-(0,0)}\sum a(k_1, k_2) z_1^{-k_1} z_2^{-k_2}. \tag{4.172}$$

The domain of the coefficients $b(k_1, k_2)$ is denoted by R_b and represents the indices in the (k_1, k_2) plane where the coefficients are nonzero. The domain for $a(k_1, k_2)$ is denoted by R_a. However, $a(0, 0) = 1$, and hence the limit for the double summation in (4.172) is $R_a - (0, 0)$. Figure 4.48 shows the domain R_a and R_b. Note that the number of a's is $M_a N_a$ and the number of b's is $M_b N_b$. Now, let $h(n_1, n_2)$ denote the impulse response of the designed filter (4.170). The error between the desired and designed impulse response is

$$e(n_1, n_2) = h_d(n_1, n_2) - h(n_1, n_2).$$

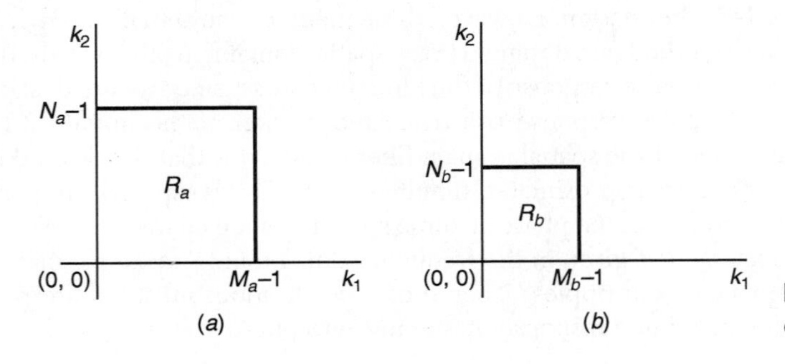

Figure 4.48 Coefficient domains for a 2-D IIR filter.

If the design is good, then we expect this error to be small for all (n_1, n_2). The design is performed by minimizing some objective function (or cost function) constructed from this error sequence. A commonly used cost function is the sum of squares error given by

$$E = \sum_{(n_1,n_2)\epsilon R_e}\sum e^2(n_1, n_2) \tag{4.173}$$

where R_e is the domain for the sum of squares. Substituting for $e(n_1, n_2)$ in (4.173) gives

$$E = \sum_{(n_1,n_2)\epsilon R_e}\sum (h_d(n_1, n_2) - h(n_1, n_2))^2. \tag{4.174}$$

It turns out that minimizing the above cost function is a nonlinear problem. This means that E is in general not a quadratic function of the coefficients. In other words, differentiating E with respect to each of the coefficients does not yield a set of linear simultaneous equations that can be solved in closed form. This is easy to show with a simple example (see Problem 59). In order to reduce the cost function to be quadratic, Shanks et al. [15] proposed the following method.

From (4.170), the impulse response of the designed filter is

$$\begin{aligned} h(n_1, n_2) &= -\sum_{(k_1,k_2)\epsilon R_a-(0,0)}\sum a(k_1, k_2)h(n_1 - k_1, n_2 - k_2) \\ &\quad + \sum_{(k_1,k_2)\epsilon R_b}\sum b(k_1, k_2)\delta(n_1 - k_1, n_2 - k_2). \end{aligned} \tag{4.175}$$

The second term in (4.175) is the convolution $b(n_1, n_2) * \delta(n_1, n_2)$ and therefore

$$h(n_1, n_2) = -\sum_{(k_1,k_2)\epsilon R_a-(0,0)}\sum a(k_1, k_2)h(n_1 - k_1, n_2 - k_2) + b(n_1, n_2). \tag{4.176}$$

If the designed filter is good, then its impulse response should closely match the desired impulse response, that is,

$$h_d(n_1, n_2) \cong h(n_1, n_2). \tag{4.177}$$

Therefore, we can replace $h(n_1-k_1, n_2-k_2)$ on the right-hand side of (4.176) by $h_d(n_1-k_1, n_2 - k_2)$ to get

$$\hat{h}(n_1, n_2) = -\sum_{(k_1,k_2)\epsilon R_a-(0,0)}\sum a(k_1, k_2)h_d(n_1 - k_1, n_2 - k_2) + b(n_1, n_2). \tag{4.178}$$

Now, we define a new error sequence:

$$\hat{e}(n_1, n_2) = h_d(n_1, n_2) - \hat{h}(n_1, n_2). \tag{4.179}$$

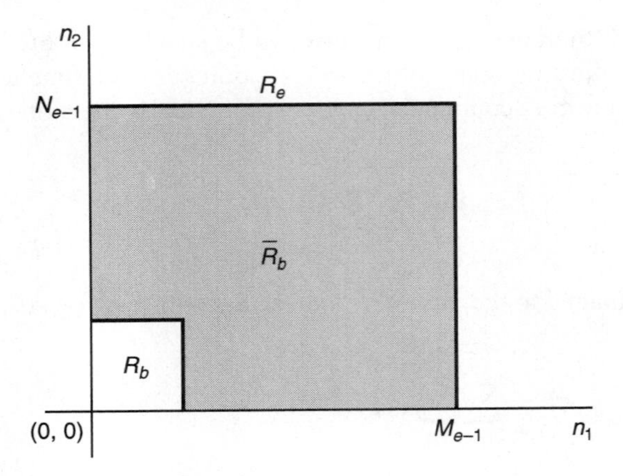

Figure 4.49 Domains for $\widehat{E}$ and R_b.

Based on this, we define a new cost function

$$\widehat{E} = \sum_{(n_1,n_2)\epsilon R_e}\sum \hat{e}^2(n_1, n_2). \tag{4.180}$$

This modified cost function is actually quadratic as we shall see shortly. The domain for this cost function is shown in Fig. 4.49, where it is subdivided into two regions as

$$\widehat{E} = \widehat{E}_{R_b} + \widehat{E}_{\overline{R}_b}. \tag{4.181}$$

The domain of $\widehat{E}_{R_b}$ is the domain given by $R_b \cap R_e$. $\overline{R}_b$ denotes the complement domain of R_b, and hence $\widehat{E}_{\overline{R}_b}$ is the squared-error in the domain $\overline{R}_b$. Notice that in the equation for $\hat{h}(n_1, n_2)$ given by (4.178), if we consider the domain $\overline{R}_b$, then $b(n_1, n_2) = 0$. Therefore,

$$\begin{aligned}\widehat{E}_{\overline{R}_b} &= \sum_{(n_1,n_2)\epsilon \overline{R}_b}\sum \left(h_d(n_1, n_2) - \hat{h}(n_1, n_2)\Big|_{\overline{R}_b}\right)^2 \\ &= \sum_{(n_1,n_2)\epsilon \overline{R}_b}\sum \left(h_d(n_1, n_2) + \sum_{(k_1,k_2)\epsilon R_a-(0,0)}\sum a(k_1, k_2) h_d(n_1 - k_1, n_2 - k_2)\right)^2. \end{aligned} \tag{4.182}$$

Equation (4.182) has $M_a N_a$ terms involving the a's which are quadratic. Differentiating (4.182) with respect to $a(k_1, k_2)$ and setting the resultant equations to zero, we can find $a(k_1, k_2)$ which minimizes the mean square error in $\overline{R}_b$. This will give us $(M_a N_a - 1)$ equations of the type

$$\begin{aligned}&\sum_{(k_1,k_2)\epsilon R_a-(0,0)}\sum a(k_1, k_2) \left[\sum_{(n_1,n_2)\epsilon \overline{R}_b}\sum h_d(n_1 - k_1, n_2 - k_2) h_d(n_1 - l_1, n_2 - l_2)\right] \\ &= -\sum_{(n_1,n_2)\epsilon \overline{R}_b}\sum h_d(n_1, n_2) h_d(n_1 - l_1, n_2 - l_2), \qquad (l_1, l_2)\epsilon R_a - (0, 0)\end{aligned} \tag{4.183}$$

where

$$\begin{aligned}
l_1 &= 0, 1, \ldots, M_a - 1 \\
l_2 &= 0, 1, \ldots, N_a - 1, (l_1, l_2) \neq (0, 0) \\
k_1 &= 0, 1, \ldots, M_a - 1 \\
k_2 &= 0, 1, \ldots, N_a - 1, (k_1, k_2) \neq (0, 0) \\
n_1 &= M_b, M_b + 1, \ldots, M_e - 1 \\
n_2 &= N_b, N_b + 1, \ldots, N_e - 1.
\end{aligned}$$

For each (l_1, l_2) pair, we get a new equation, each involving $(M_a N_a - 1)$ coefficients for a total of $(M_a N_a - 1)$ equations. By solving this set of simultaneous equations, we find all the coefficients $a(k_1, k_2)$.

Having computed the denominator $A(z_1, z_2)$ of the filter, we now compute the numerator $B(z_1, z_2)$. One way to compute $B(z_1, z_2)$ is to compute those coefficients that minimize the mean square difference between the coefficients of $H(z_1, z_2) = \frac{B(z_1,z_2)}{A(z_1,z_2)}$ and the coefficients of the desired response $H_d(z_1, z_2) = Z\{h_d(n_1, n_2)\}$. This is a discrete Wiener filtering problem in two dimensions. The concept of Wiener filtering will be addressed in Chapter 7.

A simple and less accurate method is now described. Since

$$\frac{B(z_1, z_2)}{A(z_1, z_2)} \cong H_d(z_1, z_2), \tag{4.184}$$

we can compute the coefficients of $B(z_1, z_2)$ from

$$B(z_1, z_2) \cong H_d(z_1, z_2) A(z_1, z_2).$$

This can be done by the following:

$$b(n_1, n_2) = h_d(n_1, n_2) * a(n_1, n_2), (n_1, n_2) \epsilon R_b. \tag{4.185}$$

Equations (4.183) and (4.185) are the main design equations that can be programmed as an algorithm. If the desired impulse response $h_d(n_1, n_2)$ is stable, and a good design of $H(z_1, z_2)$ is obtained, then $h(n_1, n_2)$ is likely to be stable. However, once the design is obtained, it should be tested for stability by using the algorithm described in Chapter 3. If the resulting filter is unstable, we have two options: (1) to increase the orders of $A(z_1, z_2)$ and $B(z_1, z_2)$ and redesign using Shanks's method; and (2) to use a stabilization algorithm to modify $H(z_1, z_2)$. Several stabilization algorithms have been developed, such as the complex cepstrum method and the planar least-squares inverse method [15], [16]. These methods are beyond the scope of this textbook.

Phase Response of 2-D IIR Filters

The above algorithm and most others produce filters with a first-quadrant impulse response. Because of its one-quadrant response, the filter cannot have a zero-phase response. However, we can take our one-quadrant filter and extend it symmetrically to the 2-D plane. This can be done as

$$H_{\text{zerophase}}(z_1, z_2) = H(z_1, z_2) + H(z_1, z_2^{-1}) + H(z_1^{-1}, z_2) + H(z_1^{-1}, z_2^{-1}). \tag{4.186}$$

The impulse response of the resulting filter is symmetric with respect to both the n_1 and n_2 axes and therefore has zero-phase response.

Examples

This section illustrates examples of 2-D zero-phase IIR filters designed using Shanks's algorithm. The zero-phase IIR filter is obtained using the method discussed above. First, the first-quadrant filter, whose coefficients $a(n_1, n_2)$ and $b(n_1, n_2)$ are obtained using (4.183) and (4.185), respectively, is designed. Then, the zero-phase IIR filter is obtained by using (4.186).

▶ **EXAMPLE 4.10** *The desired impulse response is that of a circular symmetric ideal lowpass filter with cut-off frequency of* 0.45π. *The circularly symmetric Hamming window with region of support* $\{-10 \leq n_1, n_2 \leq 10\}$ *is employed. The desired magnitude response is given in Fig. 4.50. Figure 4.51*

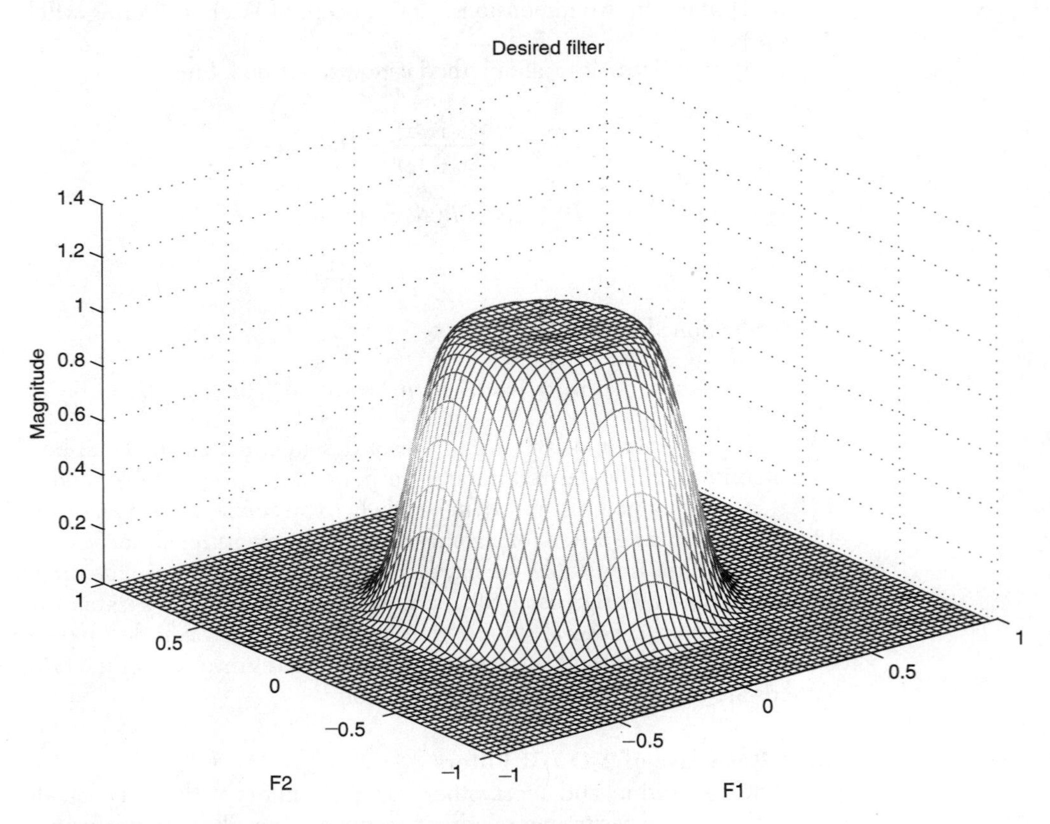

Figure 4.50 Magnitude response of ideal lowpass filter.

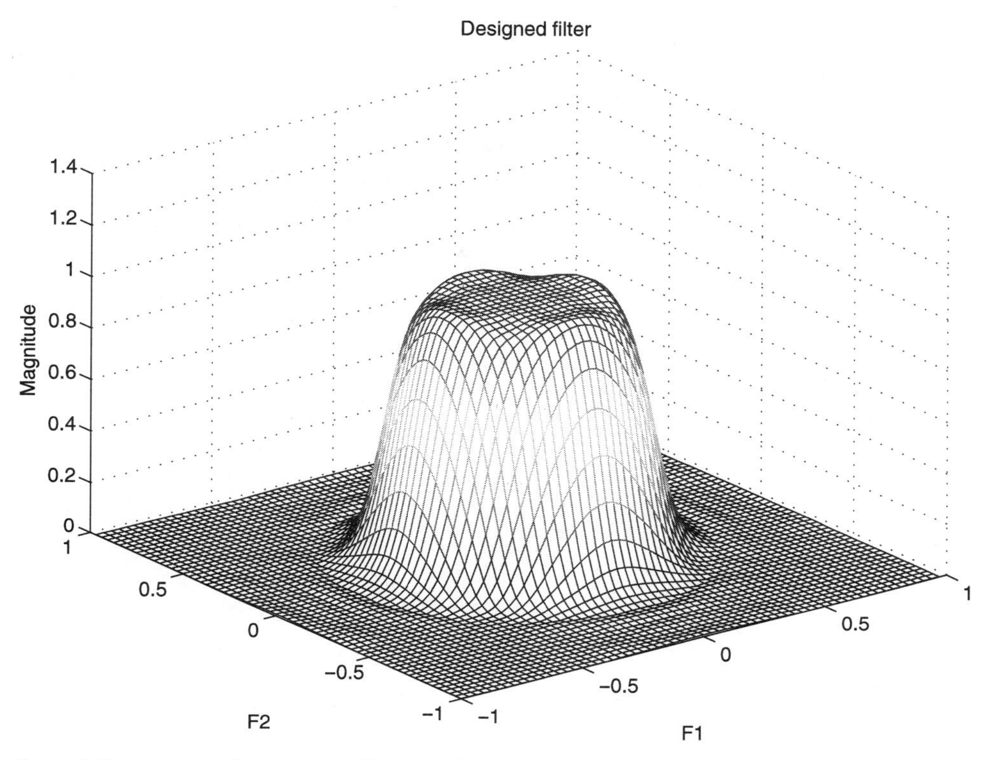

Figure 4.51 Magnitude response of lowpass filter designed using Shanks's method.

shows the magnitude response of the designed filter with a 3×3*-point* $a(n_1, n_2)$ *and* 3×3*-point* $b(n_1, n_2)$*. The magnitude responses are clearly well matched. The coefficient matrices of the designed filter are as follows:*

$$A = \begin{bmatrix} 1.000 & -0.6474 & 0.4293 & -0.0968 \\ -0.6474 & 0.4656 & -0.2786 & 0.0692 \\ 0.4293 & -0.2786 & 0.2135 & -0.0489 \\ -0.0968 & 0.0692 & -0.0489 & 0.0082 \end{bmatrix}$$

$$B = \begin{bmatrix} 0.1590 & 0.0168 & 0.0319 & 0.0001 \\ 0.0168 & 0.0062 & 0.0045 & 0.0001 \\ 0.0329 & 0.0045 & 0.0068 & 0.0000 \\ 0.0001 & 0.0001 & 0.0000 & -0.0004 \end{bmatrix}$$

EXAMPLE 4.11 *The desired impulse response is that of a circular symmetric ideal bandpass filter with low cutoff frequency of* 0.3π *and high cutoff frequency of* 0.65π*. The circularly symmetric Hamming window with region of support* $\{-10 \leq n_1, n_2 \leq 10\}$ *is employed. The desired magnitude response is illustrated in Fig. 4.52. Figure 4.53 illustrates the magnitude response of the designed filter with* 4×4*-point* $a(n_1, n_2)$ *and* 4×4*-point* $b(n_1, n_2)$*. The coefficient matrices of the designed filter are as follows:*

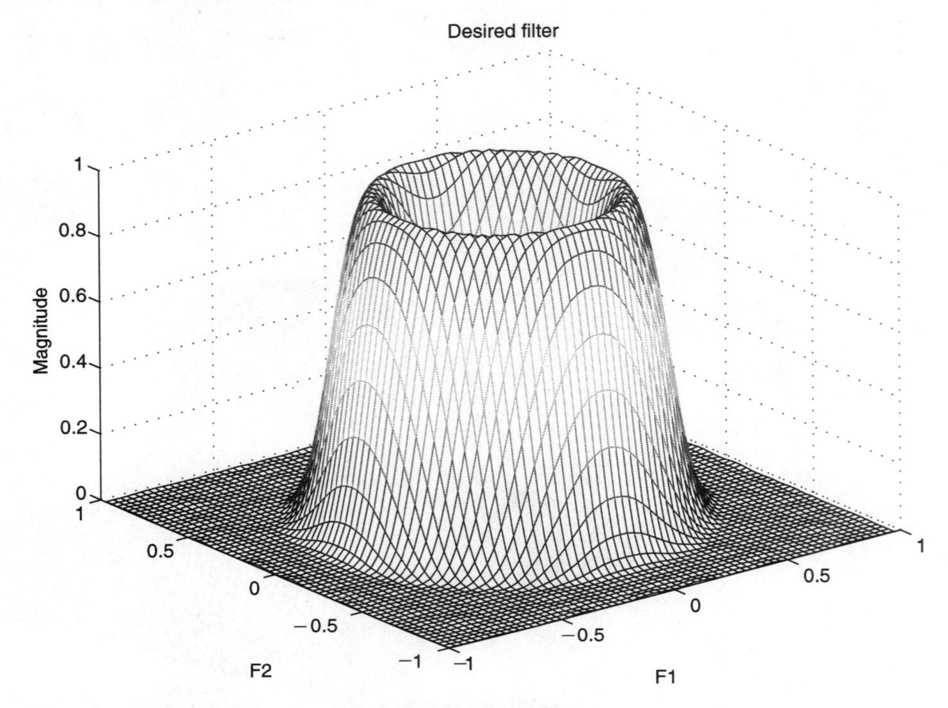

Figure 4.52 Magnitude response of ideal bandpass filter.

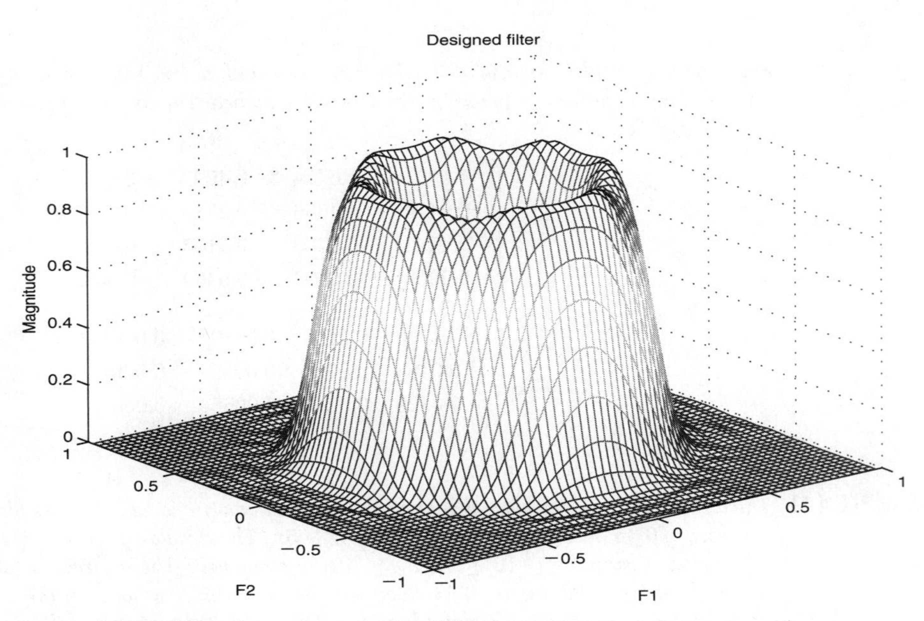

Figure 4.53 Magnitude response of bandpass filter designed using Shanks's method.

$$\mathbf{A} = \begin{bmatrix} 1.000 & -0.0036 & 0.3576 & -0.0605 & 0.0541 \\ -0.0036 & 0.0632 & 0.0478 & 0.0174 & 0.0280 \\ 0.3576 & 0.0478 & 0.2322 & -0.0111 & 0.0610 \\ -0.0605 & 0.0174 & -0.0111 & 0.0152 & 0.0181 \\ 0.0541 & 0.0280 & 0.0610 & 0.0181 & 0.0257 \end{bmatrix}$$

$$\mathbf{B} = \begin{bmatrix} 0.2611 & 0.1196 & 0.0387 & -0.0107 & 0.0011 \\ 0.1196 & 0.0452 & -0.0026 & -0.0159 & 0.0004 \\ 0.0387 & -0.0026 & -0.0140 & -0.0107 & 0.0004 \\ -0.0107 & -0.0159 & -0.0107 & -0.0025 & 0.0005 \\ 0.0011 & 0.0004 & 0.0004 & 0.0005 & 0.0000 \end{bmatrix}$$

Iterative Design Methods

As we saw in the last subsection, spatial domain algorithms for 2-D filter design attempt to minimize the p-norm (we used 2-norm) of the error between the desired response and the response of the designed filter. Two classes of algorithms are used, namely, closed form and iterative. The closed-form algorithms are suboptimal, in the sense that they slightly modify the error function so that the solution is obtained by solving a set of linear simultaneous equations. The iterative algorithms, on the other hand, use standard descent or downhill search algorithms, to minimize the nonlinear error function. The solution obtained from a closed-form algorithm such as Shanks's method is often used as the starting point for the minimization algorithms. Several different minimization algorithms [14],[16] have been used for this filter design problem, with varying degrees of success. Some of these are discussed here. One of the most successful and popular ones is the Davidon-Fletcher-Power (DFP) algorithm [17],[18], which is merely one of the two main types of the quasi-Newton method. This method has been successfully used for 2-D IIR filter design [19] but at the expense of extremely high computational expenses. The other main type of the quasi-Newton method, the Broyden-Fletcher-Goldfarb-Shanno (BFGS) method [20],[21], has also been used on the filter design problem with better performance as shown in [25].

In this section, we use the BFGS method for minimizing the error function in the spatial domain design of 2-D digital filters. The success of this method in the filter design problem is demonstrated by the fact that it converges much faster than the DFP method, thereby saving considerable computer time. We also apply a modified version of the Newton-Raphson method to this minimization problem, with significant improvement over the DFP method.

Design Algorithms

The problem of designing an IIR filter is to determine a rational and stable transfer function $H(z_1, z_2)$ that meets a given design specification. Specifically, we require $H(z_1, z_2)$, the 2-D Z-transform of the impulse response $h(n_1, n_2)$, to be a rational function of the form

$$H(z_1,z_2) = \frac{\sum\sum_{(k_1,k_2)\in R_b} b(k_1,k_2) z_1^{-k_1} z_2^{-k_2}}{1+\sum\sum_{(k_1,k_2)\in R_a-(0,0)} a(k_1,k_2) z_1^{-k_1} z_2^{-k_2}} \tag{4.187}$$

where $R_a - (0,0)$ represents the region of support of $a(k_1,k_2)$ except the origin and R_b represents the region of support of $b(k_1,k_2)$. The 2-D difference equation corresponding to (4.187) is given by

$$\begin{aligned} y(n_1,n_2) &= -\sum_{(k_1,k_2)\in R_a-(0,0)}\sum a(k_1,k_2) y(n_1-k_1,n_2-k_2) \\ &\quad + \sum_{(k_1,k_2)\in R_b-(0,0)}\sum b(k_1,k_2) u(n_1-k_1,n_2-k_2) \end{aligned} \tag{4.188}$$

where $y(n_1,n_2)$ and $u(n_1,n_2)$ represent the filter output and input, respectively. In spatial domain filter design, some desired response to a known input is assumed given. The problem is to estimate the coefficients $a(k_1,k_2)$ and $b(k_1,k_2)$ such that the designed filter response to the same input is as close as possible to the desired response in some sense. It is common to use the impulse function $\delta(n_1,n_2)$ as the input, and the desired response is then the impulse response $h_d(n_1,n_2)$. One of the commonly used error functions in filter design is the 2-norm error

$$E = \sum_{n_1,n_2\in R_c}\sum (h_d(n_1,n_2) - h(n_1,n_2))^2 \tag{4.189}$$

where R_c is the region of support of the error sequence. Minimizing (4.189) with respect to $a(k_1,k_2)$ and $b(k_1,k_2)$ is a nonlinear problem, as we mentioned earlier.

In general, an iterative descent algorithm for solving the minimization problem (P): minimize $f(\mathbf{x})$, for $\mathbf{x} \in R^n$, where f is continuously twice differentiable, can be expressed as follows. Starting at a chosen initial estimate of the solution of (P), the new estimate at kth iteration is computed as $\mathbf{x}^{k+1} = \mathbf{x}^k + \mu_k \mathbf{d}^k$, where $\mathbf{d}^k$ is a descent direction and α_k the step size. Both α_k and $\mathbf{d}^k$ are chosen so that the inequality $f(\mathbf{x}^{k+1}) < f(\mathbf{x}^k)$ is satisfied. The technique of computing the search direction $\mathbf{d}^k$ determines the iterative method. For the search direction $\mathbf{d}$ of the form $\mathbf{d} = -\mathbf{A}^{-1}\nabla f(\mathbf{x})$, some commonly used iterative methods are stated below.

1. The steepest descent method: $\mathbf{A} = \mathbf{I}$.
2. The Newton-Raphson method: $\mathbf{A} = \nabla^2 f(\mathbf{x})$, the Hessian of f at $\mathbf{x}$.
3. Quasi-Newton method: $\mathbf{A} = \mathbf{B}$, an approximation of the Hessian of f at $\mathbf{x}$.

The computation of the steepest descent method is much simpler than that of the others. Theoretically, this method has a linear rate of convergence, and it converges very slowly in practice. This method has been used on 2-D IIR filter design problem in [14], and it was noted by Cadzow that it converges in a "most disappointingly slow fashion."

The Newton-Raphson method has a locally quadratic rate of convergence and solves problem(P) efficiently if the initial estimate is chosen very close to the true solution. However, this method requires the evaluation of the true Hessian of f, which in turn requires n^2 function evaluation at each iteration, and can be computationally very expensive and time consuming. In the following, we propose a modified Newton-Raphson (MNR) method for solving the minimization problem with the objective function in (4.189), which does not require any extra function evaluations. Observe that $E(\mathbf{x})$ is of the form $\sum_{i=1}^{m} e_i^2(\mathbf{x})$. Then the Hessian of E is given by (see Problem 68):

$$\mathbf{H}(\mathbf{x}) = \mathbf{J}^T(\mathbf{x})\mathbf{J}(\mathbf{x}) + 2\sum_{i=1}^{m} e_i(\mathbf{x})H_i(\mathbf{x}) \tag{4.190}$$

where $\mathbf{J}(\mathbf{x}) = [\nabla e_1(\mathbf{x}), \ldots, \nabla e_m(\mathbf{x})]$ is the gradient vector and $\mathbf{H}_i(\mathbf{x})$ is the Hessian of e_i. If the objective function E has a zero value at its minimum point $\mathbf{x}^*$, then $\mathbf{H}(\mathbf{x}^*) = \mathbf{J}^T(\mathbf{x}^*)\mathbf{J}(\mathbf{x}^*)$. In this case, whenever $\mathbf{x} \cong \mathbf{x}^*$, $e_i(\mathbf{x}) \cong 0$, for $i = 1, \ldots m$ and then $H(\mathbf{x}) \cong \mathbf{J}^T(\mathbf{x})\mathbf{J}(\mathbf{x})$. Therefore, the MNR computes the search direction $\mathbf{d} = -\mathbf{J}^T(\mathbf{x})\mathbf{J}(\mathbf{x})\nabla f(\mathbf{x})$. If a good estimate of the solution of (P), in the sense that $E(\mathbf{x})$ is very small, is chosen, this method should work efficiently. Notice also that there is no need to form $\mathbf{J}(\mathbf{x})$ explicitly since each ∇e_i is computed and can be saved while $\nabla f(\mathbf{x})$ is evaluated. In our examples, Shanks's method is used to generate an initial point at which the function value of E is very small. Although the necessary condition (zero gradient of a minimum) may not be satisfied with this initial point, the MNR method has been shown to work very well.

In the general case, it may be more proper to use a quasi-Newton method because it does not require any information on the Hessian of the objective function E and has a locally superlinear rate of convergence. There are two types of quasi-Newton methods [22]. One updates the approximation of the Hessian, and the other updates the inverse of the approximation of the Hessian at each iteration. Here, we simply present the BFGS updating formula, which updates the approximation $\mathbf{B}$ of the Hessian. Define

$$\begin{aligned} \bar{\mathbf{x}} &= \mathbf{x} + \alpha\mathbf{d}, \\ \mathbf{s} &= \mathbf{x} - \bar{\mathbf{x}}, \text{ and} \\ \mathbf{y} &= \nabla E(\bar{\mathbf{x}}) - \nabla E(\mathbf{x}). \end{aligned} \tag{4.191}$$

The BFGS updating formula is

$$\bar{\mathbf{B}} = \mathbf{B} - \frac{\mathbf{B}\mathbf{s}\mathbf{s}^T\mathbf{B}}{\mathbf{s}^T\mathbf{B}\mathbf{s}} + \frac{\mathbf{y}\mathbf{y}^T}{\mathbf{s}^T\mathbf{y}}. \tag{4.192}$$

The updating formula satisfies the quasi-Newton equation

$$\bar{\mathbf{B}}\mathbf{s} = \mathbf{y}, \quad \bar{\mathbf{H}}\mathbf{y} = \mathbf{s}. \tag{4.193}$$

When the modified Newton-Raphson and the BFGS methods are used, the linear system of equations $\mathbf{A}\mathbf{d} = -\nabla f(\mathbf{x})$ needs to be solved at each iteration. We use the conjugate gradient method to solve this system, for the following two reasons:

1. It does not require an explicit form of the matrix $\mathbf{A}$, which we do not have when the MNR method is adapted.
2. As in a direct method, it requires at most n (the size of $\mathbf{d}$) steps to solve the system, and in many cases it needs a much smaller number of iterations.

The Conjugate Gradient Method solves the system, $\mathbf{Ad} = \mathbf{b}$, where $\mathbf{A} \in R^{n \times n}$ is symmetric positive definite and $\mathbf{b} \in R^n$. For a given initial estimate $\mathbf{d}_0$, the solution is $\mathbf{d}^* \approx \mathbf{d}_k$ if $||\mathbf{b} - \mathbf{Ad}_k||_2 \leq 10^{-8}$. At the $(k-1)$ step, we have $\mathbf{d}_{k-1}$, the direction $\mathbf{p}_{k-1}$, and gradient $\mathbf{g}_{k-1}$. Then compute $\mathbf{d}_k$, $\mathbf{p}_k$ and $\mathbf{g}_k$ as follows:

$$\alpha_k = \frac{\mathbf{g}_{k-1}^T \mathbf{g}_{k-1}}{\mathbf{p}_{k-1}^T \mathbf{A} \mathbf{p}_{k-1}}$$

$$\mathbf{d}_k = \mathbf{d}_{k-1} + \alpha_k \mathbf{p}_{k-1}$$

$$\mathbf{g}_k = \mathbf{g}_{k-1} - \alpha_k \mathbf{A} \mathbf{p}_{k-1}$$

$$\beta_k = \frac{\mathbf{g}_k^T \mathbf{g}_k}{\mathbf{g}_{k-1}^T \mathbf{g}_{k-1}}$$

$$\mathbf{p}_k = \mathbf{g}_k + \beta_k \mathbf{p}_{k-1}.$$

If $\|\mathbf{g}_k\|_2 \leq 10^{-8}$, then $\mathbf{d}^* \approx \mathbf{d}_k$ and the algorithm is terminated; otherwise let $k = k+1$ and continue the computations. In some of our examples, it requires approximately $0.4n$ iterations to approximate $\mathbf{p}$ to the accuracy of 10^{-9}.

The Armijo line search method is adopted here to determine the step size μ. This method chooses μ as the first number from the sequence: 1,0.1,0.01,...which satisfies the following inequality:

$$E(\mathbf{x} + \mu \mathbf{d}) < E(\mathbf{x}) + 0.0001 \mu \nabla f^T(\mathbf{x}) \mathbf{d}.$$

This method is used because it is easy to implement and does not require any extra information on the gradient of E, which is very time consuming to compute for our problem. In theory, it has been shown that $\mu_k = 1$ when $\mathbf{x}^k$ is close to $\mathbf{x}^*$ and $\mathbf{d}^k$ is computed by the modified Newton-Raphson method or the quasi-Newton methods. The Armijo line search method is computationally very inexpensive. The stopping criterion is set as $\mathbf{x}^k = \mathbf{x}^*$ if $\|\nabla E(\mathbf{x}^k)\|_2 < 10^{-9}$.

Comments on Stability

Digital filters designed by the well-known commonly used spatial domain techniques such as Shanks's method, the DFP algorithm, and iterative prefiltering are not guaranteed to be stable. The BFGS and the MNR techniques described may also yield filters that are unstable. However, an unstable filter is likely to have an impulse response $h(n_1, n_2)$ with large amplitude, and hence the error $e(n_1, n_2)$ will tend to be large for some region of (n_2, n_2). Since all of the above spatial domain methods attempt to minimize the total squared error, the resulting filter is likely to be stable. After designing the filter, it should be tested for stability. If it does turn out to be unstable, some stabilization method such as the Planar Least-Squares Inverse [16] can be used to stabilize the unstable filter without seriously affecting the magnitude response of the filter.

Examples

In this section, several examples of 2-D digital filters designed by the foregoing methods are presented. We will design quarterplane filters using the methods, and then we will use the method described earlier to generate a zero-phase filter.

In the following examples, we have $m = n = p = q$. The initial and the final transfer functions are represented, respectively, as

$$H_0(z_1, z_2) = (\mathbf{a}^T \mathbf{A}_0 \mathbf{b})/(\mathbf{a}^T \mathbf{B}_0 \mathbf{b}) \tag{4.194}$$

and

$$H_f(z_1, z_2) = (\mathbf{a}^T \mathbf{A}_f b)/(\mathbf{a}^T \mathbf{B}_f \mathbf{b}) \tag{4.195}$$

where $\mathbf{a} = [1\, z_1^{-1} \ldots]^T$ and $\mathbf{b} = [1\, z_2^{-1} \ldots]^T$.

EXAMPLE 4.12 *Lowpass Filter*

The desired impulse response is a lowpass circularly symmetric function with cutoff frequency of 0.4π to which a Hamming window of radius 11 points is applied. The initial filter ($m = n = p = q = 3$) is generated by applying Shanks's algorithm, which yields the following filter:

$$A_0 = \begin{bmatrix} 0.125664 & 0.028491 & 0.020191 \\ 0.028491 & 0.008307 & 0.006173 \\ 0.0201911 & 0.006173 & 0.004956 \end{bmatrix}$$

$$B_0 = \begin{bmatrix} 1.000000 & -0.573264 & 0.255029 \\ -0.573264 & 0.353636 & -0.131060 \\ 0.255029 & -0.131060 & 0.086932 \end{bmatrix}.$$

The error function value for this initial filter is $E = 6.7048 \times 10^{-5}$, and the norm of the gradient is $\|\nabla f(\mathbf{x}_0\|_2 = 0.001589$. The error function is small enough that the MNR algorithm works very efficiently and converges in 47 iterations. In this case, BFGS converges in 36 iterations to the following filter:

$$A_f = \begin{bmatrix} 0.125664 & 0.035970 & 0.024999 \\ 0.035970 & 0.011509 & 0.009040 \\ 0.024999 & 0.009040 & 0.007634 \end{bmatrix}$$

$$B_f = \begin{bmatrix} 1.000000 & -0.512021 & 0.238950 \\ -0.512022 & 0.283301 & -0.102839 \\ 0.238949 & -0.102837 & 0.083721 \end{bmatrix}.$$

For the above filter $\|\nabla f(\mathbf{x})\|_2 < 10^{-9}$, which is the stopping criterion, hence, it represents a significant improvement over the initial point. The magnitude response of the zero-phase filter is shown in Fig. 4.54.

EXAMPLE 4.13 *Bandstop Filter*

The desired impulse response is the ideal circularly symmetric bandstop function to which a Hamming window of radius 16 points is applied. The low and high cutoff frequencies are 0.5π and 0.8π, respectively. For $m = n = p = q = 4$, Shanks's method yields an initial filter with

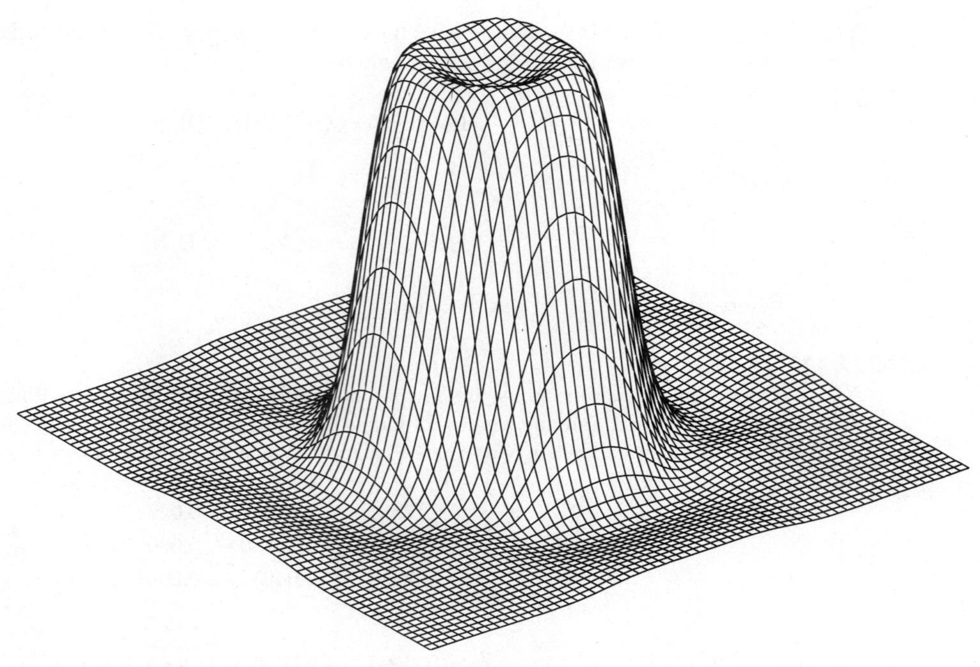

Figure 4.54 Magnitude response of lowpass filter designed using the BFGS method.

$\|\nabla f(\mathbf{x}_0)\| = 0.0076916$. The BFGS converges in 65 iterations, and MNR converges in 56 iterations to the following:

$$A_f = \begin{bmatrix} 0.306305 & 0.016072 & 0.152433 & -0.036429 \\ 0.016072 & 0.176129 & 0.162104, & 0.040054 \\ 0.152433 & 0.162103 & 0.100916 & 0.043361 \\ -0.036443 & 0.040054 & 0.043361 & 0.050239 \end{bmatrix}$$

$$B_f = \begin{bmatrix} 1.000000 & 0.232732 & 0.218761 & -0.026625 \\ 0.232773 & 0.449670 & 0.295187 & 0.040054 \\ 0.218761 & 0.295187 & 0.167098 & 0.121982 \\ -0.026625 & 0.040054 & 0.121982 & 0.109894 \end{bmatrix}.$$

The magnitude response of the final zero-phase filter is shown in Fig. 4.55. In the above examples, both the BFGS and MNR algorithms performed efficiently and were competitive with each other. However, in the next example, MNR outperforms BFGS. This is because we have increased the order of the filter, thereby reducing the error function even more.

▶ EXAMPLE 4.14 *Bandpass Filter*

The desired impulse response is the ideal circularly symmetric bandpass function with low and high cutoff frequencies of 0.5π and 0.8π, respectively. Since we are not applying a window to the ideal response, it will be rather difficult to obtain a desirable filter with low order. Therefore, we

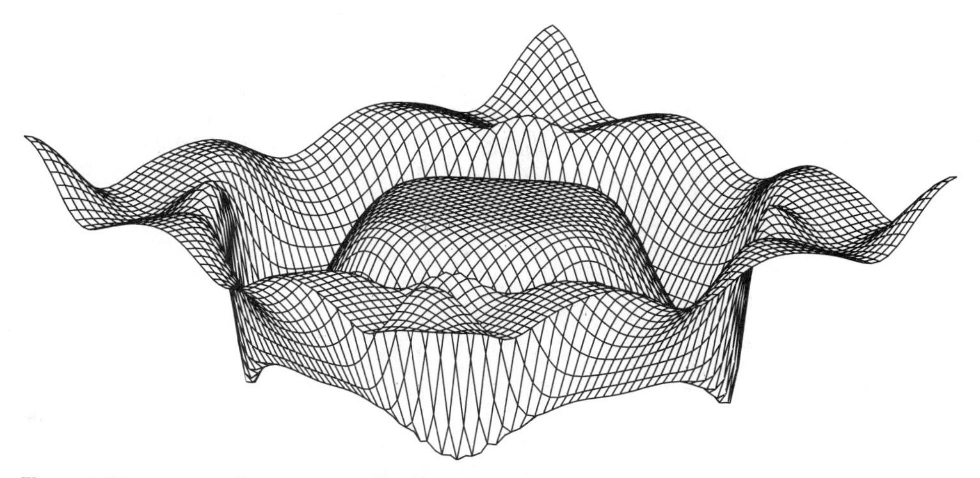

Figure 4.55 Magnitude response of bandstop filter designed by the MNR method.

try $m = n = p = q = 5$. Shanks's method yields an initial filter with $\|\nabla f(\mathbf{x}_0)\|_2 = 0.042349$. The BFGS method converged in 131 steps, and MNR converged in 65 steps to the following filter:

$$A_f = \begin{bmatrix} 0.314159 & 0.146979 & 0.074702 & 0.036228 & 0.051680 \\ 0.146979 & -0.019781 & -0.031366 & -0.021565 & 0.026786 \\ 0.074702 & -0.031366 & -0.089145 & -0.032835 & 0.035405 \\ 0.036228 & -0.021565 & -0.032835 & 0.038789 & 0.045964 \\ 0.051680 & 0.026786 & 0.035405 & 0.045964 & 0.038539 \end{bmatrix}.$$

$$B_f = \begin{bmatrix} 1.000000 & 0.047825 & 0.468307 & 0.045743 & 0.198223 \\ 0.047825 & -0.156823 & 0.057032 & -0.099338 & 0.096528 \\ 0.468307 & 0.057032 & 0.101311 & 0.025139 & 0.166621 \\ 0.045743 & -0.099338 & 0.025139 & 0.060187 & 0.063558 \\ 0.198223 & 0.096528 & 0.166621 & 0.063558 & 0.121570 \end{bmatrix}.$$

The magnitude response of the zero-phase filter is shown in Fig. 4.56.

4.6 SUMMARY

This chapter began with a discussion of the design of FIR filters. We presented the window method of design, which is the simplest method, followed by a detailed description of the frequency sampling method. One important method of design, the Parks-McClellan algorithm, has not been presented here and can be found in the references [26], [27], as well as in most popular texts on signal processing. For the design of IIR filters, we take the classical approach, in which the design is accomplished via an analog lowpass prototype. This technique is well understood and is used in the MATLAB filter design routines. We also described the analog-to-digital transformations such as the impulse invariant and bilinear transformation methods. Implementation of FIR and IIR digital filters was also discussed in detail. We covered all the basic structures such as the Direct Forms, cascade, parallel, Coupled Form, lattice, and ladder structures. The standard state-space form structure was also presented.

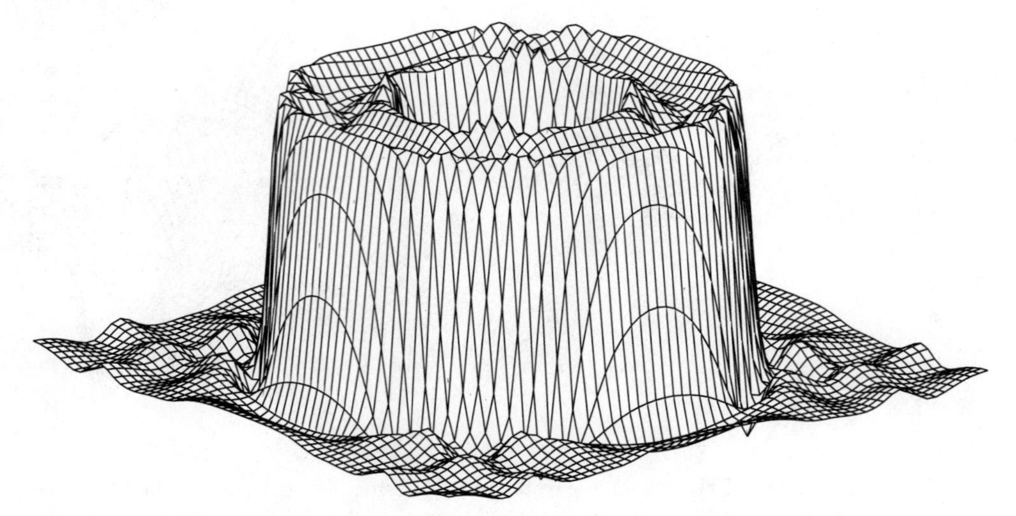

Figure 4.56 Magnitude response of bandpass filter designed by the MNR method.

Most of the design methods for FIR filters can be extended to 2-D. In this chapter, we presented the window method only for 2-D filter design. The design using the frequency sampling method has been assigned as a design project, and the design of IIR filters has been considered in the spatial domain only. The popular Shanks's method, which yields a closed-form solution, was described in detail. We also presented two iterative methods—the BFGS and the MNR method—which yield very good results for filter design.

▶ 4.7 MATLAB

The MATLAB commands will be presented with reference to the transfer function of a 1-D digital filter written as

$$H(z) = \frac{b_0 + b_1 z^{-1} + b_2 z^{-2} + \cdots + b_M z^{-M}}{1 + a_1 z^{-1} + a_2 z^{-2} + \cdots + a_N z^{-N}}. \tag{4.196}$$

The numerator and denominator are represented as the following row vectors:

$$\mathbf{b} = [b_0 \ b_1 \ \ldots \ b_M]$$

$$\mathbf{a} = [1 \ \ a_1 \ \ldots \ a_N].$$

FIR Filter Design

b = fir1(N,Wn); This function uses the window method to design an Nth order lowpass FIR digital filter and returns the filter coefficients in length $N+1$ vector **b**. The cutoff frequency Wn must be between $0 < Wn < 1.0$, with 1.0 corresponding to half the sample rate. The filter **b** is real and has linear phase, that is, even symmetric coefficients. If Wn is a two-element vector, **Wn** = [W1 W2], then firl returns an Nth order bandpass filter with the given 3-dB frequencies. **b = firl(N, Wn, 'high')** designs a highpass filter. **b = firl(N,Wn,'stop')** is a bandstop filter if **Wn** = [W1 W2]. The default window is Hamming. Other window functions can be specified as a trailing option such as **b = firl(N,Wn, 'high', bartlett (N + 1))**, where **bartlett(N + 1)** generates a $N+1$ point window. Other available windows include Boxcar (rectangular), Hanning, Blackman, Kaiser, and Chebwin.

W = kaiser(N,beta); returns the beta-valued N-point Kaiser window.

b = remez(N,F,A); uses the Parks-McClellan algorithm. It returns a length $N + 1$ linear phase (real, symmetric coefficients) FIR filter, which has the best approximation to the desired frequency response described by **F** and **A**. **F** is a vector of frequency band edges in pairs, in ascending order between 0 and 1. **A** is a real vector the same size as **F**, which specifies the desired magnitude. **b = remez(N,F,A,W)** uses the weights in **W** to weight the error. **W** has one entry per band (so it is half the length of F and A), which tells remez how much emphasis to put on minimizing the error in each band relative to the other bands.

IIR Filter Design

[b,a] = bilinear(NUM,DEN,Fs); converts an s-domain transfer function given by vectors **NUM** and **DEN** to the Z domain by bilinear transformation. The sampling frequency is Fs.

[b,a] = impinvar(NUM,DEN,Fs); creates a digital filter with numerator and denominator coefficients **b** and **a**, respectively, whose impulse response is equal to the impulse response of the analog filter with coefficients **NUM** and **DEN** sampled at a frequency of Fs Hertz.

[b,a] = butter(N,Wn); designs an Nth order lowpass digital Butterworth filter and returns the filter coefficients in length $N + 1$ vectors **b** (numerator) and **A** (denominator). The coefficients are listed in descending powers of z. The cutoff frequency Wn must be 0 < Wn < 1.0, with 1.0 corresponding to half the sample rate. If Wn is a two-element vector, **Wn** = [W1 W2], we get an order $2N$ bandpass filter. **[b,a] = BUTTER(N,Wn, 'high')** designs a highpass filter. **[b,a] = BUTTER(N,Wn,'stop')** is a bandstop filter if **Wn** = [W1 W2].

[N, Wn] = buttord(Wp, Ws, Rp, Rs); returns the order N of the lowest order digital Butterworth filter that loses no more than Rp dB in the passband and has at least Rs dB of attenuation in the stopband.

[b,a] = cheby1(N,R,Wn); designs an Nth order lowpass Chebyshev-I digital filter with R decibels of ripple in the passband and a normalized cutoff frequency of Wn. The other options are the same as in the butter function.

[N, Wn] = cheb1ord(Wp, Ws, Rp, Rs); returns the order N of the lowest order digital Chebyshev-I filter that loses no more than Rp dB in the passband and has at least Rs dB of attenuation in the stopband.

[B,A] = cheby2(N,R,Ws); designs an Nth order lowpass Chebyshev-II digital filter with the stopband ripple R decibels down and stopband edge frequency Ws.

[N, Wn] = cheb2ord(Wp, Ws, Rp, Rs); returns the order *N* of the lowest order digital Chebyshev-II filter that loses no more than Rp dB in the passband and has at least Rs dB of attenuation in the stopband.

[b,a] = ellip(N, Rp, Rs, Wn); designs an *N*th order lowpass digital elliptic filter with Rp decibels of ripple in the passband and a stopband that is Rs decibels down.

[N, Wn] = ellipord(Wp, Ws, Rp, Rs); returns the order *N* of the lowest order digital elliptic filter that loses no more than Rp dB in the passband and has at least Rs dB of attenuation in the stopband.

[A,B,C,D] = tf2ss(b,a); converts from the transfer function to the state-space model in Direct-II form. Uses a numerator polynomial vector **b** that has been padded with zeros to make it the same length as the denominator **a**.

[NUMT,DENT] = lp2lp(NUM,DEN,Wo); transforms the analog lowpass filter prototype NUM(s)/DEN(s) with unity cutoff frequency to a lowpass filter with cutoff frequency Wo.

[NUMT,DENT] = lp2hp(NUM,DEN,Wo); transforms the analog lowpass filter prototype NUM(s)/DEN(s) with unity cutoff frequency to a highpass filter with cutoff frequency Wo.

[NUMT,DENT] = lp2bp(NUM,DEN,Wo,Bw); transforms the lowpass filter prototype NUM(s)/DEN(s) with unity cutoff frequency to a bandpass filter with center frequency Wo and bandwidth Bw.

[NUMT,DENT] = lp2bs(NUM,DEN,Wo,Bw); transforms the lowpass filter prototype NUM(s)/DEN(s) with unity cutoff frequency to a bandstop filter with center frequency Wo and bandwidth Bw.

Filter Structures

[b,a] = ss2tf(A,B,C,D); converts from the state-space model to the transfer function.

[K,V] = tf2latc(b,a); finds the lattice parameters **K** and the ladder parameters **V** for an IIR lattice-ladder filter.

K = tf2latc(b); finds the lattice parameters **K** for an FIR lattice filter.

[b,a] = latc2tf(K,V); finds the transfer function from the IIR lattice coefficients **K** and ladder coefficients **V**.

b = latc2tf(K); finds the transfer function numerator from the FIR lattice coefficients specified by **K**.

[F,G] = latcfilt(K,x); filters the input signal **x** with the FIR lattice coefficients in vector **K**. Vector **F** is the upper channel output, and vector **G** is the lower channel output.

2-D Filter Design

h = fwind1(HD,win); designs a two-dimensional FIR filter **h** with frequency response HD. **fwind1** uses the one-dimensional window **win** to form an approximately circularly symmetric two-dimensional window using Huang's method. You can specify **win** using window functions in the Signal Processing Toolbox, such as BOXCAR, HAMMING, HANNING, BARTLETT, BLACKMAN, KAISER, or CHEBWIN. If length(WIN) is N, then **h** is N-by-N. **HD** is a matrix containing the desired frequency response sampled at equally spaced points between -1.0 and 1.0 (in normalized frequency, where 1.0 corresponds to half the sampling frequency, or pi radians) along the x and y frequency axes.

h = fwind1(HD,WIN1,WIN2); uses the two one-dimensional windows **WIN1** and **WIN2** to create a separable two-dimensional window.

h = fwind2(HD,WIN); produces the two-dimensional FIR filter **h** using an inverse Fourier transform of the desired frequency response **HD** and multiplication by the specified 2-D window **WIN**.

4.8 PROBLEMS

1. The ideal analog differentiator is given by

$$y(t) = \frac{dx(t)}{dt}$$

where $x(t)$ and $y(t)$ are the input and output, respectively.

(a) Find the frequency response of the differentiator. Plot the magnitude and phase responses.

(b) The digital differentiator is described by $y(n) = x(n) - x(n-1)$. Find its frequency response and plot the magnitude and phase responses using MATLAB. Compare these plots with those of (a) and comment on any differences.

(c) Compute the response of the digital differentiator to the input $x(n) = A\cos(\omega_0 n + \theta)$.

2. Find the impulse response, $h_{HP}(n)$ of an ideal highpass filter. By truncating and shifting this impulse response, find the impulse response, $\hat{h}_{HP}(n)$, of a causal highpass filter of length $M+1$. How is $\hat{h}_{HP}(n)$ related to its lowpass equivalent $\hat{h}_{LP}(n)$?

3. Find the impulse response of a zero-phase lowpass filter whose frequency response is shown in Fig. P4.1.

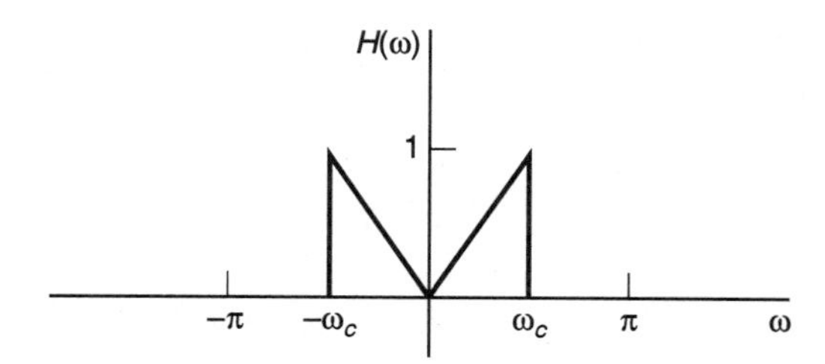

Figure P4.1

4. Consider the problem of designing FIR filters to approximate the ideal frequency response given by

$$H_I(\omega) = \begin{cases} 1, & |\omega| \le \frac{\pi}{3} \\ 0, & \frac{\pi}{3} < |\omega| < \pi. \end{cases}$$

(a) Design a 25-tap FIR filter using a rectangular window. Plot the magnitude and phase response of the designed filter using MATLAB.
(b) Repeat part (a) using a Hamming window.
(c) Repeat part (a) using a Blackman window.

5. Repeat Problem 4 using the Hanning and Barlett windows.

6. Using the Kaiser window, design a 25-tap FIR filter to approximate an ideal integrator. Use your own criteria for the passband ripples and the side lobes. Plot the magnitude and phase responses using MATLAB.

7. Design a 10-tap FIR filter to approximate an ideal digital integrator. Use the method of frequency sampling.

(a) Find the filter taps.
(b) Plot the magnitude and phase of the designed filter using MATLAB.

8. Using the frequency sampling approach, design a linear phase lowpass FIR filter of length 35 with a passband edge frequency at $\omega_p = 0.3\pi$. Assume an ideal brickwall characteristic for the desired magnitude response. Plot the magnitude and phase of the designed filter using MATLAB.

9. Using the frequency sampling approach, design a linear phase bandpass FIR filter of length 35 with passband edge frequencies at $\omega_l = 0.3\pi$ and $\omega_p = 0.7\pi$.

10. Consider an analog filter with transfer function

$$H_a(s) = \sum_{i=1}^{N} \frac{b_i}{s - p_i}.$$

Show that the impulse invariant transformation does not preserve the step response.

11. A single-pole analog filter is given by

$$H(s) = \frac{\Omega_c}{s + \Omega_c}$$

where Ω_c is the 3-dB (cutoff) frequency.

(a) Using bilinear transformation, design a lowpass digital filter with a 3-dB frequency of 0.3π.
(b) Use impulse invariant transformation to design the filter in (a).

12. An analog filter is given by

$$H(s) = \frac{s + \frac{1}{2}}{(s + \frac{1}{2})^2 + 16}.$$

(a) Use the impulse invariant method to convert the above to a digital filter. The sampling period is $T = 0.1$.
(b) Use bilinear transformation and repeat (a).

13. An ideal analog integrator is described by $H(s) = 1/s$.

(a) Use bilinear transformation to find an equivalent digital integrator.
(b) Find the frequency response of the analog integrator and plot the magnitude and phase response using MATLAB.
(c) Find the frequency response of the digital integrator and plot the magnitude and phase response using MATLAB.

(d) Compare the plots of (b) and (c) and comment on how well they match.
(e) The digital integrator has a pole at $z = 1$. What option do we have in implementing it?

14. Design the filter given in Example 4.5 using the impulse invariant transformation. Plot the magnitude response of the designed filter. Compare this magnitude response with that of the filter designed by bilinear transformation and comment on your observations.

15. Let $H(s)$ be a real coefficient causal and stable analog filter. Let the step response of this filter be denoted by $s(t)$. In the step-invariance method of digital filter design, the step response is sampled. That is, the step response of the digital filter is obtained as

$$g(n) = s(nT), \quad n = 0, 1, 2, \ldots$$

Determine the expression for $G(z)$ if

$$H(s) = \sum_{i=0}^{N} \frac{A_i}{s - p_i}.$$

16. Given the definition of the Chebyshev polynomial in equation (4.72), show that these polynomials can be generated by the recursive equation

$$C_N(x) = 2xC_{N-1}(x) - C_{N-2}(x),$$
$$\text{where } C_0(x) = 1, \quad C_1(x) = x.$$

Hint: Use $x = \cos\theta$. Then find $C_{N+1}(x)$ and $C_{N-1}(x)$.

17. Consider the allpass filter transfer function

$$H_{ap}(z) = \frac{z^{-1} - \alpha}{1 - \alpha z^{-1}}$$

with α real. Find a formula for α so that the filter has a phase of ϕ_0 at a frequency of ω_0.

18. A lowpass IIR digital filter is required to meet the following specifications:

Passband ripple less than 2 dB.

Passband edge = 2 kHz.

Stopband attenuation > 50 dB.

Stopband edge = 3 kHz.

Sample rate = 10 kHz.

The filter is to be designed using bilinear transformation on an analog prototype. Determine the filter order required for Butterworth, Chebyshev, and Elliptic filter designs.

19. A highpass IIR digital filter is required to meet the following specifications:

Passband ripple less than 5 dB.

Passband edge = 3 kHz.

Stopband attenuation > 30 dB.

Stopband edge = 2 kHz.

Sample rate = 20 kHz.

The filter is to be designed using bilinear transformation on an analog prototype. Determine the filter order required for Butterworth, Chebyshev, and elliptic filter designs.

20. Consider the digital-to-digital transformation

$$z^{-1} = \prod_{k=1}^{N} \frac{\hat{z}^{-1} - \alpha_k}{1 - \alpha_k \hat{z}^{-1}}$$

with real α.

(a) Show that by this transformation, the unit circle maps to the unit circle. Also show that the inside of the unit circle also maps into itself.

(b) Let $H(z)$ be a digital lowpass prototype filter with passband edge frequency ω_p. Find a formula for α so that this transformation yields another lowpass filter with passband edge frequency ω_p^{LP}. In other words, prove the first item in Table 4.4.

21. A second-order lowpass Butterworth filter with a 3-dB frequency of 0.4π is given by the transfer function

$$H(z) = \frac{0.207 + 0.413z^{-1} + 0.207z^{-2}}{1 - 0.369z^{-1} + 0.196z^{-2}}.$$

Design a second-order lowpass filter with 3-dB frequency at 0.5π by transforming the above using spectral transformation. Use MATLAB to plot the magnitude responses of the two lowpass filters.

22. A second-order lowpass Butterworth filter with a 3-dB frequency of 0.4π is given by the transfer function

$$H(z) = \frac{0.207 + 0.413z^{-1} + 0.207z^{-2}}{1 - 0.369z^{-1} + 0.196z^{-2}}.$$

Design a second-order highpass filter with 3-dB frequency at 0.3π by transforming the above using spectral transformation. Use MATLAB to plot the magnitude responses of the two filters.

23. A second-order highpass Butterworth filter with a 3-dB frequency of 0.4π is given by the transfer function

$$H(z) = \frac{0.391 - 0.782z^{-1} + 0.391z^{-2}}{1 - 0.369z^{-1} + 0.196z^{-2}}.$$

Design a bandpass filter with 3-dB frequencies at 0.3π and 0.6π by transforming the above using spectral transformation. Use MATLAB to plot the magnitude responses of the two filters.

24. A Chebyshev-I lowpass filter with a 3-dB frequency of 0.4π is given by the transfer function

$$H(z) = \frac{0.074 + 0.221z^{-1} + 0.221z^{-2} + 0.074z^{-3}}{1 - 0.976z^{-1} + 0.857z^{-2} - 0.292z^{-3}}.$$

Design a highpass filter with 3-dB frequency at 0.3π by transforming the above using spectral transformation. Use MATLAB to plot the magnitude responses of the two filters.

25. A Chebyshev-I lowpass filter with a 3-dB frequency of 0.4π is given by the transfer function

$$H(z) = \frac{0.074 + 0.221z^{-1} + 0.221z^{-2} + 0.074z^{-3}}{1 - 0.976z^{-1} + 0.857z^{-2} - 0.292z^{-3}}.$$

Design a bandpass filter with 3-dB frequencies at 0.3π and 0.6π by transforming the above using spectral transformation. Use MATLAB to plot the magnitude responses of the two filters.

26. Design a lowpass Chebyshev-I digital filter that meets the following specifications:

(a) 1-dB ripple in the passband, $0 \leq |\omega| \leq 0.4\pi$.
(b) At least 50-dB attenuation in the stopband $0.45\pi \leq |\omega| \leq \pi$. Plot the magnitude and the phase response using MATLAB.

27. Design a lowpass Chebyshev-II digital filter that meets the following specifications:

(a) 1-dB ripple in the passband, $0 \leq |\omega| \leq 0.4\pi$.
(b) At least 50-dB attenuation in the stopband $0.45\pi \leq |\omega| \leq \pi$. Plot the magnitude and the phase response using MATLAB.

28. The frequency response of an ideal zero-phase notch filter is defined by

$$H_N(\omega) = \begin{cases} 0, & \omega = \pm\omega_0 \\ 1, & \text{otherwise.} \end{cases}$$

Determine the impulse response of this filter and plot it using MATLAB.

29. A second-order analog notch filter has the transfer function

$$H(s) = \frac{s^2 + \Omega_0^2}{s^2 + Ks + \Omega_0^2}.$$

(a) Using bilinear transformation, show that the transfer function $G(z)$ of the digital notch filter is [13]

$$G(z) = \frac{1}{2}\,\frac{1(1+\alpha) - 2\beta(1+\alpha)z^{-1} + (1+\alpha)z^{-2}}{1 - \beta(1+\alpha)z^{-1} + \alpha z^{-2}}$$

where

$$\alpha = \frac{1 + \Omega_0^2 - K}{1 + \Omega_0^2 + K}$$

$$\beta = \frac{1 - \Omega_0^2}{1 + \Omega_0^2}.$$

(b) Show that the notch frequency ω_0 and the 3-dB notch bandwidth B of the digital notch filter are related to the constants α and β by

$$\alpha = \frac{1 - \tan(B/2)}{1 + \tan(B/2)}$$

$$\beta = \cos\omega_0.$$

30. Based on Problem 29, design a second-order digital notch filter with a notch frequency at 100 Hz and a 3-dB notch bandwidth of 10 Hz. Assume a sampling frequency of 400 Hz.

31. Given the matrices

$$\widehat{\mathbf{A}} = \begin{bmatrix} \sigma & \omega \\ -\omega & \sigma \end{bmatrix}, \quad \widehat{\mathbf{B}} \triangleq \begin{bmatrix} \hat{b}_1 & \hat{b}_2 \end{bmatrix}^T, \quad \widehat{\mathbf{C}} \triangleq \begin{bmatrix} \hat{c}_1 & \hat{c}_2 \end{bmatrix}, \quad \widehat{D} \triangleq \hat{d},$$

$$\overline{B} = \begin{bmatrix} 1 & 0 \end{bmatrix}^T, \quad \overline{\mathbf{C}} \triangleq \begin{bmatrix} \bar{c}_1 & \bar{c}_2 \end{bmatrix}.$$

Find $\bar{c}_1$ and $\bar{c}_2$ in terms of $\hat{b}_1$, $\hat{b}_2$, $\hat{c}_1$, and $\hat{c}_2$ so that the transfer function of the state-space system $\{\widehat{\mathbf{A}}, \widehat{\mathbf{B}}, \widehat{\mathbf{C}}, \widehat{D}\}$ is the same as that of $\{\widehat{\mathbf{A}}, \overline{\mathbf{B}}, \overline{\mathbf{C}}, \widehat{D}\}$.

32. Let $\{\mathbf{A}, \mathbf{B}, \mathbf{C}, D\}$ represent the state-space structure of a digital filter. The transpose filter has the structure $\{\mathbf{A}^T, \mathbf{C}^T, \mathbf{B}^T, D\}$. Show that these filter structures have the same transfer function.

33. Refer to the last problem regarding the definition of a transpose structure of a filter. Find the transposed structure of the filter shown in Fig. P4.2. Verify that the original and the transposed structures have the same transfer function.

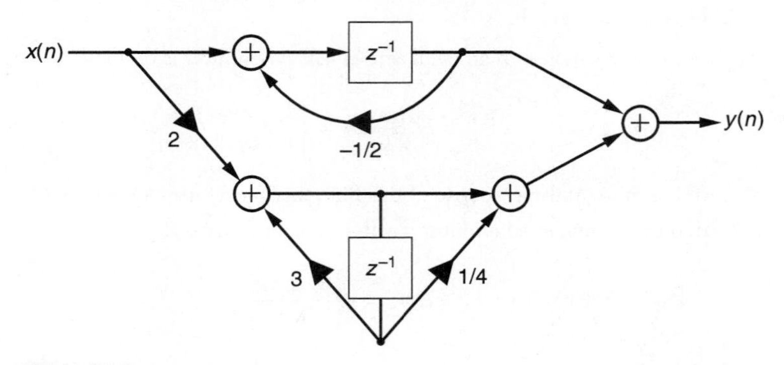

Figure P4.2

34. Given the FIR filter with transfer function

$$H(z) = 1 + 3.1z^{-1} + 5.5z^{-2} + 4.\,2z^{-3} + 2.3z^{-4},$$

(a) Sketch the Direct Form realization.
(b) Sketch the lattice realization of the filter.

35. Given the FIR filter with transfer function

$$H(z) = -0.00130 + 0.0264z^{-1} + 0.1558z^{-2} + 0.3192z^{-3} + 0.3192z^{-4} + 0.1558z^{-5} + 0.0264z^{-6} - 0.0013z^{-7}$$

(a) Sketch the Direct Form realization.
(b) Sketch the lattice realization of the filter.

36. Obtain the Direct Form II, Direct Form II$_T$, parallel, and cascade structures for the following filters:

(a)

$$y(n) + \frac{3}{4}y(n-1) + \frac{1}{8}y(n-2) = x(n) + 2x(n-1).$$

(b)

$$H(z) = \frac{z^2 + 2z + 5}{(z+0.5)(z^2+z+1)}.$$

(c)

$$H(z) = \frac{1 + z^{-1} + z^{-2}}{(1+0.3z^{-1})(1+0.4z^{-1})(1+0.5z^{-1})}.$$

(d)

$$y(n) + 0.2y(n-1) = x(n) + 3x(n-1) + x(n-2).$$

37. Find the lattice structure for the digital filter difference equation

$$y(n) = x(n) + 0.25x(n-1) + 0.5x(n-2) + 0.75x(n-3) + x(n-4).$$

38. Find the lattice-ladder structure for the filter given by the transfer function

$$H(z) = \frac{1 + z^{-1} + z^{-2}}{(1 + 0.3z^{-1})(1 + 0.4z^{-1})(1 + 0.5z^{-1})}.$$

39. Find the Coupled Form and the lattice-ladder structures for the filter given by the difference equation

$$y(n) + \frac{3}{4}y(n-1) + \frac{1}{4}y(n-2) = x(n) + 2x(n-1).$$

40. A lattice filter is specified by the reflection coefficients, $K_1 = 0.5$, $K_2 = -0.1$, $K_3 = 0.9$, $K_4 = 0.7$.

(a) Find the impulse response of the filter by tracing an impulse through the flow diagram.
(b) Find the Direct Form FIR structure for this filter.

41. A lattice filter is specified by the reflection coefficients, $K_1 = 0.5$, $K_2 = -0.1$, $K_3 = 1$. Find the Direct Form structure for this filter.

42. Given a causal IIR filter with the transfer function:

$$H(z) = \frac{1 + 3z^{-1} + 3z^{-2} + 2z^{-3}}{1 + 0.1z^{-1} + 0.2z^{-2} + 0.3z^{-3}}$$

find the lattice-ladder realization for this filter.

43. Given the following digital filter:

$$H(z) = \frac{z(z+1)}{z^2 - z + 1.06},$$

(a) Find the Coupled Form realization.
(b) Find the Direct Form II structure.
(c) Find the impulse response.
(d) Find the lattice-ladder structure.

44. The state-space model for a digital filter is given by

$$\mathbf{v}(n+1) = \begin{bmatrix} 0 & 1 \\ -0.25 & 1 \end{bmatrix} \mathbf{v}(n) + \begin{bmatrix} 0 \\ 1 \end{bmatrix} x(n)$$

$$y(n) = \begin{bmatrix} -0.25 & 1 \end{bmatrix} \mathbf{v}(n) + [1]\, x(n).$$

(a) Find the poles of the filter.
(b) Find the transfer function and the impulse response.

(c) Sketch the flow diagram.
(d) Find the state vector $\mathbf{v}(n)$ with zero input and an initial condition of $\mathbf{v}(0) = [1 - 1]^T$.

45. The state-space model for a digital filter is given by

$$\mathbf{v}(n+1) = \begin{bmatrix} 0 & 1 \\ -0.5 & -0.25 \end{bmatrix} \mathbf{v}(n) + \begin{bmatrix} 0 \\ 1 \end{bmatrix} x(n)$$

$$y(n) = \begin{bmatrix} 1 & 0 \end{bmatrix} \mathbf{v}(n) + [0]\, x(n).$$

(a) Find the poles of the filter.
(b) Find the transfer function and the impulse response.
(c) Sketch the flow diagram.
(d) Find the state vector $\mathbf{v}(n)$ with zero input and an initial condition of $\mathbf{v}(0) = [1 - 1]^T$.

46. The state-space model for a digital filter is given by

$$\mathbf{v}(n+1) = \begin{bmatrix} -0.3 & 0.8 \\ -0.8 & -0.3 \end{bmatrix} \mathbf{v}(n) + \begin{bmatrix} 1 \\ 0 \end{bmatrix} x(n)$$

$$y(n) = \begin{bmatrix} 2 & 3 \end{bmatrix} \mathbf{v}(n) + [0.1]x(n).$$

(a) Find the poles of the filter.
(b) Find the transfer function and the impulse response.
(c) Sketch the flow diagram.
(d) Find the state vector $\mathbf{v}(n)$ with zero input and an initial condition of $\mathbf{v}(0) = [1 - 1]^T$.

47. Prove Lemma 4.1.

48. Derive (4.132).

49. Derive the algorithms for converting from a lattice structure to Direct Form and from Direct Form to lattice.

50. Show that the one-multiplier lattice (Fig. P4.3) yields the same outputs as the two-multiplier lattice (upto a scale factor).

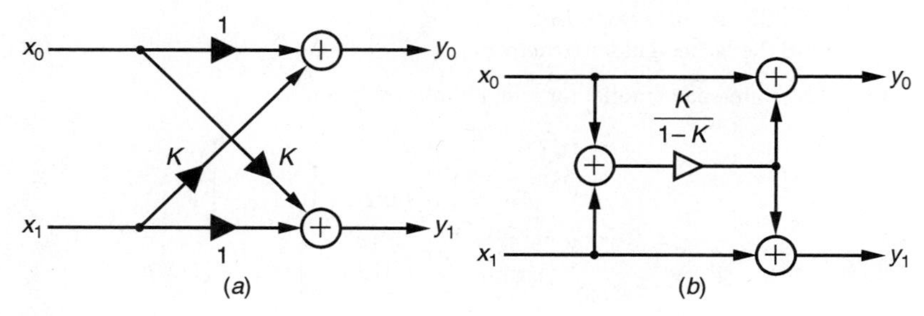

Figure P4.3

51. Derive equation (4.160).

52. Find the lattice-ladder structure for the following state-space model:

$$\mathbf{v}(n+1) = \begin{bmatrix} 0 & 1 \\ -0.5 & -0.25 \end{bmatrix} \mathbf{v}(n) + \begin{bmatrix} 0 \\ 1 \end{bmatrix} x(n)$$

$$y(n) = \begin{bmatrix} 1 & 0 \end{bmatrix} \mathbf{v}(n) + [0]\, x(n).$$

53. Find the lattice-ladder structure for the following state-space model:

$$\mathbf{v}(n+1) = \begin{bmatrix} 0 & 1 \\ -0.25 & -0.75 \end{bmatrix} \mathbf{v}(n) + \begin{bmatrix} 0.1 \\ 0.2 \end{bmatrix} x(n)$$

$$y(n) = \begin{bmatrix} 1 & 0 \end{bmatrix} \mathbf{v}(n) + [1]\, x(n).$$

54. Let $h_{LP}(n_1, n_2)$ denote the impulse response of a 2-D lowpass filter. A new impulse response is obtained as follows:

$$h(n_1, n_2) = (-1)^{n_1}(-1)^{n_2} h_{LP}(n_1, n_2).$$

What type of filter is this and why?

55. Letting $x_1(n_1, n_2)$ and $x_2(n_1, n_2)$ be zero-phase sequences, show that the product $x_1(n_1, n_2)x_2(n_1, n_2)$ is also a zero-phase sequence.

56. Design a 2-D circularly symmetric lowpass filter with a 3-dB frequency of 0.5π using the Bartlett window. Plot the magnitude response and the phase response using MATLAB.

57. Design a 2-D separable bandpass filter using the Bartlett window with 3-dB frequencies of $\omega_{c1} = 0.3\pi, \omega_{c2} = 0.6\pi$. Plot the magnitude response and the phase response using MATLAB.

58. Design a 2-D circularly symmetric lowpass filter with a 3-dB frequency of 0.5π using the Blackman window. Plot the magnitude response and the phase response using MATLAB.

59. Let a desired impulse response $h_d(n_1, n_2)$ be given. We would like to design a 2-D IIR filter to approximate this response with the transfer function

$$H(z_1, z_2) = \frac{b}{1 + a_1 z_1^{-1} + a_2 z_2^{-1}}$$

where b, a_1, and a_2 are the coefficients to be estimated. Show that the cost function given by

$$J = \sum_{(n_1,n_2)\epsilon R_d} \sum (h_d(n_1, n_2) - h(n_1, n_2))^2$$

is a nonquadradic function of the coefficients. $h(n_1, n_2)$ is the actual impulse response of the filter. You can show this by using only a few terms.

60. Let $H(z)$ be the impulse response of a lowpass 1-D causal IIR filter with a cutoff frequency of $\pi/3$. We design a 2-D filter as

$$H(z_1, z_2) = H(z_1)H(z_2).$$

What is the approximate magnitude response of $H(z_1, z_2)$? Is this a stable filter?

61. Let $H(z)$ be the impulse response of a lowpass 1-D causal IIR filter with a cutoff frequency of $\pi/3$. We design a 2-D filter as

$$H(z_1, z_2) = H(z_1 z_2)H(z_1^{-1} z_2).$$

What is the approximate magnitude response of $H(z_1, z_2)$? Is this a stable filter?

62. Consider a 2-D FIR filter with the region of support given by

$$-N \leq n_1 \leq N$$
$$-N \leq n_2 \leq N.$$

The impulse response also has symmetry given by

$$h(n_1, n_2) = h(-n_1, n_2) = h(n_1, -n_2) = h(-n_1, -n_2).$$

By exploiting this symmetry, how many multiplications and additions are required to implement this filter?

63. Let $H(z_1, z_2)$ be the transfer function of a 2-D filter given by

$$H(z_1, z_2) = \frac{b}{1 + a_1 z_1^{-1} + a_2 z_2^{-1}}.$$

Sketch the flow diagram of this filter.

64. Let $H(z_1, z_2)$ be the transfer function of a 2-D filter given by

$$H(z_1, z_2) = \frac{1 - z_2^{-1}}{(1 + a z_1^{-1})(1 + a z_2^{-1})}.$$

Sketch the flow diagram of this filter for (a) Direct Form, and (b) cascade form.

65. Consider a 2-D state-space realization given by

$$\begin{bmatrix} x^h(n_1+1, n_2) \\ x^v(n_1, n_2+1) \end{bmatrix} = \begin{bmatrix} 1 & 2 \\ 4 & 3 \end{bmatrix} \begin{bmatrix} x^h(n_1, n_2) \\ x^v(n_1, n_2) \end{bmatrix} + \begin{bmatrix} 1 \\ 0 \end{bmatrix} u(n_1, n_2)$$

$$y(n_1, n_2) = \begin{bmatrix} 7 & 8 \end{bmatrix} \begin{bmatrix} x^h(n_1, n_2) \\ x^v(n_1, n_2) \end{bmatrix} + [10]\, u(n_1, n_2)$$

where $x^h(n_1, n_2)$ is the horizontal state variable and $x^v(n_1, n_2)$ is the vertical state variable, $u(n_1, n_2)$ is the input, and $y(n_1, n_2)$ is the output. This is a special case of the well-known Givone-Roesser state-space model.

(a) Sketch the flow diagram of the filter.
(b) Find the transfer function of the filter.

66. The general form of the 2-D Givone-Roesser state-space model is given by

$$\begin{bmatrix} \mathbf{x}^h(n_1+1, n_2) \\ \mathbf{x}^v(n_1, n_2+1) \end{bmatrix} = \begin{bmatrix} \mathbf{A}_1 & \mathbf{A}_2 \\ \mathbf{A}_3 & \mathbf{A}_4 \end{bmatrix} \begin{bmatrix} \mathbf{x}^h(n_1, n_2) \\ \mathbf{x}^v(n_1, n_2) \end{bmatrix} + \begin{bmatrix} \mathbf{B}_1 \\ \mathbf{B}_2 \end{bmatrix} u(n_1, n_2)$$

$$y(n_1, n_2) = \begin{bmatrix} \mathbf{C}_1 & \mathbf{C}_2 \end{bmatrix} \begin{bmatrix} \mathbf{x}^h(n_1, n_2) \\ \mathbf{x}^v(n_1, n_2) \end{bmatrix}$$

where $\mathbf{x}(n_1, n_2)$ is the $M \times 1$ horizontal state vector, $\mathbf{x}^v(n_1, n_2)$ is the $N \times 1$ vertical state vector, $u(n_1, n_2)$ is the input, and $y(n_1, n_2)$ is the output. All the other matrices have appropriate dimensions. Find the transfer function of this system.

67. Letting $x_1(n_1, n_2, n_3)$ and $x_2(n_1, n_2, n_3)$ be zero-phase sequences, show that the product $x_1(n_1, n_2, n_3)x_2(n_1, n_2, n_3)x_3(n_1, n_2, n_3)$ is also a zero-phase sequence.

68. Derive equation (4.190).

4.9 COMPUTER PROJECTS

In the following projects, use any MATLAB command available.

Project 1
Write an interactive program for designing an FIR filter using the Kaiser window. The program should be menu controlled and have the following features.

Specifications

- LP, BP, HP, or BS
- Sampling frequency
- Passband edge(s), stopband edge(s) frequencies
- Passband ripple, stopband ripple.

Output Menu

- Filter coefficients
- Impulse response
- Absolute magnitude response
- Decibel magnitude response
- Phase response.

Run your program to design a bandpass filter with a sampling frequency of 10 kHz, edge frequencies of 0.4, 0.6, 3.8, and 4 kHz, and a ripple of 0.05 in both the passband and stopband.

Project 2 (Frequency sampling design)

Procedure

1. Design a 128-tap FIR bandpass filter using the MATLAB command fir1(), with cutoff frequencies of 0.4π and 0.6π. Plot the magnitude and phase responses of this filter.
2. Write a program (in C or MATLAB) for designing and implementing a FIR filter using the frequency sampling method. The filter should be implemented as a parallel connection of IIR resonators. The inputs to the program must include the following: The samples of the desired magnitude response of the filter in an N-point vector. These samples must consist of a set of uniformly sampled

magnitude response data in the interval $(0, 2\pi)$. The outputs of the program are: (a) impulse response of the filter; (b) magnitude response; and (c) phase response. The impulse response of the implemented filter can be found by applying an impulse at the input and "measuring" the output for as many samples as desired.

3. Use 16 data samples of the magnitude response in (1) as the input to your program written in (2). Plot these data samples and the magnitude response of the designed filter on the same graph. Also plot the impulse response and the phase response of the designed filter. Discuss your results.
4. Apply a square-wave at the input to the parallel resonator bank and measure the output. Plot the input and output and their magnitude responses. Discuss your results.

Question. How many real multiplications and real additions are required for the design and implementation of an N-tap FIR filter using the frequency sampling method with parallel resonators?

Project 3

Write a program to design IIR digital filters using the analog prototype. The program should be menu controlled and have the following features.

Specifications

- Butterworth, Chebyshev-I, Chebyshev-II, or elliptic
- LP, HP, BP, or BS
- Sampling frequency
- Passband edge(s), stopband edge(s) frequencies
- Passband ripple, stopband ripple.

Output Menu

- Filter coefficients
- Impulse response
- Absolute magnitude response
- Decibel magnitude response
- Phase response

Run your program on the test example given in Project 1. Design Butterworth, Chebyshev-I, Chebyshev-II, and elliptic filters for this example. Compare and comment on the results.

Project 4

Write a program to design 2-D FIR filters using the window method. The program should be menu controlled and have the following features.

Specifications

- Rectangular or circularly symmetric response
- LP, HP, BP, or BS
- Type of window: Rectangular, Bartlett, Hamming, or Hanning
- Normalized cutoff frequencies in the range of 0 to π.

Output Menu

- Filter coefficients
- Impulse response
- Absolute magnitude response
- Decibel magnitude response
- Phase response

Project 5

Write a program for designing and implementing a 2-D FIR filter using the frequency sampling method. The filter should be implemented as a parallel connection of separable IIR resonators. The specification should be a set of uniformly sampled magnitude response data in the interval $(0, 2\pi)$. For testing your program, use the samples of the 1-D designed magnitude response from Project 2. Of course, the specifications must be converted to 2-D using separability.

Project 6

Write a program to design a 2-D IIR digital filter using Shanks's method. The inputs to the program are as follows: (a) a 2-D impulse response given as a matrix; and (b) the orders of the numerator and denominator of the desired filter.

The program should design the filter and produce the following outputs: (a) coefficient matrices **B** and **A**; (b) 2-D magnitude response of the filter; (c) 2-D impulse response; and (d) the norm of the error between the desired and designed impulse responses.

Test your program on the following inputs:

1. Lowpass circularly symmetric zero-phase impulse response with a cutoff frequency of 0.2π, windowed by a 21×21 Hamming window.
2. Bandpass circularly symmetric zero-phase impulse response with cutoff frequencies of 0.2π and 0.6π, windowed by a 21×21 Hamming window.

Project 7

Write a program to design a 2-D IIR filter using the BFGS method of optimization. Use the specifications of Project 6 as design examples. The filters generated by Project 6 should be used as the initial conditions for the BFGS method. The outputs of the filter should be the following: (a) coefficient matrices **B** and **A**; (b) 2-D magnitude response of the filter; (c) 2-D impulse response; and (d) the norm of the error between the desired and designed impulse responses.

BIBLIOGRAPHY

[1] J.F. Kaiser, "Nonrecursive digital filter design using the I_0-sinh window function," *Proc. IEEE Intl. Symp. on Circuits and Systems*, pp. 20–23, April 1974.

[2] H.D. Helms, "Nonrecursive digital filters: Design methods for achieving specifications on frequency response," *IEEE Trans. Audio Electroacoustics*, AU-16, pp. 336–342, September 1968.

[3] A. Tustin, "A method of analyzing the behavior of linear systems in terms of time series," *Journal of IEE* 94, Part IIA, pp. 130–142, May 1947.

[4] M. Abramowitz and I.A. Stegun, *Handbook of Mathematical Functions*, Dover Publications, New York, 1964.

[5] A. Antoniou, *Digital Filters: Analysis, Design, and Applications*, McGraw-Hill, New York, 1993.

[6] A. Ambardar, *Analog and Digital Signal Processing*, PWS Publishing Company, Boston, 1995.

[7] T.W. Parks and C.S. Burrus, *Digital Filter Design*, Wiley, New York, 1987.

[8] B. Porat, *A Course in Digital Signal Processing*, Wiley, 1997.

[9] A.G. Constantinides, "Spectral transformations for digital filters," *Proc. IEE* 117, no. 8, pp. 1585–1590, August 1970.

[10] J.G. Proakis and D.G. Manolakis, *Digital Signal Processing: Principles, Algorithms, and Applications*, Prentice Hall, Upper Saddle River, NJ, 1996.

[11] S. Haykin, *Adaptive Filter Theory*, Prentice Hall, Englewood Cliffs, NJ, 1996.

[12] G. Strang and T. Nguyen, *Wavelets and Filter Banks*, Wellesley-Cambridge Press, Wellesley, MA, 1996.

[13] S.K. Mitra, *Digital Signal Processing*, McGraw-Hill, 2001.

[14] J.A. Cadzow, "Resursive digital filter synthesis via gradient based algorithms," *IEEE Transactions on Acoustic, Speech, Signal Processing*, ASSP-24, pp. 349–355, October 1976.

[15] J.L. Shanks, S. Trietel, and J.H. Justice, "Stability and synthesis of two-dimensional recursive filters," *IEEE Trans. Audio Electro-acoustics*, AU-20, pp. 115–128, June 1972.

[16] J.S. Lim, *Two-Dimensional Signal and Image Processing*, Prentice Hall, Englewood Cliffs, NJ, 1990.

[17] W.C. Davidson, "Variable metric method for minimization," *AEC Res, and Devel. Rep. ANL-5990*, 1959.

[18] R. Fletcher and M.J.D. Powell, "A rapidly convergent descent method for minimization," *Computer J.* 6, no. 2, pp. 163–168, July 1963.

[19] G.A. Shaw and R.M. Mersereau, "Design, stability, and performance of two-dimensional recursive digital filters," *Tech Report E21-B05-1, Georgia Institute of Technology*, School of Electrical Engineering, December 1979.

[20] P. Gill, W. Murray, and M. Wright, *Practical Optimization*, Academic, London, 1981.

[21] R. Fletcher, *Practical Methods of Optimization*, Wiley, New York, 1980.

[22] J. Dennis and J. Moore, "Quasi-newton methods, motivation, and theory," *SIAM Rev.* 19, pp. 46–89, 1977.

[23] M. Powell, "How bad are the BFGS and DFP methods when the objective function is quadratic?," *Rep. DAMTP* 1984/NA3, Cambridge, MA, 1984.

[24] D.E. Dudgeon and R.M. Mersereau, *Multidimensional Digital Signal Processing*, Prentice Hall, Englewood Cliffs, NJ, 1984.

[25] T. Bose and M.-Q. Chen, "Design of two-dimensional digital filters in the spatial domain," *IEEE Transactions on Signal Processing* 42, pp. 1464–1469, March 1993.

[26] T.W. Parks and J.H. McClellan, "Chebyshev approximation for nonrecursive digital filters with linear phase," *IEEE Trans. on Circuit Theory* CT-19, pp. 189–194, March 1972.

[27] T.W. Parks and J.H. McClellan, "A program for the design of linear phase finite impulse response digital filters," *IEEE Trans. on Audio Electroacoustics*, AU-20, pp. 195–199, August 1972.

[28] T.S. Huang, "Two dimensional windows," *IEEE Trans. Audio Electroacoust.*, AU-20, pp. 88–90, March 1972.

CHAPTER 5

Multirate Signal Processing

So far, this book has dealt with signal processing techniques in which the same sampling rate is used throughout the system. In Chapter 2, we briefly introduced the concept of sampling rate conversion. In many applications, the sampling rates (or data rates) are different in the various components of a system. Therefore, digital filters in these components process the signal at different rates. Hence, the sampling rate of the signal must be altered as it propagates from one component to another. The theory of processing signals at different sampling rates is called *multirate signal processing*.

The sampling rate of a signal can be altered in two different ways. First, we can use a D/A converter to transform the signal to the continuous-time domain and then resample it at the desired rate using an A/D converter. This adds significant hardware to the system and introduces quantization noise every time an A/D or D/A is used. Therefore, this is not a good option. The second method is to alter the sampling rate in the digital domain. Recall that this is done using downsamplers and upsamplers. Downsampling is usually preceded by a lowpass filter, and the combination is called *decimation*. Upsampling is also usually followed by a lowpass filter, and the combination is called *interpolation*.

In this chapter, we first review the concepts of decimation and interpolation and present the effects of these operations in the frequency domain and the Z-domain. The filters for these operations are then discussed in detail. Computationally efficient structures are derived for these filters, which are called *polyphase* structures. The design of narrowband FIR filters is then presented. This is called the Interpolated FIR (IFIR) filter design, and it can significantly reduce the number of coefficients that are needed for narrowband filters. The IFIR approach is then used for the implementation of computationally efficient decimators and interpolators in multiple stages. Next, the concept of subband coding is described with motivation in speech and image compression. This discussion leads to the theory of filter banks. The simple DFT filter bank is presented, after which Quadrature Mirror Filter (QMF) banks are described in detail. Design methodologies for these filters are also developed, and examples are given to illustrate the concepts.

5.1 DECIMATION AND INTERPOLATION

Downsampling a sequence $x(n)$ by a factor of M is the process of picking every Mth sample in the sequence and discarding the rest. The downsampled sequence is given by

$$x_M(n) \triangleq (\downarrow M)x(n) = x(Mn), \quad n = 0, \pm 1, \pm 2, \ldots \tag{5.1}$$

In Chapter 2, we have shown that when a sequence is downsampled, the resulting spectrum is the sum of the frequency-scaled and -shifted versions of the original spectrum. The spectrum of a signal downsampled by a factor of 2 is given by

$$X_{(2)}(\omega) = \frac{1}{2}\left[X\left(\frac{\omega}{2}\right) + X\left(\frac{\omega}{2} + \pi\right)\right]. \tag{5.2}$$

This operation will cause aliasing unless the highest frequency component, ω_h, of the original signal is less than or equal to $\frac{\pi}{2}$. This can be extended to a factor-of-M decimation. The following results from Chapter 2 are repeated here for convenience.

Theorem 5.1 *Let $x(n)$ and $X(\omega)$ denote a sequence and its DTFT. Let $x_M(n)$ and $X_M(\omega)$ denote the downsampled signal and its DTFT, where M is the downsampling factor. Then*

$$X_M(\omega) = \frac{1}{M}\sum_{k=0}^{M-1} X\left(\frac{\omega}{M} + \frac{2\pi k}{M}\right). \tag{5.3}$$

Theorem 5.2 *If a discrete sequence is downsampled by a factor of M, we must satisfy $\Omega_h \leq \frac{\Omega_s}{2M}$ in order to avoid aliasing, where Ω_s is the sampling frequency and Ω_h is the highest frequency component in the signal. Since $\Omega_s = \frac{2\pi}{T_s}$, this condition can also be written as $\Omega_h T_s \leq \frac{\pi}{M}$ or $\omega_h \leq \frac{\pi}{M}$.*

Theorem 5.2 states that, in order to avoid aliasing due to downsampling, the spectrum of the signal must be appropriately bandlimited by filtering. This can be done by filtering the signal with a lowpass filter with a cutoff frequency of $\frac{\pi}{M}$. This process is called decimation and is shown in Fig. 5.1.

Upsampling by a factor of M is the process of inserting $M-1$ zeros between every two consecutive sample. This can be mathematically described as

$$(\uparrow M)x(n) \triangleq x^{(M)}(n) = \begin{cases} x(\frac{n}{M}), & n = 0, \pm M, \pm 2M, \ldots \\ 0, & \text{otherwise.} \end{cases} \tag{5.4}$$

The relationship between the DTFT of an upsampled sequence and that of the original sequence was derived in Chapter 2 and is repeated here.

Theorem 5.3 *Let $x(n)$ and $X(\omega)$ denote a discrete-time sequence and its DTFT. When this sequence is upsampled by a factor of M, the DTFT of the upsampled sequence is related to that of the original sequence by*

$$X^{(M)}(\omega) = X(\omega M). \tag{5.5}$$

Figure 5.1 Decimation by a factor of M.

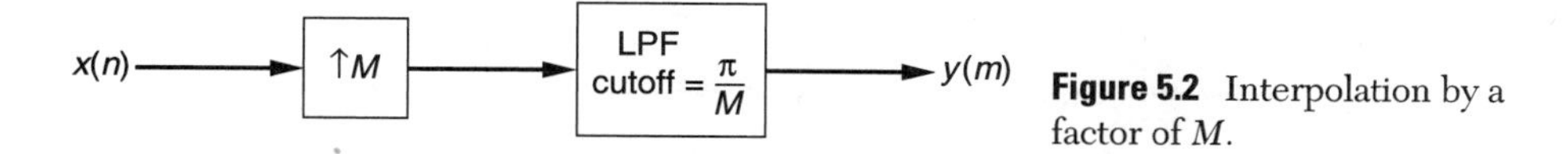

Figure 5.2 Interpolation by a factor of M.

The spectrum of the upsampled sequence is obtained from that of the original by stretching the frequency axis. Therefore, M spectral components appear in the interval $(-\pi, \pi)$. The $M-1$ extra components are called images and must be filtered out. The required filter is a lowpass filter with a cutoff frequency of $\frac{\pi}{M}$. This process of postfiltering after upsampling is called interpolation and is depicted in Fig. 5.2.

5.2 UPSAMPLING AND DOWNSAMPLING IN THE Z-DOMAIN

The results on downsampling and upsampling so far have been in the DTFT domain. Now we extend these results to the Z-transform domain. This will prove to be very useful in this chapter. The DTFT of a sequence is denoted by $X(\omega)$. However, we know that this really means $X(e^{j\omega})$. To transform to the Z-domain, we need to set $z = e^{j\omega}$. That is, we have the mappings

$$\begin{aligned} \omega &\to z \\ \frac{\omega}{M} &\to z^{\frac{1}{M}} \\ \omega M &\to z^{M} \\ \omega + \frac{2\pi}{M} &\to z e^{j\frac{2\pi}{M}}. \end{aligned}$$

Using the above mappings on the foregoing results, we have the following.

Theorem 5.4 *The Z-transform of* $x_M(n) = (\downarrow M)x(n)$ *is*

$$X_M(z) = \frac{1}{M} \sum_{k=0}^{M-1} X(z^{\frac{1}{M}} e^{j\frac{2\pi k}{M}}). \tag{5.6}$$

Corollary 5.1 *The Z-transform of* $x_2(n) = (\downarrow 2)x(n)$ *is*

$$X_2(z) = \frac{1}{2}\left[X(z^{\frac{1}{2}}) + X(-z^{\frac{1}{2}})\right]. \tag{5.7}$$

Theorem 5.5 *The Z-transform of* $x^{(M)}(n) = (\uparrow M)x(n)$ *is*

$$X^{(M)}(z) = X(z^{M}). \tag{5.8}$$

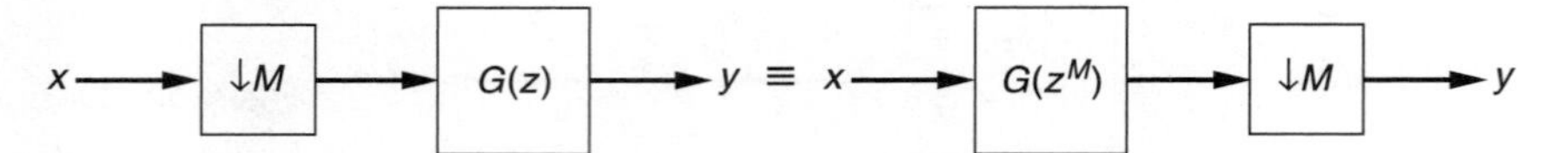

Figure 5.3 First Noble Identity.

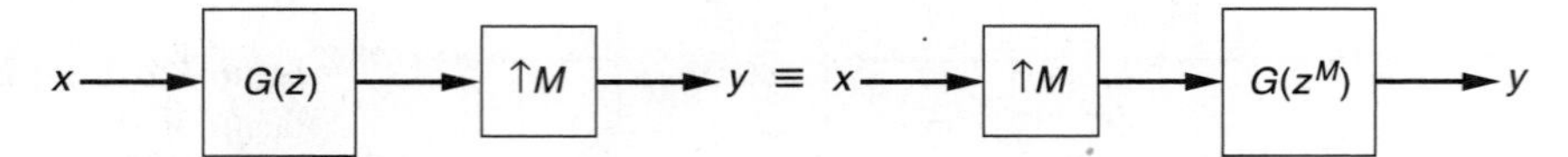

Figure 5.4 Second Noble Identity.

Corollary 5.2 *The Z-transform of* $x^{(2)}(n) = (\uparrow 2)x(n)$ *is*

$$X^{(2)}(z) = X(z^2). \tag{5.9}$$

In this chapter, we will also have situations where we need to swap the position of a filter with a downsampler or an upsampler. The nice properties that govern these operations are called the Noble identities. These identities are given in the following. The proofs are straightforward and are assigned as problems.

Theorem 5.6 *(First Noble Identity) The swapping of the positions of a filter and downsampler is governed by the identity*

$$G(z)(\downarrow M) = (\downarrow M)G(z^M).$$

This identity is depicted in Fig. 5.3. The left-hand side of the identity describes the downsampling operation followed by filtering. The right-hand side is the reverse.

Theorem 5.7 *(Second Noble Identity) The swapping of the positions of a filter and upsampler is governed by the identity*

$$(\uparrow M)G(z) = G(z^M)(\uparrow M).$$

This identity is depicted in Fig. 5.4. The Noble identities apply only when the filter has the form $G(z^M)$. This means that the filter impulse response has $M - 1$ zeros between the samples in the sequence. This is a very special filter. The purpose of polyphase filter structures is to decompose a given filter into a set of M parallel filters that have this special property. The Noble Identity will then be applicable to each of these parallel filters.

5.3 FIR FILTER POLYPHASE STRUCTURE

The polyphase structure of an FIR filter is an interesting structure in its own right. It is also useful in deriving efficient realizations of filters for decimation and interpolation. For a FIR filter of length-M, the polyphase structure consists of I FIR filters in parallel, where M is selected to be an integer multiple of I. That is, I must divide M. Therefore, each parallel FIR filter has $K = \frac{M}{I}$ coefficients. To derive the polyphase structure, consider a simple example of a length-6 FIR filter ($M = 6$) given by

$$H(z) = h(0) + h(1)z^{-1} + h(2)z^{-2} + h(3)z^{-3} + h(4)z^{-4} + h(5)z^{-5}. \tag{5.10}$$

If we let $I = 2$, then $H(z)$ can be written as

$$H(z) = (h(0) + h(2)z^{-2} + h(4)z^{-4}) + (h(1)z^{-1} + h(3)z^{-3} + h(5)z^{-5}), \tag{5.11}$$

where we have grouped even and odd terms. Now we rewrite (5.11) as

$$H(z) = (h(0) + h(2)z^{-2} + h(4)z^{-4}) + z^{-1}(h(1) + h(3)z^{-2} + h(5)z^{-4}). \tag{5.12}$$

Define the following subfilters,

$$H_0(z) = h(0) + h(2)z^{-1} + h(4)z^{-2} \tag{5.13}$$

$$H_1(z) = h(1) + h(3)z^{-1} + h(5)z^{-2}. \tag{5.14}$$

Then the original filters can be written in terms of the subfilters in (5.13) and (5.14) as

$$H(z) = H_0(z^2) + z^{-1}H_1(z^2). \tag{5.15}$$

This decomposition is shown in Fig. 5.5. Now, if we want a decomposition with three parallel filters, that is, if $I = 3$, then we pick every third term in $H(z)$ to form groups as

$$H(z) = (h(0) + h(3)z^{-3}) + (h(1)z^{-1} + h(4)z^{-4}) + (h(2)z^{-2} + h(5)z^{-5}). \tag{5.16}$$

This gives the following decomposition:

$$H(z) = H_0(z^3) + z^{-1}H_1(z^3) + z^{-2}H_2(z^3), \tag{5.17}$$

where

$$H_0(z) = h(0) + h(3)z^{-1} \tag{5.18}$$

$$H_1(z) = h(1) + h(4)z^{-1} \tag{5.19}$$

$$H_2(z) = h(2) + h(5)z^{-1}. \tag{5.20}$$

This decomposition is shown in Fig. 5.6.

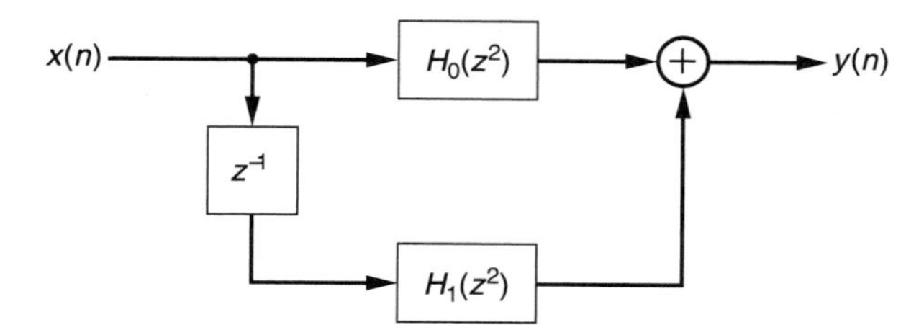

Figure 5.5 Polyphase structure of a FIR filter with two subfilters.

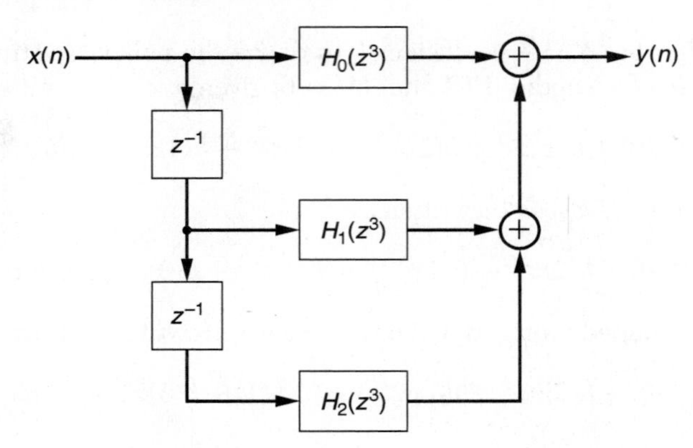

Figure 5.6 Polyphase structure of a FIR filter with three subfilters.

This analysis shows that, in general, for a FIR filter of length-M, a polyphase representation with I parallel filters is given by

$$\begin{aligned} H(z) &= H_0(z^I) + z^{-1}H_1(z^I) + z^{-2}H_2(z^I) + \cdots + z^{-(I-1)}H_{I-1}(z^I) \\ &= \sum_{i=0}^{I-1} z^{-i}H_i(z^I). \end{aligned} \tag{5.21}$$

The subfilters in (5.21) have transfer functions

$$\begin{aligned} H_0(z) &= h(0) + h(I)z^{-1} + h(2I)z^{-2} + \cdots + h((K-1)I)z^{-(K-1)} \\ H_1(z) &= h(1) + h(1+I)z^{-1} + h(1+2I)z^{-2} + \cdots + h(1+(K-1)I)z^{-(K-1)} \\ &\vdots \\ H_{I-1}(z) &= h(I-1) + h(I-1+I)z^{-1} + h(I-1+2I)z^{-2} + \cdots \\ &\quad + h(I-1+(K-1)I)z^{-(K-1)}, \end{aligned}$$

where $K = \frac{M}{I}$ is an integer. In general, we have the subfilter transfer function

$$H_i(z) = \sum_{j=0}^{K-1} h(i+jI)z^{-j}, \quad 0 \le i \le I-1. \tag{5.22}$$

This decomposition is shown in Fig. 5.7.

5.4 FILTERS FOR DECIMATION AND INTERPOLATION

Some filters required for decimation and interpolation are computationally efficient, whereas others are not. We begin with the concept of altering the sampling rate by a rational factor I/D. Clearly, we need to upsample by a factor of I and downsample by a factor of D. Two different schemes for achieving this are shown in Fig. 5.8. In Fig. 5.8(a), we first decimate and follow it by interpolation. For decimation, a prefilter

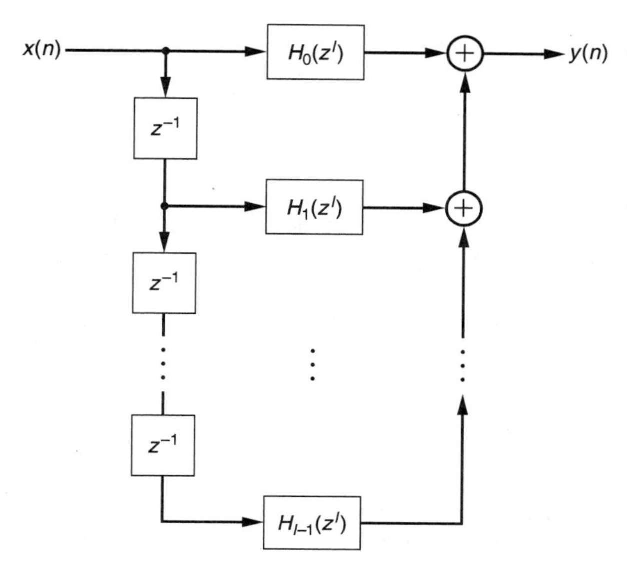

Figure 5.7 Polyphase structure of a length-M FIR filter into I parallel subfilters, where $K = M/I$ is an integer.

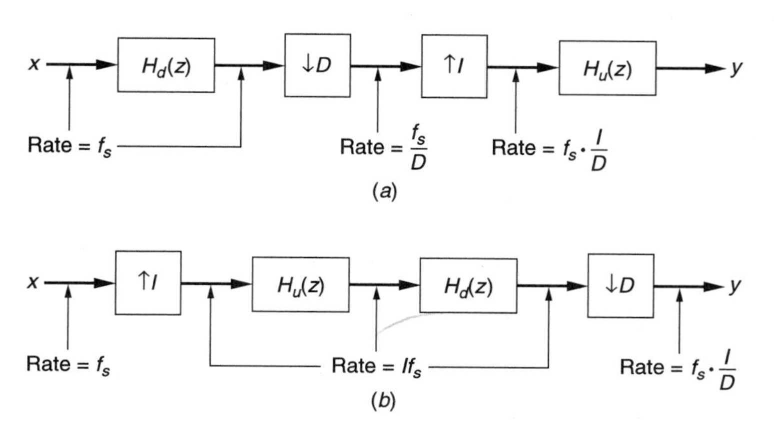

Figure 5.8 Schemes for sampling rate alteration by a rational factor of I/D.

$H_d(z)$ bandlimits the signal to $\frac{\pi}{D}$ in order to avoid aliasing. For interpolation, a postfilter $H_u(z)$ is used to remove the images introduced by the upsampling process. The natural question now is, which one of these schemes is better, if any? The second scheme (interpolation followed by decimation) is actually preferable for the following reasons: (1) In the first scheme, we filter by $H_d(z)$ initially, which can destroy some desirable spectral characteristics of the signal. This does not happen in the second scheme. (2) In the second scheme, the two filters $H_u(z)$ and $H_d(z)$ operate at the same sampling rate. Therefore, these two filters can be combined into one lowpass filter. This can potentially save computations.

This single filter, $H(z) = H_d(z)H_u(z)$, must be designed so that it has the frequency response

$$H(\omega) = \begin{cases} 1, & 0 \leq |\omega| \leq \min\left(\frac{\pi}{D}, \frac{\pi}{I}\right) \\ 0, & \text{otherwise} \end{cases} \tag{5.23}$$

where π represents the frequency of $If_s/2$. Having presented these schemes, let us now return to the problem of realizing these filters. We consider the case of decimation and interpolation separately.

For decimation, the filter can be simply realized in direct form as

$$H_d(z) = \sum_{k=0}^{M-1} h(k)z^{-k}. \tag{5.24}$$

The Direct Form decimation filter is shown in Fig. 5.9. This structure is simple but computationally inefficient. This is because we filter at a high sampling rate and then keep every Dth sample only and throw away the rest. The solution is to use a polyphase structure and then swap the decimator and the filter using the Noble Identity as we will now show. The decimator implementation using a polyphase FIR structure is shown in Fig. 5.10, where we have decomposed the downsampler into M parallel ones and inserted

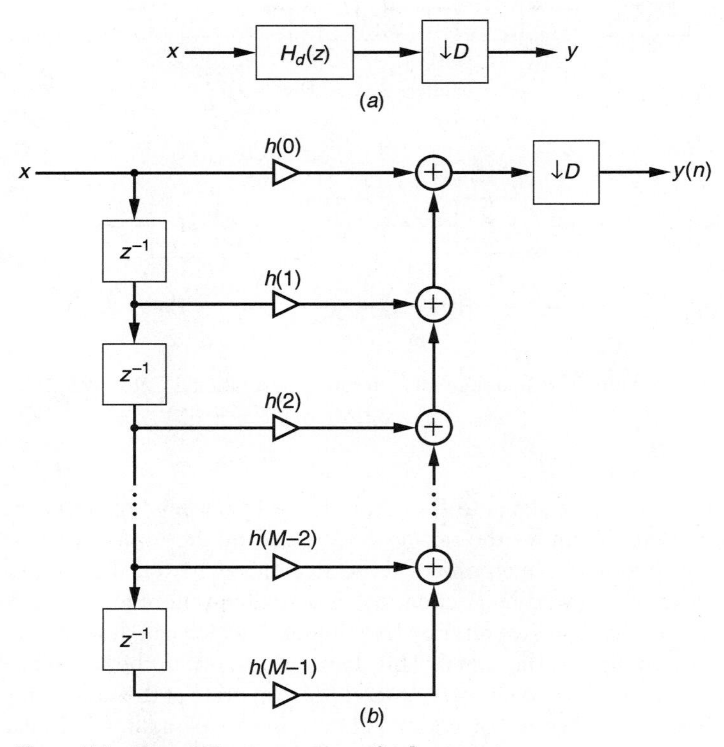

Figure 5.9 Direct Form structure of a decimator.

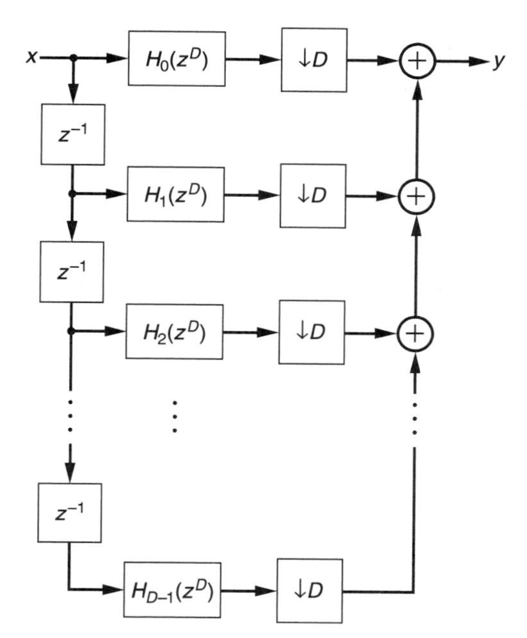

Figure 5.10 Decimator realization using a polyphase structure.

them in each parallel branch. In each parallel branch, we swap the filter and downsampler to get the structure of Fig. 5.11. The filtering in each branch is now performed after downsampling at a lower sampling rate. This structure can actually be made even more efficient if the decimation filter $H_d(z)$ is designed to have linear phase with symmetric coefficients. In this case, each subfilter turns out to be either symmetric or a mirror image of another subfilter. These properties can be used to further reduce the computational complexity. As an example, consider a linear phase $H_d(z)$ with symmetric coefficients given by

$$\begin{aligned} H_d(z) = {} & h(0) + h(1)z^{-1} + h(2)z^{-2} + h(3)z^{-3} + h(4)z^{-4} \\ & + h(3)z^{-5} + h(2)z^{-6} + h(1)z^{-7} + h(0)z^{-8}. \end{aligned} \tag{5.25}$$

The filter length is $M = 9$. Let us design the polyphase structure for $D = 3$. In this case, the subfilters are given by the following transfer function:

$$H_0(z) = h(0) + h(3)z^{-1} + h(2)z^{-2} \tag{5.26}$$

$$H_1(z) = h(1) + h(4)z^{-1} + h(1)z^{-2} \tag{5.27}$$

$$H_2(z) = h(2) + h(3)z^{-1} + h(0)z^{-2}. \tag{5.28}$$

Observe that $H_1(z)$ is symmetric and $H_2(z)$ is a mirror image of $H_0(z)$ centered around $h(3)$. By using these properties, we obtain the efficient decimator realization of Fig. 5.12. Note that in this figure, the positions of branches for $H_1(z)$ and $H_2(z)$ have been reversed in order to avoid too many crossover lines. The upper branch corresponds to $H_0(z)$,

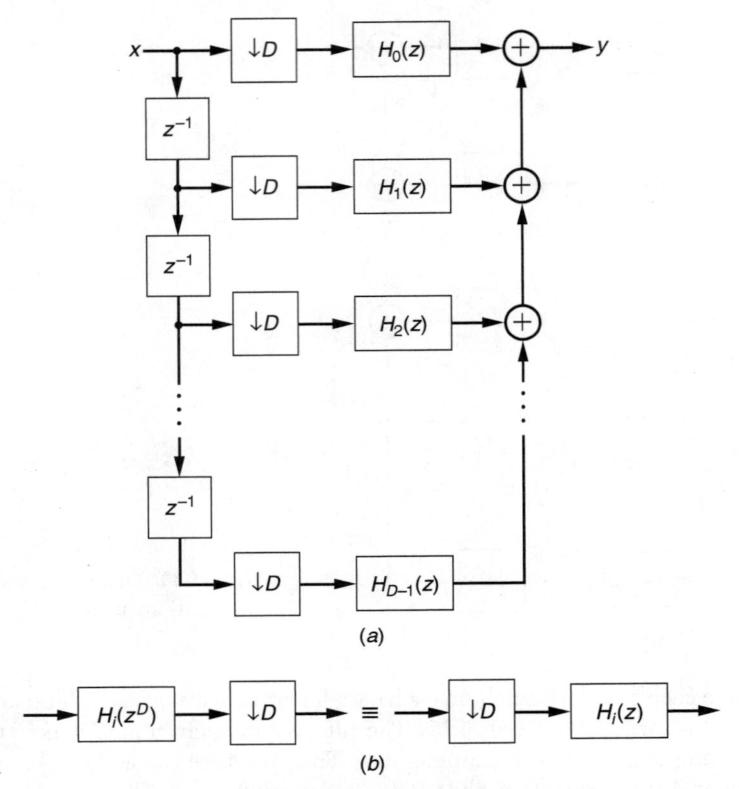

Figure 5.11 (a) Efficient decimator using a polyphase filter. (*b*) Noble Identity.

the lower branch to $H_1(z)$, and the middle branch to $H_2(z)$. This structure requires five multipliers and eight adders.

Now we extend this analysis to interpolators. In an interpolator, the filter follows the upsampler as shown in Fig. 5.13. This is an inefficient scheme because the filter operates at the higher sampling rate. Thus, the filter can be implemented as a polyphase structure as shown in Fig. 5.14. However, there is a problem here. Notice that we cannot move the upsampler to each parallel branch because of the delays. Therefore, instead of using the standard polyphase structure, we will use the transposed polyphase structure, as shown in Fig. 5.15. The transposed structure is simply the original structure with the input and output swapped, all signal directions reversed, and adders and signal tap-off points swapped. It is straightforward to show that the standard and transposed structure have the same input-output relationships. In Fig. 5.15, the upsampler can now be distributed to each parallel branch, and then the Noble Identity can be used in each branch to give the final structure of Fig. 5.16.

Now let us look at an interesting interpretation of this structure. Inspecting Fig. 5.16, we realize that each of the sequences $s_0, s_1, \ldots, s_{I-1}$ has a possibly nonzero value followed by $I-1$ zeros, followed by another (possibly) nonzero value, followed by $I-1$ zeros, and so on. Let us denote each of these sequences as

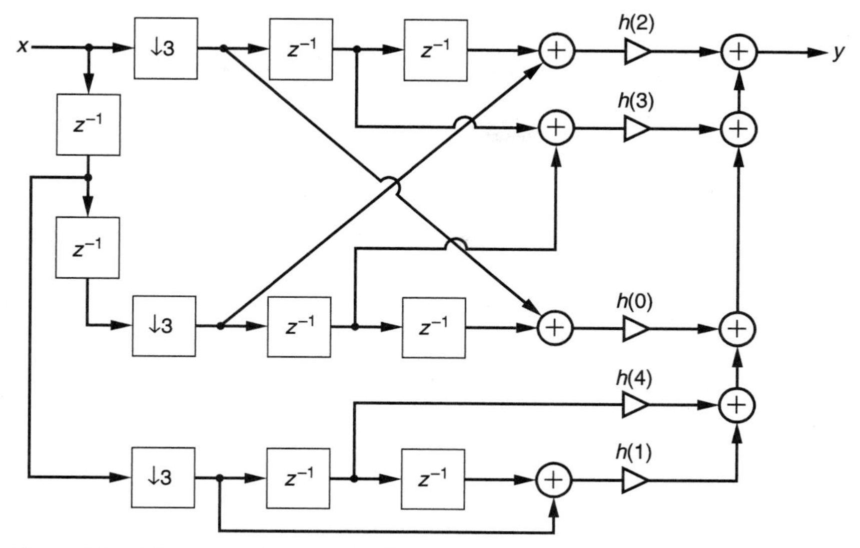

Figure 5.12 Efficient realization of a decimator with linear phase filter in a polyphase structure; $D = 3, M = 9$.

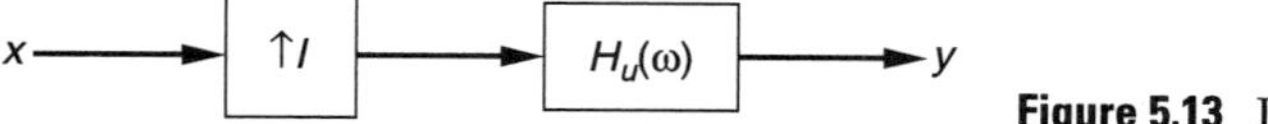

Figure 5.13 Interpolation scheme.

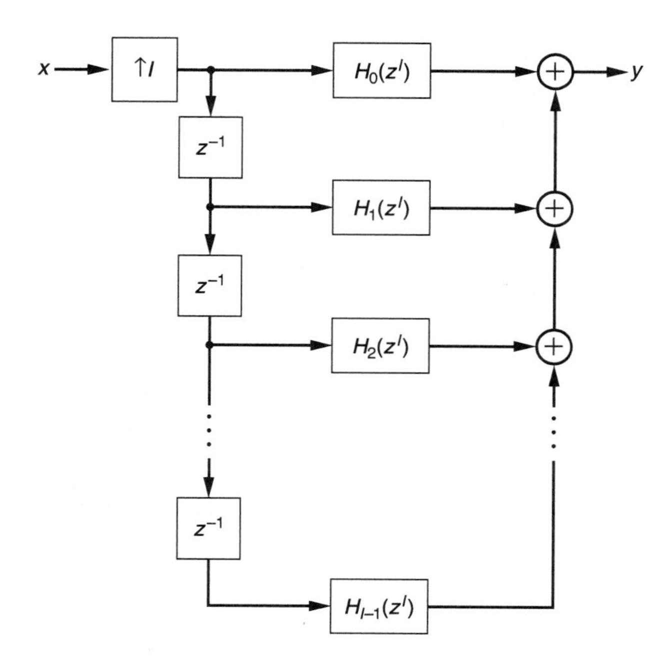

Figure 5.14 Upsampler followed by a standard polyphase filter.

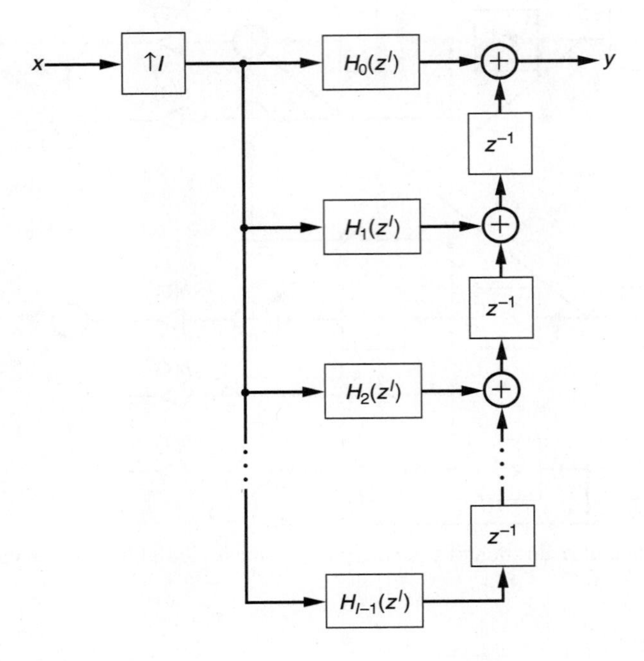

Figure 5.15 Upsampler followed by a transposed polyphase filter.

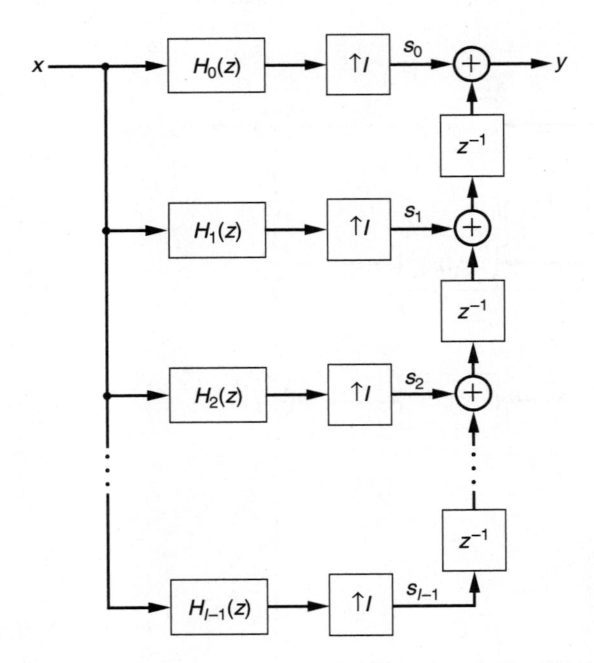

Figure 5.16 Efficient structure for interpolation using a transposed polyphase filter.

$$s_i = *, \underbrace{0, 0, \ldots, 0}_{I-1}, *, \underbrace{0, 0, \ldots, 0}_{I-1}, *, \ldots$$

where $*$ denotes a nonzero value. Also notice that s_0 is undelayed, s_1 is delayed by one sample, s_2 is delayed by two samples, and so on. The sum of these delayed sequences looks like the following (for $I = 4$ as an example):

$$y = \begin{array}{cccccccccccccccc} & *, & 0, & 0, & 0, & *, & 0, & 0, & 0, & *, & 0, & 0, & 0, & *, & & & \\ + & 0, & \blacktriangle, & 0, & 0, & 0, & \blacktriangle, & 0, & 0, & 0, & \blacktriangle, & 0, & 0, & 0, & \blacktriangle, & & \\ + & 0, & 0, & \blacksquare, & 0, & 0, & 0, & \blacksquare, & 0, & 0, & 0, & \blacksquare, & 0, & 0, & 0, & \blacksquare, & \\ + & 0, & 0, & 0, & \spadesuit, & 0, & 0, & 0, & \spadesuit, & 0, & 0, & 0, & \spadesuit, & 0, & 0, & 0, & \spadesuit, \end{array}$$

Clearly, the output at the 0th sample is from $H_0(z)$, at the 1st sample is from $H_1(z)$, at the 2nd sample is from $H_2(z)$, and so on. The output at the $(I-1)^{\text{th}}$ sample is from $H_{I-1}(z)$, the output from the Ith sample is back to $H_0(z)$, and so on. Thus, the structure of Fig. 5.16 can also be realized as in Fig. 5.17, where we have replaced the upsampler, delays, and adders by a simple commutator. The commutator is simply a rotating switch that rotates at a rate of If_s, where f_s is the sampling rate of the input. The output therefore looks like

$$y = *, \blacktriangle, \blacksquare, \spadesuit, \ldots$$

which is the same as would be obtained by Fig. 5.16. As we can see, the output of each filter is collected sequentially by the commutator to form the total output. Therefore, each sample of the output sequence is computed by a different filter, which is repeated in I samples. The parallel bank of filters can therefore be thought of as one filter with periodic coefficients, given by

$$H(z, n) = H(z, n + I). \tag{5.29}$$

This is called a periodically time-varying (PTV) filter, and its impulse response has the property

$$h(n) = h(n + I). \tag{5.30}$$

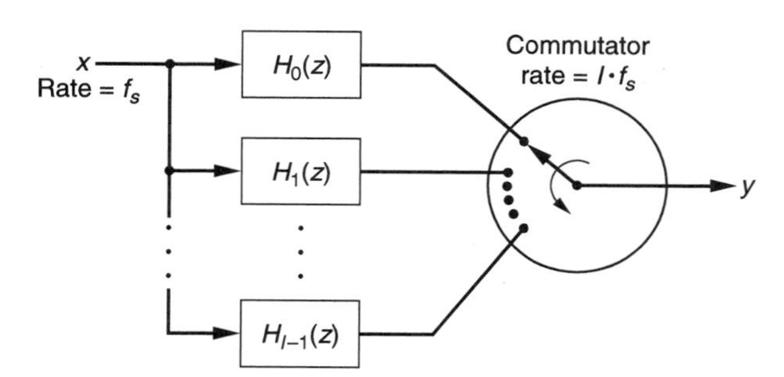

Figure 5.17 Interpolation using a polyphase filter and a commutator.

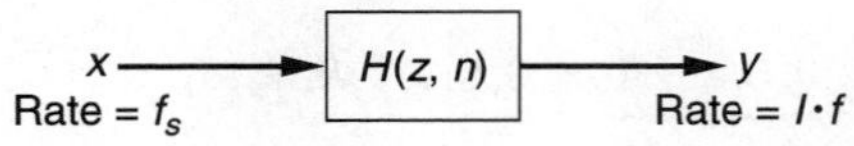

Figure 5.18 Interpolator realization using a PTV filter.

The PTV filter realization of an interpolator is shown in Fig. 5.18. The filter produces I samples for every sample of input. By buffering the right block of input samples, it is not difficult to draw a block diagram of this filter structure. This is left as an exercise.

EXAMPLE 5.1 *It is required to design a factor-of-4 decimator and realize it as a polyphase filter. It is also specified that the decimator filter must have less than* 0.12 *ripple in the passband and the stopband. Since* $D=4$*, we need a cutoff frequency of* $\frac{\pi}{4}$. *A stopband cutoff frequency is not specified, and so let us use it as* 0.35π. *This is a wide transition band and should yield a low filter order. A Kaiser window is now used to design the filter. The following MATLAB commands design the filter:*

```
c = Kaiserord([0.25*pi 0.35*pi]), [1 0], [0.1 0.1], 2*pi, 'cell');
h = fir1(c{:});
```

The length of the designed filter is $M = 16$, and the filter coefficients are the following:

$$h(0) = 0.03011$$
$$h(1) = -0.0076607$$
$$h(2) = -0.051567$$
$$h(3) = -0.063026$$
$$h(4) = -0.014227$$
$$h(5) = 0.090032$$
$$h(6) = 0.20959$$
$$h(7) = 0.28902$$
$$h(8) = 0.28902$$
$$h(9) = 0.20959$$
$$h(10) = 0.090032$$
$$h(11) = -0.014227$$
$$h(12) = -0.063026$$
$$h(13) = -0.051567$$
$$h(14) = -0.0076607$$
$$h(15) = 0.03011$$

The polyphase structure subfilters are therefore given by the following:

$$H_0(z) = h(0) + h(4)z^{-1} + h(8)z^{-2} + h(12)z^{-3}$$

$$H_1(z) = h(1) + h(5)z^{-1} + h(9)z^{-2} + h(13)z^{-3}$$

$$H_2(z) = h(2) + h(6)z^{-1} + h(10)z^{-2} + h(14)z^{-3}$$

$$H_3(z) = h(3) + h(7)z^{-1} + h(11)z^{-2} + h(15)z^{-3}.$$

In this example, the length of the filter ($M = 16$) was an integer multiple of the decimation factor ($D = 4$). Actually, in this case it is by design! If it did not turn out to be an integer multiple, then we would have had to increase the order so that it did. Increasing the order would guarantee meeting the filter specifications.

5.5 MULTISTAGE DECIMATORS AND INTERPOLATORS

In this section, we describe a technique for implementing decimators and interpolators when the factor of downsampling/upsampling M is very large. The case of decimation will be used for deriving the design methodology. We follow a derivation similar to that given in [13]. The design of interpolators is very similar.

For a factor-of-M decimation, we know that the downsampler must be preceded by a lowpass filter with a cutoff frequency of $\frac{\pi}{M}$. For large M, this is a very narrowband lowpass filter, preferably with a narrow transition band. The design of this filter will require a very high-order FIR filter and hence a high computational burden per sample. The motivation for multistate decimator design is to reduce this computational cost. To begin the derivation, consider a desired narrowband lowpass filter, $H(z)$, as shown in Fig. 5.19(a), where the ripples are not shown for simplicity. Let N_H be the number of filter coefficients required to design $H(z)$. Let us stretch this magnitude response by a factor of 2 to get another filter $S(z)$, as shown in Fig. 5.19(b). The passband and transition band of $S(z)$ are twice as wide as $H(z)$. Therefore, $S(z)$ will require half the number of coefficients, which is $\frac{N_H}{2}$. But $S(z)$ is not what we want. So, consider the interpolated filter $S(z^2)$ shown in Fig. 5.19(c). This filter has the passband and stopband edges at the desired locations. However, since this is interpolated by a factor of 2, it has an unwanted image in the range $\omega\epsilon[\pi - \omega_s, \pi]$. This image must be filtered out. A lowpass filter $L(z)$ can be used for this purpose, as shown in Fig. 5.19(d). The key to this concept is that $L(z)$ can have a large transition band and hence will require a short-length (say N_L) FIR filter. The overall system is thus a cascade of $L(z)$ followed by $S(z^2)$, as shown in Fig. 5.19(e). This would require a total of $\frac{N_H}{2} + N_L$ filter coefficients. Typically, $N_L \ll \frac{N_H}{2}$, and so there is a significant saving in computation. This technique for designing a narrowband filter is called the Interpolated FIR (IFIR) technique and was first introduced in [3]. It is straightforward to generalize this concept to interpolation of the filter $S(z)$ by a factor of M. This will result in the filter $S(z^M)$ and will have $M - 1$ unwanted images as shown in Fig. 5.20(a). The image closest to the original spectrum is centered at $\frac{2\pi}{M}$. Therefore, we need an LPF with stopband edge at $\frac{2\pi}{M} - \omega_s$, as shown in Fig. 5.20($b$). The overall IFIR filter is shown in Fig. 5.20(c).

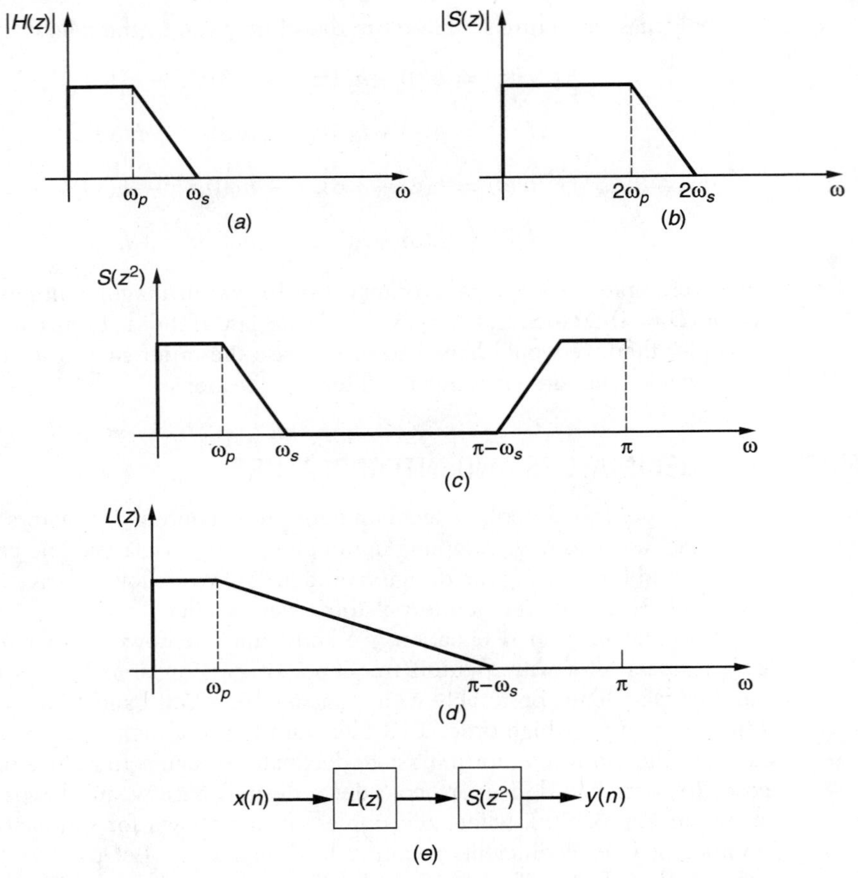

Figure 5.19 Narrow band filter design using an interpolated FIR filter.

Now we proceed to use the IFIR filter to design an efficient decimator. Let the downsampling factor M be factored as $M = M_1 M_2$, where each of the factors is an integer. The basic decimator is shown in Fig. 5.21(a), where the filter $H(z)$ has to cut off at $\frac{\pi}{M_1 M_2}$. Let us now design an IFIR filter with the interpolation factor (stretch factor) of M_1 given by $S(z^{M_1})$ and preceded by a lowpass filter $L(z)$. This is shown in Fig. 5.21(b). Now we use a Noble Identity to swap the $(\downarrow M_1)$ and $S(z^{M_1})$ to get Fig. 5.21(c). This is clearly a two-stage implementation of the desired decimator. The factors M_1 and/or M_2 may be further factored to obtain more stages. Several questions arise at this time. First, how do we factor M? Second, how many stages give the most efficient realization? Third, what should the factors be? There are no easy answers for these questions. The problem is quite nonlinear. One way to solve it is by brute force, where we write a program to try all the different combinators of factors and then use the best case where we get the smallest number of total coefficients.

Before we illustrate this design method with an example, we need the following result. The proof is left as an exercise.

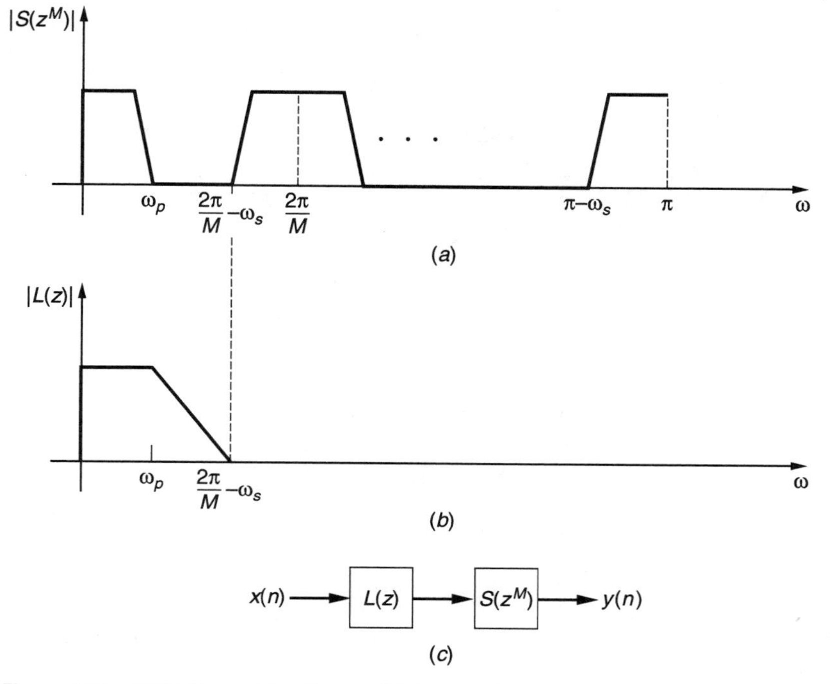

Figure 5.20 IFIR filter with a factor-of-M interpolation.

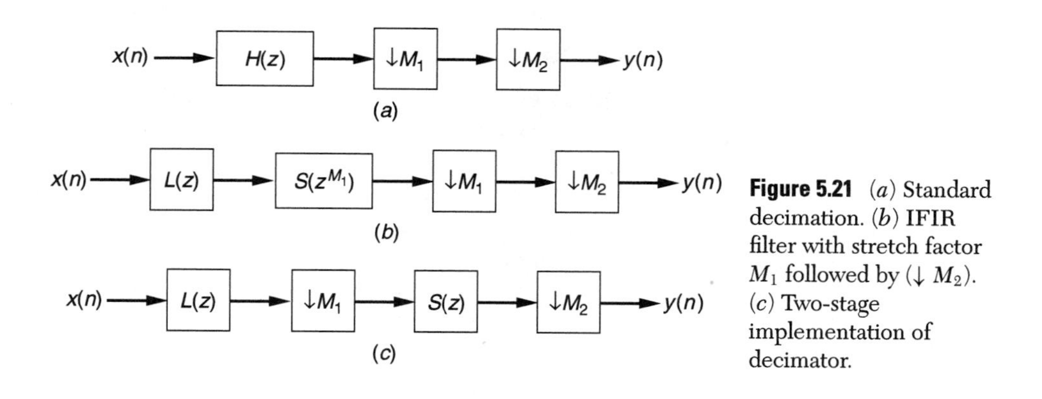

Figure 5.21 (*a*) Standard decimation. (*b*) IFIR filter with stretch factor M_1 followed by ($\downarrow M_2$). (*c*) Two-stage implementation of decimator.

Fact 1: Let $H_1(z)$ and $H_2(z)$ be two lowpass filters with peak passband ripples of δ_{p1} and δ_{p2}, respectively, and stopband edge frequencies of δ_{s1} and δ_{s2}, respectively. Assume that these ripples are much smaller than unity. Define the cascaded filter $H(z) = H_1(z)H_2(z)$ with peak passband ripple of δ_p and peak stopband ripple of δ_s. Then

$$\delta_p \leq \delta_{p1} + \delta_{p2},$$

$$\delta_s \leq \max\{\delta_{s1}, \delta_{s2}\}.$$

▶ **EXAMPLE 5.2** *A signal sampled at 100 kHz is to be decimated by a factor of $M = 50$. The decimation filter therefore must cut off at $\frac{\pi}{M} = \frac{100\,kHz/2}{50} = 1000$ Hz. Let the passband and stopband edge frequencies be 950 Hz and 1000 Hz, respectively. These correspond to $\omega_p = 0.019\pi$ and $\omega_s = 0.02\pi$. Let us assume that both the peak passband and stopband ripples are 0.005. If we design a FIR filter for these specifications using the Kaiser window, we get a filter length of $N_H = 5305$. This can be easily found by using the MATLAB function* ***kaiserord****( ).*
Now let $M = M_1M_2$, with $M_1 = 25$ and $M_2 = 2$. Next, design the filter $S(z)$ with a stretch factor of $M_1 = 25$. This filter therefore has the specifications

$$\begin{aligned}
\omega_{Sp} &= 25(0.019\pi) = 0.475\pi \\
\omega_{Ss} &= 25(0.02\pi) = 0.5\pi \\
\delta_{Sp} &= \frac{0.005}{2} = 0.0025 \text{ (see Fact 1)} \\
\delta_{Ss} &= 0.005.
\end{aligned}$$

A FIR filter designed with the Kaiser window for the above specifications yields an order of $N_S = 247$. The last filter to be designed is the lowpass filter $L(z)$ needed for image suppression. This filter has the following specifications:

$$\begin{aligned}
\omega_{Lp} &= 0.019\pi \\
\omega_{Ls} &= \frac{2\pi}{M_1} - \omega_s = \frac{2\pi}{25} - 0.02\pi = 0.06\pi.
\end{aligned}$$

Again, using a Kaiser window design, we see that this yields a filter length of $N_L = 131$. The total number of coefficients for the two-stage implementation (Fig. 5.22) is $N_s + N_L = 378$. When compared to the standard decimator ($N_H = 5305$), this represents a computational savings of a factor of about 14.

The idea of IFIR-based decimator design can also be used for designing interpolators. The steps for a $M = M_1M_2$ factor interpolator design are shown in Fig. 5.23.

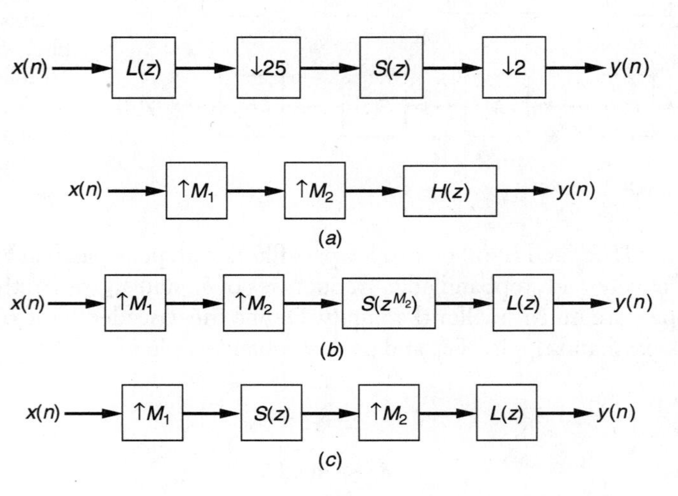

Figure 5.22 IFIR-based decimator design for a factor-of-50 decimator example.

Figure 5.23 (*a*) Standard interpolator with $M = M_1M_2$. (*b*) Upsampler followed by IFIR filter. (*c*) Two-stage design of interpolator based on IFIR approach.

▶ 5.6 FILTER BANKS

Digital filter banks have become increasingly popular during the last two decades because of their numerous applications in audio and image processing. These filters are useful in decomposing a signal into a set of subband signals that occupy different portions of the frequency spectrum. In order to perform this separation, we need a set of bandpass filters operating in parallel on the same signal as shown in Fig. 5.24. This is called an analysis filter bank. For example, the filters may be designed so that the spectrum is divided uniformly into M separate segments (see Fig. 5.25). In practice, ideal filters cannot be realized, and so the subbands will overlap. This leads to an important application. In speech signals, most of the energy is located in the lower part of the spectrum. By decomposing the signal using a filter bank into separate subbands, we can code the lower subband with more bits than the higher subband. This leads to an efficient coding and compression of the signal. These decomposed and coded signals are then transmitted or stored. These signals must eventually be combined to reconstruct the original signal. This is done by a synthesis filter bank as shown in Fig. 5.26.

At this point, one might think that by the way the analysis filter bank is set up, we are actually sending M times the total numbers of samples as the original. This is extremely inefficient. So, actually, each subband signal is decimated. Since each filter has a bandwidth of $\frac{\pi}{M}$, we decimate by a factor of M. A typical subband coding and decoding system is shown in Fig. 5.27, where Q denotes quantization to a desired number of bits. The number of bits varies from subband to subband, as mentioned earlier. In the higher subbands, the samples are often coded with zero bits in applications where there is no energy or no perceptible content in the high-frequency bands. In other words, the

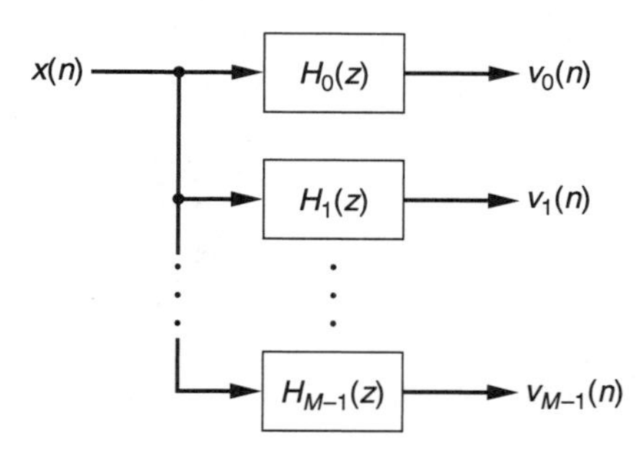

Figure 5.24 Analysis filter bank.

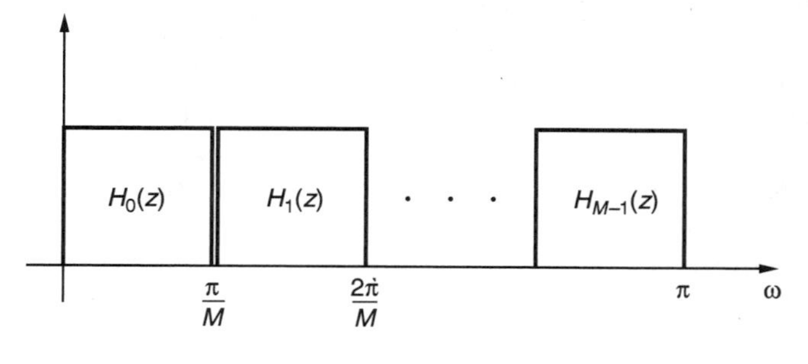

Figure 5.25 Decomposition of the spectrum by an analysis filter bank.

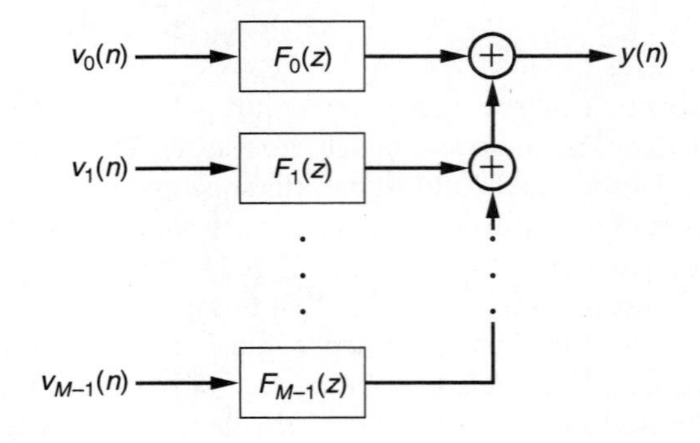

Figure 5.26 Synthesis filter bank.

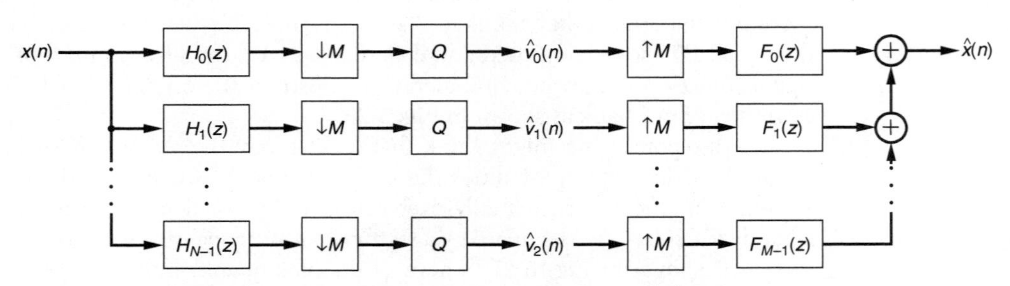

Figure 5.27 A uniform subband coding and decoding scheme using filter banks.

samples in these subbands are discarded. The subband coding scheme of Fig. 5.27 is called uniform subband coding because the spectrum is split equally by the M different filters. The spectral resolution is therefore uniform and constant. The compression ratio of speech and image signals can be increased further by using a nonuniform subband coder.

A typical nonuniform three-level coder is shown in Fig. 5.28(a). In this case, we have multiresolution of the spectrum. The original signal is split into two equal frequency bands by a lowpass filter (LPF) and a highpass filter (HPF). These are therefore half-band filters. The outputs of these filters are downsampled by a factor of 2. The lower channel produces the high-frequency ("detail") signal, and the upper channel produces the low-frequency ("smooth") signal. The detail signal is coded by a small number of bits. The smooth signal is bandlimited to $\frac{\pi}{2}$ with respect to the original signal and is called level-1 smooth signal. It is then passed through a parallel set of LPF and HPF for further decomposition of its spectrum. The outputs of these filters are again downsampled by a factor of 2. The detail signal is coded by an appropriate number of bits. The smooth signal is now a level-2 smooth signal and is bandlimited to $\frac{\pi}{4}$ with respect to original signal. This process can be continued to as many levels as desired or dictated by the application at hand.

Fig. 5.28(b) shows a three-level decomposition of the spectrum. Such a multiresolution decomposition is closely related to wavelet transforms. In fact, this is how fast wavelet transform is performed. It turns out that the design of the filters determines the type of wavelet transform. The design of these filters is therefore extremely important in the performance of this subband coding scheme. Since we cannot use ideal (brickwall) filters in practice, aliasing is inevitable in our frequency decomposition. Aliasing must therefore be canceled by proper design of the reconstruction filters. Also, distortion of the magnitude and phase can occur in the subband coder. The reconstruction filters must also take care of distortion. In fact, the filters in the coder and decoder are designed simultaneously for aliasing cancellation and distortion-free performance. Such filters are called Quadrature Mirror Filters (QMF) and will be described shortly.

The multiresolution decoder scheme is shown in Fig. 5.29 for three levels. The low-frequency (smooth) and high-frequency (detail) signals from the last level are upsampled

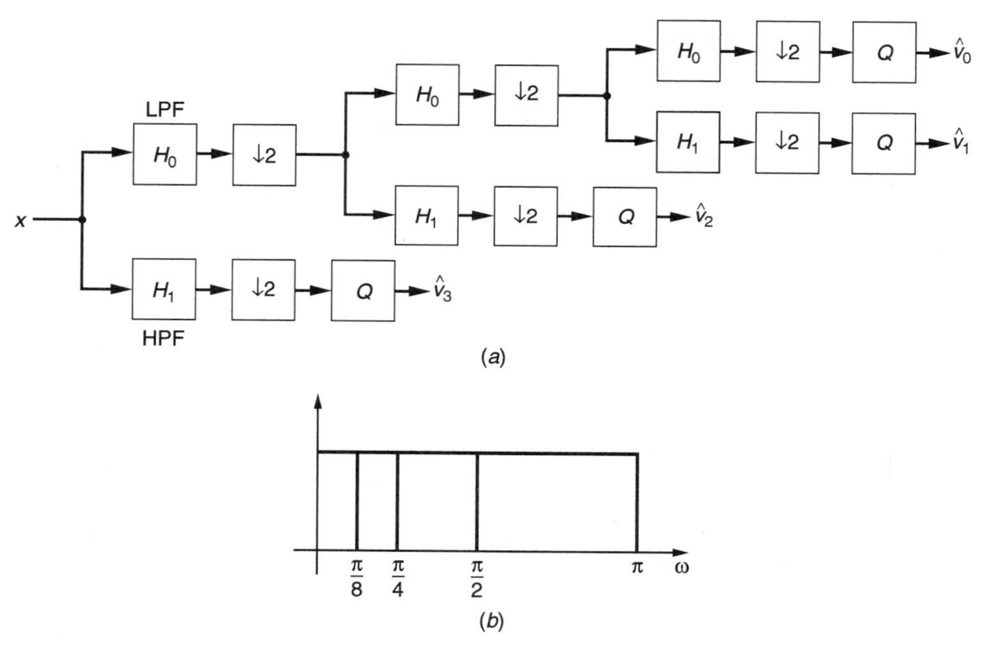

Figure 5.28 (a) A three-level multiresolution subband coder. (b) Decomposition of the frequency spectrum.

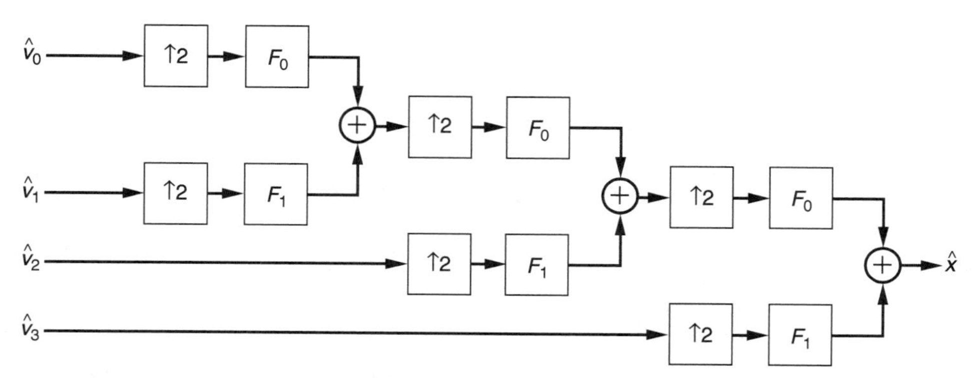

Figure 5.29 A three-level multiresolution subband decoder.

(by a factor of 2) and processed by two different filters and then summed up to produce a reconstructed smooth signal. This forms a pair with the detail signal arriving from the next lower level. These signals are then upsampled, filtered, and combined to produce the smooth signal at the next level. This process is repeated for as many levels as in the coder. Such multiresolution coding and decoding have become very popular in many speech and image compression applications. They have become standard in ITU[1] (speech) [1] and JPEG (image) compression standards [2].

5.7 UNIFORM DFT FILTER BANK

A filter bank can be used for the computation of DFT; such a scheme is called a uniform DFT filter bank. The concept is rather simple. In Figs. 5.24 and 5.25, we have seen how an analysis filter bank can be used to uniformly decompose a signal spectrum. If each of these filters has a very narrowband and we take one sample per band, then essentially we have an M-point DFT. We can think of each of the filters $H_1(z), \ldots, H_{M-1}(z)$ as having frequency responses that are shifted versions of $H_0(z)$. A shift in the frequency domain $(\omega - \frac{2\pi}{M}k)$ implies a multiplication in the time domain by $e^{j\frac{2\pi}{M}kn}$. The impulse responses of these filters are complex, which implies that these filters have complex coefficients. These filters are therefore computationally inefficient. We will get back to the computational issue later. For the moment, let us call the filter $H_0(z)$ as the prototype filter. The other filters are shifted versions of this. In the time domain, the kth filter has the impulse response

$$h_k(n) = h_0(n)e^{j\frac{2\pi}{M}kn}, \quad 1 \le k \le M-1, \tag{5.31}$$

where $h_0(n)$ is the impulse response of the filter $H_0(z)$. In the Z-domain, we have for $1 \le k \le M-1$,

$$\begin{aligned} H_k(z) &= \sum_{n=0}^{\infty} h_k(n)z^{-n} \\ &= \sum_{n=0}^{\infty} h_0(n)e^{j\frac{2\pi}{M}kn}z^{-n} \\ &= \sum_{n=0}^{\infty} h_0(n)\left(e^{-j\frac{2\pi}{M}k}z\right)^{-n} \\ &= H_0\left(ze^{-j\frac{2\pi}{M}k}\right). \end{aligned} \tag{5.32}$$

In the frequency domain, we have for $1 \le k \le M-1$,

$$H_k(\omega) = H_0\left(\omega - \frac{2\pi}{M}k\right). \tag{5.33}$$

[1] ITU(International Telecommunications Union) was called International Consultative Committee for Telephone and Telegraph (CCITT) until 1995.

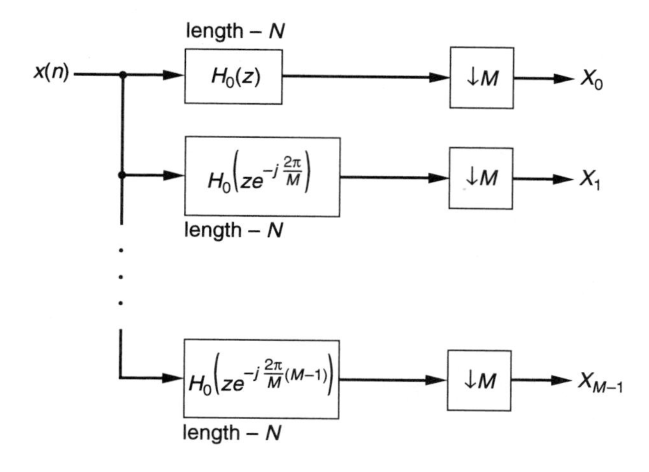

Figure 5.30 A uniform DFT filter bank; inefficient realization.

The uniform DFT filter bank is shown in Fig. 5.30 with the above notation. Note that each filter has a bandwidth of $\frac{\pi}{M}$, and it is for this reason that the output of each filter is downsampled by a factor of M. This is also logical because for every block of M samples of input $x(n)$, we get M samples of DFT, $X_0, \ldots, X_{M-1}$.

Now let us consider the computational complexity of this filter bank structure. Let each filter have N complex coefficients for FIR implementations, except possibly $H_0(z)$. Assuming $x(n)$ to be real, we see that each filter requires $2N$ real multiplications for a total of $2MN$ real multiplications per sample. Hence, this is a very inefficient realization.

Polyphase Realization of the Uniform DFT Filter Bank

The polyphase realization can significantly reduce the computational complexity of the uniform DFT filter bank. This is done by decomposing the prototype filter $H_0(z)$ in a polyphase structure as

$$H_0(z) = \sum_{i=0}^{M-1} z^{-i} E_i(z^M) \tag{5.34}$$

where $E_i(z^M)$ is the ith polyphase filter. Recall that $E_i(z)$ is formed by picking every m^{th} coefficient of $H_0(z)$ starting from the ith element. In general,

$$\begin{aligned} E_i(z) &= h_0(i) + h_0(i+M)z^{-1} + h_0(i+2M)z^{-2} + \cdots \\ &\quad + h_0(i+(M-1)M)z^{-(M-1)} \\ &= \sum_{m=0}^{M-1} h_0(i+mM)z^{-m}. \end{aligned} \tag{5.35}$$

Also, recall that the kth filter in the uniform DFT filter bank is related to the prototype filter by the relation

$$H_k(z) = H_0(ze^{-j\frac{2\pi}{M}k}). \tag{5.36}$$

Substituting (5.36) into (5.34), and using the fact that $e^{-j2\pi k} = 1$, we get for $1 \leq k \leq M-1$,

$$\begin{aligned} H_k(z) = H_0(ze^{-j\frac{2\pi}{M}k}) &= \sum_{i=0}^{M-1} (ze^{-j\frac{2\pi}{M}k})^{-i} E_i((ze^{-j\frac{2\pi}{M}k})^M) \\ &= \sum_{i=0}^{M-1} z^{-i} e^{j\frac{2\pi}{M}ki} E_i(z^M). \end{aligned} \tag{5.37}$$

Now let us expand the above for each value of k as follows:

$$\begin{aligned} H_0(z) &= E_0(z^M) + z^{-1}E_1(z^M) + z^{-2}E_2(z^M) + \cdots + z^{-(M-1)}E_{M-1}(z^M) \\ H_1(z) &= E_0(z^M) + z^{-1}E_1(z^M)e^{j\frac{2\pi}{M}.1.1} + z^{-2}E_2(z^M)e^{j\frac{2\pi}{M}.1.2} + \cdots \\ &\quad + z^{-(M-1)}E_{M-1}(z^M)e^{j\frac{2\pi}{M}.1.(M-1)} \\ &\vdots \\ H_{M-1}(z) &= E_0(z^M) + z^{-1}E_1(z^M)e^{j\frac{2\pi}{M}.(M-1).1} + z^{-2}E_2(z^M)e^{j\frac{2\pi}{M}.(M-1).2} + \cdots \\ &\quad + z^{-(M-1)}E_{M-1}(z^M)e^{j\frac{2\pi}{M}.(M-1).(M-1)} \end{aligned}$$

In matrix form, the above equations can be written as

$$\widehat{\mathbf{H}}_k(z) = \mathbf{D}\widehat{\mathbf{E}}(z^M) \tag{5.38}$$

where

$$\widehat{\mathbf{H}}_k(z) = \begin{bmatrix} H_0(z) \\ H_1(z) \\ \vdots \\ H_{M-1}(z) \end{bmatrix}, \quad \widehat{\mathbf{E}}(z^M) = \begin{bmatrix} E_0(z^M) \\ z^{-1}E_1(z^M) \\ \vdots \\ z^{-(M-1)}E_{M-1}(z^M) \end{bmatrix}$$

and

$$\mathbf{D} = \begin{bmatrix} 1 & 1 & 1 & \cdots & 1 \\ 1 & e^{j\frac{2\pi}{M}} & e^{j\frac{2\pi}{M}.2} & \cdots & e^{j\frac{2\pi}{M}.(M-1)} \\ \vdots & & \vdots & & \vdots \\ 1 & e^{j\frac{2\pi}{M}.(M-1)} & e^{j\frac{2\pi}{M}.(M-1)2} & \cdots & e^{j\frac{2\pi}{M}.(M-1)(M-1)} \end{bmatrix}. \tag{5.39}$$

Recall that the inverse DFT of a sequence is

$$x(n) = \frac{1}{M}\sum_{k=0}^{M-1} X(k)e^{j\frac{2\pi}{m}kn}, 0 \leq n \leq M-1. \tag{5.40}$$

The matrix $\mathbf{D}$ (5.39) is therefore the inverse DFT matrix kernel except for the missing $\frac{1}{M}$ factor. From (5.38), we see that if a polyphase structure is implemented using $E_0(z), \ldots, E_{M-1}(z)$, then we must perform an inverse DFT of the outputs in order to get the signals $H_k(z)$ for $k = 0, \ldots, M-1$. The inverse DFT will actually perform the multiplication $\frac{1}{M}\mathbf{D}$. Therefore, the original signal must be pre-multiplied by M as shown in Fig. 5.31, which is the complete polyphase structure using an IDFT. The IDFT is simply a set of multipliers, and so the downsamplers can be moved to precede the IDFT. Furthermore, the downsamplers can be moved before the filters by using the First Noble Identity. The result is an efficient realization of the uniform DFT filter bank, as shown in Fig. 5.32.

Now let us find the computational complexity of this realization for real input. Let $H_0(z)$ be an N-tap FIR filter. Each polyphase filter therefore has $\frac{N}{M}$ taps. For M branches,

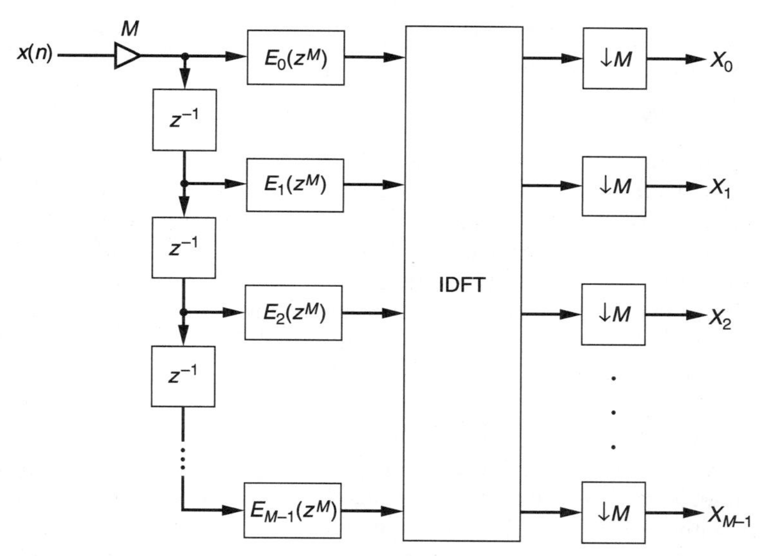

Figure 5.31 A uniform DFT filter bank implemented using a polyphase structure and IDFT.

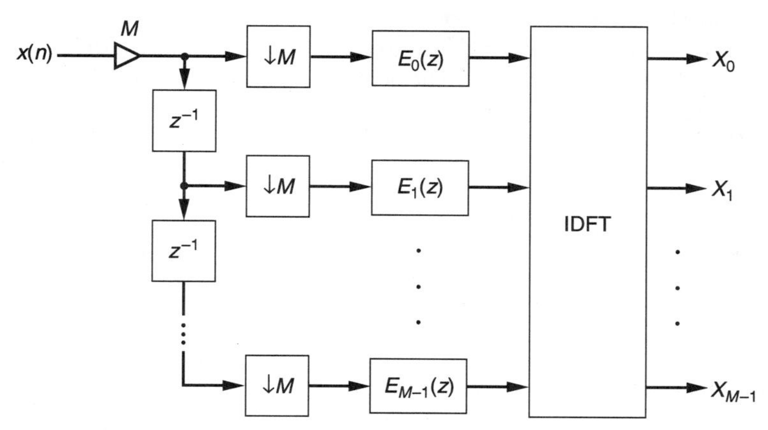

Figure 5.32 Efficient realization of a uniform DFT filter bank.

there is a total of N real multipliers required for the filters. The IDFT can be performed by a radix-2 FFT algorithm and require $2M \log_2 M$ real multipliers, for a total $N + 2M \log_2 M$ multipliers. However, there is a downsampler before every filter. Therefore, the actual number of multipliers per sample of $x(n)$ is

$$\text{total number of real multipliers} = \frac{N}{M} + 2\log_2 M + 1, \tag{5.41}$$

where we have accounted for the single multiplier at the very beginning. Let us now compare this to the $2MN$ real multipliers required for the direct realization using an example. Let $N = 160$ and $= 16$. The direct realization gives 5120 multipliers per sample. The efficient polyphase structure gives 19 multipliers; a savings of a factor of about 270.

5.8 TWO-CHANNEL QMF BANK

In a previous section, we discussed the concept of subband coding and decoding and explained a three-level subband coder-decoder in detail. The filters in such a filter bank structure must be designed carefully in order to avoid distortion and cancel aliasing. This will lead to perfect reconstruction after the synthesis filter bank. In this section, we will study a 2-channel Quadrature Mirror Filter (QMF) bank in detail. A 2-channel QMF bank is shown in Fig. 5.33. The input signal is filtered in parallel by $H_0(z)$ and $H_1(z)$, downsampled, coded, and then stored or transmitted. The coder is represented by Q, where the signal is coded into a specified number of bits. This is where quantization of the signal takes place. In general, $H_0(z)$ is a lowpass filter, and $H_1(z)$ is a highpass filter, each having a bandwidth of $\frac{\pi}{2}$. If the input signal has most of its energy in the lower part of the spectrum (as in speech and images), the upper channel coder is assigned more bits than the lower channel coder. This would lead to efficient compression of the data. At the receiver end, the synthesis filter bank is responsible for reconstructing the original signal. Of course, the transmitted data (v_o) may not be exactly the same as the received data ($\hat{v}_0$) owing to degradation in the channel. The job of the channel equalizer is to restore the signal as much as possible. The transmitted signal is also entropy coded with some redundancy built in. This is called error correction coding. The decoder has an algorithm to remove the errors that were not corrected by the channel equalizer. In this discussion, we will assume that $\hat{v}$ is exactly the same as v. The goal here is to design the filters so that the errors due to aliasing and distortion (in the filter bank) are minimized. We focus on the 2-channel QMF shown in Fig. 5.34, where the channel errors and quantizers have been removed.

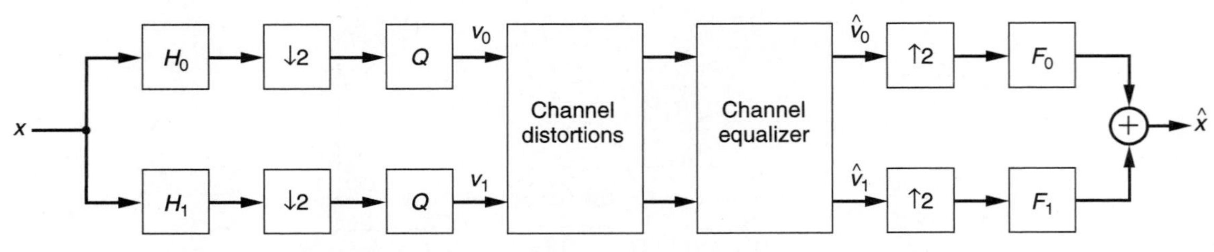

Figure 5.33 2-channel QMF bank with channel degradations.

Choices for H_0 and H_1

As mentioned earlier, the analysis filter H_0 is lowpass, and H_1 is highpass with a bandwidth of $\frac{\pi}{2}$. In practice, these filters cannot be brickwall and therefore have nonzero transition bandwidths. A couple of choices are shown in Fig. 5.35. In one case, the magnitude responses do not overlap. If the stopband attenuations of the filters are sufficiently large, then aliasing due to downsampling is negligible. However, the input signal energy is drastically reduced around $\frac{\pi}{2}$. Theoretically, this can be compensated for by designing the synthesis filters F_0 and F_1 to have high gain around $\frac{\pi}{2}$. But this will cause amplification of channel noise, coding noise, and quantization noise. Hence, this is not a good option. In the other case, the magnitude responses of H_0 and H_1 overlap. This will cause significant aliasing due to downsampling. However, this aliasing can be canceled by the synthesis filters, as will be shown shortly.

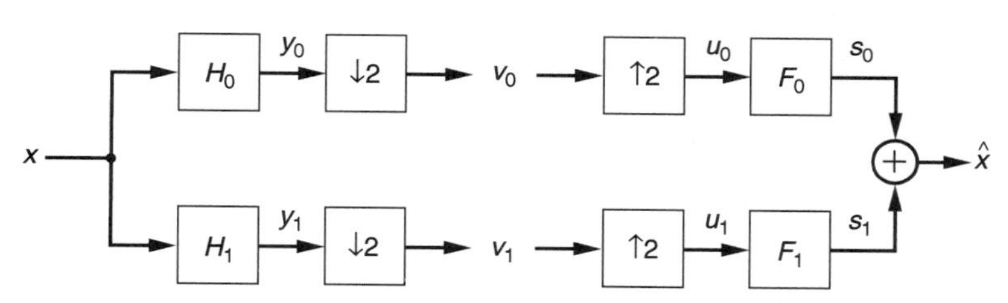

Figure 5.34 2-channel QMF bank with no channel distortion.

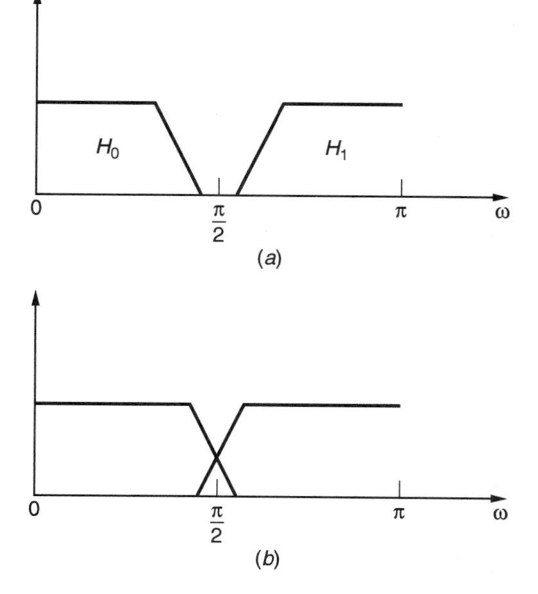

Figure 5.35 Magnitude responses for the analysis filters: (*a*) nonoverlapping and (*b*) overlapping.

Perfect Reconstruction (PR)

Let us now revisit Fig. 5.34 and analyze it using the Z-transform domain results for downsampling and upsampling obtained earlier in this chapter. We obtain the following equations for the upper channel:

$$Y_0(z) = H_0(z)X(z) \tag{5.42}$$

$$\begin{aligned} V_0(z) &= (\downarrow 2)Y_0(z) \\ &= \frac{1}{2}\left[Y_0(z^{\frac{1}{2}}) + Y_0(-z^{\frac{1}{2}})\right] \\ &= \frac{1}{2}\left[H_0(z^{\frac{1}{2}})X(z^{\frac{1}{2}}) + H_0(-z^{\frac{1}{2}})X(-z^{\frac{1}{2}})\right] \end{aligned} \tag{5.43}$$

$$\begin{aligned} U_0(z) &= (\uparrow 2)V_0(z) \\ &= V_0(z^2) \\ &= \frac{1}{2}[H_0(z)X(z) + H_0(-z)X(-z)] \end{aligned} \tag{5.44}$$

$$\begin{aligned} S_0(z) &= F_0(z)U_0(z) \\ &= \frac{1}{2}F_0(z)\;[H_0(z)X(z) + H_0(-z)X(-z)]. \end{aligned} \tag{5.45}$$

In a similar fashion, the output of the lower channel is

$$S_1(z) = \frac{1}{2}F_1(z)[H_1(z)X(z) + H_1(-z)X(-z)]. \tag{5.46}$$

The output of the synthesis filter bank is thus

$$\begin{aligned} \widehat{X}(z) &= S_0(z) + S_1(z) \\ &= \frac{1}{2}[F_0(z)H_0(z) + F_1(z)H_1(z)]\;X(z) \\ &\quad + \frac{1}{2}[F_0(z)H_0(-z) + F_1(z)H_1(-z)]\;X(-z). \end{aligned} \tag{5.47}$$

In equation (5.47), the term corresponding to $X(-z)$ represents aliasing, and the term corresponding to $X(z)$ represents distortion (amplitude and/or phase). Now we have the following theorem.

Theorem 5.8 *A 2-channel QMF bank gives perfect reconstruction if and only if*

$$F_0(z)H_0(-z) + F_1(z)H_1(-z) = 0 \tag{5.48}$$

and

$$F_0(z)H_0(z) + F_1(z)H_1(z) = 2z^{-1}, \tag{5.49}$$

where l is a nonnegative integer.

Equation (5.48) is called the alias cancellation condition, while (5.49) is the condition for no distortion. In matrix format, the PR condition is written as

$$\begin{bmatrix} F_0(z) & F_1(z) \end{bmatrix} \underbrace{\begin{bmatrix} H_0(z) & H_0(-z) \\ H_1(z) & H_1(-z) \end{bmatrix}}_{\mathbf{H(z)}} = \begin{bmatrix} 2z^{-l} & 0 \end{bmatrix}. \tag{5.50}$$

The matrix $\mathbf{H}(z)$ is called the modulation matrix.

Sufficient Condition for Alias Cancellation

From (5.48), we can choose the synthesis filters such that $F_0(z)H_0(-z)+F_1(z)H_1(-z) = 0$. A simple sufficient condition for alias cancellation is

$$F_0(z) = H_1(-z), \; F_1(z) = -H_0(-z). \tag{5.51}$$

This choice automatically satisfies the alias cancellation condition and leaves us with only two filters to design, namely, $H_0(z)$ and $H_1(z)$. It is very easy to obtain $F_0(z)$ from $H_1(z)$. If

$$H_1(z) = b_0 + b_1z^{-1} + b_2z^{-2} + b_3z^{-3} + \cdots \tag{5.52}$$

then

$$F_0(z) = H_1(-z) = b_0 - b_1z^{-1} + b_2z^{-2} - b_3z^{-3} + \cdots \tag{5.53}$$

Likewise, if

$$H_0(z) = a_0 + a_1z^{-1} + a_2z^{-2} + a_3z^{-3} + \cdots \tag{5.54}$$

then

$$F_1(z) = -a_0 + a_1z^{-1} - a_2z^{-2} + a_3z^{-3} - \ldots \tag{5.55}$$

In other words, the coefficients of $F_0(z)$ and $F_1(z)$ are obtained by appropriately alternating the signs of $H_0(z)$ and $H_1(z)$.

Distortion

Suppose we have achieved alias cancellation. Then from (5.47), we have

$$\widehat{X}(z) = D(z)X(z) \tag{5.56}$$

where

$$D(z) = \frac{1}{2}\left[F_0(z)H_0(z) + F_1(z)H_1(z)\right] \tag{5.57}$$

is called the distortion transfer function. Using the alias cancellation condition of (5.51), we get

$$D(z) = \frac{1}{2}[H_0(z)H_1(-z) - H_1(z)H_0(-z)]. \tag{5.58}$$

In the frequency domain, let $D(\omega) = |D(\omega)|e^{j\theta}$. Then from (5.56), we have

$$\widehat{X}(\omega) = |D(\omega)|X(\omega)e^{j\theta}. \tag{5.59}$$

If $|D(\omega)|$ is constant for all ω, then we do not have any amplitude distortion. In order to have no phase distortion, we need $D(\omega)$ to have linear phase; that is, $\theta(\omega) = a + b\omega$, for constants a and b. Therefore, we need $D(\omega)$ to be a linear phase all-pass filter in order to avoid any magnitude or phase distortion.

Earliest QMF

Given that we satisfy the alias cancellation condition of (5.51), we still have the problem of designing $H_0(z)$ and $H_1(z)$. In the earliest known QMF, Croisier, Esteban, and Galand [7] chose to use

$$H_1(z) = H_0(-z). \tag{5.60}$$

That is, $H_1(z)$ is obtained by alternating the signs of the coefficients of $H_0(z)$. In the frequency domain, this gives

$$|H_1(\omega)| = |H_0(\omega - \pi)|. \tag{5.61}$$

Thus, if $H_0(\omega)$ is an LPF, then $H_1(\omega)$ is an HPF. $|H_1(\omega)|$ is a mirror image of $|H_0(\omega)|$ with respect to $\frac{\pi}{2}$, the quadrature frequency—hence, the justification for the name QMF. With the choice of (5.60), we now have only one filter to design, namely, $H_0(z)$. The distortion function of (5.58) now becomes

$$D(z) = \frac{1}{2}[H_0^2(z) - H_0^2(-z)]. \tag{5.62}$$

For PR, we need $D(z) = z^{-l}$, which gives the condition

$$H_0^2(z) - H_0^2(-z) = 2z^{-l}. \tag{5.63}$$

This is a very interesting condition and requires some thought. The even coefficients of polynomials $H_0^2(z)$ and $H_0^2(-z)$ have the same sign, while the odd coefficients have the opposite sign. Therefore, the even coefficients cancel out in $H_0^2(z) - H_0^2(-z)$. The right-hand side of (5.63) has only one term, $2z^{-l}$. Therefore, $H_0^2(z)$ must have only one odd power, z^{-l}. What restriction does this place on $H_0(z)$? It forces the FIR filter $H_0(z)$ to have only two coefficients. It can be shown (Problem 31) that a FIR filter $H_0(z)$ with three or more coefficients will have at least two odd powers in $H_0^2(z)$. A two-coefficient FIR filter cannot have good stopband attenuation and sharp transition band. Therefore, PR QMF banks with FIR filters under the constraint $H_1(z) = H_0(-z)$ are not useful in practice. So, we must resort to longer (than two taps) FIR filters. We will not get PR, but we can design the filters to minimize amplitude and phase distortion.

Now let us look at realization of the QMF. Once we design $H_0(z)$, it can be decomposed into a polyphase form as

$$H_0(z) = E_0(z^2) + z^{-1}E_1(z^2). \tag{5.64}$$

Since $H_1(z) = H_0(-z)$, we have

$$H_1(z) = E_0(z^2) - z^{-1}E_1(z^2). \tag{5.65}$$

Using equations (5.64) and (5.65), the polyphase realization of the analysis bank can be obtained as in Fig. 5.36(*a*). The synthesis filters are given by the alias cancellation condition of (5.51) as

$$F_0(z) = H_1(-z) = H_0(z) = z^{-1}E_1(z^2) + E_0(z^2) \tag{5.66}$$

$$F_1(z) = -H_0(-z) = -H_1(z) = z^{-1}E_1(z^2) - E_0(z^2). \tag{5.67}$$

To get the synthesis bank, we do the following. Equation (5.66) is implemented using the upper channel signal u_0 as input and $\hat{x}$ as output. The transpose of this structure is then obtained. Equation (5.67) is then implemented with the lower channel signal u_1 as input and $\hat{x}$ as output. The transpose of this structure is then obtained. Next the two transposed structures are combined to get the synthesis bank of Fig. 5.36(*b*). The details are left as an exercise (Problem 19). In the analysis bank, the downsamplers can be moved to precede the filters by using the First Noble Identity. In the synthesis bank, the upsamplers can be moved to follow the filters by using the Second Noble Identity. As a result, we get the complete QMF bank in polyphase form as shown in Fig. 5.37. This structure is computationally very efficient. If $H_0(z)$ is a FIR filter of length N, then $E_0(z)$ and $E_1(z)$ are of length $\frac{N}{2}$. The analysis bank then requires N multiplications per sample after downsampling. In other words, it requires $\frac{N}{2}$ multiplications per sample of the input. The synthesis bank has the same complexity.

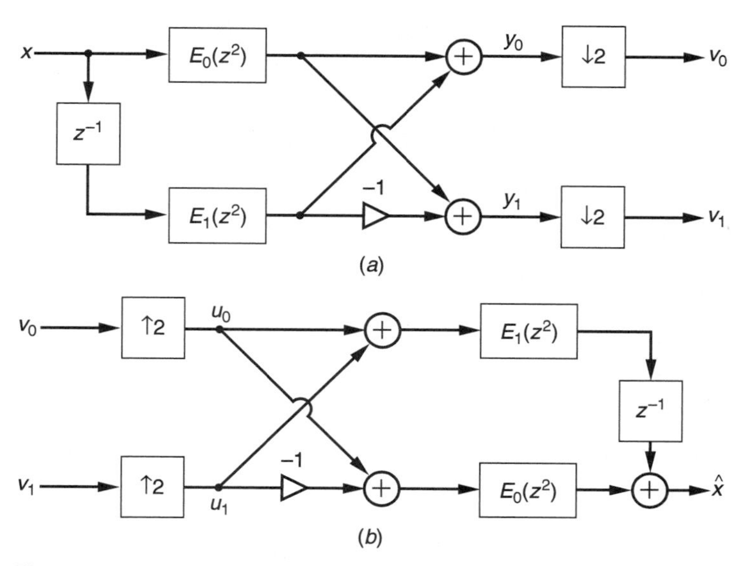

Figure 5.36 Polyphase QMF bank: (*a*) analysis bank and (*b*) synthesis bank.

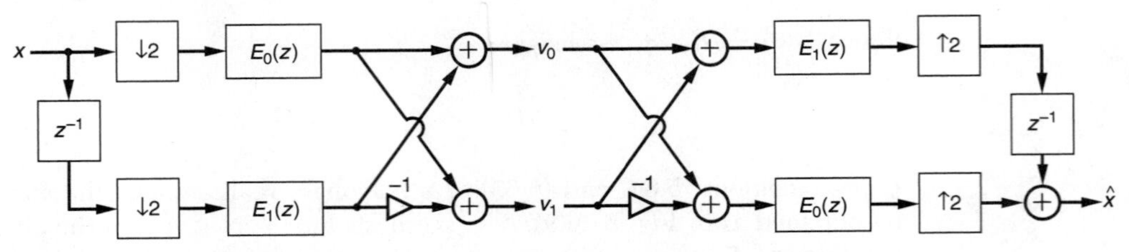

Figure 5.37 Complete QMF bank in polyphase form; efficient realization.

Phase and Amplitude Distortion

The distortion function can cause potential phase and amplitude distortion. Recall that the distortion function for the QMF bank with the constraint $H_1(z) = H_0(-z)$ is

$$D(z) = \frac{1}{2}[H_0^2(z) - H_0^2(-z)]. \tag{5.68}$$

To avoid any phase distortion, we need $D(z)$ to be linear phase, which in turn requires $H_0(z)$ to be also linear phase. Since $H_0(z)$ is a lowpass filter, its impulse response must satisfy

$$h_0(n) = h_0(N-1-n) \tag{5.69}$$

to make it linear phase. Therefore, its frequency response is

$$H_0(\omega) = H_r(\omega)e^{-j\omega\frac{N-1}{2}} \tag{5.70}$$

where $H_r(\omega)$ is a real function.

Then

$$\begin{aligned} H_0^2(\omega) &= H_r^2(\omega)e^{-j\omega(N-1)} \\ &= |H_r(\omega)|^2 e^{-j\omega(N-1)}. \end{aligned} \tag{5.71}$$

Also, the frequency response of $H_0(-z)$ is

$$\begin{aligned} H_0^2(\omega-\pi) &= H_r^2(\omega-\pi)e^{-j(\omega-\pi)(N-1)} \\ &= e^{j\pi(N-1)}|H_0(\omega-\pi)|^2 e^{-j\omega(N-1)} \\ &= (-1)^{N-1}|H_0(\omega-\pi)|^2 e^{-j\omega(N-1)}. \end{aligned} \tag{5.72}$$

Using (5.71) and (5.72), we find that the frequency response of the distortion function becomes

$$D(\omega) = \frac{1}{2}\left[|H_0(\omega)|^2 - (-1)^{N-1}|H_0(\omega-\pi)|^2\right]e^{-j\omega(N-1)}. \tag{5.73}$$

If N is an odd number, then for $\omega = \frac{\pi}{2}$, $D(\omega) = 0$. This is because $|H_0(-\frac{\pi}{2})| = |H_0(\frac{\pi}{2})|$, since $|H_0(\omega)|$ is an even function. This causes severe distortion of the signal at the frequency of $\frac{\pi}{2}$. An odd N is therefore not an option. For even N, we have

$$D(\omega) = \frac{1}{2}\left[|H_0(\omega)|^2 + |H_0(\omega-\pi)|^2\right]e^{-j\omega(N-1)}. \tag{5.74}$$

The above will avoid phase distortion but not amplitude distortion unless we have the condition

$$|H_0(\omega)|^2 + |H_0(\omega - \pi)|^2 = 1, \forall\omega. \tag{5.75}$$

This condition is referred to as *power-complementary*. As pointed out earlier, (5.75) is impossible to achieve with FIR filters unless $H_0(z)$ is a FIR filter with only two coefficients. This is not an acceptable solution for most subband coding applications. Therefore, we try to minimize distortion by making (5.75) approximately constant. This can be done by designing $H_0(z)$ by some optimization techniques in order to minimize some constructed cost function. Johnston [8] proposed the following cost function

$$J = \alpha \int_{\omega_s}^{\pi} |H(\omega)|^2 d\omega + (1-\alpha) \int_{0}^{\pi} \left(1 - |H_0(\omega)|^2 - |H_0(\omega - \pi)|^2\right) d\omega \tag{5.76}$$

where α is a weighting factor in the range $0 < \alpha < 1$. The first term assures good attenuation in the stopband. The second term assures that the magnitude response is almost flat over the entire frequency band. Tables of optimized filter coefficients are given in [6] and [8].

EXAMPLE 5.3 *In this example, we use the Johnston filter [8] labeled 48D as our analysis bank lowpass filter $H_0(z)$. This is a length-48 FIR filter. The impulse response of this filter is shown in Fig. 5.38(a). The analysis bank highpass filter is obtained as $H_1(z) = H_0(-z)$. The filter magnitude responses are shown in Fig. 5.38(b) and (c). In part (d) of the figure, we plot $|H_0(\omega)|^2 + |H_1(\omega)|^2$ in the dB power scale. It is clear that the filters are power-complementary. The dB scale shows slight variations of the power, which would be indiscernible in the absolute value scale.*

QMF Bank Using IIR Filters

IIR filters can be used to get PR in a QMF bank without any restriction on the number of filter coefficients. However, it is very hard to achieve linear phase with an IIR filter. Thus,

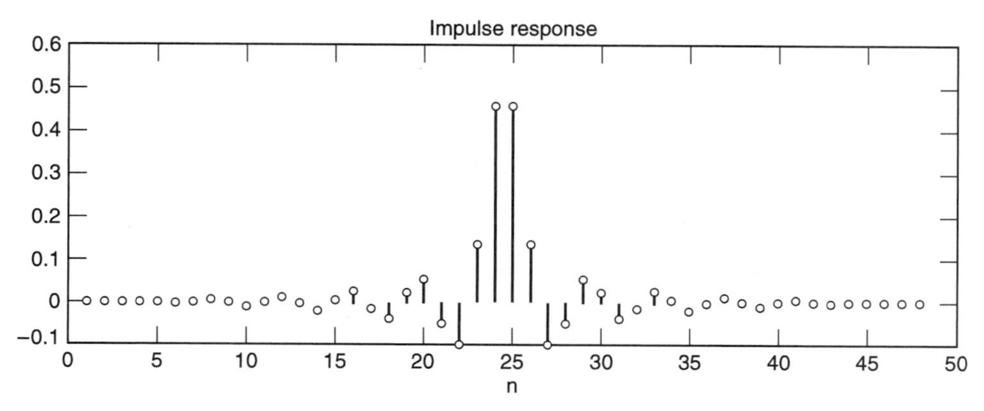

Figure 5.38 (*a*) Impulse response of the 48D Johnston filter.

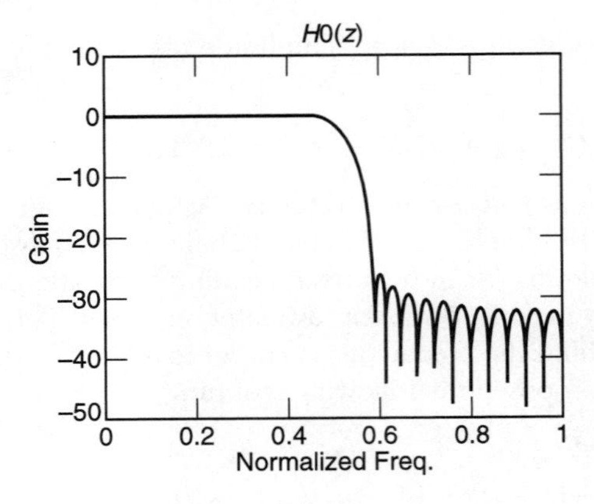

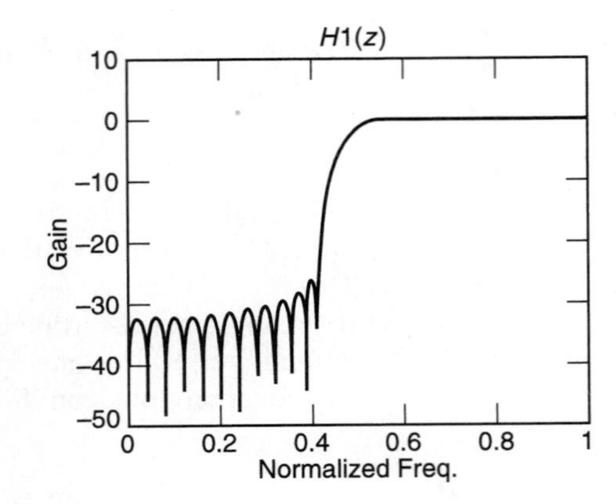

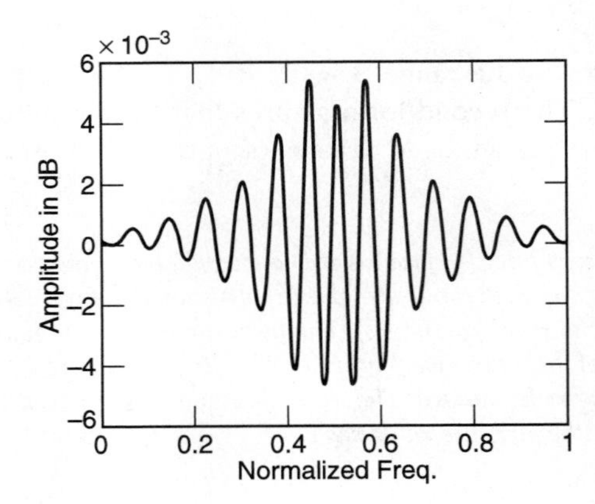

Figure 5.38 (*b*) Magnitude responses of filters $H_0(z)$. (*c*) Magnitude response of $H_1(z)$. (*d*) $10\log_{10}\left(|H_0(\omega)|^2 + |H_1(\omega)|^2\right)$ for the design example using the Croisier-Esteban-Galand filter bank and the Johnston 48D filter.

phase distortion can still remain. We just have to design $H_0(z)$ so that phase distortion is minimized.

Let $H_0(z)$ now represent an IIR filter with the polyphase representation

$$H_0(z) = E_0(z^2) + z^{-1}E_1(z^2). \tag{5.77}$$

The polyphase components are now IIR; let them be represented as

$$E_0(z) = \frac{A_0(z)}{2}, E_1(z) = \frac{A_1(z)}{2} \quad \text{and} \tag{5.78}$$

$$H_0(z) = \frac{A_0(z^2)}{2} + z^{-1}\frac{A_1(z^2)}{2}. \tag{5.79}$$

The distortion function then becomes

$$\begin{aligned} D(z) &= \frac{1}{2}\left[H_0^2(z) - H_0^2(-z)\right] \\ &= 2z^{-1}E_0(z^2)E_1(z^2) \\ &= \frac{1}{2}z^{-1}A_0(z^2)A_1(z^2). \end{aligned} \tag{5.80}$$

If $A_0(z)$ and $A_1(z)$ are constrained to be all-pass, then $D(z)$ is a simple delay and we have no amplitude distortion. This was shown in [12]. However, note that the all-pass constraint is sufficient but not necessary. For example, if $A_1(z) = \frac{1}{A_0(z)}$, then $D(z)$ is also a single delay! But this of course gives us a simple and useless $H_0(z)$ for most QMF applications. The flow diagram for the IIR QMF is essentially the same as that in Fig. 5.37, with $E_0(z)$ and $E_1(z)$ replaced by (5.78). The factors of $\frac{1}{2}$ can be all lumped into two multipliers, one at the input and the other at the output.

The question that remains now is if it is possible to design an LPF filter $H_0(z)$ so that it is in the form of (5.79), where $A_0(z)$ and $A_1(z)$ are all-pass. It has been shown in [12] that any odd-order elliptic half-band LPF $H_0(z)$ can always be expressed as (5.79) if

$$\omega_p + \omega_s = \pi \tag{5.81}$$

and

$$\delta_s^2 = 4\delta_0(1 - \delta_p) \tag{5.82}$$

where ω_p is the passband edge frequency, ω_s is the stopband edge frequency, δ_p is the passband ripple and δ_s is the stopband ripple. The poles of the designed elliptic filter satisfying these conditions lie strictly on the imaginary axis. A technique called pole-interfacing property is then used to decompose this filter into all-pass sections $A_0(z)$ and $A_1(z)$ as in (5.79). We omit this design process here. Interested readers are referred to Section 6.10 [10]. The IIR QMF thus designed will be free of amplitude distortion but will have phase distortion. The output of the QMF bank can finally be passed through an all-pass phase equalizer to minimize this distortion.

5.9 FIR QMF BANKS WITH PR

We have a total of four filters, namely, $H_0(z), H_1(z), F_0(z)$, and $F_1(z)$. We must always satisfy the alias cancellation condition given by

$$F_0(z) = H_1(-z), \; F_1(z) = -H_0(-z). \tag{5.83}$$

With alias cancellation, the output of the filter bank is the distortion term given by

$$\widehat{X}(z) = \frac{1}{2}[F_0(z)H_0(z) + F_1(z)H_1(z)]\, X(z) \tag{5.84}$$

$$= \underbrace{\frac{1}{2}[H_0(z)H_1(-z) - H_1(z)H_0(-z)]}_{D(z)}\; X(z). \tag{5.85}$$

For perfect reconstruction (PR), we need the distortion function to equal to $2z^{-l}$. The earliest filter bank by Croisier-Esteban-Galand used the condition $H_1(z) = H_0(-z)$. This does not give PR for an FIR $H_0(z)$ unless it has only two coefficients, in which case the QMF bank is not useful in practice. Smith and Barnwell [11] and Mintzer [9] proposed a better choice, which is as follows.

Alternating Flip

$$H_1(z) = -z^{-(N-1)}H_0(-z^{-1}) \tag{5.86}$$

where N is even.

So $H_1(z)$ is obtained by "flipping" or reversing the coefficients of $H_0(z)$ and then changing the signs of alternate coefficients. For example, let

$$H_0(z) = b_0 + b_1 z^{-1} + b_2 z^{-2} + b_3 z^{-3} \tag{5.87}$$

where $N - 1 = 3$. Then we have

$$\begin{aligned} H_0(-z^{-1}) &= b_0 - b_1 z + b_2 z^2 - b_3 z^3, \\ -z^{-(N-1)} H_0(-z^{-1}) &= -b_0 z^{-3} + b_1 z^{-2} - b_2 z^{-1} + b_3 \\ &= b_3 - b_2 z^{-1} + b_1 z^{-2} - b_0 z^{-3}. \end{aligned} \tag{5.88}$$

In short-hand notation, let $H_0(z) = (b_0, b_1, b_2, b_3)$. The alternating flip yields $H_1(z) = (b_3, -b_2, b_1, -b_0)$.

Now, substituting (5.86) into (5.85) gives

$$\begin{aligned} \widehat{X}(z) &= \frac{1}{2}\left[H_0(z) z^{-(N-1)} H_0(z^{-1}) + z^{-(N-1)} H_0(-z^{-1}) H_0(-z)\right] X(z) \\ &= \frac{1}{2} z^{-(N-1)} \left[H_0(z) H_0(z^{-1}) + H_0(-z) H_0(-z^{-1})\right] X(z). \end{aligned} \tag{5.89}$$

Notice that if the quantity inside the square brackets is a constant, K, then $\hat{x}(n) = \frac{K}{2} x(n - N + 1)$. That is, we have perfect reconstruction. This condition is called the *power symmetry condition*. As will be shown shortly, it is not difficult to design an FIR $H_0(z)$, so that this condition is satisfied. First, let us formalize this important result.

Definition 5.1 *A filter $H_0(z)$ is said to be power symmetric if*

$$H_0(z) H_0(z^{-1}) + H_0(-z) H_0(-z^{-1}) = K \tag{5.90}$$

where K is some constant.

Theorem 5.9 *Let the filter $H_0(z)$ in a QMF bank be designed to satisfy the power symmetric condition. Also, let*

$$F_0(z) = H_1(-z), F_1(z) = -H_0(-z), \tag{5.91}$$

so that the alias cancellation condition is satisfied. If $H_1(z)$ is the alternating flip of $H_0(z)$, that is,

$$H_1(z) = -z^{-(N-1)} H_0(-z^{-1}) \tag{5.92}$$

for even N, then the QMF bank achieves perfect reconstruction.

So, $H_1(z)$ is the "alternating flip" of $H_0(z)$; $F_0(z)$ is the "alternating sign" of $H_1(z)$; and $F_1(z)$ is the "negative alternating sign" of $H_0(z)$. An example of this procedure is shown in Fig. 5.39.

The only remaining problem is how to design $H_0(z)$ so that it is power symmetric. The design will require some knowledge of half-band filters.

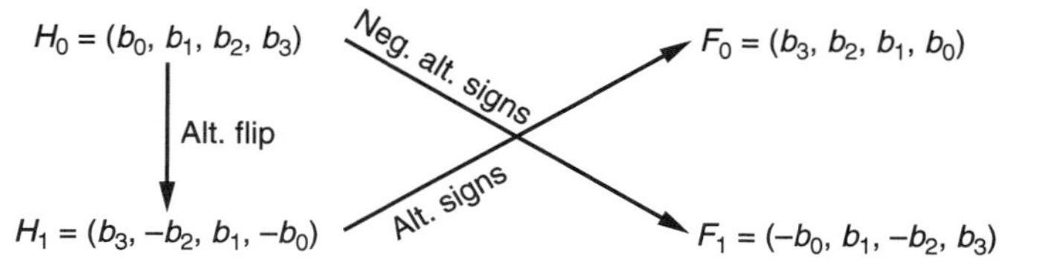

Figure 5.39 Relationship among filters in a Smith-Barnwell QMF bank.

Half-band Filters

A half-band filter is defined as an FIR filter whose impulse response is given by

$$h(2n) = \begin{cases} \beta, & n = 0 \\ 0, & n \neq 0. \end{cases} \tag{5.93}$$

Therefore, all the even coefficients are zero except $h(0) = \beta$. The odd coefficients do all the work. It is easy to see that the Z-transform of this filter satisfies the condition

$$H(z) + H(-z) = 2\beta. \tag{5.94}$$

In the frequency domain, this is equivalent to

$$H(\omega) + H(\omega - \pi) = 2\beta. \tag{5.95}$$

For a lowpass $H(\omega)$, this means that $H(\omega)$ is symmetric about $\frac{\pi}{2}$. The ripples in the passband and stopband are equal, and the edge frequencies ω_p and ω_s are equidistant from $\frac{\pi}{2}$ (see Fig. 5.40).

Now define a product filter,

$$P(z) = H_0(z)H_0(z^{-1}). \tag{5.96}$$

Using this in the power symmetric definition of (5.90), we get

$$P(z) + P(-z) = K. \tag{5.97}$$

That is, $P(z)$ is a half-band filter. Hence, we have the following result.

Theorem 5.10 *$H_0(z)$ is a power symmetric if and only if the product filter $P(z) = H_0(z)H_0(z^{-1})$ is a half-band filter.*

The implication of this theorem is that we can first design a half-band filter $P(z)$ and then factor it into $H_0(z)$ and $H_0(z^{-1})$. This process is called *spectral factorization*. Is this factorization always possible? The answer is yes if certain conditions are satisfied, as explained below.

Theorem 5.11 *Let $P(z) = H_0(z)H_0(z^{-1})$ be a half-band FIR lowpass filter. Spectral factorization of $P(z)$ into $H_0(z)$ and $H_0(z^{-1})$ is possible if $P(z)$ is a zero-phase filter and $P(z)|_{z=e^{j\omega}} \geq 0$ for all ω.*

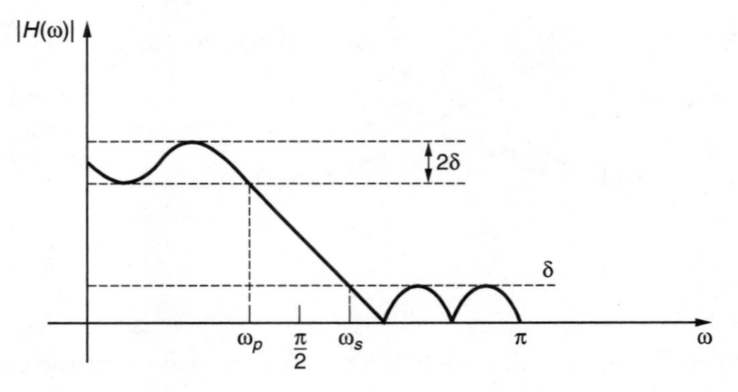

Figure 5.40 Magnitude response of half-band filter showing symmetry about $\frac{\pi}{2}$.

Proof. Recall from Chapter 3 that for a zero-phase FIR filter, $P(z)$, if z_0 is a root, then $\frac{1}{z_0}$ is also a root. In other words, the roots occur in reciprocal pairs. Also, the hypothesis, $P(z) \geq 0$, for $z = e^{j\omega}$ implies that the roots on the unit circle have even multiplicity (i.e., 2,4,6, ...). Therefore, once we know the roots of $P(z)$, we can assign a root to $H_0(z)$ and its reciprocal to $H_0(z^{-1})$. We can repeat this process for all the roots of $P(z)$. Hence, we can find $H_0(z)$. ■

Design Methodology

In the following, we outline the steps in the design of a 2-channel PR filter bank.

Step 1: Design a zero-phase half-band lowpass FIR filter $P(z)$ with $2N-1$ coefficients, where N is even.

Step 2: Let δ_s denote the maximum stopband ripple in $P(z)$. Obtain the new filter

$$\widehat{P}(z) = P(z) + \delta_s. \tag{5.98}$$

This assures that $\widehat{P}(z) \geq 0$ at $z = e^{j\omega}$, for all ω. In the time domain, this implies $\hat{p}(0) = p(0) + \delta_s$. All the other coefficients of $\widehat{P}(z)$ remain the same as those of $P(z)$.

Step 3: Perform the spectral factorization of $\widehat{P}(z) = H_0(z)H_0(z^{-1})$ as outlined in the proof of Theorem 5.11. $\widehat{P}(z)$ has $2N-1$ coefficients or $2N-2$ roots. Therefore, $H_0(z)$ has $N-1$ roots or N coefficients, where N is even.

Step 4: From $H_0(z)$, find $H_1(z), F_0(z)$, and $F_1(z)$ by the method depicted in Fig. 5.39.

An example is now in order.

▶ **EXAMPLE 5.4** *In this example, we design a complete Smith-Barnwell filter bank. The product filter is specified as a lowpass filter with passband edge frequency* $\omega_p = o.4\pi$, *stopband edge frequency* $\omega_p = 0.6\pi$, *and a maximum ripple of* 0.025 *in the passband and stopband. We design this filter* $P(z)$ *using the Remez algorithm and the following MATLAB commands:*

$$[Nb, Fo, Ao, W] = remezord([0.4\ \ 0.6], [1\ \ 0], [0.025\ \ 0.025]);$$

$$[P, err] = remez(Nb, Fo, Ao, W);$$

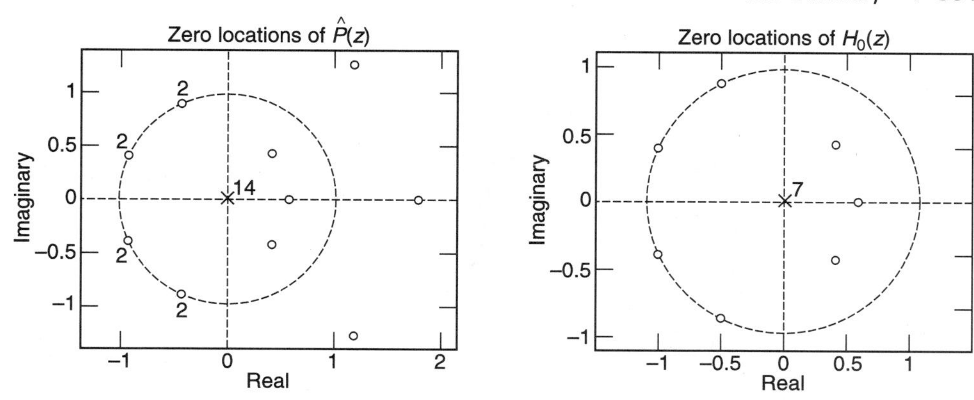

Figure 5.41 Zero locations of $\widehat{P}(z)$ and $H_0(z)$ for the design example of a Smith-Barnwell filter bank.

The parameter err returns the actual maximum ripple height. As outlined in Step 2, this ripple is used in (5.98) to find $\widehat{P}(z)$, which guarantees that $(\widehat{P}(z) \geq 0)$ at $z = e^{j\omega}$ for all ω. Now, we perform the spectral factorization of this polynomial to find $H_0(z)$. We factor it in such a way that $H_0(z)$ is a semi-minimum phase (i.e., all the zeros are on or inside the unit circle). The zero locations of $\widehat{P}(z)$ and $H_0(z)$ are shown in Fig. 5.41(a) and (b). The spectral factorization gives

$$H_0(z) = 0.3648 + 0.5051z^{-1} + 0.2668z^{-2} - 0.0865z^{-3}$$
$$-0.1339z^{-4} + 0.0342z^{-5} + 0.0955z^{-6} - 0.0692z^{-7}.$$

Using the method of Fig. 5.39, we find all the other filters in the QMF bank. The magnitude responses of all the filters are given in Fig. 5.42. The filters have quite wide transition bands as specified. The stopband ripples are very high owing to the low order of the filters.

5.10 SUMMARY

This chapter presents the fundamentals of multirate signal processing and describes decimators and interpolators in the time domain, frequency domain, and Z-domain. The design of multistage decimators and interpolators is discussed in detail, with emphasis on computational complexity. The polyphase decompostion is derived for the efficient design of filters for decimators and interpolators. The conceptual design of the DFT filter bank is also discussed. The DFT filter bank concept is very important from the point of view of computational simplicity and real-time spectral analysis. The theory of QMF banks is then developed in detail. The concepts of distortion and perfect reconstruction are important in subband coding application. We cover the early Croisier-Esteban-Galand QMF bank and the more recent Smith-Barnwell filter bank. Both FIR and IIR filter banks are discussed in connection with perfect reconstruction. The design algorithms for these filter banks are also presented and illustrated with examples.

Multirate signal processing is still a very active reseach area. Many new and exciting theories are being developed while this text is being written. Filter banks have found their way into several data compression standards. QMF banks are used for fast implementation of wavelet transforms, and this is now a standard in the modern versions of JPEG and MPEG. A QMF is used in the

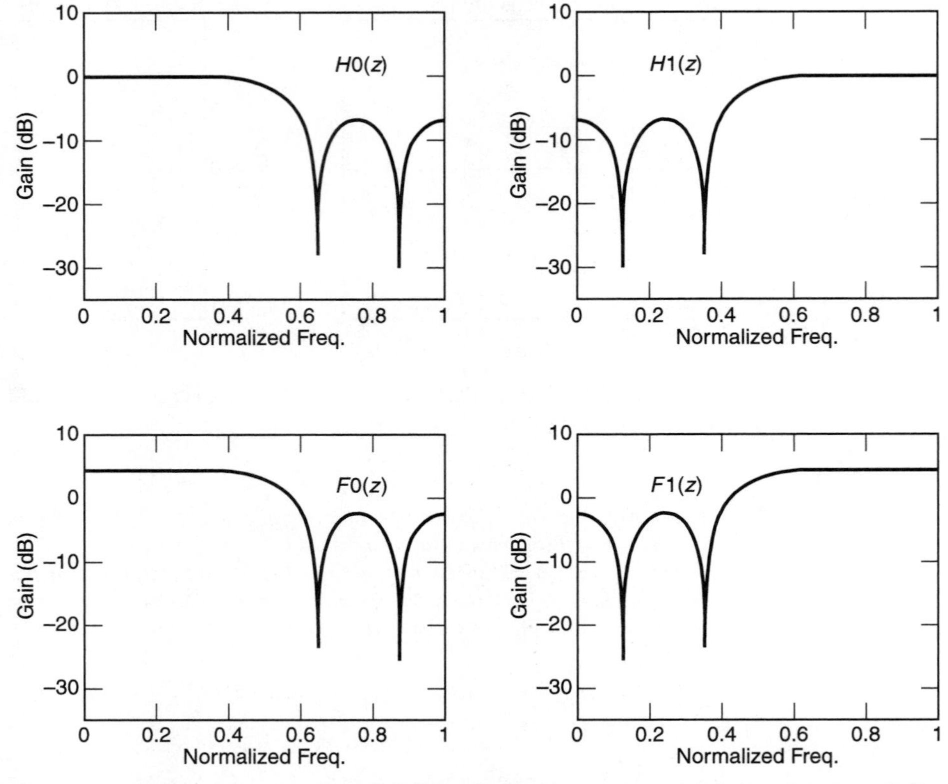

Figure 5.42 Magnitude responses of all filters for the design example of a Smith-Barnwell filter bank.

wideband standard ITU[2] G.722 [15], and in the AC-3 and DTS surround-sound standards. Details on the use of QMF in several different audio applications have been nicely documented in a tutorial paper by Painter and Spanias [14].

5.11 PROBLEMS

1. Consider the signal $x(n) = (0.5)^n u(n)$.

(a) Find the spectrum of $x(n)$ and sketch it.
(b) Determine the spectrum of $(\downarrow 2)x(n)$ and sketch it.
(c) Determine the spectrum of $(\uparrow 2)x(n)$ and sketch it.

2. Show that the upsampling operation $(\uparrow M)$ is linear and time-variant.

3. Determine if the downsampling operation $(\downarrow M)$ is linear. Is it time-variant?

4. Prove the First Noble Identity.

5. Prove the Second Noble Identity.

[2] ITU stands for International Telecommunications Union. This organization was formerly known as the CCITT.

6. Prove Fact 1.
7. Consider the filter bank of Fig. P5.1. This is a filter bank without filters. Find the output $\hat{x}(n)$.

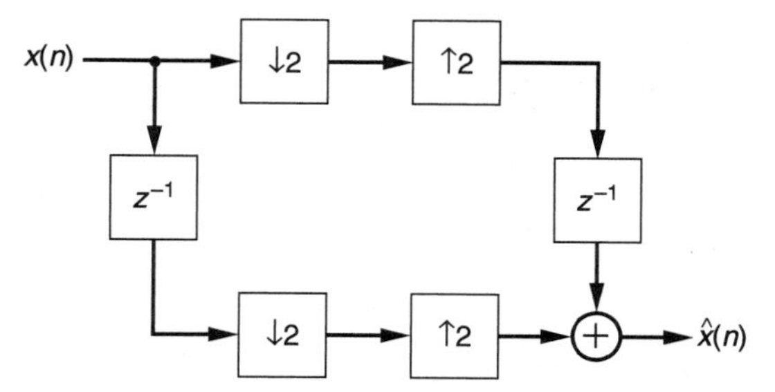

Figure P5.1

8. Following Fig. P5.1, draw the flow diagram of an M-channel filter bank without filters. Find the output of the filter bank in terms of the input.
9. Find the magnitude response of the filter in Fig. P5.2 that avoids aliasing and images.

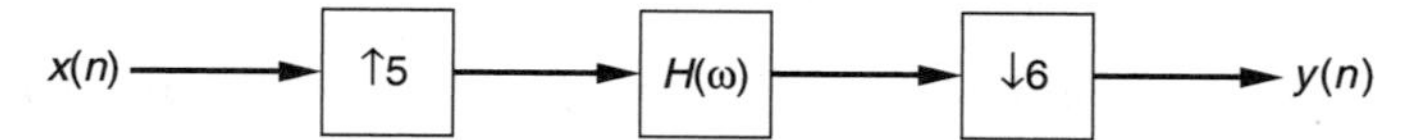

Figure P5.2

10. Find the magnitude response of the filter in Fig. P5.3 that avoids aliasing and images.

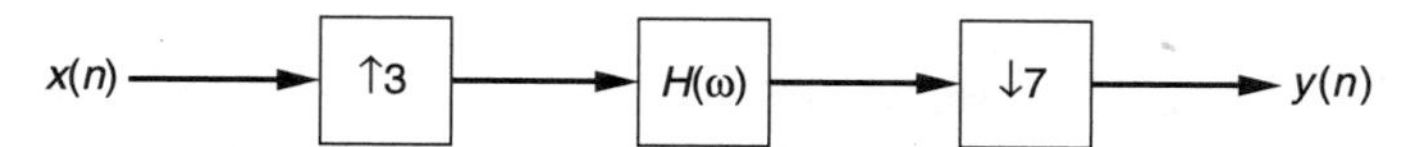

Figure P5.3

11. Show that the transpose of a 2-channel analysis filter bank is a 2-channel synthesis filter bank.
12. Show that the transpose of a M-channel synthesis filter bank is a M-channel analysis filter bank.
13. Consider the system of Fig. P5.4 with the spectrum of $x(n)$ shown. Sketch the spectrum of each of the signals $y(n)$, $v(n)$, $u(n)$, and $w(n)$.

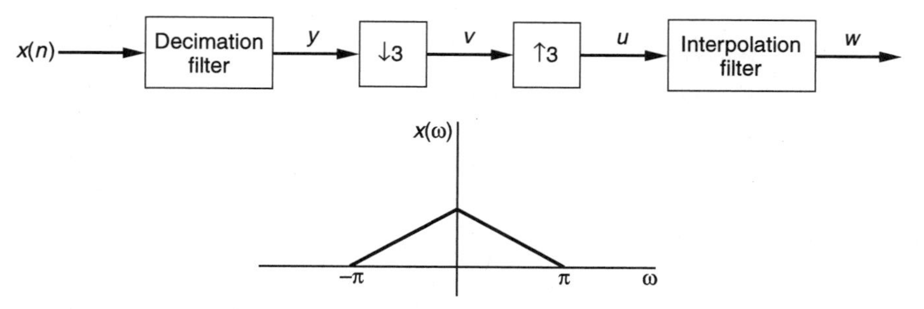

Figure P5.4

14. Prove the following: The operations $(\downarrow M)$ and $(\uparrow N)$ commute if and only if M and N are relatively prime. Using this result, simplify the system of Fig. P5.5 and find $Y(z)$ in terms of $X(z)$.

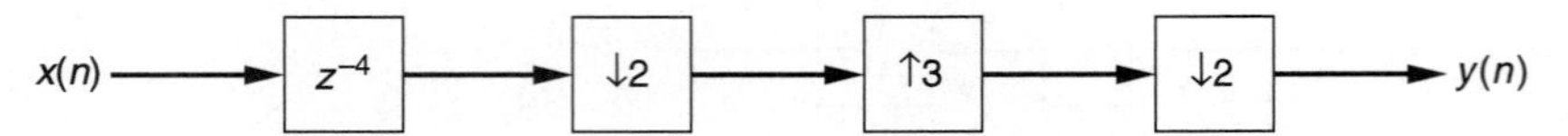

Figure P5.5

15. In the system of Fig. P5.6, find $Y(z)$ in terms of $X(z)$. Also find $y(n)$ in terms of $x(n)$.

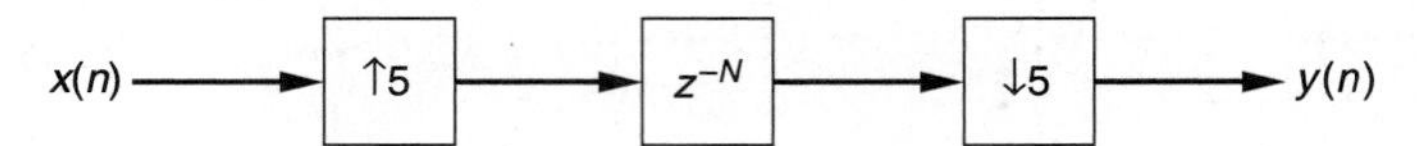

Figure P5.6

16. Draw the flow diagram of a computationally efficient factor-of-3 decimator using a length-12 linear phase FIR filter.

17. Draw the flow diagram of a computationally efficient factor-of-5 decimator using a length-15 nonlinear phase FIR filter.

18. Draw the flow diagram of a computationally efficient factor-of-6 interpolator using a length-12 linear phase FIR filter.

19. Derive the synthesis bank of Fig. 5.36(b).

20. Find the two polyphase components of a symmetric half-band FIR filter of length-N.

21. Given $H(z) = 1 + az^{-1} + bz^{-2} + cz^{-3} + dz^{-4} + dz^{-5} + cz^{-6} + bz^{-7} + az^{-8} + z^{-9}$. Find the polyphase components $H_0(z)$ and $H_1(z)$.

22. Given a FIR filter $H(z)$ with polyphase components $H_0(z)$ and $H_1(z)$, find the relationship between $H_0(z)$ and $H_1(z)$ for a symmetric $H(z)$ of odd length and an antisymmetric $H(z)$ of even length.

23. For a first-order IIR filter

$$H(z) = \frac{b}{1 - az^{-1}},$$

find the two-component polyphase decomposition.

24. Design a two-stage decimator that meets the following specifications: Factor-of-decimation = 100; input sampling rate = 100 kHz; transition bandwidth = 50 Hz; passband ripple = 0.1; stopband ripple = 0.01.

25. Find a two-component polyphase decomposition of the following IIR filter:

$$H(z) = \frac{b_0 + b_1 z^{-1}}{1 + a_1 z^{-1}}, \ |a_1| < 1.$$

26. Find a three-component polyphase decomposition of the following IIR filter:

$$H(z) = \frac{b_0 + b_1 z^{-1}}{1 + a_1 z^{-1}}, \quad |a_1| < 1.$$

27. Is the 2-channel QMF bank in general linear?

28. Show that the 2-channel QMF bank is periodically time-varying with a period of 2.

29. Consider the system of Fig. P5.7 where the spectrum of $x(n)$ is shown and $H(\omega)$ is a half-band lowpass filter. Sketch the spectra of the output $y(n)$ and all intermediate signals.

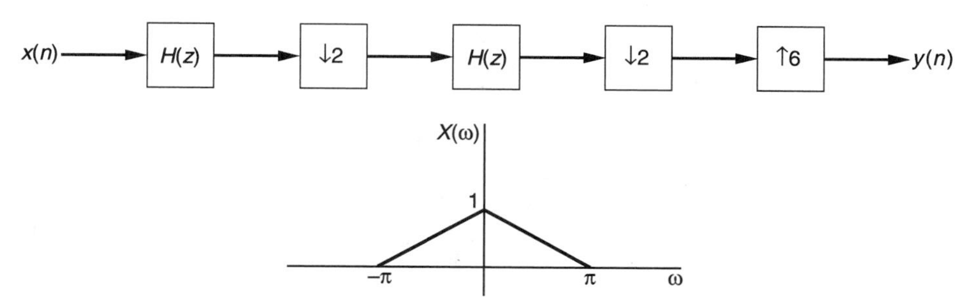

Figure P5.7

30. Consider a uniform DFT analysis bank with two polyphase components. Let the polyphase components be $E_0(z) = 1 + 3z^{-1}$ and $E_1(z) = 2 + 4z^{-1}$. Find the expressions for $H_k(z)$ for $k = 0, 1$.

31. Consider the equation $H_0^2(z) - H_0^2(-z) = 2z^{-l}$. Therefore, $H_0^2(z)$ must have only one odd power, z^{-l}. Show that a FIR filter $H_0(z)$ with three or more coefficients will have at least two odd powers in $H_0^2(z)$.

32. Consider a 2-channel Crosier-Esteban-Galand QMF bank with $H_0(z) = 1 + z^{-1}$.

(a) Find the polyphase components of the analysis and synthesis filters.
(b) Find the distortion transfer function of this filter bank.

33. Let $H_0 = 1 + az^{-1} + bz^{-2} + az^{-3} + z^{-4}$. Also, let $H_1(z) = H_0(-z)$. Sketch the QMF analysis bank in the form of a uniform DFT bank. Find all the polyphase components and the IDFT matrix.

34. Let $P(z) = H_0(z)H_0(z^{-1})$, where $H_0(z)$ is a lowpass filter.

(a) What type of filter is $P(z)$?
(b) Is $P(z)$ symmetric or antisymmetric?

35. A 2-channel QMF bank has the filters $H_0(z) = a + bz^{-1}$ and $H_1(z) = a - bz^{-1}$. Find the synthesis filters $F_0(z)$ and $F_1(z)$ for perfect reconstruction.

36. Consider the product filter given by

$$P_0(z) = \frac{1}{16}(-1 + 9z^{-2} + 16z^{-3} + 9z^{-4} - z^{-6}).$$

(a) If this filter is shifted to make it zero phase, then it is a half-band filter. Why?
(b) This half-band filter is called the binomial or maxflat filter because its frequency response is maximally flat at $\omega = 0$ and $\omega = \pi$. Show this.
(c) Find all the possible factorizations of $P_0(z)$ into two polynomials $H_0(z)$ and $F_0(z)$.
(d) If $F_0(z) = \frac{1}{2}(1+2z^{-1}+z^{-2})$, find $H_0(z)$. This is called the Daubechies 5/3 filter bank [4] with 5 coefficients for $H_0(z)$ and 3 coefficients for $F_0(z)$. Plot the magnitude response of the analysis bank filters.

37. Consider the product filter given by

$$P_0(z) = \frac{1}{16}(-1 + 9z^{-2} + 16z^{-3} + 9z^{-4} - z^{-6})$$

where $P_0(z) = H_0(z)F_0(z)$. We know that one of the conditions for alias cancellation is $F_0(z) = H_1(-z)$. If $H_0(z) = \frac{1}{8}(1 + 3z^{-1} + 3z^{-2} + z^{-3})$, find $H_1(z)$. This is called the Daubechies 4/4 filter. Plot the magnitude response of the analysis filters.

38. Consider the Croisier-Esteban-Galand filter bank $H_1(z) = H_0(-z)$. If $H_0(z)$ is a symmetric lowpass filter with even length, show that $H_1(z)$ is an antisymmetric highpass filter.

39. Given a length-N FIR filter, consider the "flip" operation: $G(z) = z^{-(N-1)}H(z^{-1})$.

(a) Find the impulse response $g(n)$ in terms of $h(n)$.
(b) If $H(z)$ is symmetric with even-length, is $G(z)$ symmetric or antisymmetric?
(c) Find $|G(\omega)|$ in terms of $|H(\omega)|$.

40. Given a length-N FIR filter, consider the "alternate flip" operation: $G(z) = z^{-(N-1)}H(-z^{-1})$.

(a) Find the impulse response $g(n)$ in terms of $h(n)$.
(b) If $H(z)$ is symmetric with even-length, is $G(z)$ symmetric or antisymmetric?
(c) Find $|G(\omega)|$ in terms of $|H(\omega)|$.

41. The Haar filter bank [5] has $H_0(z) = \frac{1}{\sqrt{2}}(1 + z^{-1})$ and $H_1(z) = H_0(-z)$. Find the synthesis filter banks for perfect reconstruction.

42. Given a perfect reconstruction filter bank $\{H_0, H_1, F_0, F_1\}$, interchange the filters in the analysis and synthesis filter banks. Show that the resulting filter bank is perfect reconstruction.

43. Given a perfect reconstruction filter bank $\{H_0, H_1, F_0, F_1\}$, define $\widehat{H}_k(z) = H_k(-z)$ and $\widehat{F}_k(z) = F_k(-z)$. Is the new filter bank perfect reconstruction? Why?

▶ 5.12 COMPUTER PROJECTS

Project 1

1. Generate a signal sampled at 8 kHz, which contains a sum of three sinusoids at frequencies 50 Hz, 500 Hz, and 1000 Hz. Plot this signal for a reasonable number of samples.
2. Design a standard factor-of-40 decimator. The edge frequencies of the decimator filter are 90 Hz and 100 Hz. The peak passband and stopband ripples are 0.001 and 0.005, respectively. Design the filter using a Kaiser window. Plot the magnitude and phase response of the filter.
3. Process the input signal by the standard decimator. Plot the output of the decimator.
4. Design a two-stage decimator based on the IFIR approach. Use 20 as the stretch factor for the IFIR filter. Plot the magnitude and phase responses of the two filters.
5. Process the input signal by the two-stage IFIR decimator. Plot the output of the decimator and compare with that of part (2).

Project 2

Repeat Project 1 for a factor-of-40 interpolation.

Project 3

1. Design a 4-channel uniform DFT analysis filter bank. The prototype filter must be an FIR filter with linear phase and with maximum passband or stopband ripple of 0.05.
2. Plot the magnitude response of each polyphase filter.
3. Implement this filter bank in MATLAB. Apply a sum of 2 sinusoids of known frequency at the input. Plot the output in a meaningful way. Discuss the results and explain if they make sense.
4. Design the synthesis filter bank as a polyphase filter. Plot the magnitude response of each polyphase component.
5. Implement the synthesis filter bank. Apply the output of the analysis bank as input to the synthesis bank. Plot the output of the synthesis bank and discuss if it makes sense.

Project 4

1. Design a lowpass power symmetric FIR filter $H_0(z)$ with a stopband edge at 0.6π. The stopband attenuation must be at least 40 dB. Plot the magnitude response of this filter.
2. Design a perfect reconstruction Smith-Barnwell QMF filter bank based on $H_0(z)$. Give the transfer function, plot the magnitude response of each analysis and synthesis filter.
3. Implement the filter in MATLAB.
4. Apply a known input to the input of the QMF bank and measure the output. Plot the input and output in a meaningful way. Discuss your results.

BIBLIOGRAPHY

[1] *ITU Draft Recommendation G.728*, "Coding of speech at 16 kbits/s using low-delay code excited linear prediction (LD-CELP)," 1992.

[2] *JPEG 2000 Image Coding System*, www.jpeg.org.

[3] Y. Neuvo, C.-Y. Dong, and S.K. Mitra, "Interpolated finite impulse response filters," *IEEE Trans. on Acoust., Speech, and Signal Process.*, ASSP-32, pp. 563–570, June 1984.

[4] I. Daubechies, "Orthonormal bases of compactly supported wavelets," *Comm. Pure Appl. Math.* 41, pp. 909–996, 1988.

[5] A. Haar, "Zur Theorie de orthogonalen Funktionen-Systeme," *Math. Ann.* 69, pp. 331–371, 1910.

[6] R.E. Crochiere, and R.L. Rabiner, *Multirate Digital Signal Processing*, Prentice Hall, Englewood Cliffs, NJ, 1983.

[7] A. Croisier, D. Esteban, and C. Galand, "Perfect channel splitting by use of interpolation/decimation/tree decomposition techniques," *Int. Symp. on Info., Circuits and Systems*, Patras, Greece, 1976.

[8] J.D. Johnston, "A filter family designed for use in quadrature mirror filter banks," *Proc. IEEE Int. Conf. Acoust. Speech and Signal Proc.*, pp. 291–294, April 1980.

[9] F. Mintzer, "Filters for distortion-free two-band multirate filter banks," *IEEE Trans. on Acoust. Speech and Signal Proc.*, ASSP-33, pp. 626–630, June 1985.

[10] S.K. Mitra, *Digital Signal Processing: A Computer-Based Approach*, 2nd ed., Irwin, IL, McGraw-Hill, 2001.

[11] M.J.T. SMITH and T.P. BARNWELL III, "A procedure for designing exact reconstruction filter banks for tree-structured subband coders," *Proc. IEEE Conf. Acoust. Speech and Signal Proc.*, pp. 27.1.1–27.1.4, March 1984.

[12] P.P. VAIDYANATHAN, P. REGALIA, and S. K. MITRA, "Design of doubly complementary IIR digital filters using a single complex allpass filter, with multirate applications," *IEEE Trans. on Circuits and Systems*, CAS-34, pp. 378–389, April 1987.

[13] P.P. VAIDYANATHAN, *Multirate Systems and Filter Banks*, Prentice Hall, Englewood Cliffs, NJ, 1993.

[14] T. PAINTER and A.S. SPANIAS, "Perceptual coding of digital audio," *Proc. of the IEEE*. 88 (4), pp. 451–513, April 2000.

[15] ITU Draft Recommendation G.722, "7 kHz audio coding within 64 kbits/s," in *Blue Book*, vol. III, Fascicle, III, October 1988.

CHAPTER 6

Finite-Wordlength Effects

All digital processors have registers and accumulators with a finite number of bits. In other words, all processors have finite wordlength. Once we design a digital filter or an algorithm, it must be implemented in a finite-wordlength machine. Digital filters and algorithms usually have many sum of product calculations. Several errors can occur in the implementation as follows. (a) All the filter coefficients must be quantized for representation with a finite number of bits, which may introduce errors in the filter frequency response. (b) Quantization of the products introduces noise into the filter, which has the effect of reducing the signal to noise ratio (SNR) of the filter. (c) Accumulator overflow can occur during addition, which can cause very large errors, thereby reducing the SNR. Overflow can also yield high-amplitude oscillations (limit cycles) in the filter; this in turn can lead to instability of an otherwise stable filter. (d) The sum of products must also be quantized after computation, and this can lead to a reduction in the SNR and also yield limit cycles.

Digital filters must therefore be designed carefully in order to minimize these effects. In addition, we can take other steps during implementation to improve the SNR and avoid instability. There are basically two types of arithmetic systems, namely, fixed point and floating point. Both of these types are useful for implementing signal processing algorithms. In the fixed point system, the binary point is assumed to be in a fixed position. On the other hand, in a floating point processor, the position of the binary point can be variable. This gives a larger dynamic range and typically lower quantization noise. However, the quantization step size is not uniform, and the noise analysis is more complicated. In addition, the hardware required for a floating point processor implementation is more complex than that of a fixed point processor. Therefore, fixed point processors are always faster than their floating point counterparts. Moreover, fixed point DSP chips are cheaper and consume less power than floating point chips. However, fixed point processors can have severe nonlinear effects in filter implementations, often causing instability, limit-cycle oscillations, and high quantization noise. Therefore, the filters must be carefully designed for fixed point implementation. In this chapter, we will study the different types of nonlinear conditions that occur in filter implementations and the ways to minimize their effects. Most of the analysis will be done for fixed point implementation.

The chapter begins with a discussion of fixed point and floating point number representations and then presents, the effects of coefficient quantization on the frequency response of IIR and FIR filters. The other sections give a detailed quantization noise analysis for IIR filters; discuss overflow effects, establish filter design constraints for the suppression of overflow limit cycles; derives scaling methods for the prevention of adder overflow; and describes product quantization effects. Several different conditions are derived for the stability of quantized filters.

6.1 NUMBER REPRESENTATIONS

Fixed Point Numbers

A number in binary, may be represented in many ways. The most common formats are called sign-magnitude and two's complement and will be presented here. Consider a number X that needs to be represented in binary. Let the number be scaled so that

$$X = X_s x \tag{6.1}$$

where X_s is the scale factor such that $|x| \leq 1$. Now, x will be represented in $(B+1)$ bits of binary and is therefore subject to quantization. Let the quantized value be represented as

$$\hat{x} = b_0.b_1 b_2 \ldots b_B \tag{6.2}$$

where b_0 represents the sign bit. The value of the above number will depend on the type of format, as described shortly. In digital filter implementations, all signals and coefficients are scaled to be binary fractions. The scale factor X_s is not explicitly represented and is implicit in the implementation of the filter. Usually, X_s is implicitly a power-of-two number such as 2^p, where p is a nonnegative integer. So, the actual signal or coefficient value is the representation of (6.2), with the binary point shifted p bits to the right. However, b_0 will still serve as the sign bit.

Sign-magnitude In sign-magnitude format, the value of the number is given by

$$\begin{aligned}\hat{x} &= sgn \cdot \{b_1 2^{-1} + b_2 2^{-2} + \cdots + b_B 2^{-B}\} \\ &= sgn \cdot \sum_{i=1}^{B} b_i 2^{-i} \end{aligned} \tag{6.3}$$

where

$$sgn = \begin{cases} 1, & \text{if } b_0 = 0 \\ -1, & \text{if } b_0 = 1. \end{cases}$$

In this representation, b_0 denotes the sign, and the rest of the bits represent the magnitude. This representation is clearly symmetric for both positive and negative numbers. For a 3-bit number, the values are represented on a circle in Fig. 6.1 for both integer and fractions. In the sign-magnitude number system, there are two formats for zero, which is a slight waste in the range of representable numbers. In digital filter implementations, all

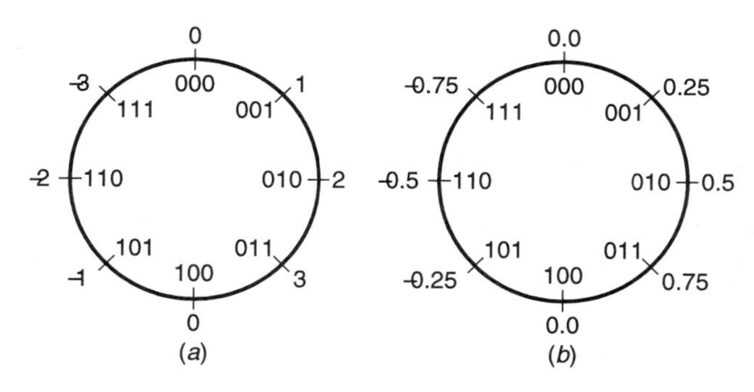

Figure 6.1 Number circles for sign-magnitude: (a) integer (b) fractions.

coefficients and signals are usually represented in the fractional format. For a $(B+1)$ bit fractional representation, the quantization step size is $q = 2^{-B}$ and the range of numbers is $-(1-2^{-B}) \leq \hat{x} \leq (1-2^{-B})$. When a higher precision number is quantized down to a lower precision (less bits), we have a couple of choices for the new number, namely, truncation and rounding. These are now explained with an example.

EXAMPLE 6.1 Consider a 4-bit number in fractional sign-magnitude format given by $\hat{a} = 0011$, which has a decimal value of 0.375. This number is now quantized down to 3 bits, which has a quantization step size of 0.25, and its possible values are given in Fig. 6.1(b). The value of 0.375 lies exactly halfway between the new possible values of 0.25 (001) or 0.50 (010). Under the *magnitude truncation* scheme, all values are quantized down to the lower quantization level. On the other hand, under the *rounding* scheme, values under the halfway point are quantized down, while values that are at halfway and above are quantized up to the next level. For negative values, quantization is done on the magnitude only, and these schemes are therefore symmetric. The transfer characteristics for magnitude truncation and rounding schemes are shown in Fig. 6.2. In this particular example, truncation results in $Q\{\hat{a}\} = 001 = 0.25$, whereas rounding gives $Q\{\hat{a}\} = 010 = 0.50$.

In the analysis of quantization, it is often useful to find the quantization error defined as

$$e = Q\{\hat{a}\} - \hat{a}. \tag{6.4}$$

From the transfer characteristics for truncation and rounding, the following are easily seen.

Magnitude truncation

$$-q < e_t < q$$

Rounding

$$-\frac{q}{2} < e_R \leq \frac{q}{2}.$$

Two's Complement Numbers In two's complement format, a $(B+1)$ bit number as in (6.2) has the value

$$\begin{aligned} \hat{x} &= -b_0 + b_1 2^{-1} + b_2 2^{-2} + \cdots + b_B 2^{-B} \\ &= -b_0 + \sum_{i=1}^{B} b_i 2^{-i}. \end{aligned} \tag{6.5}$$

In this case, b_0 also denotes the sign. However, this representation does not have two zeros. For a 3-bit number, the values are denoted on a circle in Fig. 6.3. The range of fractional numbers is $-1 \leq \hat{x} \leq (1 - 2^{-B})$. Now let us see what happens when a two's complement number is quantized to a smaller number of bits.

▶ **EXAMPLE 6.2** Consider a 4-bit number in two's complement format given by $\hat{a} = 0011$, which has a decimal value of 0.375. This is a positive number, and a simple truncation to 3 bits results in $Q\{\hat{a}\} = 001 = 0.25$. This is the same as in sign-magnitude truncation. However, consider the negative number $\hat{b} = 1101 = -0.375$. A simple truncation gives $Q\{\hat{b}\} = 110 = -0.50$. The magnitude of the number actually increased in this case. Also note that the magnitude of the quantized values of the positive number and the negative number of equal magnitude were different. Therefore, the *two's complement truncation* (TCT) scheme is not symmetric with respect to the sign of the number.

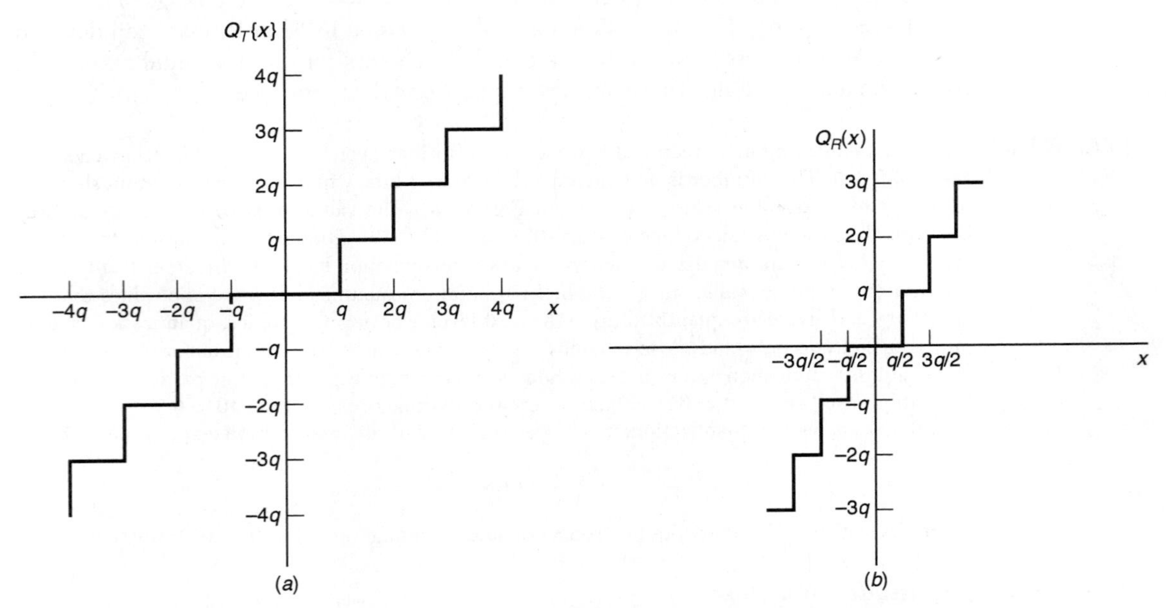

Figure 6.2 Transfer characteristics for quantization with sign-magnitude numbers: (*a*) truncation; (*b*) rounding.

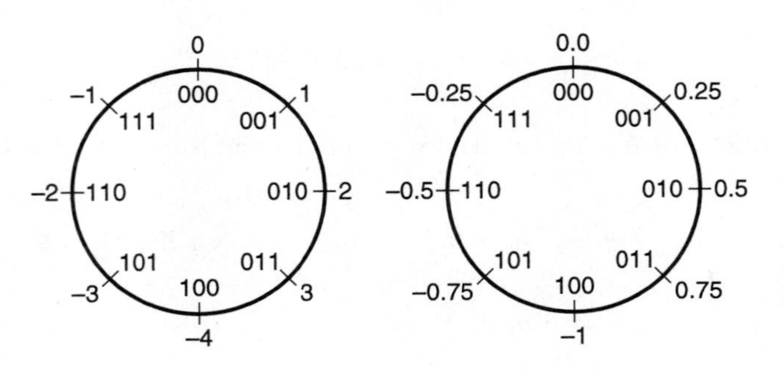

Figure 6.3 Number circles for two's complement: (a) integers, (b) fractions.

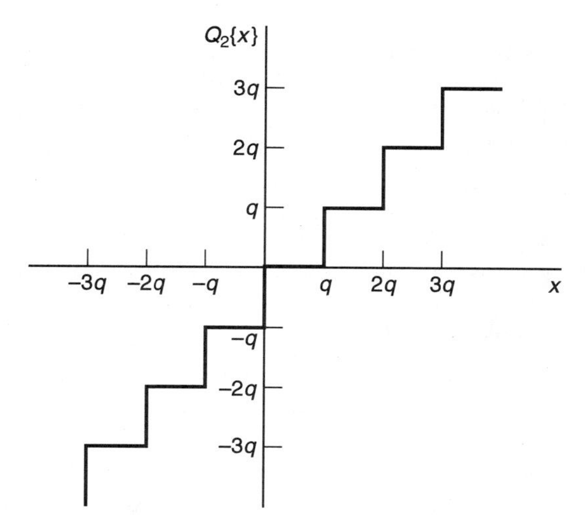

Figure 6.4 Transfer characteristic for two's complement truncation.

The transfer characteristic of the TCT scheme is shown in Fig. 6.4. The error in quantization is given by

$$-q < e_{2t} \leq 0.$$

Floating Point Numbers

In fixed point numbers, the quantization step size, q, is constant. This is not the case in floating point numbers. Because of a variable step size, the floating point representation has a much higher dynamic range. This is achieved by using a small step size (higher resolution) for smaller numbers and a large step size (coarse resolution) for larger numbers. In general, a floating point number is represented as

$$\hat{x} = M \cdot 2^E$$

where M is the mantissa and E is the exponent. The mantissa and exponent can have sign bits and can be represented in any fixed point format. In most cases, the mantissa is restricted such that $\frac{1}{2} \leq |M| < 1$. When two floating point numbers are added, their exponents must be equal. If they are not, then one of the exponents (and its mantissa) must be adjusted before addition can take place. When an exponent is adjusted in this way, it can cause a loss of precision, as will be shown shortly in an example. The multiplication of two floating point numbers are done in the following way:

$$\begin{aligned} \hat{x}_1 \cdot \hat{x}_2 &= M_1 2^{E_1} \cdot M_2 2^{E_2} \\ &= M_1 M_2 \cdot 2^{E_1+E_2}. \end{aligned}$$

Therefore, a floating point multiplication requires a fixed point multiplication as well as a fixed point addition. An example is in order.

EXAMPLE 6.3 Let us assume that the mantissa has 6 bits and the exponent has 3 bits and that both are represented in a fixed point sign-magnitude system. Consider the number $x_1 = \frac{5}{8}$. One method to represent this number in floating point is simply to write the numerator as a binary number with the least number of bits necessary, insert a preceding binary point, insert trailing zeros to satisfy the length of the mantissa (MSB reserved for the sign), and then calculate the exponent. Then $M_1 = 0.10100$, and this will require an exponent of $E_1 = 000$. Now consider the number $x_2 = 3$. By the above method, $M_2 = 0.11000$, which requires an exponent of $E_2 = 010$. The product is given by

$$\begin{aligned} x_1 x_2 &= (0.10100)(0.11000)2^{000+010} \\ &= (0.01111)2^2 \end{aligned}$$

which gives a result of $1\frac{7}{8}$, which is the correct answer. Now let us assume that the mantissa is 5 bits long and the exponent is still 3 bits. Addition of the two numbers would require the exponents to be equal. The sum can be calculated as follows:

$$\begin{aligned} x_1 + x_2 &= 0.1010 \cdot 2^{000} + 0.1100 \cdot 2^{010} \\ &= 0.0010 \cdot 2^{010} + 0.1100 \cdot 2^{010} \\ &= 0.1110 \cdot 2^{010}. \end{aligned}$$

Note that in adjusting the exponent in x_1 above, a loss in precision took place, which is not unusual in floating point additions.

The most widely used floating point format is the standard set by the Institute of Electrical and Electronic Engineers (IEEE) and is called the IEEE 754 standard. The 32-bit format is shown in Fig. 6.5, where S is the sign bit, and E and M represent the exponent and mantissa fields, respectively. The exponent is coded to be biased as $E - 127$. The floating point number has the value

$$X = (-1)^S \cdot M \cdot 2^{E-127}$$

where the mantissa is coded as $0 \leq M < 1$. The standard also specifies the following properties for the format.

If $E = 255$ and $M \neq 0$, then X is an undefined number.

If $E = 255$ and $M = 0$, then $X = (-1)^S \infty$.

If $0 < E < 255$, then $X = (-1)^S \cdot (1.M) \cdot 2^{E-127}$.

If $E = 0$ and $M \neq 0$, then $X = (-1)^S \cdot (0.M) \cdot 2^{-126}$.

If $E = 0$ and $M = 0$, then $X = (-1)^S \cdot 0$.

In these properties, $1.M$ denotes a representation with a one integer bit and 23 fractional bits, and $0.M$ denotes a pure fraction. It can be shown (Problem 1) that the IEEE-754 format has a range from 1.18×10^{-38} to 3.4×10^{38}. Therefore, the dynamic range is approximately 10^{76}, with fine resolution for small numbers and coarse resolution for large numbers. This kind of variable step size makes the analysis of roundoff noise difficult in floating point implementations.

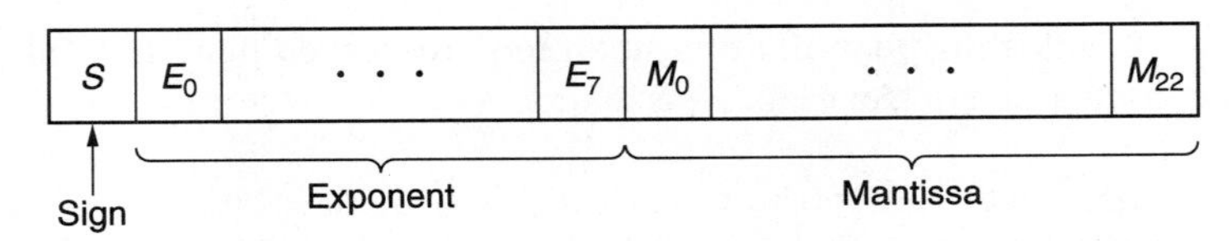

Figure 6.5 IEEE-754 format for floating point numbers.

6.2 COEFFICIENT QUANTIZATION

When a digital filter is designed, the filter coefficients are usually obtained in high precision, normally with 32-bit accuracy. If this filter is implemented in a shorter wordlength fixed point processor, the coefficients must be appropriately quantized. The frequency response of the resulting filter may differ significantly from that of the original filter. The sensitivity of the frequency response to coefficient quantization depends on the filter structure, as we will see shortly.

Consider the transfer function of a Direct Form filter given by

$$H(z) = \frac{\sum_{k=0}^{M} b_k z^{-k}}{1 + \sum_{k=1}^{N} a_k z^{-k}} \tag{6.6}$$

where the coefficients a_k and b_k are assumed to be in infinite precision. Let $\hat{a}_k$ and $\hat{b}_k$ denote the quantized coefficients such that

$$\begin{aligned} \hat{a}_k &= a_k + \Delta a_k, \quad k = 1, 2, \ldots, N \\ \hat{b}_k &= b_k + \Delta b_k, \quad k = 0, 1, \ldots, M \end{aligned}$$

where Δa_k and Δb_k represent the quantization errors. The quantized filter now has the transfer function

$$\widehat{H}(z) = \frac{\sum_{k=0}^{M} \hat{b}_k z^{-k}}{1 + \sum_{k=1}^{N} \hat{a}_k z^{-k}}. \tag{6.7}$$

In order to find a relationship between the quantization errors and the frequency response (or pole-zero positions), we consider the denominator and the numerator separately. Let the denominator be represented as

$$A(z) = 1 + \sum_{k=1}^{N} \hat{a}_k z^{-k} \tag{6.8}$$

$$= \prod_{i=1}^{N} (1 - p_i z^{-1}) \tag{6.9}$$

where p_i are the poles of the filter. The perturbation of this ith pole as a function of the coefficient quantization errors is

$$\Delta p_i = \sum_{k=1}^{N} \frac{\partial p_i}{\partial a_k} \Delta a_k. \tag{6.10}$$

To find $\frac{\partial p_i}{\partial a_k}$, use the chain rule

$$\left(\frac{\partial A(z)}{\partial a_k}\right)_{z=p_i} = \left(\frac{\partial A(z)}{\partial z}\right)_{z=p_i} \cdot \left(\frac{\partial p_i}{\partial a_k}\right). \tag{6.11}$$

Using (6.8), we obtain the left-hand side of equation (6.11) as

$$\left(\frac{\partial A(z)}{\partial a_k}\right)_{z=p_i} = \frac{\partial}{\partial z}\left\{1+\sum_{k=1}^{N}\hat{a}_k z^{-k}\right\}_{z=p_i}$$
$$= z^{-k}|_{z=p_i} = p_i^{-k}. \tag{6.12}$$

The term $\left(\frac{\partial A(z)}{\partial z}\right)_{z=p_i}$ in the right-hand side of (6.11) can be found by using (6.9) as

$$\left(\frac{\partial A(z)}{\partial z}\right)_{z=p_i} = \frac{\partial}{\partial z}\left\{\prod_{i=1}^{N}(1-p_i z^{-1})\right\}_{z=p_i}. \tag{6.13}$$

It can be shown (Problem 7) that the above derivative is

$$\left(\frac{\partial A(z)}{\partial z}\right)_{z=p_i} = \frac{1}{p_i^N}\prod_{\substack{j=1\\j\neq i}}^{N}(p_i-p_j). \tag{6.14}$$

Substituting (6.12) and (6.14) in (6.11), we get

$$\frac{\partial p_i}{\partial a_k} = \frac{p_i^{N-k}}{\prod\limits_{\substack{j=1\\j\neq i}}^{N}(p_i-p_j)}. \tag{6.15}$$

Now substituting (6.15) in (6.10), we get the desired result

$$\Delta p_i = \sum_{k=1}^{N}\frac{p_i^{N-k}\Delta a_k}{\prod\limits_{\substack{j=1\\j\neq i}}^{N}(p_i-p_j)}. \tag{6.16}$$

Equation (6.16) gives a measure of the sensitivity of the ith pole as a function of the coefficient quantization errors. The term $(p_i - p_j)$ in the denominator is a vector between the two poles. So, if the poles are clustered together, the magnitude of the denominator in (6.16) can be small, thereby resulting in a large perturbation, Δp_i. This in turn would lead to a large error in the frequency response. Closely clustered poles is a common occurrence in narrowband filters, which makes these filters highly sensitive to coefficient quantization. This problem can be circumvented by implementing the filter as a cascade or parallel combination of first- and/or second-order modules. The real poles can be implemented as first-order modules. For each of these first-order sections, there is only one pole and hence there is no sensitivity issue. Complex-conjugate pole-pairs can be implemented as second-order modules. Since the poles within a complex-conjugate pair are usually far apart, the sensitivity function (6.16) is low.

A formula analogous to (6.16) can be obtained for the zeros of the transfer function. For a cascade form realization, the zeros can be factored into first- and second-order factors. On the other hand, a parallel form realization is obtained using a partial fraction expansion. Therefore, each first-order module has no zero association, and each second-order section has one zero. In the case of either the cascade or parallel form, there is no clustering of the zeros for each individual module. Thus, cascade and parallel realizations are much less sensitive to coefficient quantization than their Direct Form counterparts.

As we know from implementing second-order systems, we have some options. Let us first consider the Direct Form filter transfer function with a denominator polynomial

$$A(z) = 1 + a_1 z^{-1} + a_2 z^{-2}.$$

Let the poles be located at $z = re^{\pm j\theta}$. By finding the roots of $A(z)$, it is easy to show that the filter coefficients are related to the poles as

$$\begin{aligned} a_1 &= -2r\cos\theta \\ a_2 &= r^2. \end{aligned}$$

If the coefficients are now quantized, the poles are located as shown in Fig. 6.6. When $a_2 = r^2$ is quantized, we get the circles, and when $a_1 = -2r\cos\theta$ is quantized, we get the vertical lines. The intersection of the vertical lines and the circles gives the pole positions of the quantized filter. In Fig. 6.6, we have a 4-bit quantization, with one bit reserved for the sign. This leads to eight circles and eight vertical lines, as shown in the first quadrant. The other quadrants are symmetric mirrored versions of the first quadrant. Notice that near the horizontal axis ($\theta = 0$ or π), there is the largest possible shift in the poles. Therefore, if the original filter has poles near $\theta = 0$ or π, the quantized filter will have high degradation in the frequency response.

Now we consider a second-order Coupled Form structure. The poles of the filter are again assumed to be at $re^{\pm j\theta} = \sigma \pm j\omega$. The filter structure was given in Fig. 4.31, where the filter coefficients are σ, σ, ω, and $-\omega$, $\sigma = r\cos\theta$, and $\omega = r\sin\theta$. When these coefficients are quantized to a 4-bit precision, the possible pole locations are shown in Fig. 6.7. The pole locations are uniformly distributed within the unit circle. Therefore, Coupled Form filters do not have severe quantization errors at certain regions within the

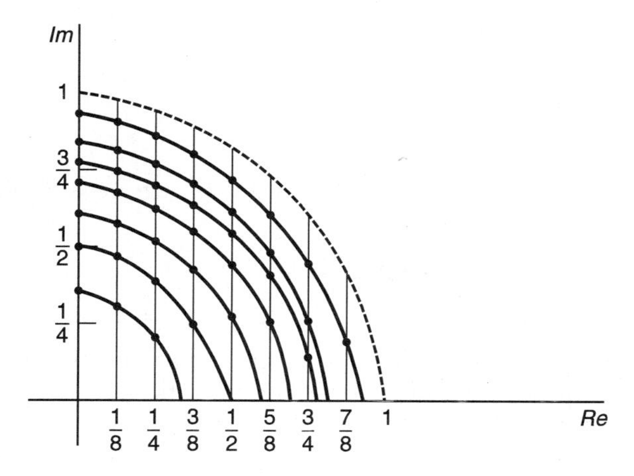

Figure 6.6 Pole locations for a second-order Direct Form filter quantized with 4 bits.

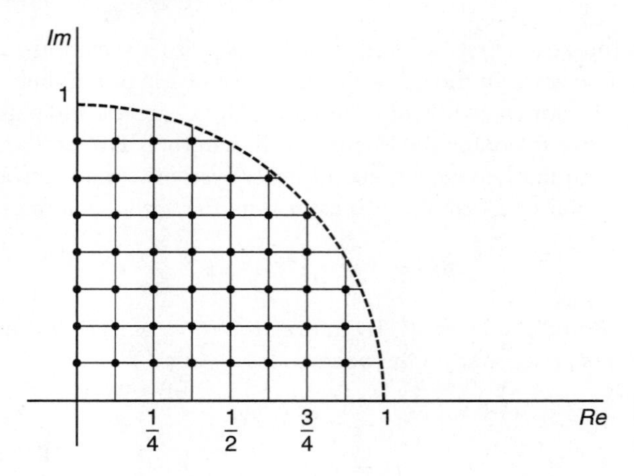

Figure 6.7 Pole locations for a second-order Coupled-Form filter quantized with 4 bits.

unit circle. However, as we know, the Coupled Form filter requires two extra multipliers than its Direct Form counterpart.

▶ EXAMPLE 6.4 In this example, we illustrate the effect of coefficient quantization on the frequency response of an IIR digital filter. Consider the design of a bandpass Butterworth filter of order $N = 10$ and cutoff frequencies at 0.5π and 0.6π. This filter can be easily designed in MATLAB. The magnitude and phase responses are shown in Fig. 6.8. Then we quantize each coefficient to a precision of 8 bits, and the resulting frequency response is given in Fig. 6.9. The degradation in the magnitude response is quite significant, and the phase response is clearly destroyed. The original filter is then converted to a cascade form, with each second-order section in Direct Form. The coefficients of the cascade form filter are then quantized to 8 bits. The resulting frequency response is shown in Fig. 6.10. Both the magnitude and phase responses are very close to those of the unquantized filter. This is expected in light of the foregoing discussion. If each second-order section is implemented as a Coupled Form and then quantized, then we expect the result to be even better. This is left as an exercise.

Coefficient Quantization in FIR Filters

Consider the unquantized FIR filter given by the transfer function

$$H(z) = \sum_{n=0}^{M} h(n)z^{-n} \tag{6.17}$$

where $h(n)$ are the filter coefficients. The filter with quantized coefficients is

$$\widehat{H}(z) = \sum_{n=0}^{M} \hat{h}(n)z^{-n} \tag{6.18}$$

where $\hat{h}(n)$ are the quantized coefficients. In our earlier discussion, we derived the sensitivity function for the poles of an IIR filter when the coefficients are quantized. The same analysis also holds for the zeros of a FIR filter. In most FIR filter designs, the zeros are not clustered. Therefore, the perturbation function for the zeros is not

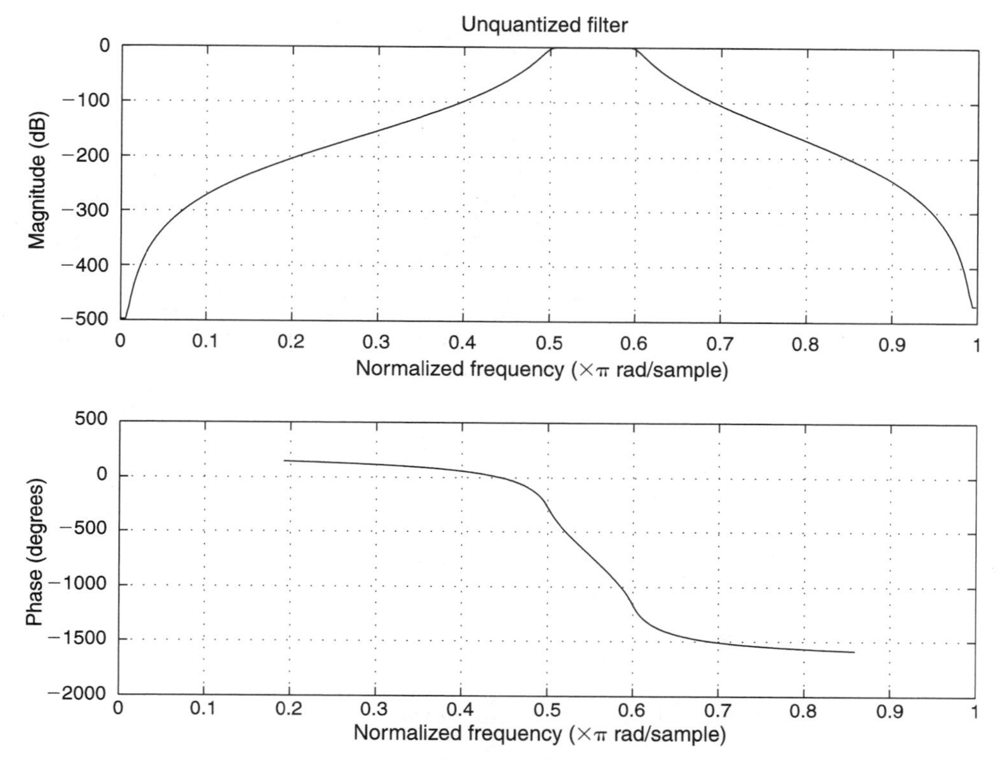

Figure 6.8 Frequency response of unquantized Butterworth filter.

very sensitive to coefficient quantization. This is the reason quantized FIR filters are often implemented in direct form and there is very little degradation in the frequency response. An attractive feature of FIR filters is that they have linear phase property if the coefficients are symmetric ($h(n) = h(M-n)$) or antisymmetric ($h(n) = -h(M-n)$). These properties are preserved even after quantization. This is another advantage of implementing FIR filters in direct form. However, in a couple of situations a cascade-form realization may be desirable. These are when the quantization is very coarse (very short-wordlengths) or when the zeros of the filter happen to be closely clustered. When the polynomial $H(z)$ is factored for cascade form implementation, each section must have linear phase for the overall filter to have linear phase. This may not be the case after the coefficients are quantized. Let us now investigate this in detail. In Chapter 3, we derived the zero position for a linear phase FIR filter. It was shown that the zeros occur in mirror image pair about the unit circle. That is, if z_0 is a zero of $H(z)$, then the reciprocal $\frac{1}{z_0}$ is also a zero. Of course, all complex zeros occur in conjugate pairs if $h(n)$ is real. The zero locations are shown in Fig. 6.11. Now consider all the possible cases for a cascade-form implementation.

Case 1: The real zeros at $z = \pm 1$ can be realized as first-order sections. Quantization of these zeros will still preserve the linear phase property.

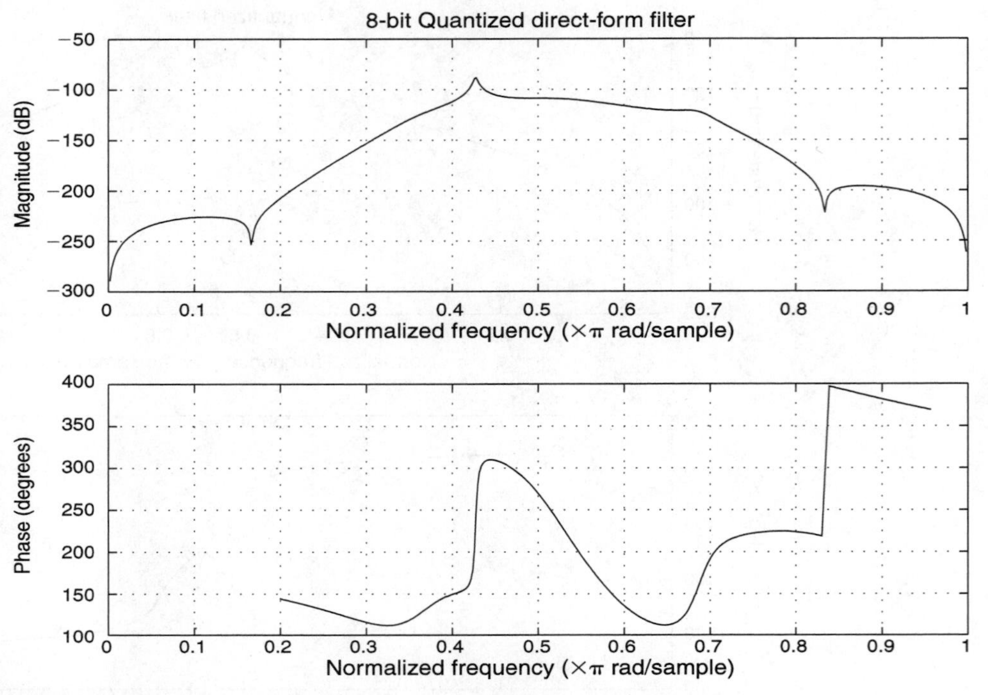

Figure 6.9 Frequency response of Direct Form filter quantized to 8 bits.

Case 2: Real zeros occur as reciprocal pairs as $z = -b, -\frac{1}{b}$. These can be realized as first-order sections $(1 + bz^{-1})$ and $(1 + b^{-1}z^{-1})$. The latter section can be realized as $\frac{1}{b}(b + z^{-1})$, where $\frac{1}{b}$ is simply a gain factor. Now if b is quantized to $\hat{b}$, the zero locations are $\hat{b}$ and $\frac{1}{\hat{b}}$, which are still reciprocal and therefore maintain linear phase.

Case 3: Complex zeros on the unit circle can be realized as a second-order section of the form $H_{uc}(z) = 1 + \alpha z^{-1} + z^{-2}$. It is easy to show that this quadratic yields roots on the unit circle. Now if we quantize α to $\hat{\alpha}$, then the roots are still on the unit circle, and linear phase is preserved.

Case 4: Complex zeros not on the unit circle occur as reciprocal pairs. If $re^{\pm j\theta}$ is a pair of zeros inside the unit circle ($r < 1$), then $r^{-1}e^{\pm j\theta}$ are the reciprocal zeros outside the unit circle. The quadratics corresponding to these pairs of roots are as follows.

$$re^{\pm j\theta} \Longrightarrow$$

$$\begin{aligned} H_1(z) &= (1 - re^{j\theta}z^{-1})(1 - re^{-j\theta}z^{-1}) \\ &= 1 - (2r\cos\theta)z^{-1} + r^2z^{-2} \\ &= 1 - \beta z^{-1} + r^2z^{-2}. \end{aligned}$$

$$r^{-1}e^{\pm j\theta} \Longrightarrow$$

$$\begin{aligned} H_2(z) &= (1 - r^{-1}e^{j\theta}z^{-1})(1 - r^{-1}e^{-j\theta}z^{-1}) \\ &= 1 - (2r^{-1}\cos\theta)z^{-1} + r^{-2}z^{-2}. \end{aligned}$$

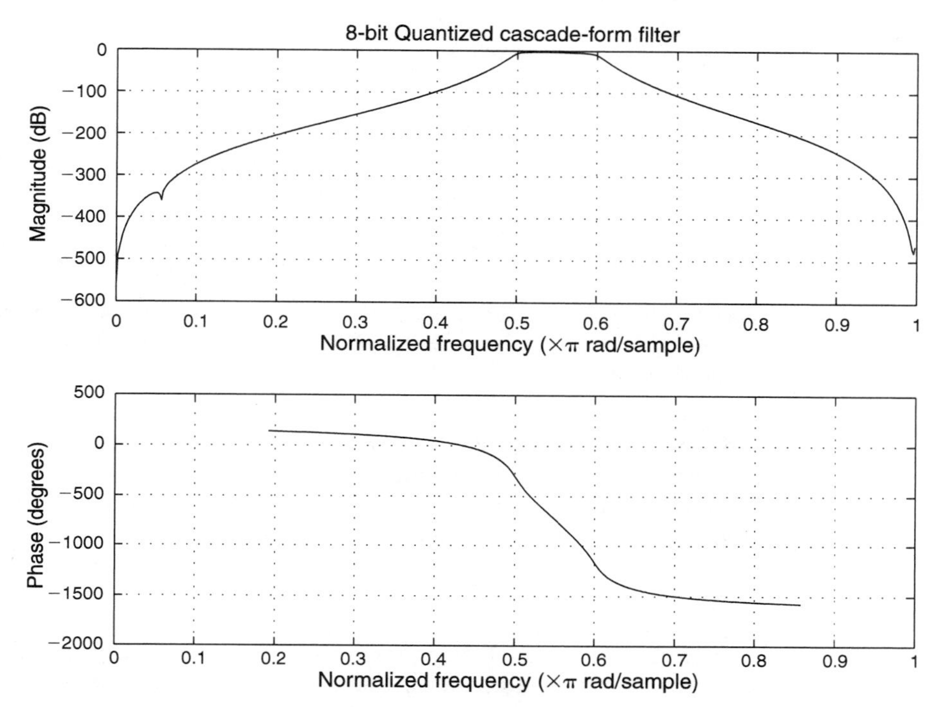

Figure 6.10 Frequency response of cascade-form filter quantized to 8 bits.

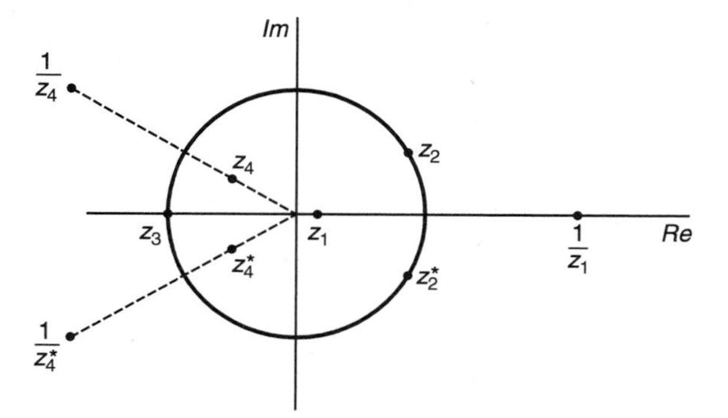

Figure 6.11 Zero locations for a linear phase FIR filter.

If $H_1(z)$ and $H_2(z)$ are quantized and implemented, then $\widehat{r^2}$ may not be the reciprocal of $\widehat{r^{-2}}$, and the same is true for the other coefficient. This means that the zeros may not be reciprocal after quantization, thereby losing the linear phase property. This can be corrected by implementing $H_2(z)$ as

$$\begin{aligned} H_2(z) &= \frac{1}{r^2}(r^2 - (2r\cos\theta)z^{-1} + z^{-2}) \\ &= \frac{1}{r^2}(r^2 - \beta z^{-1} + z^{-2}). \end{aligned}$$

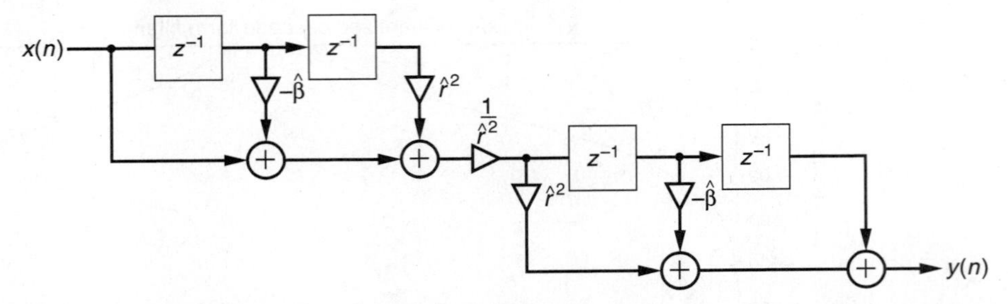

Figure 6.12 Implementation of a pair of complex conjugate poles and their reciprocals with quantization of coefficients. ^ represents coefficient quantization. This structure preserves the linear phase property.

Notice that $H_2(z)$ now has the same coefficients $(1, -\beta, r^2)$ as $H_1(z)$ except in reverse order and that $\frac{1}{r^2}$ is simply a gain factor. Quantization of $H_1(z)$ and $H_2(z)$ will now yield roots that are reciprocal and complex conjugate. This implementation is shown in Fig. 6.12, and it preserves the linear phase property.

6.3 QUANTIZATION NOISE ANALYSIS

In this section, we present techniques to find the signal to noise ratio due to quantization in the various filter structures due to the different types of quantization schemes. Such an analysis will help determine the type of filter structure, the quantization scheme, and the number of bits required in the design of a given system. The most popular method for such an analysis is by the use of linear models. In this section, we will perform the noise analysis for a first-order filter and the Direct Form II structure only. A similar technique can be used for other structures as well.

First-Order Filter

We begin our analysis with a single-pole digital filter. The linear difference equation of the filter is given by

$$v(n) = av(n-1) + u(n) \tag{6.19}$$

where $u(n)$ and $v(n)$ are the input and output, respectively.

When the filter is implemented in fixed point arithmetic, the resulting nonlinear model is

$$y(n) = Q\{ay(n-1)\} + u(n) \tag{6.20}$$

where Q is the quantization operation under consideration. Let us assume that all signal values and coefficients are represented by $(B+1)$ bit binary numbers. The product of $ay(n-1)$ is possibly $2B+1$ bits long, which is quantized down to $(B+1)$ bits by either truncation (sign-magnitude or two's complement) or rounding. The effect of this quantization can be modeled as

$$Q\{ay(n-1)\} = ay(n-1) + e(n) \tag{6.21}$$

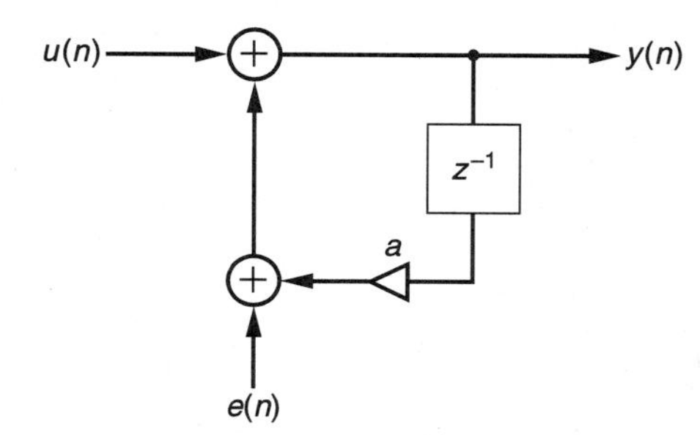

Figure 6.13 Quantization noise model for a first-order filter.

where $e(n)$ is an additive noise source injected at the quantizer. Substituting (6.21) into (6.20) gives the linear model of the quantized filter as

$$y(n) = ay(n-1) + u(n) + e(n). \tag{6.22}$$

The net effect is the addition of a noise source $e(n)$ as an input, which also propagates through the filter. The flow diagram of the quantized filter is shown in Fig. 6.13. Our task now is to find the variance of the noise at the output of the filter. To this end, we make a few assumptions in order to simplify the analysis.

Assumptions

(1) Each quantization noise source $e(n)$ is a wide-sense stationary (WSS) white noise process.

(2) The noise sequence, $e(n)$, is uniformly distributed over the quantization range.

(3) The noise sequence, $e(n)$, is uncorrelated with the input to the quantizer. It is also uncorrelated with the filter input $u(n)$ and all other quantization noise sources, if present.

Now let us examine the significance of these assumptions. Let μ_e and σ_e^2 denote the mean and variance of the noise source. From Assumption 1, the autocorrelation of the noise sequence $e(n)$ is

$$r_e(m) = \sigma_e^2 \delta(m) + \mu_e^2. \tag{6.23}$$

From Assumption 2, we know that the probability density function (pdf) of the noise sequence is flat. The quantization interval depends on the type of quantization. These intervals were derived in Section 6.2 for sign-magnitude truncation, two's complement truncation, and rounding. The corresponding pdf is shown in Fig. 6.14, where $q = 2^{-B}$ is the quantization step size.

From the pdf, the noise variance can be obtained for each of the quantization schemes. We know that the variance is given by

$$\begin{aligned}\sigma_e^2 &= E\{e^2(n)\} - (E\{e(n)\})^2 \\ &= E\{e^2(n)\} - \mu_e^2 \end{aligned} \tag{6.24}$$

The mean square value can be found using the formula

$$E\{e^2(n)\} = \int_{e_1}^{e_2} e^2 p(e)\, de \tag{6.25}$$

where e_1 and e_2, denote the interval of e in the pdf.

Using (6.24) and (6.25), we can easily obtain the variance for each quantization scheme (Problem 11) as follows.

Sign-magnitude truncation

$$\mu_t = 0$$

$$\sigma_t^2 = \frac{2^{-2B}}{3}$$

Two's complement truncation

$$\mu_{2t} = \frac{-2^{-B}}{2}$$

$$\sigma_{2t}^2 = \frac{2^{-2B}}{3}$$

Rounding

$$\mu_r = 0$$

$$\sigma_r^2 = \frac{2^{-2B}}{12}$$

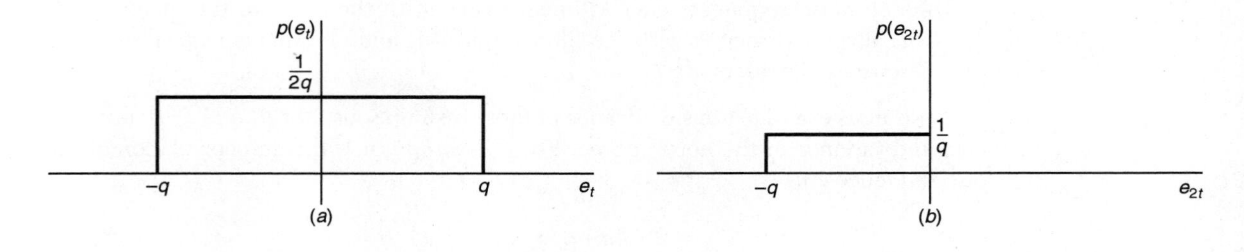

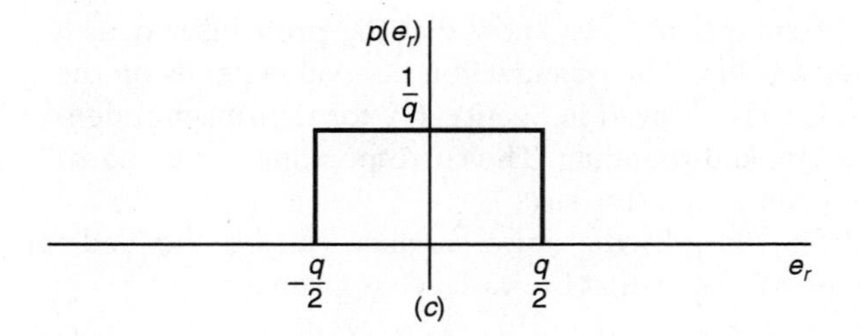

Figure 6.14 Probability density function for noise sources: (*a*) sign-magnitude; (*b*) two's complement truncation; (*c*) rounding.

Assumption 3 tells us that the variance of the input signal and that of the noise source are completely separate entities. It also states that if we have more than one quantization, then the total variance of the noise can be obtained by adding the variance of the individual noise sources.

Now we return to our quantized first-order digital filter. We have shown that the output $y(n)$ is composed of (a) the filtered version of the input $u(n)$ and (b) the filtered version of the noise $e(n)$. Now we determine what effect the filter has on the variance of the noise source. The following theorem is used for this purpose. In order to maintain the continuity of our discussion, the proof is postponed to the Appendix.

Theorem 6.1 *Consider a linear system with frequency response $H(\omega)$, with a white noise input $e(n)$ that has a mean of μ_e and a variance of σ_e^2. Then the mean and variance of the output $f(n)$ are given by*

$$\mu_f = \mu_e H(0) \tag{6.26}$$

$$\sigma_f^2 = \sigma_e^2 \sum_{n=-\infty}^{\infty} |h(n)|^2. \tag{6.27}$$

A useful result can be obtained by using Parseval's theorem on (6.27), which gives

$$\sigma_f^2 = \sigma_e^2 \frac{1}{2\pi} \int_{-\pi}^{\pi} |H(\omega)|^2 d\omega. \tag{6.28}$$

In Z-transform domain, (6.28) can be written as

$$\sigma_f^2 = \sigma_e^2 \frac{1}{2\pi j} \oint_c H(z)H(z^{-1})z^{-1}dz \tag{6.29}$$

This equation can be evaluated using Cauchy's residue theorem as discussed in Chapter 3.

For the first-order filter, the impulse response is

$$h(n) = a^n, \quad n \geq 0.$$

The total noise variance at the output is therefore

$$\sigma_f^2 = \sigma_e^2 \sum_{n=0}^{\infty} a^{2n}$$

where σ_e^2 depends on the quantization scheme used. Since for stability we need $|a| < 1$, the above can be simplified as

$$\sigma_f^2 = \sigma_e^2 \frac{1}{1-a^2}.$$

Thus, as the pole $z = a$ approaches the unit circle, the noise variance increases. So, for a prescribed SNR, the wordlength must be determined based on the pole position of the filter.

Second-Order Direct Form II

Now we consider the effect of quantization noise in a second-order Direct Form II filter. The linear model of the filter is given by the equations (see Chapter 1):

$$g(n) = -a_1 g(n-1) - a_2 g(n-2) + u(n) \tag{6.30}$$

$$y(n) = b_0 g(n) + b_1 g(n-1) + b_2 g(n-2). \tag{6.31}$$

When this filter is implemented in fixed point arithmetic, we have the following options.

Single-Length Accumulator All registers including the accumulator are $(B+1)$ bits long. Therefore, every product must be quantized to $(B+1)$ bits. The nonlinear model then becomes

$$g(n) = -Q\{a_1 g(n-1)\} - Q\{a_2 g(n-2)\} + u(n) \tag{6.32}$$

$$y(n) = Q\{b_0 g(n)\} + Q\{b_1 g(n-1)\} + Q\{b_2 g(n-2)\}. \tag{6.33}$$

As described earlier, each product quantization is modeled as a noise source. The noise model for the quantized filter is shown in Fig. 6.15.

There are five different noise sources. Sources e_3, e_4, e_5 are added directly to the output adder, whereas the sources e_1 and e_2 are actually filtered by the transfer function. The variance of each of these noise sources is the same and as derived earlier. Therefore, the total noise variance at the output is

$$\sigma_f^2 = 2\sigma_e^2 \sum_{n=-\infty}^{\infty} |h(n)|^2 + 3\sigma_e^2. \tag{6.34}$$

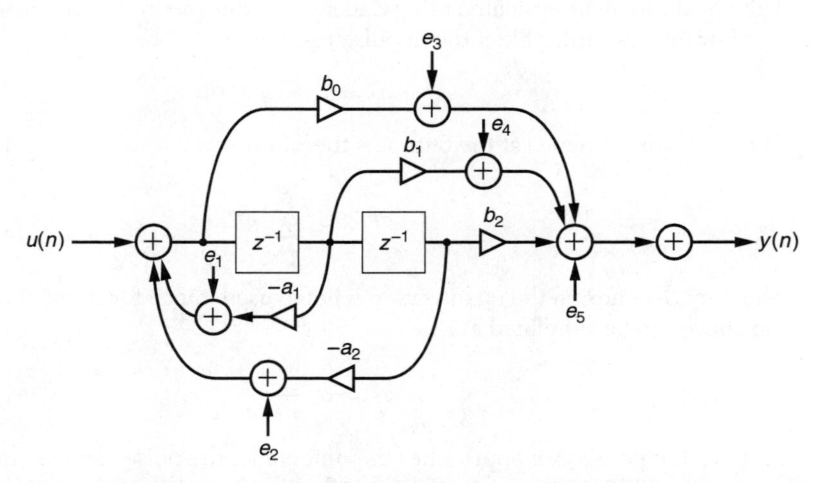

Figure 6.15 Noise model for a second-order Direct Form II filter using a single-length accumulator.

For a general order Direct Form filter, the equations are

$$g(n) = -\sum_{k=1}^{N} Q\{a_k g(n-k)\} + u(n) \tag{6.35}$$

$$y(n) = \sum_{k=0}^{M} Q\{b_k g(n-k)\}. \tag{6.36}$$

The noise variance at the output is then

$$\sigma_f^2 = N\sigma_e^2 \sum_{n=-\infty}^{\infty} |h(n)|^2 + M\sigma_e^2 \tag{6.37}$$

$$= \frac{N\sigma_e^2}{2\pi} \int_{-\pi}^{\pi} |H(\omega)|^2 d\omega + M\sigma_e^2 \tag{6.38}$$

$$= \frac{N\sigma_e^2}{2\pi j} \oint_c H(z)H(z^{-1})z^{-1}dz + M\sigma_e^2 \tag{6.39}$$

where (6.38) and (6.39) were obtained by using (6.28) and (6.29), respectively.

Double-Length Accumulator The accumulator is $(2B+1)$ bits long, while all the registers are $(B+1)$ bits long. It is also assumed that at least one product register is $(2B+1)$ bits long. This means that the intermediate products do not have to be quantized until after addition. With this type of architecture, the nonlinear model for the Direct Form II filter can be written as

$$g(n) = Q\left\{-\sum_{k=1}^{N} a_k g(n-k) + u(n)\right\} \tag{6.40}$$

$$y(n) = Q\left\{\sum_{k=0}^{M} b_k g(n-k)\right\}. \tag{6.41}$$

This nonlinear model introduces only one noise source at the input adder and one at the output adder of the filter. The total noise variance of the output is therefore immensely reduced and is given by

$$\sigma_f^2 = \sigma_e^2 \sum_{n=-\infty}^{\infty} |h(n)|^2 + \sigma_e^2. \tag{6.42}$$

For the second-order case with a double-length accumulator, let the transfer function be

$$H(z) = \frac{1}{(1-re^{j\theta}z^{-1})(1-re^{-j\theta}z^{-1})}. \tag{6.43}$$

The filter therefore has a pair of complex conjugate poles. By using (6.39) with $M = N = 1$, the total noise variance can be found (Problem 15) as

$$\sigma_f^2 = \sigma_e^2 \left[\left(\frac{1+r^2}{1-r^2} \right) \frac{1}{r^4 + 1 - 2r^2 \cos 2\theta} + 1 \right]. \tag{6.44}$$

The significance of the above lies in the fact that as $r \longrightarrow 1$; that is, as the poles approach the unit circle, σ_f^2 increases. Thus, for a prescribed SNR, longer wordlengths are required as the poles are closer to the unit circle. This was also the case for the first-order filter.

▶ **EXAMPLE 6.5** *Consider a causal filter*

$$H(z) = \frac{1 - 0.3z^{-1}}{(1 - 0.2z^{-1})(1 + 0.7z^{-1})}.$$

(a) Find the Direct Form II and cascade structures for the filter.

(b) For each of the structures, find the output noise variance for a (4 + 1)-bit two's complement truncation with a double-length accumulator.

SOLUTION

(a) The Direct Form II structure of the filter is shown in Fig. 6.16, and the cascade structure for the filter is given in Fig. 6.17.

(b) In the Direct Form II structure, there are two different noise sources: e_1 and e_2. Source e_1 is filtered by the transfer function, whereas source e_2 is added directly to the output adder. So

$$\sigma_f^2 = \sigma_e^2 + \sigma_e^2 \sum_{n=-\infty}^{\infty} |h(n)|^2.$$

In the above, we have

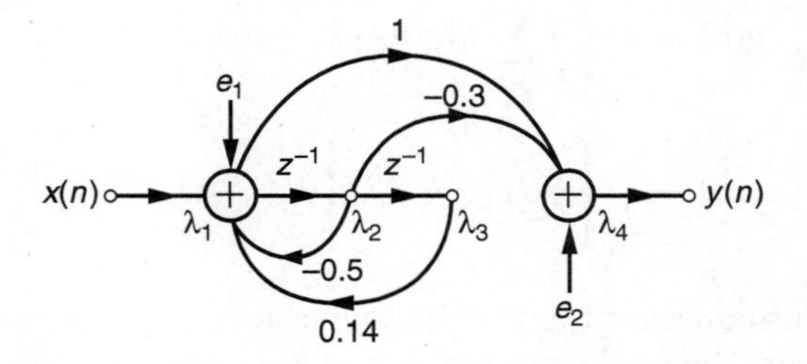

Figure 6.16 Direct Form II structure for example.

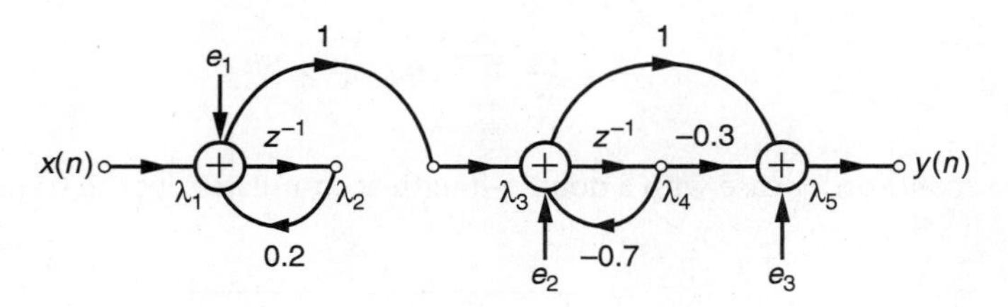

Figure 6.17 Cascade form structure, for example.

$$\sum_{n=-\infty}^{\infty} |h(n)|^2 = \frac{1}{2\pi}\int_{-\pi}^{\pi} |H(\omega)|^2 d\omega$$

$$= \frac{1}{2\pi j}\oint_C H(z)H(z^{-1})z^{-1}dz$$

$$= \sum_m Res[H(z)H(z^{-1})z^{-1}]_{Z=Z_m}$$

where Z_m are poles inside C. Since the filter is causal, the ROC is $|z| > 0.7$.

Then,

$$\sum_{n=-\infty}^{\infty} |h(n)|^2 = \sum_m Res\left[\frac{(1-0.3z^{-1})}{(1-0.2z^{-1})(1+0.7z^{-1})}\frac{(1-0.3z)}{(1-0.2z)(1+0.7z)}z^{-1}\right]_{Z=Z_m}$$

$$= \sum_m Res\left[\frac{(z-0.3)}{(z-0.2)(z+0.7)}\frac{(1-0.3z)}{(1-0.2z)(1+0.7z)}\right]_{Z=Z_m}$$

$$= \left[\frac{(z-0.3)}{(z+0.7)}\frac{(1-0.3z)}{(1-0.2z)(1+0.7z)}\right]_{Z=0.2} +$$

$$\left[\frac{(z-0.3)}{(z-0.2)}\frac{(1-0.3z)}{(1-0.2z)(1+0.7z)}\right]_{Z=-0.7}$$

$$= -0.095 + 2.312$$

$$= 2.217. \tag{6.45}$$

For two's complement truncation,

$$\sigma_e^2 = \frac{2^{-2B}}{3} = \frac{1}{3\times 2^8}.$$

So, the output noise variance for the Direct Form II structure is:

$$\sigma_f^2 = \sigma_e^2 + \sigma_e^2 \sum_{n=-\infty}^{\infty} |h(n)|^2$$

$$= \sigma_e^2(1+2.217)$$

$$= \frac{1}{3\times 2^8} \times 3.217$$

$$= 4.189\times 10^{-3}.$$

For the cascade structure, there are three different noise sources as shown in Fig. 6.17. The output noise variance is

$$\sigma_f^2 = \sigma_e^2 + \sigma_e^2 \sum_{n=-\infty}^{\infty} |h_1(n)|^2 + \sigma_e^2 \sum_{n=-\infty}^{\infty} |h_2(n)|^2$$

where

$$H_1(z) = \frac{1-0.3z^{-1}}{(1-0.2z^{-1})(1+0.7z^{-1})}$$

$$H_2(z) = \frac{1 - 0.3z^{-1}}{1 - 0.2z^{-1}}.$$

From part (a), $\sum_{n=-\infty}^{\infty} |h_1(n)|^2 = 2.217$. Also,

$$\begin{aligned}
\sum_{n=-\infty}^{\infty} |h_2(n)|^2 &= \sum_m Res\left[\frac{(z-0.3)}{(z+0.7)}\frac{(1-0.3z)}{(1+0.7z)}\frac{1}{z}\right]_{Z=Z_m} \\
&= \left[\frac{(z-0.3)}{z}\frac{(1-0.3z)}{(1+0.7z)}\right]_{Z=-0.7} + \\
&\quad \left[\frac{(z-0.3)}{(z+0.7)}\frac{(1-0.3z)}{(1+0.7z)}\right]_{Z=0} \\
&= 3.389 - 0.429. \\
&= 2.960.
\end{aligned}$$

So, the output noise variance for the cascade structure is

$$\begin{aligned}
\sigma_f^2 &= \sigma_e^2(1 + 2.217 + 2.960) \\
&= \frac{1}{3 \times 2^8} \times 6.177 \\
&= 8.043 \times 10^{-3}.
\end{aligned}$$

◀

In concluding this section, we note that the noise variance of a digital filter depends on (a) the filter structure; (b) the type of product quantization; and (c) the size of the accumulator and product register. In this section, we analyzed a first-order filter and the Direct Form II structure. Other structures, such as the transposed form, cascade, parallel, Coupled Form, and so on, can be analyzed in a similar fashion. It is not possible to say which structure has the lowest noise variance unless the coefficients of the system are known.

▶ 6.4 OVERFLOW EFFECTS IN FIXED POINT DIGITAL FILTERS

Consider two $(B + 1)$ bit numbers in some given format (sign-magnitude, two's complement, etc.). When the sum of these two numbers is greater than the dynamic range of the $(B + 1)$ bit number in the given format, then a phenomenon called *overflow* is said to occur. Let us illustrate this with some examples.

▶ **EXAMPLE 6.6** ***Sign-magnitude***

Consider a 4-bit sign-magnitude number system in which we will add two numbers, x_1 and x_2. Obviously, if the sum of these numbers is within the dynamic range $(-\frac{7}{8}, \frac{7}{8})$, then the result will be correct. Now consider the following cases where we obtain incorrect answers.

1. Let $x_1 = \frac{7}{8} = 0111$ and $x_2 = \frac{3}{8} = 0011$. The correct sum is $1\frac{1}{4}$. However, simple binary addition gives $x_1 + x_2 = 1010 = -\frac{1}{4}$, which is incorrect because there was an overflow into the sign bit.
2. Let $x_1 = \frac{7}{8} = 0111$ and $x_2 = \frac{7}{8} = 0111$. The correct sum is $1\frac{3}{4}$. Binary addition gives $x_1 + x_2 = 1110 = -\frac{3}{4}$.

3. Let $x_1 = -\frac{7}{8} = 0111$ and $x_2 = -\frac{7}{8} = 0111$. The correct sum is $-1\frac{3}{4}$. Binary addition gives $x_1 + x_2 = (1)0110 = \frac{3}{4}$, which is obtained by ignoring the overflow out of the 4 bit register.

From these examples, we can figure out the transfer characteristic for the sign-magnitude overflow. Assuming the number of bits to be large, that is, a small quantization step size, the transfer characteristic is shown in Fig. 6.18, where $\Psi\{x\}$ represents the overflow operation. Notice from the above examples that the error due to overflow can be very large. In order to avoid this problem, a popular alternative is to force the output to saturate if an overflow occurs. This is called *saturation overflow*, and its transfer characteristic is shown in Fig. 6.19.

EXAMPLE 6.7 ***Two's Complement***

Consider a 4-bit two's complement number system for adding the two numbers x_1 and x_2. The dynamic range of numbers for a 4-bit two's complement number is $(-1, \frac{7}{8})$. If the sum is within this range, then the result will be correct. Otherwise, we will get an overflow.

1. Let $x_1 = \frac{7}{8} = 0111$ and $x_2 = \frac{1}{8} = 0001$. The correct sum is 1.0. However, binary addition gives $x_1 + x_2 = 1000 = -1$, which is incorrect because there is an overflow into the sign bit.
2. Let $x_1 = \frac{5}{8} = 0101$ and $x_2 = \frac{1}{2} = 0100$. The correct sum is $1\frac{1}{8}$. However, binary addition gives $x_1 + x_2 = 1001 = -\frac{7}{8}$.

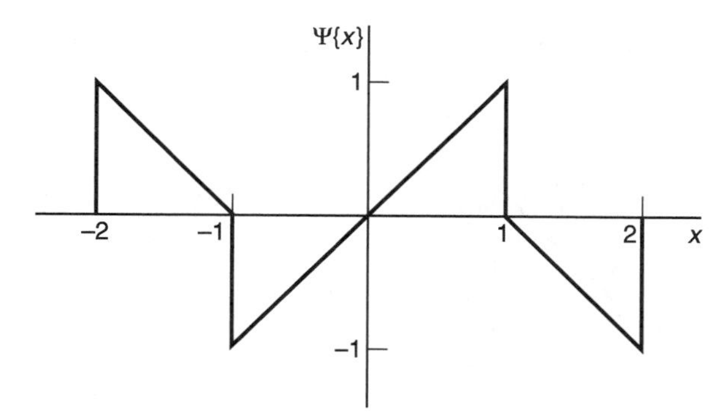

Figure 6.18 Transfer characteristic of the sign-magnitude overflow operation.

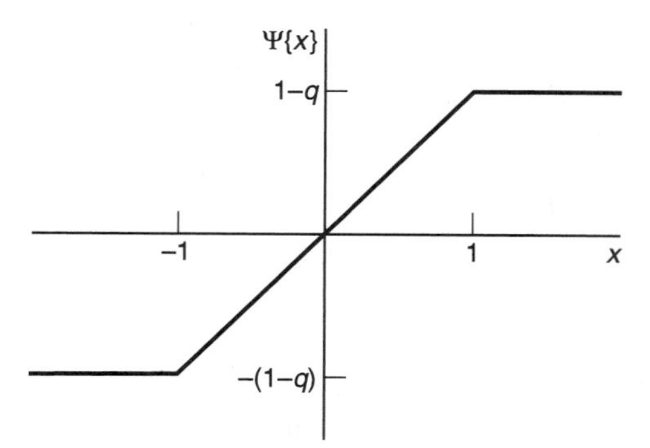

Figure 6.19 Transfer characteristic of saturation overflow.

3. Let $x_1 = -\frac{7}{8} = 1001$ and $x_2 = -\frac{7}{8} = 1001$. The correct sum is $-1\frac{3}{4}$. However, binary addition gives $x_1 + x_2 = (1)0010 = \frac{1}{4}$, which is obtained by ignoring the extra bit.

From the above examples and some similar ones, we can find the transfer characteristic of the two's complement overflow operation. This is shown in Fig. 6.20 for a large number of bits (i.e., when the quantization step size is small). Note that the unfilled circles in the figure denote that $\Psi\{x\} = 1$ is not possible. The filled circles denote that $\Psi\{m\} = -1$, for any odd integer $m \neq 0$. Again, we see that the error due to overflow can be very large. An alternative is the saturation overflow described earlier or the *zeroing overflow* characteristic shown in Fig. 6.21. From the transfer function of the two's complement overflow, we have

$$\Psi\{x\} = x + 2p \tag{6.46}$$

for some integer p and $-1 \leq \Psi\{x\} < 1$. Another equation that describes this overflow is

$$\Psi\{x + 2p\} = x \tag{6.47}$$

for $-1 \leq x < 1$. The two's complement overflow has a unique desirable property, which is not shared by any of the other overflow types. This property is stated as follows.

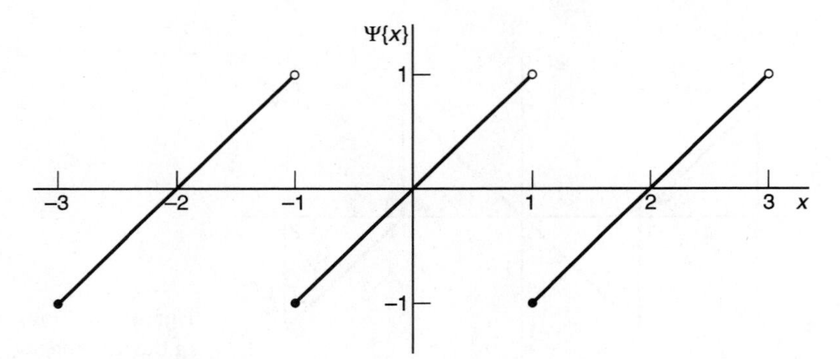

Figure 6.20 Transfer characteristic of the two's complement overflow operation.

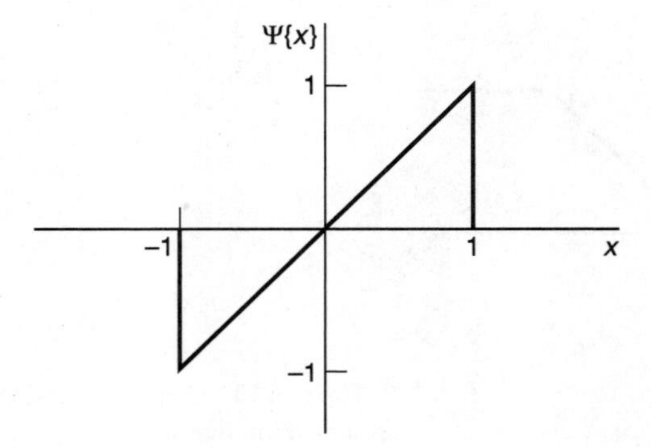

Figure 6.21 Transfer characteristic of the zeroing overflow operation.

Fact 1: Consider the sum $x_s = x_1 + x_2 + \cdots + x_n$, where $-1 \leq x_s < 1$. Then $\Psi\{x_1 + x_2 + \ldots x_n\}$ will not overflow even though intermediate sums may have an overflow.

Proof. Let us prove the above for the sum of three numbers. The general result then follows in a similar fashion. Assume that we have the sum

$$s = \Psi\{x_1 + x_2 + x_3\}$$

where we assume that $|x_1 + x_2| > 1$; that is, an overflow will occur in this intermediate sum. The complete sum is performed as

$$s = \Psi\{\Psi\{x_1 + x_2\} + x_3\}.$$

Using the property (6.46), we find that an integer p exists such that

$$s = \Psi\{(x_1 + x_2 + 2p) + x_3\}$$

where $-1 \leq (x_1 + x_2 + 2p) < 1$. The above then becomes

$$s = \Psi\{(x_1 + x_2 + x_3) + 2p\}.$$

From the hypothesis, we have $-1 \leq (x_1 + x_2 + x_3) < 1$. Therefore, we can now use the property (6.47) to get

$$s = x_1 + x_2 + x_3$$

and the conclusion follows. ■

The above property makes two's complement overflow the most desirable type of overflow. All of the overflow operations we have described can cause oscillations with zero input in digital filters, even though they are designed to be stable. These oscillations are of course caused by the nonlinear effect of the overflow and are called *limit cycles*. We now state this formally.

Definition When linear and stable digital filters are implemented in finite-wordlength fixed point processors, nonlinear effects are introduced owing to the overflow and quantization operations. These nonlinear effects can cause oscillations even with zero input. The oscillations are called limit cycles, which can render a digital filter unstable and/or increase the noise in the output signal.

Digital filters must therefore be designed carefully in order to make sure that limit cycles are not possible when they are implemented. We first give an example of a limit cycle in the familiar Direct Form II digital filter. Then we will present some design constraints for limit cycle free digital filters.

EXAMPLE 6.8 ***Limit Cycle***

Consider a Direct Form digital filter implemented in two's complement arithmetic. Let the input to the filter be zero. The difference equation can be written as

$$y(n) = \Psi\{ay(n-1) + by(n-2)\}$$

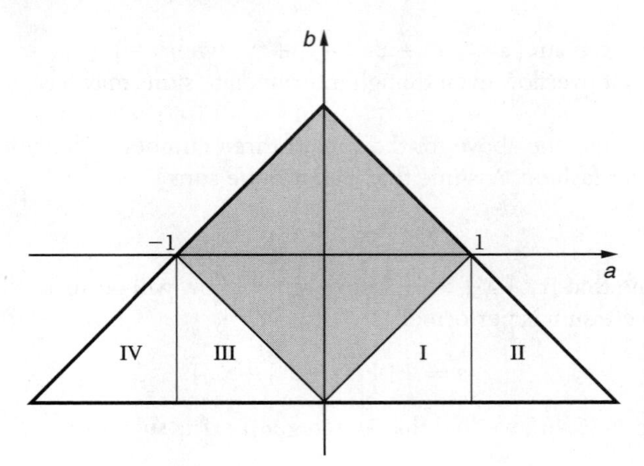

Figure 6.22 Stability triangle for a second-order Direct Form filter. Shaded region: overflow stability region.

where $\Psi\{.\}$ denotes the two's complement overflow operation. Let the coefficients and initial conditions be $a = 0.5, b = -0.95, y(0) = 0.8163, y(-1) = -0.8163$. The stability region for the linear filter is the stability triangle derived in Chapter 3 and repeated in Fig. 6.22. Note that the coefficient pair (a, b) is within the stability triangle; therefore, the system without the overflow nonlinearity is stable. In other words, the system is *linearly stable*, and $y(n) \longrightarrow 0$, as $n \longrightarrow \infty$. However, observe what happens as a result of the presence of the overflow nonlinearity. Simple recursion gives

$$y(1) = \Psi\{ay(0) + by(-1)\} = \Psi\{1.1837\} = -0.8163$$

$$y(2) = \Psi\{ay(1) + by(0)\} = \Psi\{-1.1837\} = 0.8163.$$

Thus, $y(n) = -0.8163, 0.8163, -0.8163, 0.8163, \ldots$, which is a limit cycle of period 2.

Notice that in the above example, we started with a pair of initial conditions $(y(-1), y(0))$ that gave us a limit cycle. The question is, how did we know what initial conditions we should start with to give us a limit cycle? The answer is given in a theorem in [1], where a condition is derived for the existence of a limit cycle of a given period and it also gives the actual limit cycle values. This theorem is beyond the scope of this text and is omitted.

Direct Form Filter

Now we find the stability region for a second-order Direct Form filter in the presence of overflow. Consider the Direct Form filter given by

$$y(n) = ay(n-1) + by(n-2) + u(n)$$

where $u(n)$ and $y(n)$ are the input and output, respectively. In the presence of overflow (any type) and zero input, the filter equation is

$$y(n) = \Psi\{ay(n-1) + by(n-2)\}.$$

Because of the overflow nonlinearity, the output is constrained to be $|y(n)| < 1$, except in the case of two's complement where the value of -1 is also possible. Therefore, if

$$|ay(n) + by(n-1)| < 1,$$

then overflow cannot occur. Since $|y(n)| < 1$, the above is true if $|a| + |b| < 1$. This defines the overflow stability region for a Direct Form filter and is shown as the shaded region in Fig. 6.22. In this region, zero-input limit cycles cannot occur under any type of overflow operation. This result was first obtained in [2]. It has been established in [1] that initial conditions can be found such that in regions I and II period-one limit cycles always exist. It has also been shown that in regions III and IV, period-two limit cycles can always be found. Therefore, the stability region for the Direct Form filter is greatly reduced owing to the overflow nonlinearity.

State-Space Digital Filters

In linear form, a recursive digital filter can be described as

$$\mathbf{x}(n+1) = \mathbf{A}\mathbf{x}(n) + \mathbf{B}u(n)$$
$$y(n) = \mathbf{C}x(n) + Du(n),$$

where $u(n)$ and $y(n)$ are the input and output, respectively. We know that for stability, the poles of the system or the eigenvalues of the system matrix $\mathbf{A}$ must be inside the unit circle. We also know from Chapter 3 that the above equations can be transformed by similarity transformation to another system $\{\widehat{\mathbf{A}}, \widehat{\mathbf{B}}, \widehat{\mathbf{C}}, \widehat{D}\}$, with the same transfer function. It is well known that different equivalent structures of the same system may behave differently in the presence of nonlinearities. This will be clear from this discussion.

Under zero input and overflow, the nonlinear system has the state equation $\mathbf{x}(n+1) = \Psi\{\mathbf{A}\mathbf{x}(n)\}$, where the initial condition is given by $\mathbf{x}(0)$. The overflow operation is assumed to act on each element of the vector independently of the others. In other words, for a vector $\mathbf{v}$, $\Psi\{\mathbf{v}\} = [\Psi\{v_1\}\Psi\{v_2\}\ldots\Psi\{v_p\}]^T$. We wish to find a sufficient condition for the stability of the above system. Toward that end, recall some basic definitions of vector and matrix norms. The Euclidean vector norm is defined by $\|\mathbf{x}\|_2 = \sqrt{x_1^2 + x_2^2 + \cdots + x_N^2}$. The spectral norm of a matrix is defined as

$$|||\mathbf{A}|||_2 = \sqrt{\rho(\mathbf{A}^T\mathbf{A})} \tag{6.48}$$

where $\rho(\cdot)$ is called the spectral radius and defined as the maximum modulus of the eigenvalue. That is, if λ_i, represents the eigenvalues of a matrix $\mathbf{B}$, then $\rho(\mathbf{B}) = \max_i\{|\lambda_i|\}$. We now prove the following theorem, which was originally derived in [3].

Theorem 6.2 *Consider a zero-input state-space digital filter described by*

$$\mathbf{x}(n+1) = \Psi\{\mathbf{A}\mathbf{x}(n)\} \tag{6.49}$$

where $\Psi\{\cdot\}$ *defines any kind of overflow nonlinearity. If* $|||\mathbf{A}|||_2 < 1$*, then the system is globally asymptotically stable.*

Proof. Taking the Euclidean norm of both sides of (6.49) and using the property of the overflow nonlinearity, we get

$$\begin{aligned}||\mathbf{x}(n+1)||_2 &= ||\Psi\{\mathbf{A}\mathbf{x}(n)\}||_2 \\ &\leq ||\mathbf{A}\mathbf{x}(n)||_2 \\ &\leq |||\mathbf{A}|||_2||\mathbf{x}(n)||_2 \\ &< ||\mathbf{x}(n)||_2.\end{aligned}$$

The above implies that $||\mathbf{x}(n)||_2 \longrightarrow 0$, as $n \longrightarrow \infty$ and the conclusion follows. ■

This is a very powerful result and has significance for the type of filter structures that we choose to implement. We first state the following lemmas, which are not difficult to prove (Problems 6.8 and 6.9).

Lemma 6.1 *If* $|||\cdot|||$ *is any matrix norm and* $\mathbf{A}$ *is a square matrix, then* $|||\mathbf{A}||| \geq \rho(\mathbf{A})$.

Lemma 6.2 *If* $\mathbf{A}$ *is a normal matrix, that is,* $\mathbf{A}\mathbf{A}^T = \mathbf{A}^T\mathbf{A}$, *then its spectral norm satisfies*

$$|||\mathbf{A}|||_2 = \rho(\mathbf{A}).$$

Every state-space digital filter is designed to be linearly stable, that is, $\rho(\mathbf{A}) < 1$. This implies that if the system matrix $\mathbf{A}$ is normal, then from Lemma 2 we have $|||\mathbf{A}|||_2 < 1$. Hence, all digital filters with a normal system matrix are stable in the presence of any type of overflow. These filters are called *normal form* or *minimum norm* digital filters. Let us now consider some simple examples.

First-order filters with zero-input behavior described by

$$\mathbf{x}(n+1) = \Psi\{\lambda\mathbf{x}(n)\}$$

are automatically minimum norm since a scalar always satisfies the property of a normal matrix. Thus, the above system is immune to overflow oscillations as long as $|\lambda| < 1$. If a second-order system has distinct eigenvalues, λ_1 and λ_2, then we know that the system matrix can be diagonalized by a similarity transformation. This system can then be implemented as

$$\mathbf{x}(n+1) = \Psi\left\{\begin{bmatrix}\lambda_1 & 0 \\ 0 & \lambda_2\end{bmatrix}\mathbf{x}(n)\right\}.$$

The diagonal matrix is, of course, a normal matrix, and the system is free of overflow oscillations. If a second-order system has complex eigenvalues $\sigma \pm j\omega$, then it can be implemented as a Coupled Form structure as

$$\mathbf{x}(n+1) = \Psi\left\{\begin{bmatrix}\sigma & \omega \\ -\omega & \sigma\end{bmatrix}\mathbf{x}(n)\right\}.$$

It is easy to check that this system matrix is normal. Therefore, Coupled Form structures are also stable in the presence of any type of overflow as long as they are designed to be

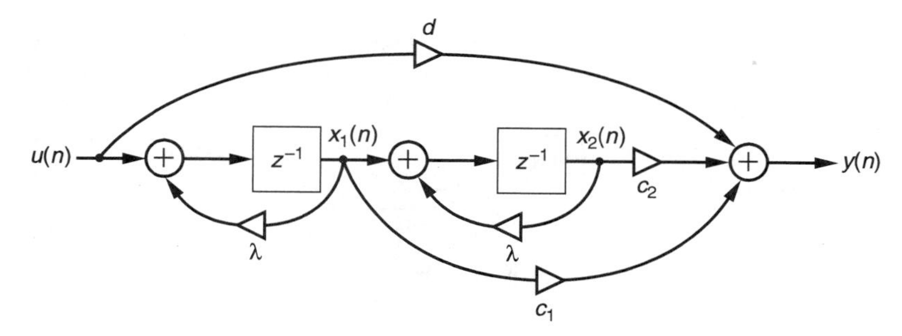

Figure 6.23 Stability realization of second-order filter with double pole at λ.

linearly stable. This is in contrast to Direct Form structures, which are stable only in a subset of the stability triangle. This desirable property of the Coupled Form structure comes at the cost of two extra multipliers. Now consider the case of a second-order filter with a double pole at λ. The system matrix cannot be diagonalized. However, it can be transformed into a Jordan form, so that the implementation becomes

$$\mathbf{x}(n+1) = \Psi \left\{ \begin{bmatrix} \lambda & 0 \\ 1 & \lambda \end{bmatrix} \mathbf{x}(n) \right\}.$$

This matrix is not normal; therefore, this implementation is susceptible to overflow. However, this system can be implemented as a cascade of two first-order systems as shown in Fig. 6.23, which is limit cycle free. In this figure, the state-space representation is given by $\mathbf{B} = [1\ 0]^T$, $\mathbf{C} = [c_1\ c_2]$, and $D = [d]$. Overflow stable filters of higher order can be realized by parallel or cascade structures of first- and second-order minimum norm systems.

The condition of Theorem 6.2 has been shown to be a special case of the one established in [4]. However, the condition of Theorem 6.2 is more suitable for applicability and was therefore presented here. Interested readers are encouraged to study the elegant results of [4].

6.5 SCALING IIR FILTERS TO PREVENT OVERFLOW

In the last section, we saw that errors caused by accumulator overflow can be very large. In fact, among all quantization effects, the error precipitated by overflow is usually the most dominant error. By scaling the input signal appropriately, it is possible to prevent overflow at each of the internal nodes where a sum of products is computed. However, there is a price to pay. Later in this section we will show that scaling has the effect of reducing the quantization SNR.

Now we proceed to derive several scaling rules. These rules depend on the assumption made about the input signal.

Bounded Input

In this case, we assume that the input is bounded such that $|u(n)| \leq 1$, for all n. Let $v_i(n)$ denote the internal nodes of the filter where an overflow can possibly occur. The impulse response of the ith node is denoted by $h_i(n)$, and its transfer function by $H_i(\omega)$. In order to prevent overflow, we must scale the input by a factor $0 < \beta \leq 1$, such that $|v_i(n)| < 1$ for all n and i. Now we can write $v_i(n)$ in terms of its impulse response as

$$v_i(n) = \sum_{m=-\infty}^{\infty} h_i(n)\beta u(n-m). \tag{6.50}$$

Taking the absolute value of both sides of (6.50) gives

$$\begin{aligned} |v_i(n)| &\leq \sum_{m=-\infty}^{\infty} |h_i(n)|\,|\beta u(n-m)| \\ &\leq \beta U_{\max} \sum_{m=-\infty}^{\infty} |h_i(n)|, \end{aligned} \tag{6.51}$$

where $U_{\max} = \max_n |u(n)|$. If the right-hand side of the above inequality is less than unity, then $|v_i(n)| < 1$. Thus, in order to prevent overflow, we must satisfy

$$\beta U_{\max} \sum_{m=-\infty}^{\infty} |h_i(n)| < 1$$

which gives the scaling rule

$$\beta U_{\max} < \frac{1}{\sum_{m=-\infty}^{\infty} |h_i(n)|}, \forall\, i.$$

The denominator in the right-hand side is a definition of the L_1 norm denoted by $||h_i||_1$. The scaling rule is therefore

$$\beta U_{\max} < \frac{1}{\max_i ||h_i||_1}. \tag{6.52}$$

Sinusoidal Input

Let the input be

$$\beta u(n) = \beta U_{\max} \cos(\omega_o n),$$

where β is again the scale factor.

The internal node signals are

$$v_i(n) = \beta U_{\max} |H_i(\omega)| \cos(\omega_o n + \theta_i)$$

where θ_i is the phase introduced by the transfer function $H_i(\omega)$. To prevent overflow, we need $|v_i(n)| < 1, \forall\, i$. This can be satisfied if

$$\beta U_{\max} < \frac{1}{|H_i(\omega)|}, \quad \forall\, i,\ |\omega| \leq \pi.$$

The scaling rule can therefore be written as

$$\beta U_{\max} < \frac{1}{\max\limits_{i, |\omega| \le \pi} |H_i(\omega)|}. \tag{6.53}$$

The L_∞ norm of $H_i(\omega)$ is defined by

$$||H_i||_\infty = \max_{|\omega| \le \pi} |H_i(\omega)|.$$

Using this definition, we find that the scaling rule becomes

$$\beta U_{\max} < \frac{1}{\max\limits_{i} ||H_i||_\infty}. \tag{6.54}$$

General Scaling Rules

The internal node signals can be written as a function of the scaled input $\beta u(n)$ as

$$\begin{aligned} v_i(n) &= F^{-1}\{H_i(\omega)\beta U(\omega)\} \\ &= \frac{\beta}{2\pi}\int_{-\pi}^{\pi} H_i(\omega)U(\omega)e^{j\omega n}d\omega \end{aligned} \tag{6.55}$$

where F^{-1} denotes the inverse DTFT. Therefore,

$$|v_i(n)| \le \frac{\beta}{2\pi}\int_{-\pi}^{\pi} |H_i(\omega)|\ |U(\omega)|\ d\omega. \tag{6.56}$$

Define

$$||H_i||_1 = \frac{1}{2\pi}\int_{-\pi}^{\pi} |H_i(\omega)|\ d\omega,$$

$$||U||_\infty = \max_{|\omega| \le \pi} |U(\omega)|.$$

Then (6.56) becomes

$$|v_i(n)| \le \beta ||U||_\infty\, ||H_i||_1.$$

To prevent overflow, we need $|v_i(n)| < 1$, which in turn requires

$$\beta < \frac{1}{||U||_\infty\, ||H_i||_1}, \forall\, i. \tag{6.57}$$

By reversing the norms in H_i and U, we can get a second condition

$$\beta < \frac{1}{||H_i||_\infty\, ||U||_1}, \forall\, i. \tag{6.58}$$

The scaling rules of (6.57) or (6.58) can be useful, depending on whether $||U||_\infty$ or $||U||_1$ is finite.

Now we derive the scaling rule for a wide-sense stationary signal. Let $S_u(\omega)$ denote the power spectral density of the input signal. In deriving the scaling rule, we will bound the expectation of the signals at the internal nodes; that is, we will bound $E\{v_i^2(n)\} < 1$. We first define the L_2 norm

$$||H_i||_2 = \left[\frac{1}{2\pi} \int_{-\pi}^{\pi} |H_i(\omega)|^2 d\omega \right]^{\frac{1}{2}}, \tag{6.59}$$

where $H_i(\omega)$ is the transfer function from the input to the ith internal node.
It is well known that

$$E\{v_i^2(n)\} = \frac{1}{2\pi} \int_{-\pi}^{\pi} |H_i(\omega)|^2 S_u(\omega) d\omega. \tag{6.60}$$

If the scaling factor is β, then (6.60) becomes

$$\begin{aligned} E\{v_i^2(n)\} &= \frac{\beta^2}{2\pi} \int_{-\pi}^{\pi} |H_i(\omega)|^2 \, S_u(\omega) d\omega \\ &\leq \beta^2 ||S_u||_\infty \, ||H_i||_2. \end{aligned} \tag{6.61}$$

This gives the scaling rule

$$\beta < \frac{1}{||S_u||_\infty \, ||H_i||_2}, \forall \, i. \tag{6.62}$$

Bounding $E\{v_i^2(n)\} < 1$ does not guarantee that $|v_i(n)| < 1$; that is why this is referred to as a probabilistic bound.

In the special case of a white input, $||S_u||_\infty = 1$, the bound becomes

$$\beta < \frac{1}{||H_i||_\infty}, \forall \, i. \tag{6.63}$$

The norms introduced so far satisfy the inequality

$$||H_i||_1 \leq ||H_i||_2 \leq ||H_i||_\infty \leq ||h||_1. \tag{6.64}$$

Thus, the most conservative scale factor is obtained by using $||h||_1$. However, the choice of the scale factor also depends on the type of input.

▶ **EXAMPLE 6.9** *Consider a causal filter:*

$$H(z) = \frac{1 - 0.3z^{-1}}{(1 - 0.2z^{-1})(1 + 0.7z^{-1})}.$$

Find the absolute bound on the input to prevent overflow in the Direct Form II and cascade structures.

SOLUTION This is the same example as in the last section. According to the most conservative scaling rule, we have for $\beta = 1$,

$$U_{\max} < \frac{1}{\max\|h_i\|_1} = \frac{1}{\max(\sum |h_i(n)|)}.$$

For the Direct Form II structure, there are four internal nodes where an overflow can possibly occur, namely, $\lambda_1(n), \lambda_2(n), \lambda_3(n), \lambda_4(n)$ in Fig. 6.16. Since $\lambda_2(n)$ and $\lambda_3(n)$ are only the delayed versions of $\lambda_1(n)$, we need simply check $\lambda_1(n)$ and $\lambda_4(n)$. So,

$$\begin{aligned}
\lambda_1(z) &= \frac{1}{(1-0.2z^{-1})(1+0.7z^{-1})} \\
&= \frac{2/9}{1-0.2z^{-1}} + \frac{7/9}{1+0.7z^{-1}} \\
\lambda_1(n) &= \frac{2}{9}(0.2)^n + \frac{7}{9}(-0.7)^n \\
&\sum_n |\lambda_1(n)| = 2.7778 \\
\lambda_4(z) &= H(z) \\
&= \frac{1-0.3z^{-1}}{(1-0.2z^{-1})(1+0.7z^{-1})} \\
&= \frac{-1/9}{1-0.2z^{-1}} + \frac{10/9}{1+0.7z^{-1}} \\
\lambda_4(n) &= \frac{-1}{9}(0.2)^n + \frac{10}{9}(-0.7)^n \\
&\sum_n |\lambda_4(n)| = 3.6111 \\
U_{\max} &< \frac{1}{\max(2.7778, 3.6111)} \\
&Umax < 0.277.
\end{aligned}$$

For the cascade structure (Fig. 6.17), there are five internal nodes. But $\lambda_2(n)$ is the delayed version of $\lambda_1(n)$, and $\lambda_4(n)$ is the delayed version of $\lambda_3(n)$. So, we need to check $\lambda_1(n), \lambda_3(n)$, and $\lambda_5(n)$ as follows:

$$\begin{aligned}
\lambda_1(z) &= \frac{1}{(1-0.2z^{-1})} \\
\lambda_1(n) &= (0.2)^n \\
\sum_n |\lambda_1(n)| &= 1.25 \\
\lambda_3(z) &= \frac{1}{(1-0.2z^{-1})(1+0.7z^{-1})} \\
\lambda_5(z) &= \frac{1-0.3z^{-1}}{(1-0.2z^{-1})(1+0.7z^{-1})}.
\end{aligned}$$

We have already calculated $\sum_n |\lambda_3(n)| = 2.7778$ and $\sum_n |\lambda_5(n)| = 3.6111$. Therefore

$$U_{\max} < \frac{1}{\max(1.25, 2.7778, 3.6111)}$$
$$= 0.277.$$

Effect on SNR

If the input is scaled down ($\beta < 1$), the quantization SNR (QSNR) at the output of the filter will be reduced. Consider the second-order Direct Form II filter with a scaled input and quantization noise as shown in Fig. 6.24. We assume a double-length accumulator; therefore we have shown only one noise source at the input adder and another one at the output adder. As the figure shows, the noise sources go through the same transfer function as the input signal. However, the input signal is now scaled by a factor $\beta < 1$. Therefore, the signal power is multiplied by β^2, thereby reducing the output SNR.

In an earlier discussion, we showed that, as the poles of a filter approach the unit circle, the QSNR decreases. Also, as the poles approach the unit circle, the system gain increases, which in turn requires the input to be scaled further in order to prevent overflow. Thus, overflow and quantization noise work against each other to decrease the filter QSNR.

6.6 EFFECTS OF PRODUCT QUANTIZATION

In this section, the effects of product quantization will be studied in detail. We will first show that product quantization can cause a filter to exhibit limit cycles, even though this filter is linearly stable. Then we will find regions of stability for each of the three types of quantization, namely, rounding, truncation, and two's complement.

Rounding

Let us begin with a first-order zero-input filter under rounding given by

$$y(n) = Q_R\{ay(n-1)\}. \tag{6.65}$$

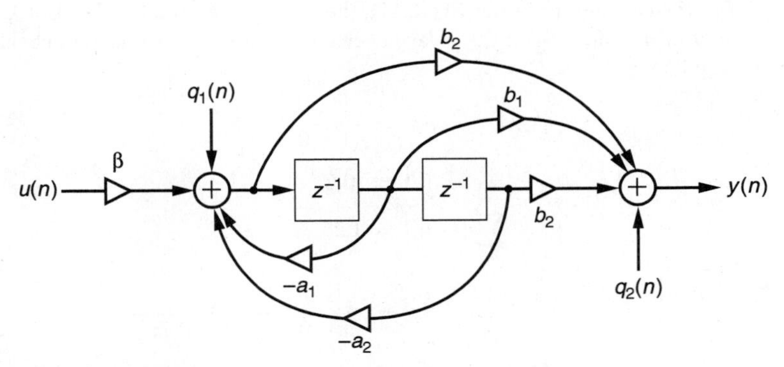

Figure 6.24 Quantization noise sources in a scaled second-order Direct Form II digital filter.

Assume that a 5-bit sign-magnitude representation is used for the implementation of the filter. Let $a(0.5)_{10} = (0.1000)_2$ and $y(0) = (\frac{15}{16})_{10} = (0.1111)_2$. In the absence of the quantizer, the filter output is $y(n) = (0.5)^n y(0)$, which goes to zero as $n \longrightarrow \infty$. Now observe what the quantizer does to the output.

$$y(1) = Q_R\left\{\left(\frac{15}{32}\right)_{10}\right\} = Q_R\{(0.01111)_2\} = (0.1000)_2 = \left(\frac{1}{2}\right)_{10}$$

$$y(2) = Q_R\left\{\left(\frac{1}{2}\cdot\frac{1}{2}\right)_{10}\right\} = Q_R\{(0.0100)_2\} = (0.0100)_2 = \left(\frac{1}{4}\right)_{10}$$

$$y(3) = Q_R\left\{\left(\frac{1}{8}\right)_{10}\right\} = Q_R\{(0.0010)_2\} = (0.0010)_2 = \left(\frac{1}{8}\right)_{10}$$

$$y(4) = Q_R\left\{\left(\frac{1}{16}\right)_{10}\right\} = Q_R\{(0.0001)_2\} = (0.0001)_2 = \left(\frac{1}{16}\right)_{10}$$

$$y(5) = Q_R\left\{\left(\frac{1}{32}\right)_{10}\right\} = Q_R\{(0.00001)_2\} = (0.0001)_2 = \left(\frac{1}{16}\right)_{10}$$

Hence, $y(n) = \frac{1}{16}$ for $n \geq 4$. This filter is said to exhibit an oscillation with a period of one. In other words, we get a period-one limit cycle. It is easy to show that if we chose a negative filter coefficient, then we would get a period-two limit cycle. The range of amplitude values of a limit cycle is referred to as the **dead band**.

In the above example, the quantization step size was $q = 2^{-4}$. We could have performed this analysis independently of q by writing the outputs as simple integers. The assumption is that these integers are implicitly multiplied by q. For example, we had $y(0) = (\frac{15}{16})_{10}$. But now we write $y(0) = 15$. Continuing the analysis, we get

$$\begin{aligned}
y(1) &= Q_R\{(0.5)(15)\} = Q_R\{7.5\} = 8 \\
y(2) &= Q_R\{(0.5)(8)\} = Q_R\{4\} = 4 \\
y(3) &= Q_R\{(0.5)(4)\} = Q_R\{2\} = 2 \\
y(4) &= Q_R\{(0.5)(2)\} = Q_R\{2\} = 1 \\
y(5) &= Q_R\{(0.5)(1)\} = Q_R\{0.5\} = 1
\end{aligned}$$

So, we get $y(n) = 1, n \geq 4$, which really means $y(n) = q, n \geq 4$, and we have the same result as before. Because of this phenomenon, we will perform the rest of the analysis independently of q, that is, with $q = 1$.

Now we proceed to show that large-amplitude limit cycles are possible owing to rounding. This was first shown in [5]. In the first-order filter given by (6.65), limit cycles of period-one and period-two occur when we have the following conditions:

$$y(n) = Q_R\{ay(n-1)\} = \begin{cases} y(n-1), & a > 0, \text{ period-one} \\ -y(n-1), & a < 0, \text{ period-two.} \end{cases} \tag{6.66}$$

The above is equivalent to the pole of the filter being effectively on the unit circle. From the property of the rounding quantizer, we have

$$|Q_R\{ay(n-1)\} - ay(n-1)| \leq \frac{q}{2}. \tag{6.67}$$

Now, equation (6.66) implies

$$|Q_R\{ay(n-1)\}| = |y(n-1)|, \tag{6.68}$$

and equation (6.67) implies

$$|Q_R\{ay(n-1\}| - |ay(n-1)| \leq |Q_R\{ay(n-1)\} - ay(n-1)| \leq \frac{q}{2}. \tag{6.69}$$

Combining (6.68) and (6.69), we get

$$|y(n-1)| - |ay(n-1)| \leq \frac{q}{2}. \tag{6.70}$$

Therefore,

$$|y(n-1) \leq \frac{q/2}{1-|a|}. \tag{6.71}$$

Equation (6.71) gives the dead band for a first-order filter. For $a = \frac{1}{2}$, the dead band is q, and for $a = \frac{3}{4}$, it increases to $2q$. If the sign-magnitude number system is used, the maximum possible value of $|a|$ is $1 - q$ for a stable filter. This gives a dead band of $\frac{1}{2}$, which is a very high-amplitude oscillation.

For second-order filters, computer simulation shows that rounding can lead to limit cycles of many different periods. It can also be shown that these limit cycles can have large amplitudes. Now we find a formula for the dead band. The second-order quantized equation is

$$y(n) = Q\{ay(n-1)\} + Q\{by(n-2)\} \tag{6.72}$$

where we have assumed that the products are rounded before addition or accumulation. This is the case when all the registers, including the accumulator, have the same length. A second-order unquantized filter with zero input is given by

$$v(n) = a_1 v(n-1) + a_2 v(n-2). \tag{6.73}$$

This filter has complex-conjugate poles if $a_1^2 < -4a_1a_2$, and the poles appear at $z = re^{\pm j\theta}$, where $a_2 = -r^2$ and $a_1 = 2rcos\theta$. Obviously, if $a_2 = -1$, the poles are on the unit circle. For limit cycles to occur in the nonlinear system of (6.72), the poles of the equivalent linear system must be on the unit circle. Comparing (6.73) and (6.72), if we want $a_2 = -1$, then in (6.72) we must have

$$Q\{by(n-2)\} = -y(n-2) \tag{6.74}$$

As in the first-order case, we also know that

$$|Q\{by(n-2)\}| - |by(n-2)| \leq |Q\{by(n-2)\} - by(n-2)| \leq \frac{q}{2}. \tag{6.75}$$

Combining (6.74) and (6.75),

$$|y(n-2)| - |b||y(n-2)| \leq \frac{q}{2}$$

which gives the bound for the dead band as

$$|y(n-2)| \leq \frac{q/2}{1-|b|}. \tag{6.76}$$

The dead band depends only on b, whereas a controls the period of the limit cycle. This was the case of limit cycles with the equivalent complex-conjugate poles on the unit circle. The above analysis can also be derived with the equivalent poles at $z = \pm 1$. This would give a dead band bound of (Problem 23)

$$|y(n-2)| \leq \frac{q}{1-|a|-b}. \tag{6.77}$$

A similar analysis can be made for filters with truncation. For higher order digital filters implemented as parallel or cascade of first- and second-order sections, our analysis holds for each section. For parallel structures, the output is simply a sum of the outputs of each section. Therefore, all the limit cycles would simply add up to create the final output. However, in cascade structures, the limit cycles get filtered by the downstream sections. The output is therefore a function of the limit cycles and the transfer function of each section. Cascade structures are therefore much more difficult to analyze for limit cycles. In any case, each second-order section must be designed to be free of limit cycles.

Stability Regions for Rounding

In this section, a deterministic analysis is presented for the stability of state-space digital filters with rounding type of quantization. A zero-input statespace filter with rounding can be described as

$$\mathbf{x}(n+1) = Q_R\{\mathbf{A}\mathbf{x}(n)\} \tag{6.78}$$

where $Q_R\{.\}$ represents the rounding operator. The quantizer is assumed to act on each element of the vector independently of the others. That is, for a vector $\mathbf{v}$, $Q_R\{\mathbf{v}\} = [Q_R\{v_1\}\ Q_R\{v_2\} \ldots Q_R\{v_N\}]^T$. Also note that in (6.78), the quantizer acts on the sum of products and not on each product. This implies the use of a double-length accumulator. If the registers are $(B+1)$-bit long, then the accumulator has $(2B+1)$ bits. Also, we assume a $(2B+1)$-bit product register where the intermediate products can be stored. Therefore, the products do not have to be quantized until after accumulation. From the characteristic of the roundoff quantizer, we can write

$$|Q_R\{v_i\}| \leq 2|v_i|, \forall\, i. \tag{6.79}$$

The above implies that

$$||Q_R\{\mathbf{v}\}||_p \leq 2||\mathbf{v}||_p,$$

for any p-norm.

Taking the p-norm norm of both sides of (6.78), we get

$$\begin{aligned} ||\mathbf{x}(n+1)||_p &= ||Q_R\{\mathbf{A}\mathbf{x}(n)\}||_p \\ &\leq 2||\mathbf{A}\mathbf{x}(n)||_p \\ &\leq 2|||\mathbf{A}|||_p\ ||\mathbf{x}(n)||_p \end{aligned} \tag{6.80}$$

where $|||\cdot|||_p$ represents the matrix norm induced by the vector p-norm. Using recursion on (6.80) gives

$$||\mathbf{x}(n)||_p \leq 2^n\ |||\mathbf{A}|||_p^n\ ||\mathbf{x}(0)||_p. \tag{6.81}$$

If $|||\mathbf{A}|||_p < \frac{1}{2}$, then $||\mathbf{x}(n)||_p \longrightarrow 0$ as $n \longrightarrow \infty$, and we have the following theorem.

Theorem 6.3 *Consider a zero-input state-space filter*

$$\mathbf{x}(n+1) = Q_R\{\mathbf{A}\mathbf{x}(n)\}$$

where $Q_R\{.\}$ represents the rounding quantization. If $|||\mathbf{A}|||_p < \frac{1}{2}$ for any matrix p-norm, then the system is globally asymptotically stable.

Theorem 6.3 gives a sufficient condition for a limit cycle-free digital filter. A second sufficient condition will now be derived. Let $|\mathbf{A}| = \{|a_{ij}|\}$, and for a vector, let $|\mathbf{v}| = [|v_i|]$. Then using (6.78) and (6.79),

$$\begin{aligned} |\mathbf{x}(n+1)| &= |Q_R\{\mathbf{A}\mathbf{x}(n)\}| \\ &= |[Q_R\{a_{i1}x_1(n) + \cdots + a_{ip}x_p(n)\}]| \\ &\leq 2|[a_{i1}x_1(n) + \cdots + a_{ip}x_p(n)]| \\ &\leq 2[|a_{i1}||x_1(n)| + \cdots + |a_{ip}|\,|x_p(n)|] \\ &\leq 2|\mathbf{A}|\,|\mathbf{x}(n)|. \end{aligned} \tag{6.82}$$

Recursing on (6.82) gives

$$|\mathbf{x}(n)| \leq 2^n|\mathbf{A}|^n|\mathbf{x}(0)|. \tag{6.83}$$

Thus, if $\rho(|\mathbf{A}|) < \frac{1}{2}$, where $\rho(\cdot)$ denotes the spectral radius of a matrix, then $|\mathbf{x}(n)| \longrightarrow 0$ as $n \longrightarrow \infty$, and the following theorem is established.

Theorem 6.4 *Consider a zero-input state-space filter*

$$\mathbf{x}(n+1) = Q_R\{\mathbf{A}\mathbf{x}(n)\}$$

where $Q_R\{\cdot\}$ represents the rounding quantization. If $\rho(|\mathbf{A}|) < \frac{1}{2}$, then the filter is globally asymptotically stable.

Since $\rho(\mathbf{A}) \leq |||\mathbf{A}|||_p$ and $\rho(\mathbf{A}) \leq \rho(|\mathbf{A}|)$, if the system matrix satisfies either Theorem 6.3 or Theorem 6.4, then $\rho(\mathbf{A}) < \frac{1}{2}$ and the system is also linearly stable. The conditions of these two theorems are not equivalent. In fact, it can be shown with examples (Problem 24) that, in general, these conditions are mutually exclusive; that is, one is not a special case of the other. Now we consider the significance of the above results to some filter structures. As seen earlier, a minimum norm filter is one that satisfies the condition

$$|||\mathbf{A}|||_2 = \rho(\mathbf{A}).$$

We also know that the above is true for any normal matrix $\mathbf{A}$. Thus, all minimum norm filter structures will be roundoff stable if the poles are restricted within the "one-half circle" as shown in Fig. 6.25. This also holds for a second-order Coupled Form structure, which has a normal system matrix given by

$$\mathbf{A} = \begin{bmatrix} \sigma & \omega \\ -\omega & \sigma \end{bmatrix},$$

where the system poles are at $\sigma \pm j\omega$. For a Direct Form filter, the system matrix is

$$\mathbf{A} = \begin{bmatrix} 0 & 1 \\ b & a \end{bmatrix}.$$

The condition of Theorem 6.3 ($|||\mathbf{A}|||_p < \frac{1}{2}$) is not satisfied for the commonly used norms $p = 1, 2, \infty$. The norm definitions are as follows. For $\mathbf{A} = [a_{ij}]$, the infinite-norm is given by

$$|||\mathbf{A}|||_\infty = \max_i \left\{ \sum_{j=1}^{p} |a_{ij}| \right\},$$

which is the maximum absolute row sum. The 1-norm is defined as

$$|||\mathbf{A}|||_1 = \max_j \left\{ \sum_{i=1}^{p} |a_{ij}| \right\},$$

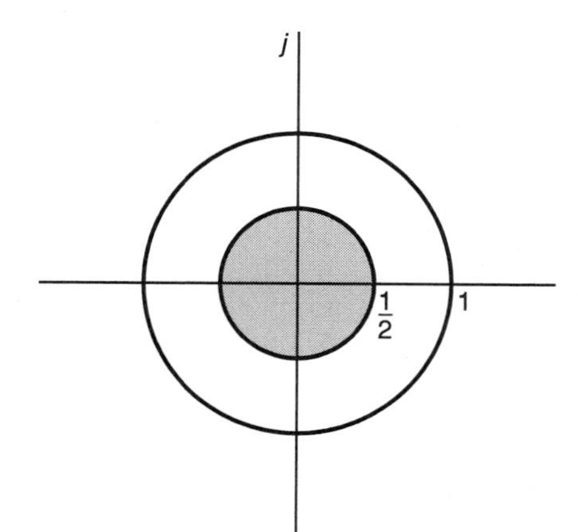

Figure 6.25 Roundoff stable region for a minimum norm filter.

which is the maximum absolute column sum. Since the condition of Theorem 6.3 does not hold for a Direct Form structure, we apply Theorem 6.4 ($\rho(|\mathbf{A}|) < \frac{1}{2}$. The resulting region of stability is shown in Fig. 6.26. The details are left to Problem 25.

These results on the stability of filters under roundoff quantization were established in [6]. Limit cycles of various amplitudes and periods have been shown to exist outside the stability region derived here. However, in some areas outside these stability regions, no limit cycle has been found in computer simulations. But the only regions guaranteed to be free of limit cycles are those shown in Figs. 6.25 and 6.26. These regions are quite restrictive. In the next subsection we show that magnitude truncation leads to a larger stability region.

Magnitude Truncation

First, consider the zero-input state-space system

$$\mathbf{x}(n+1) = Q_T\{\mathbf{A}\mathbf{x}(n)\} \tag{6.84}$$

where $Q_T\{\cdot\}$ represents magnitude truncation quantization. For a vector $\mathbf{v} \doteq [v_1, \ldots, u_N]^T$, this nonlinearity satisfies

$$|Q_T\{\mathbf{v}\}| \leq |\mathbf{v}|. \tag{6.85}$$

Since this is the same condition as for the overflow nonlinearity, if $|||\mathbf{A}|||_2 < 1$, then the system (6.84) is asymptotically stable. That is, Theorem 6.2 is also applicable to magnitude truncation. In fact, a more general condition exists for both of these nonlinearities, which is presented below without proof [11].

Theorem 6.5 *Consider the zero-input state-space filter*

$$\mathbf{x}(n+1) = \Phi\{\mathbf{A}\mathbf{x}(n)\}$$

where $\Phi\{\cdot\}$ represents magnitude truncation, any type of overflow, or a combination of magnitude truncation and overflow. If

$$\mathbf{A}^T\mathbf{D}\mathbf{A} \leq \mathbf{D} \tag{6.86}$$

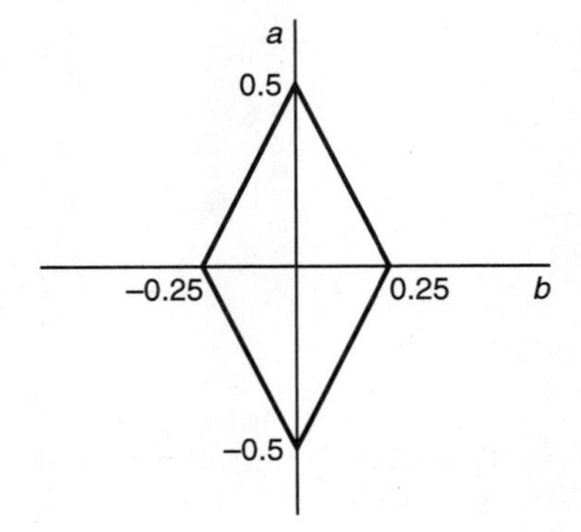

Figure 6.26 Roundoff stable region for a Direct Form filter.

for some diagonal matrix $\mathbf{D}$, *where* T *denotes transpose, then the system is free of limit cycles.*

Notice that for the special case of $\mathbf{D} = \mathbf{I}$, equation (6.86) reduces to the condition

$$\mathbf{A}^T\mathbf{A} \leq \mathbf{I}. \tag{6.87}$$

It is not difficult to prove (Problem 17) that

$$\mathbf{A}^T\mathbf{A} \leq \mathbf{I} \Longleftrightarrow |||\mathbf{A}|||_2 \leq 1. \tag{6.88}$$

Thus, condition (6.88) is sufficient for the stability of a zero-input filter with overflow and/or magnitude truncation. The condition (6.88) is an improvement over the previously established condition $|||\mathbf{A}|||_2 < 1$, where strict inequality was used.

For a normal form digital filter ($\mathbf{A}\mathbf{A}^T = \mathbf{A}^T\mathbf{A}$), we know that $|||\mathbf{A}|||_2 = \rho(\mathbf{A})$. Therefore, if the filter is designed to be linearly stable, that is, if $\rho(\mathbf{A}) < 1$, then $|||\mathbf{A}|||_2 < 1$ and the filter are free from limit cycles owing to overflow and/or magnitude truncation. In this case, the improvement in (6.88) over strict inequality did not make any difference. However, the improved condition in (6.88) explains the absence of limit cycles in Gray-Markel lattice structures. This was shown in [11]. The analysis is beyond the scope of this text and is therefore omitted.

Now, we present another sufficient condition for stability under magnitude truncation. This condition is similar to that of Theorem 6.4 for rounding but is less stringent. The proof follows in the same way and is omitted.

Theorem 6.6 *Consider the zero-input state-space filter*

$$\mathbf{x}(n+1) = \Phi\{\mathbf{A}\mathbf{x}(n)\}$$

where $\Phi\{\cdot\}$ *represents magnitude truncation, any type of overflow, or a combination of magnitude truncation and overflow. If*

$$\rho(|\mathbf{A}|) < 1,$$

then the filter is globally asymptotically stable.

Note that Theorem 6.6 works for magnitude truncation, overflow, or a combination because for all these cases the inequality (6.85) holds, which is the key in the proof. Now, we consider the simple second-order filters like the Coupled Form and Direct Form. A Coupled Form filter with eigen-values $\sigma \pm j\omega$ has the system matrix $\mathbf{A} = \begin{bmatrix} \sigma & \omega \\ -\omega & \sigma \end{bmatrix}$, which is a normal matrix. Thus, $|||\mathbf{A}|||_2 = \rho(\mathbf{A})$. So, if the filter is designed to be linearly stable, then $|||\mathbf{A}|||_2 < 1$ and the filter is also stable under magnitude truncation. The Coupled Form filter is therefore stable anywhere within the unit circle, as shown in Fig 6.27(a). For a Direct Form filter, we use Theorem 6.6, and the stability condition is $\rho(|\mathbf{A}|) < 1$, where

$$|\mathbf{A}| = \begin{bmatrix} 0 & 1 \\ |b| & |a| \end{bmatrix}.$$

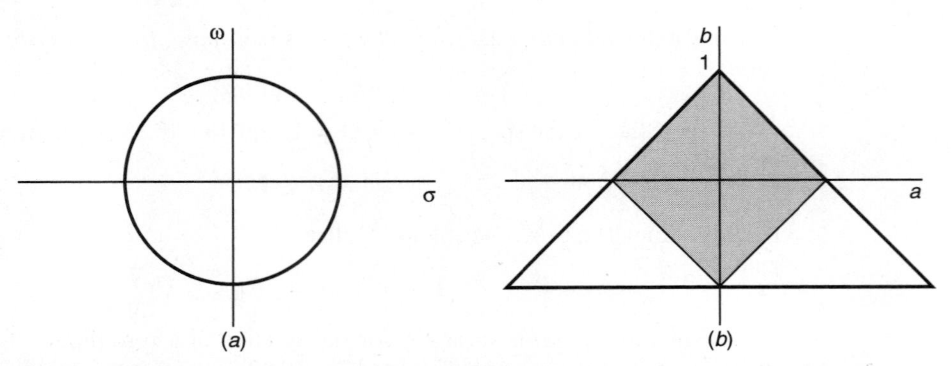

Figure 6.27 Stability region under magnitude truncation. (*a*) Coupled-Form filter, (*b*) Direct Form filter.

It is easy to show that the above condition reduces to $|a| + |b| < 1$, which is the stability region shown in Fig. 6.27(*b*). The stability regions for magnitude truncation are therefore larger than those for rounding.

Two's Complement Quantization

This section begins with the analysis of a quantized state-space model and then considers Direct Form and Coupled Form second-order structures. Again, the goal is to find conditions for the suppression of zero-input limit cycles. The state-space model of a zero-input filter under quantization is given by

$$\mathbf{x}(n+1) = Q_2\{\mathbf{Ax}(n)\}$$

when $Q_2\{.\}$ is the two's complement quantization. This assumes a double-length accumulator since the quantizer operates on a sum of products. As in the case of the roundoff quantizer, the two's complement quantizer is also assumed to operate on each element of a vector independently of the others; that is, for a vector $\mathbf{v} \in R^{N\times 1}$, $Q_2\{\mathbf{v}\} = [Q_2\{v_1\}, Q_2\{v_2\}, \ldots, Q_2\{v_N\}]^T$. From the characteristic of the quantizer, we can write for any x,

$$0 \leq x - Q_2\{x\} < q.$$

Without loss of generality, it will be assumed that the quantization step size is unity; that is, the quantized numbers are all integers. However, as explained earlier, the quantized values are integer multiples of q. For a unit step size, the quantizer satisfies the inequality

$$0 \leq x - Q_2\{x\} < 1.$$

Notice that for positive integers S and R, if

$$-S < x < R,$$

then

$$-S \leq Q_2\{x\} \leq R - 1. \tag{6.89}$$

This is the basis for the following lemma, which will be used in a subsequent theorem.

Lemma 6.3 *Let* $\mathbf{A}=[a_{ij}] \in R^{N\times N}$ *such that* $a_{ij} \leq 0\ \forall i, j$ and $|||\mathbf{A}|||_\infty < 1$. *Let* $\mathbf{W} = [w_j] \in R^{N\times 1}$ *be an integer vector with the largest nonnegative element S and smallest nonpositive element* $-R$. *Then*

$$-S\tau \leq Q_2\{\mathbf{AW}\} \leq (R-1)\tau$$

where $\tau = [1\,1 \ldots 1]^T$.

Proof. Define

$$\mathbf{Y} = \mathbf{AW} = \begin{bmatrix} a_{11}w_1 + \cdots a_{1N}w_N \\ \vdots \\ a_{N1}w_1 + \cdots a_{NN}w_N \end{bmatrix}.$$

Since $a_{ij} \leq 0\ \forall i,j$, we can write

$$\mathbf{Y} = -\begin{bmatrix} |a_{11}|w_1 + \cdots + |a_{1N}|w_N \\ \vdots \\ |a_{N1}|w_1 + \cdots + |a_{NN}|w_N \end{bmatrix}.$$

Now we consider the ith element of $-\mathbf{Y}$. Using the definition of S and $-R$ in the hypothesis, we get

$$-(|a_{i1}| + \cdots + |a_{iN}|)R \leq -y_i \leq (|a_{i1}| + \cdots + |a_{iN}|)S. \tag{6.90}$$

Since $|||\mathbf{A}|||_\infty < 1$, (6.90) gives

$$-R < -y_i < S,$$

which implies

$$-S < y_i < R.$$

Using the property of the quantization in (6.89), we find that the above gives

$$-S \leq Q_2\{y_i\} \leq R-1,$$

and the conclusion follows. ∎

Using the foregoing lemma, we can now establish the following theorem. It gives a sufficient condition for the asymptotic stability of a state-space filter under two's complement quantization.

Theorem 6.7 *Consider the zero-input state-space equation of a 1-D digital filter implemented with two's complement quantization as*

$$\mathbf{x}(n+1) = Q_2\{\mathbf{Ax}(n)\},\ \mathbf{x}(n_0)$$

where $\mathbf{x} \in R^{N\times 1}$, $\mathbf{A} \in R^{N\times N}$, *and* $\mathbf{x}(n_0)$ *is an integer representing the initial condition. If* $\mathbf{A}$ *is nonpositive and* $|||\mathbf{A}|||_\infty < 1$, *then the system is asymptotically stable.*

Proof. Let $m = ||x(n_0)||_\infty$. Since $\mathbf{A}$ is nonpositive with $|||\mathbf{A}|||_\infty < 1$ from Lemma 6.3, we have $-m\tau \leq \mathbf{x}(n_0+1) \leq (m-1)\tau$, where τ is the unit vector in $R^{N\times 1}$. Since $\mathbf{x}(n+2) = Q_2\{\mathbf{Ax}(n+1)\}$, applying Lemma 6.3 again gives

$$-(m-1)\tau \leq \mathbf{x}(n_0+2) \leq (m-1)\tau.$$

Repeating the above process leads to the following set of inequalities:

$$-(m-1)\tau \leq \mathbf{x}(n_0+3) \leq (m-2)\tau,$$

$$-(m-2)\tau \leq \mathbf{x}(n_0+4) \leq (m-2)\tau,$$

$$-(m-2)\tau \leq \mathbf{x}(n_0+5) \leq (m-3)\tau,$$

$$-(m-3)\tau \leq \mathbf{x}(n_0+6) \leq (m-3)\tau, \ldots$$

Hence, $\mathbf{x}(n) = 0$ for $k \geqslant k_0 + 2m$, and the conclusion follows. ■

Theorem 6.7 is important in its own right for checking the stability in state-space filters. However, it is easy to see that the conditions of this theorem are not satisfied for Direct Form and Coupled Form filters. Therefore, a special analysis must be performed for each of these cases for finding the regions of stability. However, before we do that, we present Theorem 6.8 for a first-order filter.

Theorem 6.8 *The first-order system*

$$x(n+1) = Q_2\{\alpha x(n)\}$$

is asymptotically stable if and only if $-1 < \alpha < 0$.

The sufficiency part in Theorem 6.8 follows from a direct application of Theorem 6.7. The necessary condition can be established by showing that if $\alpha > 0$, an initial condition can always be found so that a period-one limit cycle is obtained. The details are left as an exercise.

Direct Form

For a Direct Form second-order filter with two's complement quantization, the equations with zero input are

$$\begin{aligned} x_1(n+1) &= x_2(n) \\ x_2(n+1) &= Q_2\{bx_1(n) + ax_2(n)\}. \end{aligned} \tag{6.91}$$

In [7] and [8], some conditions were found for the existence of limit cycles of period-one and period-two within the stability triangle (Fig. 6.28). It has been shown that in regions 1 and 2, initial conditions can always be found, so that period-two limit cycles exist. In region 3, period-two limit cycles may be present. It has also been shown that $a+b>0$ is a necessary and sufficient condition for the existence of period-one limit cycles (Fig. 6.28(b)). With two's complement quantization, large-amplitude limit cycles

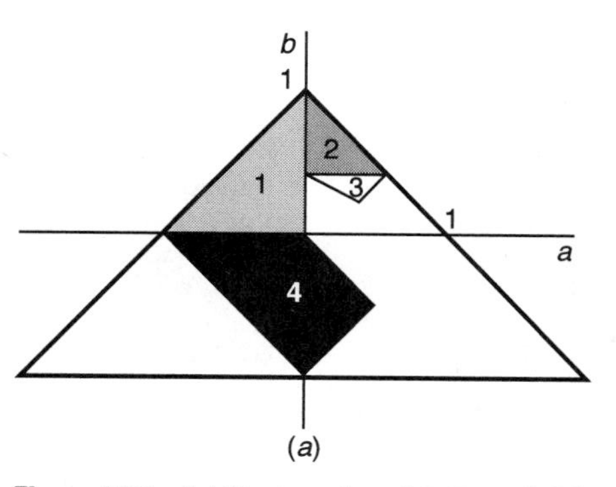

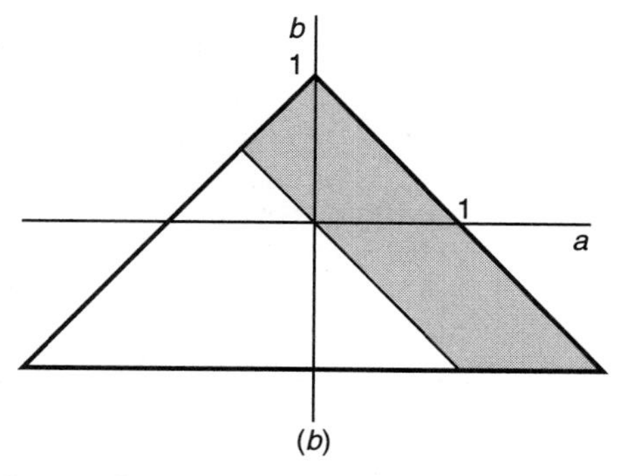

Figure 6.28 (*a*) Regions 1 and 2: Period-2 limit cycles can always be obtained; Region 3: Period-2 limit cycles may be present depending on initial conditions; Region 4: Region of global asymptotic stability; (*b*) Shaded region is the necessary and sufficient condition for the existence of period-1 limit cycles.

TABLE 6.1 Examples of Limit Cycles in a Direct Form Digital Filter

	a	b	$x(0)$	Period
(1)	−0.19	−0.999	$\begin{bmatrix} 1 \\ 5 \end{bmatrix}$	62
(2)	−1.8	−0.99	$\begin{bmatrix} -10 \\ 17 \end{bmatrix}$	7
(3)	−1.48	−0.9	$\begin{bmatrix} -3 \\ -2 \end{bmatrix}$	5
(4)	−1.48	−0.99	$\begin{bmatrix} 4 \\ -2 \end{bmatrix}$	8

can be exhibited. Some examples are given in Table 6.1. Thus, it is important to find a region in the parameter space where the filter is stable. This was done in [8], and the stability region is shown in Fig. 6.28(*a*). The derivation is not trivial and is omitted.

Clearly, second-order Direct Form structures are highly susceptible to large-amplitude limit cycles. In addition, the stability region is quite small, as shown in Fig. 6.28(*a*). This restricts the class of filters that can be designed. For this reason, Direct Form filters should be avoided for implementation with two's complement quantization.

Coupled Form Filters

These realizations are attractive because they are free of limit cycles owing to overflow and MT quantization as long as they are linearly stable. However, these filters are still susceptible to limit cycles due to two's complement quantization (TCQ). The following theorem, established in [8], gives the stability region for a Coupled Form filter under two's complement quantization.

Theorem 6.9 *Consider a second-order Coupled Form digital filter*

$$\mathbf{x}(n+1) = Q\{\mathbf{A}\mathbf{x}(n)\}$$

where

$$\mathbf{A} = \begin{bmatrix} \sigma & \omega \\ -\omega & \sigma \end{bmatrix}, \quad \mathbf{x}(n) = \begin{bmatrix} x_1(n) \\ x_2(n) \end{bmatrix},$$

and $Q\{.\}$ *represents two's complement quantization. If* $|\sigma| + |\omega| < 1$ *and* $\sigma < 0$, *then* $x(n) \longrightarrow 0$, *as* $n \longrightarrow \infty$.

The proof of this theorem is lengthy and nontrivial. Interested readers are referred to [8]. The stability region given by this theorem is shown Fig. 6.29. The region is quite restrictive in the class of filters that can be designed. In [9] and [10], the stability region has been extended by finding some bounds on the limit cycles and then using an exhaustive search. The extended region is given in Fig. 6.30. Now we use a block form realization to further extend the stability region. Consider a digital filter with state and output equations

$$\begin{bmatrix} \mathbf{x}(k+1) \\ y(k) \end{bmatrix} = \begin{bmatrix} \mathbf{A} & \mathbf{B} \\ \mathbf{C} & D \end{bmatrix} \begin{bmatrix} \mathbf{x}(k) \\ u(k) \end{bmatrix} \tag{6.92}$$

where $u(k)$, $y(k)$ are scalars representing the input and output, respectively. Instead of processing a signal input $u(k)$ to obtain signal output $y(k)$, we process the input sequence in blocks of length L. That is, the state vector is updated every L samples by using input and output buffers. Writing out (6.92) for $k+2, k+3, \ldots, k+L$ gives the following (Problem 18):

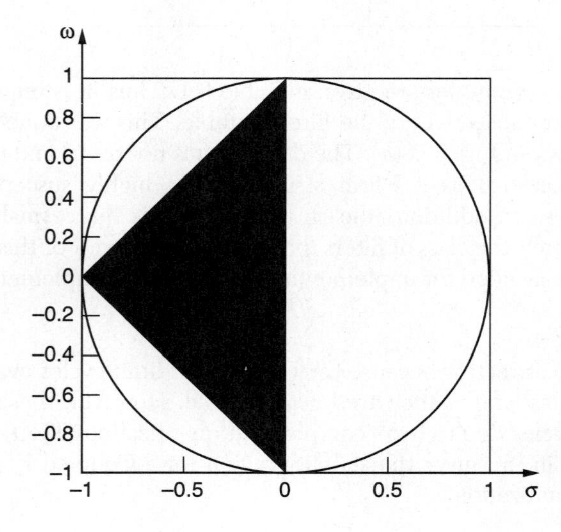

Figure 6.29 Stability region for a Coupled Form filter with TCQ given by sufficient condition.

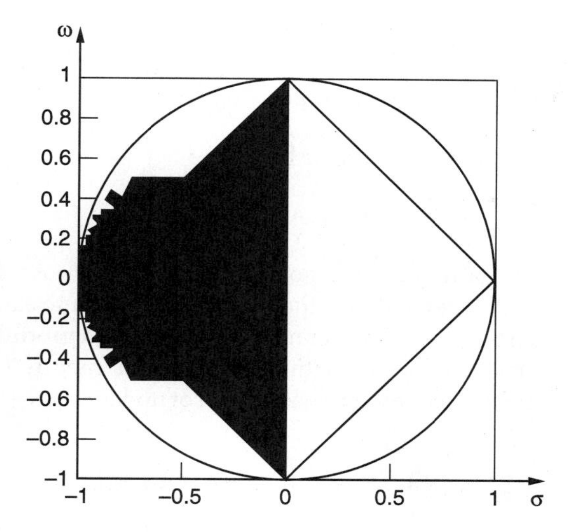

Figure 6.30 Stability region for a Coupled Form filter with TCQ, obtained by an exhaustive search.

$$
\begin{bmatrix} \mathbf{x}'(k+L) \\ y(k) \\ y(k+1) \\ \vdots \\ y(k+L-1) \end{bmatrix} = \begin{bmatrix} \mathbf{A}' & \mathbf{B}' \\ \mathbf{C}' & \mathbf{D}' \end{bmatrix} \begin{bmatrix} \mathbf{x}'(k) \\ u(k) \\ u(k+1) \\ \vdots \\ u(k+L-1) \end{bmatrix} \tag{6.93}
$$

where

$$
\begin{aligned}
\mathbf{A}' &= \mathbf{A}^L; \\
\mathbf{B}' &= [\mathbf{A}^{L-1}\mathbf{B}, \mathbf{A}^{L-2}\mathbf{B}, \ldots, \mathbf{B}]; \\
\mathbf{C}' &= [\mathbf{C}, \; \mathbf{CA}, \ldots, \mathbf{CA}^{L-1}]^T;
\end{aligned}
$$

and

$$
\mathbf{D}' = \begin{bmatrix} \mathbf{D} & \mathbf{0} & \ldots & \mathbf{0} \\ \mathbf{CB} & \mathbf{D} & \ldots & \mathbf{0} \\ & \vdots & & \\ \mathbf{CA}^{L-2}\mathbf{B} & \mathbf{CA}^{L-3}\mathbf{B} & \ldots & \mathbf{D} \end{bmatrix}.
$$

The above is the well-known block form structure. We now present a new block structure that extends the stability region.

Let $L = 2$ and $\mathbf{A}$ be a second-order normal form realization; that is

$$
\mathbf{A} = \begin{bmatrix} \sigma & \omega \\ -\omega & \sigma \end{bmatrix}.
$$

Define a diagonal matrix **E** such that

$$\mathbf{E}_{ii} = -sgn\{\mathbf{A}_{ii}\} = -sgn\{\sigma\}$$

where $sgn\{\cdot\}$ denotes the signum operation. Also define

$$\mathbf{A}\_ = \mathbf{AE}.$$

It is easy to see that $\mathbf{A}\_$ is negative in the diagonal and that $\mathbf{A}\_\mathbf{A}\_ = \mathbf{A}^2$. Please note that the above also holds for first-order systems, that is, with **A** being a scalar.

Let us now realize a first- or second-order normal form filter in **modified block form** as shown in Fig. 6.31, where the block length is $L = 2$. This structure is essentially the same as a standard block realization, except that instead of updating $\mathbf{x}(k+2) = Q\{\mathbf{A}^2\mathbf{x}(k)\}$, we perform

$$\mathbf{x}(k+2) = Q\{\mathbf{A}\_\mathbf{x}(k+1)\} = Q\{\mathbf{A}\_Q\{\mathbf{A}\_\mathbf{x}(k)\}\}. \tag{6.94}$$

Note that $\mathbf{x}(k+2)$ in (6.94) represents the samples of a state-space system in normal form realization with system matrix $\mathbf{A}\_$. Since $\mathbf{A}\_$ has negative diagonal entries, the system is stable based on Theorem 6.9. Therefore, the realization of Fig. 6.31 is globally asymptotically stable. We have proved the following theorems.

Theorem 6.10 *Consider a first-order digital filter with system matrix* $\mathbf{A} = \lambda$. *If* $|\lambda| < 1$, *then the block filter realized, as in Fig. 6.31 with two's complement quantization, is globally asymptotically stable.*

Theorem 6.11 *Consider a second-order normal form digital filter with system matrix*

$$\mathbf{A} = \begin{bmatrix} \sigma & \omega \\ -\omega & \sigma \end{bmatrix}$$

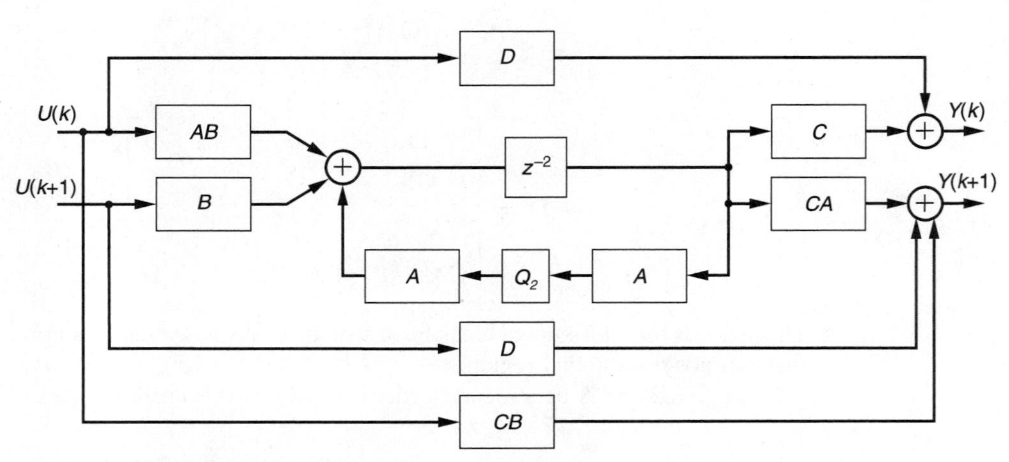

Figure 6.31 Modified block structure.

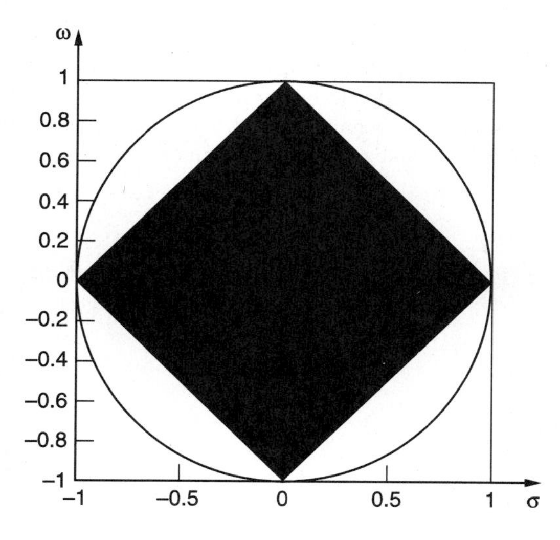

Figure 6.32 Stability region for a Coupled Form filter with TCQ implementation in a modified block form with a block size of 2.

implemented in block form as in Fig. 6.31 with two's complement quantization. If $|\sigma| + |\omega| < 1$, *then the filter is globally asymptotically stable.*

The results of Theorems 6.10 and 6.11 are significant because, as illustrated in Fig. 6.32, the stability region has been extended to the right-hand side of parameter plane. That is, the region has been doubled with a block length of only 2. The stability region can actually be extended to the entire unit circle by increasing the block length. Since the eigenvalues of the block system matrix are $\lambda' = \lambda^L$, it is possible to move the eigenvalues further inside by increasing L. A suitable search algorithm for finding the minimum block length is given in [12]. This algorithm guarantees an implementation with a minimal block size.

6.7 SUMMARY

In this chapter, we have considered many aspects of finite wordlength in the implementation of digital filters. We began with a presentation of fixed point and floating point numbers. The relative advantages and disadvantages of these representations were discussed. The rest of the chapter dealt with the effects of fixed point arithmetic in digital filters.

Overflow effects in digital filters have been discussed in detail, and the different types of overflow characteristics have been presented. It was shown that large-amplitude limit cycles can result from overflow. The stability regions in the parameter space were derived for various filter structures under overflow. The concept of scaling the input in order to prevent overflow has been described. Scaling rules have been derived for different types of input signals. It has also been shown that scaling results in a decrease in the signal to noise ratio due to quantization.

The topic of product quantization in digital filter implementations has also been considered. The common types of quantization schemes, namely, rounding, magnitude truncation, and two's complement quantization, have been studied in detail. The popular filter structures have been considered under each of these quantization schemes, and stability regions have been established. It has been shown that the Coupled Form filter structure has the largest stability region under the nonlinear effects of both overflow and product quantization.

The effects of finite wordlength can also be studied in a stochastic approach. This has been done for first-order and second-order filters, and simple formulas have been derived for the quantization SNR. These formulas can be easily extended to higher order systems. Coefficient quantization in both IIR and FIR digital filters have also been presented. Techniques have been derived in order to minimize the effect of coefficient quantization on the frequency response of the filter.

6.8 APPENDIX

Proof of Theorem 6.1: The output $f(n)$ is given by the convolution sum

$$f(n) = \sum_{k=-\infty}^{\infty} h(k)e(n-k).$$

The mean of the output is

$$\begin{aligned}\mu_f &= E\{f(n)\} = \sum_{k=-\infty}^{\infty} h(k)E\{e(n-k)\} \\ &= \mu \sum_{k=-\infty}^{\infty} h(k) = \mu H(0).\end{aligned}$$

The autocorrelation of the input white noise sequence is

$$\begin{aligned}r_w(m) &= E\{e(n)e(n-m)\} \\ &= \begin{cases}\mu^2, & m \neq 0 \\ \sigma^2 + \mu^2, & m = 0\end{cases} \\ &= \sigma^2\delta(m) + \mu^2\end{aligned}$$

The autocorrelation of the output is

$$\begin{aligned}E\{f(n)f(n-m)\} &= E\left\{\sum_{k=-\infty}^{\infty} h(k)e(n-k) \sum_{r=-\infty}^{\infty} h(r)e(n-m-r)\right\} \\ &= \sum_{k=-\infty}^{\infty} h(k) \sum_{r=-\infty}^{\infty} h(r)E\{e(n-k)e(n-m-r)\}.\end{aligned}$$

Since $w(k)$ is wide-sense stationary, we have $E\{e(n-k)e(n-m-r)\} = r_e(m+r-k)$. Then

$$E\{f(n)f(n-m)\} = \sum_{k=-\infty}^{\infty} h(k) \sum_{r=-\infty}^{\infty} h(r)r_e(m+r-k).$$

The above is dependent only on m and not on n. Therefore, the output is wide-sense stationary, and its autocorrelation can be written as

$$r_f(m) = \sum_{k=-\infty}^{\infty} h(k) \sum_{r=-\infty}^{\infty} h(r)r_e(m+r-k).$$

Let $l = k - r$. Then

$$r_f(m) = \sum_{l=-\infty}^{\infty} r_e(m-l) \sum_{r=-\infty}^{\infty} h(r)h(r+l)$$
$$= \sum_{l=-\infty}^{\infty} r_e(m-l)c(l)$$
$$= r_e(m) * c(m),$$

where $c(m) = h(m) * h(-m)$. Substituting for $r_e(m)$, the above becomes

$$r_f(m) = (\sigma_e^2 \delta(m) + \mu_e^2) * c(m)$$
$$= \sigma_e^2 c(m) + \mu_e^2 * c(m)$$
$$= \sigma_e^2 h(m) * h(-m) + \mu_e^2 * h(m) * h(-m).$$

The Fourier transform of the above is

$$R_f(\omega) = \sigma_e^2 |H(\omega)|^2 + \mu_e^2 \delta(\omega)|H(\omega)|^2$$
$$= \sigma_e^2 |H(\omega)|^2 + \mu_e^2 |H(0)|^2 \delta(\omega).$$

Taking the inverse Fourier transform gives

$$r_f(m) = \sigma_e^2 c(m) + \mu_e^2 |H(0)|^2.$$

For $m = 0$, we have the mean square value

$$r_f(0) = \sigma_e^2 c(0) + \mu_e^2 |H(0)|^2 = \sigma_f^2 + \mu_f^2.$$

Noting that $\mu_f^2 = \mu_e^2 |H(0)|^2$, and

$$c(0) = h(m) * h(-m)|_{m=0} = \sum_{n=-\infty}^{\infty} h(n)h(n+m)\Big|_{m=0} = \sum_{n=-\infty}^{\infty} |h(n)|^2,$$

we get

$$\sigma_f^2 = \sigma_e^2 \sum_{n=-\infty}^{\infty} |h(n)|^2$$

which completes the proof.

6.9 MATLAB

zp2tf: zero-pole to transfer function conversion. Given a set of zero locations in vector **Z**, a set of pole locations in vector **P**, and a gain in scalar K, the function [**num,den**] = zp2tf(**Z,P,K**) forms the transfer function $H(s)$ $\frac{\text{num(s)}}{\text{den(s)}}$. Vectors

num and **den** are returned with the numerator and denominator coefficients in descending powers of s.

tf2zp: transfer function to zero-pole conversion. The command

[Z,P,K] = tf2zp (num,den) finds the zeros, poles, and gains such that

$$H(s) = K\frac{(s - z_1)(s - z_2)\ldots}{(s - p_1)(s - p_2)\ldots}.$$

tf2sos: transfer function to second-Order section conversion. The command [**SOS**,G] = **tf2sos (B,A)** finds a matrix **SOS** in second-order section form and a gain G, which represent the same system $H(z)$ as the one with numerator **B** and denominator **A**. The poles and zeros of $H(z)$ must be in complex conjugate pairs. **SOS** is an $L \times 1$ matrix, where L is the number of second-order sections. Each row of the matrix describes a second-order transfer function, with the first three elements representing the numerator coefficients and the last three the denominator coefficients.

sos2tf: second-order section to transfer function. The command **[B,A] = sos2tf (SOS**,G) simply performs the reverse of the above function.

sos2ss: Second-order sections to state-space model conversion. The command **[A,B,C,D] = sos2ss** (**SOS**,G) returns the state-space matrices **A,B,C**, and D of the discrete-time system given by the gain G and the matrix **SOS** in second-order sections form.

ss2sos: state-space to second-order section model conversion. The command [**SOS**,G] = **ss2sos**(**A,B,C**,D) simply performs the reverse of the above function. FIX rounds toward zero.

fix(X): rounds the elements of **X** to the nearest integers toward zero.

floor (X): rounds the elements of **X** to the nearest integers toward minus infinity.

ceil (X): rounds the elements of **X** to the nearest integers toward infinity.

6.10 PROBLEMS

1. Show that the IEEE 754 standard for floating point numbers has a range from 1.18×10^{-38} to 3.4×10^{38}.

2. Consider the filter

$$y(n) = y(n-1) - 0.26y(n-2) + u(n)$$

which is required to be implemented in a 4-bit two's complement arithmetic.

(a) Compute the poles of the system in infinite precision.

(b) Compute the coefficients and poles of the quantized filter when implemented in Direct Form II. Draw the flow graph.

(c) Compute the coefficients and poles of the quantized filter when implemented in Coupled Form. Draw the flow graph.

3. Consider the filter

$$y(n) = y(n-1) - 0.26y(n-2) + u(n)$$

which is required to be implemented in a 4-bit two's complement arithmetic. Sketch and compare the frequency responses of the unquantized filter, the quantized Direct Form II filter, and the quantized Coupled Form filter.

4. A linear phase FIR filter is given by the transfer function

$$H(z) = 1 - 0.2z^{-1} + 1.85z^{-2} - 0.2z^{-3} + z^{-4}.$$

The filter is required to be implemented as a cascade structure with a short wordlength, such that the linear phase property is preserved. Find and sketch the filter structure and give the coefficients for a 4-bit two's complement representation.

5. Plot and compare the frequency responses of the unquantized and quantized filters of Problem 4.

6. Realize the filter

$$H(z) = 1 - 0.2z^{-1} + 0.5z^{-2} - 0.3z^{-3} + 0.3z^{-4}$$

as a lattice structure. Quantize the coefficients for 4-bit sign-magnitude representation. Find and compare the frequency responses of the unquantized and quantized filters.

7. Derive equation (6.14).

8. Prove Lemma 6.1.

9. Prove Lemma 6.2.

10. The transfer function of a first-order system is given by

$$H(z) = \frac{b(1+z^{-1})}{1-0.5z^{-1}}.$$

Let this filter be implemented in fixed point arithmetic, with the products rounded to $(B+1)$ bits before addition.

(a) Draw three equivalent flow graphs for this system. (Note that the gain b can be moved to different positions in the circuit.)

(b) For each of the flow graphs in (a), find the total noise power at the output.

(c) For each flow graph, find the maximum absolute value of the input to prevent overflow anywhere in the circuit.

(d) Repeat part (c) if the input is a uniformly distributed white noise sequence.

11. Using (6.24) and (6.25), find the mean and variance for the quantization schemes: (a) sign-magnitude truncation, (b) two's complement truncation, and (c) rounding.

12. (a) The flow graph of a second-order filter is given in Fig. P6.1. Find an absolute bound on the input to prevent overflow. Also, find the output noise variance if the products are quantized before addition.

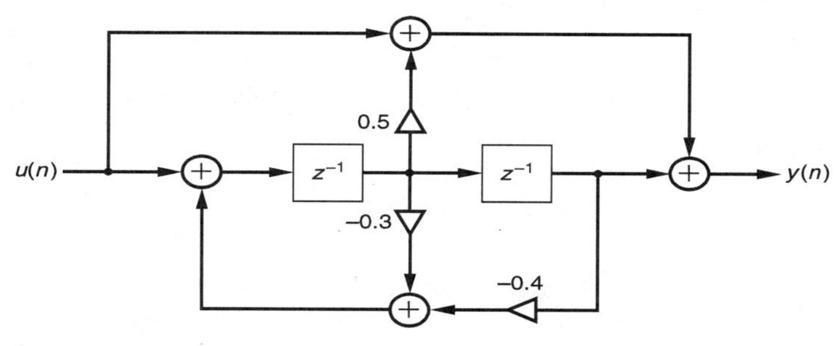

Figure P6.1

13. The transfer function of a first-order system is given by

$$H(z) = \frac{b(1+z^{-1})}{1-0.5z^{-1}}.$$

If the input is sinusoidal, find a bound on the input to prevent overflow. Also, find the output noise variance if the products are quantized before addition.

14. The transfer function of a first-order system is given by

$$H(z) = \frac{b(1+z^{-1})}{1-0.5z^{-1}}.$$

If the input is a uniformly distributed white noise sequence, find a bound on the input to prevent overflow. Also, find the output noise variance if the products are quantized before addition.

15. Derive equation (6.44).

16. The transfer function of a filter is given by

$$H(z) = \frac{z^{-1}-a}{1-az^{-1}}.$$

(a) Show that this is an all-pass filter.
(b) Find the Direct Form II realization of the filter.
(c) If the coefficients are quantized, is the filter still all-pass?
(d) Find the noise variance at the output of Direct Form II structure if $(B+1)$-bit two's complement quantization is used before addition.
(e) Now consider the filter to be implemented as $y(n) = a[y(n-1) - u(n)] + u(n-1)$. If the coefficients are quantized, is the filter still all-pass?
Find the noise variance at the output for this filter structure.

17. Prove that $\mathbf{A}^T\mathbf{A} \leq \mathbf{I}$ if and only if $|||\mathbf{A}|||_2 \leq 1$.

18. Derive equation (6.93).

19. Consider the filter

$$H(z) = \frac{1-0.5z^{-1}}{1-z^{-1}+0.3125z^{-2}}.$$

(a) Find the Direct Form II structure.
(b) Find the Coupled Form structure.
(c) For each of the structures, find the output noise variance for a $(B+1)$-bit magnitude truncation after quantization.
(d) Find the absolute bound on the input to prevent overflow in each of the structures.

20. Consider the filter

$$H(z) = \frac{1-0.5z^{-1}}{(1-0.25z^{-1})(1+0.75z^{-1})}.$$

(a) Find the Direct Form II, cascade, and parallel structures for the filter.
(b) For each structure, find the output noise variance for a $(B+1)$-bit rounding after quantization.
(c) Find the absolute bound on the input to prevent overflow in each of the structures.

21. Generate a period-two limit cycle using product rounding in a first-order digital filter. The chosen filter must be linearly stable.

22. Generate a period-two limit cycle using product rounding in a second-order Direct Form digital filter. The chosen filter must be linearly stable.

23. Derive the dead band for a second-order digital filter with rounding and equivalent poles at $z = \pm 1$; that is, derive equation (6.77).

24. Show that the two sufficient conditions for stability under rounding are mutually exclusive; that is, one is not a special case of the other. Show this by finding an example that satisfies one of the conditions and not the other and another example that is vice versa.

25. Derive the roundoff stable region for a Direct Form filter as shown in Fig. 6.26.

26. Consider a first-order filter with two's complement quantization given by $x(n+1) = Q_2\{\alpha x(n)\}$. Show that if $\alpha > 0$, an initial condition can always be found so that a period-one limit cycle is obtained.

27. The flow graph of a first-order system is given in Fig. P6.2.

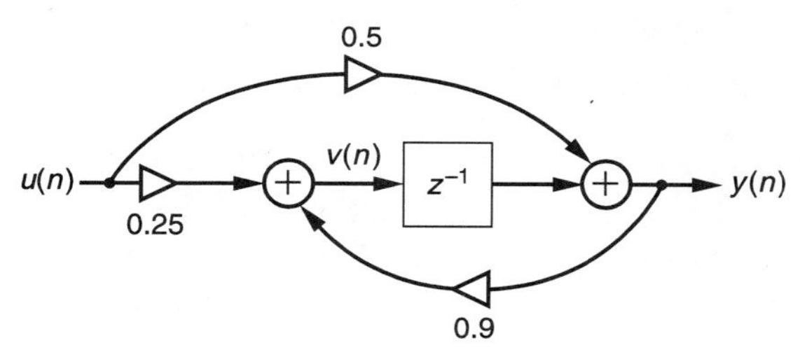

Figure P6.2

(a) Find the transfer function of the filter.
(b) Find the impulse response of the filter.
(c) The products are quantized to $(B+1)$ bits before addition. Draw a linear noise model of the quantized system. Find the total noise power at the output.
(d) Find the maximum value of $u(n)$ required to ensure that $|y(n)| < 1$ and $|v(n)| < 1$.
(e) If $u(n)$ is a uniformly white noise sequence, find the maximum value of the input to prevent overflow.

28. A first-order filter is given by

$$y(n) = Q_2\{ay(n-1) + u(n)\}$$

where Q_2 represents a 4-bit two's complement quantization. The input is $u(n) = \delta(n)$.

(a) If $a = -0.25$, does the filter go into a limit cycle? If so, what are the elements of the cycle?
(b) If $a = 0.875$, does the filter go into a limit cycle? If so, what are the element of the cycle?

29. Repeat Problem 28 for a 4-bit rounding quantizer.

30. Repeat Problem 28 for a 4-bit magnitude truncation quantizer.

6.11 COMPUTER PROJECTS

Project 1: Quantized Pole Locations

Write a MATLAB program to plot the possible pole locations for a quantized second-order filter. The program should ask the user for the following inputs:

(a) Number of bits.
(b) Type of quantization: sign-magnitude, rounding, or two's complement.
(c) Direct Form II or Coupled Form filter.

The program should then plot all the possible pole locations in the first quadrant of the unit circle.

Project 2: Scaling
Write a program to find the bound on the input for a given filter in order to prevent overflow. The input to the program is a transfer function in the form of a numerator polynomial vector and a denominator polynomial vector. The user will be prompted for the following:

(a) The numerator and denominator polynomial vectors.
(b) The type of input (bounded input, sinusoid, or white).
(c) The maximum known bound on the input.

The program will then find the scaling factor, β, for the Direct Form realization.

Project 3: Limit Cycles
You are given a first-order zero-input digital filter of the form $y(n) = Q_R\{ay(n-1)\}$, where the quantization is rounding. The user will input the coefficient a and the number of bits $B+1$. The program is then required to compute all possible limit cycles and the corresponding period.

Project 4: Coefficient Quantization
You are given a digital filter transfer function in the form of a numerator polynomial vector and a denominator polynomial vector. The inputs to the program are as follows:

(a) The numerator and denominator polynomial vectors.
(b) The type of quantization: magnitude truncation, rounding, or two's complement quantization.
(c) The number of bits for each coefficient.
(d) The filter structure: Direct Form II, cascade form, parallel form.

Also, for each second-order section, the user is prompted to decide on Direct Form II or Coupled Form.

The program then plots the frequency response (magnitude and phase) of the unquantized filter and the quantized filter on the same graph.

Project 5: Limit Cycles
Consider a second-order digital filter with the following transfer function

$$H(z) = \frac{1 - 0.5z^{-1}}{1 + 1.48z^{-1} - 0.9z^{-2}}.$$

Implement this filter with a 4-bit sign-magnitude arithmetic assuming the effect of product quantization only. In other words, assume infinite precision (or full MATLAB precision) for the coefficients. Implement in (a) Direct Form II and (b) Coupled Form. Assume the input to be zero. Use different initial conditions to show that the Direct Form II structure can exhibit limit cycles, but the Coupled Form filter cannot.

Project 6: Quantization Noise

Consider a second-order digital filter with the following transfer function:

$$H(z) = \frac{1 - 0.5z^{-1}}{1 + 1.48z^{-1} - 0.9z^{-2}}.$$

Implement the above filter in the following ways.

(a) Infinite precision (full MATLAB) precision arithmetic (any realization).

(b) Direct Form II realization: Use 4-bit two's complement arithmetic and simulate the effects of coefficient quantization, overflow, and product quantization.

(c) Coupled Form realization: Use 4-bit two's complement arithmetic and simulate the effects of coefficient quantization, overflow, and product quantization.

Let the input be a sunusoid of frequency 100 Hz sampled at 2 kHz. Apply 100 cycles of the input to each of the three implementations and measure the output. Plot the input and output for each case. Find the SNR of the outputs for (b) and (c) with respect to that of (a). Comment on your results.

BIBLIOGRAPHY

[1] T. Bose and M.-Q. Chen, "Overflow oscillations in state-space digital filters," *IEEE Trans. Circuits and Systems*, CAS-38, pp. 807–810, July 1991.

[2] P.M. Ebert, E. Mazo, and M.C. Taylor, "Overflow oscillations in digital filters," *Bell System Technical Journal*, pp. 2999–3020, November 1969.

[3] C.W. Barnes and A.T. Fam, "Minimum norm recursive digital filters that are free of overflow limit cycles," *IEEE Transactions on Circuits and Systems*, CAS-24, pp. 569–574, October 1977.

[4] W.L. Mills, C.T. Mullis, and R.A. Roberts, "Digital filter realizations without overflow oscillations," *IEEE Trans. on Acoustics, Speech, and Signal Processing*, ASSP-26, pp. 334–338, August 1978.

[5] L. Jackson, "An analysis of limit cycles due to multiplicative rounding in recursive digital filters," *Proc. 7th Allerton Conf. Circuit System Theory*, pp. 69–78, 1969.

[6] T. Bose and D.P. Brown, "Limit cycles due to roundoff in state-space digital filters," *IEEE Trans. on Acoustics, Speech, and Signal Processing*, ASSP-38, pp. 1460–1462, August 1990.

[7] T. Thong and B. Liu, "Limit cycles in the combinatorial implementation of digital filters," *IEEE Trans. on Acoustics, Speech, and Signal Processing*, ASSP-24, pp. 248–256, June 1976.

[8] T. Bose and M.-Q. Chen, "Stability of digital filters implemented with two's complement truncation quantization," *IEEE Trans. on Signal Processing*, SP-40, pp. 24–31, January 1992.

[9] T. Bose, M.-Q. Chen, and F. Brammer, "Stability of normal form digital filters with two's complement quantization," *Proc. of the IEEE Intl. Conf. Acous. Speech and Signal Processing*, vol. 3, pp. 1889–1892, May 1991.

[10] K. Premaratne, E. Kulasekere, P. Bauer, and L.-J. Leclerc, "An exhaustive search algorithm for checking limit cycle behavior of digital filters," *IEEE Trans. on Signal Processing*, SP-44, pp. 2405–2412, October 1996.

[11] P.P. Vaidyanathan and V. Liu, "An improved sufficient condition for the absence of limit cycles in digital filters," *IEEE Trans. Circuits and Systems*, CAS-34, pp. 319–322, March. 1987.

[12] G.F. Xu and T. Bose, "Elimination of limit cycles due to two's complement quantization in normal form digital filters," *IEEE Transactions on Signal Processing*, SP-45, pp. 2891–2895, December 1997.

CHAPTER 7

Adaptive Signal Processing

In many situations in signal processing, the characteristics of the input signal, the noise, and/or the dynamics of the physical system change with time. We need digital filters that adapt to these changes so that the desired output is achieved. This is done by varying the coefficients of the filter in response to the various changes. Filters with changing coefficients are called *adaptive filters*. An adaptive filter consists of the following two basic elements: (1) a digital filter, which produces an output in response to an input signal, and (2) an adaptive algorithm, which adjusts the coefficients of the digital filter. A simplified block diagram is shown in Fig. 7.1. The signal $d(n)$ is called the desired signal. The input and output of the filter are denoted by $x(n)$ and $y(n)$, respectively. The signal $e(n)$ is called the estimation error and is defined by $e(n) = d(n) - y(n)$. The adaptive algorithm is designed so that it minimizes some objective function constructed from this error signal. Many different adaptive filtering schemes are available. The basic difference among these schemes is the way the desired signal is extracted. Some basic applications are as follows.

System Identification In digital signal processing, most systems are characterized by linear or nonlinear difference equations. If an unknown system is given, it is often useful to find a model for this system, that is, the coefficients of its difference equation. This is referred to as system identification. A block diagram for system identification is given in Fig. 7.2(a). The order of the difference equation model of the unknown system must first be estimated based on some prior knowledge of the system. This difference equation with unknown coefficients is the adaptive filter. An input is applied simultaneously to the unknown system and the adaptive filter. The difference between the outputs is called the error signal. If the adaptive filter models the unknown system perfectly, then this error signal will be zero. The coefficients of the adaptive filter are calculated by an adaptive algorithm that minimizes some objective function (or cost function) based on the error signal. Note that it is conventional to denote an adaptive filter by an arrow through it as shown in the figure. Also, the block labeled adaptive algorithm is usually not shown in most adaptive filtering literature. Two examples of objective functions are as follows, where n is the present time index:

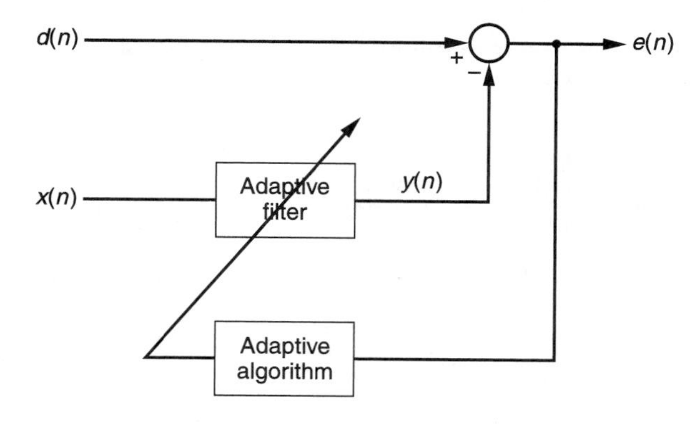

Figure 7.1 Adaptive filter block diagram.

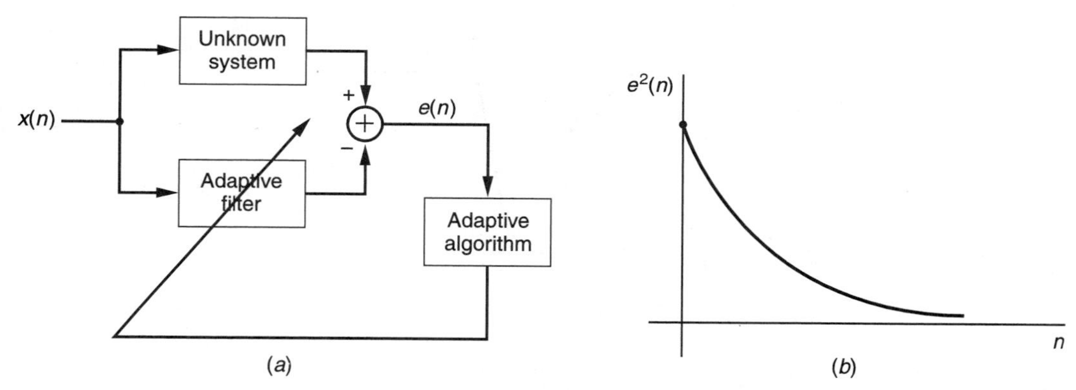

Figure 7.2 (*a*) System identification scheme; (*b*) A typical squared error plot.

1. $J_n = E\{e^2(n)\}$: The expectation of the instantaneous squared error.
2. $J_n = \sum_{i=0}^{n} e^2(i)$: Sum of squared errors up to the present sample.

The adaptive algorithm calculates the filter coefficients and provides this information to the digital filter, which in turn filters the input signal with these coefficients. The adaptive algorithm usually takes a certain amount of time before it achieves the minimum of the objective function. This is referred to as the *convergence time*. A plot of the instantaneous squared error $e^2(n)$ versus the sample index n is usually decreasing. A typical plot is shown in Fig. 7.2(b), which illustrates that the algorithm is converging. The adaptive algorithm continuously updates the filter coefficients, even after convergence. Therefore, if the parameters of the unknown system change, the filter coefficients also change in order to constantly track the unknown system.

Sinusoid Tracking Consider a signal that is composed of a sinusoid plus some noise added to it. The noise is assumed to be wideband; that is, it is composed of a wide band of frequencies. The goal is to extract the sinusoid and remove the noise even when the frequency of the sinusoid changes, that is, to track the sinusoid. A block diagram is shown

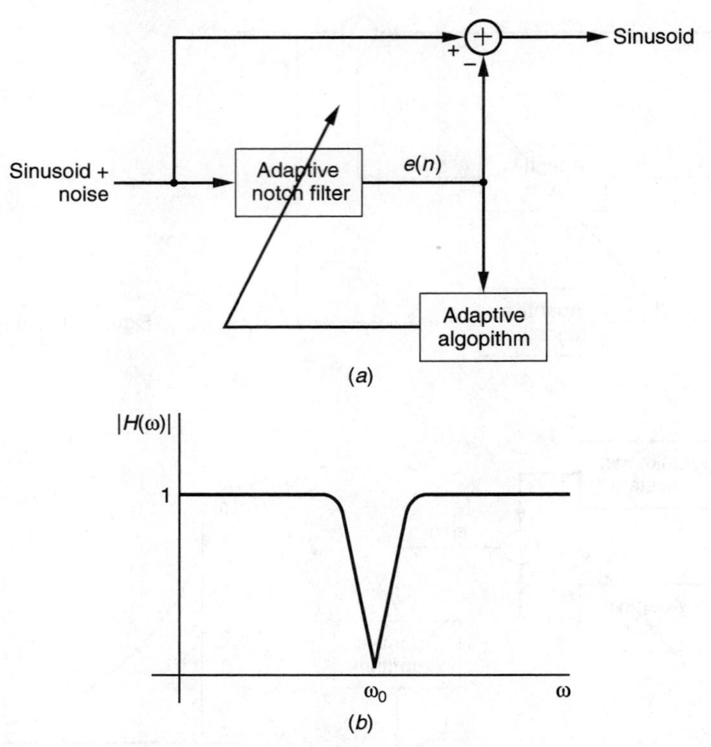

Figure 7.3 (a) Sinusoidal tracking scheme; (b) Notch filter frequency response.

in Fig. 7.3(a). An adaptive notch filter is used in this application. As we know, a notch filter blocks out a single frequency (sinusoid) and passes the rest. The frequency response of a notch filter is shown in Fig. 7.3(b), where ω_0 is the notched frequency and $|H(\omega)|$ is the gain of the filter as a function of frequency. Clearly, this filter blocks out the signal with frequency ω_0 and passes all others except those in the neighborhood of ω_0, which are attenuated to some extent. In an adaptive notch filter, the notch frequency is varied by the adaptive algorithm. In this example, the error signal is the output of the notch filter. The objective function is minimized when the notch frequency is the same as that of the sinusoid. The output of the notch filter is then nearly equal to the noise.[1] The sinusoid can be obtained by subtracting this error signal from the input signal. As the frequency of the sinusoid changes, the adaptive algorithm updates the coefficients of the filter so that its notch frequency tracks the sinusoidal frequency.

Noise Cancellation This term refers to cancellation of noise from a sum of signal and noise. Let $s(n)$ represent the signal (not necessarily a sinusoid as in the above example) and $w(n)$ be the noise. It is assumed that the signal and the noise are not correlated. This means that it is not possible to obtain one from the other by filtering it. Another noise

[1] We assume that the notch filter is zero phase. That is, it does not cause a phase shift.

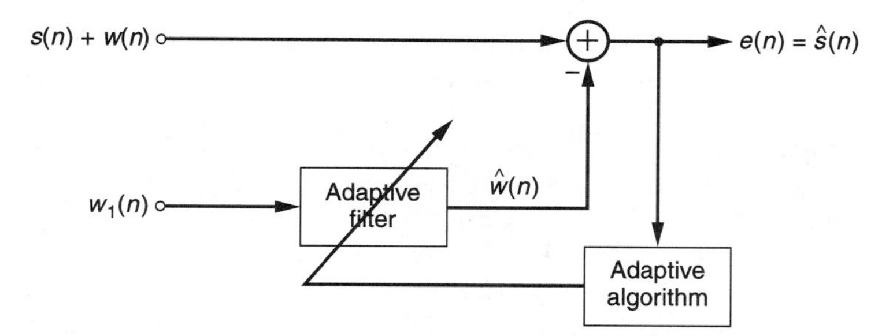

Figure 7.4 Adaptive noise cancellation.

signal denoted by $w_1(n)$ is assumed to be available. A good example of this situation is given in [4], which we shall now borrow. The schematic diagram is shown in Fig. 7.4. Consider $s(n)$ to be the speech of the pilot in the cockpit of an aircraft and $w(n)$ as the engine noise. The sum of these signals enters a microphone used by the pilot. A second microphone can be placed away from the other one somewhere else in the cockpit. This microphone also picks up the engine noise but not the pilot's voice. The engine noises picked up by the two microphones are not exactly the same but are correlated. That is, by filtering $w_1(n)$, we can obtain a close approximation of $w(n)$. Since the exact relationship between $w_1(n)$ and $w(n)$ is unknown and may also change with time, an adaptive filter is the obvious choice. Again, the adaptive algorithm minimizes an objective function constructed from the error signal $e(n)$. This minimization is achieved by canceling one of the components of $s(n) + w(n)$. Since the noise is uncorrelated with the signal, $w_1(n)$ cannot be converted to $s(n)$. Therefore, in an attempt to minimize the objective function, the filter converts $w_1(n)$ to an approximation of $w(n)$ denoted by $\hat{w}(n)$. The output of the adder is therefore an approximation of the signal denoted by $\hat{s}(n)$. In the ideal case, the output is the original signal $s(n)$.

Numerous adaptive filtering algorithms are described in the literature for many different applications. Some of these are more suitable for certain applications than others. However, a few common characteristics can be used to measure the goodness of an adaptive algorithm.

Computational Complexity Recall from Fig. 7.1 that the adaptive filter consists of a digital filter and an adaptive algorithm that calculates the coefficients of the filter. For every sample of incoming data, the adaptive algorithm must complete its calculation of the filter coefficients, and then the digital filter must perform its filtering operation (sum of products). Therefore, the number of multiplications and additions (computations) required by the adaptive algorithm is an important factor. The lower the computational burden, the higher the allowable sampling rate of the signal.

Rate of Convergence This is the speed at which the adaptive algorithm converges toward the optimal solution. The higher this rate, the better suited is the algorithm for adapting to rapid changes in the system or signal characteristics.

Numerical Robustness In practice, adaptive algorithms are implemented in processors that have finite register length and therefore are subjected to quantization errors.

Numerical robustness refers to the convergence of the algorithm under quantization effects. It is also a measure of the errors introduced as a result of quantization.

Misadjustment As mentioned earlier, the adaptive algorithm is designed to minimize some objective function constructed from the error $e(n)$. One popular function is the mean square error (MSE) denoted by $E\{e(n)\}$. Theoretically, the Wiener filter is the optimal solution for minimum MSE. The Wiener filter is not suitable for practical implementation, for reasons that we shall discuss in this chapter. Adaptive algorithms are designed that attempt to achieve the Wiener solution. Let J_{opt} denote the MSE attained by the optimum Wiener filter, and let J_{ss} denote the steady-state MSE attained by some adaptive algorithm. The difference $J_{ex} = J_{ss} - J_{opt}$ is the price paid for using an adaptive approach and is called the *Excess MSE*. The ratio J_{ex}/J_{opt} is called the misadjustment. Clearly, a low misadjustment is desirable.

7.1 THE CORRELATION MATRIX AND SOME LINEAR ALGEBRA

Let the sequence $x(n)$ denote a single realization of a random process. This sequence will be denoted by the vector

$$\mathbf{x}(n) = \left[x(n), x(n-1) \; \ldots \; x(n-N+1)\right]^T. \tag{7.1}$$

The vector is depicted by boldface and is of length N. The correlation matrix of this sequence is defined by

$$\mathbf{R} = E\{\mathbf{x}(n)\mathbf{x}^{\dagger}(n)\} \tag{7.2}$$

where $\dagger$ denotes the transpose and complex conjugation operation. Using the above and the definition of autocorrelation, we get

$$\mathbf{R} = E\left\{\begin{bmatrix} x(n) \\ x(n-1) \\ \vdots \\ x(n-N+1) \end{bmatrix} \left[x^*(n)\, x^*(n-1) \; \ldots \; x^*(n-N+1)\right]\right\}$$

$$= \begin{bmatrix} r(0) & r(1) & \ldots & r(N-1) \\ r(-1) & r(0) & \ldots & r(N-2) \\ \vdots & \vdots & & \vdots \\ r(-N+1) & r(-N+2) & \ldots & r(0) \end{bmatrix}. \tag{7.3}$$

This correlation matrix has several interesting and useful properties. A good understanding of these properties is important in the analysis and design of adaptive filtering algorithms. Some of the properties are very straightforward, whereas others require some background in linear algebra. When the need arises, we present the necessary theorems of linear algebra along with their proofs. The proofs may be omitted if desired without loss of continuity. Here is the first property of the correlation matrix.

Property 1: The correlation matrix of stationary random process is Hermitian; that is,

$$\mathbf{R}^{\dagger} = \mathbf{R}. \tag{7.4}$$

Proof. The proof follows directly from the matrix **R** given in (7.3). ∎

Definition 7.1 A matrix **A** is said to be *positive semidefinite* if for any vector $\mathbf{v} \neq 0$, $\mathbf{v}^{\dagger}\mathbf{A}\mathbf{v} \geq 0$. The function $\mathbf{v}^{\dagger}\mathbf{A}\mathbf{v}$ is called the *quadratic form*. This condition is also denoted by $\mathbf{A} \geq \mathbf{0}$. If $\mathbf{A} > \mathbf{0}$, then the matrix is said to be positive definite.

Property 2: The correlation matrix is positive semidefinite; that is,

$$\mathbf{R} \geq \mathbf{0}. \tag{7.5}$$

Proof. Let $\mathbf{R} = E\{\mathbf{x}(n)\mathbf{x}^{\dagger}(n)\}$. For some vector $\mathbf{x} \neq 0$, consider the function[2]

$$\begin{aligned} \mathbf{v}^{\dagger}\mathbf{R}\mathbf{v} &= \mathbf{v}^{\dagger}E\{\mathbf{x}(n)\mathbf{x}^{\dagger}(n)\}\mathbf{v} \\ &= E\{\mathbf{v}^{\dagger}\mathbf{x}(n)\mathbf{x}^{\dagger}(n)\mathbf{v}\} \\ &= E\{[\mathbf{x}^{\dagger}(n)\mathbf{v}]^{\dagger}[\mathbf{x}^{\dagger}(n)\mathbf{v}]\}. \end{aligned} \tag{7.6}$$

Now define a vector $\mathbf{p} \triangleq \mathbf{x}^{\dagger}(n)\mathbf{v}$. Then the above becomes

$$\mathbf{v}^{\dagger}\mathbf{R}\mathbf{v} = E\{\mathbf{p}^{\dagger}\mathbf{p}\} \geq 0 \tag{7.7}$$

and the proof is complete. ∎

Properties 3 and 4 given next will be useful for the Levinson-Durbin algorithm in Chapter 9. The proofs follow easily by rewriting the elements of the correlation matrix representation given in (7.3).

Property 3: Let the input vector be $\mathbf{x}(n) = [x(n)\, x(n-1)\, \ldots\, x(n-N+1)]^T$. We know that its correlation matrix is defined as $\mathbf{R} = E\{\mathbf{x}(n)\mathbf{x}^{\dagger}(n)\}$. Define a new vector with the elements of the input vector rearranged backward as

$$\mathbf{x}^{B}(n) \triangleq [x(n-N+1)\, \ldots\, x(n-1)\, x(n)]^T. \tag{7.8}$$

Then the correlation matrix of this new vector is the transpose of the original correlation matrix; that is,

$$E\{\mathbf{x}^{B}(n)[\mathbf{x}^{B}(n)]^{\dagger}\} = \mathbf{R}^T. \tag{7.9}$$

Property 4: Let $\mathbf{x}_N(n) = [x(n)\;\; x(n-1)\;\; \ldots\;\; x(n-N+1)]^T$, where the subscript N denotes the number of observations in the input vector. Define its correlation matrix

[2] In matrix algebra, $(\mathbf{AB})^{\dagger} = \mathbf{B}^{\dagger}\mathbf{A}^{\dagger}$.

as $\mathbf{R}_N = E\{\mathbf{x}_N(n)\mathbf{x}_N^\dagger(n)\}$. Similarly, define $\mathbf{x}_{N+1}(n) = [x(n)\ x(n-1)\ \ldots\ x(n-N+1)\ x(n-N)]^T$ and its correlation matrix $\mathbf{R}_{N+1} = E\{\mathbf{x}_{N+1}(n)\mathbf{x}_{N+1}^\dagger(n)\}$. Also, define $\mathbf{r} = [r(-1)\ r(-2)\ \ldots\ r(-N)]^T$ and $\mathbf{r}^{BT} = [r(-N)\ \ldots\ r(-2)\ r(-1)]$. Then the correlation matrix $\mathbf{R}_{N+1}$ can be represented as follows:

$$\mathbf{R}_{N+1} = \begin{bmatrix} r(0) & \mathbf{r}^\dagger \\ \mathbf{r} & \mathbf{R}_N \end{bmatrix} \tag{7.10}$$

$$= \begin{bmatrix} \mathbf{R}_N & \mathbf{r}^{B*} \\ \mathbf{r}^{BT} & r(0) \end{bmatrix}. \tag{7.11}$$

Definition 7.2 *Let* $\mathbf{A}$ *denote an* $N \times N$ *matrix. The eigenvalues of the matrix are the roots of the characteristic equation*

$$\det\{\mathbf{A} - \lambda\mathbf{I}\} = 0 \tag{7.12}$$

where $\mathbf{I}$ *denotes the identity matrix of appropriate dimension and* det *is the determinant. If* $\lambda_i,\ i = 1, 2, \ldots, N$ *denote the eigenvalues and* $v_i,\ i = 1, 2, \ldots, N$ *denote the eigenvectors, then*

$$\mathbf{A}\mathbf{v}_i = \lambda_i\mathbf{v}_i. \tag{7.13}$$

The eigenvector $\mathbf{v}_i$ *is said to be associated with the eigenvalue* λ_i.

Definition 7.3 *A matrix* $\mathbf{Q}$ *is said to be unitary if*

$$\mathbf{Q}^\dagger\mathbf{Q} = \mathbf{I} \tag{7.14}$$

which implies that $\mathbf{Q}^\dagger = \mathbf{Q}^{-1}$.

Definition 7.4 *Two vectors* $\mathbf{x}$ *and* $\mathbf{y}$ *are said to be mutually orthogonal if* $\mathbf{x}^\dagger\mathbf{y} = 0$.

Theorem 7.1 *A square matrix* $\mathbf{Q}$ *is unitary if and only if its columns (or rows) are mutually orthogonal unit vectors.*

Proof. Let $\mathbf{Q} = \begin{bmatrix} \mathbf{q}_1 & \mathbf{q}_2 & \cdots & \mathbf{q}_N \end{bmatrix}$, where $\mathbf{q}_i$ are column vectors. Then we can write

$$\mathbf{Q}^\dagger\mathbf{Q} = \begin{bmatrix} \mathbf{q}_1^\dagger \\ \mathbf{q}_2^\dagger \\ \vdots \\ \mathbf{q}_N^\dagger \end{bmatrix} \begin{bmatrix} \mathbf{q}_1 & \mathbf{q}_2 & \cdots & \mathbf{q}_N \end{bmatrix} \tag{7.15}$$

$$= \begin{bmatrix} \mathbf{q}_1^\dagger\mathbf{q}_1 & \mathbf{q}_1^\dagger\mathbf{q}_2 & \cdots & \mathbf{q}_1^\dagger\mathbf{q}_N \\ \vdots & & & \vdots \\ \mathbf{q}_N^\dagger\mathbf{q}_1 & \mathbf{q}_N^\dagger\mathbf{q}_2 & \cdots & \mathbf{q}_N^\dagger\mathbf{q}_N \end{bmatrix}.$$

First, suppose that $\mathbf{Q}$ is unitary. Then $\mathbf{Q}^\dagger\mathbf{Q} = \mathbf{I}$, which implies

$$\mathbf{q}_i^\dagger \mathbf{q}_j = \begin{cases} 1, & i = j \\ 0, & i \neq j. \end{cases} \tag{7.16}$$

Equation (7.16) then implies that the columns are orthogonal unit vectors (orthonormal). This completes the proof of necessity.

Now suppose that the columns of $\mathbf{Q}$ are mutually orthonormal. Then the matrix in (7.15) is identity, and the proof of sufficiency is complete.

The above can be proved for the rows of $\mathbf{Q}$ by considering the matrix $\mathbf{Q}\mathbf{Q}^\dagger$. ▪

Theorem 7.2 *The eigenvalues of a Hermitian matrix are all real.*

Proof. Let $\mathbf{A}$ be a Hermitian matrix. Let λ and $\mathbf{v}$ be an eigenvalue and an associated unit eigenvector, respectively. Then

$$\mathbf{A}\mathbf{v} = \lambda\mathbf{v}.$$

Pre-multiplying both sides by $\mathbf{v}^\dagger$ and using the fact that $\mathbf{v}$ is a unit vector, we get

$$\begin{aligned} \mathbf{v}^\dagger\mathbf{A}\mathbf{v} &= \mathbf{v}^\dagger\lambda\mathbf{v} \\ &= \lambda\mathbf{v}^\dagger\mathbf{v} \\ &= \lambda. \end{aligned} \tag{7.17}$$

Using the above and the fact that $\mathbf{A} = \mathbf{A}^\dagger$, we write

$$\begin{aligned} \lambda^* &= (\mathbf{v}^\dagger\mathbf{A}\mathbf{v})^* \\ &= (\mathbf{v}^\dagger\mathbf{A}\mathbf{v})^\dagger \\ &= \mathbf{v}^\dagger\mathbf{A}^\dagger\mathbf{v} \\ &= \mathbf{v}^\dagger\mathbf{A}\mathbf{v} \\ &= \lambda. \end{aligned} \tag{7.18}$$

Equation (7.18) shows that all eigenvalues of a Hermitian matrix are real. ▪

Theorem 7.3 *Matrix* $\mathbf{A}$ *and its unitary transformation* $\mathbf{Q}^\dagger\mathbf{A}\mathbf{Q}$ *have the same eigenvalues, where* $\mathbf{Q}$ *is a unitary matrix.*

The proof is assigned as a problem.

Theorem 7.4 *A matrix* $\mathbf{H}$ *is Hermitian if and only if its quadratic form* $h = \mathbf{v}^\dagger\mathbf{H}\mathbf{v}$ *is real for any* $\mathbf{v}$.

Proof. **Necessity:** Suppose H is Hermitian. Because h is a scalar, we have

$$\begin{aligned} h^* &= h^\dagger \\ &= (\mathbf{v}^\dagger\mathbf{H}\mathbf{v})^\dagger \\ &= \mathbf{v}^\dagger\mathbf{H}^\dagger\mathbf{v} \\ &= \mathbf{v}^\dagger\mathbf{H}\mathbf{v} = h. \end{aligned}$$

Thus, the quadratic form is real, and the proof of necessity is complete.

Sufficiency: Suppose h is real for all $\mathbf{v}$. Then we have $h^* = h$, which gives

$$\mathbf{v}^\dagger \mathbf{H} \mathbf{v} = \mathbf{v}^\dagger \mathbf{H}^\dagger \mathbf{v}.$$

The above then implies that $\mathbf{H} = \mathbf{H}^\dagger$, and the proof is complete. ■

Property 5: All eigenvalues of the correlation matrix are real and nonnegative.

Proof. Since $\mathbf{R}$ is Hermitian, from Theorem 7.2 we know that all its eigenvalues are real. Let λ be an eigenvalue and $\mathbf{v}$ the associated eigenvector, that is, $\mathbf{R}\mathbf{v} = \lambda\mathbf{v}$. Pre-multiplying both sides by $\mathbf{v}^\dagger$, we have

$$\mathbf{v}^\dagger \mathbf{R} \mathbf{v} = \lambda \mathbf{v}^\dagger \mathbf{v} \tag{7.19}$$

which gives

$$\lambda = \frac{\mathbf{v}^\dagger \mathbf{R} \mathbf{v}}{\mathbf{v}^\dagger \mathbf{v}}. \tag{7.20}$$

In equation (7.20), $\mathbf{v}^\dagger \mathbf{v} \geq \mathbf{0}$. Since $\mathbf{R}$ is Hermitian, from Property 2 we know that it is positive semidefinite, that is, $\mathbf{v}^\dagger \mathbf{R} \mathbf{v} \geq 0$. Therefore, $\lambda \geq 0$. ■

The following theorem is about triangularizing matrices. This theorem will be used to derive subsequent results. One of the problems in adaptive signal processing is to iteratively solve the equation $\mathbf{w} = \mathbf{R}^{-1}\mathbf{p}$, where $\mathbf{R}$ is the autocorrelation matrix, $\mathbf{p}$ is a cross-correlation vector, and $\mathbf{w}$ is the unknown weight vector. This is the well-known Wiener filter, as we shall see in the next section. The following theorem is also important in its own right because if matrix $\mathbf{R}$ is triangularizable, then it can be solved in fewer computations than that required to compute the inverse.

Theorem 7.5 *Any square matrix can be reduced by a unitary transformation to upper triangular form with its eigenvalues on the diagonal. That is, a matrix* $\mathbf{A}$ *of order* $N \times N$ *satisfies*

$$\mathbf{Q}^\dagger \mathbf{A} \mathbf{Q} = \mathbf{U}\{\lambda_1, \lambda_2, \ldots, \lambda_N\} \tag{7.21}$$

where $\lambda_1, \lambda_2, \ldots, \lambda_N$ *are the eigenvalues of* $\mathbf{A}$ *and* $\mathbf{U}\{\lambda_1, \lambda_2, \ldots, \lambda_N\}$ *denotes an upper triangular matrix with* $\lambda_1, \lambda_2, \ldots, \lambda_N$ *on the diagonal.*

Proof. The theorem is proved by mathematical induction. First consider $N = 2$. Let $\mathbf{A}$ have eigenvalues λ_1 and λ_2. Let λ_1 have the unit eigenvector $\mathbf{v}_1$. Let $\mathbf{v}_2$ be a unit vector orthogonal to $\mathbf{v}_1$, but $\mathbf{v}_2$ is not necessarily the eigenvector associated with λ_2. Define $\mathbf{Q} = [\mathbf{v}_1,\ \mathbf{v}_2]$. This matrix is unitary since $\mathbf{v}_1 \perp \mathbf{v}_2$. Also,

$$\mathbf{Q}^\dagger \mathbf{A} \mathbf{Q} = \begin{bmatrix} \mathbf{v}_1^\dagger \\ \mathbf{v}_2^\dagger \end{bmatrix} \mathbf{A}[\mathbf{v}_1\ \mathbf{v}_2] = \begin{bmatrix} \mathbf{v}_1^\dagger \mathbf{A} \mathbf{v}_1 & \mathbf{v}_1^\dagger \mathbf{A} \mathbf{v}_2 \\ \mathbf{v}_2^\dagger \mathbf{A} \mathbf{v}_1 & \mathbf{v}_2^\dagger \mathbf{A} \mathbf{v}_2 \end{bmatrix}. \tag{7.22}$$

Since $\mathbf{A}\mathbf{v}_1 = \lambda_1 \mathbf{v}_1$, $\mathbf{v}_1^\dagger \mathbf{v}_1 = 1$, and $\mathbf{v}_2^\dagger \mathbf{v}_1 = 0$, (7.22) becomes

$$\mathbf{Q}^\dagger \mathbf{A} \mathbf{Q} = \begin{bmatrix} \lambda_1 & \mathbf{v}_1^\dagger \mathbf{A} \mathbf{v}_2 \\ 0 & \mathbf{v}_2^\dagger \mathbf{A} \mathbf{v}_2 \end{bmatrix}. \tag{7.23}$$

The eigenvalues of a upper triangular matrix must be on the diagonal. Also, since $\mathbf{A}$ and $\mathbf{Q}^\dagger \mathbf{A} \mathbf{Q}$ have the same eigenvalues (Theorem 7.3), it follows that $\mathbf{v}_2^\dagger \mathbf{A} \mathbf{v}_2 = \lambda_2$, and the theorem is proved for $N = 2$. Now suppose that the theorem holds for all matrices of order $N - 1$ and let $\mathbf{A}$ be of order $N \times N$. Let λ_1 be an eigenvalue and $\mathbf{v}_1$ be its associated unit eigenvector. Let $\mathbf{Q} = [\mathbf{v}_1, \mathbf{v}_2, \ldots, \mathbf{v}_N]$ be unitary, that is, $\mathbf{v}_i$ are mutually orthogonal. Thus

$$\mathbf{Q}^\dagger \mathbf{A} \mathbf{Q} = \begin{bmatrix} \lambda_1 & \mathbf{v}_1^\dagger \mathbf{A} \mathbf{v}_2 & \ldots & \mathbf{v}_1^\dagger \mathbf{A} \mathbf{v}_N \\ 0 & \mathbf{v}_2^\dagger \mathbf{A} \mathbf{v}_2 & \ldots & \mathbf{v}_2^\dagger \mathbf{A} \mathbf{v}_N \\ \vdots & & & \\ 0 & \mathbf{v}_N^\dagger \mathbf{A} \mathbf{v}_2 & \ldots & \mathbf{v}_N^\dagger \mathbf{A} \mathbf{v}_N \end{bmatrix} \tag{7.24}$$

$$= \begin{bmatrix} \lambda_1 & \mathbf{B} \\ 0 & \mathbf{C}_{(N-1)\times(N-1)} \end{bmatrix}. \tag{7.25}$$

By the induction hypotheses, there exists a unitary matrix $\mathbf{W}$, which will upper triangularize $\mathbf{C}$. Let $\mathbf{V} = \begin{bmatrix} 1 & 0 \\ 0 & \mathbf{W} \end{bmatrix}$. Then

$$\mathbf{V}^\dagger \mathbf{V} = \begin{bmatrix} 1 & 0 \\ 0 & \mathbf{W}^\dagger \end{bmatrix} \begin{bmatrix} 1 & 0 \\ 0 & \mathbf{W} \end{bmatrix} = \begin{bmatrix} 1 & 0 \\ 0 & \underbrace{\mathbf{W}^\dagger \mathbf{W}}_{\mathbf{I}_{N-1}} \end{bmatrix} = \mathbf{I}_N. \tag{7.26}$$

Therefore, $\mathbf{V}$ is unitary. Now consider

$$\mathbf{V}^\dagger \left(\mathbf{Q}^\dagger \mathbf{A} \mathbf{Q}\right) \mathbf{V} = \begin{bmatrix} 1 & 0 \\ 0 & \mathbf{W}^\dagger \end{bmatrix} \begin{bmatrix} \lambda_1 & \mathbf{B} \\ 0 & \mathbf{C} \end{bmatrix} \begin{bmatrix} 1 & 0 \\ 0 & \mathbf{W} \end{bmatrix}$$

$$= \begin{bmatrix} \lambda_1 & \mathbf{B} \\ 0 & \mathbf{W}^\dagger \mathbf{C} \mathbf{W} \end{bmatrix}, \tag{7.27}$$

where $\mathbf{W}^\dagger \mathbf{C} \mathbf{W}$ is upper triangular. Thus, the above matrix is also upper triangular. So,

$$\mathbf{V}^\dagger \left(\mathbf{Q}^\dagger \mathbf{A} \mathbf{Q}\right) \mathbf{V} = (\mathbf{Q}\mathbf{V})^\dagger \mathbf{A} (\mathbf{Q}\mathbf{V}) = \begin{bmatrix} \lambda_1 & b_{12} & b_{13} & \ldots & b_{1n} \\ 0 & \lambda_2 & b_{23} & \ldots & b_{2n} \\ \vdots & & & & \\ 0 & 0 & 0 & \ldots & \lambda_n \end{bmatrix}. \tag{7.28}$$

The matrix $(\mathbf{QV})$ is unitary since $(\mathbf{QV})^\dagger(\mathbf{QV}) = \mathbf{V}^\dagger \underbrace{\mathbf{Q}^\dagger\mathbf{Q}}_{\mathbf{I}}\mathbf{V} = \mathbf{V}^\dagger\mathbf{V} = \mathbf{I}$. Since $(\mathbf{QV})$ is unitary, the eigenvalues of $\mathbf{A}$ are the same as those of the above upper triangular matrix. That is, the diagonal elements of the above matrix are the eigenvalues of $\mathbf{A}$, and the theorem is proved. ■

Now, we derive some properties of normal matrices. As we have seen in earlier chapters, normal matrices are very useful for digital filter implementations and have several desirable properties. The definition of a normal matrix is restated here for convenience.

Definition 7.5 *Matrix $\mathbf{A}$ is said to be normal if it satisfies $\mathbf{A}^\dagger\mathbf{A} = \mathbf{A}\mathbf{A}^\dagger$.*

It is easy to show that Hermitian and unitary matrices are special cases of a normal matrix. The following theorem will be useful in subsequent results. Its proof is straightforward and is left as an exercise.

Theorem 7.6 *If $\mathbf{Q}$ is unitary, then any matrix $\mathbf{A}$ is normal if and only if $\mathbf{Q}^\dagger\mathbf{A}\mathbf{Q}$ is normal.*

Theorem 7.7 *Let $\lambda_1, \lambda_2, \ldots \lambda_N$ denote the eigenvalues of matrix $\mathbf{A}$ and let $\mathbf{Q}$ be a unitary matrix. This matrix can be diagonalized by a unitary transformation, i.e. $(\mathbf{Q}^\dagger\mathbf{A}\mathbf{Q}) = \mathbf{D}\{\lambda_1, \lambda_2 \ldots \lambda_N\}$, if and only if $\mathbf{A}$ is normal.*

Proof. **Necessity:** Suppose $\mathbf{Q}^\dagger\mathbf{A}\mathbf{Q} = \mathbf{D}\{\lambda_1, \lambda_2, \ldots \lambda_n\}$. Since $\mathbf{Q}$ is unitary, $\mathbf{A} = \mathbf{Q}\mathbf{D}\mathbf{Q}^\dagger$, and

$$\mathbf{A}^\dagger\mathbf{A} = (\mathbf{Q}\mathbf{D}\mathbf{Q}^\dagger)^\dagger\,(\mathbf{Q}\mathbf{D}\mathbf{Q}^\dagger) = \mathbf{Q}\mathbf{D}^\dagger\,\underbrace{\mathbf{Q}^\dagger\mathbf{Q}}_{\mathbf{I}}\,\mathbf{D}\mathbf{Q}^\dagger = \mathbf{Q}\mathbf{D}^\dagger\mathbf{D}\mathbf{Q}^\dagger. \tag{7.29}$$

Also,

$$\mathbf{A}\mathbf{A}^\dagger = (\mathbf{Q}\mathbf{D}\mathbf{Q}^\dagger)\,(\mathbf{Q}\mathbf{D}\mathbf{Q}^\dagger)^\dagger = \mathbf{Q}\mathbf{D}\,\underbrace{\mathbf{Q}^\dagger\mathbf{Q}}_{\mathbf{I}}\,\mathbf{D}^\dagger\mathbf{Q}^\dagger = \mathbf{Q}\mathbf{D}\mathbf{D}^\dagger\mathbf{Q}^\dagger. \tag{7.30}$$

Since $\mathbf{D}$ is diagonal, $\mathbf{D}^\dagger\mathbf{D} = \mathbf{D}\mathbf{D}^\dagger$. Thus, equation (7.30) gives $\mathbf{A}^\dagger\mathbf{A} = \mathbf{A}\mathbf{A}^\dagger$; that is, $\mathbf{A}$ is normal. The proof of necessity is complete.

Sufficiency: Now suppose $\mathbf{A}$ is normal. Then by Theorem 7.5, there exists a unitary matrix $\mathbf{Q}$ such that $\mathbf{Q}^\dagger\mathbf{A}\mathbf{Q} = \mathbf{B}$, where $\mathbf{B}$ is upper triangular with eigenvalues of $\mathbf{A}$ on the diagonal. Also, by Theorem 7.6, $\mathbf{B}$ is normal. So, let

$$\mathbf{B} = \begin{bmatrix} \lambda_1 & b_{12} & b_{13} & \ldots & b_{1N} \\ 0 & \lambda_2 & b_{23} & \ldots & b_{2N} \\ \vdots & & & & \\ 0 & 0 & 0 & \ldots & \lambda_N \end{bmatrix}. \tag{7.31}$$

Since $\mathbf{B}^\dagger\mathbf{B} = \mathbf{B}\mathbf{B}^\dagger$, we can calculate the (1,1) entries of both sides to get

$$\lambda_1^*\lambda_1 = \lambda_1\lambda_1^* + b_{12}b_{12}^* + \cdots + b_{1N}b_{1N}^* \tag{7.32}$$

which gives

$$0 = |b_{12}|^2 + \cdots + |b_{1N}|^2 \tag{7.33}$$

or

$$b_{12} = b_{13} = \cdots = b_{1N} = 0. \tag{7.34}$$

Similarly, calculating the (2,2) entries of $\mathbf{B}^\dagger\mathbf{B}$ and $\mathbf{B}\mathbf{B}^\dagger$ and equating them, we get

$$b_{23} = b_{24} = \cdots = b_{2N} = 0 \tag{7.35}$$

and so on for the the other off-diagonal entries. Thus, if $\mathbf{A}$ is normal then $\mathbf{B}$ is diagonal. The theorem is now proved. ■

Theorem 7.8 *If a matrix* $\mathbf{A}$ *is diagonalizable by a unitary transformation, that is,* $\mathbf{Q}^\dagger\mathbf{A}\mathbf{Q} = \mathit{diag}\{\lambda_1, \lambda_2, \cdots \lambda_N\} \triangleq \mathbf{D}$*, then* $\mathbf{Q} = [\mathbf{v}_1, \mathbf{v}_2, \ldots, \mathbf{v}_N]$*, where* $\mathbf{v}_i$ *are the eigenvectors of* $\mathbf{A}$.

Proof. Let $\mathbf{Q} = [\mathbf{v}_1, \mathbf{v}_2, \ldots, \mathbf{v}_N]$, where $\mathbf{v}_i$ are arbitrary vectors. By hypothesis,

$$\mathbf{Q}^\dagger\mathbf{A}\mathbf{Q} = \mathbf{D}. \tag{7.36}$$

Equation (7.36) implies

$$\mathbf{A}\mathbf{Q} = (\mathbf{Q}^\dagger)^{-1}\mathbf{D} = \mathbf{Q}\mathbf{D} \tag{7.37}$$

or

$$\mathbf{A}[\mathbf{v}_1, \mathbf{v}_2, \ldots, \mathbf{v}_N] = [\mathbf{v}_1, \mathbf{v}_2, \ldots, \mathbf{v}_N]\begin{bmatrix} \lambda_1 & 0 & & 0 \\ 0 & \lambda_2 & \ddots & \\ & \ddots & \ddots & 0 \\ 0 & & 0 & \lambda_N \end{bmatrix} \tag{7.38}$$

or

$$[\mathbf{A}\mathbf{v}_1, \mathbf{A}\mathbf{v}_2, \ldots, \mathbf{A}\mathbf{v}_N] = [\lambda_1\mathbf{v}_1, \lambda_2\mathbf{v}_2, \ldots, \lambda_N\mathbf{v}_N] \tag{7.39}$$

which implies that $\mathbf{v}_i$ is the eigenvector associated with λ_i. ■

Property 6: The correlation matrix $\mathbf{R}$ is unitarily diagonalizable, that is, $\mathbf{Q}^\dagger\mathbf{R}\mathbf{Q} = \mathit{diag}\{\lambda_1, \lambda_2, \ldots \lambda_N\} \triangleq \mathbf{D}$ where $\mathbf{Q} = [\mathbf{v}_1, \mathbf{v}_2, \ldots, \mathbf{v}_N]$, and λ_i and $\mathbf{v}_i$ are the eigenvalues and corresponding eigenvectors of $\mathbf{R}$.

Proof. $\mathbf{R}$ is Hermitian $\Rightarrow$ $\mathbf{R}$ is normal $\overset{Thm.\ 7.7}{\Rightarrow}$ $\mathbf{Q}^\dagger\mathbf{R}\mathbf{Q} = \mathbf{D}$, where $\mathbf{Q}$ is unitary $\overset{Thm.\ 7.8}{\Rightarrow}$ $\mathbf{Q} = [\mathbf{v}_1, \mathbf{v}_2, \cdots, \mathbf{v}_N]$. ■

Property 7: Let $\mathbf{v}_1, \mathbf{v}_2, \ldots, \mathbf{v}_N$ denote the eigenvectors corresponding to the *distinct* eigenvalues $\lambda_1, \lambda_2, \ldots \lambda_N$ of the $N \times N$ correlation matrix, $\mathbf{R}$. Then the eigenvectors are orthogonal to each other.

Proof. From the hypotheses,

$$\mathbf{R}\mathbf{v}_i = \lambda_i \mathbf{v}_i \tag{7.40}$$

and

$$\mathbf{R}\mathbf{v}_j = \lambda_j \mathbf{v}_j. \tag{7.41}$$

Pre-multiply both sides of (7.40) by $\mathbf{v}_j^\dagger$ to get

$$\mathbf{v}_j^\dagger \mathbf{R}\mathbf{v}_i = \lambda_i \mathbf{v}_j^\dagger \mathbf{v}_i. \tag{7.42}$$

The Hermitian of both sides of (7.41) gives

$$\mathbf{v}_j^\dagger \mathbf{R} = \lambda_j \mathbf{v}_j^\dagger. \tag{7.43}$$

Post-multiply both sides of above by $\mathbf{v}_i$ to get

$$\mathbf{v}_j^\dagger \mathbf{R}\mathbf{v}_i = \lambda_j \mathbf{v}_j^\dagger \mathbf{v}_i. \tag{7.44}$$

Now, subtracting (7.44) from (7.42) gives

$$(\lambda_i - \lambda_j)\mathbf{v}_j^\dagger \mathbf{v}_i = 0. \tag{7.45}$$

Since $\lambda_i \neq \lambda_j$ by hypotheses, we have $\mathbf{v}_j^\dagger \mathbf{v}_i = 0$ and the property is proved. ■

Property 8: Let $\lambda_1, \lambda_2, \cdots \lambda_N$ be the eigenvalues of the $N \times N$ correlation matrix, $\mathbf{R}$. Then the sum of these eigenvalues equals the trace[3] of $\mathbf{R}$.

Proof. From Property 6, we have

$$\mathbf{Q}^\dagger \mathbf{R}\mathbf{Q} = diag\{\lambda_1, \lambda_2, \cdots \lambda_N\} \triangleq \mathbf{D}. \tag{7.46}$$

Therefore

$$tr\{\mathbf{Q}^\dagger \mathbf{R}\mathbf{Q}\} = tr\{\mathbf{D}\} = \sum_{i=1}^{N} \lambda_i. \tag{7.47}$$

It is easy to show that for two matrices $\mathbf{A}$ and $\mathbf{B}$, $tr\{\mathbf{AB}\} = tr\{\mathbf{BA}\}$. Therefore,

$$\sum_{i=1}^{N} \lambda_i = tr\{\mathbf{Q}^\dagger(\mathbf{R}\mathbf{Q})\} = tr\{\mathbf{R}\underbrace{\mathbf{Q}\mathbf{Q}^\dagger}_{\mathbf{I}}\} = tr\{\mathbf{R}\} \tag{7.48}$$

which completes the proof. ■

[3]The *trace* (or *tr* in short) of a matrix is defined as the sum of its diagonal elements.

The following result is called the *spectral property* of the correlation matrix. The proof follows from the property that matrix $\mathbf{R}$ can be diagonalized by a unitary transformation. It is left as an exercise.

Property 9: Let the correlation matrix $\mathbf{R}$ have eigenvalues $\lambda_1, \lambda_2, \ldots \lambda_N$ and eigenvectors $\mathbf{v}_1, \mathbf{v}_2, \ldots, \mathbf{v}_N$. Then the correlation matrix can be written as

$$\mathbf{R} = \sum_{i=1}^{N} \lambda_i \mathbf{v}\mathbf{v}^{\dagger}. \tag{7.49}$$

At this time, it is natural to wonder about the diagonalization of a general matrix and not one as special as the correlation matrix. The following theorem gives the conditions for diagonalizing a matrix using the similarity transformation. Recall that the transformation $\mathbf{P}^{-1}\mathbf{A}\mathbf{P}$ is called the similarity transformation of matrix $\mathbf{A}$. This theorem will not be useful to the adaptive filter theory as treated in this text and may be omitted. However, this is an important result for many aspects of signal processing and is therefore presented here for completeness.

Theorem 7.9 *Let* $\mathbf{A}$ *be an* $N \times N$ *real matrix with eigenvalues* $\lambda_1, \lambda_2, \ldots \lambda_N$ *and corresponding eigenvectors* $\mathbf{v}_1, \mathbf{v}_2, \ldots \mathbf{v}_N$. *Define* $\mathbf{P} = [\mathbf{v}_1 \ \mathbf{v}_2 \ldots \mathbf{v}_N]$. *The following results hold.*

1. *If* $\mathbf{P}^{-1}$ *exists, then* $\mathbf{P}^{-1}\mathbf{A}\mathbf{P} = diag\{\lambda_1, \lambda_2, \cdots \lambda_N\} \triangleq \mathbf{D}$.
2. *If the eigenvalues are distinct, that is,* $\lambda_i \neq \lambda_j$, *then* $\mathbf{P}^{-1}\mathbf{A}\mathbf{P} = \mathbf{D}$.

Proof. (1) By hypothesis, $\mathbf{P}^{-1}$ exists. Consider

$$\begin{aligned} \mathbf{I} &= \mathbf{P}^{-1}\mathbf{P} \\ &= \mathbf{P}^{-1}[\mathbf{v}_1 \mathbf{v}_2 \cdots \mathbf{v}_N] \\ &= [\mathbf{P}^{-1}\mathbf{v}_1 \ \mathbf{P}^{-1}\mathbf{v}_2 \ \ldots \ \mathbf{P}^{-1}\mathbf{v}_N] \end{aligned} \tag{7.50}$$

$$= \begin{bmatrix} 1 \ 0 \ \ldots \ 0 \\ 0 \ 1 \ \ldots \ 0 \\ \vdots \quad \ddots \\ 0 \ 0 \ \ldots \ 1 \end{bmatrix}. \tag{7.51}$$

Since $\mathbf{P}^{-1}\mathbf{v}_i$ is a column, from the above it is easy to see that it has a 1 in the ith position and zeros elsewhere; that is,

$$\mathbf{P}^{-1}\mathbf{v}_i = \begin{bmatrix} 0 \cdots 0 & \underbrace{\mathbf{1}}_{\text{position } i} & 0 \cdots 0 \end{bmatrix}^T. \tag{7.52}$$

Also, since $\mathbf{Av}_i = \lambda_i \mathbf{v}_i$, using the above we have

$$\begin{aligned}\mathbf{P}^{-1}\mathbf{AP} &= \mathbf{P}^{-1}[\lambda_1\mathbf{v}_1 \ \lambda_2\mathbf{v}_2 \ \cdots \ \lambda_N\mathbf{v}_N] \\ &= [\lambda_1\mathbf{P}^{-1}\mathbf{v}_1 \ \lambda_2\mathbf{P}^{-1}\mathbf{v}_2 \ \cdots \ \lambda_N\mathbf{P}^{-1}\mathbf{v}_N] \end{aligned} \tag{7.53}$$

$$= \begin{bmatrix} \lambda_1 & 0 & \dots & 0 \\ 0 & \lambda_2 & \dots & 0 \\ \vdots & & \ddots & \\ 0 & 0 & \dots & \lambda_N \end{bmatrix} \tag{7.54}$$

and the proof of the first part is complete.

(2) In this part we will prove that if the eigenvalues of $\mathbf{A}$ are distinct, then $\mathbf{P}^{-1}$ exists. Then by (1) we will conclude that the matrix is diagonalizable by a similarity transformation. The proof will use the technique of contradiction. Suppose that $\mathbf{P}^{-1}$ does not exist. This means that the columns of $\mathbf{P}$ are *linearly dependent*.[4] That is, there exists constants α_i not all zero such that

$$\alpha_1\mathbf{v}_1 + \alpha_2\mathbf{v}_2 + \ldots + \alpha_N\mathbf{v}_N = \mathbf{0}. \tag{7.55}$$

Note that the $\mathbf{0}$ on the right-hand side is a vector and hence in boldface. Multiplying the above equation throughout by $\mathbf{A}$ and realizing that $\mathbf{Av}_i = \lambda_i\mathbf{v}_i$, we get

$$\alpha_1\lambda_1\mathbf{v}_1 + \alpha_2\lambda_2\mathbf{v}_2 + \ldots + \alpha_N\lambda_N\mathbf{v}_N = \mathbf{0}. \tag{7.56}$$

Repeatedly multiplying by $\mathbf{A}$ gives the following:

$$\alpha_1\lambda_1^2\mathbf{v}_1 + \alpha_2\lambda_2^2\mathbf{v}_2 + \ldots + \alpha_N\lambda_N^2\mathbf{v}_N = \mathbf{0} \tag{7.57}$$

$$\vdots$$

$$\alpha_1\lambda_1^{N-1}\mathbf{v}_1 + \alpha_2\lambda_2^{N-1}\mathbf{v}_2 + \ldots + \alpha_N\lambda_N^{N-1}\mathbf{v}_N = \mathbf{0}. \tag{7.58}$$

Equations (7.56), (7.57), and (7.58) can be written in the form

$$\begin{bmatrix} 1 & 1 & \dots & 1 \\ \lambda_1 & \lambda_2 & \dots & \lambda_N \\ \lambda_1^2 & \lambda_2^2 & \dots & \lambda_N^2 \\ \vdots & & & \\ \lambda_1^{N-1} & \lambda_2^{N-1} & & \lambda_N^{N-1} \end{bmatrix} \begin{bmatrix} \alpha_1\mathbf{v}_1^T \\ \alpha_2\mathbf{v}_2^T \\ \alpha_3\mathbf{v}_3^T \\ \vdots \\ \alpha_N\mathbf{v}_N^T \end{bmatrix} = \mathbf{0}. \tag{7.59}$$

[4] Linear dependency means that one vector can be represented in terms of a linear combination of one or more vectors. This implies that the determinant of the matrix is zero.

The matrix on the left in the above equation is called the *Vandermonde matrix*. This matrix has some nice properties, one of which is that its determinant is [2]

$$\det(\mathbf{V}) = \prod_{i=1\,j=1}^{N,N} (\lambda_i - \lambda_j). \tag{7.60}$$

Defining $\mathbf{V}$ as the Vandermonde matrix and $\mathbf{X}$ as the other matrix, we have

$$\mathbf{VX} = \mathbf{0}. \tag{7.61}$$

Since by hypothesis, the eigenvalues are distinct, from (7.60) we have $\det(\mathbf{V}) \neq 0$. Therefore

$$\mathbf{X} = \mathbf{V}^{-1}\mathbf{0} = \mathbf{0} \tag{7.62}$$

or

$$\begin{bmatrix} \alpha_1 \mathbf{v}_1^T \\ \alpha_2 \mathbf{v}_2^T \\ \alpha_3 \mathbf{v}_3^T \\ \vdots \\ \alpha_N \mathbf{v}_N^T \end{bmatrix} = \mathbf{0} \tag{7.63}$$

which in turn implies that $\alpha_i = 0$ for all i. This is a contradiction to our assumption. Hence $\mathbf{P}^{-1}$ exists, and then by (1), matrix $\mathbf{A}$ is diagonalizable. ■

We now present examples to illustrate some of these concepts.

EXAMPLE 7.1 *Consider the matrix*

$$\mathbf{A} = \begin{bmatrix} 0 & 1 \\ -3 & -4 \end{bmatrix}.$$

The eigenvalues are found as the solution of the characteristic equation $\det(\lambda\,\mathbf{I} - \mathbf{A}) = 0$, *as*

$$\det \begin{bmatrix} \lambda & -1 \\ 3 & \lambda + 4 \end{bmatrix} = \lambda^2 + 4\lambda + 3 = 0$$

which gives $\lambda_1 = -1, \lambda_2 = -3$. *The eigenvectors can be found as follows.*

$$\mathbf{Av}_1 = \lambda_1 \mathbf{v}_1,$$

$$(\mathbf{A} - \lambda_1 \mathbf{I})\mathbf{v}_1 = 0, \quad \mathbf{v}_1 = [a\,b]^T,$$

$$\begin{bmatrix} 1 & 1 \\ -3 & -3 \end{bmatrix} \begin{bmatrix} a \\ b \end{bmatrix} = \begin{bmatrix} 0 \\ 0 \end{bmatrix}.$$

The above gives $a = -b$. Thus, we have a choice in the eigenvector. We may choose $v_1 = [1\ -1]^T$. If we want a normalized unit eigenvector, we can simply divide the eigenvector by $\frac{1}{\sqrt{2}}$. Similarly, we can find the other eigenvector, $v_2 = [1\ -3]^T$. The eigenvalues are distinct and therefore by Theorem 7.9, part(2), matrix $\mathbf{A}$ is diagonalizable by a similarity transformation. By setting $\mathbf{P} = [\mathbf{v}_1\ \mathbf{v}_2]$, it is easy to verify that $\mathbf{P}^{-1}\mathbf{A}\mathbf{P} = diag\{-1, -3\}$.

EXAMPLE 7.2 *Now consider the matrix*

$$\mathbf{A} = \begin{bmatrix} 1 & 1 & 0 \\ 0 & 2 & 0 \\ 0 & 0 & 1 \end{bmatrix}.$$

The eigenvalues of this matrix are $\lambda_1 = \lambda_2 = 1$, $\lambda_3 = 2$. The eigenvalues are not distinct, and therefore the condition of Theorem 7.9, part (2) fails. But this condition is only a sufficient condition. Therefore, we cannot say that $\mathbf{A}$ is not diagonalizable. To find out for sure, we need to find the eigenvectors corresponding to the repeated eigenvalues and see if we can make them linearly independent. Let $\mathbf{v}_{1,2} = [a\ b\ c]^T$ represent the eigenvector corresponding to the eigenvalue of $\lambda_1 = \lambda_2 = 1$. By following the method of the last example, we get

$$(\mathbf{A} - \mathbf{I})\mathbf{v}_{1,2} = 0,$$

$$\begin{bmatrix} 0 & 1 & 0 \\ 0 & 1 & 0 \\ 0 & 0 & 0 \end{bmatrix} \begin{bmatrix} a \\ b \\ c \end{bmatrix} = \begin{bmatrix} 0 \\ 0 \\ 0 \end{bmatrix}$$

which gives $b = 0$. The elements a and c are arbitrary. So we can choose $a = 1$, $c = 0$, for eigenvector $\mathbf{v}_1$ and $a = 0$, $c = 1$ for $\mathbf{v}_2$. This will make the eigenvectors linearly independent. Since λ_3 is distinct from $\lambda_1 = \lambda_2$, the eigenvector v_3 will be linearly independent with $\mathbf{v}_1$ and $\mathbf{v}_2$. Therefore, matrix $\mathbf{A}$ is diagonalizable.

7.2 THE WIENER FILTER

The Wiener filter is the basis of adaptive filter theory. It is the optimal filter that most adaptive filtering algorithms attempt to achieve. In this section and subsequently, we will assume that all the signals and filter coefficients are real. This is done for notational simplicity and because this is true of most signals encountered in practice. The generalization to the complex case is straightforward. Consider the situation in Fig. 7.5 where $d(n)$ is the desired signal and $x(n)$ is the input signal. The input signal is processed by a filter so

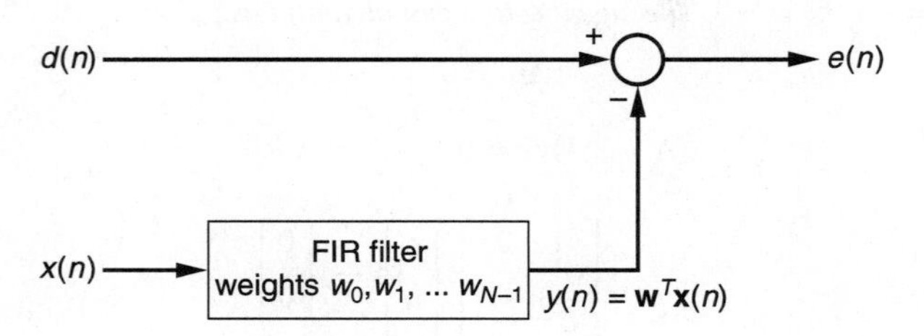

Figure 7.5 The Wiener filtering problem.

that the output is $y(n)$. The goal is to find the impulse response coefficients of this filter so that the expected value of the squared error, $E\{e^2(n)\}$ is minimized. For simplicity we will assume an FIR filter given by

$$y(n) = w_0x(n) + w_1x(n-1) + \ldots + w_{N-1}x(n-N+1) \tag{7.64}$$

$$= \mathbf{w}^T\mathbf{x}(n) \tag{7.65}$$

where $\mathbf{w}^T = [w_0\, w_1 \ldots w_{N-1}]$ and $\mathbf{x}(n) = [x(n)\, x(n-1) \ldots x(n-N+1)]^T$. The function to be minimized is called the *cost function* or *objective function*, and for the Wiener filtering problem it is given by

$$J(\mathbf{w}) = E\{e^2(n)\}$$

$$= E\{[d(n) - y(n)]^2\} \tag{7.66}$$

$$= E\{d^2(n) - 2d(n)y(n) + y^2(n)\}. \tag{7.67}$$

Now we can substitute for $y(n)$ from (7.64). Note that since $y(n)$ is a sum of products, $y^2(n)$ can be written as $\mathbf{w}^T\mathbf{x}(n)\mathbf{x}^T(n)\mathbf{w}$. With this substitution and using the fact that the expectation operation is linear, the above equation becomes

$$J(\mathbf{w}) = E\{d^2(n) - 2d(n)y(n) + y^2(n)\}$$

$$= E\{d^2(n)\} - 2E\{d(n)\mathbf{w}^T\mathbf{x}(n)\} + E\{\mathbf{w}^T\mathbf{x}(n)\mathbf{x}^T(n)\mathbf{w}\}. \tag{7.68}$$

Since the filter weight vector, $\mathbf{w}$, is not a random variable, the cost function reduces to

$$J(\mathbf{w}) = E\{d^2(n)\} - 2\mathbf{w}^TE\{d(n)\mathbf{x}(n)\} + \mathbf{w}^TE\{\mathbf{x}(n)\mathbf{x}^T(n)\}\mathbf{w}. \tag{7.69}$$

Now we assume that $d(n)$ is zero mean. Then the first term in (7.69) is equal to σ_d^2, the variance of $d(n)$. Also define $\mathbf{p} \triangleq E\{d(n)\mathbf{x}(n)\} = [p(0)\, p(1) \ldots p(N-1)]^T$ as the cross correlation between the desired signal and the input signal vector. In the third term in (7.69), we easily recognize the correlation matrix $\mathbf{R} = E\{\mathbf{x}(n)\mathbf{x}^T(n)\}$. The cost function now becomes

$$J(\mathbf{w}) = \sigma_d^2 - 2\mathbf{w}^T\mathbf{p} + \mathbf{w}^T\mathbf{R}\mathbf{w}. \tag{7.70}$$

The second term in equation (7.70) is linear in terms of the filter coefficients, and the third term is quadratic. Hence, the overall cost function is quadratic (also called convex). A convex function has a unique minimum point, which can be easily solved for by taking the gradient with respect to $\mathbf{w}$ and setting it to zero. To see this process, it is instructive to study the case of a 2-tap Wiener filter. The cost function in this case is

$$\begin{aligned}
J(w_0, w_1) &= E\left\{d^2(n)\right\} - 2E\left\{d(n)y(n)\right\} + E\left\{y^2(n)\right\} \\
&= \sigma_d^2 - 2E\left\{d(n)[w_0x(n) + w_1x(n-1)]\right\} + E\left\{[w_0x(n) + w_1x(n-1)]^2\right\} \\
&= \sigma_d^2 - 2w_0E\left\{d(n)x(n)\right\} - 2w_1E\left\{d(n)x(n-1)\right\} \\
&\quad + w_0^2E\left\{x^2(n)\right\} + 2w_0w_1E\left\{x(n)x(n-1)\right\} + w_1^2E\left\{x^2(n-1)\right\} \\
&= \sigma_d^2 - 2w_0p(0) - 2w_1p(1) + w_0^2r(0) + 2w_0w_1r(1) + w_1^2r(0).
\end{aligned} \tag{7.71}$$

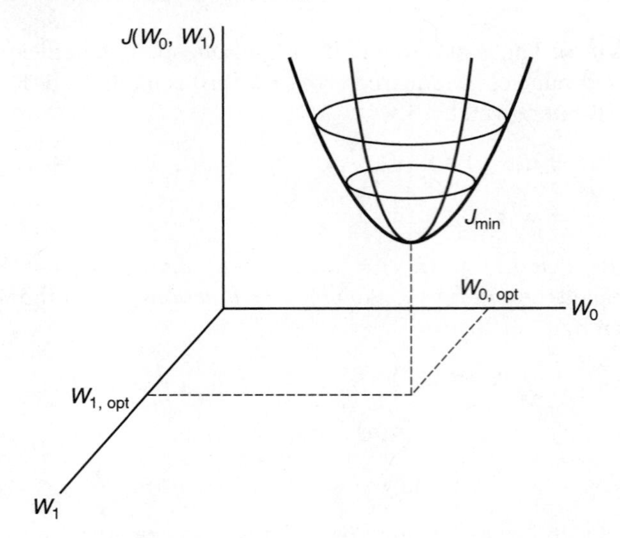

Figure 7.6 Quadratic cost function.

With the cross-correlation and autocorrelation as constants, this cost function is clearly a quadratic function in w_0 and w_1. A typical plot of this function is shown in Fig. 7.6. The minimum is at the bottom of the "bowl." The derivatives of the cost function with respect to the two weights are

$$\frac{\partial J}{\partial w_0} = -2p(0) + 2w_0 r(0) + 2w_1 r(1) \tag{7.72}$$

$$\frac{\partial J}{\partial w_1} = -2p(1) + 2w_0 r(1) + 2w_1 r(0). \tag{7.73}$$

The gradient, $\mathbf{\nabla J}(w_0, w_1)$ can then be written in matrix format as

$$\mathbf{\nabla J}(w_0, w_1) = \begin{bmatrix} \dfrac{\partial J}{\partial w_0} \\ \dfrac{\partial J}{\partial w_1} \end{bmatrix} \tag{7.74}$$

$$= -2 \begin{bmatrix} p(0) \\ p(1) \end{bmatrix} + 2 \begin{bmatrix} r(0) & r(1) \\ r(1) & r(0) \end{bmatrix} \begin{bmatrix} w_0 \\ w_1 \end{bmatrix}. \tag{7.75}$$

For the general case of the $(N-1)$-tap filter, we can apply the above process to the cost function in (7.70) to get the gradient

$$\mathbf{\nabla J}(\mathbf{w}) = -2\mathbf{p} + 2\mathbf{R}\mathbf{w}. \tag{7.76}$$

By setting this gradient to zero, we get the Wiener filter

$$\mathbf{w}_{\text{opt}} = \mathbf{R}^{-1}\mathbf{p} \tag{7.77}$$

where we have used the subscript "opt" to denote the optimal weight vector. The Wiener filter is the optimal solution to the described problem, but it is not suitable to implement in practice for a couple of reasons. First, computing the inverse of matrix **R** is computationally intensive and prohibitive for most real-time applications. Second, in practice we deal with single realizations of all the signals, and therefore **R** and **p** are not available and must be somehow estimated. Therefore, most adaptive filtering algorithms attempt to achieve the Wiener solution in an iterative manner, with the least number of computations possible per iteration. The goodness of the algorithm then depends on its computational simplicity, speed of convergence, numerical stability, and so on. In subsequent sections, we develop some popular adaptive filtering algorithms.

7.3 GRADIENT SEARCH ALGORITHMS

Let $J(\mathbf{w})$ denote the cost function to minimize, where $\mathbf{w}$ is the weight vector. Let $\mathbf{w}(n)$ be the estimate of the weights at the nth sample or iteration. The weight vector is updated as follows:

$$\mathbf{w}(n+1) = \mathbf{w}(n) + \mathbf{g}(n) \tag{7.78}$$

where $\mathbf{g}(n)$ is the direction vector. The goal now is to calculate this direction vector so that the function value at the next iteration, $J(\mathbf{w}(n+1)) = J(\mathbf{w}(n) + \mathbf{g}(n))$ is a minimum. To do this, we write the Taylor series of $J(\mathbf{w}(n+1))$, differentiate it with respect to $\mathbf{g}$, and set it equal to zero. The Taylor series is [4]

$$J(\mathbf{w}+\mathbf{g}) = J(\mathbf{w}) + \sum_i g_i \frac{\partial J(\mathbf{w})}{\partial w_i} + \frac{1}{2}\sum_i \sum_j g_i g_j \frac{\partial^2 J(\mathbf{w})}{\partial w_i \partial w_j} + \cdots \tag{7.79}$$

where we have omitted the arguments of $\mathbf{w}$ and $\mathbf{g}$ for convenience. It is useful to study the case of a 2-tap filter for which the Taylor series becomes

$$J(\mathbf{w}+\mathbf{g}) = J(\mathbf{w}) + g_0 \frac{\partial J(\mathbf{w})}{\partial w_0} + g_1 \frac{\partial J(\mathbf{w})}{\partial w_1} +$$

$$\frac{1}{2}[g_0^2 \frac{\partial^2 J(\mathbf{w})}{\partial w_0 \partial w_0} + g_0 g_1 \frac{\partial^2 J(\mathbf{w})}{\partial w_0 \partial w_1} + \tag{7.80}$$

$$g_0 g_1 \frac{\partial^2 J(\mathbf{w})}{\partial w_1 \partial w_0} + g_1^2 \frac{\partial^2 J(\mathbf{w})}{\partial w_1 \partial w_1}]. \tag{7.81}$$

Differentiating the above with respect to g_0 and g_1 gives

$$\frac{\partial J(\mathbf{w}+\mathbf{g})}{\partial g_0} = \frac{\partial J(\mathbf{w})}{\partial w_0} + g_0 \frac{\partial^2 J(\mathbf{w})}{\partial w_0 \partial w_0} + g_1 \frac{\partial^2 J(\mathbf{w})}{\partial w_0 \partial w_1} \tag{7.82}$$

$$\frac{\partial J(\mathbf{w}+\mathbf{g})}{\partial g_1} = \frac{\partial J(\mathbf{w})}{\partial w_1} + g_0 \frac{\partial^2 J(\mathbf{w})}{\partial w_0 \partial w_1} + g_1 \frac{\partial^2 J(\mathbf{w})}{\partial w_1 \partial w_1}. \tag{7.83}$$

Now define the following:

$$\frac{\partial J(\mathbf{w}+\mathbf{g})}{\partial \mathbf{g}} \triangleq \begin{bmatrix} \dfrac{\partial J(\mathbf{w}+\mathbf{g})}{\partial g_0} \\ \dfrac{\partial J(\mathbf{w}+\mathbf{g})}{\partial g_1} \end{bmatrix} \tag{7.84}$$

$$\mathbf{\nabla J} \triangleq \begin{bmatrix} \dfrac{\partial J(\mathbf{w})}{\partial w_0} \\ \dfrac{\partial J(\mathbf{w})}{\partial w_1} \end{bmatrix}. \tag{7.85}$$

Equations (7.82) and (7.83) can then be written in matrix-vector format as

$$\frac{\partial J(\mathbf{w}+\mathbf{g})}{\partial \mathbf{g}} = \mathbf{\nabla J} + \begin{bmatrix} \dfrac{\partial^2 J(\mathbf{w})}{\partial w_0 \partial w_0} & \dfrac{\partial^2 J(\mathbf{w})}{\partial w_0 \partial w_1} \\ \dfrac{\partial^2 J(\mathbf{w})}{\partial w_0 \partial w_1} & \dfrac{\partial^2 J(\mathbf{w})}{\partial w_1 \partial w_1} \end{bmatrix} \begin{bmatrix} g_0 \\ g_1 \end{bmatrix}. \tag{7.86}$$

The matrix in (7.86) is called the Hessian and is denoted by $\mathbf{H}$. Its generalization to $N \times N$ is straightforward. To find the value of $\mathbf{g}$ for which $J(\mathbf{w}+\mathbf{g})$ is minimum, we set equation (7.86) to zero. This gives

$$\mathbf{g} = -\mathbf{H}^{-1}\mathbf{\nabla J}. \tag{7.87}$$

Substituting (7.87) in (7.78), we get the update equation

$$\mathbf{w}(n+1) = \mathbf{w}(n) - \mathbf{H}^{-1}\mathbf{\nabla J}. \tag{7.88}$$

The above is called *Newton's algorithm*. We derived it for the second-order case, but it can just as easily be done for the general case. As seen by the above derivation, Newton's algorithm converges in one step if the Hessian $\mathbf{H}$ and the gradient $\mathbf{\nabla J}$ are known. Now consider Newton's algorithm as applied to the adaptive filtering problem, with the cost function given in (7.70). The gradient is given in (7.76) as $\mathbf{\nabla J}(\mathbf{w}) = -2\mathbf{p} + 2\mathbf{R}\mathbf{w}$. The Hessian is therefore $\mathbf{H} = 2\mathbf{R}$. Substituting these in Newton's update equation gives $\mathbf{w}(n+1) = \mathbf{R}^{-1}\mathbf{p} = \mathbf{w}_{\text{opt}}$. Thus, Newton's algorithm achieves the Wiener solution in one step.

The computation of the inverse of the Hessian is computationally intensive, and so Newton's algorithm is not feasible for most real-time applications. There are many ways of approximating $\mathbf{H}^{-1}$, which lead to various derivatives of Newton's algorithm.

Steepest Descent Algorithm

The steepest descent algorithm uses the approximation $\mathbf{H} = 2\mathbf{I}$ in Newton's update equation. A constant $\mu > 0$, is introduced to control the rate of convergence. The general update equation is

$$\mathbf{w}(n+1) = \mathbf{w}(n) - \frac{\mu}{2}\mathbf{\nabla J}. \tag{7.89}$$

For the adaptive filtering problem at hand, we substitute for the gradient in (7.89) to get

$$\mathbf{w}(n+1) = \mathbf{w}(n) + \mu[\mathbf{p} - \mathbf{R}\mathbf{w}(n)]. \tag{7.90}$$

The algorithm is summarized in Table 7.1. Since this is an approximation of Newton's algorithm, its convergence must be established. To do so, we first define a weight error vector

$$\boldsymbol{\epsilon}(n) = \mathbf{w}(\mathbf{n}) - \mathbf{w}_{\text{opt}} \tag{7.91}$$

where $\mathbf{w}_{\text{opt}}$ is the optimal Wiener solution as before. Subtracting $\mathbf{w}_{\text{opt}}$ from both sides of (7.90), we get

$$\mathbf{w}(n+1) - \mathbf{w}_{\text{opt}} = \mathbf{w}(n) - \mathbf{w}_{\text{opt}} + \mu[\mathbf{p} - \mathbf{R}\mathbf{w}(n)]. \tag{7.92}$$

Using the definition of the weight error vector and the fact that $\mathbf{p} = \mathbf{R}\mathbf{w}_{\text{opt}}$ gives

$$\begin{aligned} \boldsymbol{\epsilon}(n+1) &= \boldsymbol{\epsilon}(n) + \mu[\mathbf{R}\mathbf{w}_{\text{opt}} - \mathbf{R}\mathbf{w}(n)] \\ &= (\mathbf{I} - \mu\mathbf{R})\boldsymbol{\epsilon}(n). \end{aligned} \tag{7.93}$$

Now use the property that $\mathbf{R}$ can be represented as $\mathbf{R} = \mathbf{Q}\mathbf{D}\mathbf{Q}^T$ where $\mathbf{D}$ is a diagonal matrix with the eigenvalues of $\mathbf{R}$ on the diagonal and $\mathbf{Q}$ is a unitary matrix with the corresponding eigenvectors as columns. This gives

$$\boldsymbol{\epsilon}(n+1) = (\mathbf{I} - \mu\mathbf{Q}\mathbf{D}\mathbf{Q}^T)\,\boldsymbol{\epsilon}\,(n). \tag{7.94}$$

Pre-multiply the above equation by the matrix $\mathbf{Q}^T$, to get

$$\begin{aligned} \mathbf{Q}^T\boldsymbol{\epsilon}(n+1) &= \mathbf{Q}^T\boldsymbol{\epsilon}(n) - \mu\mathbf{Q}^T\mathbf{Q}\mathbf{D}\mathbf{Q}^T\boldsymbol{\epsilon}(n) \\ &= \mathbf{Q}^T\boldsymbol{\epsilon}(n) - \mu\mathbf{D}\mathbf{Q}^T\boldsymbol{\epsilon}(n). \end{aligned} \tag{7.95}$$

Define

$$\hat{\boldsymbol{\epsilon}}(n) \triangleq \mathbf{Q}^T\boldsymbol{\epsilon}(n) \tag{7.96}$$

where $\hat{\boldsymbol{\epsilon}}$ is used to denote the transformed (or rotated) version of $\boldsymbol{\epsilon}$. Equation (7.95) then becomes

$$\hat{\boldsymbol{\epsilon}}(n+1) = (\mathbf{I} - \mu\mathbf{D})\hat{\boldsymbol{\epsilon}}(n). \tag{7.97}$$

TABLE 7.1 Steepest Descent Algorithm

Initialization

$$\mathbf{w}(0) = \mathbf{0}$$

Algorithm

For $n = 0, 1, 2, \ldots$

$$\mathbf{w}(n+1) = \mathbf{w}(n) + \mu[\mathbf{p} - \mathbf{R}\mathbf{w}(n)]$$

Equation (7.97) represents a set of first-order difference equations. If $\epsilon_i(n)$ represents the ith element of the vector, then (7.94) represents the set of equations

$$\hat{\epsilon}_i(n+1) = (1 - \mu\lambda_i)\hat{\epsilon}_i(n) \tag{7.98}$$

where λ_i are the eigenvalues of $\mathbf{R}$. It is easy to verify that $\hat{\epsilon}_i(n) \longrightarrow 0$ as $n \longrightarrow \infty$ if

$$|1 - \mu\lambda_i| < 1. \tag{7.99}$$

Given the definition in (7.96), if for all i, $\hat{\epsilon}_i(n) \longrightarrow 0$ then $\epsilon_i(n) \longrightarrow 0$. This in turn implies that $\mathbf{w}(n) \longrightarrow \mathbf{w}_{\text{opt}}$; that is, the algorithm converges. Since $\lambda_i \geq 0$ (Property 5) and $\mu > 0$, the above condition is equivalent to

$$0 < \mu < \frac{2}{\lambda_i}, \forall i. \tag{7.100}$$

This must be satisfied for all the eigenvalues. Therefore, the condition for the convergence of the steepest descent algorithm is

$$0 < \mu < \frac{2}{\lambda_{\max}} \tag{7.101}$$

where $\lambda_{\max}$ is the largest eigenvalue of $\mathbf{R}$.

Computer Experiment

Let us assume that the correlation matrix of an input signal, the cross-correlation of the input and the desired signal, and the variance of the desired signal are given, respectively, by

$$\mathbf{R} = \begin{bmatrix} a & 1 \\ 1 & a \end{bmatrix}, \quad \mathbf{p} = \begin{bmatrix} 0.9 \\ 0.2 \end{bmatrix}, \quad \sigma_d^2 = 1$$

where $-1 \leq a \leq 1$. This condition ensures that the eigenvalues are nonnegative, which is true for all correlation matrices (Property 5). The eigenvalues are $1 \pm a$. The eigenvalue spread of matrix $\mathbf{R}$ is defined as the ratio of the largest eigenvalue to the smallest one. In this case, it is $\xi(\mathbf{R}) = \frac{1+a}{1-a}$. In this experiment, we will investigate the effects of varying the step size, μ and the eigenvalue spread $\xi(\mathbf{R})$ on the convergence rate of the steepest descent algorithm. The cost function for a 2-tap filter is given by (7.71) and is repeated here for convenience:

$$J_n(w_0, w_1) = \sigma_d^2 - 2w_0(n)p(0) - 2w_1(n)p(1) + w_0^2(n)r(0) + \tag{7.102}$$

$$2w_0(n)w_1(n)r(1) + w_1^2(n)r(0). \tag{7.103}$$

Notice that we have introduced the argument (n) for the weights w_0 and w_1 since they are updated every sample instant n. Consequently, we have introduced a subscript to the cost function to read J_n, in order to convey its time-varying nature. For some n, equation (7.103) gives the contour of the surface at that instant. As n progresses, these contours should get smaller and smaller as the weights approach the bottom of the bowl. In this example, by varying a, the eigenvalue spread can be varied. The update equation for the steepest descent is repeated here for convenience,

$$\mathbf{w}(n+1) = \mathbf{w}(n) + \mu[\mathbf{p} - \mathbf{R}\mathbf{w}(n)]. \tag{7.104}$$

Using a zero initial condition, the above can be easily simulated. In Fig. 7.7, we have plotted the progression of the weights $(w_0(n), w_1(n))$ with increasing n. In these two figures, we have the same eigenvalue spread of 2 but different step sizes. With $\mu = 0.1$, a lot of contours are evident, which means that the weights are converging at a slow pace. On the other hand, with $\mu = 0.8$, we see fewer contours, which implies a faster rate of convergence. This is also illustrated in Fig. 7.10. In Fig. 7.8, the two figures have the same step size but different eigenvalue spreads. The larger eigenvalue spread bends the trajectory and seems to create more contours, implying a slower rate of convergence. This is also illustrated by the learning curves in Fig. 7.9. Notice that with a smaller eigenvalue spread, the steady-state MSE is reached more quickly.

7.4 THE LMS ALGORITHM

The least mean square or LMS algorithm is simply an approximate version of the steepest descent method. For the adaptive filtering problem with a mean square cost function, recall that the steepest descent algorithm is given by

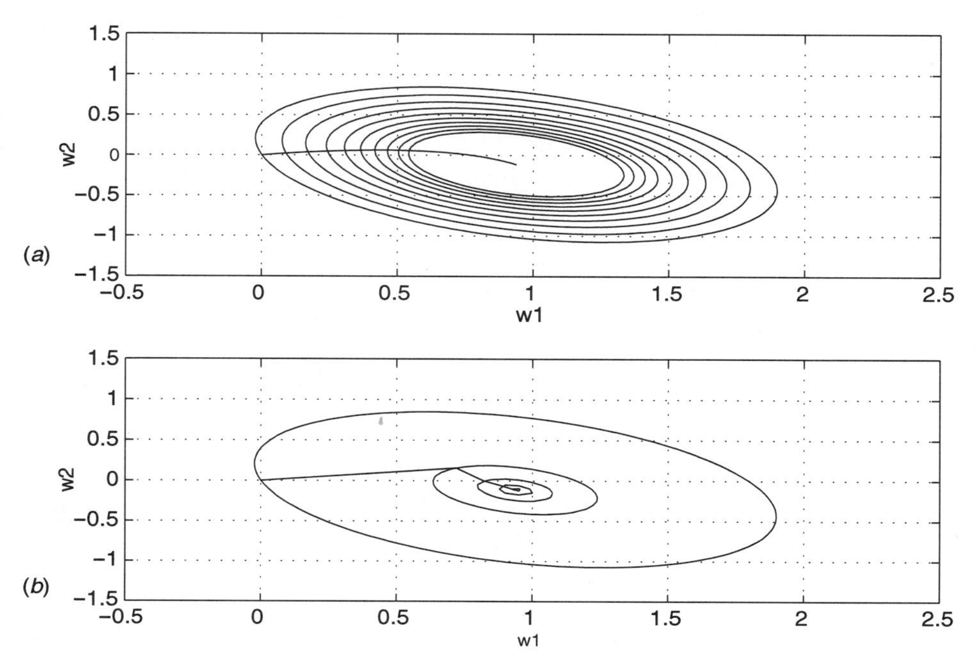

Figure 7.7 (*a*) Loci of (w_1, w_2) for $\mu = 0.1$ and eig. spread $= 2$; (*b*) loci of (w_1, w_2) for $\mu = 0.8$ and eig. spread $= 2$.

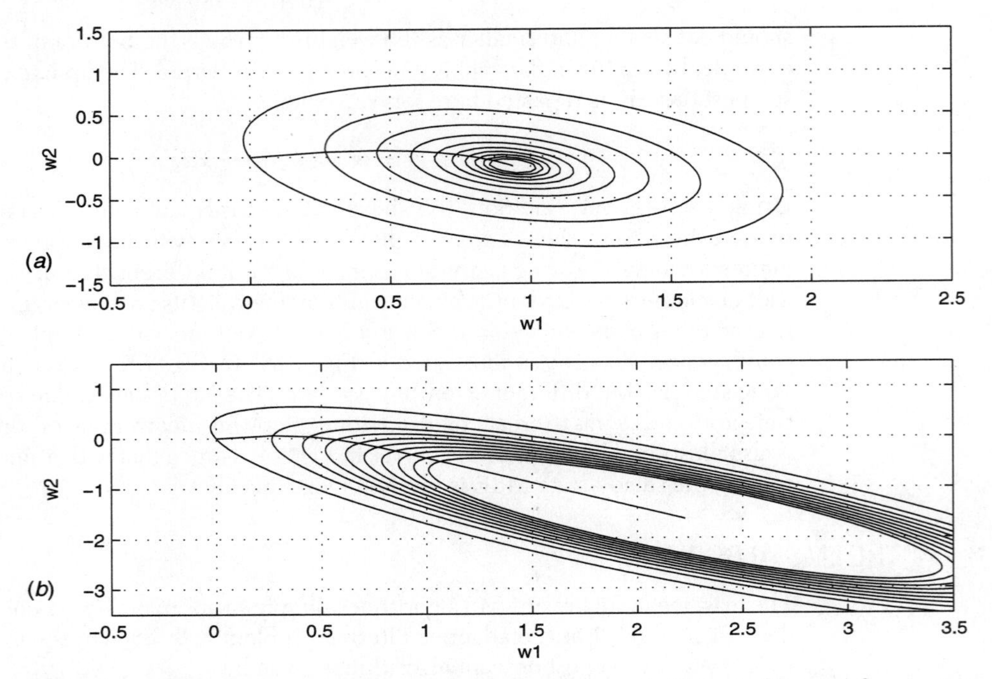

Figure 7.8 (*a*) Loci of (w_1, w_2) for $\mu = 0.3$ and eig. spread = 2; (*b*) Loci of (w_1, w_2) for $\mu = 0.3$ and eig. spread = 10.

$$\mathbf{w}(n+1) = \mathbf{w}(n) + \mu[\mathbf{p} - \mathbf{R}\mathbf{w}(n)]. \tag{7.105}$$

For most real-time applications, the parameters $\mathbf{R}$ and $\mathbf{p}$ are not known and therefore must be estimated. The simplest way to do this is the following:

$$\begin{aligned} \mathbf{R} &= E\{\mathbf{x}(n)\mathbf{x}^T(n)\} \\ &\simeq \mathbf{x}(n)\mathbf{x}^T(n); \end{aligned} \tag{7.106}$$

$$\begin{aligned} \mathbf{p} &= E\{d(n)\mathbf{x}(n)\} \\ &\simeq d(n)\mathbf{x}(n). \end{aligned} \tag{7.107}$$

Substituting the above in the steepest descent algorithm gives

$$\begin{aligned} \mathbf{w}(n+1) &= \mathbf{w}(n) + \mu[d(n)\mathbf{x}(n) - \mathbf{x}(n)\mathbf{x}^T(n)\mathbf{w}(n)] \\ &= \mathbf{w}(n) + \mu\mathbf{x}(n)[d(n) - \mathbf{x}^T(n)\mathbf{w}(n)]. \end{aligned} \tag{7.108}$$

Now, the terms inside the square brackets on the right-hand side is the error signal. Making this substitution, we get

$$\mathbf{w}(n+1) = \mathbf{w}(n) + \mu\mathbf{x}(n)e(n). \tag{7.109}$$

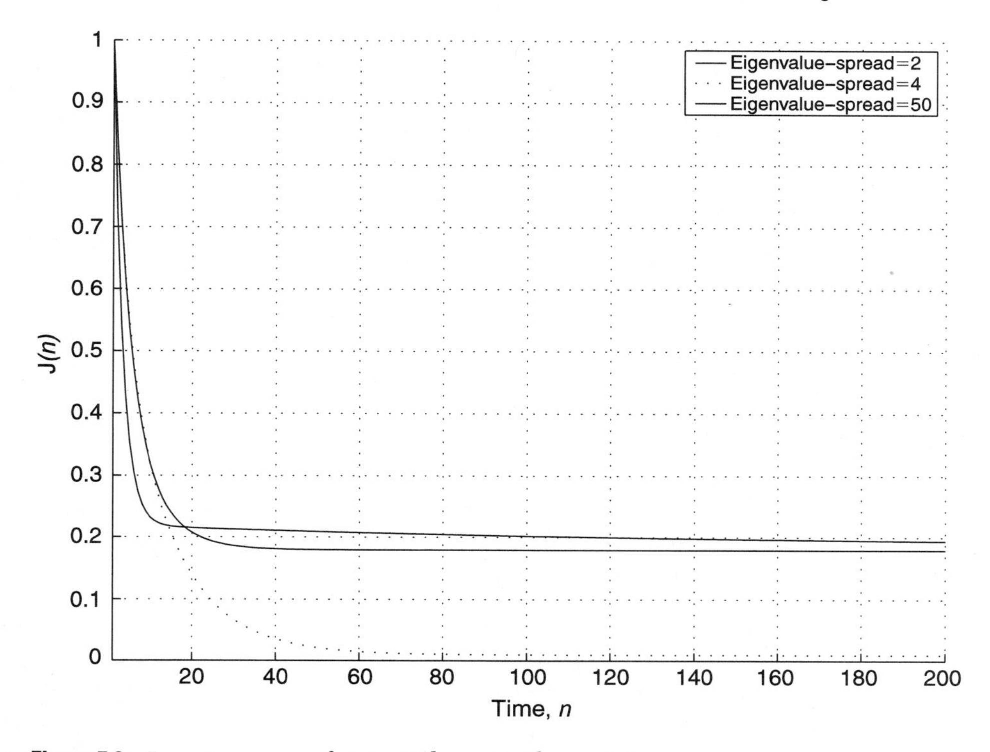

Figure 7.9 Learning curves of steepest descent with $\mu = 0.1$.

The above is the celebrated LMS algorithm, which can be traced back to Widrow and Hoff [3]. Table 7.2 summarizes the algorithm. The popularity of the LMS algorithm is due to its simplicity in implementation and computation. For a filter with N taps, the update equation (7.109) requires $N + 1$ multiplications and N additions. The filter equation $y(n) = \mathbf{x}^T(n)\mathbf{w}(n)$ requires N multiplications and $N - 1$ additions.[5] The error $e(n) = d(n) - y(n)$ requires one addition. Therefore, the LMS algorithm requires $2N + 1$ multiplications and $2N$ additions. For large N, we can say that the algorithm requires $2N$ multiplications and $2N$ additions, or $2N$ computations.[6] If an algorithm requires $(k_1 N + k_2)$ computations, where k_1 and k_2 are constants, it is said to have an "Order N computational complexity," denoted by $O(N)$. The LMS algorithm has been used in many different applications with considerable success. Numerous variations of the LMS have been reported in order to improve its performance in several applications. Computer simulations of the LMS in some applications will be presented later in this chapter. In the next two subsections, we present some theoretical results on the performance of the LMS algorithm.

[5]The accumulator in a digital processor adds two numbers at a time. Therefore, in the calculation of computation complexity, we consider only two-item additions. Also, a subtraction is considered an addition.
[6]As defined in Chapter 3, one computation equals one multiplication and one addition.

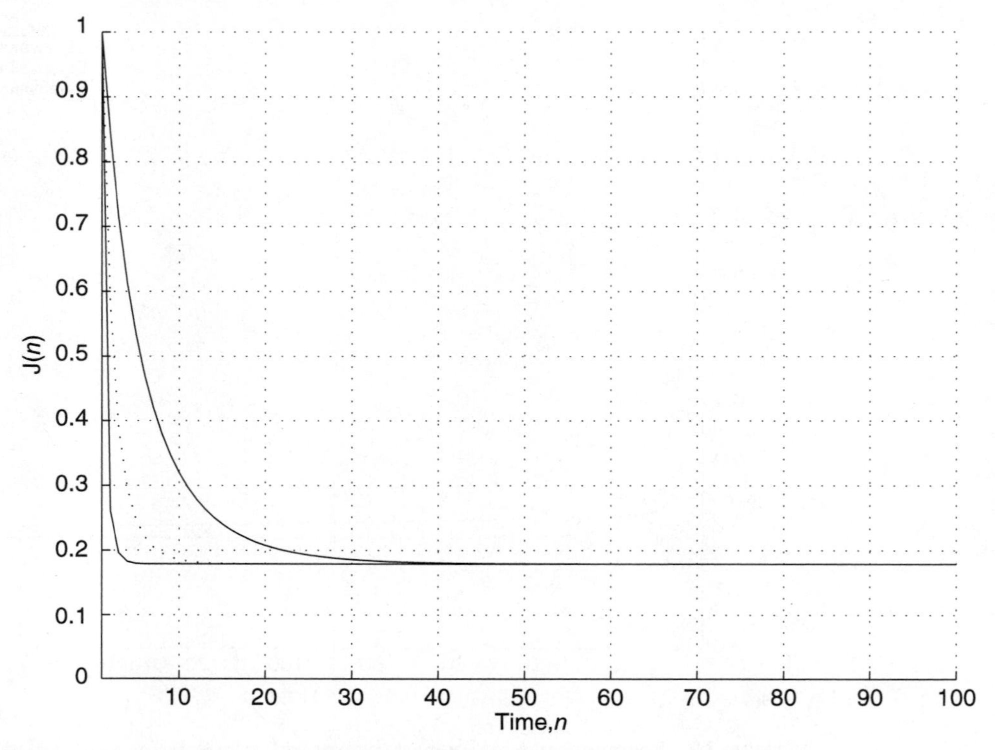

Figure 7.10 Learning curves of steepest descent with eig. spread = 2, and step sizes of 0.1, 0.3, and 0.8.

TABLE 7.2 LMS Algorithm

Initialization

$$\mathbf{w}(0) = \mathbf{0}$$

Algorithm

$$\begin{aligned} \text{For } n &= 0, 1, 2, \ldots \\ y(n) &= \mathbf{w}^T(n)\mathbf{x}(n) \\ e(n) &= d(n) - y(n) \\ \mathbf{w}(n+1) &= \mathbf{w}(n) + \mu e(n)\mathbf{x}(n) \end{aligned}$$

Convergence

Since the LMS algorithm is an approximation of the steepest descent method, its convergence must be proven. In doing so, we used the so-called *independence assumption*. The different aspects of this assumption are as follows.

1. The input vectors $\mathbf{x}(1), \mathbf{x}(2), \ldots, \mathbf{x}(n)$, are statistically independent of each other.

2. The weight vector $\mathbf{w}(n)$ is independent of the input vector $\mathbf{x}(n)$.
3. The input vector $\mathbf{x}(n)$ is independent of the previous samples of the desired signal $d(n-1), d(n-2), \ldots$

For the FIR filter structures we have been using, these assumptions do not hold because the vector $\mathbf{x}(n)$ is a shifted version of $\mathbf{x}(n-1)$ with just one new entry. This makes items (1) and (2) not valid. However, item (3) is valid. In spite of its theoretical invalidity for the FIR case, the independence assumption is still used in the the analysis of the LMS algorithm because it simplifies matters significantly. In addition, the results obtained using this simplified analysis match closely with experiments. There are certain applications such as *beamforming* [6], where the independence assumption is theoretically valid. The independence assumption can be traced back to Widrow [5]. There are proofs of convergence of the LMS algorithm without the independence assumption [7], but these are quite involved and beyond the scope of this book.

In proceeding with a simplified analysis of the LMS algorithm, we use the definition of the weight error vector defined earlier as

$$\boldsymbol{\epsilon}(n) = \mathbf{w}(n) - \mathbf{w}_{\text{opt}} \tag{7.110}$$

where $\mathbf{w}_{\text{opt}}$ is the optimal or Wiener solution. Now, subtract $\mathbf{w}_{\text{opt}}$ from both sides of the LMS update equation to get

$$\mathbf{w}(n+1) - \mathbf{w}_{\text{opt}} = \mathbf{w}(n) - \mathbf{w}_{\text{opt}} + \mu\mathbf{x}(n)\left[d(n) - \mathbf{x}^T(n)\mathbf{w}(n)\right]. \tag{7.111}$$

Taking the expectation of the above gives

$$E\left\{\boldsymbol{\epsilon}(n+1)\right\} = E\left\{\boldsymbol{\epsilon}(n)\right\} + \mu E\left\{\mathbf{x}(n)d(n)\right\} - \mu E\left\{\mathbf{x}(n)\mathbf{x}^T(n)\mathbf{w}(n)\right\}. \tag{7.112}$$

Because of item (2) in the independence assumption, we can write $E\{\mathbf{x}(n)\mathbf{x}^T(n)\mathbf{w}(n)\} = E\{\mathbf{x}(n)\mathbf{x}^T(n)\}E\{\mathbf{w}(n)\}$. Equation (7.112) then reduces to

$$E\{\boldsymbol{\epsilon}(n+1)\} = E\{\boldsymbol{\epsilon}(n)\} + \mu\left[\mathbf{p} - \mathbf{R}E\{\mathbf{w}(n)\}\right]. \tag{7.113}$$

From Wiener filter theory, we have $\mathbf{w}_{\text{opt}} = \mathbf{R}^{-1}\mathbf{p}$. Using this in (7.113), we get

$$E\{\boldsymbol{\epsilon}(n+1)\} = E\{\boldsymbol{\epsilon}(n)\} + \mu\left[\mathbf{R}\mathbf{w}_{\text{opt}} - \mathbf{R}E\{\mathbf{w}(n)\}\right] \tag{7.114}$$

or

$$E\{\boldsymbol{\epsilon}(n+1)\} = (\mathbf{I} - \mu\mathbf{R})\,E\{\boldsymbol{\epsilon}(n)\}. \tag{7.115}$$

This equation is essentially the same as equation (7.93) obtained during the proof of convergence of the steepest descent method. The only difference is that here we have an expectation operator on the weight error vectors. Using the same analysis as before, we can show that $E\{\boldsymbol{\epsilon}(n)\} \longrightarrow 0$ as $n \longrightarrow \infty$ if

$$0 < \mu < \frac{2}{\lambda_{\max}} \tag{7.116}$$

where $\lambda_{\max}$ is the largest eigenvalue of $\mathbf{R}$. This implies that $E\{\mathbf{w}(n) - w_{\text{opt}}\} \longrightarrow 0$; that is, $E\{\mathbf{w}(n)\} \longrightarrow w_{\text{opt}}$. We have thus proved that under the condition of (7.116), the expected value of the weight vector converges to the Wiener solution. This is called convergence in the mean and is a simplified form of the convergence analysis.

In practice, the eigenvalues of the correlation matrix are not usually known. Therefore, μ cannot be calculated using the formula in (7.116). A practically useful formula is now derived. Let $\mathbf{D} = \mathbf{Q}^T\mathbf{R}\mathbf{Q}$, be the unitary transformation of $\mathbf{R}$ as before, where $\mathbf{D} = \text{diag}\{\lambda_1, \ldots, \lambda_N\}$. Obviously, $\lambda_{\max} \leq tr\{\mathbf{D}\}$. From Property 8, we know that $tr\{\mathbf{R}\} = tr\{\mathbf{D}\}$. Thus, we have

$$\begin{aligned} \lambda_{\max} &\leq tr\{\mathbf{D}\} \\ &= tr\{\mathbf{R}\}. \end{aligned} \tag{7.117}$$

In the LMS algorithm, the correlation matrix is approximated as $\mathbf{R} \simeq \mathbf{x}(n)\mathbf{x}^T(n)$. Therefore

$$\begin{aligned} tr\{\mathbf{R}\} &\simeq tr\{\mathbf{x}(n)\mathbf{x}^T(n)\} \\ &= tr\left\{\begin{bmatrix} x^2(n) & x(n)x(n-1) & \ldots\ x(n)x(n-N+1) \\ x(n-1)x(n) & x^2(n-1) & \ldots\ x(n-1)x(n-N+1) \\ \vdots & & \ddots \\ x(n-N+1)x(n) & x(n-N+1)x(n-1) & \ldots\ x^2(n-N+1) \end{bmatrix}\right\} \\ &= x^2(n) + x^2(n-1) + \cdots + x^2(n-N+1) \\ &\triangleq ||\mathbf{x}(n)||^2. \end{aligned}$$

The term $||\mathbf{x}(n)||$ defined above is called the Euclidean norm of the vector. This is really the power of the signal and is usually known or can be estimated a priori. We now have

$$\lambda_{\max} \leq ||\mathbf{x}(n)||^2. \tag{7.118}$$

This condition implies that $\frac{2}{||\mathbf{x}(n)||^2} \leq \frac{2}{\lambda_{\max}}$. Using this in (7.116), another sufficient condition for the convergence of the LMS algorithm is

$$0 < \mu < \frac{2}{||\mathbf{x}(n)||^2}. \tag{7.119}$$

This condition is stricter than (7.116) but is more practical.

Misadjustment

As discussed earlier in the chapter, the misadjustment of an adaptive algorithm is a measure of its steady-state error with respect to that of the Wiener solution. It is defined as

$$\begin{aligned} M &\triangleq \frac{J_{\text{ex}}}{J_{\min}} \\ &= \frac{J_{\text{ss}} - J_{\min}}{J_{\min}} \end{aligned} \tag{7.120}$$

where J_{ex} is called the excess MSE, J_{min} is the MSE of the optimum Wiener filter, and J_{ss} is the steady-state MSE of the algorithm under consideration. It is useful to derive a formula for the misadjustment of the LMS algorithm in order to study its dependence on μ and the eigenvalues of $\mathbf{R}$. Since the derivation is lengthy, we first present a couple of lemmas that will be used in the derivation of the misadjustment formula.

Lemma 7.1 *Consider the LMS algorithm given in (7.109). Define the optimal error as* $e_{opt}(n) \triangleq d(n) - \mathbf{w}_{opt}^T\mathbf{x}(n)$. *In steady state, that is, after the filter has converged,*

$$E\{e_{opt}(n)\mathbf{x}(n)\} = 0. \tag{7.121}$$

Proof. Taking the expectation of both sides of the LMS update equation in steady state gives

$$E\{\mathbf{w}(n+1)\} = E\{\mathbf{w}(n)\} + \mu E\{e(n)\mathbf{x}(n)\}. \tag{7.122}$$

In the steady state, the means of the filter weights have converged to $\mathbf{w}_{opt}$ and $e(n)$ is replaced by $e_{opt}(n)$ in (7.122). We now have

$$\mathbf{w}_{opt} = \mathbf{w}_{opt} + \mu E\{e_{opt}(n)\mathbf{x}(n)\} \tag{7.123}$$

which leads to the conclusion. ■

The above result states that after the means of the weights have converged, the term $\mu e(n)\mathbf{x}(n)$ is a zero-mean noise vector added to the weights.

Lemma 7.2 *Consider the LMS algorithm given in (7.109). Let the definitions of weight error vector,* $\boldsymbol{\epsilon}(n)$ *and its rotated counterpart,* $\hat{\boldsymbol{\epsilon}}(n)$ *be as in (7.91) and (7.96), respectively. The excess MSE is given by*

$$J_{ex} = \sum_{k=1}^{N} \lambda_k E\{\hat{\delta}_k^2(n)\} \tag{7.124}$$

where $\lambda_1, \ldots, \lambda_N$ are the eigenvalues of the input correlation matrix, and $\hat{\delta}_k(n)$ is the kth element of the vector $\hat{\boldsymbol{\epsilon}}(n)$.

Proof. The error signal in the LMS algorithm is given by

$$e(n) = d(n) - \mathbf{w}^T\mathbf{x}(n). \tag{7.125}$$

Adding and subtracting $\mathbf{w}_{opt}^T x(n)$ on the right-hand side and defining $e_{opt}(n) \triangleq d(n) - \mathbf{w}_{opt}^T x(n)$, we get

$$\begin{aligned} e(n) &= d(n) - \mathbf{w}_{opt}^T\mathbf{x}(n) - \mathbf{w}^T\mathbf{x}(n) + \mathbf{w}_{opt}^T\mathbf{x}(n) \\ &= e_{opt}(n) - (\mathbf{w}^T(n) - \mathbf{w}_{opt}^T)\mathbf{x}(n) \\ &= e_{opt}(n) - \boldsymbol{\epsilon}^T(n)\mathbf{x}(n). \end{aligned} \tag{7.126}$$

Recall that the cost function is $J_n(\mathbf{w}) = E\{e^2(n)\}$. Squaring and taking the expectation of both sides of the above, we get

$$\begin{aligned} J_n(\mathbf{w}) &= E\left\{[e_{\text{opt}}(n) - \boldsymbol{\epsilon}^T(n)\mathbf{x}(n)]^2\right\} \\ &= E\left\{e^2_{\text{opt}}(n) - 2e_{\text{opt}}(n)\boldsymbol{\epsilon}^T(n)\mathbf{x}(n) + \boldsymbol{\epsilon}^T(n)\mathbf{x}(n)\boldsymbol{\epsilon}^T(n)\mathbf{x}(n)\right\} \\ &= E\left\{e^2_{\text{opt}}(n)\right\} - 2E\left\{e_{\text{opt}}(n)\boldsymbol{\epsilon}^T(n)\mathbf{x}(n)\right\} + E\{\boldsymbol{\epsilon}^T(n)\mathbf{x}(n)\mathbf{x}^T(n)\boldsymbol{\epsilon}(n)\}. \end{aligned} \quad (7.127)$$

Because of the independence assumption, the expectation operator in (7.127) equation can be distributed as

$$J_n(\mathbf{w}) = E\left\{e^2_{\text{opt}}(n)\right\} - 2E\left\{\boldsymbol{\epsilon}^T(n)\right\} E\left\{e_{\text{opt}}(n)\mathbf{x}(n)\right\} + E\{\boldsymbol{\epsilon}^T(n)\mathbf{x}(n)\mathbf{x}^T(n)\boldsymbol{\epsilon}(n)\}. \quad (7.128)$$

By Lemma 7.1, $E\{e_{\text{opt}}(n)\mathbf{x}(n)\} = 0$, and so the second term in the above vanishes. The first term is really $J_{\min}$. We therefore get

$$J_n(\mathbf{w}) = J_{\min} + E\left\{\boldsymbol{\epsilon}^T(n)\mathbf{x}(n)\mathbf{x}^T(n)\boldsymbol{\epsilon}(n)\right\}. \quad (7.129)$$

If we assume that n is sufficiently large that the filter has reached steady state, then $J(\mathbf{w})|_{n \text{ large}} = J_{ss}$. Also recall that $J_{ex} = J_{ss} - J_{\text{opt}}$. Thus, equation (7.129) becomes

$$J_{ex} = E\left\{\boldsymbol{\epsilon}^T(n)\mathbf{x}(n)\mathbf{x}^T(n)\boldsymbol{\epsilon}(n)\right\}. \quad (7.130)$$

Now we will perform some manipulations on the above to get the desired result. Clearly, J_{ex} is a scalar. The trace of a scalar is the scalar itself. In addition, for two matrices of compatible dimensions, it is known that $tr\{\mathbf{AB}\} = tr\{\mathbf{BA}\}$. Using these facts, we manipulate (7.130) as follows:

$$\begin{aligned} J_{ex} &= E\left\{tr\left\{[\boldsymbol{\epsilon}^T(n)][\mathbf{x}(n)\mathbf{x}^T(n)\boldsymbol{\epsilon}(n)]\right\}\right\} \\ &= E\left\{tr[\mathbf{x}(n)\mathbf{x}^T(n)\boldsymbol{\epsilon}(n)][\boldsymbol{\epsilon}^T(n)]\right\} \\ &= tr\left\{E\left\{\mathbf{x}(n)\mathbf{x}^T(n)\boldsymbol{\epsilon}(n)\boldsymbol{\epsilon}^T(n)\right\}\right\} \\ &= tr\left\{E\left\{\mathbf{x}(n)\mathbf{x}^T(n)\right\}E\{\boldsymbol{\epsilon}(n)\boldsymbol{\epsilon}^T(n)\}\right\}. \end{aligned} \quad (7.131)$$

In the last step above, the independence assumption was used to distribute the expectation operator. Now write $E\{\mathbf{x}(n)\mathbf{x}^T(n)\} = \mathbf{R} = \mathbf{QDQ}^T$, where $\mathbf{Q}$ is the unitary transformation matrix and $\mathbf{D}$ is a diagonal matrix of eigenvalues, as before. Equation (7.131) then becomes

$$\begin{aligned} J_{ex} &= tr\{\mathbf{R}E\{\boldsymbol{\epsilon}(n)\boldsymbol{\epsilon}^T(n)\}\} \\ &= tr\{E\{\mathbf{R}\boldsymbol{\epsilon}(n)\boldsymbol{\epsilon}^T(n)\}\} \\ &= tr\{E\{\mathbf{QDQ}^T\boldsymbol{\epsilon}(n)\boldsymbol{\epsilon}^T(n)\}\} \\ &= E\{tr\{[\mathbf{QDQ}^T\boldsymbol{\epsilon}(n)][\boldsymbol{\epsilon}^T(n)]\}\} \\ &= E\{tr\{\boldsymbol{\epsilon}^T(n)\mathbf{QDQ}^T\boldsymbol{\epsilon}(n)\}\} \\ &= E\{tr\{[\mathbf{Q}^T\boldsymbol{\epsilon}(n)]^T\mathbf{D}[\mathbf{Q}^T\boldsymbol{\epsilon}(n)]\}\}. \end{aligned} \quad (7.132)$$

Note that $\mathbf{Q}^T\boldsymbol{\epsilon}(n) = \hat{\boldsymbol{\epsilon}}(n)$, as defined earlier, the rotated version of $\boldsymbol{\epsilon}(n)$. The quantity within the trace is a scalar, and so the trace can be removed. The excess MSE is therefore

$$\begin{aligned} J_{\text{ex}} &= E\{\hat{\boldsymbol{\epsilon}}^T(n)\mathbf{D}\hat{\boldsymbol{\epsilon}}^T(n)\} \\ &= \sum_{k=1}^{N} \lambda_k E\{\hat{\delta}_k^2(n)\} \end{aligned} \tag{7.133}$$

and the proof is complete. ■

The above lemmas will now be used to derive the formula for misadjustment given in the following theorem.

Theorem 7.10 *Let $\lambda_1, \ldots, \lambda_N$ denote the eigenvalues of the correlation matrix of the input, and let μ denote the step size. The misadjustment of the LMS algorithm is given by*

$$M = \sum_{k=1}^{N} \frac{\mu\lambda_k}{2 - \mu\lambda_k}. \tag{7.134}$$

Proof. Consider the LMS algorithm, and subtract $\mathbf{w}_{\text{opt}}$ from both sides to get

$$\mathbf{w}(n+1) - \mathbf{w}_{\text{opt}} = \mathbf{w}(n) - \mathbf{w}_{\text{opt}} + \mu e(n)\mathbf{x}(n). \tag{7.135}$$

Recalling the definition of the weight error vector, $\boldsymbol{\epsilon}(n)$, we get

$$\begin{aligned} \boldsymbol{\epsilon}(n+1) &= \boldsymbol{\epsilon}(n) + \mu e(n)\mathbf{x}(n) \\ &= \boldsymbol{\epsilon}(n) + \mu\mathbf{x}(n)[d(n) - \mathbf{x}^T(n)\mathbf{w}(n)] \\ &= \boldsymbol{\epsilon}(n) + \mu\mathbf{x}(n)d(n) - \mathbf{x}(n)\mathbf{x}^T(n)\mathbf{w}(n). \end{aligned} \tag{7.136}$$

Taking the expectation of both sides of the above and using the independence assumption, we get

$$E\{\boldsymbol{\epsilon}(n+1)\} = E\{\boldsymbol{\epsilon}(n)\} + \mu\mathbf{p} - \mu\mathbf{R}E\{\mathbf{w}(n)\} \tag{7.137}$$

where $\mathbf{R}$ and $\mathbf{p}$ are as usual the autocorrelation matrix and cross-correlation vector, respectively. Substituting $\mathbf{p} = \mathbf{R}\mathbf{w}_{\text{opt}}$ in the above and after some simple manipulations, we get

$$E\{\boldsymbol{\epsilon}(n+1)\} = E\{\boldsymbol{\epsilon}(n)\} - \mu\mathbf{R}E\{\boldsymbol{\epsilon}(n)\}. \tag{7.138}$$

Now we multiply throughout by the unitary transformation matrix, $\mathbf{Q}^T$ and use the fact that $\mathbf{Q}^T E\{\boldsymbol{\epsilon}(n)\} = E\{\mathbf{Q}^T\boldsymbol{\epsilon}(n)\} \triangleq E\{\hat{\boldsymbol{\epsilon}}(n)\}$, which is the rotated weight error vector. We also substitute $\mathbf{R} = \mathbf{Q}\mathbf{D}\mathbf{Q}^T$. Then we have

$$\begin{aligned} E\{\hat{\boldsymbol{\epsilon}}(n+1)\} &= E\{\hat{\boldsymbol{\epsilon}}(n)\} - \mu\mathbf{Q}^T\mathbf{Q}\mathbf{D}\mathbf{Q}^T E\{\boldsymbol{\epsilon}(n)\} \\ &= (\mathbf{I} - \mu\mathbf{D})E\{\hat{\boldsymbol{\epsilon}}(n)\}. \end{aligned} \tag{7.139}$$

The above is a set of first-order difference equations given by

$$E\{\hat{\delta}_k(n+1)\} = (1-\mu\lambda_k)E\{\hat{\delta}_k(n)\}, \quad k=1,2,\ldots,N. \tag{7.140}$$

If the step size satisfies the condition $0 < \mu < \frac{2}{\lambda_{max}}$, then all of the above equations converge to zero, that is, in equation (7.139), $E\{\hat{\boldsymbol{\epsilon}}(n)\} \longrightarrow 0$. Now, if we take the expectation operators out in this equation, then we must add a zero-mean noise vector on the right-hand side as follows.

$$\hat{\boldsymbol{\epsilon}}(n+1) = (\mathbf{I}-\mu\mathbf{D})\hat{\boldsymbol{\epsilon}}(n) + \boldsymbol{\Upsilon}(n) \tag{7.141}$$

Recall from earlier discussions that the error on the weight vector is the zero-mean noise vector $\mu e(n)\mathbf{x}(n)$. However, in equation (7.141), the weight error vector is rotated. Thus, we have

$$\boldsymbol{\Upsilon}(n) = \mu\mathbf{Q}^T e(n)\mathbf{x}(n). \tag{7.142}$$

Now we will show that the elements of $\boldsymbol{\Upsilon}(n)$ are not correlated with each other. This will assure that (7.141) is a set of independent first-order difference equations. The correlation matrix of $\boldsymbol{\Upsilon}(n)$ is

$$\begin{aligned} E\{\boldsymbol{\Upsilon}(n)\boldsymbol{\Upsilon}^T(n)\} &= \mu^2 E\{e^2(n)\mathbf{Q}^T\mathbf{x}(n)[\mathbf{Q}^T\mathbf{x}(n)]^T\} \\ &= \mu^2 E\{e^2(n)\mathbf{Q}^T\mathbf{x}(n)\mathbf{x}^T(n)\mathbf{Q}\} \\ &= \mu^2 E\{e^2(n)\}\mathbf{Q}^T\mathbf{R}\mathbf{Q}. \end{aligned} \tag{7.143}$$

Since we have assumed that the weights have converged to the optimal values, $E\{e^2(n)\} = J_{\min}$. So, the above becomes

$$E\{\boldsymbol{\Upsilon}(n)\boldsymbol{\Upsilon}^T(n)\} = \mu^2 J_{\min}\mathbf{D}. \tag{7.144}$$

The above matrix is clearly diagonal, with the diagonal elements being $\mu^2 J_{\min}\lambda_k$. Thus, we have proved that the elements $\boldsymbol{\Upsilon}_k(n)$ of the rotated noise vector $\boldsymbol{\Upsilon}(n)$ are uncorrelated. That is, the first-order equations in (7.141) are completely uncoupled. The diagonal elements of the correlation matrix are the variances of the elements of $\boldsymbol{\Upsilon}(n)$, that is,

$$\sigma_k^2 = \mu^2 J_{\min}\lambda_k. \tag{7.145}$$

The first-order equations in (7.141) are

$$\hat{\delta}_k(n+1) = (1-\mu\lambda_k)\hat{\delta}_k(n) + \Upsilon_k(n), \quad k=1,2,\ldots,N. \tag{7.146}$$

The impulse response of the above first-order filter is

$$g(n) = (1-\mu\lambda_k)^n, \quad n \geq 0. \tag{7.147}$$

The output of the filter in response to an input of $\Upsilon_k(n)$ is simply its convolution with $g(n)$ and is given by

$$\hat{\delta}_k(n) = \sum_{i=0}^{\infty}(1-\mu\lambda_k)^i \Upsilon_k(n-i). \tag{7.148}$$

Now we use equation (7.124) in Lemma 7.2 to find the excess MSE. Toward that end, we first find

$$E\{\hat{\delta}_k^2(n)\} = E\left\{\sum_{i=0}^{\infty}\sum_{j=0}^{\infty}(1-\mu\lambda_k)^i(1-\mu\lambda_k)^j\Upsilon_k(n-i)\Upsilon_k(n-j)\right\}$$

$$= \sum_{i=0}^{\infty}\sum_{j=0}^{\infty}(1-\mu\lambda_k)^i(1-\mu\lambda_k)^j E\{\Upsilon_k(n-i)\Upsilon_k(n-j)\}. \quad (7.149)$$

Since we have already shown that the elements of $\mathbf{\Upsilon}(n)$ are uncorrelated, we have

$$E\{\Upsilon_k(n-i)\Upsilon_k(n-j)\} = \begin{cases} 0, & i \neq j \\ \sigma_k^2, & i = j. \end{cases} \quad (7.150)$$

Substituting (7.150) in (7.149), we get

$$E\{\hat{\delta}_k^2(n)\} = \sum_{i=0}^{\infty}(1-\mu\lambda_k)^i(1-\mu\lambda_k)^i\sigma_k^2$$

$$= \sum_{i=0}^{\infty}(1-\mu\lambda_k)^{2i}\sigma_k^2$$

$$= \frac{\sigma_k^2}{1-(1-\mu\lambda_k)^2}. \quad (7.151)$$

Now, we substitute (7.151) in the formula for the excess MSE given in Lemma 7.2, to get

$$J_{ex} = \sum_{k=1}^{N}\lambda_k E\{\hat{\delta}_k^2(n)\}$$

$$= \sum_{k=1}^{N}\frac{\lambda_k\sigma_k^2}{1-(1-\mu\lambda_k)^2}. \quad (7.152)$$

Next we substitute the formula for σ_k^2 given in (7.145) in equation (7.152) to get

$$J_{ex} = \mu^2 J_{\min}\sum_{k=1}^{N}\frac{\lambda_k^2}{1-(1-\mu\lambda_k)^2}. \quad (7.153)$$

This gives

$$M = \frac{J_{ex}}{J_{\min}}$$

$$= \mu^2\sum_{k=1}^{N}\frac{\lambda_k^2}{1-(1-\mu\lambda_k)^2}$$

$$= \sum_{k=1}^{N}\frac{\mu\lambda_k}{2-\mu\lambda_k} \quad (7.154)$$

and the proof is complete. ■

If we examine the formula for misadjustment, we see that the denominator is always positive because of the convergence condition of the LMS algorithm. We also know that μ is positive and λ_k are nonnegative. Therefore, the larger the step size, the higher the misadjustment. We also know that the larger the step size (within its limits), the faster the rate of convergence. So, the step size must be chosen in order to trade off between the misadjustment and convergence rate. It is always nice to illustrate the theoretical results with experimental simulations. Hence, we will present two experiments based on the LMS algorithm. However, a little background on noise cancellation is first necessary.

Noise Cancellation

A block diagram of a basic noise cancellation scheme is shown in Fig. 7.11. The primary source is the signal of interest, denoted by $s(n)$ plus a noise sequence, $\Upsilon_1(n)$. The secondary source is another noise sequence, $\Upsilon_2(n)$. It is assumed that all of these signals are zero-mean and stationary. It is also assumed that the noise sources are correlated with each other but not with $s(n)$. The error signal is given by

$$e(n) = s(n) + \Upsilon_1(n) - y(n). \tag{7.155}$$

Squaring both sides and taking the expectation, we get

$$e^2(n) = s^2(n) + [\Upsilon_1(n) - y(n)]^2 - 2s(n)[\Upsilon_1(n) - y(n)], \tag{7.156}$$

$$E\{e^2(n)\} = E\{s^2(n)\} + E\{[\Upsilon_1(n) - y(n)]^2\} - 2E\{s(n)[\Upsilon_1(n) - y(n)]\}. \tag{7.157}$$

Since $s(n)$ is not correlated with any of the noise sequences, the last term on the right-hand side vanishes, and we have

$$E\{e^2(n)\} = E\{s^2(n)\} + E\{[\Upsilon_1(n) - y(n)]^2\}. \tag{7.158}$$

In (7.158), $E\{s^2(n)\}$ is independent of the filter weights. So, minimizing the MSE is essentially the same as minimizing $E\{[\Upsilon_1(n) - y(n)]^2\}$. This means that the error signal $e(n)$ becomes equal to an estimate of the desired signal $s(n)$.

In the above scheme, we assumed the availability of a primary source (signal plus noise) and a secondary source (correlated noise). However, this is not always true in practice. Often, we simply have one source which is composed of the signal plus noise. Consider the block diagram shown in Fig. 7.12. The input is a sum of a narrowband signal, $N(n)$, and a broadband signal, $B(n)$, where these signals are not correlated with each other.

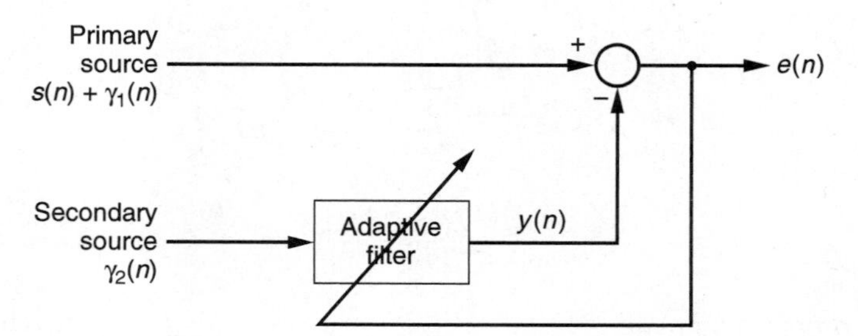

Figure 7.11 A basic noise cancellation scheme.

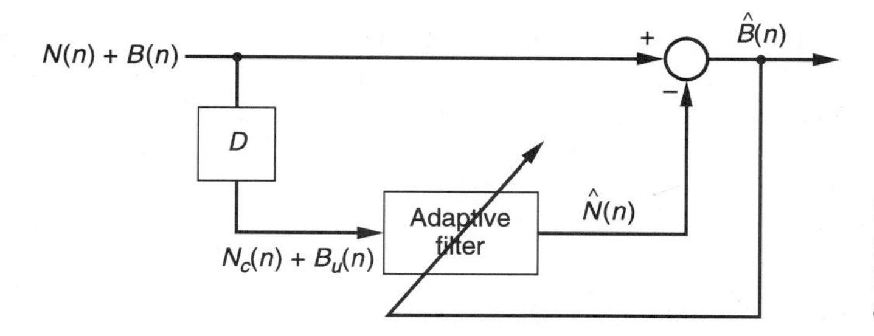

Figure 7.12 Adaptive Line Enhancer (ALE) or Narrowband Interference Canceler (NIC).

The sum of these signals is delayed and applied to the input of an adaptive filter. The amount of delay should be sufficient to decorrelate the broadband signal, and it depends on its spectrum. If $B(n)$ is white, then a delay of one sample will decorrelate it. Since $N(n)$ is narrowband, simply delaying it does not decorrelate it. Now, the input to the adaptive filter is $N_c(n)+B_u(n)$, where $N_c(n)$ is correlated with $N(n)$ and $B_u(n)$ is uncorrelated with $B(n)$. We have used the subscript c for correlated and u for uncorrelated. By using the same technique as in equations (7.155)–(7.158), it is not difficult to show that by minimizing $E\{e^2(n)\}$, the output of the adaptive filter becomes an estimate of the narrowband signal $N(n)$. The mathematical proof of this is left as an exercise. This also makes sense intuitively. The filter will not be able to convert $B_u(n)$ to $B(n)$ since these signals are not correlated. So, the filter converts $N_c(n)$ to an estimate of $N(n)$, denoted by $\widehat{N}(n)$, thereby canceling the narrowband signals in the error. The adaptive filter thus predicts the current sample of $N(n)$ from its past samples and also attenuates the broadband signal. This scheme is therefore called *Linear Prediction*. It is now easy to see that the frequency response of the filter should be bandpass in order to pull out the narrowband signal from the broadband signal. If the narrowband signal is the useful signal, then the output is taken from the output of the filter and the system is called an *Adaptive Line Enhancer* (ALE). If the broadband signal is the useful signal and the narrowband signal is the noise, then the output is taken from the adder, and the system is called a *Narrowband Interference Canceler* (NIC).

Computer Experiments

System Identification

In this example, we will consider the identification of an FIR system given by the difference equation

$$g(n) = 0.1x(n) + 0.2x(n-1) + 0.3x(n-2).$$

Let the input signal be a white Gaussian sequence with unit variance. The output is then generated using the above difference equation. White Gaussian noise is then added to this output to form the desired signal, $d(n)=g(n)+v(n)$. The SNR of the desired signal is 10 dB; that is,

$$10\log\left(\frac{\text{var}\{d(n)\}}{\text{var}\{v(n)\}}\right) = 10 \text{ dB}.$$

The desired signal and the input signal are then used to identify the system with a 3-tap FIR filter and the LMS algorithm. A step size of $\mu=0.1$ is chosen. The experiment is repeated 100 times to get an ensemble of data. The instantaneous squared error for one run and the ensemble average (MSE) are shown in Fig. 7.13(a). Obviously, the MSE does not vary as much as the instantaneous squared error. The weights are plotted in Fig. 7.13(b) and clearly converge in less than 50 iterations. In Fig. 7.13(c), we show the MSE for step sizes of 0.1 and 0.01. The MSE for the smaller step size converges more slowly, as expected.

Adaptive Line Enhancer (ALE)
Consider the block diagram of Fig. 7.12. The useful signal is $N(n) = \sin(2\pi f_0/f_s)$, with $f_0 = 25$ Hz and $f_s = 500$ Hz. The discrete sine wave therefore has 20 samples per cycle. The sequence $B(n)$ is a white Gaussian noise with variance such that the SNR of $N(n) + B(n)$ is 10 dB. The noise therefore has very high amplitude as shown in Fig. 7.14(a). The signal plus noise combination is delayed by one sample and used as an input to a 128-tap adaptive FIR filter. The LMS algorithm is used with $\mu = 0.00005$. The output of the filter after 3000 iterations is shown in Fig. 7.14(b), which is clearly the sine wave. The magnitude response of the filter after convergence is shown in Fig. 7.14(c). It is clearly a narrowband bandpass filter. This filter passes the normalized frequency 0.1. In this plot, the unit normalized frequency is equal to $\frac{f_s}{2} = 250$ Hz. Therefore, the filter passes the frequency of 25 Hz and attenuates the rest, as desired.

Normalized LMS Algorithm

The normalized LMS (NLMS) algorithm is a special case of the LMS where the step size is optimized for faster convergence. This comes at the price of higher computational complexity. In order to derive the algorithm, we will use the line search idea used in [8].

Recall the cost function for the LMS algorithm as

$$J(\mathbf{w}) = \sigma_d^2 - 2\mathbf{w}^T\mathbf{p} + \mathbf{W}^T\mathbf{R}\mathbf{w}, \tag{7.159}$$

where the correlation matrix and the cross-correlation vector are approximated as

$$\begin{aligned} \mathbf{R} &= E\{\mathbf{x}(n)\mathbf{x}^T(n)\} \\ &\simeq \mathbf{x}(n)\mathbf{x}^T(n); \end{aligned} \tag{7.160}$$

$$\begin{aligned} \mathbf{p} &= E\{d(n)\mathbf{x}(n)\} \\ &\simeq d(n)\mathbf{x}(n). \end{aligned} \tag{7.161}$$

The gradient is then

$$\nabla \mathbf{J} = -2\mathbf{x}(n)d(n) + 2\mathbf{x}(n)\mathbf{x}^T(n)\mathbf{w}(n). \tag{7.162}$$

If we use a time-varying step size $\mu(n)$, then the LMS algorithm can be written as

$$\mathbf{w}(n+1) = \mathbf{w}(n) + \mu(n)e(n)\mathbf{x}(n). \tag{7.163}$$

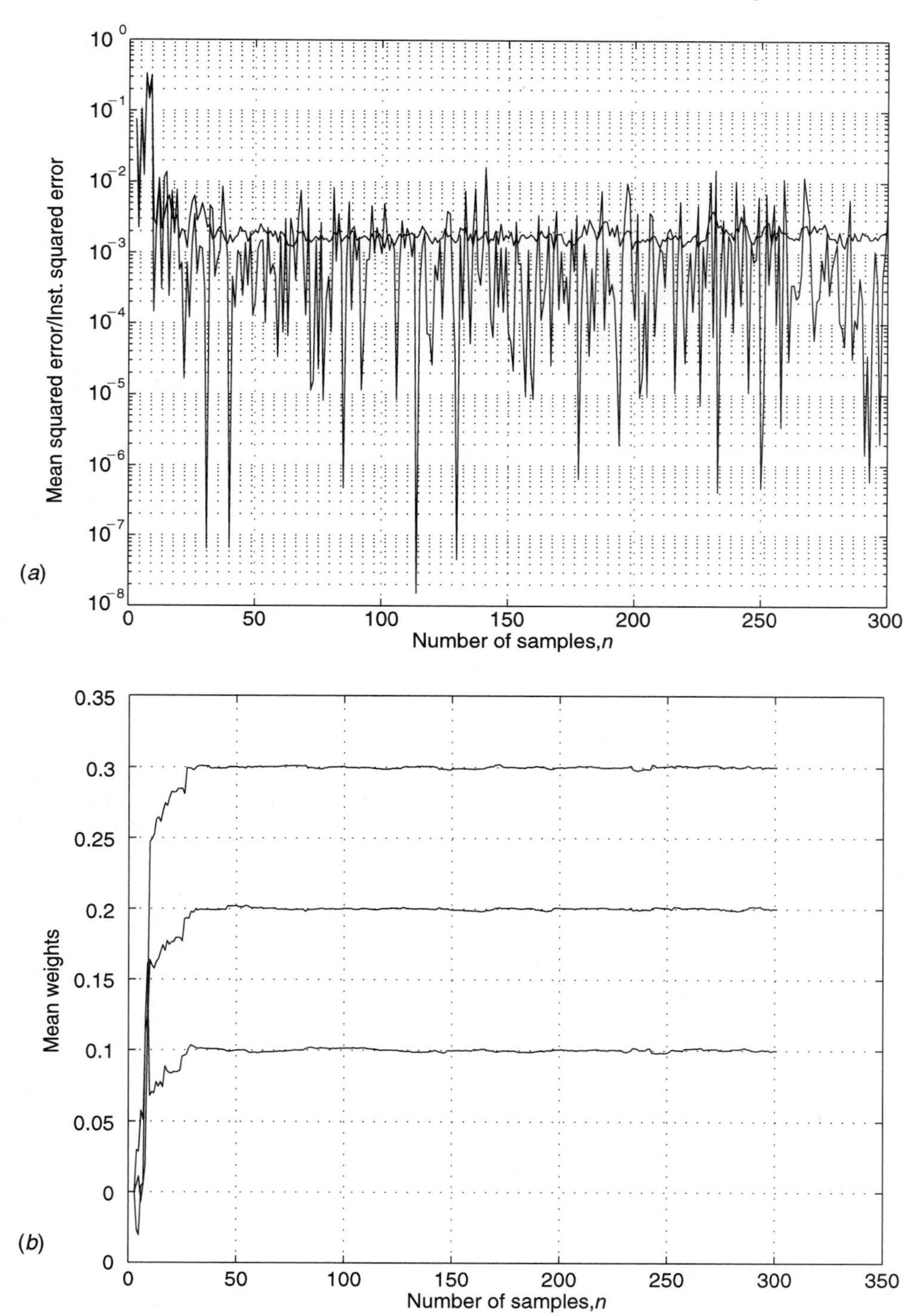

Figure 7.13 (continued)

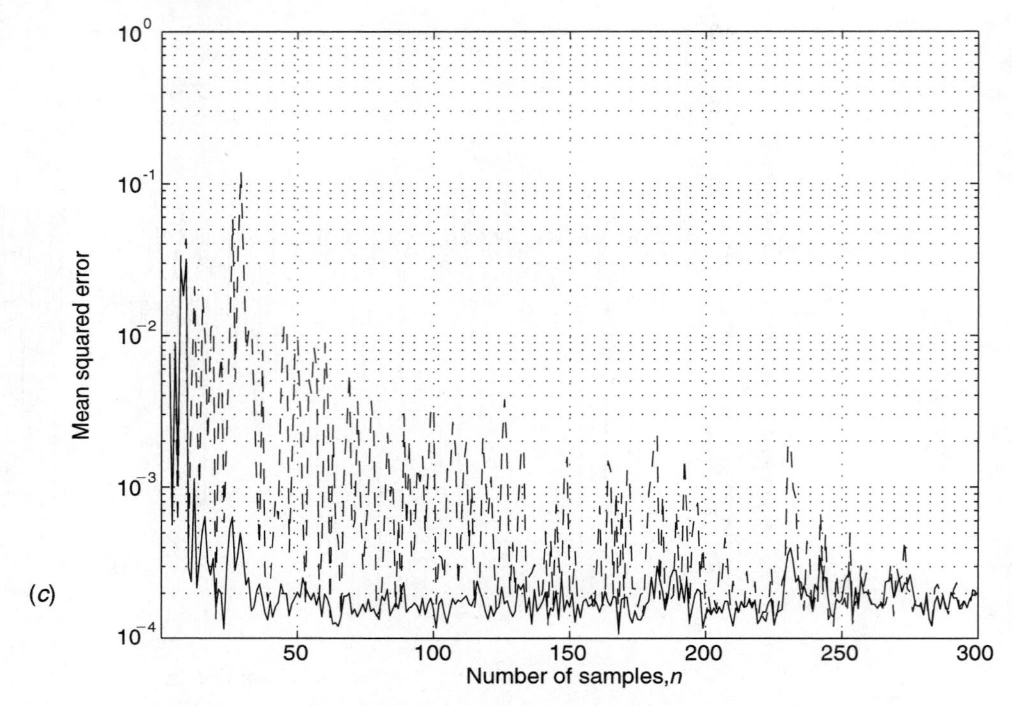

Figure 7.13 System identification example: (*a*) MSE and instantaneous squared error; (*b*) convergence of filter weights; (*c*) solid line for $\mu = 0.1$; dashed line for $\mu = 0.01$.

The concept of line search is that we find the optimum value of μ at every iteration, so that the cost function at the next iteration $J(\mathbf{w}(n+1))$ is minimized. From the steepest descent method, we know that the next iteration is

$$\mathbf{w}(n+1) = \mathbf{w}(n) - \frac{\mu}{2}\nabla\mathbf{J} \tag{7.164}$$

where we have used $\frac{\mu}{2}$ so that the final result is consistent with the most common form of the NLMS. The cost function at the next iteration is

$$\begin{aligned} J(\mathbf{w}(n+1)) &= J(\mathbf{w}(n) + \mu(n)e(n)\mathbf{x}(n)) \\ &= \sigma_d^2 - 2\mathbf{w}^T(n+1)\mathbf{p} + \mathbf{w}^T(n+1)\mathbf{R}\mathbf{w}(n+1). \end{aligned} \tag{7.165}$$

Now substitute for $\mathbf{R}$,$\mathbf{p}$, and $\mathbf{w}(n+1)$ from (7.160), (7.161), and (7.164), respectively, to get

$$\begin{aligned} J(\mathbf{w}(n+1)) &= \sigma_d^2 - 2[\mathbf{w}(n) - \frac{\mu}{2}\nabla\mathbf{J}]^T d(n)\mathbf{x}(n) + \\ &\quad [\mathbf{w}(n) - \frac{\mu}{2}\nabla\mathbf{J}]^T\mathbf{x}(n)\mathbf{x}^T(n)[\mathbf{w}(n) - \frac{\mu}{2}\nabla\mathbf{J}]. \end{aligned} \tag{7.166}$$

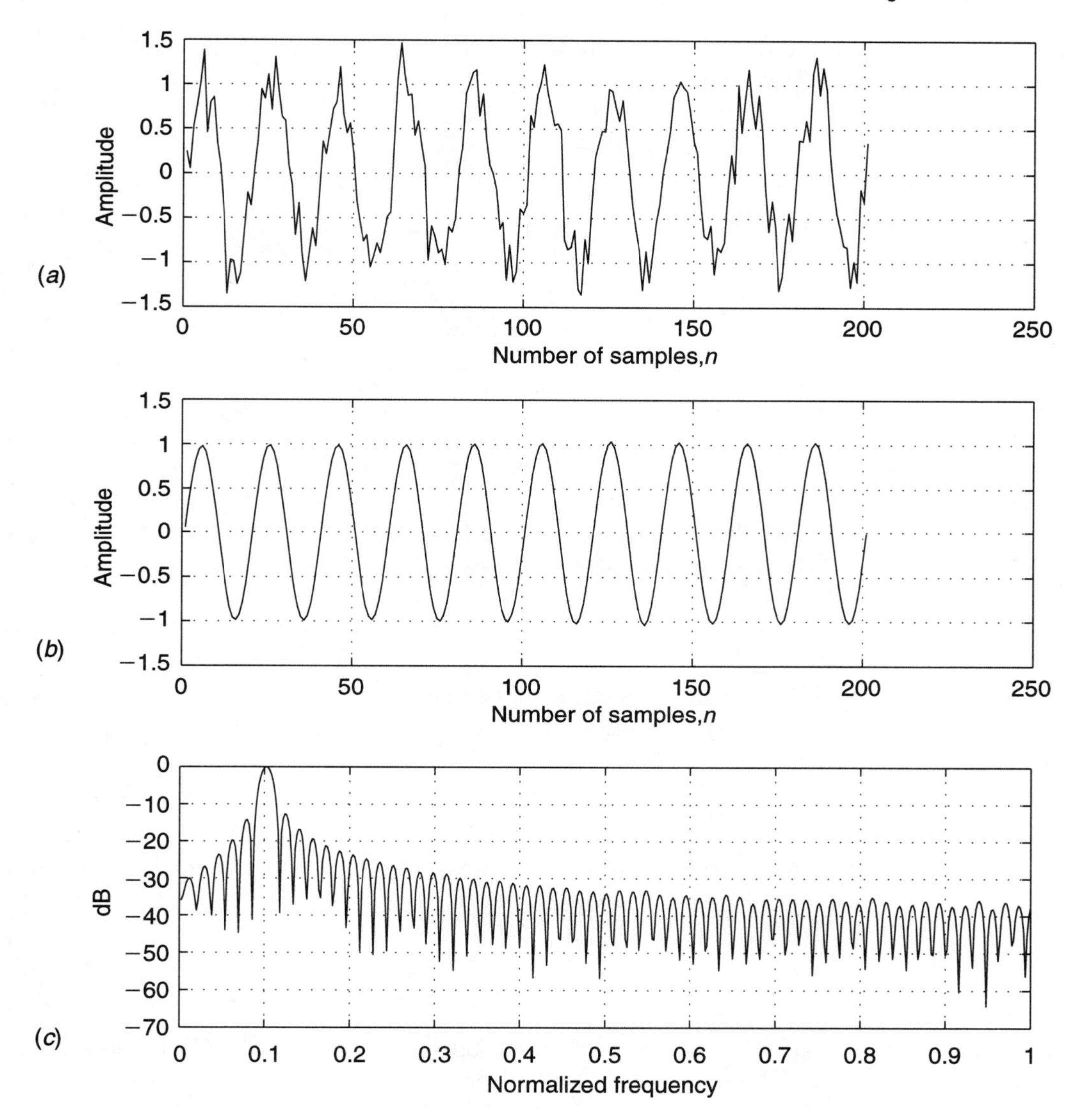

Figure 7.14 ALE example: (*a*) signal plus noise; (*b*) output of adaptive filter; (*c*) magnitude response of filter after convergence.

In order to find the value of μ that will minimize the cost function at the next step, we differentiate the above with respect to μ and set it equal to zero. The result is

$$\mu = -\frac{2[d(n) - \mathbf{w}^T(n)\mathbf{x}(n)]}{\nabla\mathbf{J}^T\mathbf{x}(n)}$$

$$= -\frac{2e(n)}{\nabla\mathbf{J}^T\mathbf{x}(n)}. \tag{7.167}$$

Substituting (7.167) in the LMS update equation and using the expression for $\nabla\mathbf{J}$ from (7.162), we get

$$\begin{aligned}\mathbf{w}(n+1) &= \mathbf{w}(n) - \frac{2e(n)}{\nabla\mathbf{J}^T\mathbf{x}(n)}e(n)\mathbf{x}(n)\\ &= \mathbf{w}(n) - \frac{2e^2(n)\mathbf{x}(n)}{[-2\mathbf{x}(n)d(n) + 2\mathbf{x}(n)\mathbf{x}^T(n)\mathbf{w}(n)]^T\mathbf{x}(n)}\\ &= \mathbf{w}(n) + \frac{e^2(n)\mathbf{x}(n)}{[\mathbf{x}(n)e(n)]^T\mathbf{x}(n)}\\ &= \mathbf{w}(n) + \frac{e(n)\mathbf{x}(n)}{\mathbf{x}^T(n)\mathbf{x}(n)}\\ &= \mathbf{w}(n) + \frac{1}{||\mathbf{x}(n)||^2}e(n)\mathbf{x}(n).\end{aligned} \tag{7.168}$$

For flexibility in the control of convergence speed, a positive constant α may be introduced. The final NLMS algorithm is

$$\mathbf{w}(n+1) = \mathbf{w}(n) + \frac{\alpha}{||\mathbf{x}(n)||^2}e(n)\mathbf{x}(n). \tag{7.169}$$

In order to find the condition for convergence of the NLMS algorithm, we use one of the convergence conditions of the LMS algorithm given by

$$0 < \mu < \frac{2}{||\mathbf{x}(n)||^2}. \tag{7.170}$$

The step size in the NLMS is $\frac{\alpha}{||\mathbf{x}(n)||^2}$. Applying the above condition on this step size, we get the convergence condition

$$0 < \alpha < 2. \tag{7.171}$$

At this time, some comments on the NLMS algorithm are in order.

1. The NLMS converges faster than the standard LMS because the step size is optimized at each iteration.
2. If the input signal is very small for some length of time, then $||\mathbf{x}(n)||^2$ is also small. This may cause numerical instability problems in the division. This can be overcome by adding a small positive number, β to $||\mathbf{x}(n)||^2$. However, this makes the line search not optimal any more. The NLMS algorithm is summarized in Table 7.3.
3. The computational complexity of the NLMS is higher than the LMS due to the computation of $||\mathbf{x}(n)||^2$. However, note that $||\mathbf{x}(n)||^2 = x^2(n) + x^2(n-1) + \ldots + x^2(n-N+1)$. At each iteration, the complete sum does not need to be calculated. We can simply add the square of the current input and subtract the square of the oldest input. This is equivalent to two additions. There is one more addition due to the constant β. In terms of multiplication, we need an extra one

TABLE 7.3 Normalized LMS

Initialization
$\mathbf{w}(0) = \mathbf{0}$
Algorithm
For $n = 0, 1, 2, \ldots$
$y(n) = \mathbf{w}^T(n)\mathbf{x}(n)$
$e(n) = d(n) - y(n)$
$\mathbf{w}(n+1) = \mathbf{w}(n) + \frac{\alpha}{\lVert\mathbf{x}(n)\rVert^2 + \beta} e(n)\mathbf{x}(n)$

to form $x^2(n)$. Also, there is a division (considered a multiplication) by $||\mathbf{x}(n)||^2$. Therefore, the NLMS requires a total of $2N + 3$ multiplications and $2N + 3$ additions per iteration. The computational complexity is still $O(N)$.

It would seem that the increase in computations over the LMS is minimal and that this should be the preferred algorithm owing to its faster convergence rate. However, the division required by the NLMS is a problem in many real-time implementations. Moreover, the introduction of the constant β makes the line search suboptimal, which may slow the convergence rate in many cases. For these reasons, the standard LMS is still very popular. Computer experiments involving the NLMS algorithm are assigned as computer projects.

7.5 FREQUENCY-DOMAIN ADAPTIVE FILTERING

In Frequency-Domain Adaptive Filtering (FDAF), the input signal $u(n)$ and the desired signal $d(n)$ are first transformed to the frequency domain by FFT. Then filtering is done on the transformed signal $u(n)$ to produce the transformed output signal. Next, an inverse FFT produces the time-domain output signal $y(n)$. This frequency-domain approach has the following advantages.

1. As we know, the fast adaptive algorithms have a computational complexity of $O(N)$, where N is the number of filter taps. If N is large, the total number of computations per sample can be very high. This can be circumvented by using an IIR filter. However, adaptive IIR filters can have instability problems. Therefore, the FDAF offers an alternate solution where the number of computations can be significantly reduced, as we shall see shortly.
2. In Time-Domain Adaptive Filtering, the rate of convergence is dictated by the eigenvalue spread of the input autocorrelation matrix. By frequency transformation, it is possible to remove or reduce eigenvalue disparity problems, thereby improving the convergence rate of the algorithm.
3. In many applications, an FFT of the input or output signal is necessary. In the case of FDAF, the FFTs are readily available at no extra computational cost.

The block diagram of the FDAF is shown in Fig. 7.15. The first step is to take the serial input data $u(n)$ and form blocks of data. If L is the block size, then the data can be segmented as

$$u_i(n) = u(iL + n),\ n = 0, 1, \ldots, L - 1,\ i = 0, 1, 2, \ldots \tag{7.172}$$

where $u_i(n)$ denotes the samples within the ith block. The same is done to the desired signal $d(n)$ so that

$$d_i(n) = d(iL + n),\ n = 0, 1, \ldots, L - 1,\ i = 0, 1, 2, \ldots \tag{7.173}$$

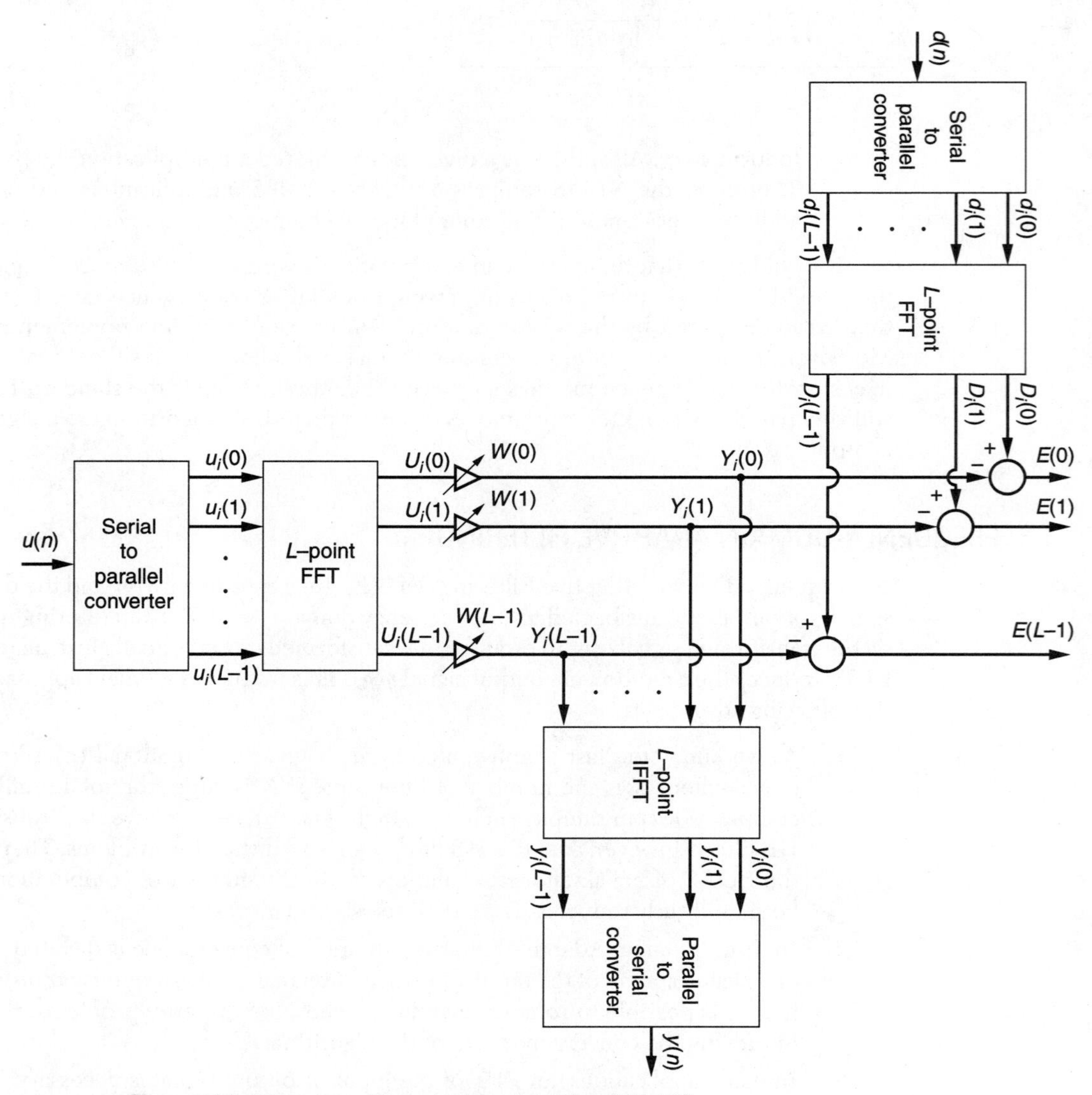

Figure 7.15 Frequency Domain Adaptive Filter (FDAF).

The data blocks for the input and desired signals are transformed by FFT into their respective frequency domains. Filtering and weight update are performed in the frequency domain. The transformed data $U_i(n)$ and $D_i(n)$ are possibly complex. The input signal, desired signal, and the weights $W(l), l = 0, 1, \ldots, L-1$ are assumed to be real. The weight update can be done by any algorithm, but we will now use the LMS for simplicity. The filtered signal is obtained as

$$Y_i(k) = W_i(k)U_i(k), \; k = 0, 1, \ldots, L-1 \tag{7.174}$$

where $Y_i(k)$ denotes the ith block of output data. By analogy with the time-domain LMS algorithm, the weight update equation becomes

$$W_{i+1}(k) = W_i(k) + \mu_f E_i(k)U_i^*(k), \; k = 0, 1, \ldots, L-1 \tag{7.175}$$

where

$$E_i(k) = D_i(k) - Y_i(k) \tag{7.176}$$

and the complex conjugate of $U_i(k)$ is used since the transformed data are possibly complex. Define the following vectors and matrix:

$$\mathbf{W}_i = [W_i(0), W_i(1), \ldots, W_i(L-1)]^T$$

$$\mathbf{E}_i = [E_i(0), E_i(1), \ldots, E_i(L-1)]^T$$

$$\mathbf{U}_i = \begin{bmatrix} U_i(0) & 0 & 0 \ldots & 0 \\ 0 & U_i(1) & 0 \ldots & 0 \\ \vdots & & & \\ 0 & 0 & 0 \ldots & U_i(L-1) \end{bmatrix}.$$

The update equation in vector-matrix format then becomes

$$\mathbf{W}_{i+1} = \mathbf{W}_i + \mu_f \mathbf{U}_i^* \mathbf{E}_i \tag{7.177}$$

for $i = 0, 1, 2, \ldots$ Comparing this equation to the time-domain LMS, we make the following observations.

1. The weight vector is updated only once for every block of data, that is, every L data points. In the time-domain LMS, the weight vector is updated every sample.
2. In the FDAF, a complex conjugate of the transformed data is used.
3. In the FDAF, the $\mathbf{U}_i^*$ matrix is diagonal. This means that the update equation (7.177) is decoupled and can be written as L independent scalar equations, as given by (7.175). This is not the case for the time-domain LMS algorithm.

Computational Complexity

For a filter length of L, the time-domain LMS requires $2L+1$ multiplications per sample. For a block of L samples, this translates to $2L^2+L$ multiplications per block. In the FDAF, the block size is the same as the filter length. It requires three L-point FFTs (including an IFFT), each requiring $2L \log_2 L$ real multiplications. Since the input, desired signal, and output are all real, the transforms are all symmetric. Therefore, only half as many multiplications are required. This gives a total of $3L \log_2 L$ multiplications for the three FFTs. The filtering operation (7.174) requires $2L$ real multiplications. This is from the fact that the weights $W_i(k)$ are real and the transformed input data $U_i(k)$ are complex. The update equation (7.177) requires $4L$ real multiplications. Summing it all up, we find that the FDAF requires a total of $3L \log_2 L + 6L$ real multiplications per block. When this is compared to the time-domain LMS, the computational advantage is significant, especially for long filters. The computational advantage for the FDAF is approximately a factor $\frac{2L+1}{3\log_2 L+6}$. For example, if $L = 256$, the number of multiplications required by the time-domain LMS is about 17 times that of the FDAF.

Convergence

The simplest notion of convergence is that in the mean, as we saw earlier for the time-domain LMS. A similar analysis can be done for the FDAF. Combining equations (7.174) and (7.175), we get

$$W_{i+1}(k) = (1 - \mu_f |U_i(k)|^2)\, W_i(k) + \mu_f D_i(k) U_i^*(k). \tag{7.178}$$

Taking the expectation of both sides gives

$$E\{W_{i+1}(k)\} = (1 - \mu_f E\{|U_i(k)|^2\})\, E\{W_i(k)\} + \mu_f E\{D_i(k) U_i^*(k)\}, \tag{7.179}$$

where we have assumed the independence of $U_i(k)$ and $W_i(k)$ as before. Now define

$$V_i(k) = E\{W_i(k)\}.$$

Then we have

$$V_{i+1}(k) = (1 - \mu_f E\{|U_i(k)|^2\}) V_i(k) + \mu_f E\{D_i(k) U_i^*(k)\}. \tag{7.180}$$

For a given k, the above is a first-order difference equation of the form

$$V_{i+1}(k) = \beta V_i(k) + p \tag{7.181}$$

where

$$\beta = 1 - \mu_f E\{|U_i(k)|^2\} \tag{7.182}$$

$$p = \mu_f E\{D_i(k) U_i^*(k)\}. \tag{7.183}$$

Assuming that p is bounded,

$$\lim_{i\to\infty} V_{i+1} = \frac{p}{1-\beta}$$

provided $|\beta| < 1$. Putting the index k back, and using (7.182) and (7.183), we have

$$\lim_{i\to\infty} V_{i+1}(k) = \frac{E\{D_i(k)U_i^*(k)\}}{E\{|U_i(k)|^2\}}. \tag{7.184}$$

Thus, the mean of the weights converges as

$$E\{W_i(k)\} \stackrel{i\to\infty}{\longrightarrow} \frac{E\{D_i(k)U_i^*(k)\}}{E\{|U_i(k)|^2\}}, \quad k = 0, 1, \ldots, L. \tag{7.185}$$

The sufficient condition for convergence is $-1 < \beta < 1$, which translates to

$$0 < \mu_f < \frac{2}{E\{|U_i(k)|^2\}}. \tag{7.186}$$

Note that the expectation operation is on the index i since it stands for the ensemble average. The convergence condition must therefore be satisfied for all $k = 0, 1, \ldots, L-1$. The above condition is similar to that of the time-domain LMS, except that it is now dependent on the power of the input in the frequency domain.

7.6 BLOCK LMS ALGORITHM

The block LMS (BLMS) algorithm is a time-domain algorithm designed to reduce the computational complexity of the LMS. As in the FDAF, this is done by updating the coefficients only once per block of data instead of at every sample. The BLMS is also important because of its relationship to the Fast LMS (FLMS) algorithm, which is a frequency-domain implementation of the block LMS. The FLMS will be discussed in the next section.

The block diagram of the BLMS algorithm is shown in Fig. 7.16. The serial input signal $u(n)$ is segmented into blocks of length L. Let i denote the block index and N denote the length of the filter. Then the actual time index n can be written as

$$n = iL + j, \quad j = 0, 1, \ldots, L-1, \quad i = 0, 1, 2, \ldots \tag{7.187}$$

For each block, the weight vector is

$$\mathbf{w}(i) = [w_0(i)\, w_1(i) \ldots w_{N-1}(i)]^T, \quad i = 0, 1, \ldots \tag{7.188}$$

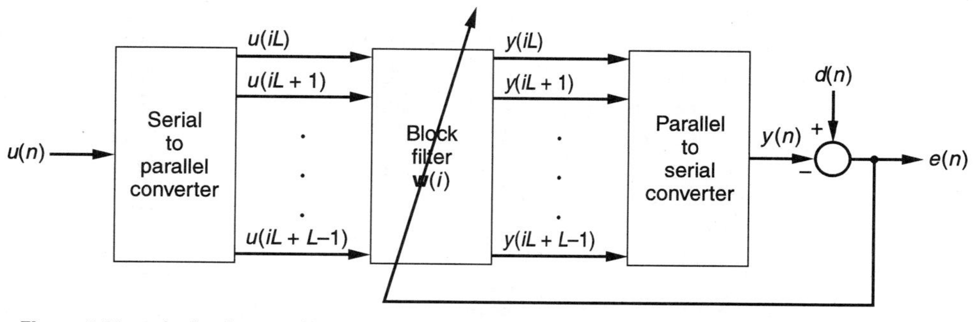

Figure 7.16 Block adaptive filter.

In each block, there are L input data samples, and therefore we need to calculate L output data samples. The filtering is done using the same weight vector $\mathbf{w}(i)$ for the entire ith block. We know that filtering is done as

$$y(n) = \mathbf{w}(i)^T \mathbf{u}(n) \tag{7.189}$$

where

$$\mathbf{u}(n) = [u(n)\, u(n-1) \ldots u(n-N+1)]^T. \tag{7.190}$$

Substituting (7.187), we get

$$\mathbf{u}(iL+j) = [u(iL+j)\, u(iL+j-1) \ldots u(iL+j-N+1)]^T \tag{7.191}$$

and

$$y(iL+j) = \mathbf{w}^T(i)\mathbf{u}(iL+j), \tag{7.192}$$

where $j = 0, 1, \ldots, L-1, i = 0, 1, 2, \ldots$ The error signal is now calculated as

$$e(iL+j) = d(iL+j) - y(iL+j). \tag{7.193}$$

Recall that in the LMS algorithm, the error signal is used in the weight update equation. In the BLMS algorithm, the weight vector $\mathbf{w}(i)$ is only updated once per block, but we have L different error signals. So, the gradient estimate is obtained by averaging the products $\mathbf{u}(iL+j)e(iL+j)$ over all the values of $j = 0, 1, \ldots, L-1$. The BLMS update equation is then given by

$$\mathbf{w}(i+1) = \mathbf{w}(i) + \frac{\mu_B}{L} \sum_{j=0}^{L-1} \mathbf{u}(iL+j)e(iL+j). \tag{7.194}$$

Equations (7.191)–(7.194) completely describe the BLMS algorithm and are summarized in Table 7.4.

Computational Complexity

For a given block of data, the BLMS requires N multiplications for equation (7.192) and approximately $NL + L$ multiplications for (7.194)—that is, a total of $(L+1)N + L$ multiplications per block. This compares to $(2N+1)L$ multiplications per block for the time-domain LMS. Is is easy to show that if $L \geq 2$, the BLMS is more efficient than the time-domain LMS. The choice of the block size L has been studied in [6]. If $L < N$, there is not enough data in the block to use all the tap weights, and some are therefore wasted. On the other hand, if $L > N$, the gradient estimate is computed over more points than the length of the filter, thereby introducing redundancy. Therefore, $L = N$ has been suggested as the most practical choice.

Convergence

The condition for convergence can be formed in a similar fashion as for the LMS algorithm. We start with the BLMS equation

TABLE 7.4 BLMS Algorithm

Given:

Input data sequence $u(n)$

Desired data sequence $d(n)$

$N =$ filter length

$L =$ block size

Initialization:

$$\mathbf{w}(0) = \mathbf{0}_{N\times 1}$$

Algorithm:

Segment the data into blocks of length L with i denoting the block number.

$$\begin{aligned}
&\text{For } i = 0, 1, 2, \ldots \\
&\text{For } j = 0, 1, \ldots, L-1 \\
\mathbf{u}(iL+j) &= [u(iL+j), u(iL+j-1), \ldots, u(iL+j-N+1)]^T \\
y(iL+j) &= \mathbf{w}^T(i)\mathbf{u}(iL+j) \\
e(iL+j) &= d(iL+j) - y(iL+j) \\
&\text{end } j \\
\mathbf{w}(i+1) &= \mathbf{w}(i) + \frac{\mu_B}{L}\sum_{j=0}^{L-1}\mathbf{u}(iL+j)e(iL+j)
\end{aligned}$$

$$\begin{aligned}
\mathbf{w}(i+1) &= \mathbf{w}(i) + \frac{\mu_B}{L}\sum_{j=0}^{L-1}[d(iL+j) - \mathbf{w}^T(i)\mathbf{u}(iL+j)]\mathbf{u}(iL+j) \\
&= \left(1 - \frac{\mu_B}{L}\sum_{j=0}^{L-1}\mathbf{u}^T(iL+j)\mathbf{u}(iL+j)\right)\mathbf{w}(i) \\
&\quad + \frac{\mu_B}{L}\sum_{j=0}^{L-1}d(iL+j)\mathbf{u}(iL+j).
\end{aligned} \tag{7.195}$$

Now, we take the expectation of both sides of (7.195) and use the usual stationarity and independence assumptions to get an equation in terms of the input autocorrelation matrix $\mathbf{R}$ and the cross-correlation vector p. Then we define an error vector $\mathbf{v}(i) = \mathbf{w}(i) - \mathbf{w}_0$, where $\mathbf{w}_0$ is the optimal weight vector. This will lead to another equation, which will ultimately give the sufficient condition for convergence as

$$0 < \mu_B < \frac{2}{\lambda_{max}} \tag{7.196}$$

where λ_{max} is the largest eigenvalue of the input autocorrelation matrix. The details are left as a problem.

7.7 FAST LMS ALGORITHM

The Fast LMS (FLMS) algorithm is a frequency domain-implementation of the BLMS algorithm, with a further reduction in computational complexity. Referring to the BLMS algorithm, we repeat two key equations:

$$y(iL+j) = \mathbf{w}^T(i)\mathbf{u}(iL+j) \tag{7.197}$$

$$\mathbf{w}(i+1) = \mathbf{w}(i) + \frac{\mu_B}{L}\sum_{j=0}^{L-1} e(iL+j)\mathbf{u}(iL+j). \tag{7.198}$$

Note that (7.197) is a convolution and that the second term in (7.198) is a correlation operation. Let us rewrite the correlation operation as

$$\mathbf{c}(i) = \sum_{j=0}^{L-1} e(iL+j)\mathbf{u}(iL+j). \tag{7.199}$$

Then (7.199) becomes

$$\mathbf{w}(i+1) = \mathbf{w}(i) + \frac{\mu_B}{L}\mathbf{c}(i). \tag{7.200}$$

The computational burden of the convolution of (7.197) can be reduced by performing the convolution in the FFT domain. However, the operation IFFT {FFT{$\mathbf{w}$}. FFT{$\mathbf{u}$}} yields a circular convolution rather than linear convolution. In Chapter 2, we pointed out that by zero-padding the vectors appropriately, we can obtain linear convolution. However, in a real-time adaptive filtering application, we have a continuous flow of data and cannot transform the entire data set. Therefore, the filtering is done as blocks of data, as seen in the BLMS. In the FFT domain, we must recognize that the outputs from successive blocks will overlap. So care must be taken to put together the output block in order to get the correct results. There are two principal methods for handling this, namely, the overlap-add [8] and the overlap-save [6], [7]. The overlap-save method requires less computation and will be presented here. It is essentially the same analysis as was done in Chapter 3.

In the FLMS based on overlap save, the filter weight vector is padded with an equal number of zeros, and then a $2N$ point FFT is performed. The transformed weight vector is therefore $2N$ points long and is given by

$$\mathbf{W}(i) = FFT\left\{\begin{bmatrix}\mathbf{w}(i)\\ \mathbf{0}\end{bmatrix}_{2N\times 1}\right\}. \tag{7.201}$$

The input vector must also be $2N$ points long and is segmented as

$$\mathbf{U}(i) = FFT\left\{\left[\underbrace{u(iN-N),\ldots,u(iN-1)}_{(i-1)^{\text{th}}\text{block}}\underbrace{u(iN),\ldots,u(iN+N-1)}_{i^{\text{th}}\text{block}}\right]^T_{2N\times 1}\right\}. \quad (7.202)$$

Notice that the block size has been chosen to be the same as the filter length. By virtue of the overlap-save method, the linear convolution is obtained as

$$\begin{aligned}\mathbf{y}(i) &= [y(iN),\ y(iN+1),\ldots,\ y(iN+N-1)]^T \\ &= \text{last } N \text{ elements of } IFFT\{[\mathbf{W}(i)\bullet\mathbf{U}(i)]_{2N\times 1}\}\end{aligned} \quad (7.203)$$

where $\bullet$ represents an element-by-element multiplication of the two vectors. The overlap-save method states that the IFFT of the product yields a $2N \times 1$ vector whose first N elements correspond to circular convolution and the last N elements correspond to linear convolution. To compute the correlation function of (7.199), we form the error block vector as

$$\begin{aligned}\mathbf{e}(i) &= [e(iN),\quad e(iN+1),\ldots,\quad e(iN+N-1)]^T_{N\times 1} \\ &= \mathbf{d}(i) - \mathbf{y}(i)\end{aligned} \quad (7.204)$$

where

$$\mathbf{d}(i) = [d(iN),\quad d(iN+1),\ldots,\quad d(iN+N-1)]^T_{N\times 1}. \quad (7.205)$$

The implementation of the correlation function is nothing but a reversed convolution given by $\mathbf{e}(-n) * \mathbf{u}(n)$. Therefore, this can also be done by the overlap-save method as follows:

$$\mathbf{E}(i) = FFT\left\{\begin{bmatrix}\mathbf{0}_{N\times 1} \\ \mathbf{e}(i)_{N\times 1}\end{bmatrix}\right\} \quad (7.206)$$

$$\mathbf{c}(i) = \text{first } N \text{ elements of } IFFT\{[\mathbf{E}(i)\bullet\mathbf{U}(i)]_{2N\times 1}\}. \quad (7.207)$$

Notice that in (7.206) we should have $\mathbf{E}^*(i)$. But we dropped the conjugation because we assume all real signals. In this case, the first N elements are used because this is a reversed convolution. The weight vector can now be updated in the FFT domain as

$$\mathbf{W}(i+1) = \mathbf{W}(i) + \mu_f FFT\left\{\begin{bmatrix}\mathbf{c}(i)_{N\times 1} \\ \mathbf{0}_{N\times 1}\end{bmatrix}\right\} \quad (7.208)$$

where μ_f is the step size. This completes the FLMS algorithm and is given by equations (7.201) to (7.208). The algorithm is summarized in Table 7.5.

TABLE 7.5 FLMS Algorithm Using Overlap-save

Given:

Input data block sequence $\mathbf{u}(n)$
Desired data block sequence $\mathbf{d}(n)$

Initialization:

$$\mathbf{W}(0) = \mathbf{0}_{2N\times 1}$$

Algorithm:

Segment incoming data into blocks of length $= N$ with i denoting the block number.

For $i = 0, 1, 2, \ldots$

$$U(i) = \text{FFT}\left\{[u(iN-N), \ldots, u(iN-1), u(iN), \ldots, u(iN+N-1)]^T\right\}$$

$$\mathbf{y}(i) = \text{last } N \text{ elements of IFFT } \{[\mathbf{W}(i) \cdot \mathbf{U}(i)]\}$$

$$\boldsymbol{e}(i) = \mathbf{d}(i) - \mathbf{y}(i)$$

$$\boldsymbol{E}(i) = \text{FFT}\left\{\begin{bmatrix} \mathbf{0}_{N\times 1} \\ \boldsymbol{e}(i) \end{bmatrix}\right\}$$

$$\mathbf{c}(i) = \text{first } N \text{ elements of IFFT } \left\{[\mathbf{E}(i) \cdot \mathbf{U}(i)]\right\}$$

$$\mathbf{W}(i+1) = \mathbf{W}(i) + \mu_f \cdot \text{FFT}\left\{\begin{bmatrix} \mathbf{c}(i) \\ \mathbf{0}_{N\times 1} \end{bmatrix}\right\}$$

Computational Complexity

Again, we consider the number of multiplications only. For a block of N data samples, the FLMS requires five $2N$-point FFTs (including IFFTs), each requiring $N \log_2 2N$ complex multiplications. Since the data are real, each FFT requires $2N \log_2 2N$ real multiplications, for a total of $10N \log_2 2N$ real multiplications for all the FFTs. The product in equation (7.203) requires $8N$ real multiplications since $\mathbf{W}(i)$ and $\mathbf{U}(i)$ are both complex. The product in equation (7.207) requires $8N$ real multiplications. Finally, the product with μ_f in (7.208) requires $4N$ multiplications. Therefore,

$$\begin{aligned}\text{Total number of real multiplications / block} &= 10N \log_2 2N + 12N \\ &= 10N \log_2 N + 22N.\end{aligned}$$

In contrast, the BLMS algorithm requires $N^2 + 2N$ real multiplications per block. The complexity ratio for the BLMS to the FLMS is given by

$$\frac{\text{BLMS}}{\text{FLMS}} \text{ complexity} = \frac{N^2 + 2N}{10N \log_2 N + 22N}.$$

TABLE 7.6 Computational Complexity

Algorithm	No. of Real Multiplications/Block
LMS	$2N^2 + N$
BLMS	$N^2 + 2N$
FLMS	$10N \log_2 N + 22N$

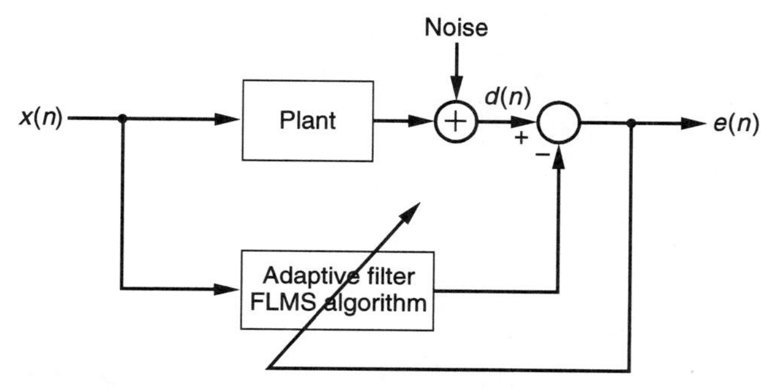

Figure 7.17 System identification scheme using the FLMS algorithm.

For $N = 128$, the number of multiplications required by the BLMS is about five times that of the FLMS. For $N = 256$, this factor increases to 11 times. The computational complexity for the LMS, BLMS, and FLMS is tabulated in Table 7.6.

Example

Let us consider the case of system identification using the FLMS method. The overall scheme is shown in Fig. 7.17, where the "Plant" is a 16-tap bandpass FIR filter with normalized cutoff frequencies at 0.4 and 0.5. The filter is designed using the window method with a Chebyshev window. The frequency response of this Plant is shown in Fig. 7.18. The input to the plant is a zero-mean Gaussian white noise with unit variance. The desired signal is obtained by adding a zero-mean white Gaussian noise of variance 0.0001 to the output of the plant. The FLMS algorithm is then used to identify the plant. A step size of 0.005 is used. The results are ensemble averaged over 200 independent runs. The MSE is shown in Fig. 7.19, where the steady-state error settles down to the level of the noise, as expected. For comparison, the result of using the LMS algorithm is shown in Fig. 7.20. The results are comparable. The FLMS achieves the same convergence rate at a lower computational cost.

7.8 SUMMARY

We reviewed some fundamental concepts from linear algebra that are useful in adaptive filtering. These include eigenvalues and eigenvectors of matrices, normal and Hermitian matrices, unitary transformation, similarity transformation, and diagonalizations. These concepts were used to derive some important properties of the correlation matrix. Some useful references on linear algebra are

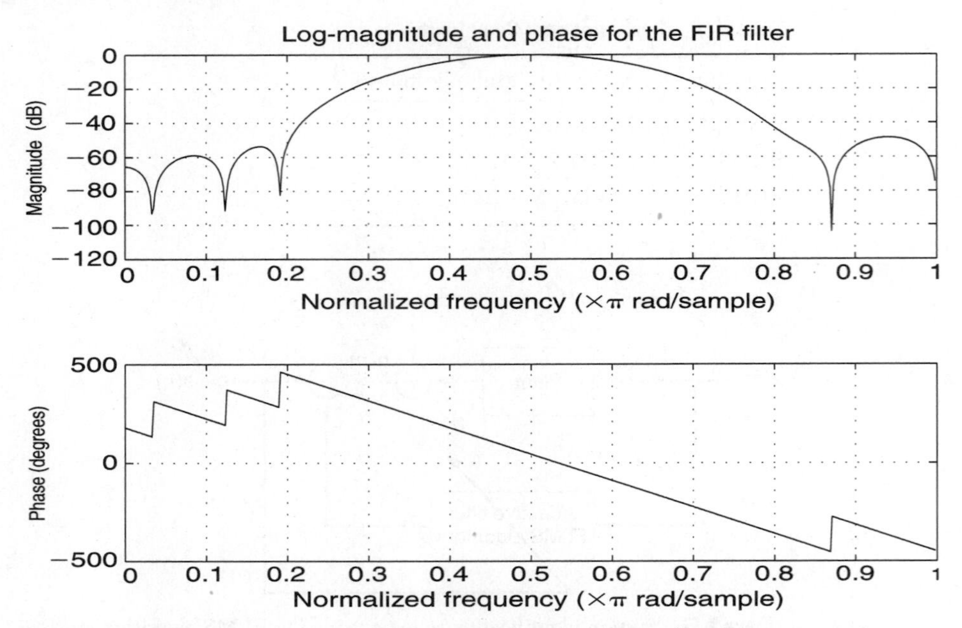

Figure 7.18 Frequency response of the plant used in the system identification example.

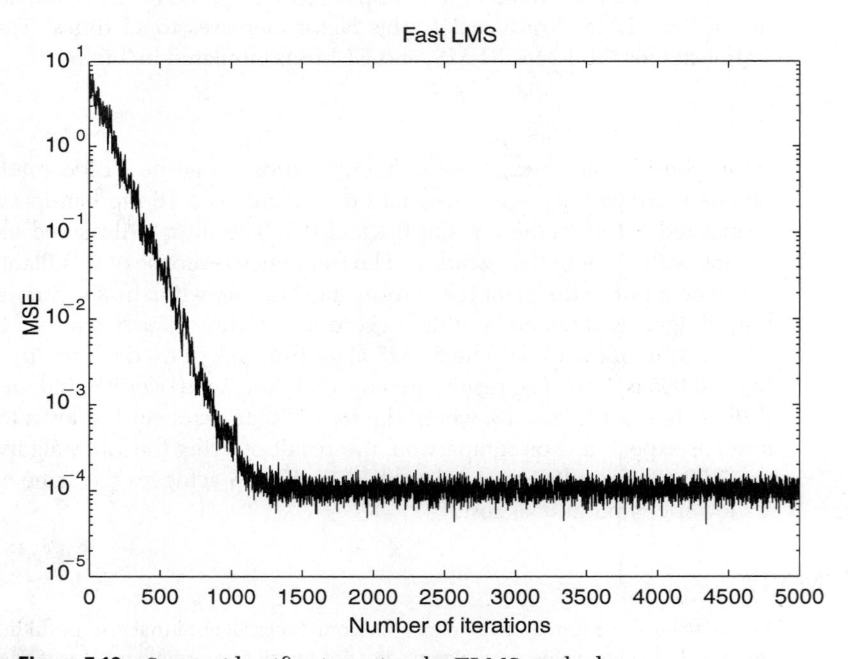

Figure 7.19 System identification using the FLMS method.

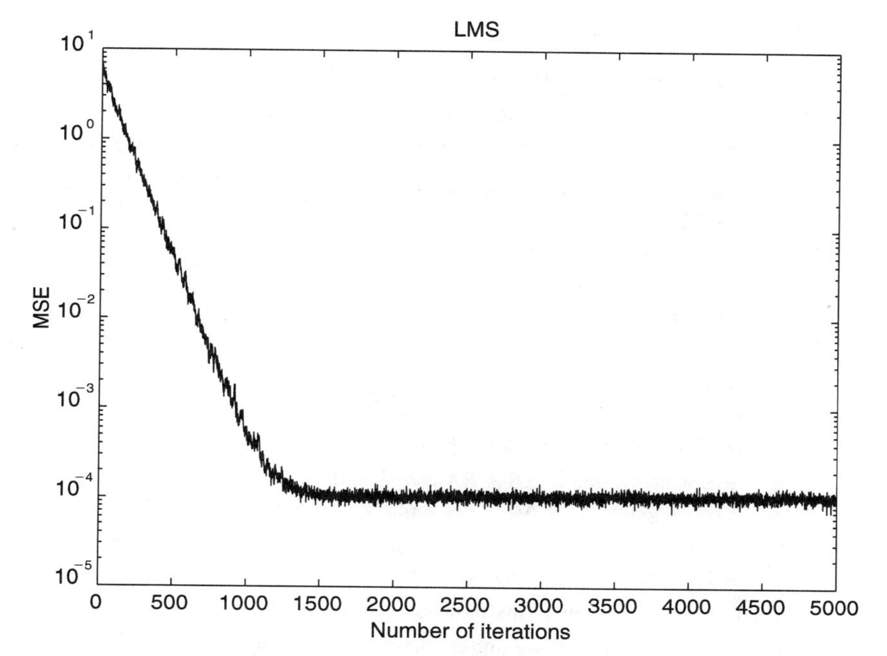

Figure 7.20 System identification using the LMS algorithm.

the texts by Horn and Johnson [9], Gantmacher [10], Goldberg [11], Hohn [12], and Strang [2], among many others.

The Wiener filter was presented next. This is the optimal filter that all adaptive algorithms try to achieve in a recursive fashion. To solve the Wiener filtering problem recursively, we introduced the method of steepest descent, which is one of the many gradient search algorithms. The simplicity of the steepest descent makes it appealing in adaptive filtering. The LMS algorithm is derived from the steepest descent by approximating the correlation matrix and the cross-correlation vector. The convergence and misadjustment properties of the LMS were discussed in detail. Some computer experiments were then presented to illustrate the algorithm. The LMS algorithm has become immensely popular in the last three decades. Literally thousands of papers have appeared in the literature on the analysis and applications of this method. It has been used in numerous signal processing algorithms in many areas of engineering and has become the industry standard for several applications. Several useful variations of the LMS algorithm have been developed such as in [13]–[16]. This is by no means a comprehensive list.

Next, the normalized LMS (NLMS) was derived. By using the technique of optimal line search, the derivation of the NLMS is considerably simplified. Numerous papers have also appeared on the analysis and applications of this algorithm. Several block-filtering algorithms have also been presented in this chapter. These algorithms are especially useful for frequency-domain adaptive filtering and are computationally more efficient. There are numerous references and variations of these block algorithms. A few citations on the FDAF, BLMS, and FLMS algorithms can be found in [17] and [18]. A fast frequency-domain optimal block algorithm has been proposed in [19], where a formula for an optimal step size has been established. Another algorithm called the Fast Transversal Filter (FTF) is also quite popular; it is a fast version of the RLS algorithm. Numerical stability problems have been reported on the original version of the algorithm. Some numerically stable FTF algorithms were since published. Interested readers are referred to [20]–[24].

▶ 7.9 PROBLEMS

1. Let $\mathbf{A}$ be a real matrix with $\mathbf{A}^T = -\mathbf{A}$.

(a) Show that $\mathbf{I} + \mathbf{A}$ is nonsingular.
(b) Show that the Cayley transform $\mathbf{Q} = (\mathbf{I} - \mathbf{A})(\mathbf{I} + \mathbf{A})^{-1}$ is orthogonal.

2. Show that matrix $\mathbf{A}$ and its unitary transformation $\mathbf{Q}^\dagger\mathbf{A}\mathbf{Q}$ have the same eigenvalues, where $\mathbf{Q}$ is a unitary matrix.

3. Let $\mathbf{A}$ be a $N \times N$ Hermitian matrix and let $\rho(\mathbf{A})$ denote the spectral radius.

(a) Show that $||\mathbf{A}||^m = \rho\{\mathbf{A}^m\}$.
(b) Show that the limiting relation

$$\mathbf{A}^m\mathbf{x} \to 0 \text{ as } m \to \infty$$

holds for each vector $\mathbf{x}$ if and only if $\rho(\mathbf{A}) < 1$.

4. Let $\mathbf{A}$ and $\mathbf{B}$ be Hermitian matrices. Show that

(a) every eigenvalue of $\mathbf{AB} + \mathbf{BA}$ is real; and
(b) every eigenvalue of $\mathbf{AB} - \mathbf{BA}$ is purely imaginary.

5. Let $\mathbf{A}$ be a square matrix. Prove that $\mathbf{A}$ is positive definite if and only if there is a positive definite matrix $\mathbf{S}$ such that $\mathbf{A} = \mathbf{S}^2$.

6. Let $\mathbf{A}$ be a Hermitian matrix whose inverse exists. Prove that $\mathbf{A}^{-1}$ is also Hermitian.

7. Given that $\mathbf{A}$ and $\mathbf{B}$ are unitary matrices such that $\mathbf{AB} = \mathbf{BA}$. Show that

$$\frac{1}{\sqrt{2}}\begin{bmatrix} \mathbf{A} & -\mathbf{B} \\ \mathbf{B}^\dagger & \mathbf{A}^\dagger \end{bmatrix}$$

is also unitary.

8. Show that if $\mathbf{U}$ is a unitary matrix, so are $\mathbf{U}^\dagger$ and $\mathbf{U}^*$.

9. Prove Theorem 7.6.

10. Diagonalize the following matrix by similarity transformation, if possible:

$$A = \begin{bmatrix} 1 & 1 \\ -3 & -4 \end{bmatrix}.$$

11. Diagonalize the following matrix by similarity transformation, if possible:

$$A = \begin{bmatrix} 1 & 0 & -1 \\ 0 & 1 & 0 \\ 0 & 0 & 2 \end{bmatrix}.$$

12. Diagonalize the following matrix by similarity transformation, if possible:

$$A = \begin{bmatrix} 1 & 1 & 2 \\ 0 & 1 & 3 \\ 0 & 0 & 2 \end{bmatrix}.$$

13. Consider a Wiener filtering problem with

$$\mathbf{R} = \begin{bmatrix} 1 & 0.3 \\ 0.3 & 1 \end{bmatrix}$$

$$\mathbf{P} = \begin{bmatrix} 0.2 \\ 0.3 \end{bmatrix}.$$

(a) Find a suitable range for the step size μ so that the steepest descent method would converge.
(b) Let the tap weight elements be w_1 and w_2. Starting with zero initial conditions, plot one element against the other as $n = 0, 1, 2, \ldots$
(c) Investigate the effect of varying μ on part (b).

14. Consider the steepest descent algorithm for a scalar weight $w(n)$. Find the mean square error $J(n)$, the optimal weight w_o, and the minimum mean square error $J_{\min}$.

15. Formulate the LMS algorithm for a one-step linear predictor. Such a predictor is a filter that predicts the sample $x(n)$ from past samples $x(n-1), x(n-2), \ldots, x(n-N+1)$.

16. Repeat the above problem for the steepest descent algorithm.

17. The input to an LMS algorithm is a white noise with mean μ and variance σ^2.

(a) Find the condition for convergence of the algorithm in the mean.
(b) Find the excess mean squared error.

18. **Least Mean Fourth (LMF) Algorithm:** Let the cost function for the LMS algorithm be modified as

$$J(\mathbf{w}) = E\{e^{2K}\}$$

where K is an integer. Derive an LMS-like algorithm by using the instantaneous gradient vector. Show that the update equation is

$$\mathbf{w}(n+1) = \mathbf{w}(n) + \mu K \mathbf{u}(n)[e(n)]^{2K-1}. \tag{7.209}$$

19. **Leaky LMS Algorithm:** The cost function for this algorithm is

$$J(\mathbf{w}) = e^2(n) + a\mathbf{w}^T\mathbf{w}, \tag{7.210}$$

where $0 < a < 1$.

(a) Using a similar approach as in the derivation of the LMS algorithm, show that the update equation for the leaky LMS algorithm is

$$\mathbf{w}(n+1) = (1 - \mu a)\mathbf{w}(n) + \mu e(n)\mathbf{u}(n). \tag{7.211}$$

(b) Show that the mean of the weight vector converges if

$$0 < \mu < \frac{2}{\lambda_{\max} + a}$$

where $\lambda_{\max}$ is the largest eigenvalue of the input correlation matrix.
(c) Assuming that the input and desired signals are zero mean, use the independence assumption to show that the mean of the weight vector converges to

$$\lim_{n\to\infty} E\{\mathbf{w}(n)\} = (\mathbf{R} + a\mathbf{I})^{-1}\mathbf{p} \tag{7.212}$$

where $\mathbf{R}$ is the correlation matrix of the input and $\mathbf{p}$ is the cross-correlation vector between the input and the desired signal.

20. **Sign LMS Algorithm:** The cost function for this algorithm is

$$J(\mathbf{w}) = E\{|e(n)|\}. \tag{7.213}$$

Using a similar approach as the LMS, show that the Sign LMS algorithm has the update equation

$$\mathbf{w}(n+1) = \mathbf{w}(n) + \mu \operatorname{sgn}\{e(n)\}\mathbf{u}(n). \tag{7.214}$$

21. **Variable step-size LMS Algorithm:** The update equation for this algorithm is

$$\mathbf{w}(n+1) = \mathbf{w}(n) + e(n)\mathbf{M}(n)\mathbf{u}(n) \tag{7.215}$$

where $\mathbf{M}(n)$ is a diagonal matrix of adaptation factors μ_i given by

$$\mathbf{M}(n) = \operatorname{diag}\{\mu_1(n), \mu_1(n), \ldots, \mu_{N-1}(n)\}.$$

Assuming zero-mean stationary inputs and using the independence assumption, find a condition so that the algorithm converges in the mean.

22. For the Block LMS algorithm, prove that the condition for the convergence in the mean is given by

$$0 < \mu_B < \frac{2}{\lambda_{\max}}$$

where $\lambda_{\max}$ is the largest eigenvalue of the input autocorrelation matrix.

23. Consider the cost function

$$J = \frac{1}{L} E\left\{\sum_{j=0}^{L-1} e^2(iL+j)\right\}$$

where

$$e(iL+j) = d(iL+j) - \mathbf{w}(i)^T\mathbf{u}(n),$$

where $j = 0, 1, \ldots, L-1$, and $i = 0, 1, 2, \ldots$. Assume zero-mean stationary inputs $x(n)$ and $d(n)$.

(a) By directly minimizing the above cost function, derive the usual normal equations given by

$$\mathbf{R}\mathbf{w} = \mathbf{p}.$$

(b) Derive the Block LMS algorithm from the above by using the steepest descent method and then replacing the $E\{\,\}$ by its instantaneous value.

24. Consider the cost function

$$J_n = \sum_{j=0}^{n} \lambda^{n-j} e^2(j)$$

where

$$e(j) = d(j) - \mathbf{w}^T(n)\mathbf{x}(n).$$

Show that the direct minimization of this cost function leads to the normal equations

$$\mathbf{R}_n\mathbf{w}_n = \mathbf{p}_n$$

where

$$\mathbf{R}_n = \sum_{j=0}^{n} \lambda^{n-j}\mathbf{x}(j)\mathbf{x}^T(j)$$

and

$$\mathbf{p}_n = \sum_{j=0}^{n} \lambda^{n-j} d(j)\mathbf{x}(j).$$

25. Show that the excess mean squared error of the Block LMS algorithm is given by

$$J_{ex} = \frac{\mu_B}{2L} \text{ trace } \{\mathbf{R}\}.$$

Use the same technique as used for the LMS algorithm.

26. Consider the second-order IIR filter

$$H(z) = \frac{b_0 + b_1 z^{-1}}{1 + a_1 z^{-1} + a_2 z^{-2}}.$$

Derive a LMS like algorithm for updating the parameters of the above filter. Assume that the parameters are slowly time-varying.

27. Repeat the above problem for the Block LMS algorithm.

7.10 COMPUTER PROJECTS

Project 1: Adaptive Line Enhancer (ALE) with LMS

Generate a signal of the form

$$s(n) = A \sin(2\pi f_1 n T_s) + B \sin(2\pi f_2 n T_s),$$

where T_s is the sampling frequency and the frequencies f_1 and f_2 can be chosen as desired. Choose T_s so that the sinusoid appears smooth. A noise $w(n)$ is added to the signal to produce the noisy signal

$$d(n) = s(n) + w(n).$$

The noise is a white uniformly distributed sequence with zero mean. The SNR of this signal should be about -7 dB. The desired signal (or primary signal) is $d(n)$. The reference signal is $d(n-1)$ and is applied to the input of an N-tap FIR filter that produces the output $y(n)$, which should be the estimate of the signal. Use the LMS algorithm for adaptation. Choose N and the step size appropriately for good results. Submit the following for your best case.

(a) Plot the MSE for an ensemble of 100 runs.

(b) Plot a few cycles of $s(n)$ for one run.

(c) Plot a few cycles of $y(n)$ after convergence (one run).

(d) Plot the impulse response and the frequency response of the filter after convergence (one run).

(e) Find the SNR of the filtered signal. Make sure you correct for phase if necessary.

Provide all block diagrams, equations, and algorithms in your report. Discuss your results. Are they as expected? Any variations from your expectations? What are the effects of changing the step size? What are the effects of changing N? Comment on the convergence rate in your experiments.

Project 2: Variable Step-size LMS for the ALE
The variable step-size LMS algorithm has the update equation

$$\mathbf{w}(n+1) = \mathbf{w}(n) + e(n)\mathbf{M}(n)\mathbf{u}(n)$$

where $\mathbf{M}(n)$ is a diagonal matrix of adaptation factors μ_i given by

$$\mathbf{M}(n) = \text{diag}\{\mu_1(n), \mu_1(n), \ldots, \mu_{N-1}(n)\}.$$

Solve (Problem 7.21) and find the condition for convergence in the mean in order to find μ_{max} and μ_{min}. All adaptation factors are constrained such that

$$\mu_{\text{min}} \leq \mu_i(n) \leq \mu_{\text{max}}.$$

Starting with $\mu_i(0) = \mu_{\text{max}}$ for all i, the adaptation parameters are calculated by the formula

$$g_i(n) = e(n)u(n-i)$$

$$\mu_i(n) = \mu_i(n-1) + \rho\text{sign}\{g_i(n)\}\text{sign}\{g_i(n-1)\}.$$

Repeat Project 1 using the variable step-size algorithm.

Project 3: System Identification Using LMS
An FIR system is given by

$$y(n) = x(n) - 0.5x(n-1) + 0.25x(n-2)$$

where $x(n)$ and $y(n)$ are the input and output, respectively. Let $x(n)$ be a uniformly distributed white input sequence of unit variance. Obtain a desired signal by adding white Gaussian noise, $w(n)$ as

$$d(n) = y(n) + w(n).$$

Once we have $d(n)$ and $x(n)$, we assume that the system is unknown and we perform system identification.

(a) Use SNR of 10 dB and 20 dB and identify the system using the LMS algorithm. Assume no modeling error; that is, the filter is the same length as the original system. Comment on the effects of the noise.

(b) Experiment with different values of the step size and comment on your results.

(c) Average the results for an ensemble of 100 runs and plot the squared error and filter coefficients.

Project 4: System Identification Using Normalized LMS
Repeat Project 3 using the normalized LMS algorithm. Discuss the relative performance of the two algorithms in terms of convergence rate (best case), steady-state error, and parameter estimation accuracy.

Project 5: ALE with FLMS
Repeat Project 1 using the FLMS algorithm.

(a) Experiment with different block lengths of $L = 100, 200, 400, 800$, and 1000. Discuss the effect of the block length on convergence, if any. How does the computational complexity change with the block length?

(b) Discuss the performance of the FLMS with respect to the LMS algorithm.

(c) In the FLMS, contiguous blocks of data are used. One variation is to use partially overlapping blocks in order to gain convergence speed. Use overlaps of $L/4$, $L/2$, and $3L/4$ and repeat your experiments. Discuss the performance of the algorithm. Find a formula for the computational complexity for an overlap of M points within the block.

Project 6: System Identification Using FLMS
Repeat Project 3 using the FLMS algorithm.

(a) Experiment with different block lengths of $L = 100, 200, 400, 800$, and 1000. Discuss the effect of the block length on convergence, if any. How does the computational complexity change with the block length?

(b) Discuss the performance of the FLMS with respect to the LMS algorithm.

BIBLIOGRAPHY

[1] J.G. PROAKIS and D.G. MANOLAKIS, *Digital Signal Processing: Principles, Algorithms, and Applications*, Prentice Hall, Upper Saddle River, NJ, 1996.

[2] G. STRANG, *Linear Algebra and Its Applications*, 2d ed., Academic Press, New York.

[3] B. WIDROW and M.E. HOFF Jr., "Adaptive switching circuits," IRE WESCON Conv, Rec., pt. 4, pp.96–104, 1960.

[4] C.-E. FROBERG, *Numerical Mathematics*, The Benjamin/Cummings Publishing Co., 1985.

[5] B. WIDROW et al., "Stationary and nonstationary learning characteristics of the LMS adaptive filter," *Proc. IEEE* 64, pp. 1151–1162, 1976.

[6] B.D. VAN VEEN and K.M. BUCKLEY, "Beamforming: a versatile approach to spatial filtering," *IEEE ASSP Magazine* 5, pp. 4–24, 1988.

[7] O. MACCHI, *Adaptive Processing: The LMS Approach with Applications in Transmission*, Wiley, New York, 1995.

[8] Pi SHENG CHANG,"On the Analysis of Adaptive Filtering Algorithms and Their Application to Spectral and Direction-of-Arrival Estimation," Ph.D. Diss., University of California, Los Angeles, 1998.

[9] R.A. HORN and C.R. JOHNSON, *Matrix Analysis*, Cambridge University Press, 1990.

[10] F.R. GANTMACHER, *Matrix Theory* (volume 1), Chelsea Publishing Co., 1990.

[11] JACK L. GOLDBERG, *Matrix Theory with Applications* (International Series in Pure and Applied Mathematics), McGraw-Hill, New York, 1988.

[12] F.E. HOHN, *Elementary Matrix Algebra,* Macmillan Publishing Co., New York, 1964.

[13] P.S.R. DINIZ and L.W. BISCAINHO, "Optimal variable step size for the LMS/Newton algorithm with application to subband adaptive filtering," *IEEE Trans. on Signal Processing*, SP-40, pp. 2825–2829, November 1992.

[14] W.B. Mikhael, F.H. Fu, L.G. Kazovsky, G.S. Kang, and L.J. Fransen, "Adaptive filter with individual adaptation of parameters," *IEEE Trans. on Circuits and Systems*, CAS-33, pp. 677–686, July 1986.

[15] R.W. Harris, D.M. Chabries, and F.A. Bishop, "A variable step (VS) adaptive filter algorithm," *IEEE Trans. on Acous., Speech, and Signal Processing*, ASSP-34, pp. 309–316, April 1986.

[16] J.J. Shynk and S. Roy, "Analysis of the momentum updating LMS algorithm," *IEEE Trans. on Acous., Speech, and Signal Processing*, ASSP-38, pp. 2088–2098, December 1990.

[17] E.R. Ferrara, " Frequency domain adaptive filtering," in *Adaptive Filters*, C.F.N. Cowan and P.M. Grant, eds., pp. 145–179, Prentice-Hall, Englewood Cliffs, NJ, 1985.

[18] J.J. Shynk, "Frequency domain and multirate adaptive filtering," *IEEE ASSP Magazine* 6, pp. 4–21, 1989.

[19] W.B. Mikhael and A.S. Spanias, "A fast frequency-domain adaptive algorithm," *Proc. of the IEEE* 76 (1), pp. 80–82 January 1988.

[20] D.T.M. Slock and T. Kailath, "Numerically stable fast transversal filters," *IEEE Trans. Signal Processing*, SP-39, pp. 92–114, 1991.

[21] A.L. Sayed and T. Kailath, "A state-space approach to adaptive RLS filtering," *IEEE Trans. Signal Processing Magazine* 11, pp. 18–60, 1994.

[22] J.M. Cioffi and T. Kailath, "Fast recursive least squares transversal filters for adaptive filtering," *IEEE Trans. Acous., Speech, and Signal Process.*, ASSP-32, pp. 304–337, 1984.

[23] D.T.M. Slock, "The backward consistency concept and roundoff error propagation dynamics in RLS algorithms,"*Optical Engineering* 31, pp. 1153–1169, 1992.

[24] P.A. Regalia, "Numerical stability issues in fast least-squares adaptation algorithms," *Optical Engineering* 31, pp. 1144–1152, 1992.

CHAPTER 8

Least-Squares Adaptive Algorithms

In the last chapter, we presented algorithms such as the LMS and NLMS which minimized the mean squared error. These algorithms are called mean square algorithms. In this chapter, we present algorithms that minimize the sum of squared error. These are called least-squares algorithms, the most powerful of which is known as the Recursive Least Squares (RLS) algorithm. The RLS algorithm is not popular for high-speed applications because of its high computational complexity. An alternative is the recently developed Euclidean Direction Search (EDS) algorithm. A modified version of the EDS algorithm is called the Fast EDS, which has a low computational complexity. This algorithm is also presented in this chapter.

The chapter presents the derivation of the RLS algorithm followed by a thorough performance evaluation. The application of the RLS is then demonstrated with an example of a channel equalizer. Also derived is the EDS algorithm, which performs a cyclical direction search among all of the Euclidean directions–hence the name. The relationship of the EDS algorithm to the Gauss-Seidel method and Newton's method is explained. The EDS has a computational complexity of $O(N^2)$. With a simple modification of the cost function and some other minor adjustments, we derive the Fast EDS algorithm, which has an $O(N)$ computational complexity. The channel equalizer example is used to illustrate the performance of the Fast EDS algorithm. The topic of Adaptive Pulse Code Modulation (ADPCM), which is important for data compression, is then presented. The ADPCM system can be used in conjunction with any of the adaptive algorithms developed in this or Chapter 7. Computer simulations of ADPCM are given for 1-D signals and images. Finally, this chapter deals with 2-D adaptive filters. The LMS and the EDS algorithms are extended for 2-D filters.

8.1 THE RLS ALGORITHM

Let $d(n)$ be a desired signal. Let the input and output of an FIR filter be $x(n)$ and $y(n)$, respectively. If $\mathbf{w} = [w_0\, w_1 \ldots w_{N-1}]^T$ denotes the filter coefficients, then $y(n) = \mathbf{w}^T(n)\mathbf{x}(n)$ is the filter output, where $\mathbf{x}(n) = [x(n)\, x(n-1) \ldots\, x(n-N+1)]^T$. As in the

Wiener filtering problem, the goal is to find the filter coefficients so that $y(n)$ matches $d(n)$ in some sense. This is achieved by minimizing some objective function.

In the least-squares (LS) method, the objective function is defined as

$$\begin{aligned}\mathbf{F}_n(\mathbf{w}) &\triangleq \sum_{i=1}^{n} \beta(n,i)e^2(i) \\ &= \sum_{i=1}^{n} \beta(n,i)[d(i) - \mathbf{w}^T(n)\mathbf{x}(i)]^2. \end{aligned} \tag{8.1}$$

Use of the weighting factor $\beta(n,i)$ is intended to ensure that past data are "forgotten" in order to track the statistical variations of the data in a nonstationary environment. The cost function can be written as

$$\mathbf{F}_n(\mathbf{w}) = \sum_{i=1}^{n} \beta(n,i)\{d^2(i) - 2d(i)\mathbf{w}^T(n)\mathbf{x}(i) + \mathbf{w}^T(n)\mathbf{x}(i)\mathbf{x}^T(i)\mathbf{w}(n)\}. \tag{8.2}$$

Define

$$\Phi(n) \triangleq \sum_{i=1}^{n} \beta(n,i)\mathbf{x}(i)\mathbf{x}^T(i)$$

$$\Psi(n) \triangleq \sum_{i=1}^{n} \beta(n,i)d(i)\mathbf{x}(i)$$

and

$$\gamma_d^2(n) \triangleq \sum_{i=1}^{n} \beta(n,i)d^2(i).$$

The least-squares objective function then becomes

$$\mathbf{F}_n(\mathbf{w}) = \gamma_d^2(n) - 2\mathbf{w}^T(n)\Psi(n) + \mathbf{w}^T(n)\Phi(n)\mathbf{w}(n). \tag{8.3}$$

A popular choice of the weighting function is $\beta(n,i) = \lambda^{n-i}, 0 < \lambda \leq 1$. For this choice, we rename matrix $\Phi(n)$, vector $\Psi(n)$, and scalar $\gamma_d^2(n)$, by $\mathbf{Q}(n)$, $\mathbf{r}(n)$, and $\overline{\sigma}_d^2$, respectively. The cost function is now

$$\mathbf{F}_n(\mathbf{w}) \triangleq \sum_{i=1}^{n} \lambda^{n-i}e^2(i) \tag{8.4}$$

$$= \sum_{i=1}^{n} \lambda^{n-i}[d(i) - \mathbf{w}^T(n)\mathbf{x}(i)]^2 \tag{8.5}$$

$$= \overline{\sigma}_d^2(n) - 2\mathbf{w}^T(n)\mathbf{r}(n) + \mathbf{w}^T(n)\mathbf{Q}(n)\mathbf{w}(n), \tag{8.6}$$

where

$$\mathbf{Q}(n) = \sum_{i=1}^{n} \lambda^{n-i}\mathbf{x}(i)\mathbf{x}^T(i) \tag{8.7}$$

$$\mathbf{r}(n) = \sum_{i=1}^{n} \lambda^{n-i}d(i)\mathbf{x}(i) \tag{8.8}$$

$$\overline{\sigma}_d^2(n) = \sum_{i=1}^{n} \lambda^{n-i} d^2(i). \tag{8.9}$$

If we expand equations (8.4), (8.7), and (8.8), and set $\lambda = 0$, then we get $\mathbf{F}_n(\mathbf{w}) = e^2(n)$, $\mathbf{Q}(n) = \mathbf{x}(n)\mathbf{x}(n)^T$, and $\mathbf{r}(n) = d(i)\mathbf{x}(i)$. That is, the least-squares objective function degenerates to that of the LMS algorithm as given in the last chapter. If $\lambda = 1$, the objective function is the sum of the total error squares, and it is said to have infinite memory. This is suitable only for time-invariant systems.

The minimum of the quadratic cost function given in (8.6) can be obtained by setting its gradient $\nabla \mathbf{F}_n(\mathbf{w}) = 0$. This gives

$$\begin{aligned} \nabla \mathbf{F}_n(\mathbf{w}) &\triangleq \frac{\partial \mathbf{F}_n(\mathbf{w})}{\partial \mathbf{w}} \\ &= 2\mathbf{Q}(n)\mathbf{w}(n) - \mathbf{2r}(n) \\ &= 0, \end{aligned} \tag{8.10}$$

which yields the normal equation

$$\mathbf{Q}(n)\mathbf{w}(n) = \mathbf{r}(n). \tag{8.11}$$

Because of the outer product of the input vector, $\mathbf{Q}(\mathbf{n})$ is nonsingular in general, and the time-varying optimal solution is

$$\mathbf{w}(n) = \mathbf{Q}^{-1}(n)\mathbf{r}(n). \tag{8.12}$$

In order to avoid a matrix inversion, an algorithm must be designed in order to solve the above equation recursively. To that end, we first note that there is no need to compute $\mathbf{Q}(n)$ and $\mathbf{r}(n)$ at each iteration by the computationally intensive formulas in (8.7) and (8.8). These parameters can be updated recursively as follows:

$$\begin{aligned} \mathbf{Q}(n) &= \sum_{i=1}^{n} \lambda^{n-i} \mathbf{x}(i)\mathbf{x}^T(i) \\ &= \lambda \sum_{i=1}^{n-1} \lambda^{n-1-i} \mathbf{x}(i)\mathbf{x}^T(i) + \mathbf{x}(n)\mathbf{x}^T(n) \\ &= \lambda \mathbf{Q}(n-1) + \mathbf{x}(n)\mathbf{x}^T(n); \end{aligned} \tag{8.13}$$

$$\begin{aligned} \mathbf{r}(n) &= \sum_{i=1}^{n} \lambda^{n-i} d(i)\mathbf{x}(i) \\ &= \lambda \sum_{i=1}^{n-1} \lambda^{n-1-i} d(i)\mathbf{x}(i) + d(n)\mathbf{x}(n) \\ &= \lambda \mathbf{r}(n-1) + d(n)\mathbf{x}(n). \end{aligned} \tag{8.14}$$

Now, in order to avoid inverting the matrix $\mathbf{Q}(n)$, we use the well-known Matrix Inversion Lemma given next. This lemma is also referred to in the literature as Woodbury's identity. Its usage in signal processing can be traced back to Kailath [1].

Matrix Inversion Lemma

Let **A**,**B**, *and* **D** *be positive definite matrices related by*

$$\mathbf{A} = \mathbf{B}^{-1} + \mathbf{C}\mathbf{D}^{-1}\mathbf{C}^T \tag{8.15}$$

for any matrix **C**. *Then*

$$\mathbf{A}^{-1} = \mathbf{B} - \mathbf{B}\mathbf{C}\,(\mathbf{D} + \mathbf{C}^T\mathbf{B}\mathbf{C})^{-1}\mathbf{C}^T\mathbf{B} \tag{8.16}$$

Proof. Multiplying **A** by $\mathbf{A}^{-1}$ and simplifying, it is easy to show that the result is identity. ■

Consider the update equation for matrix $\mathbf{Q}(n)$,

$$\mathbf{Q}(n) = \lambda\mathbf{Q}(n-1) + \mathbf{x}(n)\mathbf{x}^T(n). \tag{8.17}$$

Compare this equation to (8.15) and set $\mathbf{B}^{-1} = \lambda\mathbf{Q}(n-1)$, $\mathbf{C} = \mathbf{x}(n)$, and $\mathbf{D}^{-1} = 1$. Then the Matrix Inversion Lemma (8.16) yields

$$\begin{aligned}\mathbf{Q}^{-1}(n) &= \lambda^{-1}\mathbf{Q}^{-1}(n-1) - \lambda^{-1}\mathbf{Q}^{-1}(n-1)\mathbf{x}(n) \\ &\quad \times(1 + \mathbf{x}^T(n)\lambda^{-1}\mathbf{Q}^{-1}(n-1)\mathbf{x}(n))^{-1}\mathbf{x}^T(n)\lambda^{-1}\mathbf{Q}^{-1}(n-1) \\ &= \lambda^{-1}\mathbf{Q}^{-1}(n-1) - \frac{\lambda^{-2}\mathbf{Q}^{-1}(n-1)\,\mathbf{x}(n)\,\mathbf{x}^T(n)\mathbf{Q}^{-1}(n-1)}{1 + \lambda^{-1}\mathbf{x}^T(n)\mathbf{Q}^{-1}(n-1)\mathbf{x}(n)}.\end{aligned} \tag{8.18}$$

Define

$$\mathbf{P}(n) \triangleq \mathbf{Q}^{-1}(n). \tag{8.19}$$

Then (8.18) becomes

$$\mathbf{P}(n) = \lambda^{-1}\mathbf{P}(n-1) - \frac{\lambda^{-2}\mathbf{P}(n-1)\,\mathbf{x}(n)\,\mathbf{x}^T(n)\mathbf{P}(n-1)}{1 + \lambda^{-1}\mathbf{x}^T(n)\mathbf{P}(n-1)\mathbf{x}(n)} \tag{8.20}$$

$$= \lambda^{-1}\mathbf{P}(n-1) - \frac{\lambda^{-1}\mathbf{P}(n-1)\,\mathbf{x}(n)\,\mathbf{x}^T(n)\mathbf{P}(n-1)}{\lambda + \mathbf{x}^T(n)\mathbf{P}(n-1)\mathbf{x}(n)}. \tag{8.21}$$

Equation (8.21) is a recursive equation for updating $\mathbf{P}(n)$ in terms of $\mathbf{P}(n-1)$ and the input signal vector. Now we need an equation for updating the weight vector. The weight vector is given by

$$\mathbf{w}(n) = \mathbf{Q}^{-1}(n)\mathbf{r}(n)$$

$$= \mathbf{P}(n)\mathbf{r}(n) \tag{8.22}$$

$$= \mathbf{P}(n)[\lambda\mathbf{r}(n-1) + \mathrm{d}(n)\mathbf{x}(n)]. \tag{8.23}$$

Substituting for $\mathbf{P}(n)$ from (8.20), we get

$$\mathbf{w}(n) = \left(\lambda^{-1}\mathbf{P}(n-1) - \frac{\lambda^{-2}\mathbf{P}(n-1)\mathbf{x}(n)\mathbf{x}^T(n)\mathbf{P}(n-1)}{1+\lambda^{-1}\mathbf{x}^T(n)\mathbf{P}(n-1)\mathbf{x}(n)}\right)(\lambda\mathbf{r}(n-1) + d(n)\,\mathbf{x}(n))$$
$$= \mathbf{P}(n-1)\mathbf{r}(n-1) + \lambda^{-1}\mathbf{P}(n-1)\,d(n)\mathbf{x}(n)$$
$$- \frac{\lambda^{-1}\mathbf{P}(n-1)\mathbf{x}(n)\mathbf{x}^T(n)\mathbf{P}(n-1)\mathbf{r}(n-1)}{1+\lambda^{-1}\mathbf{x}^T(n)\mathbf{P}(n-1)\mathbf{x}(n)}$$
$$- \frac{\lambda^{-2}\mathbf{P}(n-1)\mathbf{x}(n)\mathbf{x}^T(n)\mathbf{P}(n-1)\,d(n)\mathbf{x}(n)}{1+\lambda^{-1}\mathbf{x}^T(n)\mathbf{P}(n-1)\mathbf{x}(n)}. \tag{8.24}$$

The first term in (8.24) can be replaced by using equation (8.22) as $\mathbf{P}(n-1)\mathbf{r}(n-1) = \mathbf{w}(n-1)$. (8.24) then becomes

$$\mathbf{w}(n) = \mathbf{w}(n-1) + \lambda^{-1}\mathbf{P}(n-1)\,d(n)\mathbf{x}(n)$$
$$- \frac{\lambda^{-1}\mathbf{P}(n-1)\mathbf{x}(n)\mathbf{x}^T(n)\mathbf{P}(n-1)\mathbf{r}(n-1)}{1+\lambda^{-1}\mathbf{x}^T(n)\mathbf{P}(n-1)\mathbf{x}(n)}$$
$$- \frac{\lambda^{-2}\mathbf{P}(n-1)\mathbf{x}(n)\mathbf{x}^T(n)\mathbf{P}(n-1)d(n)\mathbf{x}(n)}{1+\lambda^{-1}\mathbf{x}^T(n)\mathbf{P}(n-1)\mathbf{x}(n)}. \tag{8.25}$$

Straightforward simplification of equation (8.25) gives

$$\mathbf{w}(n) = \mathbf{w}(n-1) + \frac{\mathbf{P}(n-1)\mathbf{x}(n)[d(n) - \mathbf{w}^T(n-1)\mathbf{x}(n)]}{\lambda + \mathbf{x}^T(n)\mathbf{P}(n-1)\mathbf{x}(n)}. \tag{8.26}$$

Now, define

$$\hat{e}(n) = d(n) - \mathbf{w}^T(n-1)\mathbf{x}(n). \tag{8.27}$$

This is called the a priori estimation error, that is, the error at n given the weight vector at iteration $(n-1)$. This is certainly different from the error used in the last chapter, which was the "current" estimation error, $e(n)$. Substituting (8.27) in (8.26), we get

$$\mathbf{w}(n) = \mathbf{w}(n-1) + \frac{\mathbf{P}(n-1)\mathbf{x}(n)\hat{e}(n)}{\lambda + \mathbf{x}^T(n)\mathbf{P}(n-1)\mathbf{x}(n)}. \tag{8.28}$$

The RLS algorithm is essentially complete and is given by the equations for $\hat{e}(n)$, $\mathbf{w}(n)$ and $\mathbf{P}(n)$ described by (8.27), (8.28), and (8.21), respectively. Note that the term $\frac{\mathbf{P}(n-1)\mathbf{x}(n)}{\lambda+\mathbf{x}^T(n)\mathbf{P}(n-1)\mathbf{x}(n)}$ is present in both equation (8.28) and equation (8.21). In order to save computations, we define

$$\mathbf{K}(n) = \frac{\mathbf{P}(n-1)\mathbf{x}(n)}{\lambda + \mathbf{x}^T(n)\mathbf{P}(n-1)\mathbf{x}(n)} \tag{8.29}$$

and compute it only once during each update cycle. The complete RLS algorithm is given in Table 8.1. An operational count reveals that this algorithm has a computational complexity of $O(N^2)$. The high computational cost makes the RLS unsuitable for many high-speed applications. However, the RLS has several appealing properties such as fast convergence and zero misadjustment.

TABLE 8.1 The RLS Algorithm

Initialization

$$\mathbf{P}(0) = \delta^{-1}\mathbf{I}, \quad \delta = \text{small positive constant}$$
$$\mathbf{w}(0) = \mathbf{0}$$

Algorithm

For $n = 1, 2, \ldots$, compute

$$\mathbf{K}(n) = \frac{\mathbf{P}(n-1)\mathbf{x}(n)}{\lambda + \mathbf{x}^T(n)\mathbf{P}(n-1)\mathbf{x}(n)}$$
$$\hat{e}(n) = d(n) - \mathbf{w}^T(n-1)\mathbf{x}(n)$$
$$\mathbf{w}(n) = \mathbf{w}(n-1) + \mathbf{K}(n)\hat{e}(n)$$
$$\mathbf{P}(n) = \lambda^{-1}\left(\mathbf{P}(n-1) - \mathbf{K}(n)\mathbf{x}^T(n)\mathbf{P}(n-1)\right)$$

Performance Evaluation

First, it is important to establish the convergence of the RLS algorithm. The time-varying solution of the algorithm is given by

$$\mathbf{w}(n) = \mathbf{Q}^{-1}(n)\mathbf{r}(n), \quad n \geq N. \tag{8.30}$$

The condition $n \geq N$ is necessary because it allows the formation of a complete matrix $\mathbf{Q}(n)$ and vector $\mathbf{r}(n)$. The optimal error is

$$e_{\text{opt}}(n) = d(n) - \mathbf{w}_{\text{opt}}^T\,\mathbf{x}(n) \tag{8.31}$$

where $\mathbf{w}_{\text{opt}}$ is the optimal weight vector. We substitute for $d(n)$ from the above in equation (8.8) to get

$$\begin{aligned}
\mathbf{r}(n) &= \sum_{i=1}^{n} \lambda^{n-i}\mathbf{x}(i)[e_{\text{opt}}(i) + \mathbf{w}_{\text{opt}}^T\mathbf{x}(i)] \\
&= \sum_{i=1}^{n} \lambda^{n-i}\mathbf{x}(i)\mathbf{x}^T(i)\mathbf{w}_{\text{opt}} + \sum_{i=1}^{n} \lambda^{n-i}\mathbf{x}(i)e_{\text{opt}}(i) \\
&= \mathbf{Q}(n)\mathbf{w}_{\text{opt}} + \sum_{i=1}^{n} \lambda^{n-i}\mathbf{x}(i)e_{\text{opt}}(i).
\end{aligned} \tag{8.32}$$

Now we substitute (8.32) into the formula for the solution to get

$$\begin{aligned}
\mathbf{w}(n) &= \mathbf{Q}^{-1}(n)\mathbf{r}(n) \\
&= \mathbf{Q}^{-1}(n)\mathbf{Q}(n)\mathbf{w}_{\text{opt}} + \mathbf{Q}^{-1}(n)\sum_{i=1}^{n} \lambda^{n-i}\mathbf{x}(i)e_{\text{opt}}(i), \quad n \geq N.
\end{aligned} \tag{8.33}$$

Taking the expectation of both sides gives

$$E\{\mathbf{w}(n)\} = \mathbf{w}_{\text{opt}} + E\left\{\mathbf{Q}^{-1}(n)\sum_{i=1}^{n}\lambda^{n-i}\mathbf{x}(i)e_{\text{opt}}(i)\right\}. \tag{8.34}$$

Now we use a property of conditional expectation. It says that

$$E\{a\} = E\{E\{a|b\}\} \tag{8.35}$$

where the first expectation on the right-hand side is with respect to the random variable a, and $E\{a|b\}$ is the expectation of random variable a given the random variable b. This property can be used on (8.34) to get

$$E\{\mathbf{w}(n)\} = \mathbf{w}_{\text{opt}} + E\left\{E\left\{\mathbf{Q}^{-1}(n)\sum_{i=1}^{n}\lambda^{n-i}\mathbf{x}(i)e_{\text{opt}}(i)\,|\mathbf{x}(i)\right\}\right\}, \quad n \geq N. \tag{8.36}$$

Recall that $\mathbf{Q}(n)$ is formed from $\mathbf{x}(n)$. In the innermost expectation since $\mathbf{x}(i)$ is given, $\mathbf{Q}(n)$ is also given. Therefore

$$E\{\mathbf{w}(n)\} = \mathbf{w}_{\text{opt}} + E\left\{\mathbf{Q}^{-1}(n)\sum_{i=1}^{n}\lambda^{n-i}\mathbf{x}(i)E\{e_{\text{opt}}(i)\,|\mathbf{x}(i)\}\right\}, \quad n \geq N. \tag{8.37}$$

If the desired signal and the input signal have zero mean, then $E\{e_{\text{opt}}(i)\}$ is zero, which in turn implies that

$$E\{\mathbf{w}(n)\} = \mathbf{w}_{\text{opt}}, \quad n \geq N. \tag{8.38}$$

This establishes convergence in the mean. The mean of the weight vector is equal to the optimal weight vector in just N iterations. This is significantly better than in the LMS, where the mean of the weights converges to the optimal weights as $n \to \infty$. The importance of this result deserves a formal statement.

Theorem 8.1 *Let N be the number of filter coefficients in the RLS algorithm. For the number of iterations $n \geq N$, the mean of the weight vector is equal to the optimal solution.*

Now we establish a relationship between the least-squares solution and the Wiener filter. It is given by the following theorem.

Theorem 8.2 *If $\lambda = 1$, and the input signal is stationary and mean-ergodic, then the least-squares solution is equal to the Wiener solution.*

Proof. Recall that a stationary process is said to be mean-ergodic if the time-average is equal to its ensemble average. Since the input signal is mean-ergodic, we have

$$\mathbf{R} = \lim_{n\to\infty}\frac{1}{n}\sum_{i=1}^{n}\mathbf{x}(i)\mathbf{x}^T(i) \tag{8.39}$$

$$= \lim_{n\to\infty}\frac{1}{n}\mathbf{Q}(n) \tag{8.40}$$

$$\mathbf{p} = \lim_{n\to\infty} \frac{1}{n} \sum_{i=1}^{n} \mathbf{x}(i)d(i) \tag{8.41}$$

$$= \lim_{n\to\infty} \frac{1}{n}\mathbf{r}(n). \tag{8.42}$$

Now consider the solution of the least-squares problem in the limit as $n \to \infty$,

$$\begin{aligned}
\lim_{n\to\infty} \mathbf{w}(n) &= \lim_{n\to\infty} \mathbf{Q}^{-1}(n)\mathbf{r}(n) \\
&= \lim_{n\to\infty} \left(\sum_{i=1}^{n} \mathbf{x}(i)\mathbf{x}^T(i)\right)^{-1} \left(\sum_{i=1}^{n} \mathbf{x}(i)d(i)\right) \\
&= \lim_{n\to\infty} \left(\frac{1}{n}\sum_{i=1}^{n} \mathbf{x}(i)\mathbf{x}^T(i)\right)^{-1} \lim_{n\to\infty} \left(\frac{1}{n}\sum_{i=1}^{n} \mathbf{x}(i)d(i)\right) \\
&= \mathbf{R}^{-1}\mathbf{p} \\
&= \mathbf{w}_{\text{wiener}}.
\end{aligned} \tag{8.43}$$

This completes the proof. ■

The following two theorems give further insight into the RLS algorithm. The proofs are lengthy and not insightful and are therefore omitted. Interested readers are referred to [3].

Theorem 8.3 *Let the weight error vector be defined as* $\boldsymbol{\epsilon}(n) = \mathbf{w}(n) - \mathbf{w}_{\text{opt}}$, *and the minimum mean square error be* $J_{\min} = E\{e^2_{\text{opt}}(n)\}$. *Then*

$$E\{\boldsymbol{\epsilon}^T(n)\boldsymbol{\epsilon}(n)\} = \frac{J_{\min}}{n-N-1}\sum_{i=1}^{N}\frac{1}{\lambda_i}, \quad n > N+1 \tag{8.44}$$

where N *is the length of the weight vector,* n *is the discrete time sample, and* λ_i *are the eigenvalues of the input correlation matrix.*

Theorem 8.4 *Let the minimum mean square error be* $J_{\min} = E\{e^2_{\text{opt}}(n)\}$ *and the a priori mean square be* $\widehat{J}_n(\mathbf{w}) = E\{\hat{e}^2(n)\}$ *where* $\{\hat{e}(n)\}$ *is as defined in (8.27). Then*

$$\widehat{J}_n(\mathbf{w}) = J_{\min} + \frac{N \cdot J_{\min}}{n-N+1}, \quad n > N+1. \tag{8.45}$$

Based on the above results, we make the following observations for the RLS algorithm.

1. The mean of the weights converges to the optimal solution in N iterations. This is a significant improvement over the LMS, where the mean approaches the optimal weight vector as the number of iterations tend to infinity (Theorem 8.1).

2. For a stationary and ergodic input signal, the forgetting factor λ can be set equal to unity. The system is then said to have infinite memory. In this case, the mean square solution is then equal to the least-squares solution (Theorem 8.2).
3. The mean of the Euclidean norm of the weight error vector, $E\{\epsilon^T(n)\epsilon(n)\}$, is inversely proportional to the eigenvalues of the input correlation matrix. That is, for a large eigenvalue spread, $E\{\epsilon^T(n)\epsilon(n)\}$ can be very high. This in turn means that, even though the mean of the weights converges, the actual weights may vary significantly, implying a bad convergence property (Theorem 8.3).
4. The a priori MSE tends to $J_{\min}$ as $n \to \infty$. This means that the excess MSE tends to zero; that is, the RLS algorithm achieves a zero misadjustment (Theorem 8.4). This is not possible for the LMS algorithm. The zero misadjustment for the RLS is only possible for a stationary system with the forgetting factor $\lambda = 1$.

Channel Equalization

In digital communications, much effort has been devoted to the study of data transmission systems that efficiently utilize the available channel bandwidth. Adaptive channel equalization was first proposed and analyzed by Lucky in 1965 [12]–[14], building on earlier work in adaptive filtering by Widrow and Hoff in 1960 [15]. The simplest block diagram of an adaptive equalizer is shown in Fig. 8.1 where the adaptive filter is realized by an FIR filter. The input to the channel is typically a binary sequence, say ± 1. This signal is corrupted by the channel. Typical channel corruptions are lowpass filter in nature. In addition, some additive white noise is introduced in the channel. The output of the channel is the received signal, which forms the input to the adaptive filter, $x(n)$. What about the desired signal, $d(n)$, for the adaptive filter? Actually, there are two modes of operation for an adaptive equalizer, namely, the training phase and the working phase. A predetermined signal called the training signal is stored in the receiver. This signal is transmitted periodically, and this period is known as the training phase. During the training phase, the stored training signal is used as the desired signal, $d(n)$, and the filter coefficients are adapted. The working phase is the period when actual unknown data are transmitted instead of the training signal. During this phase, we have three options. First is to freeze the filter coefficients and assume that the channel characteristics do not change that rapidly. The second option is to first quantize the filter output into the actual symbols, which are ± 1 in this case. This is called *slicing*. A delayed version of the sliced

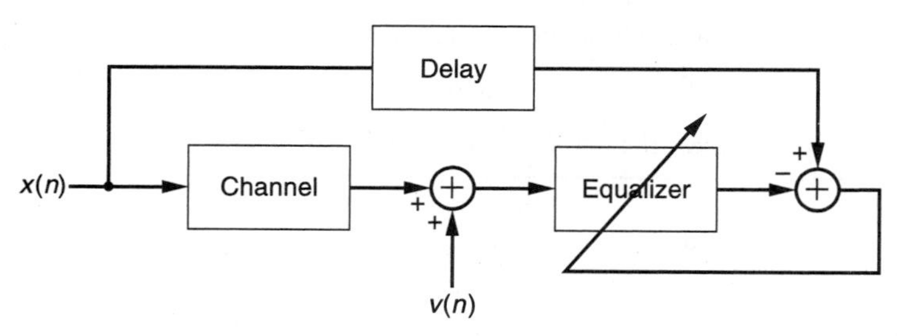

Figure 8.1 Adaptive channel equalizer.

signal is then used as the desired signal. This method works well if the channel characteristics change only slightly during this phase. The third option is to use a technique called *blind equalization*. This is a challenging technique because there is no desired signal. In this case, some additional information must be available about the system. Interested readers are referred to [3] and [16] for further study. The training phase only will be considered in our experiment.

For the numerical example, we will use the popular example from the text by Haykin [3]. The channel impulse response is assumed to be a raised cosine as

$$h(n) = \begin{cases} \frac{1}{2}(1 + \cos(2\pi(n-2)/b)), & n = 1, 2, 3 \\ 0, & \text{otherwise} \end{cases} \tag{8.46}$$

where b is a parameter that determines the eigenvalue spread of the input correlation matrix. This is proportional to the amount of amplitude distortion produced by the channel. Let the channel input be a random binary sequence $x(n) = \pm 1$ with equal probability. The desired output of the equalizer is the appropriately delayed input. The additive white noise $v(n)$ has zero mean and variance $\sigma_v^2 = 0.001$, which leads to a signal to noise ratio $SNR = 30$ dB because the channel input has zero mean and unit variance. The equalizer is chosen to be a 11-tap FIR filter. Since the channel has a symmetric impulse response about $n = 2$, it follows that the optimum weights $\mathbf{w}^*$ of the equalizer are symmetric about its center $n = 5$ as well. Accordingly, the channel input signal should be delayed by seven samples to form the desired signal. The initial weights $\mathbf{w}_0$ may be chosen as

$$w(n) = \begin{cases} 1, & n = 5, \\ 0, & n \neq 5. \end{cases} \tag{8.47}$$

The experiment is conducted with $b = 2.9, 3.1$, and 3.5. As listed in Table 8.2, the eigenvalue spread $\chi(\mathbf{Q}) = \lambda_{max}/\lambda_{min}$ of the correlation matrix $\mathbf{Q}$ ranges from 6.0782 (for $b = 2.9$) to 46.8216 (for $b = 3.5$).

First, we use the LMS algorithm to adapt the equalizer weights. The results of an ensemble average of 100 runs is shown in Fig. 8.2. Note that as the eigenvalue spread increases, the rate of convergence becomes slower. In the best case, the algorithm converges in about 200 iterations. The misadjustment is not identifiable from the figure. Recall from Chapter 7 that the misadjustment is sensitive to the eigenvalue spread and increases with it.

Now we use the RLS algorithm for adaptation and perform the ensemble average of 100 runs. The MSE is shown in Fig. 8.3 for different eigenvalue spreads. Based on this figure, we may make the following observations. First, the rate of convergence in the RLS

TABLE 8.2 Eigenvalue Spreads of Q in Channel Equalization

b	2.9	3.1	3.5
λ_{min}	0.3339	0.2136	0.0656
λ_{max}	2.0295	2.3761	3.0707
$\chi(\mathbf{Q})$	6.0782	11.1238	46.8216

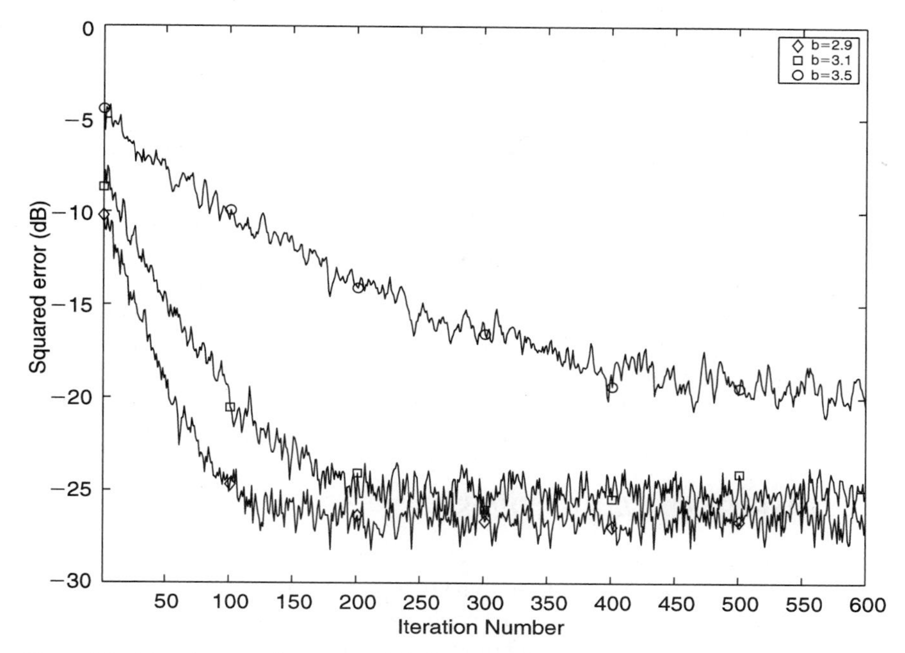

Figure 8.2 Equalizer performance with the LMS algorithm.

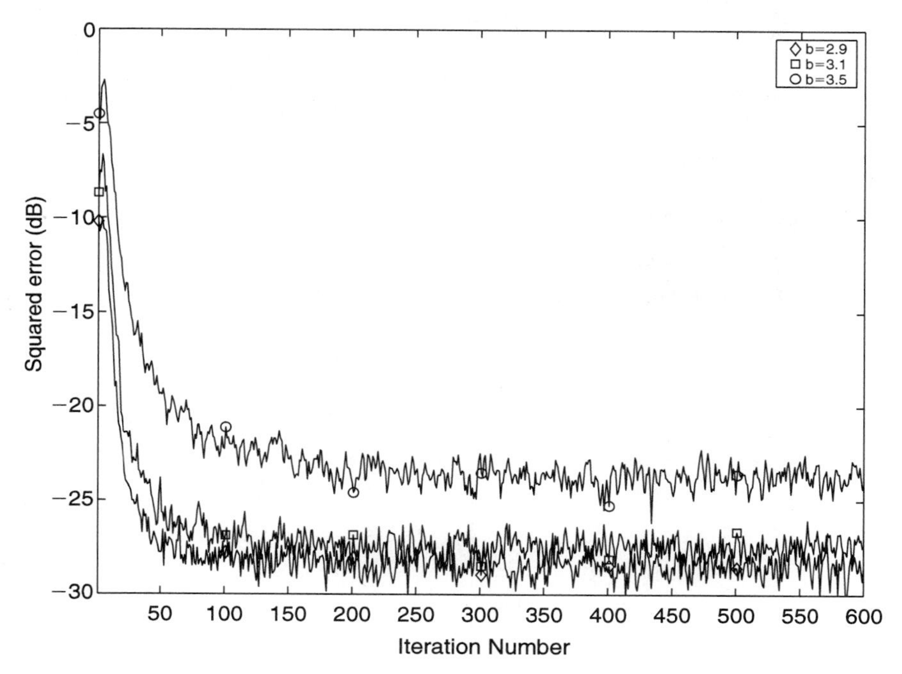

Figure 8.3 Equalizer performance with the RLS algorithm.

is not very sensitive to the variations of the eigenvalue spread $\chi(\mathbf{Q})$. Second, increasing the eigenvalue spread will increase the steady-state value of the average squared error. However, this should be the minimum MSE because theoretically the RLS gives zero misadjustment. Further experiments with the same equalizer will be given later in the chapter.

There have been numerous advances recently in the design of equalizers for communication channels. Many different algorithms have been developed for numerous applications. These algorithms include maximum likelihood estimator, blind channel estimation, decision feedback equalizer, and the constant modulus algorithm, among many others. A good tutorial on adaptive equalization is given by Tugnait, Tong, and Ding [4]. Some of these algorithms have become very useful and have found their way into several communication standards. For example, the Time Division Multiple Access (TDMA) Gaussian Synchronous Modulation (GSM) cellular system in Europe uses an FIR adaptive decision feedback equalizer with five coefficients and a training sequence of 26 symbols.

8.2 EDS AND FAST-EDS ALGORITHMS

From the earlier discussion, we know that the LMS algorithm is simple to implement and has an $O(N)$ computational complexity. However, it has a slow rate of convergence. On the other hand, the RLS algorithm has a very fast rate of convergence and zero misadjustment but has a computational complexity of $O(N^2)$. The Euclidean Direction Search (EDS) algorithm was developed recently by Xu and Bose [5]–[7] in order to combine the benefits of the RLS and LMS. The recursive version of the EDS algorithm is called the Fast EDS (FEDS) algorithm. In this section, we will show that the FEDS has an $O(N)$ computational complexity and a convergence rate that is comparable to that of the RLS. We will also show that the EDS method is closely related to the Gauss-Seidel method. First, we review the well-known Jacobi and Gauss-Seidel methods for solving a set of simultaneous equations.

Jacobi and Gauss-Seidel Methods

Recall that in the least-squares problem, we try to solve the set of equations given by $\mathbf{Qw} = \mathbf{r}$. The solution is obtained iteratively by deriving a recursive formula such as

$$\mathbf{w}(n+1) = \mathbf{Tw}(n) + \mathbf{c} \tag{8.48}$$

where $\mathbf{T}$ is a $N \times N$ matrix and $\mathbf{c}$ is a $N \times 1$ vector. To get $\mathbf{T}$ for our iterative methods, we set $\mathbf{Q} = \mathbf{L} + \mathbf{D} + \mathbf{U}$, where

$$\mathbf{D} = \begin{bmatrix} q_{11} & 0 & \cdots & & 0 \\ 0 & q_{22} & & & \vdots \\ & & \ddots & & \\ \vdots & \vdots & & & \\ & & & q_{N-1,N-1} & 0 \\ 0 & & \cdots & 0 & q_{N,N} \end{bmatrix}, \tag{8.49}$$

$$\mathbf{L} = \begin{bmatrix} 0 & 0 & & \cdots & & 0 \\ q_{21} & 0 & & & & \vdots \\ q_{31} & q_{32} & \ddots & \ddots & \ddots & \\ \vdots & \vdots & & 0 & & \\ & & & & 0 & 0 \\ q_{N1} & q_{N2} & & \cdots & q_{N,N-1} & 0 \end{bmatrix}, \tag{8.50}$$

and

$$\mathbf{U} = \begin{bmatrix} 0 & q_{12} & q_{13} & \cdots & & q_{1N} \\ & 0 & q_{23} & & & \vdots \\ & & \ddots & \ddots & \ddots & \\ \vdots & \vdots & & & q_{N-2,N-1} & q_{N-2,N} \\ & & & & 0 & q_{N-1,N} \\ 0 & & \cdots & & & 0 \end{bmatrix}. \tag{8.51}$$

If $q_{ii} \neq 0$ for each $i = 1, \ldots, N$, then $\mathbf{D}$ is nonsingular and invertible. Thus we can write

$$\begin{aligned} \mathbf{r} &= \mathbf{Q}\mathbf{w} \\ &= \mathbf{D}\mathbf{w} + (\mathbf{L} + \mathbf{U})\mathbf{w}. \end{aligned} \tag{8.52}$$

On the right-hand side of equation (8.52), the first $\mathbf{w}$ is replaced by $\mathbf{w}(n+1)$, and the second $\mathbf{w}$ is replaced by $\mathbf{w}(n)$. This implies the iterative method

$$\mathbf{w}(n+1) = -\mathbf{D}^{-1}(\mathbf{L} + \mathbf{U})\mathbf{w}(n) + \mathbf{D}^{-1}\mathbf{r}. \tag{8.53}$$

This is consistent with the desired iterative formula in (8.48). Now, we define the following notations.

$(\mathbf{w}(n))_j$: jth element of vector $\mathbf{w}(n)$;

$(\mathbf{q}^{(i)})T$: ith row of matrix $\mathbf{Q}$; therefore, $\mathbf{q}^{(i)}$ is a column vector.

By substituting the values of $\mathbf{D}$, $\mathbf{L}$, and $\mathbf{U}$, the matrix equation (8.53) can then be written as

$$(\mathbf{w}(n+1))_i = -\left(\sum_{j=1}^{i-1} q_{ij}(\mathbf{w}(n))_j + \sum_{j=i+1}^{N} q_{ij}(\mathbf{w}(n))_j - (\mathbf{r})_i\right) / q_{ii}, \quad i = 1, 2, \ldots, N. \tag{8.54}$$

This is called the *Jacobi method*. Notice that $\mathbf{w}(n+1)$ is found one element at a time and that these elements depend on the elements of $\mathbf{w}(n)$. In finding $(\mathbf{w}(n+1))_i$ for some i, note that we already have available the values $(\mathbf{w}(n+1))_j$, where $j < i$. So, the Jacobi method can be modified in order to use the latest available values of the weights. The result is

$$(\mathbf{w}(n+1))_i = -\left(\sum_{j=1}^{i-1} q_{ij}(\mathbf{w}(n+1))_j + \sum_{j=i+1}^{N} q_{ij}(\mathbf{w}(n))_j - (\mathbf{r})_i\right) / q_{ii}, \quad i = 1, 2, \ldots, N. \tag{8.55}$$

This is called the *Gauss-Seidel method*. For numerical examples of the Jacobi and Gauss-Seidel methods, refer to [2]. In the next subsection, we derive the EDS algorithm for a least-squares problem and establish its relationship to the Gauss-Seidel method.

The EDS Algorithm

Let us reconsider our least-squares cost function

$$J_n(\mathbf{w}) = \min_{\mathbf{w} \, \epsilon R^N} \{\mathbf{w}^T\mathbf{Q}(n)\mathbf{w} - 2\mathbf{w}^T\mathbf{r}(n) + \overline{\sigma}(n)^2\},$$

where $\mathbf{Q}(n)$, $\mathbf{r}(n)$, and $\mathbf{w}$ are given by (8.7), (8.8), and (8.9), respectively.

We know that an iterative algorithm is of the form

$$\mathbf{w}(n+1) = \mathbf{w}(\mathbf{n}) + \alpha(n)\mathbf{g}(n), \tag{8.56}$$

where $\alpha(n)$ is the step size and $\mathbf{g}(n)$ is the search direction at iteration n. Given an initial estimate of $\mathbf{w}$ and the search direction $\mathbf{g}$, the process of finding the best step size α for minimizing the next step $J_n(\mathbf{w}+\alpha\mathbf{g})$ is called the line search, as mentioned in Chapter 7. The optimal step size can be obtained by setting $\nabla_\alpha = 0$, where

$$\begin{aligned}\nabla_\alpha &\triangleq \frac{\partial J_n(\mathbf{w}+\alpha\mathbf{g})}{\partial\alpha} \\ &= 2\mathbf{g}^T\mathbf{Q}(n)(\mathbf{w}+\alpha\mathbf{g}) - 2\mathbf{g}^T\mathbf{r}(n).\end{aligned}$$

This gives

$$\alpha = -\frac{\mathbf{g}^T(\mathbf{Q}(n)\mathbf{w} - \mathbf{r}(n))}{\mathbf{g}^T\mathbf{Q}(n)\mathbf{g}^T}. \tag{8.57}$$

The matrix-vector multiplication involved in (8.57) requires $O(N^2)$ computational complexity in general. However, this complexity can be reduced if we have some special direction $\mathbf{g}$. The simplest of these directions is the Euclidean direction, defined as $\mathbf{g}^{(i)} = [0 \ldots 0\ 1\ 0 \ldots 0]$, where the $\mathbf{1}$ appears in the ith position. In Euclidean Direction Search (EDS), we use $\mathbf{g}^{(i)}$ for finding the ith element of $\mathbf{w}$. Now, let us see what happens when we use these directions in the line search formula of (8.57). In the numerator, for some i, $(\mathbf{g}^{(i)})^T(\mathbf{Q}(n)\mathbf{w}-\mathbf{r}(n))$ simply picks out the ith element of the vector $\mathbf{Q}(n)\mathbf{w}-\mathbf{r}(n)$. Therefore, it is not necessary to calculate all the elements of this vector. We need to compute only the ith element, which can be done in N multiplications. The denominator needs no computations because $\mathbf{g}^T\mathbf{Q}(n)\mathbf{g}$ is simply q_{ii}. Now we can explicitly write the EDS algorithm, where for each n, we sequentially use the directions $\mathbf{g}^{(i)}$, $i = 1, 2, \ldots, N$

TABLE 8.3 The EDS Algorithm

Initialization

$$\mathbf{w}(0) = \mathbf{0}$$
$$\mathbf{Q} = \mathbf{0}$$
$$\mathbf{r} = \mathbf{0}$$

Algorithm

For $n = 0, 1, 2, \ldots$

For $i = 1, 2, \ldots, N$

(1) $\varepsilon = (\mathbf{q}^{(i)})^T \mathbf{w} - (\mathbf{r})_i$

(2) $a = (\mathbf{q}^{(i)})_i$

(3) if $a \neq 0, \quad \alpha = -\varepsilon / a$

(4) $(\mathbf{w})_i = (\mathbf{w})_i + \alpha$

(5) $\mathbf{q}^{(i)} = \lambda \mathbf{q}^{(i)} + (\mathbf{x}(n+1))_i \, \mathbf{x}^T(n+1)$

(6) $(\mathbf{r})_i = \lambda (\mathbf{r})_i + (\mathbf{x}(n+1))_i \, d(n+1)$

end i

---

in order to update all the elements of the weight vector. The algorithm is obtained by using these directions and the equations (8.56) and (8.57). It is explicitly described in Table 8.3.

Let us go through the steps of the algorithm and see where they come from. For each i, we update one element of the weight vector with an Euclidean Direction Search. Items (1), (2) and (3) are for computing α in equation (8.57) with an Euclidean direction. Item (1) is the computation of the ith element of the vector $\mathbf{Q}(n)\mathbf{w} - \mathbf{r}(n)$. Item (2) is the denominator of the equation (8.57) as mentioned earlier. Item (3) completes the calculation of the step size for the ith Euclidean direction. We have used the condition $a \neq 0$ in order to avoid a division by zero. Item (4) in the algorithm is the update of the ith element of the weight vector. Items (5) and (6) are for updating matrix $\mathbf{Q}(n)$ and vector $\mathbf{r}(n)$ for the next n (not i). These items actually implement equations (8.13) and (8.14). Instead of implementing these equations directly, item (5) updates each row of $\mathbf{Q}(n)$ at a time for each i, and item (6) updates each element of vector $\mathbf{r}(n)$. The computations for updating $\mathbf{Q}(n)$ and $\mathbf{r}(n)$ are distributed during one *search cycle*; that is, $i = 1, \ldots, N$. At the end of this search cycle, we have the new and updated $\mathbf{Q}(n)$ and $\mathbf{r}(n)$.

Now let us go through the exercise of finding the computational complexity of this algorithm. We will consider the number of multiplications only. Item (1) requires N multiplications. Item (2) simply picks out the ith of the vector $\mathbf{q}^{(i)}$ and therefore needs no computations. Item (3) requires one division (multiplication), and item (4) needs none. Items (5) and (6) require $2N$ and 2 multiplications, respectively. For each i, the total number of multiplications needed is $3N + 3$. However, for each incoming data sample (i.e., for each n), we need $N(3N + 3)$ multiplications. The EDS algorithm thus has an

$O(N^2)$ computational complexity. The EDS algorithm is very effective in adaptive filtering applications, as we will illustrate shortly with an example. However, its high computational complexity may be prohibitive in many high-speed applications. This EDS algorithm is the precursor to the Fast EDS (FEDS) algorithm, which has an $O(N)$ computational complexity. This algorithm will be derived later in the chapter.

EDS and the Gauss-Seidel Method

Now we show the relationship between the EDS algorithm and the Gauss-Seidel method. The Gauss-Seidel update equation is repeated here for convenience.

$$(\mathbf{w}(n+1))_i = -\left(\sum_{j=1}^{i-1} q_{ij}(\mathbf{w}(n+1))_j + \sum_{j=i+1}^{N} q_{ij}(\mathbf{w}(n))_j - (\mathbf{r})_i\right) / q_{ii}, \quad i = 1, 2, \ldots, N. \tag{8.58}$$

Adding and subtracting the term $q_{ii}(\mathbf{w}(n))_i$ on the right-hand side gives

$$\begin{aligned}(\mathbf{w}(n+1))_i = -\Bigg[&\sum_{j=1}^{i-1} q_{ij}(\mathbf{w}(n+1))_j + q_{ii}(\mathbf{w}(n))_i + \sum_{j=i+1}^{N} q_{ij}(\mathbf{w}(n))_j \\ &-(\mathbf{r})_i - q_{ii}(\mathbf{w}(n))_i\Bigg] / q_{ii}, \quad i = 1, 2, \ldots, N.\end{aligned} \tag{8.59}$$

Define vector $\overline{\mathbf{w}}(n)$ as the weight vector with the elements being the most recent weights, that is,

$$(\overline{\mathbf{w}}(n))_j = \begin{cases} (\mathbf{w}(n+1))_j, & j < i \\ (\mathbf{w}(n))_j, & j \geq i \end{cases}, \quad i = 1, 2, \ldots, N. \tag{8.60}$$

With this definition, equation (8.60) becomes

$$(\mathbf{w}(n+1))_i = -\frac{\sum_{j=1}^{i-1} q_{ij}(\overline{\mathbf{w}}(n))_j - (\mathbf{r})_i}{q_{ii}} + (\overline{\mathbf{w}}(n))_i, \quad i = 1, 2, \ldots, N. \tag{8.61}$$

Recalling that $(\mathbf{q}^{(i)})^T$ denotes the ith row of $\mathbf{Q}(n)$, equation (8.61) can be written as

$$(\mathbf{w}(n+1))_i = -\frac{(\mathbf{q}^{(i)})^T \overline{\mathbf{w}}(n) - (\mathbf{r})_i}{q_{ii}} + (\overline{\mathbf{w}}(n))_i, \quad i = 1, 2, \ldots, N. \tag{8.62}$$

If $\overline{\mathbf{w}}(n)$ is simply redefined as $\mathbf{w}$, then we see that the numerator in the first term on the right-hand side is the same as ε in item (1) in the EDS algorithm (Table 8.3). The denominator is a in item (2). The first term is therefore $-\epsilon/a$, which is equal to α, as given in item (3) of the algorithm. So, equation (8.62) computes items (1) through (4) in the EDS. Except for the updating of $\mathbf{Q}(n)$ and $\mathbf{r}(n)$ in items (5) and (6), the EDS equations are exactly the same as the Gauss-Seidel method. It has been proven that the Gauss-Seidel method converges for any initial condition when $\mathbf{Q}$ is time-invariant, symmetric, and positive definite [8]. The convergence rate is approximately twice as fast as the Jacobi method [9].

EDS and the Newton's Method

Now we show the relationship of the EDS method to Newton's method. Recall from Chapter 7 that Newton's method for solving the problem

$$\mathbf{Q}\mathbf{w} = \mathbf{r} \tag{8.63}$$

is given by

$$\mathbf{w}(n+1) = \mathbf{w}(n) - \frac{\mu}{2}\mathbf{Q}^{-1}\nabla\mathbf{J}, \tag{8.64}$$

where $\nabla\mathbf{J}$ denotes the gradient and a convergence factor μ is introduced with $0 < \mu < 2$. We know from earlier discussion that for our cost function, $\nabla\mathbf{J} = 2(\mathbf{Q}\mathbf{w} - \mathbf{r})$. We also know that for $\mu = 1$, Newton's method converges in one step. Convergence in one step is a source of satisfaction to a numerical analyst who likes to minimize the number of iterations necessary to accomplish a surface search. However, to an adaptive system designer, it is generally too fast and not really desirable. A reasonable rate of adaptation provides a filtering process that ameliorates the effects of measurement noise [11]. Also, computing the inverse of $\mathbf{Q}$ is computationally unrealistic for real-time applications. Therefore, $\mathbf{Q}$ is usually estimated or approximated for most adaptive algorithms. Recall that for the steepest descent algorithm, we used the identity matrix as an estimate of $\mathbf{Q}$.

Now let us turn to the EDS method. By the matrix splitting concept introduced before, we can write

$$\mathbf{Q} = \mathbf{L} + \mathbf{D} + \mathbf{U} \tag{8.65}$$

where $\mathbf{D}$ is the diagonal part of $\mathbf{Q}$, $\mathbf{L}$ is the lower triangle of $\mathbf{Q}$ with zeros on the diagonal, and $\mathbf{U}$ is the upper triangle of $\mathbf{Q}$ with zeros on the diagonal, as given in equations (8.49), (8.50), and (8.51). Substituting (8.65) in (8.63), we get the Gauss-Seidel (or EDS) iterative method

$$(\mathbf{L} + \mathbf{D})\mathbf{w}(n+1) = -\mathbf{U}\mathbf{w}(n) + \mathbf{r}. \tag{8.66}$$

Subtracting $(\mathbf{L} + \mathbf{D})\mathbf{w}(n)$ from both sides of (8.66) gives

$$\mathbf{w}(n+1) = \mathbf{w}(n) - (\mathbf{L} + \mathbf{D})^{-1}(\mathbf{Q}\mathbf{w}(n) - \mathbf{r}). \tag{8.67}$$

Realizing that $\mathbf{Q}\mathbf{w}(n) - \mathbf{r}$ is the gradient for our cost function, we have

$$\mathbf{w}(n+1) = \mathbf{w}(n) - (\mathbf{L} + \mathbf{D})^{-1}\nabla\mathbf{J}. \tag{8.68}$$

The matrix $\mathbf{L} + \mathbf{D}$ in (8.68) is the lower triangular part of $\mathbf{Q}$. Thus, in the EDS method, we are using the lower triangular part of $\mathbf{Q}$ as an approximation. In the LMS (steepest descent), we used $\mathbf{I}$ as an approximation of $\mathbf{Q}$ and obtained a computational complexity of $2N + 1$. In the RLS algorithm, we used the actual $\mathbf{Q}$ but used the Matrix Inversion Lemma to compute its inverse, and we obtained a complexity of $3N^2 + 7N$. The EDS is therefore between the LMS and the RLS, where we use the lower triangular part. So, we expect the convergence rate and the computational complexity of the EDS to be somewhere between that of the LMS and RLS. The computational complexity of the EDS is $3N^2 + 3N$. A computer simulation example of the EDS will be deferred until we have developed the Fast EDS algorithm. In [17], it has been shown that for a time-invariant system, the EDS algorithm always converges to the optimal solution. In addition, the convergence rate of the EDS is mathematically proven to be faster than that of the steepest descent. We will not pursue these mathematical derivations here.

The Fast EDS Algorithm

The Fast EDS algorithm is developed in order to reduce the computational complexity of the EDS. It has the same basic philosophy as that of the EDS, whereby the N weights are updated by cyclically searching through the N Euclidean directions. In order to achieve a computational complexity of $O(N)$, we first modify the conventional exponentially weighted least-squares objective function to a *block* exponentially weighted least-squares form. Then, we only perform one Euclidean direction search for every sample of data.

As given earlier, the conventional exponentially weighted least-squares objective function is

$$J_n(\mathbf{w}) = \min_{\mathbf{w}\epsilon R^N} \left\{ \sum_{i=1}^{N} \lambda^{n-i}(d(i) - \mathbf{w}^T(n)\mathbf{x}(i))^2 \right\}. \tag{8.69}$$

The weights applied to the squared error decrease exponentially with discrete time, (see Fig. 8.4). Actually, the weights do not have to exponentially decrease with every sample. The new cost function is designed so that the weights decrease exponentially with every block as shown in Fig. 8.5. Let $n = kN + l$, where $1 \leq l \leq N$, where the block size N is the same as the number of filter coefficients. So, k is the number of full blocks within n, and l is the number of samples remaining. The block weighted cost function can then be written as

$$\begin{aligned} \min_{\mathbf{w}\epsilon R^N} J_n(\mathbf{w}) = & \sum_{i=0}^{k-1} \lambda^{k-i} \left(\sum_{j=1}^{N} (d(iN+j) - \mathbf{w}^T(n)\mathbf{x}(iN+j))^2 \right) \\ & + \sum_{j=1}^{l} (d(kN+j) - \mathbf{w}^T(n)\mathbf{x}(kN+j))^2. \end{aligned} \tag{8.70}$$

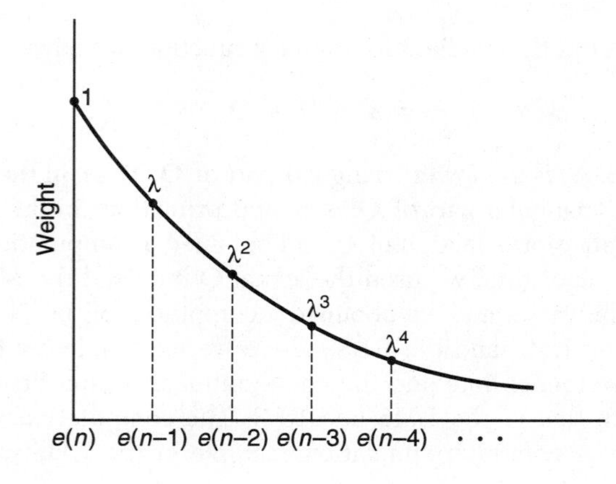

Figure 8.4 Weight for conventional least-squares cost function.

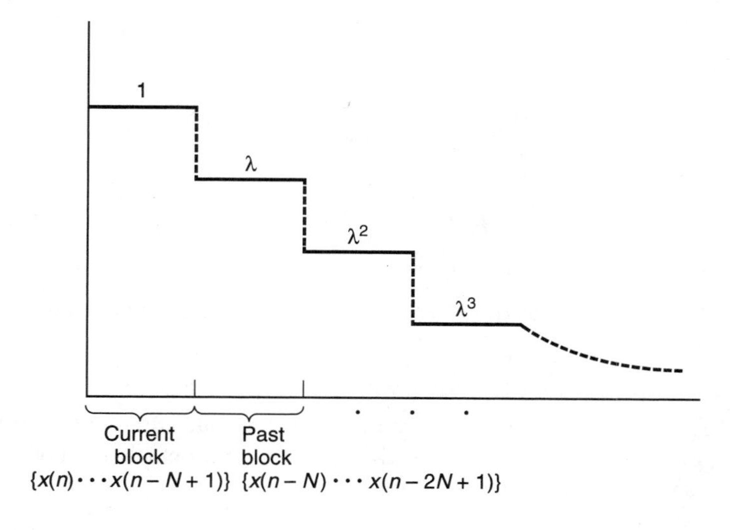

Figure 8.5 Weight for modified least-squares cost function used for the Fast EDS algorithm.

Since both (8.69) and (8.70) have the flexibility in choosing λ, the above modification should not affect the essential optimal solution. A similar modification can be found in [10]. The new cost function of (8.70) can be written in the form of the general quadratic function (8.6) with appropriate modifications in $\mathbf{Q}$, $\mathbf{r}$, and $\overline{\sigma}^2$. These are given by (Problem 7),

$$\begin{aligned}\mathbf{Q}(n) &= \mathbf{Q}(kN+l)\\ &= \sum_{i=0}^{k-1}\lambda^{k-i}\left(\sum_{j=1}^{N}\mathbf{x}(iN+j)\mathbf{x}^T(iN+j)\right)+\sum_{j=1}^{l}\mathbf{x}(kN+j)\mathbf{x}^T(kN+j); \quad (8.71)\end{aligned}$$

$$\begin{aligned}\mathbf{r}(n) &= \mathbf{r}(kN+l)\\ &= \sum_{i=0}^{k-1}\lambda^{k-i}\left(\sum_{j=1}^{N}d(iN+j)\mathbf{x}(iN+j)\right)+\sum_{j=1}^{l}d(kN+j)\mathbf{x}(kN+j); \quad (8.72)\end{aligned}$$

$$\begin{aligned}\overline{\sigma}^2(n) &= \overline{\sigma}^2(kN+l)\\ &= \sum_{i=0}^{k-1}\lambda^{k-i}\left(\sum_{j=1}^{N}d^2(iN+j)\right)+\sum_{j=1}^{l}d^2(kN+j). \quad (8.73)\end{aligned}$$

As in the RLS method, the above parameters can be computed recursively. However, because of the exponentially weighted relationship between adjacent blocks, the recursion of the above parameters hold in a block form (Problem 8) as

$$\begin{aligned}\mathbf{Q}(kN+l) &= \mathbf{Q}(n)\\ &= \lambda\mathbf{Q}(kN)+\widetilde{\mathbf{Q}}(kN+l), \quad (8.74)\end{aligned}$$

$$\begin{aligned}\mathbf{r}(kN+l) &= \mathbf{r}(n)\\ &= \lambda\mathbf{r}(kN)+\widetilde{\mathbf{r}}(kN+l), \quad (8.75)\end{aligned}$$

where

$$\widetilde{\mathbf{Q}}(kN+l) = \sum_{j=1}^{l} \mathbf{x}(kN+j)\mathbf{x}(kN+j)^T, \tag{8.76}$$

$$\widetilde{\mathbf{r}}(kN+l) = \sum_{j=1}^{l} d(kN+j)\mathbf{x}(kN+j). \tag{8.77}$$

The FEDS is the same as the EDS algorithm except in the following ways.

1. Parameters $\mathbf{Q}$,$\mathbf{r}$, and $\overline{\sigma}^2$ are updated as given by equations (8.74)–(8.77). These are updated for every sample of data as in the EDS.
2. For every sample of data, only one Euclidean Direction Search is performed instead of N searches in the EDS. That is, only one element of the weight vector is updated per sample of data.

The Fast EDS algorithm is based on the above update equations for $\mathbf{Q}$ and $\mathbf{r}$ and the equation for the step size as in the EDS. The complete Fast EDS algorithm is given in Table 8.4. Now, we will examine each equation of the algorithm and see where it comes from.

Recall that the discrete-time index is $n = kN + l$. The equations in Table 8.4 are executed for each block index, k. Within each block there are N data. For each data (index i), items (1) through (8) are performed. Then items (A) and (B) are performed at the end of each block. As seen in the EDS algorithm, for each Euclidean Direction Search (i.e., for each i), we update only one row of $\mathbf{Q}$, one element of $\mathbf{r}$, and one element of $\mathbf{w}$. Item (1) is the first part of equation (8.74) for one row. Item (2) is the first part of equation (8.75) for one element. Note that equations (8.76) and (8.77) are accumulations of $\widetilde{\mathbf{Q}}$ and $\widetilde{\mathbf{r}}$ within one block. These are done in items (3) and (4). In items (5) and (6), $\mathbf{q}^{(i)} + \widetilde{\mathbf{q}}^{(i)}$ is the update of one row of $\mathbf{Q}$, and $(\mathbf{r} + \widetilde{\mathbf{r}})_i$ is the update of one element of $\mathbf{r}$. Then in items (5), (6), and (7), the step size α is calculated in the same way as in the EDS method. Item (8) updates one element of the weight vector. At the end of each block, we must update $\mathbf{Q}$ and $\mathbf{r}$, and reset $\widetilde{\mathbf{Q}}$ and $\widetilde{\mathbf{r}}$. These are done in items (A) and (B). This completes the algorithm.

For the computational complexity we will count multiplications only.

- Item (1): N multiplications.
- Item (2): 1 multiplication.
- Item (3): Direct computation would require N^2 multiplications. However, note that given the matrix $\mathbf{x}(n)\mathbf{x}^T(n)$, the matrix $\mathbf{x}(n+1)\mathbf{x}^T(n+1)$has only N new elements (Problem 9). Therefore, item (3) requires only N multiplications.
- Item (4): N multiplications.
- Item (5): N multiplications.
- Item (6): None.
- Item (7): 1 multiplication.
- Item (8): None.

TABLE 8.4 The Fast EDS Algorithm

Initialization

$$\boldsymbol{w}(0) = \boldsymbol{0}$$
$$\widetilde{\boldsymbol{Q}} = \boldsymbol{0}$$
$$\boldsymbol{Q} = \boldsymbol{0}$$
$$\widetilde{\boldsymbol{r}} = \boldsymbol{0}$$
$$\boldsymbol{r} = \boldsymbol{0}$$

Algorithm

Time index, $n = kN + l$

For $k = 1, 2, \ldots$

For $i = 1, 2, \ldots, N$

(1) $\boldsymbol{q}^{(i)} = \lambda \mathbf{q}^{(i)}$

(2) $(\boldsymbol{r})_i = \lambda \boldsymbol{r}_i$

(3) $\widetilde{\boldsymbol{Q}} = \widetilde{\boldsymbol{Q}} + \mathbf{x}(kN+i)\mathbf{x}^T(kN+i)$

(4) $\widetilde{\boldsymbol{r}} = \widetilde{\boldsymbol{r}} + d(kN+i)\mathbf{x}(kN+i)$

(5) $\boldsymbol{\varepsilon} = (\boldsymbol{q}^{(i)} + \widetilde{\boldsymbol{q}}^{(i)})^T \mathbf{w} - (\boldsymbol{r} + \widetilde{\boldsymbol{r}})_i.$

(6) $a = q^{(i)}{}_i + (\widetilde{\mathbf{q}}^{(i)})_i$

(7) if $a \neq 0, \quad \alpha = \dfrac{-\varepsilon}{a}$

(8) $(\boldsymbol{w})_i = (\boldsymbol{w})_i + \alpha$

end i

(A) $\mathbf{Q} = \mathbf{Q} + \widetilde{\mathbf{Q}}; \quad \widetilde{\boldsymbol{Q}} = \mathbf{0}$

(B) $\boldsymbol{r} = \boldsymbol{r} + \widetilde{\boldsymbol{r}}; \quad \widetilde{\boldsymbol{r}} = \mathbf{0}$

So, items (1) through (8) require a total of $4N + 2$ multiplications; that is, the Fast EDS algorithm has a computational complexity of $O(N)$.

Fast EDS and EDS-Based Channel Equalization

Now reconsider the channel equalization example used in the last section. We first use the Fast EDS algorithm for adaptation and average the results over an ensemble of 100 runs, for the same eigenvalue spreads as before. The MSE is given in Fig. 8.6. For the best case, the steady state is reached in about 100 samples. The rate of convergence is sensitive to the eigenvalue spread as in the LMS algorithm. It is instructive to compare the performances of the LMS, RLS, and Fast EDS algorithms for this channel equalizer. We perform this comparison for one eigenvalue spread, namely, with $b = 3.5$. The MSEs for the different algorithms are plotted in Fig. 8.7. The RLS converges much faster than the LMS, and the Fast EDS is somewhere between the two. In fact, the Fast EDS

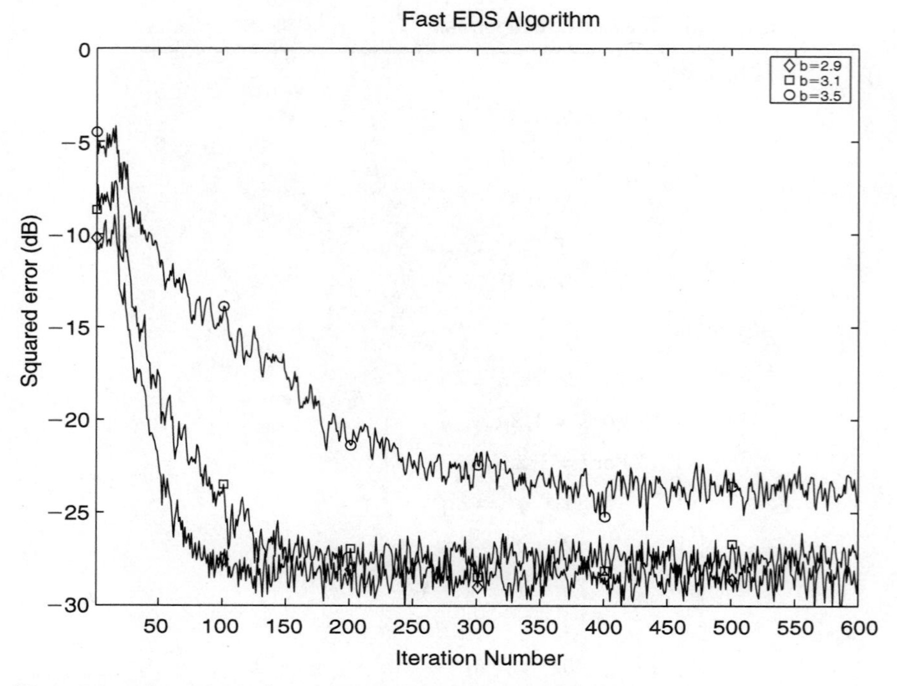

Figure 8.6 Channel equalization with the Fast EDS algorithm.

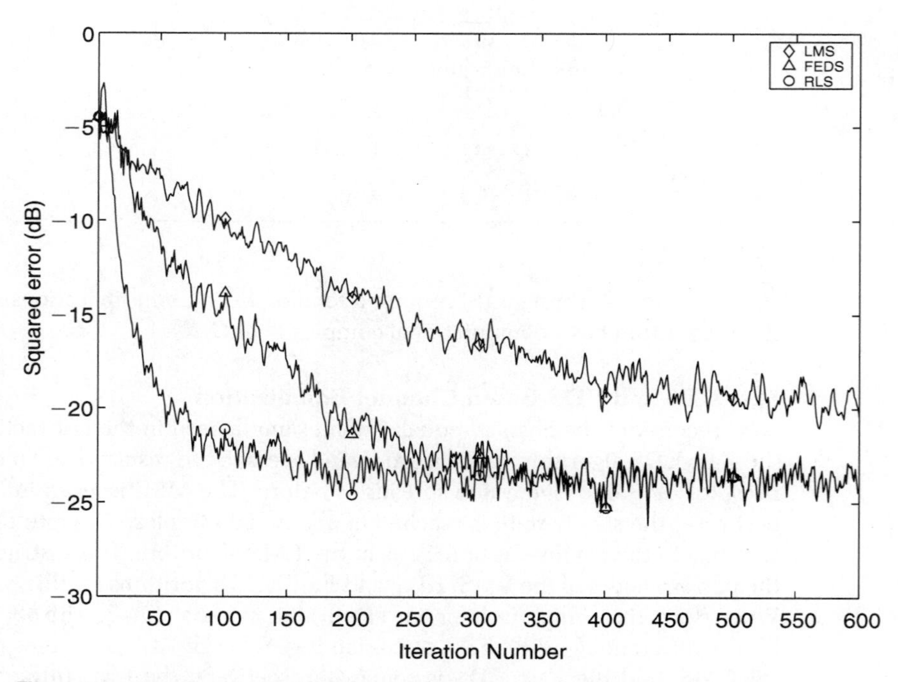

Figure 8.7 Comparison of different algorithms for channel equalization.

converges in about 300 iterations and the RLS in about 200, whereas the LMS takes more than 600. The RLS converges a little faster than the Fast EDS but at the cost of $O(N^2)$ computational complexity.

At this point, it is natural to compare the performances of the different EDS-based algorithms. The first one is the standard EDS algorithm, where N directions are searched for every sample. This algorithm has $O(N^2)$ computational complexity as discussed before. Then we have the Fast EDS algorithm, or FEDS. This is derived by slightly modifying the cost function, but it has $O(N)$ complexity $(4N+2)$. In this algorithm, only one directional search is performed for each sample of data. There is another algorithm that has not been discussed, where a two-directional search is performed for each sample. This is called the FEDS-2 algorithm; it is also an $O(N)$ algorithm with $(6N+10)$ multiplications. The derivation is complicated and is therefore omitted. Interested readers are referred to [7] and [17]. Now we use these three algorithms for the channel equalizer. The results are given in Fig. 8.8. It is clear that the EDS algorithm converges the fastest, whereas the FEDS algorithm is the slowest of the three. The reason is obvious; the more directions are searched for every sample, the faster the convergence, but the higher is the computational complexity. The EDS and the FEDS-2 have comparable convergence rates. They converge in less than 200 iterations, which is about the same rate as the RLS. Interestingly, then, the FEDS-2 algorithm with a complexity of $O(N)$ converges as fast as the RLS.

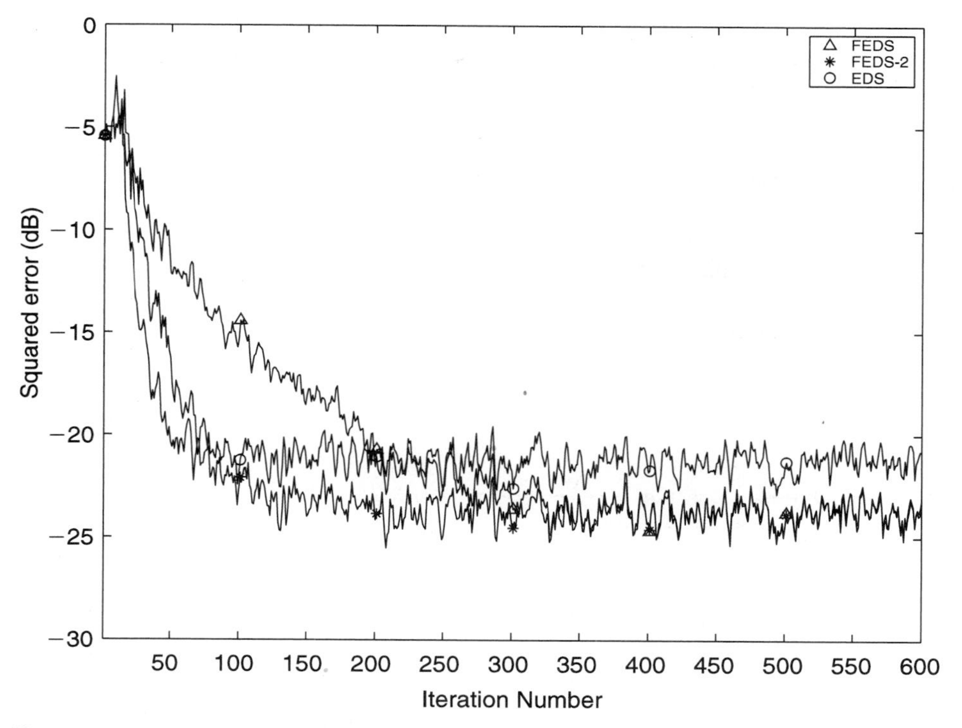

Figure 8.8 Comparison of different EDS based algorithms for channel equalization.

8.3 ADPCM

In this section, we use adaptive algorithms for Differential Pulse Code Modulation (DPCM), a popular technique for data compression that uses the concept of linear prediction. When the predictor (filter) is adaptive, we have an ADPCM system. The basic philosophy of DPCM is the following. Let $\{x(n)\}$ denote the sequence of data to be transmitted (or stored). Using some past data values $x(n-1), x(n-2), \ldots, x(n-N)$, we predict the value of the current data and call it $\hat{x}(n)$. The difference between the actual data and the predicted value is $e(n) = x(n) - \hat{x}(n)$. If the prediction is reasonably good, then $e(n)$ is much smaller than $x(n)$. Therefore, $e(n)$ requires fewer bits and is the sequence that is transmitted. The receiver has the same predictor and can reconstruct $x(n)$ from $e(n)$.

The actual DPCM scheme is slightly different from the above owing to the necessity of a quantizer and is given in Fig. 8.9(*a*) and (*b*) for the transmitter and receiver, respectively. The various signals are as follows.

$s(n)$: Signal to be transmitted or stored. It is theoretically infinite precision. In practice, it has a large number of bits per sample, typically 16 or 32.

$\tilde{e}(n)$: Quantized error signal to be actually transmitted or stored. It has B bits/sample, where B is a small number, typically 3 or 4. This is therefore a low-precision signal and is used as the error signal for the adaptive algorithm.

$\hat{s}(n)$: The output of the adaptive predictor. The predictor coefficients are high precision, and therefore this signal is high precision also.

$\tilde{s}(n)$: The input to the adaptive predictor. This signal is the sum of $\tilde{e}(n)$ (low precision) and $\hat{s}(n)$ (high precision). The number of bits is usually set to be high.

The main elements to be designed are the predictor and the quantizer. The literature presents many theories on the design of these elements for a variety of applications. Here we will consider the most fundamental designs. For the predictor, we will use an N-tap FIR adaptive filter that will be adapted by the algorithms we have learned so far. The same predictor model and adaptive algorithm are used in the receiver so that $\hat{s}(n)$ can be reconstructed.

The quantizer will be assumed to be a B-bit uniform quantizer with variable step size. The quantizer step size is defined as $\Delta = A_{fs}/2^{B-1}$, where A_{fs} is the maximum magnitude of the signal, or full-scale value. The full-scale value A_{fs} is not exactly known and must be estimated. In addition, it changes all the time. In order to get the most resolution of the quantizer, the step size (or the full-scale value) must be constantly adapted according to some algorithm. Although many such algorithms are useful for specific applications,

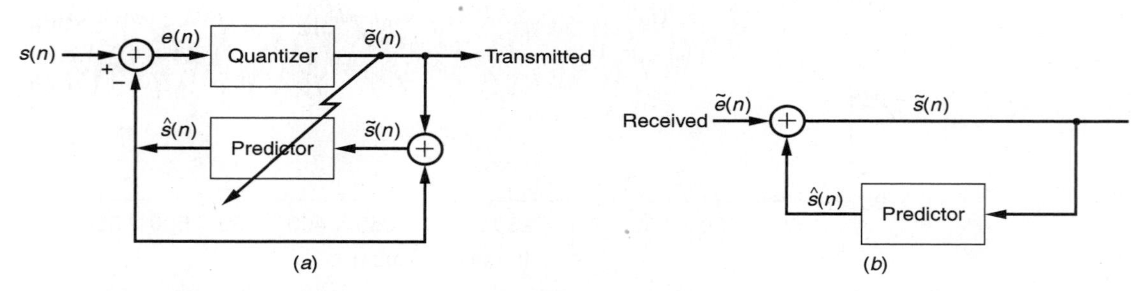

Figure 8.9 ADPCM: (a) transmitter, (b) receiver.

here we will use a simple recursive adaptive quantizer [22] where the quantizer step size is computed as

$$\Delta(n+1) = M(\beta)\Delta(n) \tag{8.78}$$

where $M(\beta)$ is a multiplier that can be optimized for different applications. The variable β is an integer that represents the quantization level. We will use simple magnitude truncation for quantization. For 4-bit magnitude truncation, the values for β are shown in Fig. 8.10. Therefore, we can write a formula for β as

$$\beta = \left\lfloor \frac{e(n)}{\Delta(n)} \right\rfloor \tag{8.79}$$

where $\lfloor x \rfloor$ represents the floor function, that is, the largest integer not greater than x. One popular set of multiplication factors is given by Jayant [22] and is optimized for speech encoding. This is called the Jayant quantizer and will be used here. The Jayant multipliers are given in Table 8.5 for 2-, 3-, and 4-bit quantization.

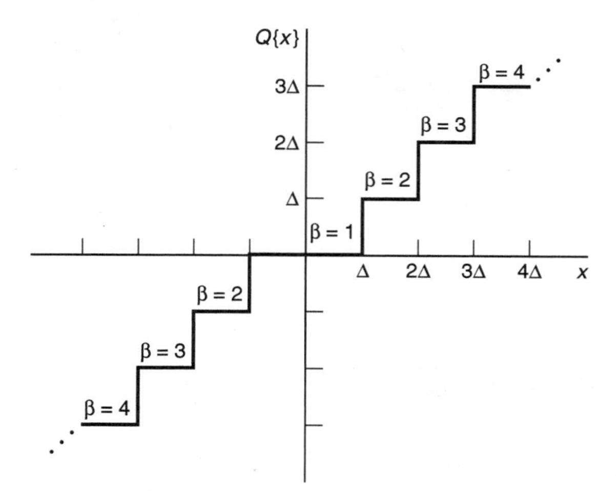

Figure 8.10 Magnitude truncation quantizer.

TABLE 8.5 Jayant Quantizer Multiplication Factors

	B = 2	B = 3	B = 4
M (1)	0.8	0.90	0.90
M (2)	1.6	0.90	0.90
M (3)		1.25	0.90
M (4)		1.70	0.90
M (5)			1.20
M (6)			1.60
M (7)			2.00
M (8)			2.40

Now we illustrate the concept of ADPCM with an example: Let $s(n) = \sin(n/33)$ be the original signal at a precision of 32 bits per sample (standard MATLAB format). This is shown in Fig. 8.11(a). If this signal is directly quantized to a 4-bit precision, the result is as shown in Fig. 8.11(b). This directly quantized signal has an SNR of about 18 dB. Now we

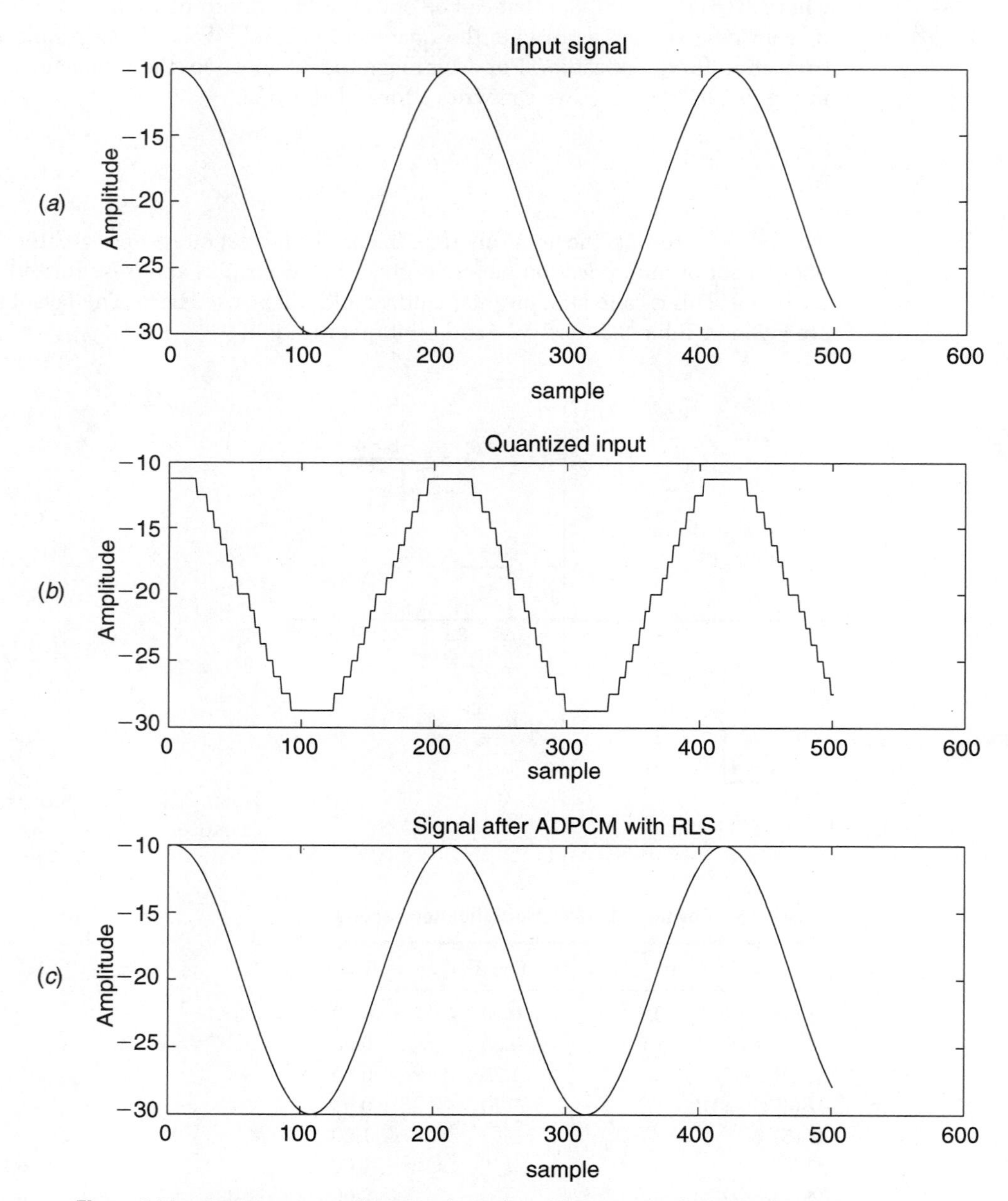

Figure 8.11 ADPCM using the RLS algorithm: (a) input signal; (b) directly quantized input signal to 4 bits resolution; (c) recovered signal after ADPCM.

perform ADPCM with a 5-tap adaptive FIR filter with the RLS algorithm. The forgetting factor is chosen as $\lambda = 0.98$. The signal $\tilde{s}(n)$ is shown in Fig. 8.11(*c*) after convergence of the adaptive filter. The SNR of this signal is about 191 dB. This is a remarkable improvement over direct quantization in spite of a compression ratio of $32/4 = 8$. This experiment is also conducted with the LMS algorithm. The SNR of $\tilde{s}(n)$ is then found to be about 38 dB. The SNR is not as high because of the nonzero misadjustment of the LMS algorithm. The experiment with the FEDS algorithm is assigned as a computer project.

▶ 8.4 2-D ADAPTIVE ALGORITHMS

In this section, we derive the 2-D counterparts of 1-D adaptive algorithms, and then we use them in image processing applications. The 2-D versions of the LMS and the FEDS are derived, whereas the 2-D versions of other adaptive algorithms are left as exercises.

The 2-D LMS Algorithm

As we know from previous chapters, a 2-D signal is represented as a matrix. If we concatenate the rows (or columns), then we have one long vector, that is, a 1-D sequence. Then we can use a 1-D filter and a 1-D adaptive algorithm. However, this approach usually leads to unsatisfactory results because in an image signal or matrix, there is strong correlation all around the neighborhood of a pixel. Therefore, a 2-D filter is necessary. In addition, if we convert the image matrix to a 1-D vector by concatenating rows (or columns), then this correlation is destroyed. Recall that a 2-D FIR filter is described by

$$y(n_1, n_2) = \sum_{i=0}^{K-1} \sum_{j=0}^{L-1} b_{ij} x(n_1 - i, n_2 - j) \tag{8.80}$$

where $x(n_1, n_2)$ is the input, $y(n_1, n_2)$ is the output, and K and L are positive integers. The 2-D FIR filter is therefore of order (K, L). In the development of 1-D adaptive algorithms, the filter output was written as $y(n) = \mathbf{w}^T \mathbf{x}(n)$, where $\mathbf{w}$ is the latest available weight vector and $\mathbf{x}(n)$ is the input data vector. In the 2-D case, we can do the same thing, but the weight vector and the input vector must be reformulated. Let us define

$$\mathbf{w} = [b_{00}\, b_{01} \ldots b_{0,L} b_{10}\, b_{11} \ldots b_{K,L}]^T \tag{8.81}$$

$$\begin{aligned} \mathbf{x}(n_1, n_2) = {} & [x(n_1, n_2)\, x(n_1, n_2 - 1) \ldots x(n_1, n_2 - L)\, x(n_1 - 1, n_2) \\ & x(n_1 - 1, n_2 - 1) \ldots x(n_1 - K, n_2 - L)]^T. \end{aligned} \tag{8.82}$$

If we define $N = (K + 1) \times (L + 1)$, the lengths of the above vectors are clearly equal to N. The 2-D filter can now be written as

$$y(n_1, n_2) = \mathbf{w}^T \mathbf{x}(n_1, n_2). \tag{8.83}$$

The LMS algorithm essentially remains the same, except that the input vector must be reformulated as above. Since we have a 2-D data sequence, we update the weight vector as we traverse the data matrix pixel by pixel, one row (or column) at a time (see Fig. 8.12). In this figure, note that each update of the weight vector **w** is represented by incrementing its argument. The weight vector is therefore written as **w**(l), where $l = n_1N_1 + n_2$. For a desired signal of $d(n_1, n_2)$, the complete 2-D LMS algorithm is given in Table 8.6.

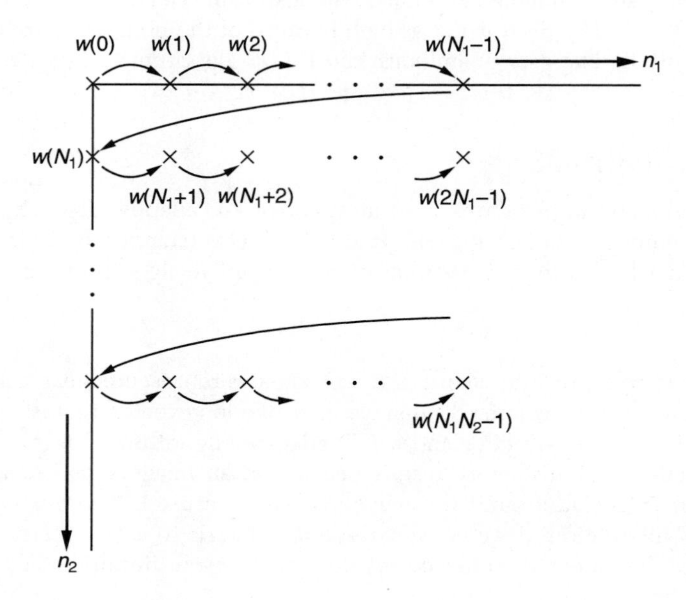

Figure 8.12 Weight vector update for a 2-D adaptive algorithm.

TABLE 8.6 The 2-D LMS

Initialization:

$$\mathbf{w} = \mathbf{0}$$

$$x(n_1, n_2) = 0, \quad n_1 \leq -1, \quad n_2 \leq -1$$

Algorithm:

For $n_2 = 0, 1, 2, \ldots, M_2$

For $n_1 = 0, 1, 2, \ldots, M_1$

$$\mathbf{x}(n_1, n_2) = [x(n_1, n_2)\, x(n_1, n_2 - 1) \ldots\, x(n_1, n_2 - M_2)$$
$$x(n_1 - 1, n_2)\, x(n_1 - 1, n_2 - 1) \ldots,$$
$$x(n_1 - M_1, n_2 - M_2)]^T$$

$$e(n_1, n_2) = d(n_1, n_2) - \mathbf{w}^T(l)\, \mathbf{x}(n_1, n_2)$$

$$\mathbf{w}(l + 1) = \mathbf{w}(l) + \mu \mathbf{x}(n_1, n_2) e(n_1, n_2)$$

end n_1

end n_2

The 2-D LMS algorithm was first reported by Hadhoud and Thomas [18] who formally derived the algorithm from 1-D and applied it to 2-D line enhancement. They showed that the 2-D LMS is effective in adapting to local statistical variations within the image and in preserving the edges. As in the 1-D case, the choice of the convergence factor μ affects the convergence rate, tracking ability, and steady-state MSE. In fact, this factor is chosen to trade off among these parameters. This choice often depends on the application at hand. Some techniques used in finding the optimal step size for different applications have been reported in [19] and [20]. These techniques require knowledge of the eigenvalues of the image autocorrelation matrix in addition to other statistical information of the image. In [21], a technique has been reported for calculating the step size depending on the input signal. This technique is similar to the normalized LMS and is useful when we do not have a priori knowledge of the image statistics. A discussion of these techniques is beyond the scope of this book; here we will simply use values of μ that are based on trial and error.

The 2-D FEDS Algorithm

Recall that in the 1-D FEDS algorithm, the data are processed within blocks of length N, where N is the number of filter coefficients. The N-element weight vector is updated only once per block, whereas the correlation matrix ($\mathbf{R}$) and the cross-correlation vector ($\mathbf{p}$) are updated for every sample within the block. By reexamining the FEDS algorithm in Table 8.4, we see that steps (1) through (8) are performed for every data sample, whereas steps (A) and (B) are computed at the end of each block. In the 2-D case, the same concept is followed except that we need to formulate it in the 2-D signal space. Let the 2-D signal be of size $N_1 \times N_2$, and the filter of size (K, L); the number of filter coefficients is then $N = K \times L$. Now consider the 2-D signal space shown in Fig. 8.13. The steps

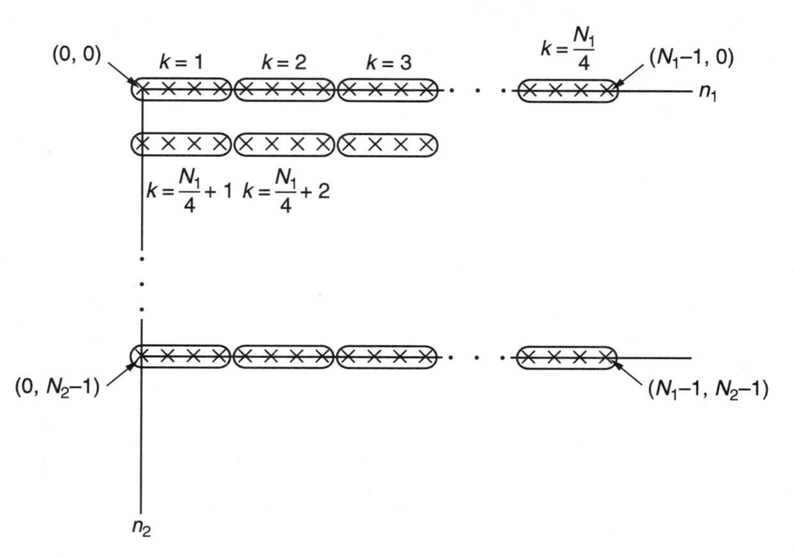

Figure 8.13 2-D space for FEDS algorithm and path of processing.

of the FEDS algorithm are now performed as follows. The image will be processed one row (or column) at a time. The pixels in each row are subdivided into blocks of length N as shown in the figure. It is assumed that N divides N_1.[1] This assumption is made so that the 2-D version of the FEDS algorithm can be described in a simple form. As we traverse each pixel, steps (1) through (8) of the FEDS algorithm are performed. Note that the input vector $\mathbf{x}$ must be formulated at each pixel so that a 2-D filtering operation is performed. At the end of each block, steps (A) and (B) are computed. When we complete a row, we go to the beginning of the next row and repeat the process. The complete 2-D FEDS algorithm is given in Table 8.7

TABLE 8.7 The 2-D FEDS Alogrithm

For $n_2 = 0, \ldots, N_2 - 1$

For $j = 0, \ldots, \dfrac{N_1}{N} - 1$

Step (I) : For $k = 0, \ldots, N-1; \quad n_1 = jN + k$

(1)$\mathrm{x} = [x(n_1, n_2)\, x(n_1, n_2 - 1) \ \ldots \ x(n_1, n_2 - L)\, x(n_1 - 1, n_2)\, x(n_1 - 1, n_2 - 1)$

$\ldots x(n_1 - K, n_2 - L)]^T$;

(2) $q^{(k)} = \lambda \mathbf{q}^k$;

(3) $(\mathbf{r})_k = \lambda (r)_k$;

(4) $\widetilde{\mathbf{Q}} = \widetilde{\mathbf{Q}} + \mathbf{x}\mathbf{x}^T$;

(5) $\widetilde{\mathbf{r}} = \widetilde{\mathbf{r}} + d(n_1, n_2)\mathbf{x}$;

(6) $\varepsilon = (\mathbf{q}^{(k)} + \widetilde{\mathbf{q}}^{(k)})^T \mathbf{w} - (\mathbf{r} + \widetilde{\mathbf{r}})_k$;

(7) $a = (\mathbf{q}^{(k)})_k + (\widetilde{\mathbf{q}}^{(k)})_k$;

(8) If $a \neq 0, \ \alpha = -\dfrac{\varepsilon}{a}; \qquad (\mathbf{w})_k = (\mathbf{w})_k + \alpha$;

end k

Step (II)

(A) $\mathbf{Q} = \mathbf{Q} + \widetilde{\mathbf{Q}}; \quad \widetilde{\mathbf{Q}} = 0$;

(B) $\mathbf{r} = \mathbf{r} + \widetilde{\mathbf{r}}; \quad \widetilde{\mathbf{r}} = 0$;

end j

end n_2

---

[1] If N does not divide N_1, then the remaining pixels in that row can be used with pixels in the next row to form the next block, and so on. If there are just a few remainder pixels, then we may choose not to perform FEDS on these pixels at all. In this case, we would simply filter these pixels with the latest available filter weights.

8.5 IMAGE RESTORATION

The 2-D adaptive algorithms derived earlier are now applied to image restoration by line enhancement as shown in Fig. 8.14. We will choose a test image and add noise to it to form the image $d(n_1, n_2)$. This is the desired signal for the adaptive algorithm. This signal is delayed by (1, 1) in order to decorrelate the noise, and this forms the input signal $x(n_1, n_2)$ to the filter. The filter attenuates the noise and produces $y(n_1, n_2)$, which should be the restored image. Two test images have been chosen: "Lena" and "Mandrill." The SNR used are -1.8 dB and -7 dB. The filter order is chosen as (4,4), that is, $K = L = 4$. The 2-D LMS (with $\mu = 10^{-6}$) and 2-D FEDS (with $\lambda = 0.99$) algorithms are applied for adaptation of the filter coefficients. The SNR of the filtered signal is calculated for each experiment. The SNR improvement (SNRI) is defined as the final SNR minus the original SNR. The results of the simulations are given in Table 8.8. From this table, we see that the SNRI figures obtained for the FEDS algorithm are higher than those obtained by the LMS. The reason for this difference can be attributed to the fast convergence rate and the zero misadjustment of the FEDS. The simulation results for "Lena" with 1.8 dB SNR are shown in Fig. 8.15(a)–(d).

8.6 SUMMARY

In this chapter, we presented the method of least-squares optimization and then described some algorithms for adaptive filtering. The popular RLS algorithm has been derived. Then a thorough performance analysis was done. The advantages of this algorithm are fast convergence rate and zero misadjustment. However, its $O(N^2)$ complexity may be a deterrent for many real-time applications. An $O(N)$ version of the RLS developed by Cioffi and Kailath [23] has received considerable

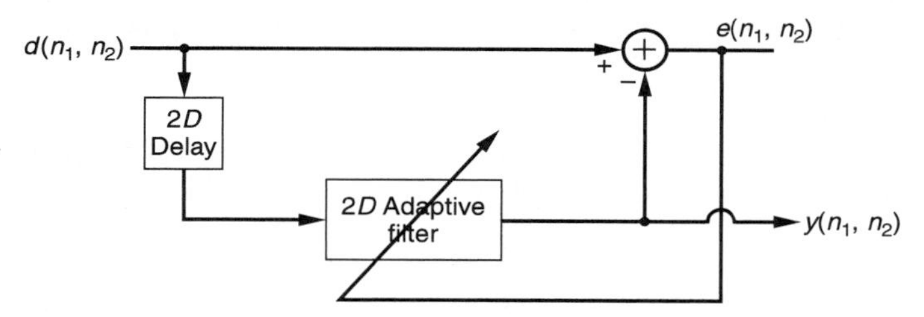

Figure 8.14 2-D line enhancer for image restoration.

TABLE 8.8 SNRI (dB)

	Lena	Mandrill
$SNR = -1.8$ dB		
LMS	8.6	5.0
FEDS	9.5	5.6
$SNR = -7$ dB		
LMS	10.6	8.5
FEDS	11.9	9.5

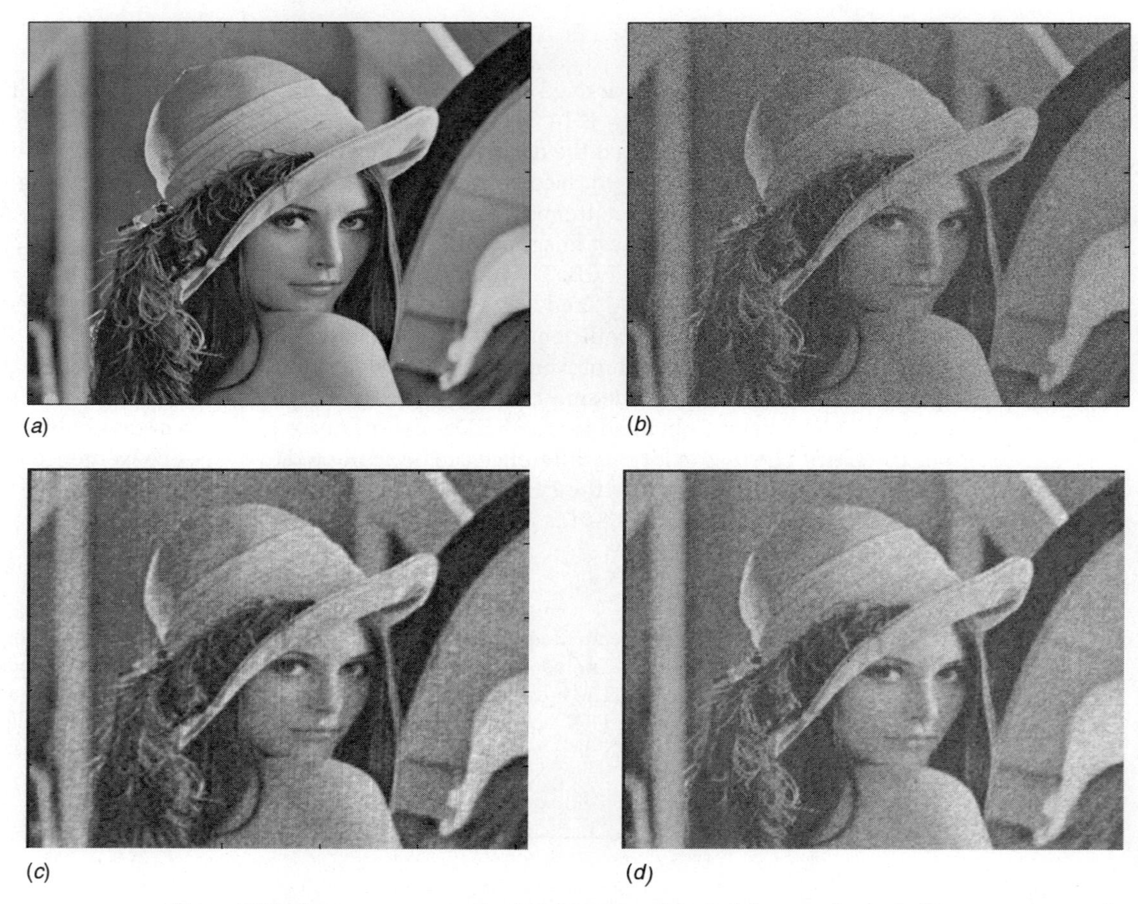

Figure 8.15 Image restoration by 2-D linear prediction: (*a*) original image; (*b*) image corrupted by 1.8 SNR with Gaussian noise; (*c*) image restored by 2-D LMS; (*d*) image restored by 2-D FEDS.

attention. This is called the Fast Transversal Filter (FTF). This algorithm has some numerical stability problems. Some stabilized versions of the FTF have since appeared in the literature [24]. Much has been written about the RLS and its fast counterparts, but we have only presented the fundamentals in this chapter. There are many good books that cover this topic in more detail and also point to numerous references. These books include those by Haykin [3], Widrow and Stearns [11], Alexander [25], Farhang-Bourjeny [26], Clarkson [27], Treichler et al. [28], Proakis et al. [29], Diniz [30], and Jenkins et al. [31].

We also presented a recently developed algorithm called the Euclidean Direction Search (EDS) algorithm. This algorithm is closely related to the Gauss-Seidel method of solving a set of linear simultaneous equations. The fast version of this algorithm is called the Fast EDS or FEDS. This algorithm has an $O(N)$ computational complexity, zero misadjustment and a convergence rate comparable to that of the RLS. A thorough performance evaluation of this method can be found in [17].

Two-dimensional versions of the LMS and the FEDS algorithms are useful for image processing applications. Several applications have been presented throughout the chapter, including channel equalization, adaptive DPCM, and image restoration.

8.7 PROBLEMS

1. Prove the Matrix Inversion Lemma.
2. Formulate the RLS algorithm for a one-step forward predictor. Such a predictor is a filter that predicts the sample $x(n)$ from the past samples $x(n-1), x(n-2), \ldots, x(n-N+1)$.
3. Formulate the EDS algorithm for a one-step forward predictor. Such a predictor is a filter that predicts the sample $x(n)$ from the past samples $x(n-1), x(n-2), \ldots, x(n-N+1)$.
4. Formulate the Fast EDS algorithm for a one-step forward predictor. Such a predictor is a filter that predicts the sample $x(n)$ from the past samples $x(n-1), x(n-2), \ldots, x(n-N+1)$.
5. Consider a modified least-squares cost function defined by

$$C(w,n) = \sum_{i=1}^{n} \lambda^{n-i} e^2(i) + \lambda^n K ||\mathbf{w}(n)||_2^2$$

where K is a constant.

(a) Using this cost function, derive a modified RLS algorithm.
(b) Derive the convergence condition for the algorithm in terms of K.

Discuss the convergence behavior for small and large K.

6. Show that the misadjustment of the RLS algorithm for $\lambda = 1$ and n large is

$$M_{\text{RLS}} = \frac{N}{n}.$$

7. Prove equations (8.71), (8.72), and (8.73).
8. Prove equations (8.74) and (8.75).
9. Show that if we have the $N \times N$ matrix $\mathbf{x}(n)\mathbf{x}^T(n)$, then computation of $\mathbf{x}(n+1)\mathbf{x}^T(n+1)$ can be done in N multiplications.
10. Derive the FEDS-2 algorithm, where the search is performed in two directions at a time.

8.8 COMPUTER PROJECTS

Project 1: Adaptive Line Enhancer (ALE)
Generate a signal of the form

$$s(n) = A\sin(2\pi f_1 n T_s) + B\sin(2\pi f_2 n T_s),$$

where T_s is the sampling frequency and the frequencies f_1 and f_2 can be chosen as desired. Choose T_s so that the sinusoid appears smooth. A noise $w(n)$ is added to the signal to produce the noisy signal

$$d(n) = s(n) + w(n).$$

The noise is a white uniformly distributed sequence with zero mean. The SNR of this signal should be about -7 dB. The desired signal (or primary signal) is $d(n)$. The reference signal is $d(n-1)$ and is applied to the input of an N-tap FIR filter that produces the output $y(n)$, which should be the estimate of the signal. Use the EDS algorithm for adaptation. Experiment with different values of N and the forgetting factor λ. Submit the following for your best case.

(a) Plot the MSE for an ensemble of 100 runs.
(b) Plot a few cycles of $s(n)$ for one run.
(c) Plot a few cycles of $y(n)$ after convergence (one run).
(d) Plot the impulse response and the frequency response of the filter after convergence (one run).
(e) Find the SNR of the filtered signal. Make sure you correct for phase, if necessary.
(f) Discuss the effects of N and λ on the performance of the algorithm.

Provide all block diagrams, equations, and algorithms in your report. Discuss your results. Are they as expected? Are there any variations from your expectations? Comment on the convergence rate in your experiments.

Project 2

Repeat Project 1 using the Fast EDS algorithm. Submit all of the items required in Project 1. In addition, discuss the rate of convergence of the Fast EDS in relation to the EDS algorithm.

Project 3

Repeat Project 1 using the RLS algorithm. Discuss the comparative performances of the EDS, Fast EDS, and the RLS algorithms.

Project 4: System Identification

An FIR system is given by

$$y(n) = x(n) + 0.1x(n-1) + 0.2x(n-2) + 0.3x(n-3)$$

where $x(n)$ and $y(n)$ are the input and output, respectively. Let $x(n)$ be a uniformly distributed white input sequence of unit variance. Obtain a desired signal by adding white Gaussian noise, $w(n)$ as

$$d(n) = y(n) + w(n).$$

Once we have $d(n)$ and $x(n)$, we assume that the system is unknown and we perform system identification. Assume no modeling error; that is, the filter is the same length as the original system.

Use SNR of infinity, 10 dB, and 20 dB, and identify the system using the EDS algorithm. Experiment with different values of N and the forgetting factor λ. Submit the following for your best case.

(a) Plot the MSE for an ensemble of 100 runs.
(b) Plot a few cycles of $d(n)$ for one run.
(c) Plot a few cycles of $y(n)$ after convergence (one run).
(d) Plot the impulse response and the frequency response of the filter after convergence (one run).
(e) Discuss the effects of N and λ on the performance of the algorithm.

Provide all block diagrams, equations, and algorithms in your report. Discuss your results. Are they as expected? Are there any variations from your expectations? Comment on the convergence rate in your experiments.

Project 5

Repeat Project 4 using the Fast EDS algorithm. Submit all of the items required in Project 3. In addition, discuss the performance of the Fast EDS in relation to the EDS algorithm.

Project 6

Repeat Project 4 using the RLS algorithm. Discuss the comparative performances of the EDS, Fast EDS, and the RLS algorithms.

Project 7: ADPCM

The objective of this project is to gain an understanding of the ADPCM encoding and decoding operations. For simulation purposes, we will use the following inputs:

(a) $s(n) = \sin(n/33)$.

(b) Colored noise generated by passing uniformly distributed zero-mean white noise through a lowpass filter of your choice. The cutoff frequency of this filter is at 0.2π. Scale the generated colored signal so that it is within $(+1, -1)$ in amplitude.

Simulate an ADPCM encoder with a feedback (Jayant) adaptive quantizer. The feedback quantizer step size is given by

$$\Delta(n+1) = \Delta(n)M(n)$$

where $M(n)$ is given by the standard Jayant quantizer. Use a 4-bit quantizer (including the sign bit) and an N-tap FIR predictor.

Use the RLS algorithm and experiment with the filter length and the forgetting factor. For your best case, submit the following:

(a) Plot $e^2(n)$.

(b) Plot a window of the input signal.

(c) Plot a window of a directly quantized (4-bit) version of the input.

(d) Find the SNR of the direct-quantized version of the input.

(e) Plot a window of the reconstructed signal after the filter has converged.

(f) Find the SNR of the reconstructed signal.

Provide all block diagrams, equations, and algorithms, Discuss your results. Discuss the effects of varying the filter length and the forgetting factor.

Project 8

Repeat Project 7 using the EDS algorithm. Discuss its performance in relation to the RLS algorithm.

Project 9

Repeat Project 7 using the Fast EDS algorithm. Discuss the relative performances of the Fast EDS, EDS, and the RLS algorithms.

Project 10

Repeat Project 7 using the LMS algorithm. Discuss the relative performances of the LMS, Fast EDS, EDS, and RLS algorithms.

▶ BIBLIOGRAPHY

[1] T. KAILATH, "Estimating filters for linear time-invariant channels," *Quarterly Progress Report 58, MIT Research Lab for Electronics*, Cambridge, MA, pp. 185–197, 1960.

[2] J. GREGORY and D. REDMOND, *Numerical Analysis*, Jones and Bartlett Publishers, Inc., 1994.

[3] S. HAYKIN, *Adaptive Filter Theory*, Prentice Hall, Englewood Cliffs, NJ, 1996.

[4] J.K. TUGNAIT, L. TONG, and Z. DING, "Single-user channel estimation and equalization," *IEEE Signal Processing Magazine*, 17 (3), pp. 17–28, May 2000.

[5] G.F. XU, T. BOSE, W. KOBER, and J. THOMAS, "A fast adaptive algorithm for image restoration," *IEEE Transactions on Circuits and Systems -I*, January 1999.

[6] G.F. XU and T. BOSE, "Analysis of the Euclidean direction set adaptive algorithm," *Proc. of the International Conference on Acoustics, Speech and Signal Processing*, pp. 1689–1692, April 1998.

[7] G.F. XU, T. BOSE, and J. SCHROEDER, "Channel equalization using an Euclidean Direction Search based adaptive algorithm, *Proc. of the Globecom*, November 1998.

[8] G.H. GOLUB and C. F. VAN LOAN, *Matrix Computations*, Johns Hopkins University Press, Baltimore, MD, 1996.

[9] W.H. PRESS, S. A. TEUKOLSKY, W. T. VETTERLING, and B. P. FLANNERY, *Numerical Recipes in Fortran*, Press Syndicate of the University of Cambridge, 1992.

[10] C.E. DAVILA, "Line search algorithm for adaptive filtering," *IEEE Trans. on Signal Processing*, 41, pp. 2490–2494, July 1993.

[11] B. WIDROW and S. D. STEARNS, *Adaptive Signal Processing*, Prentice Hall, Englewood Cliffs, NJ, 1985.

[12] R.W. LUCKY, "Automatic equalization for digital communications," *Bell Syst. Tech. Journal*, no. 44, pp. 547–588, April 1965.

[13] R.W. LUCKY and H.R. RUDIN, "An automatic equalizer for general-purpose communication channels," *Bell Sys. Tech. J.* 46, pp. 2178, November 1967.

[14] R.W. LUCKY, J. SALZ, and E.J. WELDON Jr., *Principles of Data Communication*, McGraw-Hill., New York, 1968.

[15] B. WIDROW and M.E. HOFF Jr., "Adaptive switching circuits," *IRE WESCON Conf. Rec.*, pp. 96–104, 1960.

[16] R. GODFREY and F. ROCCA, "Zero memory nonlinear deconvolution," *Geophys. Prospect*. 29, pp. 189–228, 1981.

[17] G.F. XU, *Fast Algorithms for Digital Filtering: Theory and Applications*, Ph.D. disse., University of Colorado, Boulder, 1999.

[18] M. HADHOUD and D.W. THOMAS, "The two-dimensional adaptive LMS (TDLMS) algorithm," *IEEE Trans. on Circuits and Systems*, CAS-35, pp. 485–494, May 1988.

[19] S.T. ALEXANDER and S.A. RAJALA, "Image compression results using the LMS adaptive algorithm," *IEEE Trans. Acous., Speech, and Signal Processing*, ASSP-33, June 1985.

[20] S.T. ALEXANDER and S.A. RAJALA, "Optimal gain derivation for the LMS algorithm using a visual fidelity criterion," *IEEE Trans. Acous., Speech, and Signal Processing*, ASSP-32, pp. 434–437, April 1984.

[21] W.B. MIKHAEL, F.H. WU, L.G. KAZONVSKY, G.S. KANG, and L.J. Fransen, "Adaptive filters with individual adaptation of parameters," *IEEE Trans. on Circuits and Systems*, CAS-33, pp. 677–686, July 1986.

[22] N.S. JAYANT, "Adaptive delta modulation with one-bit memory," *Bell System Technical Journal*, pp. 321–342, March 1970.

[23] J.M. CIOFFI and T. KAILATH, "Fast recursive least squares transversal filters for adaptive filtering," *IEEE Trans. Acous., Speech, and Signal Processing*, ASSP-32, pp. 304–337, 1984.

[24] P.A. REGALIA, "Numerical stability properties of a QR-based fast least squares algorithm," *IEEE Trans. Signal Processing*, SP-41, pp. 2096–2109, June 1993.

[25] S.T. ALEXANDER, *Adaptive Signal Processing Theory and Applications*, Springer-Verlag, 1986.

[26] B. FARHANG-Bourjeny, *Adaptive Filters: Theory and Applications*, Wiley, New York, 1998.

[27] P. CLARKSON, *Optimal and Adaptive Signal Processing*, CRC Press, Boca Raton, FL, 1993.

[28] J.R. TREICHLER, C.R. JOHNSON Jr., and M.G. LARIMORE, *Theory and Design or Adaptive Filters*, Wiley, & New York, 1987.

[29] J.G. PROAKIS, C.M. RADER, F. LING, and C.L. NIKIAS, *Advanced Digital Signal Processing*, MacMillan, 1992.

[30] P.S.R. DINIZ, *Adaptive Filtering: Algorithms and Practical Implementation*, Kluwer, 1997.

[31] W.K. JENKINS, et al., *Advanced Concepts in Adaptive Signal Processing*, Kluwer, 1996.

CHAPTER 9

Linear Prediction

The concept of linear prediction—the process of predicting a sample $u(n)$ from some of its past samples—has found many applications in signal processing. This process is also referred to as forward prediction. Similarly, there is the concept of backward prediction, in which, given the samples $u(n)$, $u(n-1), \ldots, u(n-M+1)$, we predict the past sample $u(n-M)$. In other words, we go backward in time. The difference between the predicted value of a sample and its actual value is called the prediction error. Interestingly enough, the lattice filter can generate both the forward prediction error and the backward prediction error at no extra computational cost. That is why lattice predictors are very popular structures.

The chapter begins with the concepts of forward and backward prediction and then presents the Levinson-Durbin algorithm. This is a powerful algorithm, providing an efficient way of calculating the coefficients of the lattice predictor. Several properties of the prediction error filter are also presented, after which the Gradient Adaptive Lattice (GAL) is derived. The advantage of this algorithm and its computational complexity are also discussed. Several techniques of speech coding, which is an important application of linear prediction, are discussed, including the waveform coder, transform coder, subband coder, and linear predictive coder.

9.1 FORWARD PREDICTION

Forward prediction is the process of predicting the current sample of a wide-sense stationary random process from a linear combination of past samples. Figure 9.1 shows the schematic of a forward predictor where the estimate of the nth sample, $\hat{u}(n)$, is predicted from past samples $u(n-1), u(n-2), \ldots, u(n-M)$. This is called one-step prediction. The predictor can be nonlinear, linear, recursive, or nonrecursive. The most commonly used predictor is a linear FIR filter given by

$$\hat{u}(n) = \sum_{k=1}^{M} w_k^f u(n-k) \tag{9.1}$$

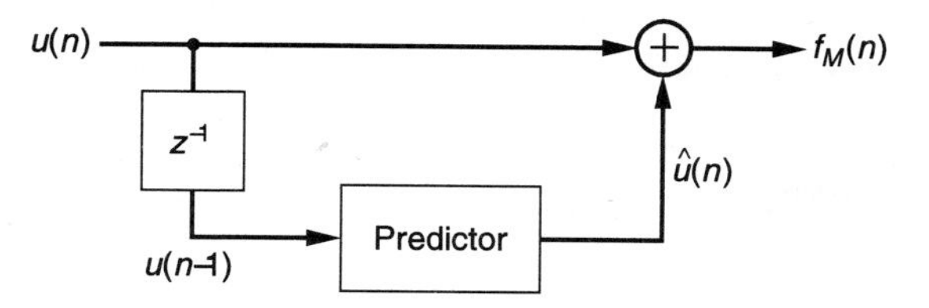

Figure 9.1 Forward linear prediction scheme.

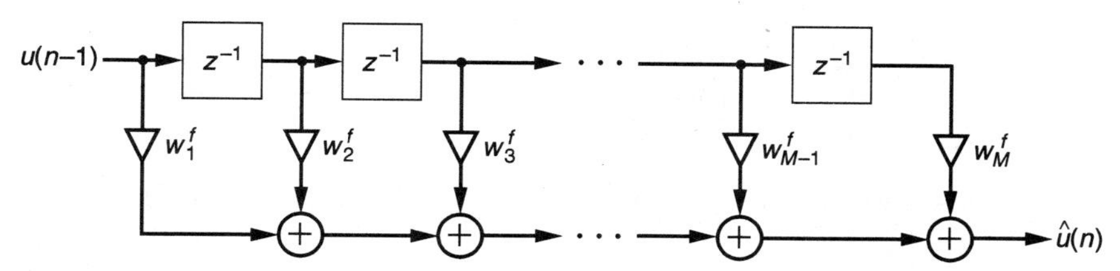

Figure 9.2 One-step linear predictor.

where w_k^f represents the filter coefficient. The superscript f is used to denote forward prediction. The error signal of its forward predictor is

$$f_M(n) = u(n) - \hat{u}(n) \tag{9.2}$$

where the subscript M denotes the order of the predictor. The forward linear predictor of order M is shown in Fig. 9.2. Now define the following:

$$\mathbf{w}^f \triangleq \left[w_1^f w_2^f \dots w_M^f\right]^T$$

$$\mathbf{u}(n-1) = [u(n-1)\, u(n-2) \cdots u(n-M)]^T.$$

Using these definitions, we can write the predicted signal and the prediction error as

$$\hat{u}(n) = \mathbf{u}^T(n-1)\mathbf{w}^f = \mathbf{w}^{f^T}\mathbf{u}(n-1)$$

and

$$f_M(n) = u(n) - \mathbf{u}^T(n-1)\mathbf{w}^f.$$

The prediction error can also be written as

$$\begin{aligned} f_M(n) &= u(n) - \sum_{k=1}^{M} w_k^f u(n-k) \\ &= \sum_{k=0}^{M} a_{M,k} u(n-k) \end{aligned} \tag{9.3}$$

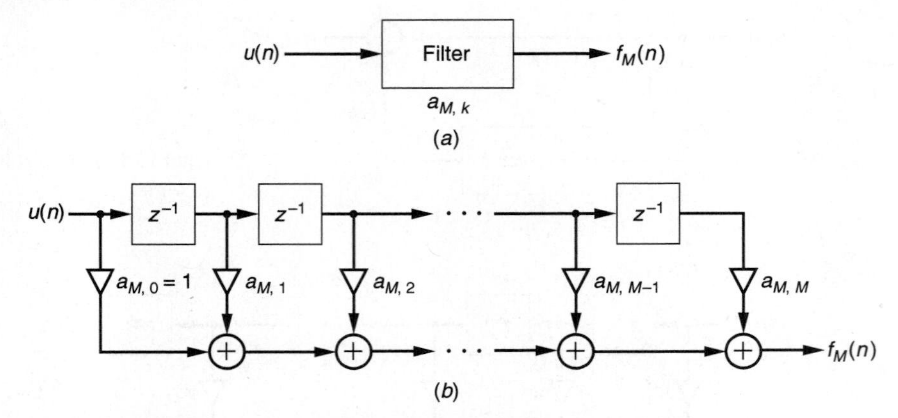

Figure 9.3 Prediction error filter: (*a*) block diagram; (*b*) flow graph.

Equation (9.3) is an FIR filter with the forward prediction error as the output. It is therefore called the forward prediction error filter; its flow graph is shown in Fig. 9.3. From (9.3), the relationship between w_k^f and $a_{M,k}$ is

$$a_{M,0} = 1,$$

$$a_{M,k} = -w_k^f, \quad k = 1, 2, \ldots, M.$$

In more compact form, we can write

$$\mathbf{a}_M = \begin{bmatrix} 1 \\ -\mathbf{w}^f \end{bmatrix}. \tag{9.4}$$

Now that the preliminaries are out of the way, let us revisit the linear prediction scheme shown in Fig. 9.1. Comparing this with the Wiener filtering problem in Chapter 7, we see that the desired signal is $u(n)$ and the filter input is $u(n-1)$. The Wiener-Hopf equations are derived by minimizing $E\{f_M^2(n)\}$ and are given by

$$\mathbf{R}\mathbf{w}^f = \mathbf{r} \tag{9.5}$$

where $\mathbf{R}$ is the correlation matrix for the filter input and $\mathbf{r}$ is the cross-correlation vector between the input and the desired signal. That is,

$$\begin{aligned} \mathbf{R} &= E\{\mathbf{u}(n-1)\mathbf{u}^T(n-1)\} \\ &= \begin{bmatrix} r(0) & r(1) & \cdots & r(M-1) \\ r(1) & r(0) & \cdots & r(M-2) \\ \vdots & & & \vdots \\ r(M-1) & r(M-2) & \cdots & r(0) \end{bmatrix} \end{aligned} \tag{9.6}$$

and

$$\mathbf{r} = E\{\mathbf{u}(n-1)u(n)\}$$

$$= \begin{bmatrix} r(1) \\ r(2) \\ \vdots \\ r(M) \end{bmatrix}. \tag{9.7}$$

Note that we have assumed a real input signal and therefore $r(k) = r(-k)$. From Chapter 7, we also know that the mean square error (MSE) is given by

$$E_M^f = E\{f_M^2(n)\}$$

$$= \sigma_d^2 - \mathbf{r}^T\mathbf{w}^f, \tag{9.8}$$

where σ_d^2 is the variance of the desired signal. In this context, the desired signal is $u(n)$ with zero mean, and so we have

$$E_M^f = r(0) - \mathbf{r}^T\mathbf{w}^f. \tag{9.9}$$

If we substitute the optimum weight vector, $\mathbf{w}^f = \mathbf{R}^{-1}\mathbf{r}$ into (9.9), we get the minimum MSE.

Once we know the optimum weight vector $\mathbf{w}^f$ of the predictor, we can find the optimum weight vector of the prediction error filter $\mathbf{a}_M$ by using equation (9.4). This filter can also be represented as a lattice structure as described in Chapter 4. The flow graph of the lattice structure is shown in Fig. 9.4. The order-recursive equations for the mth stage are given by

$$f_m(n) = f_{m-1}(n) + K_m g_{m-1}(n-1) \tag{9.10}$$

$$g_m(n) = K_m f_{m-1}(n) + g_{m-1}(n-1) \tag{9.11}$$

for $m = 1, 2, \ldots, M$. The upper channel output, $f_m(n)$ is the forward prediction error. The lower channel output, $g_m(n)$, is the backward prediction error, which will be discussed in the next section. The relationship between the lattice coefficients K_m and the Direct Form coefficients $\mathbf{a}_M$ were derived in Chapter 4. The lattice structure for the prediction error filter was introduced because the lattice coefficients K_m can be directly found from the elements in $\mathbf{R}$ in a recursive manner. This algorithm is called the Levinson-Durbin algorithm and will be described in a subsequent section. In closing this section, we write the Wiener-Hopf equation (9.5) and the MSE of the prediction error (9.8) in a compact augmented form as

$$\begin{bmatrix} r(0) & \mathbf{r}^T \\ \mathbf{r} & \mathbf{R} \end{bmatrix} \begin{bmatrix} 1 \\ -\mathbf{w}^f \end{bmatrix} = \begin{bmatrix} E_M^f \\ \mathbf{0} \end{bmatrix} \tag{9.12}$$

where $\mathbf{0}$ is a $M \times 1$ vector of zeros.

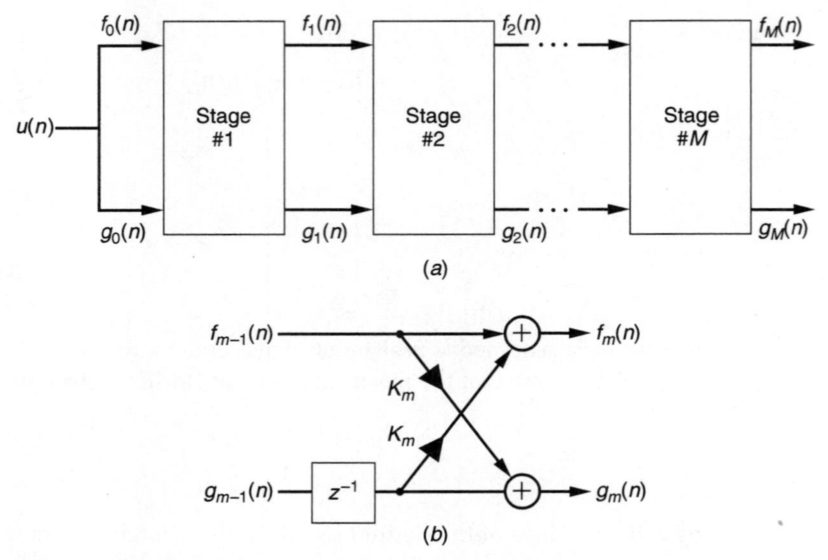

Figure 9.4 (*a*) Lattice structure; (*b*) a single stage.

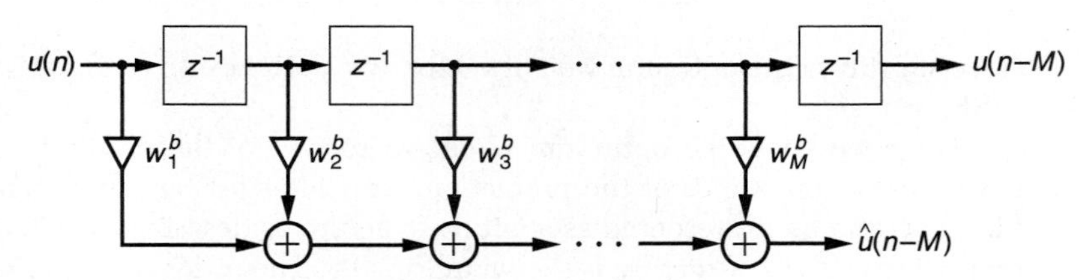

Figure 9.5 Backward linear predictor.

▶ 9.2 BACKWARD PREDICTION

The backward predictor operates on a time series in the backward direction. Given the samples $u(n)$, $u(n-1)$, ... ,$u(n-M+1)$, the backward predictor makes a prediction of the sample $u(n-M)$. This is called a one-step backward prediction. The predicted value is given by

$$\hat{u}(n-M) = \sum_{k=1}^{M} w_k^b u(n-k+1) \tag{9.13}$$

where w_k^b represents the filter coefficients and the superscript b is used to denote backward prediction. The flow graph of this prediction is given in Fig. 9.5. The backward prediction error is now

$$g_M(n) = u(n-M) - \hat{u}(n-M) \tag{9.14}$$

where the subscript M denotes the order of the predictor.

Substituting (9.13) into (9.14) gives

$$g_M(n) = u(n-M) - \sum_{k=1}^{M} w_k^b u(n-k+1)$$

$$= \sum_{k=0}^{M} b_{M,k} u(n-k) \tag{9.15}$$

where $b_{M,0} = -w_1^b$, $b_{M,1} = -w_2^b, \ldots,$ $b_{M,M-1} = -w_M^b$ and $b_{M,M} = 1$. In other words, we have

$$\mathbf{b}_M = \begin{bmatrix} -\mathbf{w}^b \\ 1 \end{bmatrix}. \tag{9.16}$$

The flow graph of the backward prediction error filter is given in Fig. 9.6. For the backward prediction, the desired signal is

$$d(n) = u(n-M). \tag{9.17}$$

In this case, the input signal vector is again

$$\mathbf{u}(n) = [u(n)\, u(n-1) \ldots u(n-M+1)]^T.$$

The correlation matrix $\mathbf{R} = E\{\mathbf{u}(n)\mathbf{u}^T(n)\}$, which is the same as given in (9.6). The cross-correlation vector is now

$$E\{\mathbf{u}(n)d(n)\} = E\{\mathbf{u}(n)u(n-M)\}$$

$$= \begin{bmatrix} r(M) \\ r(M-1) \\ \vdots \\ r(1) \end{bmatrix} = \mathbf{r}^B \tag{9.18}$$

where $\mathbf{r}^B$ denotes a reverse arrangement of vector $\mathbf{r}$ as given in (9.7). The Wiener-Hopf equations are thus

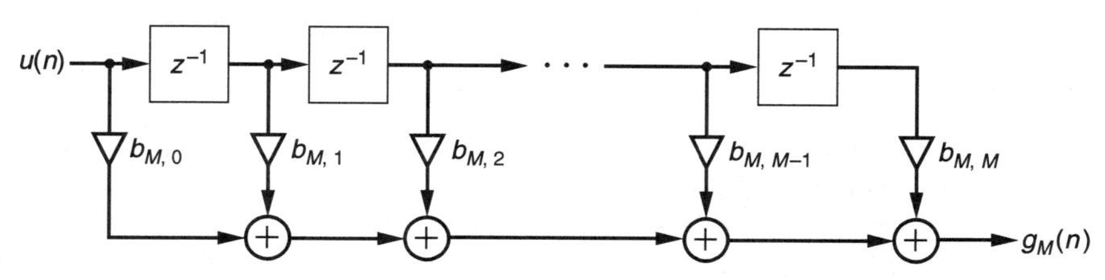

Figure 9.6 Backward prediction error filter.

$$\mathbf{R}\mathbf{w}^b = \mathbf{r}^B. \tag{9.19}$$

The MSE for the backward predictor is

$$E_M^b = r(0) - \mathbf{r}^{B^T}\mathbf{w}^b. \tag{9.20}$$

Now let us explore the relationship between the forward and backward predictors. If we compare equations (9.19) and (9.5), we see that the right-hand side of one is the reverse arrangement of the other. So, we reverse the order of the elements on both sides of (9.19) as

$$(\mathbf{R}\mathbf{w}^b)^B = (\mathbf{r}^B)^B = \mathbf{r}.$$

It is not difficult to show that $(\mathbf{R}\mathbf{w}^b)^B = \mathbf{R}\mathbf{w}^{bB}$, since $\mathbf{R}$ is a symmetric matrix. Therefore, for backward prediction, we have

$$\mathbf{R}\mathbf{w}^{bB} = \mathbf{r}. \tag{9.21}$$

The corresponding forward predictor equation is

$$\mathbf{R}\mathbf{w}^f = \mathbf{r}. \tag{9.22}$$

Comparing (9.21) and (9.22), we get

$$\mathbf{w}^f = \mathbf{w}^{bB}. \tag{9.23}$$

The coefficients for the two predictors are exactly the same, except that they are in reverse order. Now we relate the weight vectors of the forward and backward prediction error filters. From (9.4) and (9.16),

$$\mathbf{a}_M = \begin{bmatrix} 1 \\ -\mathbf{w}^f \end{bmatrix} = \begin{bmatrix} 1 \\ -\mathbf{w}^{bB} \end{bmatrix}$$

and

$$\mathbf{b}_M = \begin{bmatrix} -\mathbf{w}^b \\ 1 \end{bmatrix}.$$

Therefore, we have

$$\mathbf{a}_M = \mathbf{b}_M^B. \tag{9.24}$$

Using an analysis similar to the above, we can show (Problem 4) that the MSEs of the forward and backward predictors are identical. That is,

$$E_M^f = E_M^b. \tag{9.25}$$

The MSEs are identical because the correlation sequence is symmetric. If least-squares prediction is attempted with the use of covariances, then this equality does not hold. In that case, one may attempt to minimize the forward and backward prediction errors simultaneously as in modified covariance linear prediction. Further details can be found in [1].

9.3 LEVINSON-DURBIN ALGORITHM

The coefficient vector of the forward prediction error filter (**a**) can be calculated as follows. The correlation matrix **R** and the cross-correlation vector **r** are formed. Then the coefficient vector of the forward predictor is calculated as $\mathbf{w}^f = \mathbf{R}^{-1}\mathbf{r}$. The vector **a** is then found as $\mathbf{a} = \begin{bmatrix} 1 \\ -\mathbf{w}^f \end{bmatrix}$. This computation requires a matrix inversion that requires a computational complexity of $O(M^3)$. The Levinson-Durbin algorithm has two advantages: it finds all the coefficients in $O(M^2)$ computations, and it finds the reflection coefficients of the lattice structure directly. As a result, the order of the filter can be expanded by simply adding a stage without recomputing the coefficients of the existing stages. We will derive the algorithm for the forward predictor filter.

To set the stage for the derivation, we first recall a few earlier equations and make some new observations. Recall that an mth order prediction error filter has $m + 1$ coefficients, which are $a_{m,0} = 1,\ a_{m,k} = -w_k^f, k = 1, 2, \ldots, m$.

We know from Chapter 4 that these coefficients can be order updated by the following equations:

$$a_{m,m} = K_m \tag{9.26}$$

$$a_{m,k} = a_{m-1,k} + K_m a_{m-1,m-k}, \quad k = 1, 2, \ldots, m. \tag{9.27}$$

Note that we are not updating $a_{m,0}$ since it is equal to 1. The above equations are explicitly written as

$$\begin{bmatrix} a_{m,1} \\ a_{m,2} \\ \vdots \\ a_{m,m} \end{bmatrix} = \begin{bmatrix} a_{m-1,1} \\ a_{m-1,2} \\ \vdots \\ 0 \end{bmatrix} + \begin{bmatrix} x_{m-1,1} \\ x_{m-1,2} \\ \vdots \\ K_m \end{bmatrix} \tag{9.28}$$

where the unknown elements $x_{m-1,k}$ and K_m are to be found.

Define

$$\mathbf{x}_{m-1} = [x_{m-1,1}\ x_{m-1,2} \ldots x_{m-1,m-1}]^T. \tag{9.29}$$

Then (9.28) can be written as

$$-\mathbf{w}_m^f = \begin{bmatrix} -\mathbf{w}_{m-1}^f \\ 0 \end{bmatrix} + \begin{bmatrix} \mathbf{x}_{m-1} \\ K_m \end{bmatrix} \tag{9.30}$$

where we have introduced a subscript in $\mathbf{w}^f$ to denote the order of the predictor. A subscript will also be introduced for the correlation matrix and cross-correlation vector. So, for an mth order predictor, we have

$$\mathbf{R}_m = \begin{bmatrix} r(0) & r(1) & \dots & r(m-1) \\ r(1) & r(0) & \dots & r(m-2) \\ \vdots & & & \vdots \\ r(m-1) & r(m-2) & \dots & r(0) \end{bmatrix}, \tag{9.31}$$

and

$$\mathbf{r}_m = [r(1)\, r(2) \dots r(m)]^T. \tag{9.32}$$

The reverse arrangement of (9.32) is

$$\mathbf{r}_m^B = [r(m)\, r(m-1) \dots r(0)]^T. \tag{9.33}$$

The mth order correlation matrix can be partitioned as

$$\mathbf{R}_m = \begin{bmatrix} \mathbf{R}_{m-1} & \mathbf{r}_{m-1}^B \\ \mathbf{r}_{m-1}^{B^T} & r(0) \end{bmatrix}. \tag{9.34}$$

The Wiener-Hopf equations for the mth order predictor are

$$\mathbf{R}_m \mathbf{w}_m^f = \mathbf{r}_m. \tag{9.35}$$

Substituting (9.34) and (9.30) in (9.35), we get

$$\begin{bmatrix} \mathbf{R}_{m-1} & \mathbf{r}_{m-1}^B \\ \mathbf{r}_{m-1}^{B^T} & r(0) \end{bmatrix} \left\{ \begin{bmatrix} \mathbf{w}_{m-1}^f \\ 0 \end{bmatrix} - \begin{bmatrix} \mathbf{x}_{m-1} \\ K_m \end{bmatrix} \right\} = \begin{bmatrix} \mathbf{r}_{m-1} \\ r(m) \end{bmatrix}. \tag{9.36}$$

Equation (9.36) gives the two equations

$$\mathbf{R}_{m-1}\mathbf{w}_{m-1}^f - \mathbf{R}_{m-1}\mathbf{x}_{m-1} - \mathbf{r}_{m-1}^B K_m = \mathbf{r}_{m-1} \tag{9.37}$$

$$\mathbf{r}_{m-1}^{B^T}\mathbf{w}_{m-1}^f - \mathbf{r}_{m-1}^{B^T}\mathbf{x}_{m-1} - r(0)K_m = r(m). \tag{9.38}$$

Note that in (9.37), the first term on the left is

$$\mathbf{R}_{m-1}\mathbf{w}_{m-1}^f = \mathbf{r}_{m-1}. \tag{9.39}$$

Using this equation, (9.37) becomes

$$\begin{aligned} \mathbf{x}_{m-1} &= -K_m \mathbf{R}_{m-1}^{-1}\mathbf{r}_{m-1}^B \\ &= -K_m \mathbf{w}_{m-1}^{fB}. \end{aligned} \tag{9.40}$$

Substituting (9.40) in (9.38) and solving for K_m gives

$$\mathbf{r}_{m-1}^{B^T}\mathbf{w}_{m-1}^f + \mathbf{r}_{m-1}^{B^T} K_m \mathbf{w}_{m-1}^{fB} - r(0)K_m = r(m)$$

which simplifies to

$$K_m = \frac{-r(m) + \mathbf{r}_{m-1}^{B^T}\mathbf{w}_{m-1}^{f}}{r(0) - \mathbf{r}_{m-1}^{B^T}\mathbf{w}_{m-1}^{fB}}. \tag{9.41}$$

This completes the Levinson-Durbin algorithm, which is summarized in Table 9.1. This algorithm produces the reflection coefficients, K_m, for the lattice structure as well as the Direct Form FIR filter coefficients, $a_{m,k}$. Now let us find the computational complexity of the algorithm. Referring to equation (9.41), we see that in order to find K_m for the mth stage, the denominator requires $m-1$ multiplications and $m-1$ additions. The numerator requires the same, for a total of $2m-2$ computations (multiplications and additions) for the mth stage. Therefore, an m-order prediction error filter requires

$$2 \cdot 1 - 2 + 2 \cdot 2 - 2 + \cdots + 2 \cdot M - 2 = \frac{2M(M+1)}{2} - 2M$$

computations for computing K_m for all the stages. The coefficients $a_{m,k}$ must also be computed as in Table 9.1. These require m multiplications and m additions for each stage. For M stages, we therefore need $\frac{M(M+1)}{2}$ multiplications and the same number of additions. Combining these with the requirement for finding K_m, we have a total of

TABLE 9.1 Levinson-Durbin Algorithm

Given: $r(0)\ r(1), \ldots r(M)$

To compute: $K_1,\ K_2, \ldots K_M$

$a_{M,1},\ a_{M,2}, \ldots,\ a_{M,M}$

$$K_1 = \frac{r(1)}{r(0)}$$

$$a_{1,1} = K_1$$

For $m = 2, 3, \ldots, M$

$$\mathbf{r}_{m-1} = [r(1)\ r(2) \ldots r\ (m-1)]$$

$$\mathbf{w}_{m-1,k}^{f} = -a_{m-1,k}\ k = 1, 2, \ldots, m-1$$

$$\mathbf{w}_{m-1} = [\mathbf{w}_1^f \mathbf{w}_2^f \ldots \mathbf{w}_{m-1}^f]$$

$$K_m = \frac{-r(m) + \mathbf{r}_{m-1}^{B^T}\ \mathbf{w}_{m-1}^{f}}{r(0) - r_{m-1}^{B^T}\ \mathbf{w}_{m-1}^{fB}}$$

$$a_{m,k} = a_{m-1,k} + K_m a_{m-1,m-k},\ k = 1, 2, \ldots, m-1$$

$$a_{m,m} = K_m$$

end.

$\frac{3}{2}M(M+1) - 2M$ multiplications and the same number of additions. In other words, the computational complexity of the Levinson-Durbin algorithm is $O(M^2)$. The computational complexity can be further reduced by using parallel processors. The computation of $a_{m,k}$ will then require only one computation per stage, thereby requiring a total of $O(M)$ computations. In the calculation of K_m, the multiplications in the inner products can be computed in parallel yielding $O(M)$ computations for all the M stages. However, the additions for these inner products cannot be done simultaneously. It can be shown that the M stages require $O(M \log M)$ additions, which is the final computational complexity. There is another algorithm called the Schur algorithm [2],[3], which is structured so that it does not require the computations of inner products and therefore has a computational complexity of $O(M)$. The Schur algorithm is not described in this text.

Now we find an expression for the minimum MSE of the predictor. The MSE for an m-order predictor is

$$\begin{aligned} E_m^f &= r(0) - \mathbf{r}_m^T \mathbf{w}_m^f \\ &= r(0) - \mathbf{r}_{m-1}^{BT} \mathbf{w}_m^B \\ &= r(0) + \sum_{k=1}^{m} a_{m,k} r(k). \end{aligned} \tag{9.42}$$

We know that $a_{m,k}$ are updated as

$$a_{m,k} = a_{m-1,k} + K_m a_{m-1,m-k}, \; k = 1, 2, \ldots, m \text{ and } m = 1, 2, \ldots, M.$$

Substituting the above in (9.42) gives

$$\begin{aligned} E_m^f &= r(0) + \sum_{k=1}^{m} (a_{m-1,k} + K_m a_{m-1,m-k}) r(k) \\ &= r(0) + \sum_{k=1}^{m-1} a_{m-1,k} r(k) + \sum_{k=1}^{m-1} K_m a_{m-1,m-k} r(k) \\ &\quad + (a_{m-1,m} + K_m a_{m-1,0}) r(m). \end{aligned} \tag{9.43}$$

The first two terms on the right equal E_{m-1}^f in accordance with equation (9.42). The term $a_{m-1,m} = 0$ and $a_{m-1,0} = 1$. Thus, we have

$$\begin{aligned} E_m^f &= E_{m-1}^f + \sum_{k=1}^{m-1} K_m a_{m-1,m-k} r(k) + K_m r(m) \\ &= E_{m-1}^f - K_m \mathbf{r}_{m-1}^{BT} \mathbf{w}_{m-1}^f + K_m r(m) \\ &= E_{m-1}^f - K_m \left[-r(m) + \mathbf{r}_{m-1}^{BT} \mathbf{w}_{m-1}^f \right]. \end{aligned} \tag{9.44}$$

Now rewrite (9.41) as

$$K_m = \frac{-r(m) + \mathbf{r}_{m-1}^{BT} \mathbf{w}_{m-1}^f}{E_{m-1}^f}. \tag{9.45}$$

Using (9.45) in (9.44), we obtain

$$\begin{aligned} E_m^f &= E_{m-1}^f - K_m^2 E_{m-1}^f \\ &= (1 - K_m^2) E_{m-1}^f. \end{aligned} \tag{9.46}$$

We know that the predictor error power E_m^f is nonnegative and that as the predictor order m increases, the corresponding value of E_m^f decreases or remains the same. That is,

$$0 \leq E_m^f \leq E_{m-1}^f, \quad m \geq 1. \tag{9.47}$$

From (9.46) and (9.47), we have the important result

$$|K_m| \leq 1, \forall m. \tag{9.48}$$

9.4 PROPERTIES OF THE PREDICTION ERROR FILTER

The first two properties are direct consequences of the principle of orthogonality (Chapter 7).

Property 1 The m-order forward prediction error $f_m(n)$ and the input $u(n)$ are orthogonal. That is,

$$E\{f_m(n)\, u(n-k)\} = 0, \quad k = 1, 2, \ldots, m.$$

Property 2 The m-order backward prediction error $g_m(n)$ and the input $u(n)$ are orthogonal. That is,

$$E\{g_m(n)\, u(n-k)\} = 0, \quad k = 1, 2, \ldots, m.$$

Property 3 The backward prediction errors for different predictor orders are orthogonal. That is,

$$E\{g_m(n) g_l(n)\} = \begin{cases} 0 & 0 \leq l \leq m-1 \\ E_m^b & l = m. \end{cases} \tag{9.49}$$

This property can be proved by substituting for $g_m(n)$ and $g_l(n)$ from equation (9.15) and then comparing with the normal equation in (9.19). The details are left to Problem 5.

Property 4 (whitening filter) If the order of the prediction error filter is high enough, it whitens any stationary process applied to its filter.

The predictor predicts the estimate of $u(n)$ from the samples $u(n-1), u(n-2), \ldots$. As the order of the predictor increases, the difference between the predicted value $\hat{u}(n)$ and the actual value $u(n)$ decreases. The stronger the correlation between the adjacent samples of the input, the smaller is the error, $e(n) = \hat{u}(n) - u(n)$. If the input is white, the predictor does nothing and the error is high. If the samples are correlated, the error keeps decreasing as the order of the predictor increases. So, the predictor essentially removes the correlation between the samples. For sufficiently high order, any correlation is completely removed and the input signal is thus whitened.

Property 5 Let $H_m^f(z)$ denote the transfer function of a forward prediction error filter of order m such that

$$H_m^f(z) = \sum_{k=0}^{m} a_{m,k} z^{-k}$$

where $a_{m,0} = 1$. If $H_m^b(z)$ denotes the transfer function of a backward prediction error filter, then the following relationships hold:

$$H_m^b(z) = \sum_{k=0}^{m} a_{m,m-k} z^{-k} \tag{9.50}$$

$$H_m^f(z) = H_{m-1}^f(z) + K_m z^{-1} H_{m-1}^b(z). \tag{9.51}$$

The proof is left to Problem 6.

In conclusion, the Levinson-Durbin algorithm yields a minimum-phase polynomial for the prediction error filter if the autocorrelation sequence is estimated properly. This is an important property for speech coding applications.

9.5 JOINT PROCESS ESTIMATOR

The lattice structure developed earlier is a prediction error filter that generates both the forward prediction error and the backward prediction error for a given input signal $u(n)$ and a desired signal $d(n)$. For a joint process estimator, the desired signal is the same, but the input signal is now the backward prediction error $e^b(n)$. In other words, we now minimize the mean square error given by

$$J_J = E\{d(n) - f(e^b(n))\}^2, \tag{9.52}$$

where $f(e^b(n))$ denotes the filtered version of $e^b(n)$.

This can be done by using a transversal filter operating on $e^b(n)$. The flow graph for the joint process estimation is given in Fig. 9.7. The reason for the name is now obvious. This system jointly minimizes two cost functions. The lattice coefficients are formed such that $E\{e^b(n)\}^2$ is minimized and the FIR filter coefficients c_i are found to minimize (9.52). The estimated signal can now be written as

$$\hat{d}(n) = \sum_{i=0}^{M} c_i e_i^b(n) \tag{9.53}$$

$$= \mathbf{c}^T \mathbf{e}^b \tag{9.54}$$

where $\mathbf{c} = [c_0\, c_1 \ldots c_M]^T$ and $\mathbf{e}^b = [e_0^b\, e_1^b \ldots e_M^b]$.

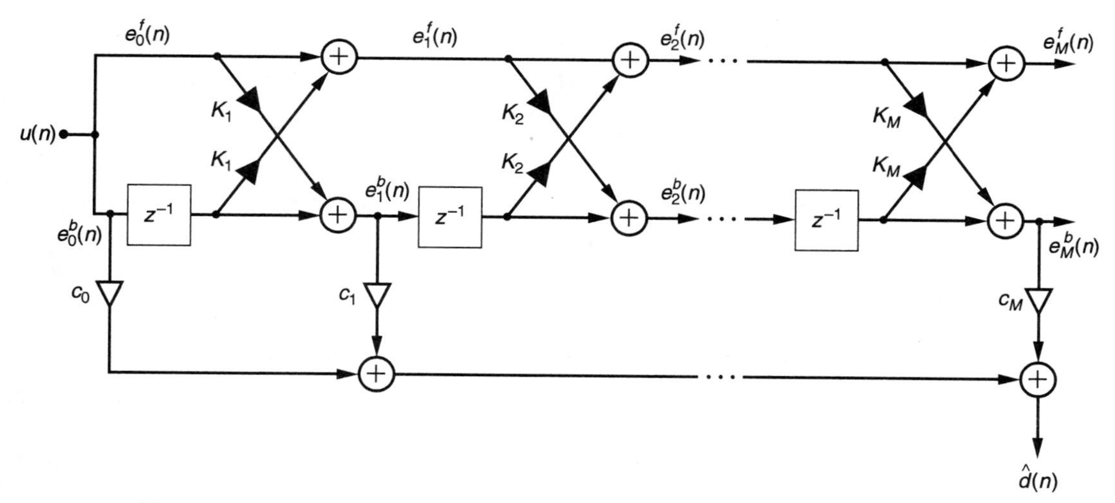

Figure 9.7 Joint process estimator in a lattice structure.

Let $\mathbf{B}$ denote the correlation matrix of $\mathbf{e}^b(n)$ and $\mathbf{S}$ denote the cross-correlation vector of $d(n)$ and $\mathbf{e}^b(n)$. That is, let

$$\mathbf{B} = E\{\mathbf{e}^b(n)\mathbf{e}^{b^T}(n)\} \tag{9.55}$$

and

$$\mathbf{S} = E\{\mathbf{e}^b(n)d(n)\}. \tag{9.56}$$

Thus, the Wiener-Hopf equations are given by

$$\mathbf{Bc} = \mathbf{S},$$

which gives the solution

$$\mathbf{c} = \mathbf{B}^{-1}\mathbf{S}. \tag{9.57}$$

Since the backward prediction errors are orthogonal (Property 3), matrix $\mathbf{B}$ is diagonal. Therefore, the computation in (9.57) is rather simple.

The coefficients of the forward (transversal) section, c_i, can actually be adapted in real time as an adaptive FIR filter. The adaptive structure is shown in Fig. 9.8. The LMS algorithm can be easily derived (as in Chapter 7) to update the coefficients, c_i. The update coefficients are

$$\hat{d}(n) = \mathbf{c}^T(n)\mathbf{e}^b(n) \tag{9.58}$$

$$e(n) = d(n) - \hat{d}(n) \tag{9.59}$$

$$\mathbf{c}(n+1) = \mathbf{c}(n) + \beta\mathbf{e}^b(n)e(n). \tag{9.60}$$

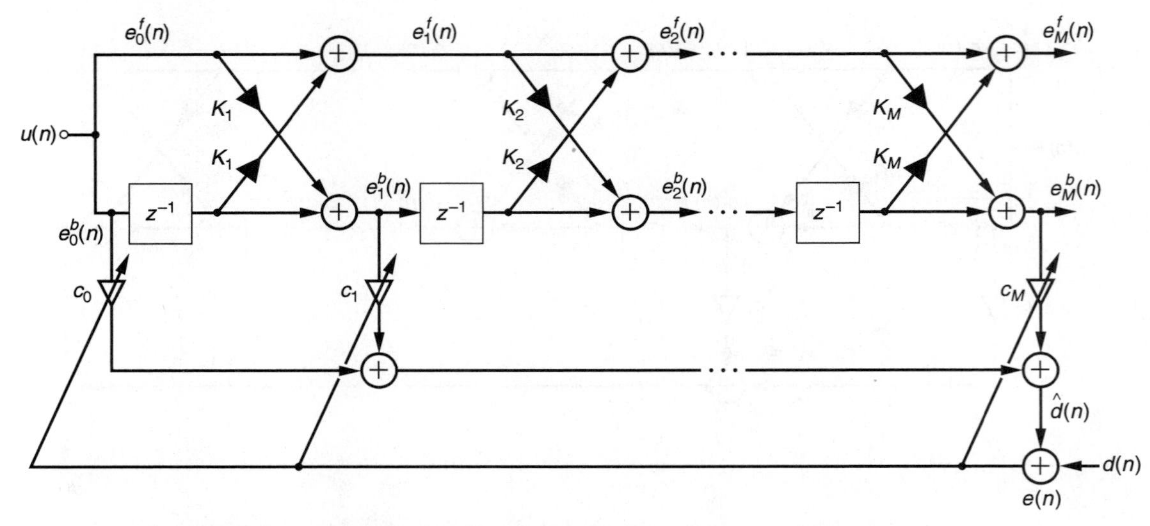

Figure 9.8 Adaptive joint process estimator.

This adaptive lattice structure has a tremendous advantage. Since the backward prediction errors are orthogonal, the transversal section of the filter operates on decoupled inputs. This means that the coefficients, c_i, will converge very rapidly since the autocorrelation matrix cannot suffer from eigenvalue disparity problems. If the lattice coefficients, K_i, are also adapted (see next section), then orthogonalization of the backward prediction errors is achieved after the coefficients converge. If the input signal is white, the reflection coefficients are zero and no orthogonalization takes place. The overall structure then reduces to a transversal filter.

9.6 ADAPTIVE LATTICE FILTER

In this section, we develop an algorithm that will continuously update the reflection coefficients. The simplest way to do so is to define the cost function to be the expectation of the sum of forward and backward prediction errors, and then use the steepest descent method of minimization. The resulting algorithm is called the Gradient Adaptive Lattice (GAL), which was initially proposed in [4] and [5]. The cost function is

$$J_i = E\left\{\left[e_i^f(n)\right]^2 + \left[e_i^b(n)\right]^2\right\}, \quad i = 1, 2, \ldots, M. \tag{9.61}$$

If we use the steepest descent method, the update equation for the reflection coefficient at the ith stage is

$$K_i(n+1) = K_i(n) - \frac{\mu}{2}\nabla J_i, \quad i = 1, 2, \ldots, M, \tag{9.62}$$

where μ is the convergence factor and ∇J_i is the gradient. As in the case with the LMS algorithm, we approximate the objective function as

$$\widehat{J}_i = \left[e_i^f(n)\right]^2 + [e_i^b(n)]^2. \tag{9.63}$$

The approximate gradient then becomes

$$\begin{aligned}\nabla\widehat{J}_i &= \frac{\partial\widehat{J}_i}{\partial K_i(n)} \\ &= 2e_i^f(n)\frac{\partial e_i^f(n)}{\partial K_i(n)} + 2e_i^b(n)\frac{\partial e_i^b(n)}{\partial K_i(n)}.\end{aligned} \tag{9.64}$$

From Fig. 9.8, the forward and backward prediction errors are related as

$$e_i^f(n) = e_{i-1}^f(n) + K_i(n)e_{i-1}^b(n-1) \tag{9.65}$$

$$e_i^b(n) = e_{i-1}^b(n-1) + K_i(n)e_{i-1}^f(n). \tag{9.66}$$

The derivatives in (9.64) can now be formed as

$$\frac{\partial e_i^f(n)}{\partial K_i(n)} = e_{i-1}^b(n-1) \tag{9.67}$$

$$\frac{\partial e_i^b(n)}{\partial K_i(n)} = e_{i-1}^f(n). \tag{9.68}$$

The gradient then becomes

$$\nabla\widehat{J}_i = 2e_i^f(n)e_{i-1}^b(n-1) + 2e_i{}^b(n)e_{i-1}^f(n). \tag{9.69}$$

Substituting (9.69) in (9.62), we get the update equation

$$K_i(n+1) = K_i(n) - \mu\left[e_i^f(n)e_{i-1}^b(n) + e_i^b(n)e_{i-1}^f(n)\right]. \tag{9.70}$$

This completes the derivation of the LMS- type algorithm for the GAL filter. When this is combined with the update equation (9.60) for the transversal section of the joint process estimation, we have a fully adaptive system. The update equations are summarized in Table 9.2.

The rest of this chapter is devoted to applications that are closely related to linear prediction and adaptive filtering.

9.7 SPEECH CODING

The subject of speech coding has received tremendous attention in the last two decades in response to the demand and spread of telecommunication networks and services in all parts of the world. The literature on this topic is extensive, and government agencies have developed and adopted numerous speech coding standards. In addition, good tutorial papers on the topic have been written by Spanias [19] and Cox [20]. Several good books on the subject are also available.

A speech signal is usually processed and coded in order to reduce the amount of data needed to represent it for transmission or storage. This technique is referred to as *speech*

TABLE 9.2 Adaptive Joint Process Estimator

Given: $u(n), d(n)$

To compute: $K_i(n),\ i = 1, 2, \ldots, M$

$c_i(n),\ i = 0, 1, 2, \ldots, M$

Initialization: $K_i(0) = 0,\ i = 1, 2, \ldots, M$

$c_i(n) = 0,\ i = 0, 1, \ldots, M$

For $n = 0, 1, 2, \ldots$

$e_o^f(n) = e_o^b(n) = u(n)$

For $i = 1, 2, \ldots, M$

$e_i^f(n) = e_{i-1}^f(n) + K_i e_{i-1}^b(n-1)$

$e_i^b(n) = e_{i-1}^b(n-1) + K_i e_{i-1}^f(n)$

end i

$\mathbf{c}(n) = [c_0(n) c_1(n)\ c_M(n)]^T$

$\mathbf{e}^b(n) = [e_o^b(n) e_1^b(n) \cdots e_M^b(n)]^T$

$\hat{d}(n) = \mathbf{c}^T(n) e^b(n)$

$\mathbf{c}(n+1) = \mathbf{c}(n) + \beta \mathbf{e}^b(n)[d(n) - \hat{d}(n)]$

coding, and it obviously results in speech compression. Speech coding involves sampling, processing, and quantization. The bandwidth of speech is under 4 kHz; therefore, speech signals are typically sampled at 8 kHz. In the simplest form of speech coding, the sampled speech signal is quantized to 8 bits per sample using a Pulse Code Modulator (PCM). The PCM simply quantizes the amplitudes of the discrete signal. The resulting data rate is therefore 64 kb/sec. Such a PCM coded signal is actually considered to be uncompressed and is used as a benchmark for rating compressed signals. Signals coded in the range of 8 to 16 kb/s are said to be *medium rate,* those 2.4 to 8 kb/s are called *low rate,* and those under 2.4 kb/s are called *very-low-rate* signals. The device that performs speech coding is called a *vocoder,* which is short for voice coder.

The performance of a vocoder is measured by several attributes, namely, bit rate, quality, complexity, and delay. The bit rate attribute does not need much explanation. The lower the bit rate, the better the vocoder. The quality attribute can be objective or subjective. One of the common objective measures is the SNR (with respect to the PCM coded signal). However, this measure can sometimes be misleading. Speech is usually composed of periods of voice and periods of silence. We may have a high SNR, but the speech may be unintelligible or of poor quality if the noise is concentrated during the voice period. More importantly, the SNR measure does not account for the perceptual properties of the ear. Therefore, the SNR measure is not always useful. One of the subjective measures is called the absolute category rating (ACR), a technique in which human subjects listen to pairs of sentences, and they rate the quality on a scale of 1 (worst) to 5 (best). The mean of the ratings is used as a score for the vocoder; this score is called the mean opinion score (MOS). The acronyms ACR and MOS are used interchangeably.

Another subjective quality measure is the Diagnostic Rhyme Test (DRT), a test that measures intelligibility. The subject is asked to identify one of two possible words in a set of rhyming pairs, such as pot/hot and pick/tick. The DRT and the MOS have become the most popular metrics of quality. The complexity attribute refers to the hardware, software, size, and power requirements of the vocoder. For VLSI implementations, complexity typically refers to the computational complexity (number of multiplications and number of additions per sample) of the algorithm in the vocoder. The lower the computational complexity, the faster, smaller, and cheaper the chip. Speed and cost are important considerations in the competitive telecommunications marketplace. This is why most vocoders are implemented in fixed point arithmetic. The last attribute is delay. If the speech signal is delayed by the vocoder, then it might cause difficulty in communication between the talkers. Delays of 300 ms or greater are very perceptible and objectionable. The vocoder algorithm must be designed so that it minimizes the amount of signal delay.

There are several important types of speech coders, especially the waveform coder, transform coder, subband coder, Linear Predictive Coder (LPC), and Linear Prediction Analysis-by-Synthesis (LPAS).

Waveform Coder

The more common types of waveform coding techniques include Linear Pulse Code Modulation (PCM), μ-law and A-law PCM, Adaptive PCM (APCM), Differential Pulse Code Modulation (DPCM), and Adaptive DPCM (ADPCM). All of these methods attempt to approximate the waveform with a small number of bits. As the number of bits increases, the approximation gets closer to the actual waveform. This is why these coders are also called waveform-following coders.

Linear PCM

Linear PCM is simply the quantization of the waveform in both time and amplitude. It happens naturally and simultaneously in the A/D converter. Quantization in the time domain converts the signal $x(t)$ to the discrete domain $x(n)$. The amplitude is also quantized to get $\hat{x}(n)$. In Chapter 6, we discussed different types of amplitude quantizers such as magnitude truncation, rounding, and two's complement truncation. The reader is encouraged to review those sections at this time. All of these quantizers are uniform quantizers, which means that they have a fixed step size. Let q represent the step size and B the number of bits. The total number of levels is then $N = 2^B$. Now we proceed to find a relationship between the number of bits and the quantization SNR (QSNR). For this purpose, let us choose a type of quantizer, say, rounding. Define the quantization noise (or error) as

$$e(n) = \hat{x}(n) - x(n). \tag{9.71}$$

From Chapter 6, we know that for the rounding quantizer, the quantization noise is in the range

$$-\frac{q}{2} \leq e(n) \leq \frac{q}{2}. \tag{9.72}$$

The following assumptions can be made on the quantization noise:

1. The quantization noise $e(n)$ is white.
2. $x(n)$ and $e(n)$ are uncorrelated.
3. $e(n)$ is uniformly distributed in the interval $(-\frac{q}{2}, \frac{q}{2})$. It is easy to show that this implies that the variance is $\sigma_e^2 = \frac{q^2}{12}$.

Let $x(n)$ be such that it satisfies

$$|x(n)| \leq X_m. \tag{9.73}$$

Then the quantization step size is

$$q = \frac{2X_m}{2^B}. \tag{9.74}$$

The variance of the quantization noise then becomes

$$\sigma_e^2 = \frac{q^2}{12} = \frac{X_m^2}{3 \cdot 2^{2B}}. \tag{9.75}$$

The quantization signal to noise ratio is

$$\begin{aligned} QSNR &= 10 \log_{10}\left(\frac{\sigma_x^2}{\sigma_e^2}\right) \\ &= 10 \log_{10}\left(\frac{\sigma_x^2 \cdot 3 \cdot 2^{2B}}{X_m^2}\right) \\ &= 20 \log_{10}(2)B + \left[10 \log_{10} 3 - 20 \log_{10}\left(\frac{X_m}{\sigma_x}\right)\right] \\ &\simeq 6B + \left[10 \log_{10} 3 - 20 \log_{10}\left(\frac{X_m}{\sigma_x}\right)\right]. \end{aligned} \tag{9.76}$$

In this equation, the term within square brackets is a constant for some given $x(n)$. Therefore, each bit contributes to an additional 6 dB of QSNR. This is an important result and can be used to determine the number of bits required in a vocoder.

Nonlinear PCM

When we add noise to a high-amplitude signal, it is not as noticeable as if we were to add the same noise to a low-amplitude signal. This is because, if the signal is large, we do not tend to notice the noise as much. From the previous discussion, we know that amplitude quantization will add noise to the signal uniformly. Therefore, in order to make

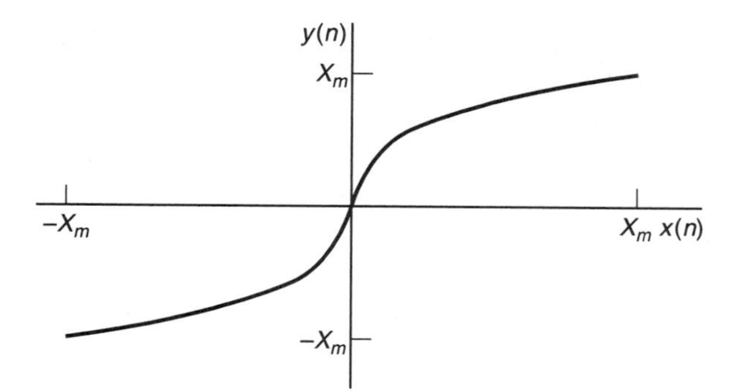

Figure 9.9 μ-law compressor characteristic.

the noise less noticeable in low-level signals, we can simply "boost" the low-level signals. The device that performs this is called a *compressor*, and it has a transfer characteristic of the form shown in Fig. 9.9 (μ−law). Clearly, the low-amplitude signals are amplified more than the high-amplitude signals. This compression can be performed before or after quantization. The two standard compressor characteristics are as follows.

μ−Law

$$y(n) = X_m \frac{\log\left[1 + \mu \frac{|x(n)|}{X_m}\right]}{\log[1+\mu]} sgn[1 + x(n)]. \tag{9.77}$$

This function is approximately logarithmic for large $x(n)$ and approximately linear for small $x(n)$.

A-Law

$$y(n) = X_m \frac{1 + \log\left[\frac{A|x(n)|}{X_m}\right]}{1 + \log A} sgn[x(n)]. \tag{9.78}$$

Both of these compressors offer similar quality of QSNR. In North America and Japan, the μ-law is used with $\mu = 255$. The rest of the world uses the A-law with $A = 87.56$. For 8 bits/sample, both compressors yield about 35 dB of QSNR.

Adaptive PCM (APCM)

In the two compressor schemes we have described, the signal levels were adjusted by using a nonlinear function, but the quantization step size was still constant. The concept of APCM is to adapt the step size in response to the level of the input signal. This is done by making the step size a function of the standard deviation of the signal as

$$q(n) = q_0 \sigma_x(n) \tag{9.79}$$

where q_0 is a constant initial step size. One way to compute the signal variance is to lowpass filter $x^2(n)$ as

$$\sigma_x^2(n) = a\sigma_x^2(n-1) + (1-a)\hat{x}^2(n-1) \tag{9.80}$$

where $0 < a < 1$ is a constant chosen to suit the application at hand. Another way to estimate $\sigma_x(n)$ is to calculate it for the past M samples as

$$\sigma_x^2(n) = \frac{1}{M} \sum_{i=n-M}^{n-1} \hat{x}^2(i). \tag{9.81}$$

Notice that in both (9.80) and (9.81), we have used the quantized value of the input, $\hat{x}(n)$ instead of the input $x(n)$. In that way, the receiver can re-create the step size $q(n)$, since it has $\hat{x}(n)$ and not $x(n)$. APCM exhibits a QSNR improvement of 4 to 8 dB over linear PCM.

DPCM and Adaptive DPCM

The concept of DPCM was described in Chapter 8. A linear predictor and a quantizer are used in a unique way to predict the current sample for a number of past samples. When the quantizer and/or the predictor are adaptive, the system is called adaptive DPCM or ADPCM. The 24, 32, and 40 kb/s ADPCM has been adopted for the G.726 ITU[1] standard. It uses a pole-zero adaptive predictor with two poles and six zeros. The LMS algorithm is used to update the coefficients of this IIR filter. Stability is checked by finding the magnitude of the two poles. A 4-bit quantizer is used. Further details of this standard can be found in [21] and [22]. ADPCM is also used to encode the subband signals in the ITU G.722 standard [24].

Transform Coder

In transform coding, the input signal is first segmented into blocks of length-N, using a serial-to-parallel converter. The blocks are then transformed by an energy-compacting transform. Many such transforms are available, including the DFT, Discrete Cosine Transform (DCT), Discrete Sine Transform (DST), Discrete Wavelet Transform (DWT), Karhunen-Loeve Transform (KLT), and Discrete Hadamard Transform (DHT). All of these transforms perform energy compaction into a few of the transform coefficients. Therefore, the coefficients with large magnitudes can be allotted a higher number of bits

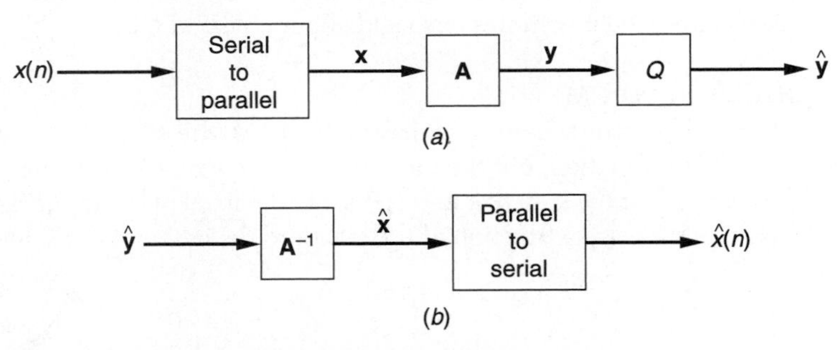

Figure 9.10 (*a*) Transform coder; (*b*) Decoder.

[1] International Consultative Committee for Telephone and Telegraph. It is currently called the International Telecommunications Union—Telecommunications Standardization Sector (ITU-TSS).

than those with smaller magnitudes. Since much of the energy of speech is typically concentrated in the lower frequencies, the transform coefficients at the higher frequencies are usually very small and can be discarded. This would obviously lead to compression. A simplified block diagram for a transform coder and decoder are shown in Fig. 9.10. The transform kernel is an $N \times N$ matrix $\mathbf{A}$, and the transform is given by

$$\mathbf{y} = \mathbf{A}\mathbf{x} \tag{9.82}$$

where $\mathbf{x}$ is a vector of N input data samples and $\mathbf{y}$ is the transformed vector. The transformed vector is then quantized and transmitted. The receiver performs the inverse transform or an approximation thereof in order to reconstruct the signal. This is given by

$$\hat{x} = \mathbf{A}^{-1}\hat{\mathbf{y}} \tag{9.83}$$

where $\hat{\mathbf{y}}$ is the vector of received samples and $\hat{\mathbf{x}}$ is the vector of reconstructed samples.

The above-mentioned transforms have varying degrees of energy compaction properties and other relative pros and cons. A comprehensive treatment of these transforms can be found in [23]. Several of these transforms are discussed in Chapter 11. The following are the important attributes of a transform.

1. *Energy compaction*: The transform must have minimal redundancy within the transformed samples. This means that the transform coefficients are not highly correlated, and therefore there is high-energy compaction into a small number of coefficients. KLT has the highest energy compaction property of all. However, it is not a computationally efficient transform.
2. *Computational complexity*: The transform should be computable with as little computation as possible. Lower computational complexity implies that a higher sampling rate can be achieved for the same processor. In other words, a lower computational complexity reduces hardware complexity, cost, and power requirements. For example, all the elements of the kernel matrix of the DHT are ± 1. Therefore, no multiplications are required in computing the transform. However, the DHT has the worst energy compaction property of all the above-mentioned transforms.
3. *Receiver complexity*: The properties of the transform kernel matrix dictate how easily $\mathbf{A}^{-1}$ can be found. For example, the DFT and the DCT are unitary transforms, that is $\mathbf{A}^{-1} = \mathbf{A}^{\dagger}$. Therefore, the inverse can be found by simply taking the conjugate transpose of the matrix.

Subband Coder

The subband coder (SBC) was briefly discussed in Chapter 5 in the context of multirate filter banks. In the SBC, the signal is divided into frequency bands by an LPF and a set of BPFs. The output of each filter is then downsampled and encoded. This is shown in Fig. 9.11 and is called a uniform filter bank. The signal in each band is encoded using a different number of bits, depending on the statistics of the signal. In speech coding applications, the higher frequency bands will require fewer bits than the lower frequency bands. As we know from Chapter 5, the filters in the coder and decoder must be designed carefully so that perfect reconstruction is achieved.

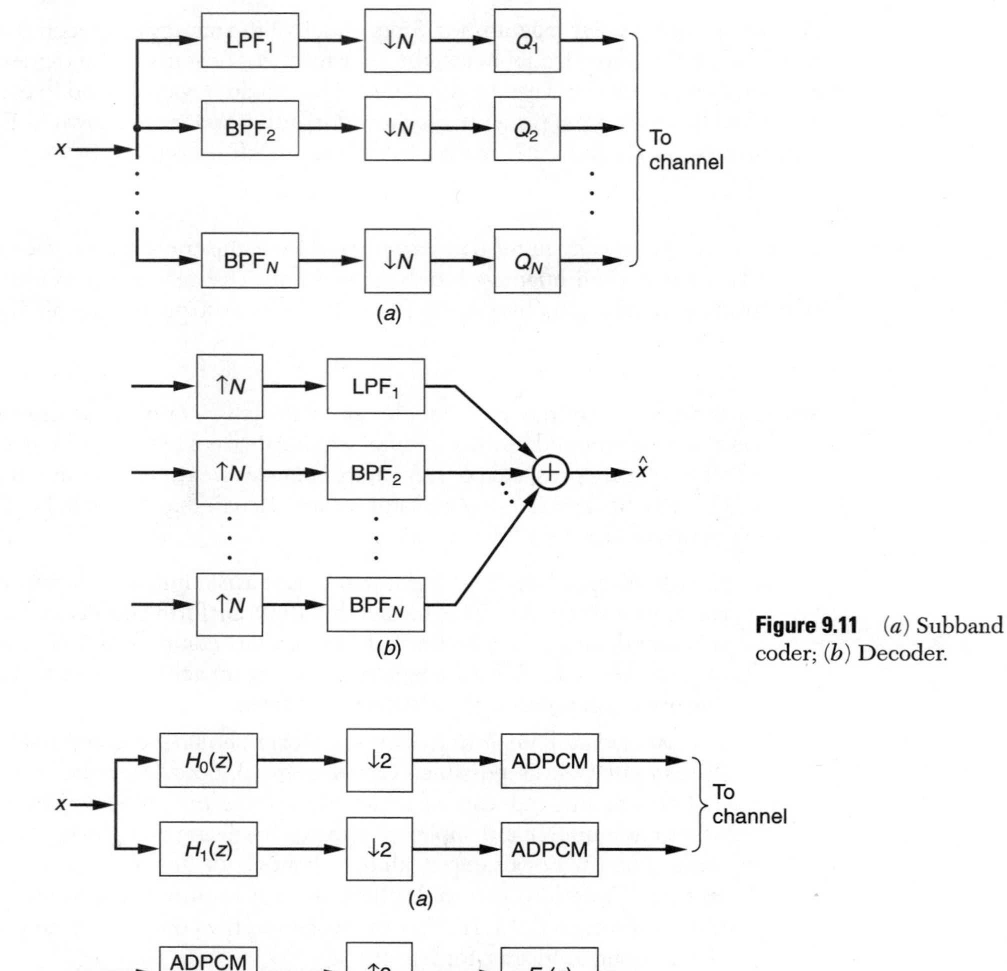

Figure 9.11 (*a*) Subband coder; (*b*) Decoder.

Figure 9.12 CCITT G.722 standard: (*a*) coder; (*b*) decoder.

The ITU standard G.722 [24] for 7 kHz audio at 64 kb/s for ISDN teleconferencing uses a 2-band SBC followed by ADPCM. The block diagram is shown in Fig. 9.12. The filter bank is a QMF bank and therefore achieves perfect reconstruction (PR). Of course, PR is only possible without the quantizers. The signal is sampled at 16 kHz. The subbands are equally divided in frequency. The upper channel (lower frequency band) after downsampling has the rate of 8 kHz. It is then coded with 6 bits/sample. This gives a bit rate of 48 kb/s. The lower channel (upper frequency band) is coded at 2 bits/sample for a bit rate of 16 kb/s. The MOS at 64 kb/s is greater than 4 for speech and slightly less for music. The signal delay due to coding is less than 3 ms. The G.722 standard also has an adaptive bit allocation scheme. Provisions have been made for quantizing the upper channel at 40 kb/s or at 32 kb/s.

The ATT voice standard [25] also uses the SBC and is used for voice transmission and storage at 16 or 24 kb/s. It consists of a 5-band tree-structured QMF bank followed by ADPCM coders. The conceptual structure is shown in Fig. 9.13. The diagram of the QMF tree is left as an exercise. The frequency ranges for each band are the following: 0–0.5, 0.5–1, 1–2, 2–3, and 3–4 kHz. For the 16 kb/s scheme, the bit allocations are {4/4/2/2/0}, and for the 24 kb/s scheme the bit allocations are {5/5/4/3/0}. The signal delay for this coder is less than 18 ms.

Linear Predictive Coder (LPC)

The LPC vocoder is a well-studied method and has received a lot of attention in the last two decades. In fact, it has been proven very effective for speech coding and has found its way into many government standards. The LPC vocoder is a model-based coder. Model-based coders are based on identifying a model for speech production. If this identification is good, then the input speech is reconstructed well. The parameters of the model are quantized and transmitted to the receiver. Unlike in waveform coders,

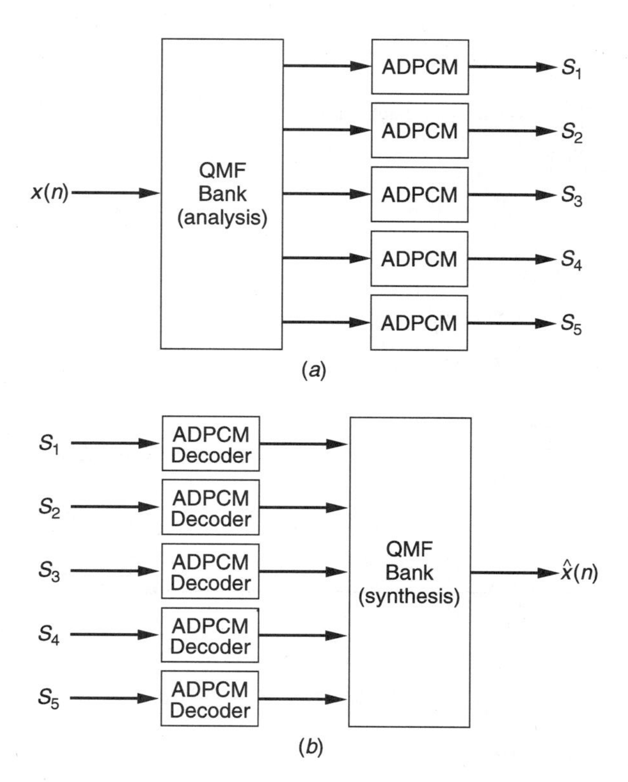

Figure 9.13 ATT subband coder: (*a*) coder, (*b*) decoder.

if there is no quantization error, the reproduced speech will still not be the same as the original speech. A block diagram of the LPC vocoder is shown in Fig. 9.14. Based on a frame of speech, two sets of parameters are quantized and transmitted, namely, the linear prediction (LP) parameters and excitation model parameters. These parameters are for the general speech production model.

Voiced speech is produced by exciting the vocal tract with periodic glottal air pulses. These pulses are generated by the vibrating vocal chords. The frequency of these pulses is called the fundamental frequency or *pitch*. This frequency is an important parameter in the synthesis of speech. On the other hand, unvoiced speech is produced by forcing air through a constriction in the vocal tract.

A typical model for speech production, shown in Fig. 9.15, is the so-called Fant model [18] and is clearly an all-pole model. The transfer function of this model is

$$S(z) = \frac{G}{1 - A(z)} X(z) \tag{9.84}$$

where G is a constant gain factor and $A(z)$ is given by

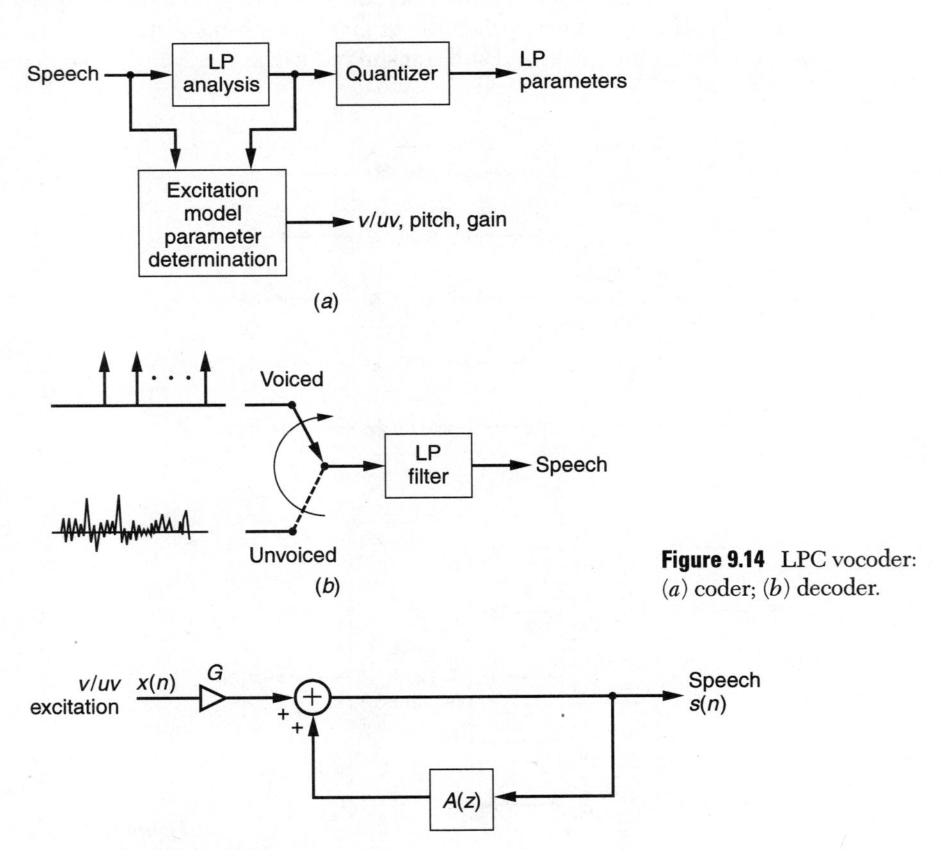

Figure 9.14 LPC vocoder: (*a*) coder; (*b*) decoder.

Figure 9.15 Speech production model.

$$A(z) = \sum_{k=1}^{M} w_k^f z^{-k}. \tag{9.85}$$

The transfer function $A(z)$ is in the same form as in equation (9.1). The coefficients of $A(z)$ are obtained using linear prediction. The speech production model of (9.84) is based on careful modeling of the human vocal tract. The parameters of $A(z)$ change depending on whether the input is voiced (v) or unvoiced (uv). For voiced speech, we need to know the pitch. The input is then modeled as an impulse train at this fundamental frequency. For unvoiced speech, the input is modeled as white noise. The excitation model parameters consist of (a) voiced/unvoiced (v/uv), (b) gain (G), and (c) pitch. The LP parameters are the coefficients of $A(z)$. The linear predictor is therefore given by

$$\hat{s}(n) = \sum_{k=1}^{M} w_k^f s(n-k). \tag{9.86}$$

The LP coefficients can be found by minimizing the following cost function:

$$J = E[(s(n) - \hat{s}(n))^2]. \tag{9.87}$$

This problem was solved earlier in this chapter by solving a set of normal equations. We have seen that the Levinson-Durbin algorithm is an efficient method because it yields the Direct Form coefficients w_k^f, as well as the reflection coefficients K_m of the lattice structure. The algorithm is summarized in Table 9.1. The autocorrelation sequence $r(m)$ has to be estimated in practice. It is typically estimated from the last N samples of the speech signal $s(n)$, using one of the following methods:

Unbiased estimator

$$\hat{r}(m) = \frac{1}{N - |m|} \sum_{i=0}^{N-|m|-1} s(n + |m|)s(n) \tag{9.88}$$

Biased estimator

$$\hat{r}(m) = \frac{1}{N} \sum_{i=0}^{N-|m|-1} s(n + |m|)s(n). \tag{9.89}$$

A biased estimator is used because such a condition guarantees semi-positiveness of the corresponding covariance matrix and, consequently, a unique solution for the set of Wiener-Hopf equations. Each frame window is typically 20 to 30 ms long, at the end of which the filter coefficients are updated. In some applications, the frame is subdivided in subframes of 5 ms, and the subframe parameters are obtained by linear interpolation of adjacent frame parameters. This is done so that frequent updates can be performed at a lower computational cost.

Now the question is, do we transmit the Direct Form coefficients or the lattice coefficients? As we have seen in Chapter 6, quantization of the Direct Form coefficients leads to high quantization noise. In addition, Direct Form filters are more susceptible to overflow oscillations and limit cycles owing to product quantization. These lead to instability and also add to quantization noise. For this reason, quantization of Direct Form coefficients is avoided. On the other hand, lattice structures are known to be robust under quantization. In addition, it is easy to check the stability of lattice structures. All the reflection coefficients must be less than unity for stability. There are also some transformations of the lattice coefficients that make them even less sensitive to quantization, notably:

Log Area Ratio (LAR)

$$LAR(m) = \log\left(\frac{1+K_m}{1-K_m}\right) \tag{9.90}$$

Inverse Sine Transform (IST)

$$\mathrm{Si}(m) = \sin^{-1}(K_m) \tag{9.91}$$

These transformations have been shown to improve speech quality [26].

Excitation Model Parameter Estimation

As pointed out, the three parameters for the excitation model are gain (G), pitch, and voice/unvoiced (v/uv). The v/uv condition is easily determined by measuring the energy in the frame. For a voiced frame, the energy will be much higher than for unvoiced. A predetermined threshold is used for this purpose. The gain is typically determined at the receiver. The power of the current analysis frame is transmitted. The receiver compares this power to that of the (LPC) synthesized speech. The ratio of the two powers then gives the gain (G). Estimation of the pitch (fundamental frequency) is not that simple. Several different algorithms for pitch estimation have been designed; these can be found in [27] and [28]. The simplest way is to suppress all harmonics and measure the lowest frequency component. This can be done using adaptive notch filters and many other means. The quality of the LPC vocoder is very sensitive to the estimated pitch. An accurate pitch tracker is needed for reasonable quality. Inaccurate pitch estimation can result in a mechanical sound for voice sounds and a buzzy quality for unvoiced sounds.

The LPC vocoder has nevertheless become quite popular and found its way in the Federal Standard 1015 [29], a standard proposed for secure speech communication at 2.4 kb/s. This standard uses an LPC with a tenth-order lattice predictor. A pitch extractor called the Average Magnitude Pitch Extractor (AMPE) is used. The block diagram of the coder and decoder is shown in Fig. 9.16. Most noise tends to have high frequency. In order to reduce the effect of noise, we can amplify the high-frequency content of the signal before processing and transmission. This process is called preemphasis, and it is simply a first-order FIR highpass filter with cutoff frequency at 100 Hz. The segmentation block segments the signal in frames. After segmentation, the linear predictor coefficients are computed. In a parallel branch, the signal is lowpass filtered, with a cutoff frequency of 800 Hz. This is the only useful band for pitch estimation, gain estimation, and v/uv determination. The receiver performs the speech synthesis, and the block diagram is self-explanatory. Further details of this standard can be found in [29].

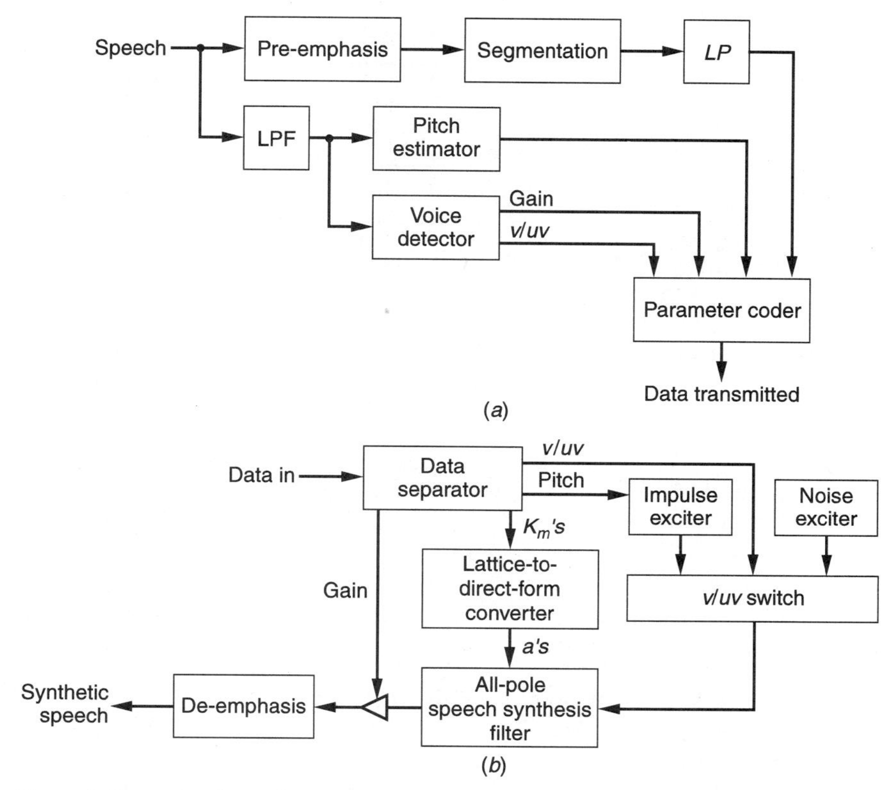

Figure 9.16 Federal Standard FS1015 LPC vocoder: (*a*) transmitter; (*b*) receiver.

Linear Prediction Analysis-by-Synthesis (LPAS) Coder

As in LPC, the LPAS obtains the predictor coefficients by linear prediction. In addition, this type of coder determines an excitation sequence by an optimization technique. Recall that the excitation sequence is used as input to the linear predictor in order to synthesize the speech signal. The LPC coefficients are quantized and transmitted along with the excitation sequence. This is done for every frame or subframe of data. A conceptual block diagram of a typical LPAS coder is shown in Fig. 9.17. First, an LPC analysis is performed on the speech signal in order to compute the predictor coefficients, and then these coefficients are quantized. Next, the quantized coefficients are used on an LPC synthesizer in order to obtain the synthesized speech. The input to this synthesizer comes from an excitation sequence generator. The error between the actual speech and the synthesized speech is then used to construct an objective function. This is a function of the excitation sequence. Minimization of this objective function gives the best excitation sequence. A deeper discussion of the LPAS coder requires a background of vector quantization and is beyond the scope of this book. It suffices to say that many different types of LPAS coders are available, including the Multipulse Excited Linear Prediction (MELP), Regular Pulse Excitation Coders (RPEC), and Code Excited Linear Prediction (CELP) coders. The difference lies in the types of codes generated for the excitation sequence

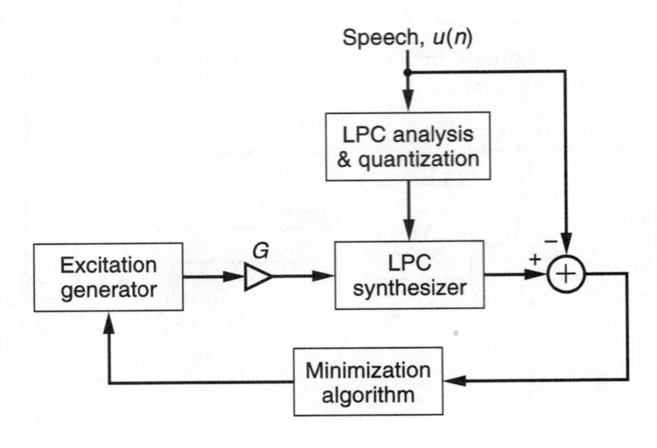

Figure 9.17 Linear prediction analysis-by-synthesis coder.

and in the type of optimization algorithm used. Interested readers are referred to the tutorial paper by Spanias [19] and the numerous references therein. Federal Standard 1016 (1991) uses the CELP coder. It has been adopted by the Department of Defense for use in third-generation secure telephone communication. Further details on this standard can be found in [30]. Several CELP standards have emerged since 1991, a detailed discussion of which can be found in the recent tutorial chapter by Spanias in [31]. Some excerpts are given below.

Second-Generation Near-Toll Quality CELP

These algorithms have high computational complexity and are used to deliver near-toll quality speech for applications in second-generation cellular phones and Internet streaming. The second-generation CELP algorithms were developed between 1993 and 1998. The high-quality speech in these algorithms is due to several factors, including the use of algebraic codebooks in the coding of excitation, line spectral frequencies in the coding of LPC parameters, and perceptually optimized split vector quantization. Algorithms in this class include the ITU G.729 [32] ACELP, the G.723.1 coder [33], the GSM EFR [34], and the IS-127 RCELOP [35].

Third-Generation CELP for 3G Cellular Standards

A lot of research has been conducted in order to find algorithms that work at multiple rates and are suitable for wideband wireless cellular standards. These third-generation (3G) algorithms are said to be multimodal because they accommodate multiple bit rates. In Europe, GSM[2] is considering the adaptive multirate coder [36] for a standard. In the United States, the TIA[3] has tested the Selectable Mode Vocoder (SMV) [37], developed by Connexant.

▶ 9.8 SUMMARY

In this chapter, we first presented the basic concepts of linear prediction and the Levinson- Durbin algorithm, which was first given by Levinson [9] and then modified by Durbin [10]. As pointed

[2] Gaussian Synchronous Modulation (GSM) in the European standardization organization.
[3] Telecommunications Industry Association.

out, the Levinson-Durbin algorithm is computationally efficient and also finds the coefficients of the lattice predictor at no extra computational cost. Some modifications of these algorithms have also been developed [11]–[12], which are computationally even more efficient. Another algorithm for linear predictor, the so-called Schur algorithm, has not been presented here. It is designed for fast implementation of a linear predictor using parallel processors. Details of this algorithm can be found in [13]–[16]. The chapter also discussed speech in detail and presented many types of speech coding schemes and some federal standards.

9.9 PROBLEMS

1. Consider the linear prediction scheme in Fig. P9.1, where $M \geq 2$. The problem is to find $\mathbf{w}$ so that $E\{e^2(n)\}$ is minimized. Find an expression for the optimal value of $\mathbf{w}$.

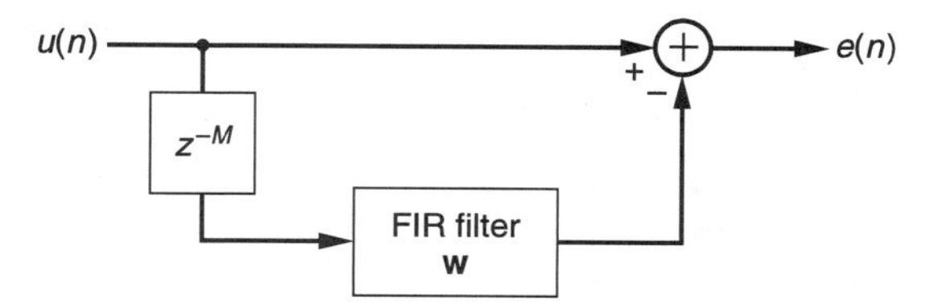

Figure P9.1

2. Consider a wide-sense stationary process with the autocorrelation values: $r(0) = 1$, $r(1) = 0.7$, $r(2) = 0.5$, $r(3) = 0.3$.

(a) Using the Levinson-Durbin algorithm, find the reflection coefficients.
(b) Sketch the lattice predictor for this process.
(c) Find the average power of the prediction error at the output of each stage of the lattice.

3. Consider a wide-sense stationary process with the autocorrelation values: $r(0) = 1$, $r(1) = 0.5$, $r(2) = 0.4$, $r(3) = 0.3$, $r(4) = 0.2$.

(a) Using the Levinson-Durbin algorithm, find the reflection coefficients.
(b) Sketch the lattice predictor for this process.
(c) Find the average power of the prediction error at the output of each stage of the lattice.

4. Prove that the MSEs of forward and backward predictors are identical.

5. Prove that the backward prediction errors for different predictor orders are orthogonal. That is, prove equation (9.49).

6. Prove Property 5 of the prediction error filter, that is, equations (9.50) and (9.51).

7. Consider the second-order autoregressive process given by

$$y(n) = y(n-1) - 0.3y(n-2) + w(n)$$

where $w(n)$ is a white noise of zero mean and unit variance.

(a) Find the reflection coefficients.
(b) Find the average power of $y(n)$.

8. Consider the third-order autoregressive process given by

$$y(n) = y(n-1) - 0.3y(n-2) - 0.2y(n-3) + w(n)$$

where $w(n)$ is a white noise of zero mean and unit variance.

(a) Find the reflection coefficients.
(b) Find the average power of $y(n)$.

9. Derive an RLS-based algorithm for the Adaptive Lattice Filter.

10. Sketch the 5-band QMF tree structure for the ATT voice standard.

11. For the ITU G.722 standard, design all four filters for the 2-channel QMF bank. Plot the magnitude response for each of these filters.

12. Consider the LAR algorithm for transformation of lattice coefficients. Show that the transformed coefficients are less sensitive to quantization than the lattice coefficients.

13. Consider the IST algorithm for transformation of lattice coefficients. Show that the transformed coefficients are less sensitive to quantization than the lattice coefficients.

9.10 COMPUTER PROJECTS

Project 1

Repeat Project 1 of Chapter 7 using the GAL algorithm. Comment on the performance with respect to the LMS algorithm.

Project 2

Repeat Project 3 of Chapter 7 using the GAL algorithm. Comment on the performance with respect to the LMS algorithm.

BIBLIOGRAPHY

[1] A.S. Spanias, "A block time and frequency modified covariance algorithm for spectral analysis," *IEEE Trans. on Signal Processing*, SP-41, no. 11, pp. 3138–3153, November 1993.

[2] I. Schur, "On power series which are bounded in the interior the unit circle," *J. Reine Angew. Math.* 147, pp. 205–232, Berlin, 1917.

[3] I. Gohberg, "Schur methods in operator theory and signal processing," Birkhauser Verlag, Stuttgart, Germany, 1986.

[4] L.J. Griffiths, "A continuously adaptive filter implemented as a lattice structure," *Proc. IEEE Intl. Conf. Acous., Speech, and Signal Processing*, pp. 683–686, 1977.

[5] L.J. Griffiths, "An adaptive lattice structure for noise cancellation applications," *Proc. IEEE Intl. Conf. Acous., Speech, and Signal Processing*, pp. 87–90, 1978.

[6] G.A. Clark, S.K. Mitra, and S.R. Parker, "Block implementation of adaptive digital filters," *IEEE Trans. Circuits and Systems*, CAS-28, pp. 584–592, 1981.

[7] G.A. Clark, S.R. Parker, and S.K. Mitra, "A unified approach to time and frequency domain realization of FIR adaptive digital filters," *IEEE Trans. Acous. Speech and Signal Processing*, ASSP-31, pp. 1073–1083, 1983.

[8] P.C.W. Sommen and J.A.K.S. Jayasinghee, "On frequency domain adaptive filters using the overlap add method," *Proc. Intl. Symp. on Circuits and Systems*, Espoo, Finland, pp. 27–30, 1988.

[9] N. Levinson, "The Wiener RMS error criterion in filter design and prediction," *J. Math. Phys.* 25, pp. 261–278, 1947.

[10] J. Durbin, "Efficient estimation of parameters in moving-average models," *Biometrika* 46, pp. 306–316, 1959.

[11] P. Delsarte and Y. Genin, "The split-Levinson algorithm," *IEEE Trans. Acous., Speech, and Signal Process.*, ASSP-34, pp. 470–478, June 1986.

[12] H. Krishna, "New split-Levinson, Schur, and lattice algorithms," *IEEE Proc. Intl. Conf. Acous., Speech, and Signal Process.*, ASSP-28, pp. 1640–1642, April 1988.

[13] J.G. PROAKIS and D.G. MANOLAKIS, *Digital Signal Processing: Principles, Algorithms, and Applications*, Prentice Hall, Upper Saddle River, NJ, 1996.

[14] T. KAILATH, "Linear estimation of stationary and near-stationary processes," *Modern Signal Processing*, T. Kailath, ed., Hemisphere Publishing Corporation, Washington, DC, 1985.

[15] T. KAILATH, "A theorem of I. Schur and its impact on modern signal processing," *I. Schur Methods in Operator Theory and Signal Processing*, I. Gohberg, ed., Birkhauser Verlag, Stuttgart, Germany, 1986.

[16] J-G. CHUNG and K.K. PARHI, *Pipelined Lattice and Wave Digital Recursive Filters*, Kluwer Academic Publishers, Norwell, MA, 1996.

[17] E.R. FERRARA, "Fast implementation of LMS adaptive filters," *IEEE Trans. Acous., Speech, and Signal Process.*, ASSP-28, pp. 470–475, 1980.

[18] G. FANT, *Acoustic Theory of Speech Production*, Mounton and Co., Gravenhage, The Netherlands, 1960.

[19] A. SPANIAS, "Speech coding: a tutorial review," *Proceedings of the IEEE*, pp. 1541–1582, 1994.

[20] R. COX, "Speech coding," *The Digital Signal Processing Handbook*, V. Madisetti and D.B. Williams, eds., CRC Press, Boca Raton, FL, 1998.

[21] ITU Recommendation G.726, "24,32, 40kb/s adaptive differential pulse code modulation (ADPCM)," in *Blue Book*, vol. III, Fascicle III.3, October 1988.

[22] ITU Draft Recommendation G.723, "Extensions of recommendation G.721 ADPCM to 24 and 40 kbits/s for DCME application," in *Blue Book*, vol. III, Fascicle III, October 1988.

[23] A.K. JAIN, *Fundamentals of Digital Image Processing*, Prentice Hall, Englewood Cliffs, NJ, 1989.

[24] ITU Draft Recommendation G.722, "7 kHz audio coding within 64 kbits/s," in *Blue Book*, vol. III, Fascicle III, October 1988.

[25] J. JOSENHANS *et al.*, "Speech processing applications standards," *AT & T Tech. J.* 65 (5), p. 23, September–October 1986.

[26] P. VARY *et al.*, "Speech codec for the European mobile radio system," *Proc. ICASSP*, p. 227, April 1988.

[27] W. HESS, *Pitch Determination of Speech Signals*, Springer Verlag, New York, 1983.

[28] R. SCHAFER and J. MARKEL, *Speech Analysis*, IEEE Press, New York, 1979.

[29] *Federal Standard 1015*, "Telecommunications: analog to digital conversion of radio voice by 2400 bit/second linear predictive coding," National Communication System—Office of Technology and Standards, November 1984.

[30] *Federal Standard 1016*, "Telecommunications: analog to digital conversion of radio voice by 4800 bit/second code excited linear prediction (CELP)," National Communication System—Office of Technology and Standards, February 1991.

[31] A.S. SPANIAS, "Vocoders," *Encyclopedia on Telecommunications*, J. Proakis, ed., Wiley, New York, 2003.

[32] ITU Study Group 15 Draft Recommendation G.729, "Coding of speech at 8 kbits/s using conjugate-structure algebraic-code-excited linear-prediction (CS-ACELP)," *International Telecommunication Union*, 1995.

[33] ITU Recommendation G.723.1, "Dual rate speech coder for multimedia communications transmitting at 5.3 and 6.3 kbits/s," *International Telecommunication Union*, 1995.

[34] GSM 06.60, "GSM digital cellular communication standards: Enhanced full-rate transcoding," *ETSI/GSM*, 1996.

[35] TIA/EIA/IS-127, "Enhanced variable rate codec," Speech Service Option 3 for Wideband Spread Spectrum Digital Systems," *TIA*, 1997.

[36] R. EKUDDEN, R. HAGEN, I. JOHANSSON, and J. SVEDBURG, "The adaptive multirate speech coder," *Proc. IEEE Workshop on Speech Coding*, pp. 117–119, 1999.

[37] Y.GAO, E. SCHLOMOT, A. BENYASSINE, J. THYSSEN, H. SU, and C. MURGIA, "The SMV algorithm selected for TIA and 3GPP2 for CDMA applications," *Proc. ICASSP*, vol. 2, Salt Lake City, UT, May 2002.

CHAPTER

10

Image Processing Fundamentals

Image processing encompasses the mathematical and engineering methods for manipulating digital images. Some of the most fundamental manipulations are depicted in Fig. 10.1. An acquisition device generates a digital image. Examples of such devices include a digital camera, an electronic microscope, an MRI scanner, and a hyperspectral camera on a spacecraft. The acquired image usually suffers from additive noise, linear distortions, nonlinear distortions, and possibly some lost samples. The image restoration process involves the removal of the noise and the distortions. It also involves filtering methods and statistical techniques. Geometric distortions may require the application of geometric transformations. These geometric transformations rely on sophisticated interpolation schemes. The visualization of the restored image can require some enhancement in order to improve the perception of the image. Enhancement techniques include grayscale modification, edge enhancement, and contrast modification. Finally, there may be a need to understand the structure of the image in more detail. The image understanding process requires the extraction of important features using automatic procedures. Edge detection is one of the most fundamental problems in image understanding and will be discussed in detail here.

Applications of image processing are ubiquitous today. The following applications are related to the topics covered in this chapter.

- Biomedical applications: noise removal, enhancement of biomedical images, and contour extractions for the quantitative analysis of biomedical images.
- Satellite imagery: noise removal and distortion correction of satellite images; analysis of colored and hyperspectral images.
- Image databases: feature extraction and search of image databases.
- Exploration and geophysics: noise removal, enhancement, and analysis of seismic data.

This chapter first describes several methods for image enhancement and then discusses image interpolation, including linear and polynomial techniques. Edge detection

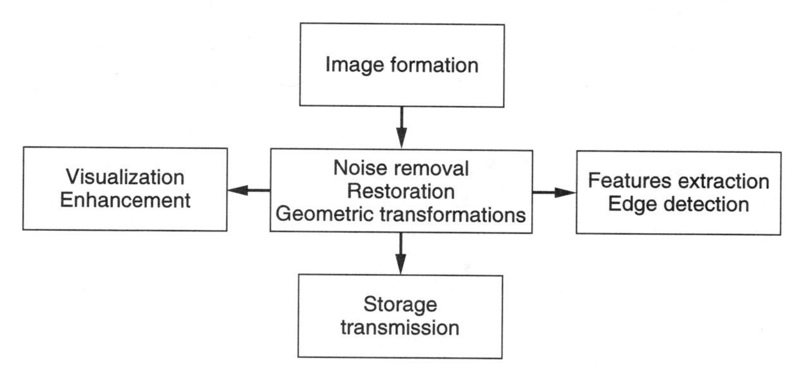

Figure 10.1 Fundamental stages in image processing from image acquisition to higher level interpretation.

is then presented, and several different methods are discussed in detail. The next section covers topics in image restoration, including inverse filtering, lowpass filtering, and median filtering. Finally, the basics of color image processing are presented.

10.1 IMAGE ENHANCEMENT

The goal of image enhancement is to improve the perception of the image through modification of intensity functions, image spectral content, or a combination of these functions. Several different classes of methods are available to enhance the image. We first present standard gray-level mapping techniques that allow us to change the perception of the image by a careful remapping of the intensity. The second class of methods involves modification of the histogram of the image. The goal is to apply a pointwise transformation of the intensity in such a way that the histogram of the transformed image has a predefined shape. Finally, the third class of methods includes linear filtering techniques that can enhance features by applying a combination of low- and highpass filters. The selection of a particular method depends on the features that one needs to detect or measure in the image.

Grayscale Transformation

The grayscale transformation function maps the intensity range onto itself. Let $[0, M]$ be the intensity range of the considered image. For example, $M = 255$ for 8-bit images, and $M = 4095$ for 12-bit images. A grayscale transformation is a function T that gives a new value $T(u)$ for each value u of the intensity; that is,

$$\begin{aligned} T\colon [0, M] &\longrightarrow [0, M] \\ u &\longrightarrow T(u). \end{aligned} \tag{10.1}$$

The function T can be used to modify the image intensity f by applying T at any pixel (n_1, n_2) as

$$(n_1, n_2) \longrightarrow T(f(n_1, n_2)). \tag{10.2}$$

A grayscale transformation is often called a point operation, because $T(f(n_1, n_2))$ only depends on the value of the intensity at the point (n_1, n_2). A linear transformation will in general not be a point operation. The output usually depends on the value of the intensity in a *neighborhood* around the point (n_1, n_2).

Grayscale transformations can increase or decrease the intensity range. They are often used to recover valuable visual information from images that are overexposed, underexposed, or have a very small dynamic range. They can also be used when the intensity range is too large to be displayed on a certain medium such as a terminal, printer, or film. Grayscale transformation algorithms are often a part of visualization software. We now present some of the popular pointwise operators.

Contrast Stretching

The intensity of the image is mapped to a new range using the following mapping, which includes a clipping (or thresholding) operation:

$$T(u) = \begin{cases} v_{\min} & \text{if } u < u_{\min} \\ v_{\min} + \dfrac{(v_{\max} - v_{\min})(u - u_{\min})}{u_{\max} - u_{\min}} & \text{if } u_{\min} < u < u_{\max} \\ v_{\max} & \text{if } u > u_{\max} \end{cases} \tag{10.3}$$

The mapping is shown in Fig. 10.2. The small values of intensity are thresholded, the large values are clipped, and the rest are linearly mapped. Figure 10.3 illustrates the effect of contrast stretching. The contrast has clearly been improved.

MATLAB
This function is implemented with the function **imadjust**. **imadjust** (**I**, $[u_{\min}\ u_{\max}]$, $[v_{\min}\ v_{\max}]$) maps the intensity values in the image **I** to new values in **J** according to equation (10.3). The following lines of code generate Fig. 10.3:

```
>> I = imread ('valley. tif');
>> imshow (I);
```

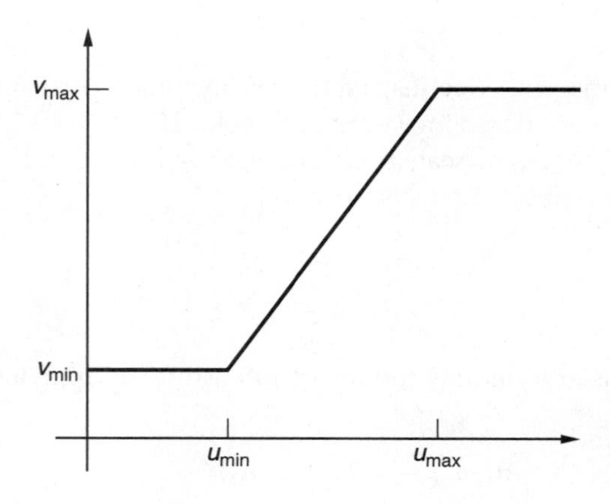

Figure 10.2 Contrast stretching with clipping and thresholding.

Figure 10.3 Intensity stretching. Left: Original valley image. Right: Transformed image after intensity stretching.

```
>> J = imadjust (I, [0 0.39], [0 1]);
>> imshow (J);
```

Values between 0 and 0.39 are mapped to values between 0 and 1. Values use above 0.39 are mapped to 1. The effect of stretching the intensity is illustrated in Fig. 10.4, where the histograms of the images are shown. The horizontal axis represents the pixel intensity values in the range [0,255], and the vertical axis shows the number of pixels for each intensity level. The histogram of the original image has most of the pixels under 50. The modifed histogram is clearly stretched. This transform simply stretches the histogram rather than modify its shape.

Gamma Correction

The gamma correction operation is defined by the following nonlinear mapping of the intensity function:

$$v = \frac{u^{\gamma}}{M^{\gamma}} \tag{10.4}$$

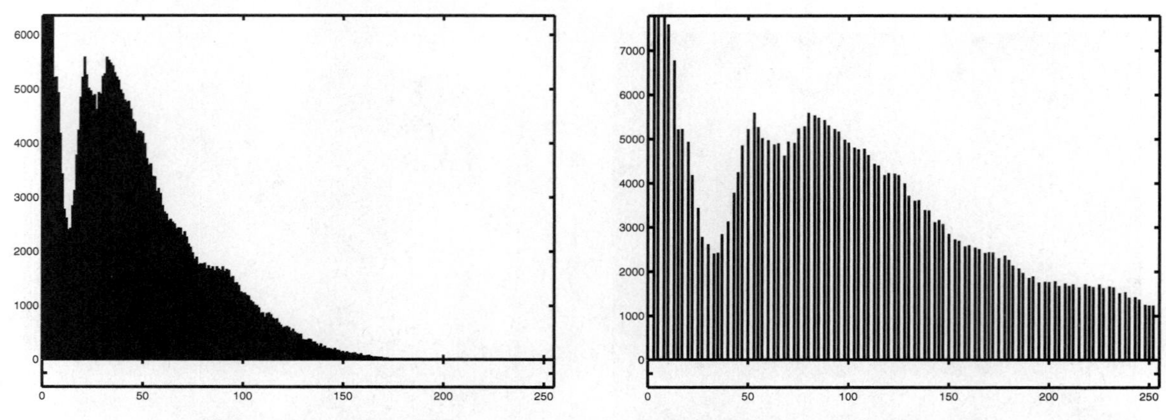

Figure 10.4 Effect of the intensity stretching. Left: Histogram of the original valley image. Right: Histogram of the transformed image after intensity stretching.

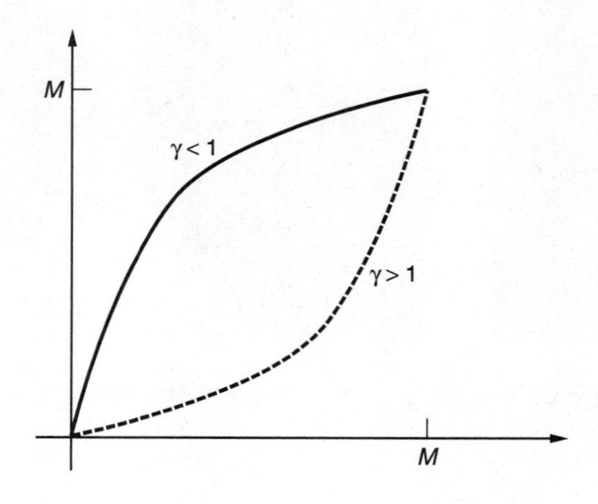

Figure 10.5 Gamma correction: μ is the input intensity value in $[0,M]$ and v is the output value.

where u is the initial intensity function in the range $[0,M]$. If γ is less than 1, all intensities are increased, and the image becomes brighter. Smaller intensity values are increased more than the larger ones. If γ is greater than 1, the image becomes darker, and the larger values are reduced more than the smaller ones. Figure 10.5 shows the gamma correction transfer characteristic, and Fig. 10.6 illustrates the effect of gamma correction with $\gamma = 2$. Notice that in the original image, the statue is underexposed. In the processed image, the details of the statue are now visible. The histograms are shown in Fig. 10.7. The original histogram has a large number of pixels under 50. The modified histogram has been stretched in a nonlinear manner toward the large-intensity values. As a result, the contrast has been increased.

MATLAB

Gamma correction is also implemented with the function **imadjust** as follows:

```
>> imadjust(I,[ ],[ ],gamma)
```

Figure 10.6 Top: Original "Neptune" image. Bottom: Transformed image after gamma correction.

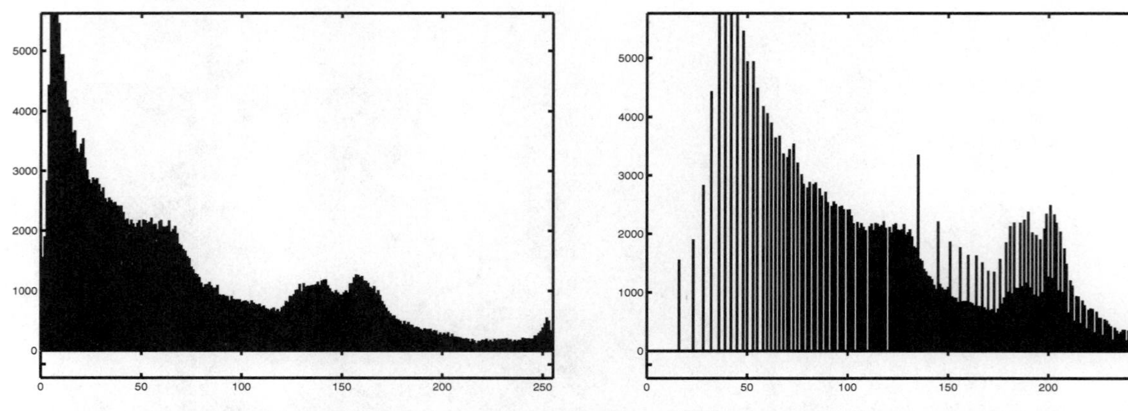

Figure 10.7 Effect of the gamma correction to the histogram. Left: Histogram of the original "Neptune" image. Right: Histogram of the transformed image after gamma correction.

where the parameter **gamma** is actually $1/\gamma$ in equation (10.4). The following code was used to generate Fig. 10.6:

```
>> I = imread ('neptune.tif');
>> imshow (I);
>> J = imadjust (I,[ ],[ ],2);
>> J = imshow (J);
>> imhist (J,256);
```

Histogram Equalization

If the image intensity is distributed over a narrow range of values, $[u_0, u_1]$, then the histogram of the intensity will be very narrow. One can automatically remap the grayscale values in order to "stretch" the histogram and utilize all the gray values that are available. This procedure is called *histogram equalization*. After histogram equalization, the histogram of the transformed image is uniform. Theoretically, then, an equal number of pixels in the image have any given intensity value. In the discrete-time case, the histogram is not exactly equalized but is nearly so.

Let us now derive the expression of the transform T such that the histogram of the transformed image is uniform. We first consider the case where the image intensity is distributed with a continuous-probability density function $p_U(u)$. We will later extend it to the discrete case.

Let $F_U(u)$ be the cumulative distribution function (cdf) of the original image U, and let T be a one-to-one mapping over the intensity range

$$T[0, M] \longrightarrow [0, M]. \tag{10.5}$$

If we apply the transform T on the image U, then the cdf of the transformed image $T(U)$ is given by

$$\text{Prob}(T(U) \leq u) = \text{Prob}(U \leq T^{-1}(u)) = F_U(T^{-1}(u)). \tag{10.6}$$

If T is the transform such that $T(U)$ has uniform distribution, then

$$\text{Prob}(T(U) \leq u) = u \tag{10.7}$$

and thus

$$F_U(T^{-1}(u)) = u. \tag{10.8}$$

If $T = F_U$, then equation (10.8) will be satisfied. Thus, for histogram equalization the transformation is nothing but the cumulative distribution function.

Now we extend this to the discrete-time case. In continuous time, the cdf is given by

$$F_U(u) = \int_0^u p_U(x)dx. \tag{10.9}$$

For the discrete case, we replace (10.9) by the cumulative normalized histogram

$$\sum_{k=u_0}^{u} \text{Prob}(U = k) \tag{10.10}$$

where u_0 is the lowest nonzero intensity value and u_1 is the largest intensity value. The discrete transformation for histogram equalization is then approximated as follows:

$$u \in [0, M], \quad T(u) = \left\lfloor M \cdot \frac{\sum_{k=u_0}^{u} \text{Prob}(U = k) - \text{Prob}(U = u_0)}{1 - \text{Prob}(U = u_0)} \right\rfloor, \tag{10.11}$$

where $\lfloor x \rfloor$ denotes the floor operation, that is, the largest integer not greater than x. The floor operation is used because the term within the floor function may not be an integer, but we need the pixel intensity to be an integer in [0,255]. Note that instead of floor, the ceiling operation $\lceil . \rceil$ can be used just as well. An illustration of histogram equalization is given in Figs. 10.8 and 10.9. The original "prison" image and its histogram are shown

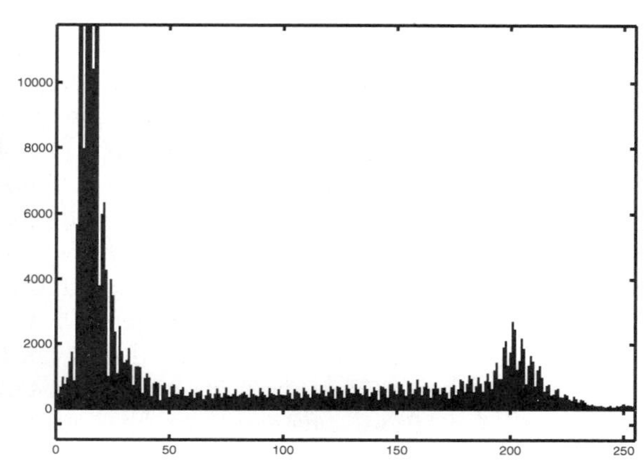

Figure 10.8 Original "prison" image and histogram.

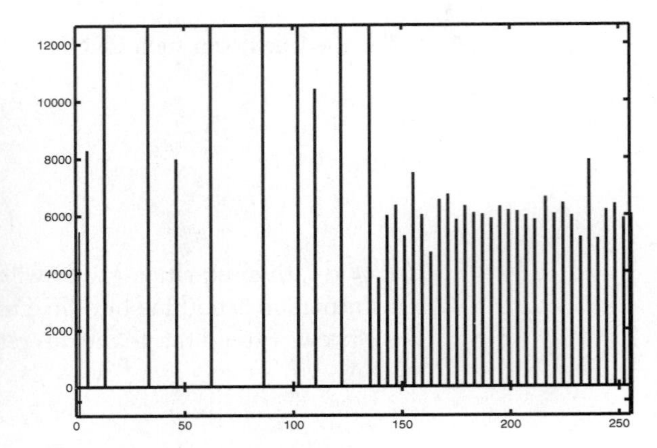

Figure 10.9 "Prison" image after histogram equalization and its histogram.

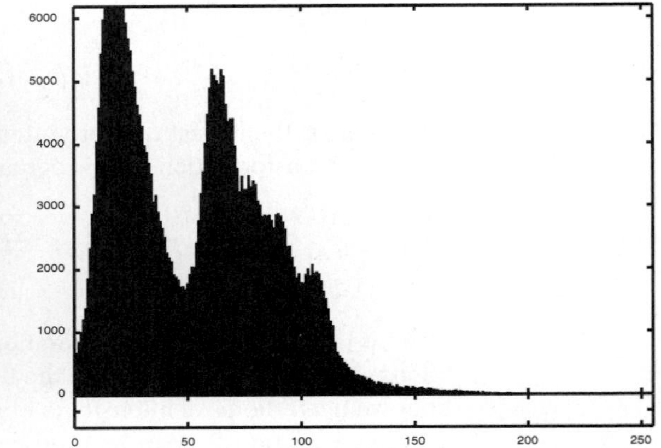

Figure 10.10 Left: original "ocean" image. Right: histogram of the image.

in Fig. 10.8. The prison wall is entirely black, which gives rise to a peak in the histogram around 25. The equalized image is shown in Fig. 10.9, where the details on the prison wall are now visible. The histogram is also much more uniform than that of the original image. For another example, the original image is shown in Fig. 10.10 along with its histogram. The histogram is bimodal; the first mode corresponds to the dark pixels in the mountain, and the second mode corresponds to the lighter pixels in the ocean. The equalized image and its histogram are shown in Fig. 10.11. The details of the reef are now visible, and the histogram is quite uniform.

Now the natural question is: Is it possible to transform an image so that it has a specified histogram that is not necessarily uniform? The answer is, of course, yes. Let $F_U(u)$ be the cdf of the given image, and let $G_V(v)$ be the cdf that we want after transformation. It can be shown that the required transformation function is

$$T(u) = G_V^{-1}(F_U(u)). \tag{10.12}$$

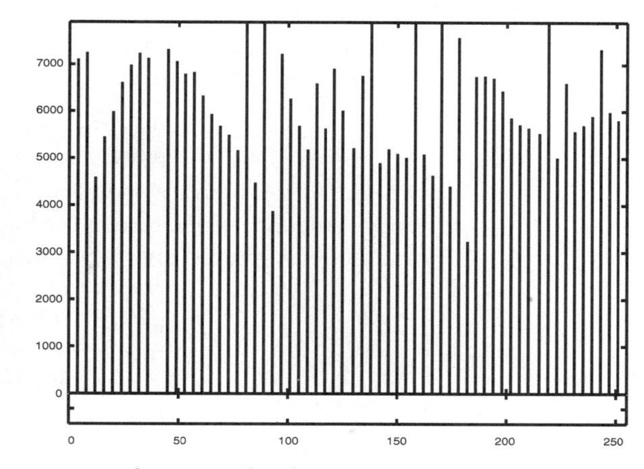

Figure 10.11 "Ocean" image after histogram equalization and its histogram.

These functions are invertible in the continuous time. In discrete time, the inverses have to be approximated. With this transformation, the cdf of the transformed image becomes $G_V(v)$. The proof is left as an exercise (see Problem 2).

MATLAB Commands

Histogram equalization is implemented with the function **histeq**.

The following lines will generate Figs. 10.8 through 10.11:

```
>> I = imread ('image.tif');
>> imshow (I);
>> imhist(I, 256);
>> J = histeq(I);
>> imshow (J);
>> imhist(J, 256);
```

Unsharp Masking

Edges provide critical information for the visual processing of images. The perception of an image can be improved by sharpening the edges: the images will be perceived as "crisper" than the original. This process can be achieved by applying a linear filter on the image that will "boost" the high frequencies. This filter acts in a similar way as an audio equalizer, which can boost high frequencies.

Edges correspond to rapid changes or discontinuities of the intensity function. These rapid changes create large high-frequency components in the Fourier transform of the signal. The discontinuity can therefore be detected by convolving the image with a highpass filter with impulse response, G as

$$f * G. \tag{10.13}$$

We now write the image as the sum of two components: a lowpass filtered image and a highpass filtered image, as

$$f = (I - G) * f + G * f. \tag{10.14}$$

where I is the identity operator. The first term above is the low-frequency component image, and the second term is the high-frequency component image. We can now amplify the highpass filtered image by a factor $\lambda > 1$. This gives

$$f_{\text{sharp}} = (I - G) * f + \lambda G * f = [I + (\lambda - 1)G] * f. \tag{10.15}$$

Define $\alpha = (\lambda - 1)$ and also define the unsharp masking filter as

$$H = I + \alpha G. \tag{10.16}$$

A standard choice for the highpass filter G is the Laplacian. The Laplacian operator is defined by

$$\Delta f = \frac{\partial^2 f}{\partial x^2} + \frac{\partial^2 f}{\partial y^2}. \tag{10.17}$$

A discrete implementation of the Laplacian is derived in Section 10.3. The DFT of Δf is given by

$$\widehat{\Delta f}(k_1, k_2) = (k_1^2 + k_2^2) f(k_1, k_2). \tag{10.18}$$

This equation represents a highpass filtering operation because as (k_1, k_2) increases, the magnitude response of the above filter increases. An example of unsharp masking is given in Fig. 10.12. In the processed image, the tiles on the roofs and the cobblestones on the street are now clearly visible. Another example is given in Fig. 10.13. In the filtered image, the anatomical contours of the brain and the cerebellum are much crisper and easy to follow.

MATLAB Commands

The unsharp masking filter can be designed with the function **fspecial**.

The following lines of code will generate Figs. 10.12 and 10.13.

```
>> G = f special('unsharp');
>> J = imfilter(I,G);
>> imshow (J);
```

10.2 IMAGE INTERPOLATION

A digital image can be obtained by sampling a continuous space image f on a regular grid. Sampling is performed with a rate sufficient to avoid aliasing. The sampling period in the horizontal and vertical directions, Δx, and Δy, characterizes the physical distances

Figure 10.12 Top: Original image. Bottom: Filtered image with the unsharp masking filter.

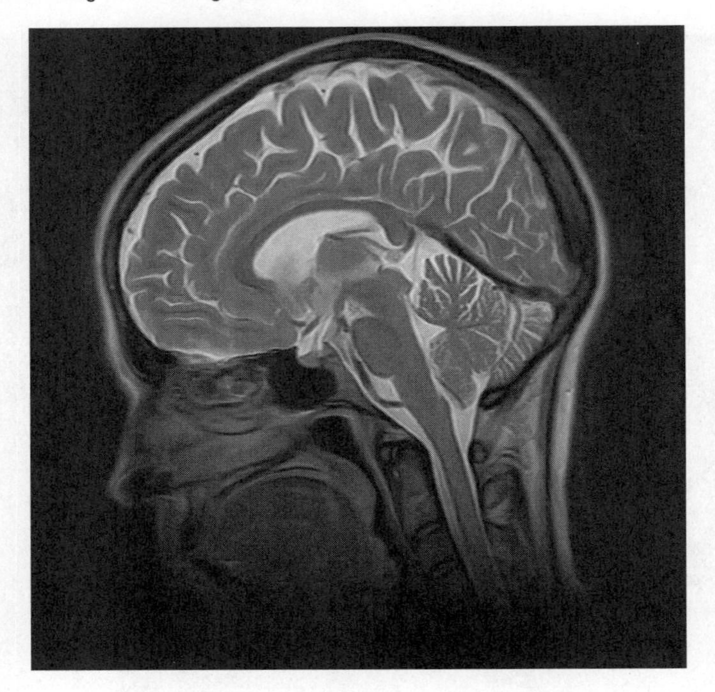

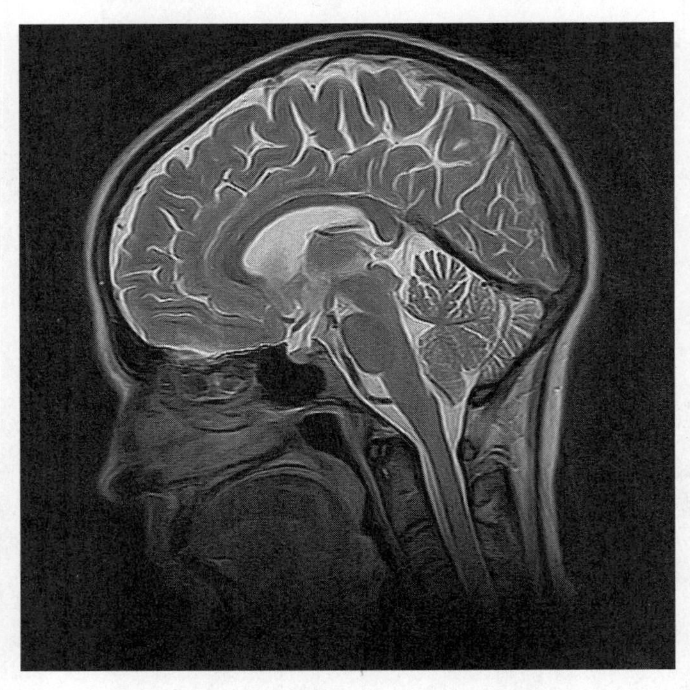

Figure 10.13 Top: Original image. Bottom: Filtered image with the unsharp masking filter.

between adjacent pixels. We know from the Whitaker-Shannon theorem that we can reconstruct the original image f from its samples using the sinc kernel:

$$f(x,y) = \sum_{n_1,n_2} f(n_1\Delta x, n_2\Delta y)k(x/\Delta x - n_1, y/\Delta y - n_2) \tag{10.19}$$

where k is the sinc kernel given by

$$k(x,y) = \frac{\sin \pi x}{\pi x}\frac{\sin \pi y}{\pi y}. \tag{10.20}$$

A large number of geometric transformations, such as translation, rotation, contraction/dilation, and shearing, will map (warp) pixels to a new position that is no longer on the original sampling grid. Figure 10.14 illustrates that a rotation of the image requires the evaluation of intensity at points that were not on the original sampling grid. Computing values at points between pixels is called *interpolation*.

Let us consider the problem of interpolating the original digital image f on a new regular grid defined by the sampling periods δx and δy. At first sight, interpolation appears to be a straightforward operation. First, the original image f can be reconstructed from its samples using the sinc kernel. One can then sample the continuous space image at the new location $m_1\delta x, m_2\delta y$. Let $g(m_1,m_2)$ be the image interpolated on the new grid,

$$g(m_1,m_2) = f(m_1\delta x, m_2\delta y). \tag{10.21}$$

Then we have

$$g(m_1,m_2) = \sum_{n_1,n_2} f(n_1,n_2)k(m_1\delta_x/\Delta x - n_1, m_1\delta_y/\Delta y - n_2), \tag{10.22}$$

where $f(n_1,n_2) = f(n_1\Delta x, n_2\Delta y)$ is the original discrete image and k is the sinc kernel. We note that k is indeed an interpolation kernel in the sense that

$$k(n_1,n_2) = \begin{cases} 1 & \text{if } n_1 = n_2 = 0 \\ 0 & \text{otherwise.} \end{cases} \tag{10.23}$$

Therefore, interpolation reproduces exactly the image f on the original sampling grid.

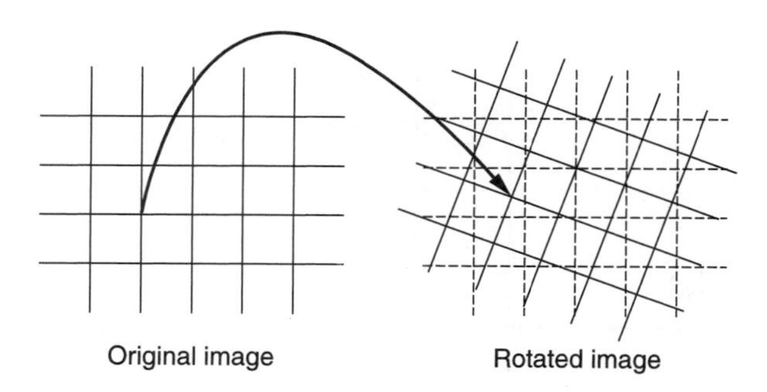

Figure 10.14 Illustration that a rotation of the image requires interpolation.

The sinc kernel has an infinite support region and therefore cannot be used in practice. The goal of interpolation is to construct kernels that have a finite support and approximate the ideal sinc kernel. In the following, we will assume that the interpolation kernel is separable, and we will limit our discussion to one-dimensional interpolation kernels. For a given 1-D interpolation kernel, we can then construct a 2-D kernel from the product of the one-dimensional kernels as done earlier in Chapter 2.

As we know from Chapter 2, if the original image was critically sampled, then interpolating on a grid with a larger sampling interval is likely to create aliasing. For this reason, one should apply a lowpass filter on the image prior to interpolating on a coarser grid. Interpolation on a coarser grid corresponds to a reduction of the image size (contraction). Reciprocally, decreasing the sampling interval will never create any aliasing, and therefore no lowpass filtering is required prior to expanding the image size (dilation or zooming).

Linear Interpolation

Linear interpolation corresponds to the case where the kernel is the "hat" function, also known as the Bartlett window, given by

$$k(x) = \begin{cases} 1 - |x| & \text{if } |x| \leq 1 \\ 0 & \text{otherwise.} \end{cases} \tag{10.24}$$

The Fourier transform of k is given by

$$K(\omega) = \frac{\sin^2 \pi\omega/2}{(\pi\omega/2)^2}. \tag{10.25}$$

As opposed to the ideal kernel, this kernel will attenuate the high frequencies and therefore will smooth the image. In practice, interpolation is computed from the linear combination of the intensity at adjacent pixels. In two dimensions, this gives rise to the bilinear interpolation scheme (see Fig. 10.15) given by

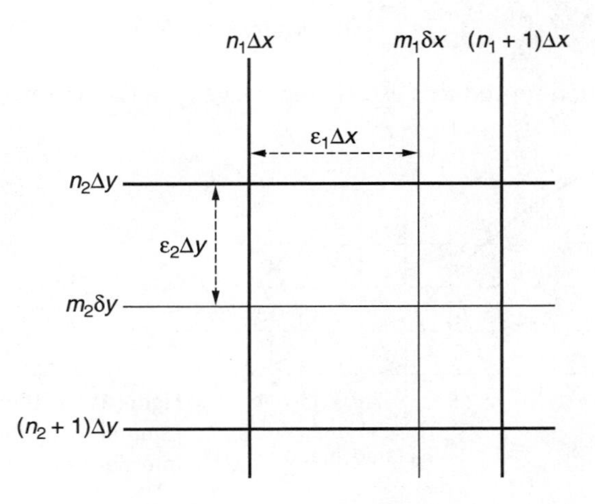

Figure 10.15 Bilinear interpolation scheme.

$$g(m_1, m_2) = \varepsilon_1\varepsilon_2 f(n_1, n_2) + (1-\varepsilon_1)\varepsilon_2 f(n_1+1, n_2)$$
$$+\varepsilon_1(1-\varepsilon_2) f(n_1, n_2+1) + (1-\varepsilon_1)(1-\varepsilon_2) f(n_1+1, n_2+1) \quad (10.26)$$

where n_1 and n_2 are integers such that

$$n_1\Delta x \le m_1\delta x < (n_1+1)\Delta x,$$
$$n_2\Delta y \le m_2\delta y < (n_2+1)\Delta y, \quad (10.27)$$

and $\varepsilon_1, \varepsilon_2$ are defined by

$$\varepsilon_1 = m_1\delta x/\Delta x - n_1,$$
$$\varepsilon_2 = m_2\delta y/\Delta y - n_2,$$

The value of the new pixel in the new grid is interpolated from the values at the four neighboring pixels on the original grid.

Polynomial Interpolation

As explained earlier, linear interpolation offers a poor approximation of the sinc interpolation kernel. A natural idea is to replace the first-order piecewise polynomial kernel with higher order interpolating polynomials. Increasing the order of the polynomial results in faster decay of the Fourier transform of the kernel. Cubic spline interpolation and Lagrange interpolation polynomials can be used. These methods better approximate the ideal interpolation kernel and do not smooth the interpolated image as much as the bilinear interpolation. Obviously, their numerical implementation requires longer filters since the higher powers require more terms in their finite difference representations.

MATLAB Commands

MATLAB provides the following three interpolation methods:

- Nearest neighbor: This is a nonlinear method, where $g(m_1, m_2)$ is equal to the intensity value at the nearest pixel.
- Bilinear interpolation.
- Bicubic interpolation: This method is based on a cubic spline interpolation.

These interpolations are performed as part of the functions **imresize** and **imrotate**, which, respectively, resize and rotate a given image. For example, the following lines will rotate an image by 12° and will generate Fig. 10.16 using bicubic interpolation.

```
>> I = imread ('bbc.tif');
>> J = imrotate(I, 12, 'bicubic');
>> imshow(J);
```

In the next example, we expand an image by a factor of 2.67 using the nearest neighbor and the bilinear methods. As is obvious from Fig. 10.17, the image created with the nearest neighbor interpolation method is of very poor quality. The bilinear

Figure 10.16 Top: Original image. Bottom: Rotated image by an angle of 12°.

Figure 10.17 Top: Original image. Bottom: Image expanded by a factor 2.67 using nearest neighbor interpolation.

Figure 10.18 Image expanded by a factor of 2.67 using bilinear interpolation.

interpolation method creates an image that has fewer artifacts (Fig. 10.18) but is clearly much smoother than the original image.

```
>> I = imread('necklace.tif');
>> J = imresize(I, 2.67, 'bilinear');
>> figure ('name', 'bilinear'); imshow (J, 'truesize');
>> J = imresize(I, 2.67, 'nearest');
>> figure ('name', 'nearest'); imshow (J, 'truesize');
```

10.3 EDGE DETECTION

The detection of object boundaries is one of the most fundamental problems in image understanding and computer vision. The importance of high-contrast lines that outline objects was clearly stated by David Hubel [6, page 86]: "light-dark contours are the most important components of our perception." In fact, the visual cortex uses 2-D filters that are designed to specifically detect black lines against a white background, or white lines against a dark background, or simply light-dark discontinuities in the intensity. Contours provide such important visual cues that the brain often makes interpolation between fragments of contours in order to reconstruct an object. As shown in the Kanizsa [8] triangle illusion in Fig. 10.19, we tend to perceive a connected contour even when only edge termination exists. An illusory triangle appears to be in front of the actual triangle and the three disks.

Figure 10.19 Kanizsa illusion.

High-contrast contours do not always correspond to object boundaries. They can be created by a variety of reasons, such as variation in the texture, variation in the reflectance, shadow boundary, and variation in depth. More than twenty years ago, David Marr suggested that computer vision and image understanding algorithms should rely on a simple sketch of the image [11]. He proposed the concept of a primal sketch that would be composed of edges and yield a "cartoon-like" representation of the image. As a result, edge detection has become a tenet of most pattern recognition and computer vision algorithms. Finally, a mathematical motivation for the importance of edges is provided by the fact that one can reconstruct the intensity at any pixel inside an image from the positions of the edges, and the value of the intensity along them [10]. These results can be summarized with the words of Hubel [6, page 87]: "What happens at the border [of an evenly lit object] is the only information that you need to know: the interior is boring."

In the following section, we propose a definition of an edge which allows us to present some classic edge detectors. We then describe more complex edge detection operators, which yield edge maps that are both less noisy and more precise.

Definition of an Edge

As explained in the previous section, many phenomena can occur at the origin of what we perceive as an edge. Unfortunately, without a precise definition of an edge, one cannot build a rigorous theory of edge detection. We choose to define edges as large values of the gradient.

Definition 10.1 *An edge is a boundary between two regions of relatively uniform intensity, indicated by a strong gradient or a discontinuity in the intensity function.*

According to our definition, the intensity profile in a direction perpendicular to the edge will behave as a step function. In the direction along the edge, the intensity can vary smoothly. Edges that correspond to a strong second-order derivative, such as "peaks," will be detected as two edges according to our definition. A simple algorithm can be derived directly from the definition of an edge as follows:

1. Compute the gradient $\nabla f(n_1, n_2)$ at all pixels.
2. Choose a threshold $\lambda > 0$, and keep only the pixels (n_1, n_2) such that $||\nabla f(n_1, n_2)|| > \lambda$. These pixels can be called edges.

Unfortunately, this algorithm presents several problems. First, most images are likely to contain some noise. Even if the power of the noise is small, taking the derivative will enhance the noise. Indeed, the frequency response of the derivative operator in 1-D is $j\omega$, and therefore the high-frequency part of the noise will be amplified. This is illustrated in Fig. 10.20, where $n(\omega)$ is the frequency response of the noise. The second problem associated with this simple algorithm is the choice of λ. For a small λ, most of the pixels will be detected as edges, and for a large λ, only very few pixels will be detected as edges. How do we choose λ? Finally, the set of pixels (n_1, n_2) such that $||\nabla f(n_1, n_2)|| > \lambda$ is likely to be composed of blobs, or regions, and not lines. Because we would like to interpret edges as outlines of objects, we expect that the edge detector algorithm will generate "lines," and not regions.

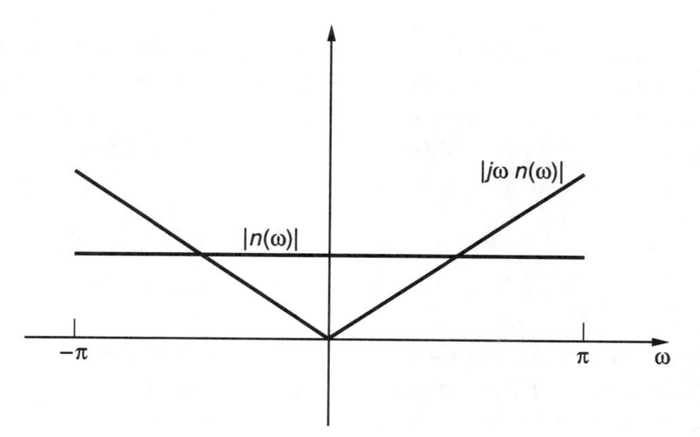

Figure 10.20 Frequency response of the derivative operator.

Finite Differences

Most edge detection algorithms require the computation of the first- or second-order derivatives of the image intensity function. Because the intensity function is sampled on a grid, one needs to approximate the derivatives with a linear combination of the intensity values on the grid. The corresponding filter is usually called a "finite difference." We assume that the distance between two samples on the grid is the same along the horizontal and vertical directions and is equal to Δ. The standard approximations of the first-order derivative of a function are given by the following difference operators:

$$
\begin{aligned}
&\tfrac{1}{\Delta}\{f(n\Delta+\Delta)-f(n\Delta)\} && \text{Forward difference}\\
&\tfrac{1}{\Delta}\{f(n\Delta)-f(n\Delta-\Delta)\} && \text{Backward difference} \qquad (10.28)\\
&\tfrac{1}{\Delta}\{f(n\Delta+\Delta)-f(n\Delta-\Delta)\} && \text{Central difference.}
\end{aligned}
$$

The forward and backward differences are asymmetric and can be useful at the boundaries of the grid where the image intensity is known only on one side. The central difference is symmetric and provides a more precise approximation of the first derivative. Expanding f into a Taylor series around $n\Delta$, we obtain

$$f(n\Delta+\Delta)=f(n\Delta)+\Delta f'(n\Delta)+\frac{\Delta^2}{2}f''(n\Delta)+O(\Delta^3), \tag{10.29}$$

where the notation $O(\Delta^3)$ means that this term can be bounded by a term of the form $C\Delta^3$, C being a constant. All the higher order terms are buried inside $O(\Delta^3)$. Similarly, we can obtain

$$f(n\Delta-\Delta)=f(n\Delta)-\Delta f'(n\Delta)+\frac{\Delta^2}{2}f''(n\Delta)+O(\Delta^3). \tag{10.30}$$

From (10.28) and (10.29), the forward difference can be written as

$$\frac{1}{\Delta}\left[f(n\Delta+\Delta)-f(n\Delta)\right]=f'(n\Delta)+O(\Delta). \tag{10.31}$$

In other words, the forward difference approximates $f'(n\Delta)$, with an error of order $O(\Delta)$. Similarly, one can check that the backward difference approximates $f'(n\Delta)$ with an error of order $O(\Delta)$. The central difference yields a better approximation of $f'(n\Delta)$, with an error of order $O(\Delta^2)$. The proof of this result is left as an exercise.

Several choices exist for the second-order derivative. The second-order central difference is given by

$$\frac{1}{\Delta^2}\left(f(n\Delta+\Delta)-2f(n\Delta)+f(n\Delta-\Delta)\right). \tag{10.32}$$

This approximates the second-order derivative $f''(n\Delta)$ with an error of order $O(\Delta^2)$.

Finite difference can be defined in a similar way in two dimensions. An example that will appear in the edge detectors is the directional derivative along the diagonal direction $\left(\frac{1}{\sqrt{2}}, \frac{1}{\sqrt{2}}\right)$, given by

$$\frac{1}{\sqrt{2}}\left(\frac{\partial f}{\partial x}+\frac{\partial f}{\partial y}\right). \tag{10.33}$$

The following forward difference operator

$$\frac{1}{\Delta}\left\{f(n_1\Delta+\Delta, n_2\Delta)-f(n_1\Delta, n_2\Delta+\Delta)\right\} \tag{10.34}$$

approximates $\frac{\partial f}{\partial x}+\frac{\partial f}{\partial y}$ with an error of order $O(\Delta)$. Similarly, the difference operator

$$\frac{1}{\Delta}\left\{f(n_1\Delta+\Delta, n_2\Delta+\Delta)-f(n_1\Delta, n_2\Delta)\right\} \tag{10.35}$$

approximates $\frac{\partial f}{\partial x}-\frac{\partial f}{\partial y}$, the partial derivative in the direction $\left(\frac{1}{\sqrt{2}}, -\frac{1}{\sqrt{2}}\right)$ with an error of order $O(\Delta)$. Finally, we recall that the Laplacian ∇^2 is defined by

$$\frac{\partial^2 f}{\partial x^2}+\frac{\partial^2 f}{\partial y^2}. \tag{10.36}$$

We can construct a finite difference approximation to the Laplacian using the second-order central difference (10.35). Consequently, the finite difference

$$\begin{aligned} &f(n_1\Delta-\Delta,\ n_2\Delta)+f(n_1\Delta+\Delta, n_2\Delta)+\\ &f(n_1\Delta, n_2\Delta-\Delta)+f(n_1\Delta, n_2\Delta+\Delta)-4f(n_1\Delta, n_2\Delta) \end{aligned} \tag{10.37}$$

approximates the Laplacian with an error of order $O(\Delta^2)$. This finite difference can be implemented with the following 2-D filter:

$$\begin{bmatrix} 0 & 1 & 0 \\ 1 & -4 & 1 \\ 0 & 1 & 0 \end{bmatrix} \tag{10.38}$$

Roberts Edge Detector

The Roberts edge detector [13] computes an approximation to

$$\left(\frac{\partial f}{\partial x}+\frac{\partial f}{\partial y}\right)^2+\left(\frac{\partial f}{\partial x}-\frac{\partial f}{\partial y}\right)^2 \tag{10.39}$$

with the first-order finite difference operators presented in the previous subsection. The 2-D filter kernel for this detector can be easily derived and is left as an exercise. The operator will detect edges that are oriented along the diagonal $y = x$ and the antidiagonal $y = -x$. An example is given in Fig. 10.21. The image on the top-right corner shows an edge map generated by the Roberts edge detector using $\lambda = 5.1$ for the threshold. As expected, the image is extremely noisy. The two images in the bottom of the figure illustrate the effect of increasing the threshold. As λ increases, spurious edges are removed, but actual weak edges are lost. Thresholding is not the solution to the problem. Rather, the fundamental problem with this edge detector is that no effort is made to smooth the image before computing the derivatives.

Sobel and Prewitt Edge Detectors

The Sobel [14] and Prewitt [12] edge detectors are obtained by smoothing the image before computing the derivatives, in the direction perpendicular to the derivative. The following filter is used for smoothing the image before computing the partial derivative in the x direction (with a forward finite difference):

$$h=\begin{bmatrix}1 & 1\\ c & c\\ 1 & 1\end{bmatrix} \tag{10.40}$$

where

$$c = 1 \quad \text{for Prewitt, and} \quad c = 2 \quad \text{for Sobel.} \tag{10.41}$$

Since the convolution and the finite difference operators are both linear operations, they can be combined, and the resulting filter is given by

$$\begin{bmatrix}1 & 0 & -1\\ c & 0 & -c\\ 1 & 0 & -1\end{bmatrix}. \tag{10.42}$$

This filter computes the partial derivative in the x (horizontal) direction. Similarly, the filter that computes the partial derivative in the y (vertical) direction is given by

$$\begin{bmatrix}1 & c & 1\\ 0 & 0 & 0\\ -1 & -c & -1\end{bmatrix}. \tag{10.43}$$

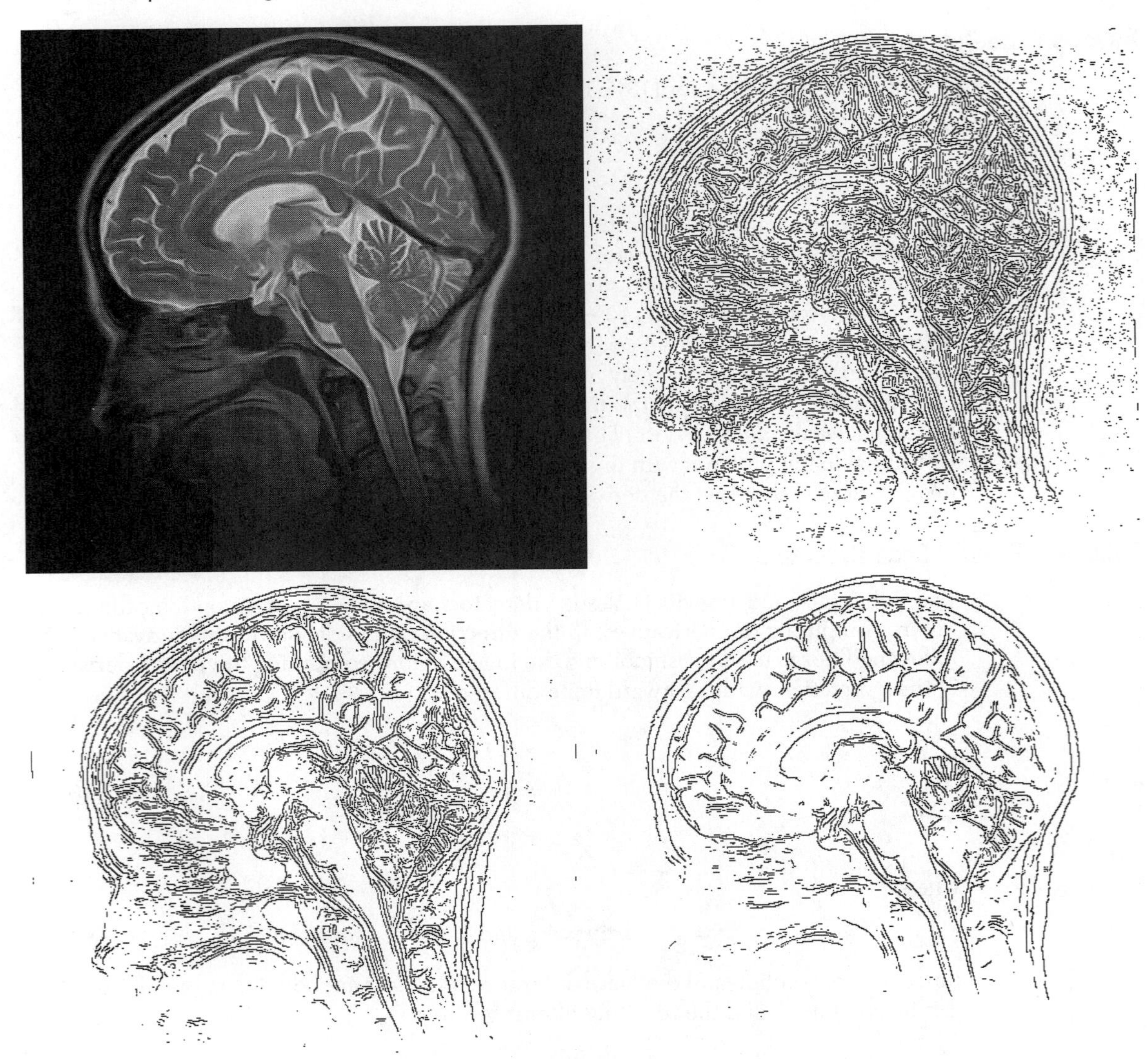

Figure 10.21 Edge map generated by the Roberts edge detector. Top left: Original image. Top right: Edge image with $\lambda = 5.1$. Bottom left: $\lambda = 10.28$. Bottom right: $\lambda = 20.4$.

Again, no smoothing is performed in the direction where the derivative is taken, and we expect these filters to be very sensitive to noise. The image on the top of Fig. 10.22 shows an edge map generated by the Sobel edge detector using $\lambda = 5.1$ for the threshold. The two images at the bottom of the figure were obtained with $\lambda = 10.2$ and 20.4. As expected, this edge detector performs poorly, and the visual results are very similar to those of the Roberts edge detector.

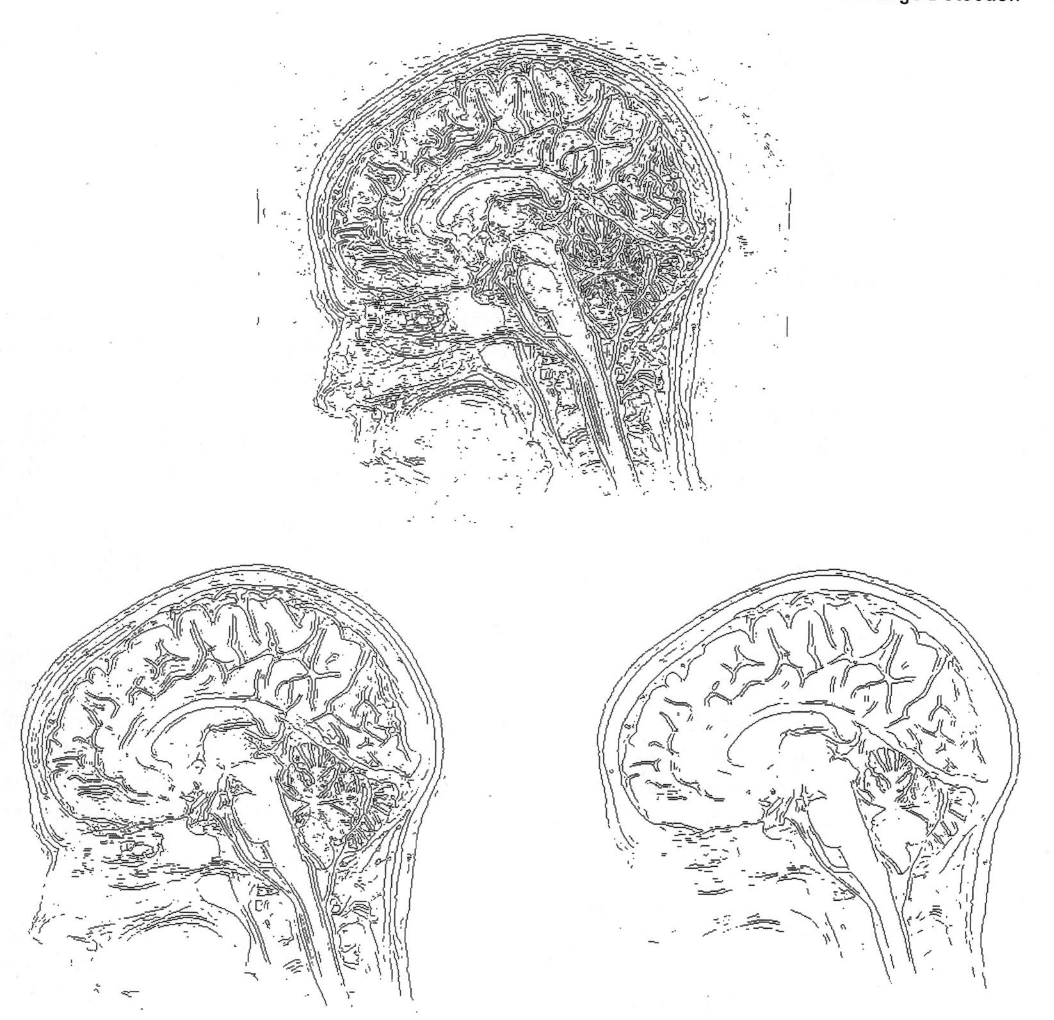

Figure 10.22 Edge maps generated by the Sobel edge detector. Top: $\lambda = 5.1$. Bottom left: $\lambda = 10.28$. Bottom right: $\lambda = 20.48$.

MATLAB Commands

The following lines will generate Figs. 10.21 and 10.22.

```
>> I = imread('mr8.tif');
>> E = edge (I,'roberts',0.020);
>> E = edge (I,'roberts',0.040);
>> E = edge (I,'roberts',0.080);
>> E = uint8(255*(1-double(E)));
>> figure('name','Roberts'),imshow(uint8(E),'truesize');
```

```
>> E = edge (I,'roberts',0.020);
>> E = edge (I,'roberts',0.040);
>> E = edge (I,'sobel',0.08);
>> E = uint8(255*(1-double(E)));
>> figure('name','Sobel'),imshow(uint8(E),'truesize');
```

Marr-Hildreth Edge Detection

Experiments with the Roberts and Sobel edge detectors indicate that we need to combine lowpass filtering (or smoothing) with a derivative operator if we want to generate edge maps that do not contain noisy spurious edges. The effect of lowpass filtering in the spatial domain is to blur the sharp edges, and therefore increase the uncertainty about the location of the edges. In fact, recall that we used a highpass filtering technique (unsharp masking) to increase the contrast of the edges. As is often the case in the design of optimal filters, edge detection involves two conflicting goals: cancellation of noise and minimization of spatial smoothing. It turns out that the Gaussian kernel provides the optimal tradeoff between these two requirements. We will assume that the image has been lowpass filtered with a 2-D Gaussian kernel of the form

$$G(x,y) = \frac{1}{2\pi\sigma^2} e^{-\frac{x^2+y^2}{2\sigma^2}}. \tag{10.44}$$

The second problem with the Roberts and Sobel edge detectors was the problematic choice of the threshold λ. The threshold was necessary to remove all edge points with a weak gradient. This problem can be solved by requiring that we keep only those edge points where the gradient is locally maximum. In 1-D, if the first derivative is maximum at a point, then the second derivative is necessarily going to change sign (provided the function is smooth enough). In other words, points where the second derivative is zero are the only candidates for local maxima.

In 2-D, there are many directions θ along which one can compute the first derivative. For each direction θ, one can look for the maximum of the derivative in that direction. If the direction is chosen to be perpendicular to an edge, then the maximum of the directional derivative will locate an edge point. Let $D_{u_\theta} f(x_0, y_0)$ be the directional derivative along the direction defined by the unit vector $u_\theta = (\cos\theta, \sin\theta)$ at the point (x_0, y_0). The restriction of the function f along the line passing through (x_0, y_0) with the direction $(\cos\theta, \sin\theta)$ is given by

$$\rho \longrightarrow f(x_0 + \rho\cos\theta, y_0 + \rho\sin\theta). \tag{10.45}$$

The directional derivative $D_{u_\theta} f(x_0, y_0)$ is then given by

$$\begin{aligned} D_{u_\theta} f(x_0, y_0) &= \frac{\partial}{\partial p} f(x_0 + \rho\cos\theta, y_0 + \rho\sin\theta)(\rho) \\ &= \frac{\partial}{\partial x} f(x_0, y_0)\cos\theta + \frac{\partial}{\partial y} f(x_0, y_0)\sin\theta \\ &= \nabla f \cdot u_\theta \end{aligned}$$

A point given by

$$\begin{cases} x = x_0 + \rho \cos\theta \\ y = y_0 + \rho \sin\theta \end{cases} \tag{10.46}$$

will be a maximum of the directional derivative of f in the direction θ if the second derivative of f in the direction $\theta, D^2_{u_\theta}$ is zero at this point. The expression for the second derivative of f in the direction θ is given by

$$\begin{aligned} D^2_{u_\theta} f(x,y) &= \frac{\partial^2}{\partial \rho^2} f(x_0 + \rho\cos\theta, y_0 + \rho\sin\theta)(\rho) \\ &= \frac{\partial^2 f}{\partial x^2}(x,y)\cos^2\theta + 2\frac{\partial^2 f}{\partial x \partial y}(x,y)\cos\theta\sin\theta + \frac{\partial^2 f}{\partial y^2}(x,y)\sin^2\theta \\ &= u_\theta^T \mathbf{H} u_\theta \end{aligned} \tag{10.47}$$

where $\mathbf{H}$ is the Hessian given by

$$\mathbf{H} = \begin{bmatrix} \frac{\partial^2 f}{\partial x^2} & \frac{\partial^2 f}{\partial x \partial y} \\ \frac{\partial^2 f}{\partial x \partial y} & \frac{\partial^2 f}{\partial x^2} \end{bmatrix}.$$

Therefore, one could detect edges by searching for the zero-crossings of $D^2_{u_\theta} f(x,y)$, where θ is aligned with the direction normal to the edge. Unfortunately, one does not know the direction of the edges, since we do not know their locations! One solution to this intricate problem is provided by the orientation of the gradient of the intensity, ∇f. If we consider a small neighborhood around a pixel in the image (Fig. 10.23), ∇f will be aligned with the direction where the function varies most rapidly in that neighborhood. If that neighborhood contains an edge, the gradient should be a good indicator of the direction perpendicular to the edge. If we follow this idea, we need to compute the directional derivative of the intensity in the direction along the gradient. As shown in Fig. 10.24, the profile of the intensity along the gradient allows us to clearly identify the edge. The points where the derivative of this directional derivative changes sign will be detected as edge points. This approach was proposed by Canny [2] and is described in the next section.

Instead of detecting the zero-crossing of $D^2_{u_\theta} f$ for all possible directions θ (hoping that one of them will be orthogonal to an edge), one can choose to replace $D^2_{u_\theta} f$ with a second-order derivative operator that would be invariant under any rotation of the image. As a result, this derivative operator could detect edges regardless of their directions. Since the only isotropic second-order derivative operator in 2-D is the Laplacian, Marr and Hildreth [11] proposed detecting edge points by finding the zero-crossings of the Laplacian of the image, after it had been convolved with a Gaussian kernel. This is denoted by the following:

$$(x,y) \in \text{edge} \iff \nabla^2(G * f)(x,y) = (\nabla^2 G) * f(x,y) = 0. \tag{10.48}$$

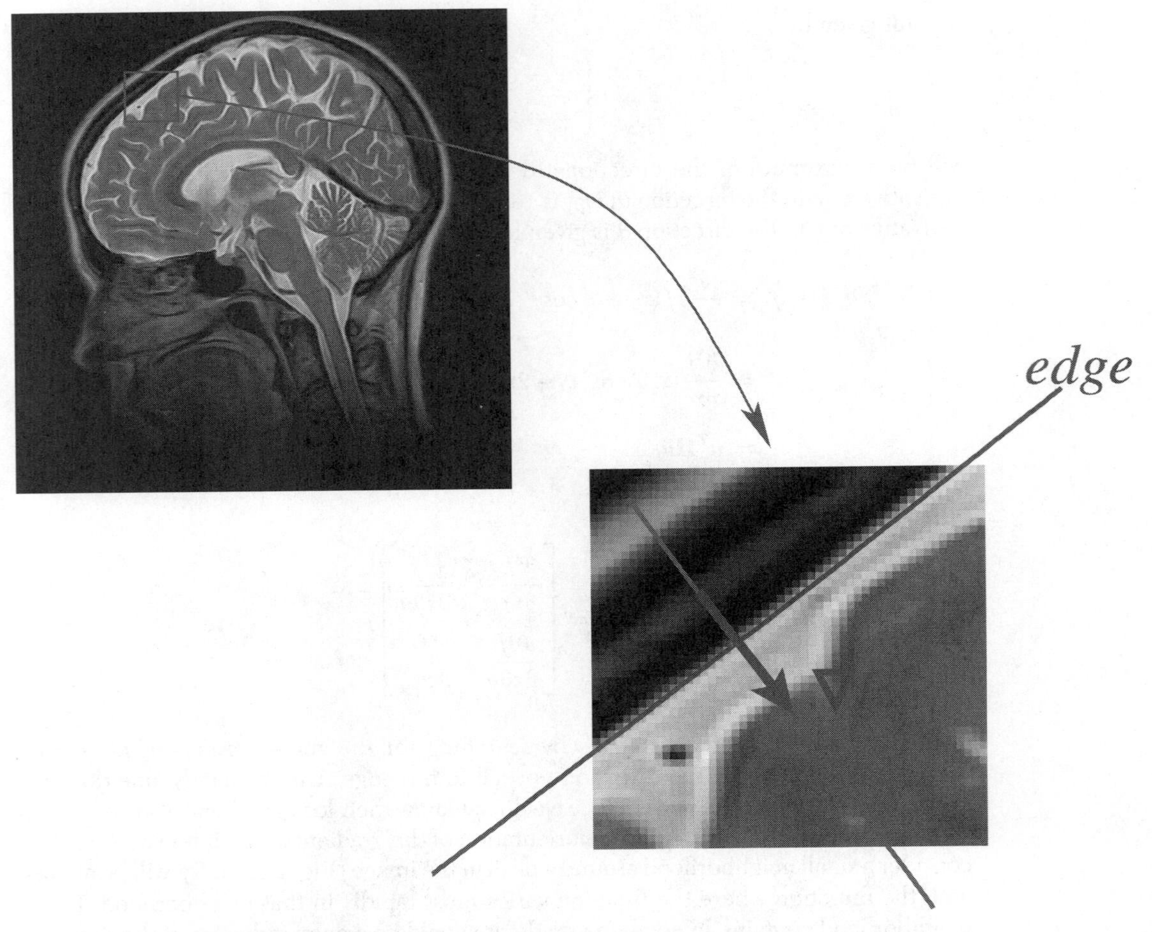

Figure 10.23 Neighbor around an edge.

The second equality holds because the Laplacian is a linear operator. The 2-D function $\nabla^2 G$ is isotropic and is shaped like a Mexican hat. The 1-D profile of the filter is shown in Fig. 10.25.

In summary, the Marr-Hildreth edge detection algorithm is given by the following steps:

1. Convolve the image with a Gaussian kernel G.
2. Compute the Laplacian $\nabla^2(G * f)$.
3. Detect the pixels where $\nabla^2(G * f)$ changes sign; these pixels are edge points.

MATLAB Commands

The following lines will generate Fig. 10.26:

```
>> I = imread ('mr8.tif');
```

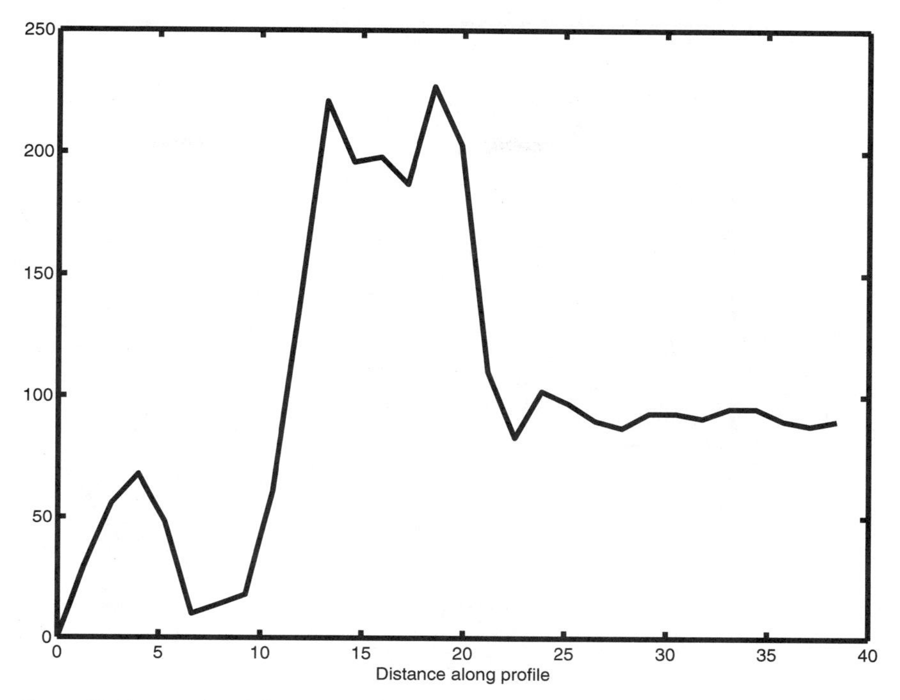

Figure 10.24 Profile of the intensity along the direction of the gradient shown in the previous figure.

```
>> E = edge (I,'zerocross', 0.003);
>> E = uint8(255*(l-double(E)));
>> figure('name','zerocross') ,imshow(uint8(E),'truesize');
```

Figure 10.26 shows an edge map generated by the Marr-Hildreth edge detector. There is no longer a need to select a threshold λ. Many small edges are now visible in the edge map, and it also contains much less noise than in the maps generated with the Sobel and Roberts edge detectors.

Canny Edge Detector

The Marr-Hildreth edge detector filters the image with a bandpass filter, which is the Laplacian of the Gaussian. Edges are the zero-crossings of the output of the filter. Although the detector performs quite well, we can ask whether the Laplacian of the Gaussian is really the optimal filter for the detection of edges. In order to talk about optimality, one needs to define some criteria to measure the quality of the filter. Canny [2] suggested the following three criteria:

1. *Ability to handle noise*: The output of the filter should not be overly sensitive to noise.

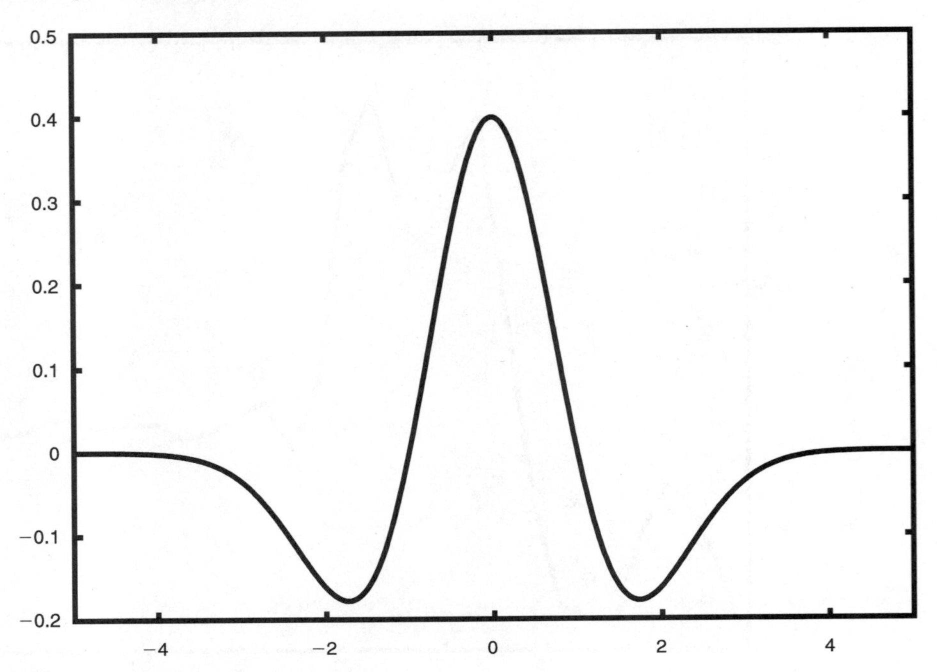

Figure 10.25 Second derivative of the Gaussian.

2. *Accurate location of the true edge*: The filter should not introduce a phase shift that would modify the location of the zero-crossings.
3. *Uniqueness of the response*: If there is only one edge in a given neighborhood, then there should be only one zero-crossing.

Canny proposed a very simple model of an edge, namely, the step edge. Let us define a step edge in a neighborhood of a point (n_1, n_2) as a continuous line along which the intensity is discontinuous. The edge separates the neighborhood into two regions within which the intensity is locally smooth. This is the ideal model of a step edge and is shown in Fig. 10.27. Given this model of the ideal edge, Canny then searched for the optimal filter according to his three criteria. He found an analytic expression for the optimal filter and realized that it was very similar to the Gaussian derivative. In fact, the performance of the filter measured in terms of the three criteria was similar to the performance of the Gaussian derivative. Instead of using the Laplacian, Canny proposed searching for the zero-crossings of the second derivative of the intensity in the direction of the gradient. One can compute the expression for $D_{\frac{\nabla f}{||\nabla f||}} f$. Using equation (10.47), we have

$$D_{\frac{\nabla f}{||\nabla f||}} f = \frac{1}{||\nabla f||^2} \left\{ \left(\frac{\partial f}{\partial x}\right)^2 \frac{\partial^2 f}{\partial x^2} + 2 \left(\frac{\partial f}{\partial x}\right)^2 \left(\frac{\partial f}{\partial y}\right)^2 + \left(\frac{\partial f}{\partial y}\right)^2 \frac{\partial^2 f}{\partial y^2} \right\}. \tag{10.49}$$

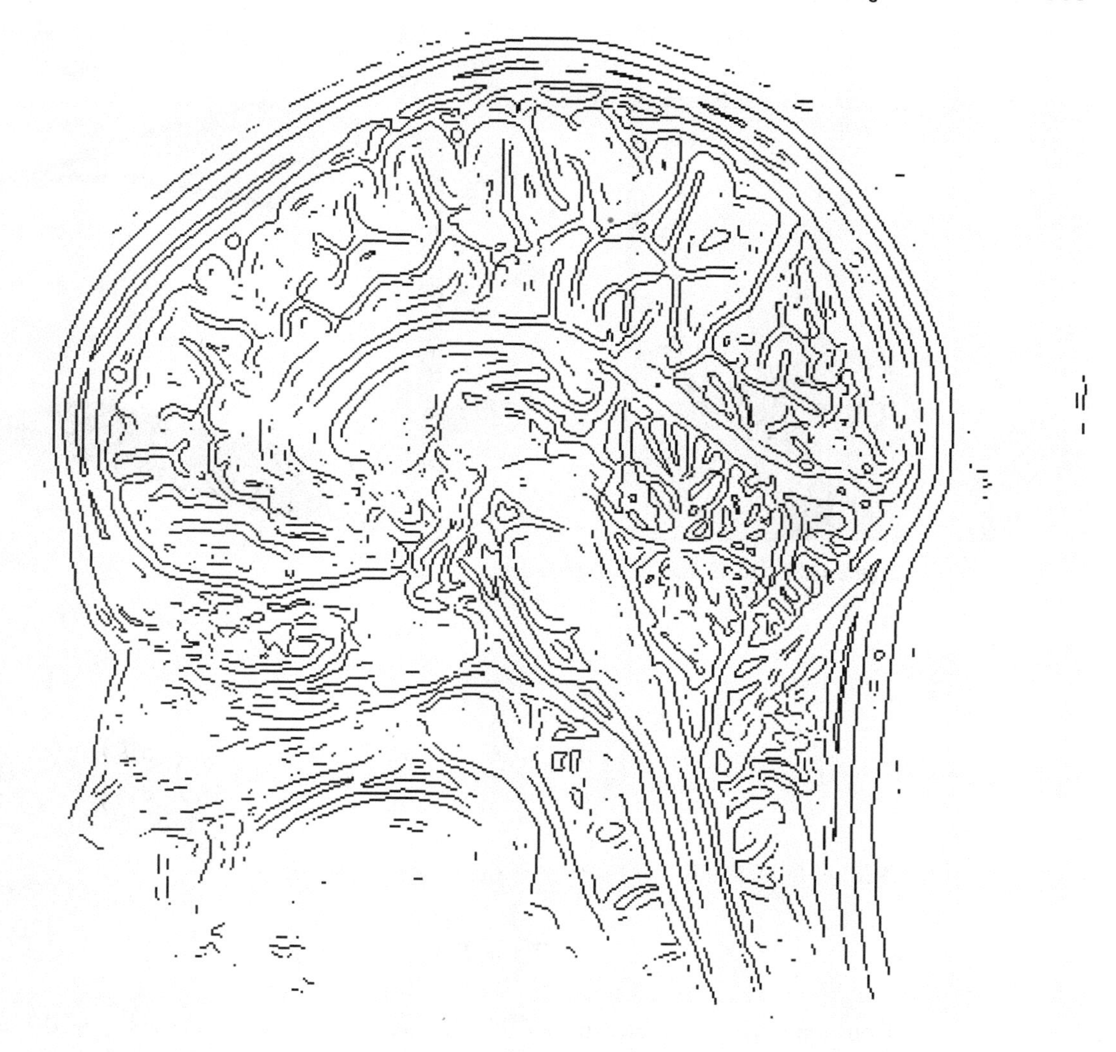

Figure 10.26 Edge map generated by the Marr-Hildreth edge detector.

Using (10.47), one can also compute the second derivative of f in the direction normal to the gradient. The direction normal to the gradient is

$$\nabla f_{\perp} = \begin{bmatrix} -\dfrac{\partial f}{\partial y} \\ \\ \dfrac{\partial f}{\partial x} \end{bmatrix}. \tag{10.50}$$

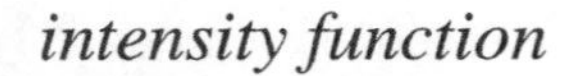

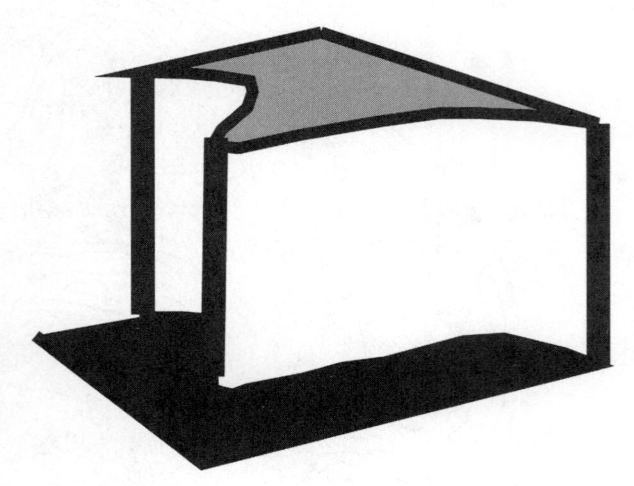

Figure 10.27 Ideal model of a step edge.

This gives

$$D_{\frac{\nabla f_\perp}{||\nabla f||}} f = \frac{1}{||\nabla f||^2}\left\{\left(\frac{\partial f}{\partial x}\right)^2 \frac{\partial^2 f}{\partial x^2} - 2\left(\frac{\partial f}{\partial x}\right)^2\left(\frac{\partial f}{\partial y}\right)^2 + \left(\frac{\partial f}{\partial y}\right)^2 \frac{\partial^2 f}{\partial y^2}\right\}. \tag{10.51}$$

Adding $D_{\frac{\nabla f}{||\Delta f||}} f$ and $D_{\frac{\nabla f_\perp}{||\Delta f||}} f$ gives us the Laplacian operator

$$\nabla^2 f = D_{\frac{\nabla f}{||\nabla f||}} f + D_{\frac{\nabla f_\perp}{||\nabla f||}} f. \tag{10.52}$$

Consequently, a zero-crossing of the Laplacian will be a zero-crossing of the second derivative in the direction of the gradient if and only if the second derivative in the direction perpendicular to the gradient is zero. This means that an edge detected by the Marr-Hildreth algorithm will coincide with an edge detected by the Canny algorithm, if and only if the intensity in the direction perpendicular to the gradient (along the edge) is locally linear.

In summary, the Canny edge detection algorithm is given by the following steps:

1. Convolve the image with a Gaussian kernel G.
2. Compute the gradient $\nabla(G * f)$ at each pixel.
3. If the gradient is nonzero, compute $D_{\frac{\nabla G * f}{||\nabla G * f||}} f$, the second derivative of the smoothed image in the direction of the gradient.
4. Detect the pixels where $D_{\frac{\nabla G * f}{||\nabla G * f||}} f$ changes sign; these pixels are edge points.

MATLAB Commands

An example of using the Canny edge detector is shown in Fig. 10.28. This is the best of all edge maps that we have seen so far; there are no spurious edges, and all edges appear to be present. The following lines will generate this figure:

```
>> I = imread ('mr8.tif');
>> %Canny edge detector;
>> th = [0.05 0.1];
```

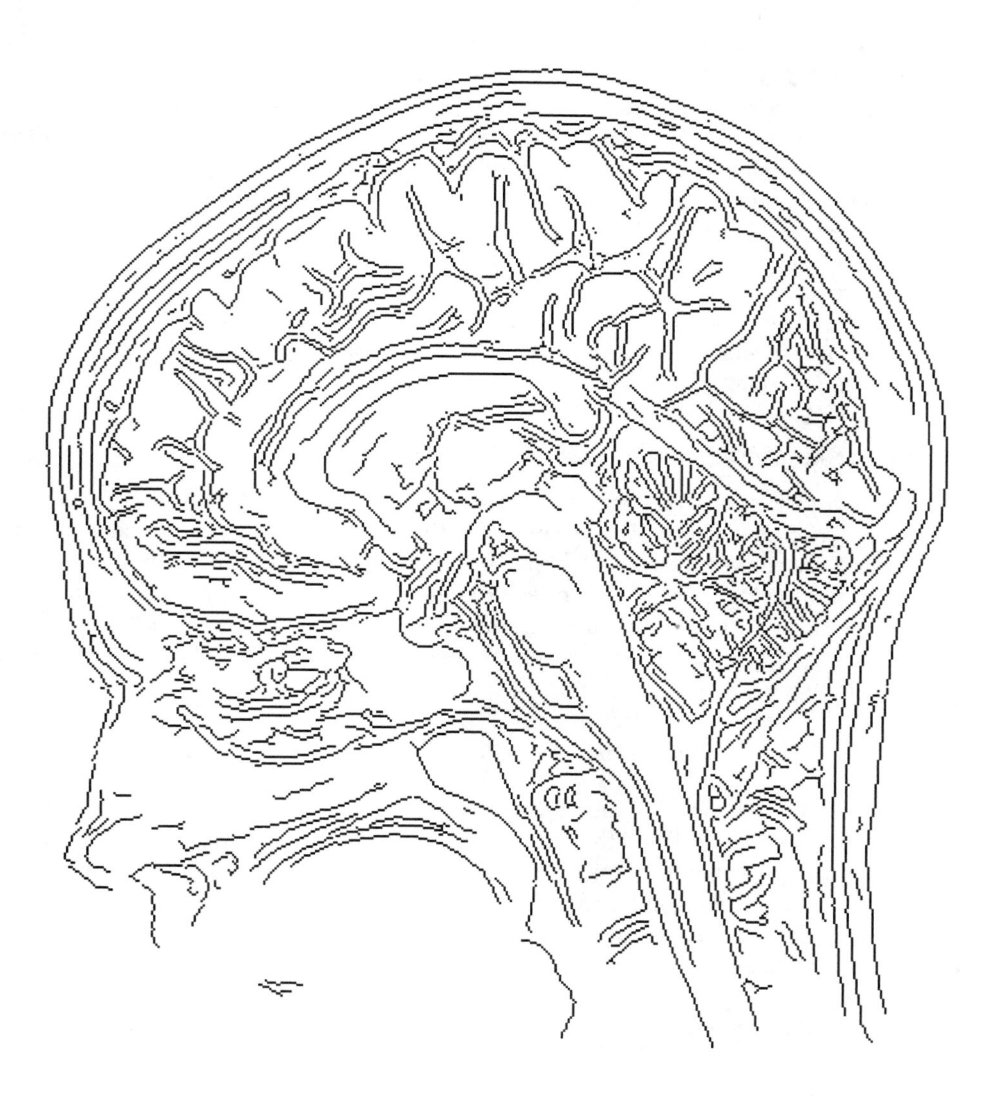

Figure 10.28 Edge map generated by the Canny edge detector.

```
>> E = edge (I,'canny', th);
>> E = uint8(255*(1-double(E)));
>> figure('name','canny'),imshow(uint8(E),'truesize');
```

10.4 IMAGE RESTORATION

Digital images often suffer from systematic distortions and addition of noise. Distortions can be caused by the defocusing of a camera or by the blurring caused by the imperfection of the optical system, such as the aberration in the optics of the Hubble telescope. If a model of the distortion is available, then one can try to invert the systematic distortion and recover an image of higher quality. The addition of random noise during the acquisition is almost inevitable. In an optical system, the noise can be caused by the quantum nature of the light; the photons hitting a CCD cell can be modeled with a Poisson distribution. Other sources of noise include thermal noise in electronic circuits and quantization noise. The goal of restoration is to correct the degradation and remove the noise. The quality of the restored image can be evaluated visually or in a quantitative manner by measuring the signal to noise ratio. The success of the restoration process relies on the knowledge of the degradation process and the design of the filter. In our discussion of restoration, we will first study systematic distortions that "blur" the image. We will present several "deblurring" and inverse filtering techniques. We then study the problem of noise removal, using linear and nonlinear filters.

Choices of Norms for Measuring Image Quality

In our discussion of degradation, we will assume that the ideal image x has been blurred or corrupted by noise. We measure a noisy or blurred image z, and we build an estimate $\hat{x}$ of x given z. Even the most elaborate restoration techniques will not be able to completely recover the ideal image x. The restored image $\hat{x}$ is always an approximation to the ideal image x. In order to measure the performance of a restoration algorithm, one needs to measure the quality of the recovered image. A standard approach consists of considering that the image

$$x(n_1, n_2), \quad n_1 = 0, \ldots, N_1 - 1, n_2 = 0, \ldots, N_2 - 1 \tag{10.53}$$

is a vector in $R^{N_1} \times R^{N_2}$. One can then use various norms to measure the distance between the restored image $\hat{x}$ and the original image x. This approach turns out to be very fruitful for the design of optimal restoration algorithms. One potential limitation of this approach is that the norm could be chosen only because a good algorithm exists to minimize this norm. Yet the norm may bear very little connection with the visual appearance of the restored image. It is therefore crucial to always visually inspect the restored image and look for visual artifacts.

A standard family of norms are the l^p norms, $0 < p < \infty$, given by

$$|x - \hat{x}|_p = \left(\sum_{n_1=0}^{N_1-1} \sum_{n_2=0}^{N_2-1} |x(n_1, n_2) - \hat{x}(n_1, n_2)|^p \right)^{1/p}. \tag{10.54}$$

If $p = 2$, then $|x - \hat{x}|_2^2$ is often called the sum of squared error (SSE). The quadratic norm will emphasize large errors and will be oblivious to small errors, even if there are many of them. In many situations, this norm is not ideal. For instance, a small error spread at every pixel in the image can result in the same SSE as a few isolated wild impulse errors. However, these impulse errors may be unnoticeable if they appear in a region with a lot of texture, whereas the small errors may be very noticeable and visually annoying. The choice $p = 1$ deemphasizes the wild impulse errors and penalizes the small variations in the errors.

In this chapter, we will only consider the case $p = 2$. The quadratic case is much simpler, and many good algorithms are available to derive optimal restoration algorithms for this norm.

Inverse Filtering and De-convolution

We assume that the image available to us is the output of a linear filter characterized by its impulse response, h. This filter has blurred the ideal image x according to the equation

$$z = h * x. \tag{10.55}$$

The image x is an ideal image that is not available to us but that we would like to estimate from its blurred version z. Clearly, the problem (10.55) involves inverse-filtering or de-convolving z in order to recover x. In this chapter, we limit ourselves to the case where h is available to us. If h is unknown, the problem of estimating x from z is called blind de-convolution. This latter problem is much more involved and will not be discussed here.

Inverse Filtering: A Naive Approach

We first describe the naive solution to our problem and explain how to modify it in order to obtain a better solution in the next section. In the Fourier domain, equation (10.55) becomes

$$Z(k_1, k_2) = H(k_1, k_2)X(k_1, k_2). \tag{10.56}$$

where X is the Fourier transform of x, Z is the Fourier transform of z, and $H(k_1, k_2)$ is the frequency response of the filter h. When $H(k_1, k_2)$ is nonzero, we have

$$X(k_1, k_2) = \frac{Z(k_1, k_2)}{H(k_1, k_2)}. \tag{10.57}$$

When $H(k_1, k_2)$ is zero, this approach does not work. Since h is typically a bandpass filter, there is a large region of the spectrum where $H(k_1, k_2)$ is zero (or close to it) and where X in (10.57) is no longer defined.

Another problem stems from the fact that if $H(k_1, k_2)$ becomes very small for some frequency (k_1, k_2), then equation (10.57) will amplify $X(k_1, k_2)$ at this frequency. This can become a problem if x contains some noise at these frequencies. This is the case, for instance, if x is contaminated by white noise or high-frequency noise. In such cases, the noise will be greatly amplified, and the image obtained from (10.57) will be extremely noisy.

An Improved Inverse Filter: The Pseudo-inverse Filter

A simple way to correct the instability created by the inverse filter is to set the inverse filter to 0 as soon as $H(k_1, k_2)$ becomes too small. The regularized pseudo-inverse filter $\tilde{h}^{-1}$ is defined by its Fourier transform as

$$\widetilde{H}^{-1}(k_1, k_2) = \begin{cases} H^{-1}(k_1, k_2) & \text{if } |H(k_1, k_2)| \geq \lambda \\ 0 & \text{if } |H(k_1, k_2)| < \lambda. \end{cases} \tag{10.58}$$

Figure 10.29 depicts the magnitude of the pseudo-inverse filter in 1-D. The pseudo-inverse filter can be applied in the Fourier domain as

$$\widehat{X}(k_1, k_2) = \widetilde{H}^{-1}(k_1, k_2) Z(k_1, k_2). \tag{10.59}$$

The deblurred image is then obtained by an inverse Fourier transform. Alternatively, one can directly compute the filter $\tilde{h}^{-1}$ from its Fourier transform $\widetilde{H}^{-1}(k_1, k_2)$ and implement the filter in the spatial domain with a convolution.

Examples and MATLAB Commands

An example of inverse filtering is shown in Fig. 10.30. The image on the left has been blurred with a rotationally invariant Gaussian lowpass filter of size 9×9 and standard deviation 5. Notice how the texture of the hat has been smeared by the lowpass filter. The output of the pseudo-inverse filter is shown in the right-hand side of Fig. 10.30. Although some high frequencies have been lost, the crispness of the hat and the face of the woman have been very well restored. The inverse filtering or de-convolution can be implemented with the function **deconvlucy**. The following lines of code generate Fig. 10.30.

```
>> I = imread('04.tif');
>> figure('name', 'original') , imshow(uint8(I));
>> h = f special('gaussian',9,5);
>> N = conv2(I,h,'same');
```

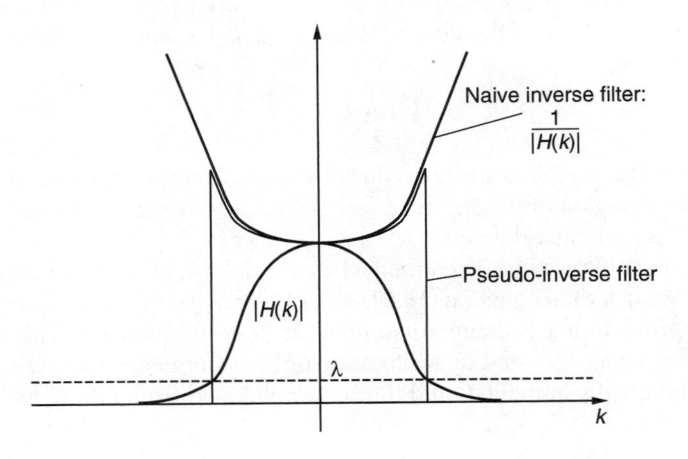

Figure 10.29 Degradation filter $H(k)$ and the pseudo-inverse filter, in one dimension.

Figure 10.30 Left: Blurred image. Right: Output of the pseudo-inverse filter.

```
>> figure ('name','blurred'), imshow(uint8(N));
>> N = edgetaper (N,h);
>> z = deconvlucy (N,h); imshow(uint8(z));
>> figure ('name','deconv'), imshow(uint8(z));
```

Because the **deconvlucy** function uses a DFT, it tends to create ringing artifacts around the borders of the image. Indeed, the DFT assumes that the image intensity is periodic along the vertical and horizontal directions, and discontinuities at the border will create the Gibbs phenomenon. To reduce this ringing, the borders of the image are first smoothed with the Gaussian filter h using the **edgetaper** function.

Figure 10.31 shows the same image after having been blurred with a rotationally invariant Gaussian lowpass filter of size 19×19 and standard deviation 10. The image has been severely degraded, and it is not surprising that it can be only partially recovered, as shown on the right. Some ringing artifacts have been introduced by the design of the filter in the Fourier domain.

Figure 10.31 Left: Blurred image. Right: Output of the pseudo-inverse filter.

Removal of Additive Noise

We first present the removal of additive noise using Wiener filtering. The Wiener filtering is first derived in 1-D and then extended to 2-D. Let $x(n)$ denote the signal and $v(n)$ denote the noise. The noisy signal is then

$$z(n) = x(n) + v(n). \tag{10.60}$$

The goal is to filter $z(n)$ with a linear filter with impulse response $h(n)$ so that the output $\hat{x}(n)$ is an estimate of the clear signal $x(n)$. Thus

$$\hat{x}(n) = z(n) * h(n), \tag{10.61}$$

where $*$ denotes convolution. Let J denote the cost function to be minimized. For Wiener filtering, the cost function is the expectation of the squared error given by

$$\begin{aligned} J &= E\{e^2(n)\} \\ &= E\{(x(n) - \hat{x}(n))^2\}. \end{aligned} \tag{10.62}$$

Assuming that the filter impulse response is causal and of possibly infinite length, we have

$$\hat{x}(n) = \sum_{k=0}^{\infty} h_k z(n-k) \tag{10.63}$$

where h_k denotes the impulse response coefficients. The cost function then becomes

$$J = E\left\{\left(x(n) - \sum_{k=0}^{\infty} h_k z(n-k)\right)^2\right\}. \tag{10.64}$$

Differentiating the above with respect to h_k and setting the result to zero, we get

$$\begin{aligned}\frac{\partial J}{\partial h_k} &= E\left\{2\left(x(n) - \sum_{k=0}^{\infty} h_k z(n-k)\right) z(n-k)\right\} \\ &= 2E\{e(n)z(n-k)\} = 0.\end{aligned} \tag{10.65}$$

Therefore

$$E\{x(n)z(n-k)\} = E\{\hat{x}(n)z(n-k)\}. \tag{10.66}$$

The above is the well-known orthogonality principle, derived in Chapter 7. The left-hand side is the cross-correlation function $R_{xz}(k)$. Of course, the assumption is that $x(n)$ and $v(n)$ are wide-sense stationary random processes. Then from (10.66), we have

$$R_{xz}(k) = E\{\hat{x}(n)z(n-k)\} \tag{10.67}$$

$$\begin{aligned} &= E\left\{\sum_{m=0}^{\infty} h_m z(n-m) z(n-k)\right\} \\ &= \sum_{m=0}^{\infty} h_m E\{z(n-m)z(n-k)\} \\ &= \sum_{m=0}^{\infty} h_m R_{zz}(k-m) \\ &= h_k * R_{zz}(k),\end{aligned} \tag{10.68}$$

where $R_{zz}(k)$ is the autocorrelation function of $z(n)$. Taking the DTFT of the above gives

$$H(\omega) = \frac{S_{xz}(\omega)}{S_{zz}(\omega)} \tag{10.69}$$

where the power spectral densities (PSDs) are

$$S_{xz}(\omega) = DTFT\{R_{xz}(k)\} \tag{10.70}$$

$$S_{zz}(\omega) = DTFT\{R_{zz}(k)\}. \tag{10.71}$$

Equation (10.68) is the general Wiener filter. Now we make some assumptions and simplify it. We first assume that the signal and the noise are uncorrelated, that is, $E\{x(n)v(n-k)\} = 0$. The cross-correlation function is

$$\begin{aligned} R_{xz}(k) &= E\{x(n)z(n-k)\} \\ &= E\{x(n)[x(n-k) + v(n-k)]\} \\ &= E\{x(n)x(n-k)\} + E\{x(n)v(n-k)\} \\ &= R_{xx}(k). \end{aligned} \tag{10.72}$$

The autocorrelation of the corrupted signal is

$$\begin{aligned} R_{zz}(k) &= E\{z(n)z(n-k)\} \\ &= E\{(x(n) + v(n))\,(x(n-k) + v(n-k))\} \\ &= E\{x(n)x(n-k)\} + E\{x(n)v(n-k)\} \\ &\quad + E\{v(n)x(n-k)\} + E\{v(n)v(n-k)\} \\ &= R_{xx}(k) + R_{vv}(k), \end{aligned} \tag{10.73}$$

where $R_{vv}(k)$ is the autocorrelation of the noise. Now, using (10.72) and (10.73) in the general Wiener filter (10.69), we get

$$H(\omega) = \frac{S_{xx}(\omega)}{S_{xx}(\omega) + S_{vv}(\omega)}. \tag{10.74}$$

Next, we make the assumption that the noise $v(n)$ is zero-mean with variance σ_v^2. Then the Wiener filter becomes

$$H(\omega) = \frac{S_{xx}(\omega)}{S_{xx}(\omega) + \sigma_v^2}. \tag{10.75}$$

In 2-D, equations (10.74) and (10.75), respectively, become

$$H(\omega_1, \omega_2) = \frac{S_{xx}(\omega_1, \omega_2)}{S_{xx}(\omega_1, \omega_2) + S_{vv}(\omega_1, \omega_2)} \tag{10.76}$$

and

$$H(\omega_1, \omega_2) = \frac{S_{xx}(\omega_1, \omega_2)}{S_{xx}(\omega_1, \omega_2) + \sigma_v^2}. \tag{10.77}$$

In equation (10.77), to find the filter $H(\omega_1, \omega_2)$, the PSD of the signal and the noise must be known a priori or estimated. In the space-variant Wiener filter, these parameters are estimated from small windows of the image, and a different filter is obtained for every window as it slides across the image. The details are omitted and can be found in [9]. The space-variant Wiener filter is performed by the MATLAB command **wiener2**.

Figure 10.32 illustrates the principle of the Wiener smoothing filter in 1-D. We assume that the noise v is white, and therefore the power spectral density $S_{vv}(\omega)$ is

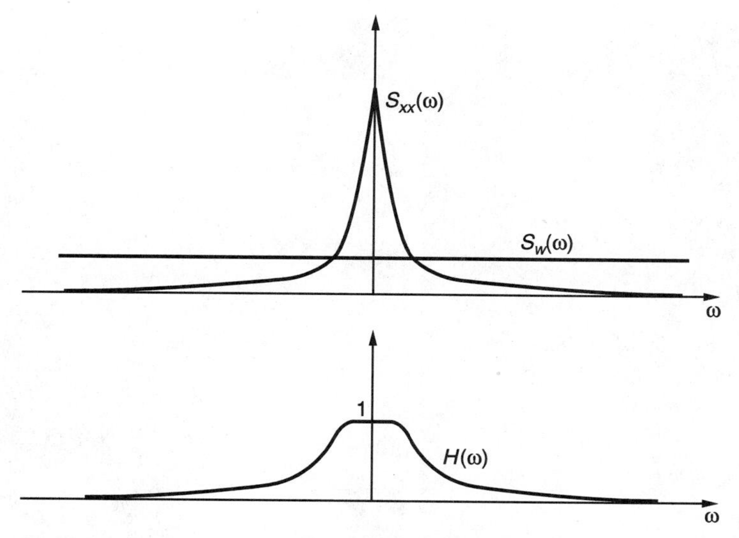

Figure 10.32 The Wiener filter concept in one dimension.

constant. In other words, the energy of the noise is spread across all frequencies. The center of mass of the power spectral density of the image $S_{xx}(\omega)$ is around 0, with a tail that decreases exponentially fast with the frequency. The smoothing Wiener filter is also shown in the figure. At low frequencies, the energy of the noise is much lower than the energy of the image, that is, $S_{xx}(\omega) \gg S_{vv}(\omega)$. Therefore, from (10.74), we have $H(\omega) \approx 1$. At low frequencies, the filter $H(\omega)$ passes all of $z(n)$. At high frequencies, the energy of the noise is much larger than the energy of the image, that is, $S_{xx}(\omega) \ll S_{vv}(\omega)$. Therefore, we have $H(\omega) \approx 0$. At high frequencies, the filter $H(\omega)$ attenuates most of $z(n)$. In this particular example, the smoothing Wiener filter is a lowpass filter. However, it is the optimal lowpass filter.

Lowpass Filtering

The Wiener filter is always the optimal linear filter for removing noise. Because most of the energy of a typical image is located at the low frequencies, and because the energy of the noise is often spread across all frequencies (white noise), the smoothing Wiener filter is often a lowpass filter. For this reason, one should expect to obtain good noise removal properties from lowpass filters.

Figure 10.33 shows an original image (left) and a corrupted image (right). The additive noise used is Gaussian white noise with a standard deviation equal to 10.2. The mean square error (MSE) between the noisy image and the original image is found to be 104.09. Figure 10.34 (left) shows the result of applying a lowpass filter to the image. The lowpass filter is a rotationally invariant Gaussian-shaped filter of variance 1 and size 7×7. The MSE between the filtered image and the original image is equal to 66.5.

The lowpass filtering algorithm can be implemented with the function **imfilter**. The following lines will generate Figs. 10.33 and 10.34 (left).

```
>> I = imread('18.tif');
>> figure('name','original'), imshow(uint8(I),'truesize');
```

Figure 10.33 Left: Original image. Right: Image corrupted by Gaussian white noise.

```
>> %% add noise
>> N = imnoise (I,'gaussian',0,5e-3);
>> E = imsubtract (I,N);
>> mse = (std2(E))^2
>> figure('name','noisy'), imshow(uint8(N),'truesize');
>> % lowpass filtering
>> h = f special('gaussian',7,1);
>> L = imfilter(N,h,'same','conv');
>> E = imsubtract (I,L);
>> mse = (std2(E))^2
>> figure('name','Lowpass'), imshow(uint8(L),'truesize');
```

We now compare the performance of the lowpass filter with that of the Wiener filter. Figure 10.34 (right) shows the result of applying the Wiener filter to the image z. The MSE between the filtered image and the original image is now 58.58.

Figure 10.34 Left: Image after lowpass filtering. Right: Image after Wiener filtering.

The Wiener filter can be implemented with the function **wiener2**. The following lines will generate Fig. 10.34 (right):

```
>> I = imread('18.tif');
>> figure('name','original'),imshow(uint8(I),'truesize');
>> %% add noise
>> N = imnoise (I,'gaussian',0,5e-3);
>> E = imsubtract (I,N);
>> mse = (std2(E))^2
>> figure('name','noisy'),imshow(uint8(N),'truesize');
>> %Wiener filtering
>> W = wiener2(N,[3 3],5e-2);
>> E = imsubtract (I,W);
>> mse = (std2(E))^2
>> figure('name','Wiener'),imshow(uint8(W),'truesize');
```

The performance of the Wiener filter is comparable to the lowpass filter in terms of MSE. A visual inspection confirms that the lowpass filtered and Wiener filtered images are very similar. Lowpass filtering works well if the spectrum of the noise is spread across all frequencies. It does not work well with impulse noise that creates a "salt and pepper" appearance on the image. Lowpass filtering will also smear the edges in the image. Because edges provide such important visual cues, it would be desirable to have some means of removing noise while preserving the edges. As is explained in the next section, this is possible with a nonlinear filter.

The Median Filter

We first recall the concept of the median of a sequence of numbers. Let $x_0, x_1, \cdots, x_{N-1}$ be a sequence of N numbers. We first rank the sequence x_k. Let $k_0, k_1, \cdots, k_{N-1}$ be the indices of x_k once they are ordered as

$$x_{k_0} \leq x_{k_1} \leq \cdots \leq x_{k_{N-1}}. \tag{10.78}$$

The median of the sequence is defined as

$$\begin{cases} x_{k_n} & \text{if } N = 2n - 1 \\ \frac{1}{2}(x_{k_{n-1}} + x_{k_n}) & \text{if } N = 2n. \end{cases} \tag{10.79}$$

In other words, the median is the number x_k that lies in the middle of the sequence, after all numbers have been ranked. As is obvious from this definition, the computation of the median requires an algorithm that can select the $\frac{N}{2}$th largest number in an array of numbers. Such an algorithm has a complexity very similar to a sorting algorithm.

We can now define the median filtering operation. We consider an image x. Around each pixel (n_1, n_2), we consider a window of size $(2\varepsilon_1 + 1) \times (2\varepsilon_2 + 1)$. The output of the median filter at position (n_1, n_2) is the median of the intensity values in the window. The output can be written as

$$y(n_1, n_2) = \text{median}\{x(m_1, m_2);\ n_1 - \varepsilon_1 \leq m_1 \leq n_1 + \varepsilon_1; n_2 - \varepsilon_2 \leq m_2 \leq n_2 + \varepsilon_2 \tag{10.80}$$

The median filter is computed by sliding the window over the entire image.

▶ **EXAMPLE 10.1** *Figure 10.35 shows a 3×3 window, centered around a pixel where the intensity is 195. First, all intensity values inside the window are ranked. The median value is then found and is equal to 185. This value is assigned to the center of the window.*

We now describe two important properties of the median filter: its ability to preserve edges and to remove impulses.

Proposition 10.1 *The intensity in a neighborhood of a step edge, between two regions where the intensity is constant, is left unchanged by a median filter.*

The proof is left as an exercise. It is important to realize that this property does not hold for a lowpass filter. Now let us define an impulse in a neighborhood as a region

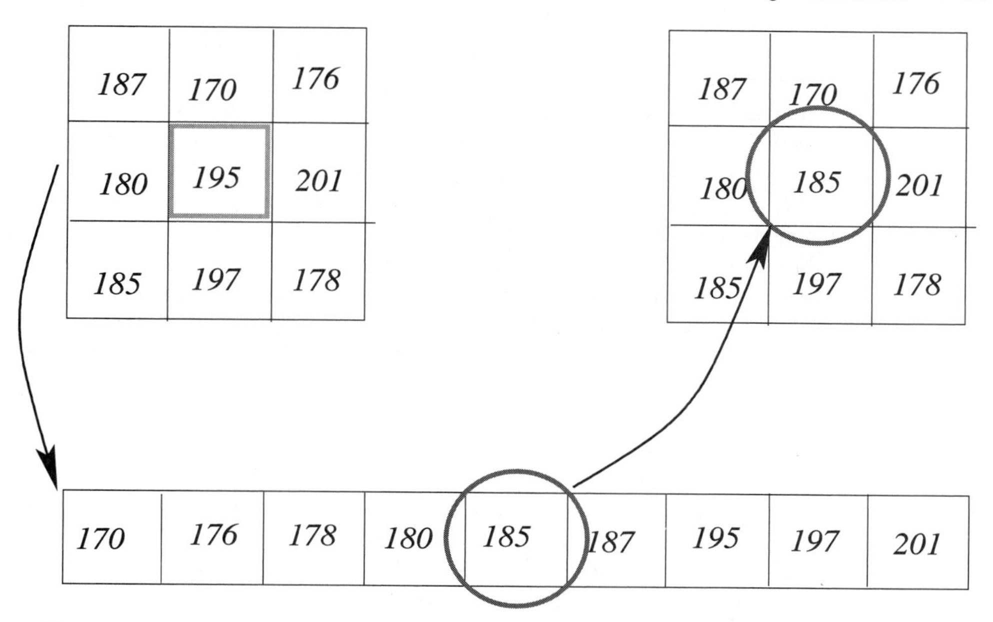

Figure 10.35 Example of the median filtering concept.

where the intensity is smooth everywhere except at a pixel where the intensity function is discontinuous, that is, significantly smaller or larger than the neighboring pixels. The median filter also has the following property.

Proposition 10.2 *An impulse is removed by the application of a median filter.*

The proof is left as an exercise. The reader can in fact generalize the result to a small number of impulses, and find the largest number of impulses that can be removed inside a window of size $(2\varepsilon_1 + 1) \times (2\varepsilon_2 + 1)$. The median filter can be implemented with the function **medfilt2**.

Figure 10.37 illustrates the properties of the median filter in one dimension. The figure shows the graph of the intensity in the image in Fig. 10.36 along the horizontal line defined by $y = 289, x \epsilon [250, 315]$. The intensity of the original image is shown on both graphs as the same dashed line. The graph on the left shows the plot of the intensity of the image after lowpass filtering with a Gaussian kernel of size 9×9. The graph on the right shows the plot of the intensity of the image after median filtering with a window of size 3×3. The lowpass filter has smoothed the edges around the points 29 and 35. It has also removed all the dips in the signal, and consequently the baseline intensity has dropped. The median filter has preserved the edges. It has removed the first and last dips but has preserved the dip around 25. A key element of the median filter is the size of the window. As the window gets larger, the filter can remove impulses that are "wider." It will also smooth more and remove more high-frequency details of the image.

We first examine the performance of the median filter on the noisy image that was corrupted by white Gaussian noise and was shown in Fig. 10.33. Figure 10.38 shows the image after median filtering. The mean square error is 65.95. This is comparable to the lowpass filtered image, and as expected, is not as good as the Wiener filter. A plot of the intensity along a horizontal line in the region of the necklace (defined by $\{y = 289,\ x \in [250, 315]\}$) reveals that the median filter has better recovered the contrast between the necklace and the dress.

A completely different type of noise is the binary (or impulse) noise that randomly selects a few pixels in the image and sets the intensity of those pixels to 0 or M (the largest intensity value) in a random fashion. Binary noise is often coined "salt and pepper" because of its visual appearance on images. Lowpass filtering performs poorly with binary noise. The spikes and small values cannot be detected by the filter, but rather are smeared in the image. As explained earlier, the median filter can efficiently remove this type of noise.

We examine the performance of the median filter on a noisy image, shown in Fig. 10.39, that was corrupted by binary noise. The mean square error between the original image and the noisy image is equal to 598.27.

Figure 10.39 (right) shows the lowpass filtered image. The mean square error has dropped to 101.79. Most of the noise has been removed, but the filtered image is blurred. The sharp edges of the texture on the rock have been smoothed. The Wiener filter performs very poorly, and the binary noise has not been entirely removed as shown in Fig. 10.40 (left). The MSE is 158.83.

Figure 10.36 The profile of the intensity in a region around the necklace is shown in a subsequent figure.

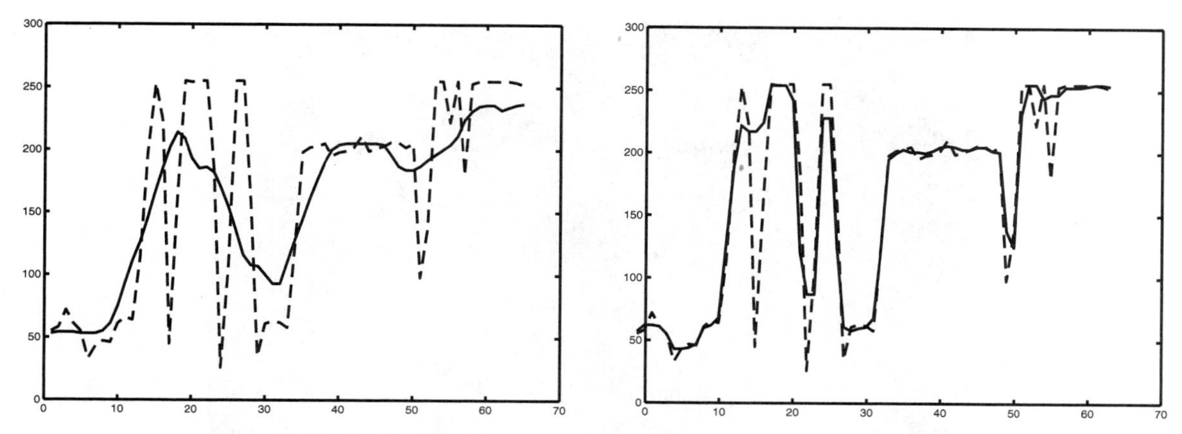

Figure 10.37 Left: Plot of the intensity after lowpass filtering. Right: Plot of intensity after median filtering.

A visual inspection of the median filtered image shows that the filter has removed most of the binary noise and has preserved the sharpness of the edges everywhere in the image. The filter used a window of size 3×3. Some fine details such as the fine gaps between the beads on the necklace, or the texture of the grass, have been removed. The restored image is very sharp, and the MSE has dropped to 82.94.

10.5 COLOR IMAGING

The subject of color imaging requires an understanding of the physics of light as well as the physiology of color perception.

Light and Pigments

Visible light is composed of electromagnetic waves with a wavelength that ranges from 400 to 700 nanometers. As was demonstrated by Isaac Newton in 1704, white light can be split into its monochromatic components with a prism. The spectrum of white light ranges from infrared to ultraviolet light. A body that is illuminated by a light source absorbs part of the light and reflects the rest of the light. A body illuminated by a white light source appears red if the *pigments* on the surface of the object absorb the middle and high frequencies and reflect the low frequencies (red light). The spectrum of the light reflected back is at the origin of our perception of color. We can already see that there are two ways to characterize the color of an object: the color of the pigments (i.e., the color of the light that is reflected) or the color of the light that is absorbed (i.e., the complementary of the pigment color). These two points of view lead to two different color models in image processing.

Color Perception

The principles underpinning the representation of color in digital imaging are directly related to our perception of color. In order to understand the rationale behind these

Figure 10.38 Median filtered image.

principles, we need to describe how the human visual system measures and processes color information. Our retina is composed of a dense array of four types of receptors, namely, *rods* and three types of *cones*. Rods are sensitive to very dim light but do not provide any color information. Rods are responsible for our ability to see at night with a very dim light. Cones contain pigments that are sensitive to three different types of wavelength. Long before anyone observed cones and studied their properties, Thomas Young (in 1802) formulated the hypothesis that the retina must possess very small red,

Figure 10.39 Left: Image corrupted by binary noise, SSE = 598.27. Right: Lowpass filtered image, SSE = 101.79.

green, and violet detectors. The peak absorption for each type of cone occurs at the following wavelengths (see Fig. 10.41).

- *430 nanometers*: The light is blue-purple at this wavelength, and the cones are called blue.
- *530 nanometers*: This light is yellow-green, and the cones are called green cones.
- *560 nanometers*: The light at this wavelength is yellow-orange, and the cones are called red cones.

The actual color of the cones, given by the color of their pigments, is the color complementary to their name. For instance, the blue cones would look red, the green cones would look yellow, and the red cones would look magenta. Although the measurement of color at the level of the retina utilizes three separate channels calibrated for the detection of red, green, and blue light, the processing of the color in the central nervous system relies on a completely different representation. It appears that in the brain the color is represented with the following three components [6]:

Figure 10.40 Left: Wiener filtering, MSE = 158.83. Right: Median filtering, MSE = 82.94.

1. A component that measures the grayscale, or luminance, from black to white.
2. A component that measures a "red to green" component.
3. A component that measures a "yellow to blue" component.

This theory was first suggested by Ewald Hering (1834–1918) but was only confirmed much later [6]. The rationale for this theory is that blue and yellow are mutually antagonistic. They correspond to opposite extremes of the same axis, and they cancel each other to give white when mixed in equal quantity. This is illustrated in Fig. 10.42. Similarly, magenta and green are complementary. The mixture of primary colors can be obtained by mixing the lights from two different flashlights that are covered with, for instance, red and green pieces of cellophane. On the other hand, the mixture of pigments can be obtained by mixing paints of different colors.

Trichromacy and the Reproduction of Color

Let $a_i(\Omega)$ be the absorption of the cone i at frequency Ω. As shown in Fig. 10.41, the cones have absorption or sensitivity curves a_i that overlap. If a blue-purple light hits the retina,

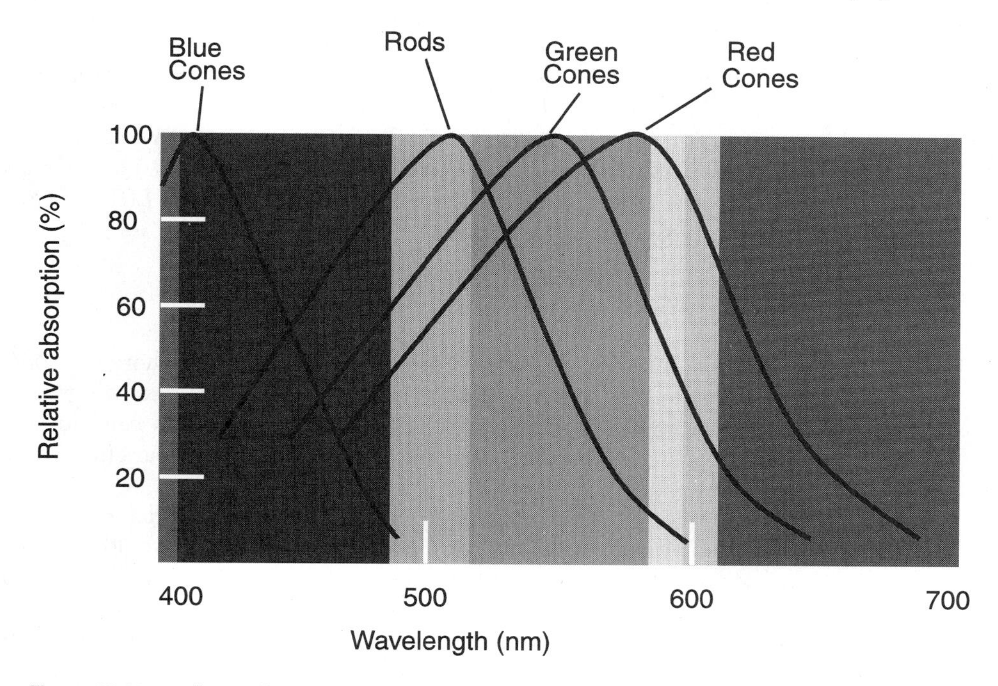

Figure 10.41 Relative absorption (or sensitivity) of the blue, green, and red cones as a function of the wavelength.

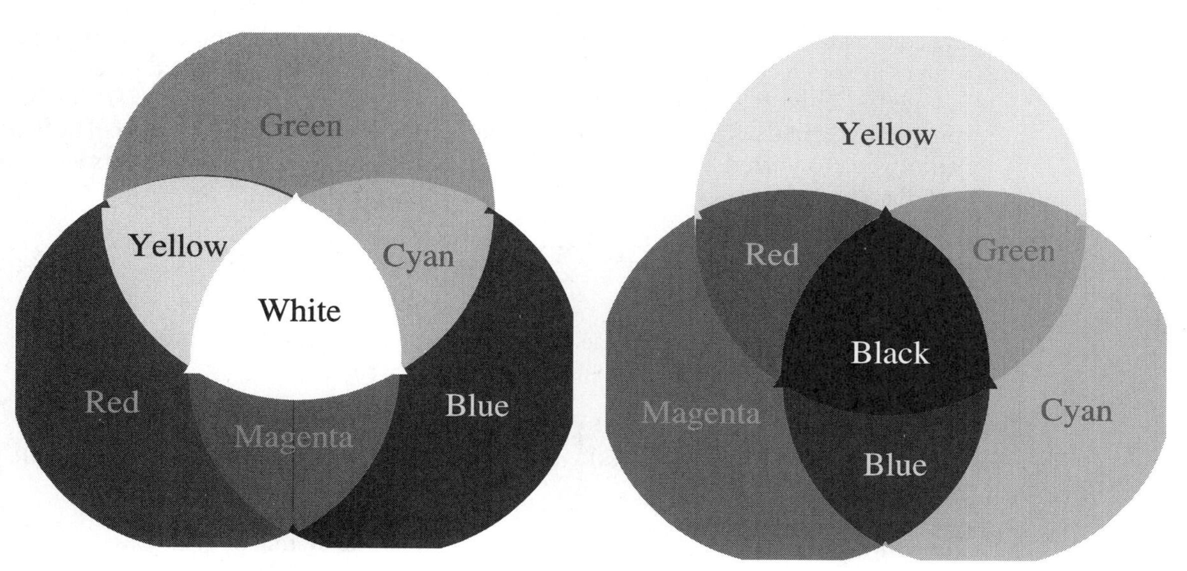

Figure 10.42 Left: The three primary colors and their combinations. Right: The primary colors of pigments and their combinations.

most of the light will be absorbed by the blue cones, and the result will be a sensation of blue-purple. If a yellow light hits the retina, then both the green and red cones will absorb the yellow light. The combination of the stimuli coming from these two types of cones will result in a sensation of yellow. Our sensation of color is the consequence of the unequal stimulation of the three types of cones. Let $r_i(L)$ be the response generated by the cone i when irradiated by a light with a spectrum $L(\Omega)$. If we assume a linear relationship, we can then write

$$r_i(L) = \int a_i(\Omega)L(\Omega)d\Omega, \tag{10.81}$$

where the integral is computed over all the frequencies where the cone i absorbs light. This relationship assumes that the intensity of the response will grow linearly with the amplitude of the incoming light. In reality, the response can saturate after the light intensity reaches a certain level. The saturation is a nonlinear effect that is not described by (10.81).

Our visual system is equipped with just three types of detectors to resolve the entire spectrum of visible light. Obviously, many more detectors would be necessary to perform a fine frequency analysis of the light received on the retina. In fact, our visual system is not even able to analyze the different components of a mixture of lights. Purple is never perceived as a mixture of red and blue. Because the visual system needs to process two large two-dimensional arrays of receptors (the two retinas), it has been designed with rudimentary frequency analysis capabilities. One can compare the visual system to the auditory system that performs a very detailed frequency analysis of sounds, and one can resolve with great precision the different frequency components of a combination of sounds. Because our sensation of color depends only on three detectors, one can generate any color with a combination of three *primary* color sources. The three primary color sources are often chosen to have their peak spectrum in the red, green, and blue.

Let $c_k(\Omega)$ be the spectrum of one of the primary color sources. Let $L(\Omega)$ be the spectrum of light that irradiates the retina, and let $r_i(L)$ be the response generated by the cone i when irradiated by the light L. We need to find three positive coefficients $w_k(L)$, such that the mixture

$$\sum_{k=1}^{3} w_k(L)c_k(\Omega) \tag{10.82}$$

reproduces the light L, in the sense that the mixture generates the same response $r_i(L)$ on the cone i. The response to the mixture of primary color is

$$\int a_i(\Omega)\left\{\sum_{k=1}^{3} w_k(L)c_k(\Omega)\right\} d\Omega = \sum_{k=1}^{3} w_k(L) \int a_i(\Omega)c_k(\Omega)d\Omega. \tag{10.83}$$

Therefore, the weights $w_k(L)$ are the solution of the linear system of equations

$$\sum_{k=1}^{3} w_k(L) \int a_i(\Omega)c_k(\Omega)d\Omega = r_i(L), \quad i = 1, 2, 3. \tag{10.84}$$

The Commission Internationale de l'Éclairage (CIE) chose the following light sources to be the three primary colors:

$$\begin{aligned} c_1(\Omega) &= \delta(\Omega - \Omega_1) \text{ with } \Omega_1 = 435.8 \text{ nm (blue)} \\ c_2(\Omega) &= \delta(\Omega - \Omega_2) \text{ with } \Omega_2 = 546.1 \text{ nm (green)} \\ c_3(\Omega) &= \delta(\Omega - \Omega_3) \text{ with } \Omega_3 = 700 \text{ nm (red)}, \end{aligned} \tag{10.85}$$

where δ represents the impulse function. For any given light spectrum $L(\Omega)$, we can find the weights by substituting (10.85) in (10.84). This gives the following set of equations that can be solved:

$$\sum_{k=1}^{3} w_k(L) a_i(\Omega_k) = r_i(L), \quad i = 1, \ldots, 3. \tag{10.86}$$

The three coefficients c_1, c_2, and c_3 are often called the R, G, and B components. The CIE's three primary colors form the basis of the RGB coordinate system. The reference white corresponds to $c_1 = c_2 = c_3 = 1$. If the coefficients c_i are restricted to be positive, then the mixture of primary sources can only generate a subset of all visible colors.

Figure 10.43 shows a color image[1] of two parrots. This image is stored in MATLAB as a $768 \times 512 \times 3$ array that contains the three coefficients c_1, c_2, and c_3. We then obtain three

Figure 10.43 Top (from left to right): Color image of two parrots; Red components of the image. Bottom (from left to right): Green and blue components of the image.

[1]The image is printed in black and white here. The MATLAB code will however display the color images.

images that quantify the blue, green, and red components with the following convention: a large coefficient is coded as white, and a small coefficient is coded as black.

The R, G, and B images shown in the figure can be obtained with the following lines of code:

```
>> I = imread ('parrot.rgb');
>> R = I(:,:,1);
>> G = I(:,:,2);
>> B = I(:,:,3);
>> figure ('name','Red'), imshow(R);
>> figure ('name','Green'), imshow(G);
>> figure ('name','Blue'), imshow(B);
```

The white in the original image (on the face of the left parrot and on the beak of the right parrot) gives rise to large coefficients in the red, green, and blue images. The head of the red parrot yields large red coefficients, and very small green and blue coefficients. The yellow color on the left parrot gives rise to large red and green components. Indeed, as is clear on the absorption curves of the cones (see Fig. 10.41), the yellow color will be absorbed by the green and red cones.

Color Models

Several color models have been defined for the purpose of measuring or reproducing color.

RGB

Most imaging devices (CCD cameras, CRT monitors, etc.) mimic the retina and utilize three separate channels calibrated for the detection of red, green, and blue colors. CRT monitors use red, green, and blue phosphors to render colors at every pixel on the screen. Unfortunately, the RGB model is quite inefficient for the coding of color images. Indeed, the red, green, and blue components have in general the same bandwidth and require the same spatial resolution. An equivalent representation is provided by applying a rotation to the RGB model and computing one luminance (or gray-level) component and two chrominance components. The CIE's X, Y, Z system provides such a representation. The Y component is the luminance component, and the other components are chrominance components. The X, Y, Z system is derived from the R, G, B system with the following linear transform

$$\begin{bmatrix} X \\ Y \\ Z \end{bmatrix} = \begin{bmatrix} 0.49 & 0.31 & 0.2 \\ 0.177 & 0.813 & 0.011 \\ 0 & 0.01 & 0.99 \end{bmatrix} \begin{bmatrix} R \\ G \\ B \end{bmatrix}.$$

It turns out that the visual system processes the chrominance information, with a spatial resolution coarser than the spatial resolution of the luminance information. It then becomes possible to downsample the chrominance components without significantly losing any color information. Most color-coding schemes used for the transmission and storage of images rely on a "luminance + chrominance" representation.

CMY or CMYK

Printers that deposit color pigments on a white paper use a color model that is complementary to the red-green-blue model. In this model, the three primary colors are the colors of the pigments that absorb the red, green, and blue lights, namely, cyan, magenta, and yellow (*CMY*), respectively. The *CMY* system is a subtractive system based on the following relationship to the *RGB* system:

$$\begin{bmatrix} C \\ M \\ Y \end{bmatrix} = \begin{bmatrix} 1 \\ 1 \\ 1 \end{bmatrix} - \begin{bmatrix} R \\ G \\ B \end{bmatrix}. \tag{10.87}$$

In the *CMY* system, the addition of cyan, magenta, and yellow components in equal proportion yields black. In practice, it is easier to have a separate coding scheme (a separate ink cartridge) for the black component. In the *CMYK* system, the black color (*K*) is encoded separately.

YIQ

The *YIQ* color model is used by the NTSC, PAL, and SECAM television standards. *Y* specifies the (grayscale) luminance component. It is primarily responsible for perception of image's brightness. It can also be used as a black-and-white representation of the image. The color information is encoded by the two other coordinates *I* and *Q*, called the chrominance components. These components are primarily responsible for the hue and saturation of the image. The *YIQ* components are obtained from *RGB* by the following linear transformation:

$$\begin{bmatrix} Y \\ I \\ Q \end{bmatrix} = \begin{bmatrix} 0.299 & 0.587 & 0.114 \\ 0.569 & -0.275 & -0.321 \\ 0.212 & -0.523 & 0.311 \end{bmatrix} \begin{bmatrix} R \\ G \\ B \end{bmatrix}. \tag{10.88}$$

Figure 10.44 shows the luminance and chrominance components of the parrot image. The luminance image appears as a grayscale rendering of the original image. Note that the high-frequency features, such as the edges and texture on the feather, are present in the luminance image. On the other hand, the *I* and *Q* images do not contain any high-frequency features. These images could easily be downsampled by a factor of 2 in each direction. This would be very useful for coding images or video.

The matrix transformation above can be implemented with the function **rgb2ntsc**. The following lines will generate Fig. 10.44:

```
>> P = mread('parrot.rgb');
>> YIQ = rgb2ntsc (P);
>> Y = YIQ (:, :, 1);
>> I = YIQ (:, :, 2);
>> Q = YIQ (:, :, 3);
>> figure('name','Y'), imshow(Y);
```

Figure 10.44 Top: Y (luminance) component of the parrot image. Bottom: (from left to right) I and Q (chrominance) components.

```
>> figure('name','I'), imshow(I);
>> figure('name','Q'), imshow(Q);
```

HSV[2]

Other color models rely on the concept of *hue* which specifies the color (as one of the many possible colors of a crayon). The color defined by the hue is then modified according to the following two parameters:

- The saturation that specifies the amount of white that is added to the hue. For instance, one can add white to red to obtain pink.
- The brightness that measures the amount of light that the color emits.

The transformation from RGB to HSV is nonlinear and can be implemented in MATLAB with the function **rgb2hsv**.

[2]A color model that is in terms of hue (or "tint"), saturation (or "shade") and value (or luminance).

Color Enhancement

Color contrast can be enhanced by using the enhancement techniques presented for grayscale images. A direct extension of these methods to color images consists of enhancing each color channel independently of the others. This offers the advantage of controlling the balance between the different channels. An example is shown in Fig. 10.45. The original appears "too blue." In order to enhance the red and green components, the original image is first split into its R, G, and B components. The R and G components are enhanced with gamma correction ($\gamma = 2$), while the blue component is not modified. The enhanced image is reconstructed from the enhanced color components. Notice that the statue of Neptune has recovered the green color of the bronze.

MATLAB Commands

The following lines will generate Fig. 10.45. The images are printed in black and white. However, the following code will display the images in color.

```
>> I = imread('neptunecolor.tif');
>> % split into RGB components
>> R = I (:, :, 1);
>> G = I (:, :, 2);
>> B = I (:, :, 3);
>> % apply gamma correction
>> Rhat = imadjust (R, [], [], .5);
>> Ghat = imadjust (G, [], [], .6);
>> % rebuild color image from RGB components
>> Ihat = Rhat;
>> Ihat (:, :, 2) = Ghat;
>> Ihat (:, :, 3) = B;
```

Color Image Filtering

The restoration and filtering algorithms described in the previous sections can be extended to multichannel color images by processing each channel independently. Figure 10.46 shows a noisy image that was corrupted by binary noise. The mean square error between the original image and the noisy image is equal to 1832. The noisy image was first split into its three R, G, and B components. The noise removal was then performed independently on each component with a median filter. Next, the filtered image was reconstructed from the filtered components. The MSE has dropped to 107.2. The median filter was capable of removing most of the noise.

The approach we have described fails to exploit the correlations between the three channels. If we consider, for instance, the parrot image, we see in Fig. 10.43 that the R, G, and B components are highly correlated. Vector approaches to color filtering consider each color pixel as a three-dimensional vector and process its magnitude and orientation. An example of such filters is discussed in Computer Project 10.4.

Figure 10.45 Top: Original image. Bottom: Enhanced image after gamma correction of the red and green components.

Figure 10.46 Top: image corrupted by binary noise. MSE = 1,832. Bottom: median filtered image. MSE = 107.2.

MATLAB Commands

The following lines will generate Fig. 10.46. The images are printed in black and white. However, the following code will display the images in color.

```
>> I = imread('parrot.tif');
>> % add noise
>> N = imnoise(I,'salt & pepper',3e-1);
>> % into split RGB components
>> R = N (:, :, 1);
>> G = N (:, :, 2);
>> B = N (:, :, 3);
>> % apply median filter
>> Rhat = medfilt2(R,[3 3]);
>> Ghat = medfilt2(G,[3 3]);
>> Bhat = medfilt2(B,[3 3]);
>> % rebuild color image from RGB components
>> Ihat = Rhat;
>> Ihat (:, :, 2) = Ghat;
>> Ihat (:,:,3) = Bhat;
```

10.6 SUMMARY

This chapter began with image enhancement techniques. The goal of these methods is to enhance the perception of the image. These techniques rely either on a pointwise transformation of the intensity or on a combination of low and highpass filters. In particular, we described the unsharp masking filter that permits enhancing the edges and other high-frequency features in the image.

We then addressed the problem of edge detection. The detection of edges, or object boundaries, remains one of the most fundamental problems in image understanding and computer vision. In order to solve the edge detection problem, we need to take into account the noise that is present in the image and to balance two conflicting goals: canceling the noise and minimizing the spatial smoothing. For further details on edge detection, the reader is referred to the excellent book by Olivier Faugeras [4].

Our third topic in this chapter was image restoration. All sensors introduce some amount of noise and suffer from systematic distortions. The goal of restoration is to correct the degradation and remove the noise. Our discussion of restoration started with a presentation of several "deblurring" and inverse filtering techniques. We then studied the problem of noise removal, using linear and nonlinear filters. Additional material on image restoration can be found in the two classic books by Jain [7] and Lim [9]. The topic of nonlinear filtering is extensively covered with many examples in the book by Astola and Kuosmanen [1].

Finally, we gave a short exposition of color imaging. Although the detailed understanding of color perception by the human visual system remains an active area of research, we demonstrated that color images could be represented using three primary colors. The book edited by Gegenfurtner [5] presents the current state of understanding regarding human color vision. A comprehensive reference on color models is provided in the book by Fairchild [3].

▶ 10.7 PROBLEMS

1. Suppose that the probability density function of an image can be modeled by an exponential density

$$P(x) = \begin{cases} \lambda e^{-\lambda x} & \text{if } x \geq 0 \\ 0 & \text{otherwise} \end{cases} \quad (10.89)$$

with $\lambda > 0$. What is the transform T that yields a uniform distribution?

2. Prove equation (10.12).

3. The impulse response of a two-dimensional Gaussian filter is given by

$$h(x, y) = \frac{1}{2\pi\sigma^2} e^{-\frac{x^2+y^2}{2\sigma^2}} \quad (10.90)$$

Show that the filter defined by h is isomorphic; that is, the output of the filter is independent of any rotation applied to the image. Show that the frequency response of the Gaussian filter is given by

$$H(\Omega, \eta) = e^{-\sigma^2(\Omega^2+\eta^2)/2} \quad (10.91)$$

4. The intensity function of an image is defined by

$$h(x, y) = e^{-(x-x_0)^2-(y-y_0)^2}(1 + \cos(\Omega_x x + \Omega_y y)) \quad (10.92)$$

Describe in plain English what the image looks like.

5. Propose an algorithm to characterize the periodic texture in an image.

6. A Gaussian filter is applied to an image. Explain what happens to the spectrum of the filtered image.

7. A 13×13 image is encoded with 4 bits (the intensity of each pixel is an integer in $(0, \cdots, 15)$. The histogram of the image is given by

$$h = [5, 13, 38, 17, 13, 5, 1, 1, 1, 3, 6, 19, 31, 11, 2, 3] \quad (10.93)$$

Compute the equalized histogram.

8. A Magnetic Resonance Image (MRI) is encoded with 14 bits, with pixel values ranging from 0 to 16,383. The image needs to be displayed on a CRT monitor that can only display 256 gray levels: $0, \cdots, 255$.

(a) The first idea is to map the interval [0,16,383] onto [0,255]. Describe the linear transformation that achieves this contrast stretching.

(b) The previous transform does not allow the radiologist to detect the subtle changes in the intensity required for a diagnostic. The manufacturer designs a new system described by the contrast stretching transform shown in Fig. P10.1.

The radiologist can specify C, the center of the interval $[A, B]$, as well as w the length of the interval $[A, B]$, using two knobs. Specify the contrast stretching transform in terms of C and w.

9. A noisy image z is composed of an image f that has been corrupted by some noise v as

$$z(n_1, n_2) = f(n_1, n_2) + v(n_1, n_2). \quad (10.94)$$

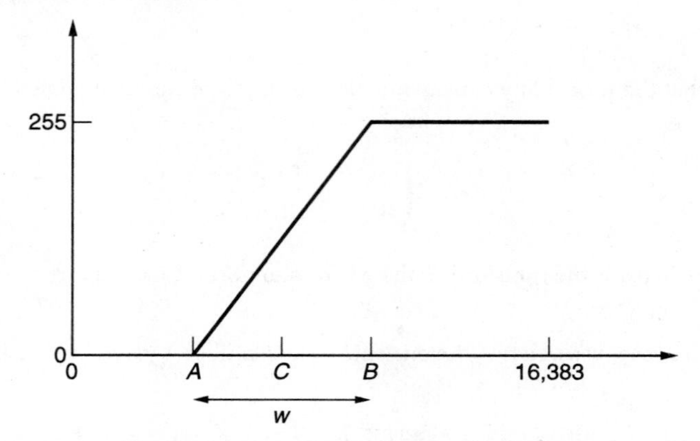

Figure P10.1 Contrast stretching that maps the interval $[A,B]$ onto $[0,255]$.

The spectrum of the noise v is flat. You measure the noisy image z and would like to enhance its features. Analyze the effect of the unsharp masking filter on the image z. *Hint*: You can use the frequency response of the unsharp masking filter (10.18).

10. A transform T is linear if both of the following properties hold:

(a) For any images f and g, $T(f+g) = Tf + Tg$.

(b) For any image f and any real number λ,

$$T(\lambda f) = \lambda T f. \tag{10.95}$$

Prove that the median filter is nonlinear by constructing one counterexample for property a). Prove that property b) holds for the median filler if $\lambda > 0$.

11. Determine if a median filter is shift-invariant or shift-variant.

12. An image is composed of a white square on a black background.

(a) Describe the image formed by applying a Gaussian filter iteratively on the image an infinite number of times. You can ignore the problems on the borders.

(b) Describe the image formed by applying a median filter iteratively on the image an infinite number of times.

13. Let $g_\theta(t)$ be the integral of $f(x, y)$ along the line

$$-\sin\theta x + \cos\theta y = t \tag{10.96}$$

The line is at an angle θ with the x axis and lies at a distance $|t|$ from the origin (see Fig. P10.2). We have

$$g_\theta(t) = \int_{\infty}^{\infty} f(-t\sin\theta + \rho\cos\theta, t\cos\theta + \rho\sin\theta)d\rho \tag{10.97}$$

Prove that the Fourier transform of $g_\theta(t)$ is given by

$$\hat{g_\theta}(\xi) = \hat{f}(-\xi\sin\theta\,, \xi\cos\theta) \tag{10.98}$$

How can we recover $f(x, y)$ from the tomographic projections $g_\theta(t), 0 \leq \theta \leq 2\pi$?

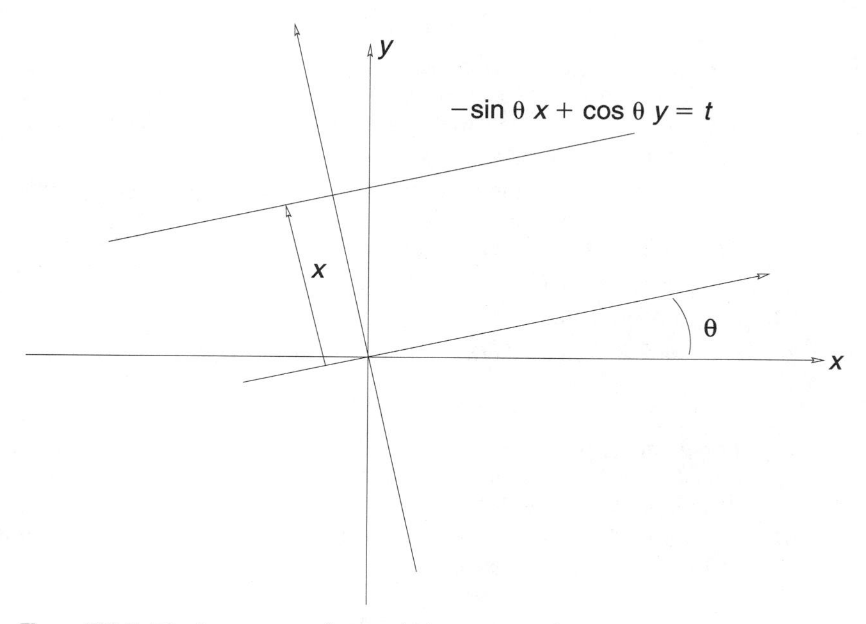

Figure P10.2 The line at an angle θ with the x-axis, and at a distance t from the origin.

14. Prove Proposition 10.1.

15. Prove Proposition 10.2.

16. We consider a one-dimensional median filter. Find the largest number of impulses that can be removed by the median filter inside a window of size ε.

17. The median filter is computed by sliding a window over the image and replacing the center of the window with the median of the intensity inside the window. Other important nonlinear operators can be defined in a similar fashion. We define two *morphological operators* that can be used to filter an image. Let x be an image. Around each pixel (n_1, n_2), we consider a window of size $(2\varepsilon_1 + 1) \times (2\varepsilon_2 + 1)$ that is moved throughout the image. The dilation operator D replaces the intensity at (n_1, n_2) by the maximum of the intensity values inside the window centered around (n_1, n_2) as

$$Dx(n_1, n_2) = \max\{x(m_1, m_2);\ n_1 - \varepsilon_1 \le m_1 \le n_1 + \varepsilon_1;\ n_2 - \varepsilon_2 \le m_2 \le n_2 + \varepsilon_2. \quad (10.99)$$

The erosion operator E replaces the intensity at (n_1, n_2) by the minimum of the intensity values inside the window centered around (n_1, n_2) as

$$Ex(n_1, n_2) = \min\{x(m_1, m_2);\ n_1 - \varepsilon_1 \le m_1 \le n_1 + \varepsilon_1;\ n_2 - \varepsilon_2 \le m_2 \le n_2 + \varepsilon_2 \quad (10.100)$$

Assume that the image is composed of several white objects with intensity 255 on a black background with intensity 0.

(a) Explain in detail the effects of the dilation operator. How does the output of the dilation operator change as a function of the shape of the window (replace the rectangle with a quincunx, for instance).

(b) Explain in detail the effects of the erosion operator.

18. The morphological operators defined in the last problem can be combined to define new operators. Two important operators are defined as follows. The *opening* of an image is obtained by applying an erosion followed by a dilation. The *closing* of an image is obtained by applying an dilation followed by an erosion.

(a) Describe the effect of the opening operator. What happens to the features of the image that are smaller than the window?

(b) Describe the effect of the closing operator. What is the effect of the operator on the contour and area of the object?

19. Suggest some applications of the two morphological operators in Problem 10.17.

20. Suppose that in a neighborhood B of an image the intensity is locally linear as:

$$f(x, y) = a + bx + cy, \quad \forall (x, y) \in B. \tag{10.101}$$

Prove that the response to the Marr-Hildreth edge detector will be zero in this region. You should neglect the response on the border of B. Note that the result holds even if the slope of the plane defined by (10.101) is very steep.

21. The Marr-Hildreth requires the convolution of the image with the Laplacian of the Gaussian. If the Gaussian filter is given by

$$h(x, y) = \frac{1}{2\pi\sigma^2} e^{-\frac{x^2+y^2}{\sigma^2}}, \tag{10.102}$$

compute the Laplacian of the Gaussian.

22. Construct a discrete two-dimensional finite impulse response filter for detecting corners.

23. The power spectrum of an image x is given by

$$S_{xx}(\Omega_1, \Omega_2) = Ae^{-(|\Omega_1| + |\Omega_2|)}. \tag{10.103}$$

The image x is corrupted by some independent white noise with constant power spectrum B. Give the expression of the Wiener filter.

24. Rotation of an image requires the interpolation of its intensity at almost every pixel in the image. The quality of the result depends on the interpolation methods. In order to get accurate results, large 2-D interpolating kernels need to be used.

(a) Prove that any 2-D rotation matrix can be decomposed as follows:

$$\begin{bmatrix} \cos\theta & -\sin\theta \\ \sin\theta & \cos\theta \end{bmatrix} = \begin{bmatrix} 1 & -\tan\theta/2 \\ 0 & 1 \end{bmatrix} \begin{bmatrix} 1 & 0 \\ \sin\theta & 1 \end{bmatrix} \begin{bmatrix} 1 & -\tan\theta/2 \\ 0 & 1 \end{bmatrix}. \tag{10.104}$$

(b) Explain the effect of applying the matrix

$$\begin{bmatrix} 1 & -\tan\theta/2 \\ 0 & 1 \end{bmatrix} \tag{10.105}$$

to an image. Describe a fast implementation of this matrix multiplication.

(c) Explain the effect of applying the matrix

$$\begin{bmatrix} 1 & 0 \\ \sin\theta & 1 \end{bmatrix} \tag{10.106}$$

to an image. Describe a fast implementation of this matrix multiplication.

25. The goal of segmentation is to divide an image into regions within which the pixels have a similar intensity. Color segmentation can be used to divide an image into regions of similar color. Describe a method to perform the color segmentation of an image.

26. The human visual system can only distinguish very few gray levels (about 25) but can discern thousands of different colors. Pseudo-colors are often used to ease the interpretation of a grayscale image by mapping the intensity of a grayscale image to three color components. Describe the map from the gray-level intensity to the R, G, and B components in order to map 2000 gray levels to a spectrum of colors ranging from purple to red.

10.8 COMPUTER PROJECTS

Project 1: The Two-Dimensional Fourier Transform

This project explores the two-dimensional Fourier transform. You will use the image available from the web site.

1. Write a function that computes the two-dimensional Fourier transform of an $N \times N$ image. What symmetries do you expect to find in the matrix that contains the Fourier coefficients when the image is real?
2. Write a function that computes the inverse 2-D Fourier transform. Verify that you can reconstruct exactly the image after a Fourier and an inverse Fourier transform.

This part explores the effect of the phase and the modulus on the Fourier coefficients. You will compute the ρ, θ representation of each complex Fourier coefficient $X[k_1, k_2]$:

$$X[k_1, k_2] = \rho[k_1, k_2] e^{j\theta[k_1, k_2]}. \tag{10.107}$$

1. Display the modulus $\rho[k_1, k_2]$ and the phase $\theta[k_1, k_2]$.
2. Set the phase θ to 0 and reconstruct the image. What do you observe?
3. Set the modulus ρ to 1 and reconstruct the image. What do you observe?

Project 2: Two-Dimensional Orientation Selective Filters

This project explores the design of linear filters in the Fourier domain. We will use a polar representation of the Fourier domain (see Fig. P10.3). You will experiment with several different values of K_0, K_1, K_2.

1. Set all the coefficients with frequencies (k_1, k_2) that are in the two shaded quadrants to zero. Reconstruct the image. What do you observe on the reconstructed image? Describe the features that have been lost in the process.
2. Set all the coefficients whose frequencies are at a distance K_1 from the origin to zero (see Fig. P10.3). Reconstruct the image. What do you observe on the reconstructed image? Describe the features that have been lost in the process. Do you observe any artifacts?
3. The previous filter corresponds to a lowpass filter. We now construct a bandpass filter. Set all the coefficients whose frequencies are in the corona defined by $K_1 \leq \sqrt{k_1^2 + k_2^2} \leq K_2$ (see Fig. P10.3). Reconstruct the image. What do you

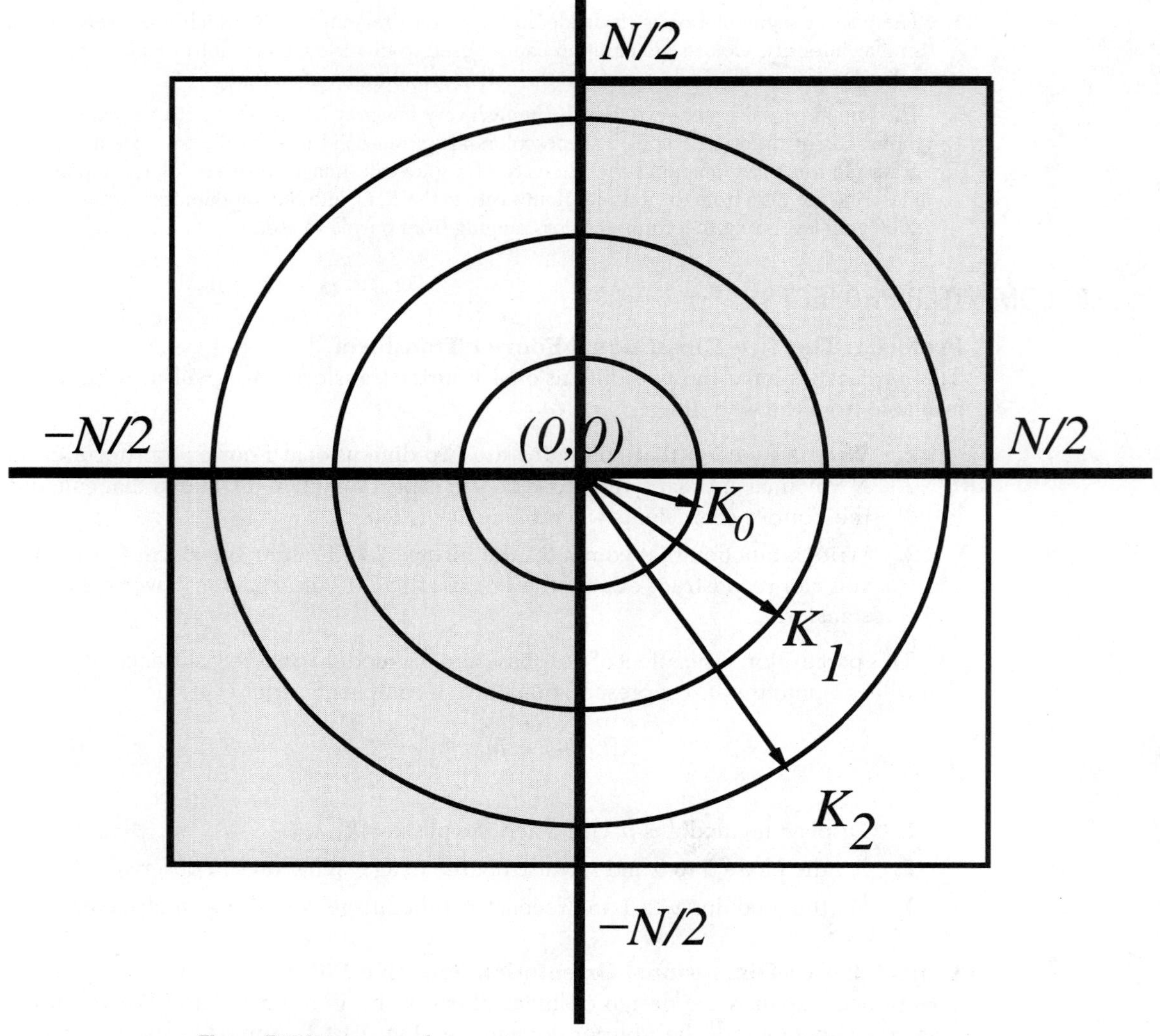

Figure P10.3 Fourier plane.

observe on the reconstructed image? Describe the features that have been lost in the process.

Project 3: The Hough Transform

An edge detector generates an image composed of pixels that are likely to belong to edges in the image. This collection of isolated edge-pixels provides only a first step toward the extraction of edges. The edge-pixels should be linked together to reconstruct curves, or simple geometric primitives such as lines or circles. If the image contains many edges that lie along straight lines, one should be able to characterize each of these lines with two coefficients. This project explores a method to detect straight lines in an image from the collection of isolated edge-pixels generated by an edge detector algorithm.

A straight line in the plane can be characterized by its distance to the origin d, and its slope measured by the angle between the normal to the line, and the x-axis (see Fig. P10.4):

$$x \cos\theta + y \sin\theta = d. \tag{10.108}$$

The angle θ can be chosen in $[0, 2\pi)$ and the distance d in $[0, +\infty)$. Reciprocally, any $(\theta, d) \in [0, 2\pi) \times [0, +\infty)$ uniquely characterizes a straight line defined by (10.108).

Given a single pixel $M_0 = (x_0, y_0)$, we can characterize in the $(\theta,\ d)$ plane the set of all lines that pass through M_0. This set is composed by the graph of the sinusoidal function:

$$d = x_0 \cos\theta + y_0 \sin\theta, \quad \theta \in [0, 2\pi) \tag{10.109}$$

If N pixels $M_i = (x_i, y_i), i = 0, \cdots, N-1$ are aligned along a straight line, then the N sinusoidal functions:

$$d = x_i \cos\theta + y_i \sin\theta, \quad \theta \in [0, 2\pi) \tag{10.110}$$

all intersect at a single point (θ, d) that defines the line that passes through the M_i (see Fig. P10.5).

The principle of the Hough transform is to detect the intersection of several sinusoidal functions. Theoretically, the intersection should be a unique point. In practice, the effect of noise combined with the discretization of θ and d will create some uncertainty regarding the location of the intersection. We consider a discrete array H to represent the (θ, d), plane:

$$H = \{(\theta_k, d_l), \quad k, l = 0, \cdots, K-1\} \tag{10.111}$$

where we can use $\theta_k = 2\pi k/K$, and $d_l = lN/K$.

1. Develop an algorithm that implements the Hough transform, according to the following principles:

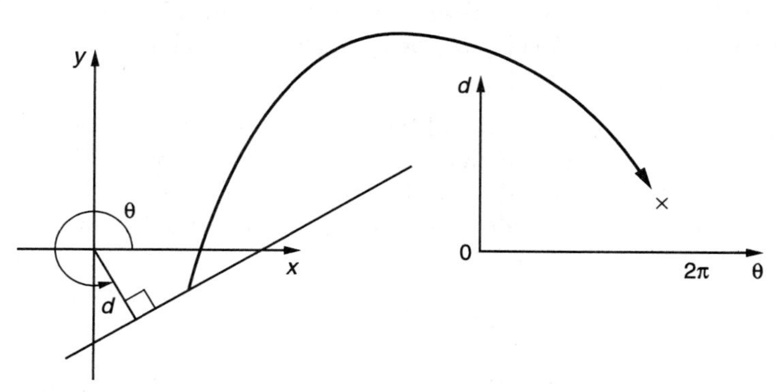

Figure P10.4 Representation of a straight line.

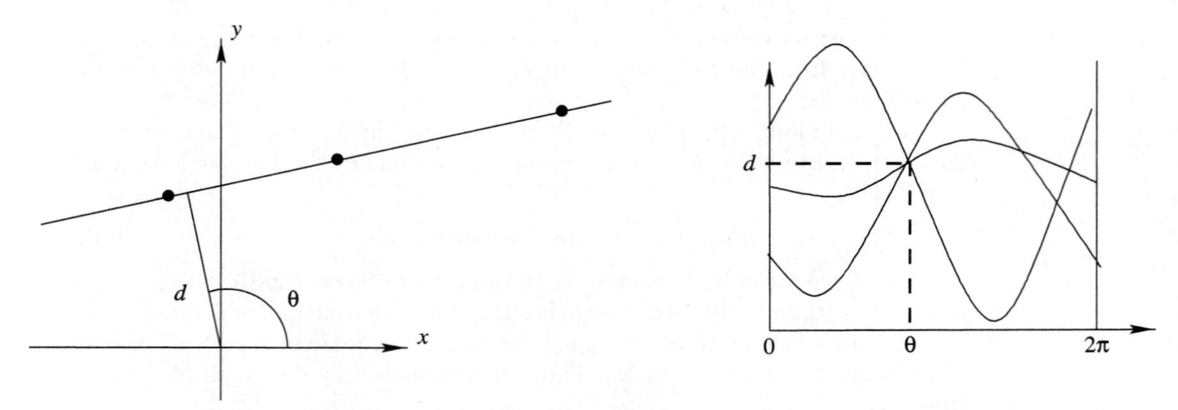

Figure P10.5 The three sinusoidal functions defined by the three aligned points intersect at a single point θ that characterizes the straight line.

For a given point (x,y), consider the set of discrete values (θ_k, d_l) from H that lie on the graph defined by (10.108). For each (θ_k, d_l) that satisfies (10.108), increment a counter for the index (k,l). In practice, the point (θ_k, d_l) may not exactly be on the graph defined by (10.108). Therefore, we replace condition (10.108) by

$$|x \cos\theta_k + y \sin\theta_k - d_l| < \varepsilon, \tag{10.112}$$

where ε is a fixed small number.

2. Verify that applying this algorithm to a set of pixels that are aligned along a straight line yields a large value of the counters at the entry (k,l) that characterizes the straight line.
3. Describe an algorithm to detect all straight lines in an image.
4. Implement and test this algorithm on an image of your choice. Make sure you choose an image so that the result is quite dramatic.

Project 4: Vector Median Filters for Color Image Filtering

In this project, we will develop vector-valued filtering techniques, and we will apply these new filters to color images.

1. We consider a sequence of N scalars $x_0, x_1, \cdots, x_{N-1}$, where N is an odd number. We recall that we defined the median of this sequence to be the number x_{med} that lies in the middle of the sequence, after all numbers have been ranked. Prove that the median can also be defined as the scalar x_{med} in the sequence $x_0, x_1, \cdots, x_{N-1}$ such that

$$x_{\text{med}} \in \{x_0, \cdots, x_{N-1}\},$$

$$\forall k = 0, 1, \ldots, N-1, \sum_{n=0}^{N-1} |x_{\text{med}} - x_n| \leq \sum_{n=0}^{N-1} |x_k - x_n|. \tag{10.113}$$

In other words, x_{med} minimizes the average distance to all numbers in the sequence.

2. We now consider a sequence of N vectors of dimension d, $\mathbf{x}_0, \cdots, \mathbf{x}_{N-1}$, where each $\mathbf{x}_n$ has d components

$$\mathbf{x}_n = \begin{bmatrix} x_n[0] \\ \vdots \\ x_n[d-1] \end{bmatrix}. \tag{10.114}$$

We propose to extend the definition of the median filter to vectors. We define the vector median as follows.

Definition 10.2 *The vector median $\mathbf{x}_{med}$ of $\mathbf{x}_0, \ldots, \mathbf{x}_{N-1}$, is defined by*

$$x_{med} \in \{x_0, \ldots, X_{N-1}\},$$

$$\forall k = 0, 1, \ldots, N-1, \sum_{n=0}^{N-1} ||x_{med} - x_n|| \leq \sum_{n=0}^{N-1} ||x_k - x_n|| \tag{10.115}$$

where the norm $||\cdot||$ is either the l^2 or l^1 norm.

We can now extend the median filtering to vectors. The vector median filtering can be defined by the following algorithm.

A window of size N moves over the image. The window is centered around $\mathbf{x}_{N/2}$. For each position of the window, we perform the following operations:

(a) For each index k, compute the sum of the distances from $\mathbf{x}_k$ to all other vectors $\mathbf{x}_n$ as

$$\sum_{n=0}^{N-1} ||\mathbf{x}_k - \mathbf{x}_n||. \tag{10.116}$$

(b) Find the index^{med} that minimizes the above sum; the vector $\mathbf{x}_{\text{med}}$ is the median.

(c) Replace the vector $\mathbf{x}_{N/2}$ with $\mathbf{x}_{\text{med}}$.

We want to apply the vector median filtering to a color image. In this case, each vector $\mathbf{x}_n$ is a color pixel with three coordinates R, G, and B. We have $d = 3$, and N is the size of the window.

Develop an algorithm in MATLAB to compute the vector median filtering. Compare the performance of the algorithm to the independent median filtering of each component separately. You can plot the performance in terms of MSE as a function of the strength of the noise.

3. What is the optimal choice for the norm $||\cdot||$?

4. A vector is characterized by its magnitude and its orientation. Our definition of vector median relies only on the magnitude differences. Could you extend this definition in such a way that the angles between vectors are also taken into account?

5. Analyze the performance of the vector median filtering equipped with your new definition of the vector median.

▶ BIBLIOGRAPHY

[1] J. Astola and P. Kuosmanen, *Fundamentals of Nonlinear Digital Filtering*, CRC Press, Boca Raton, FL, 1997.

[2] J. Canny, "A computational approach to edge detection," *IEEE Trans. on Pattern Anal. and Machine Intell*, **8** (6) pp. 679–698, 1986.

[3] M.D. Fairchild, *Color Appearance Models*, Addison-Wesley, Reading, MA, 1997.

[4] O. Faugeras, *Three-Dimensional Computer Vision: A Geometric Viewpoint*, MIT Press, Cambridge, MA, 1993.

[5] K.R. Gegenfurtner (ed.), *Color Vision: From Genes to Perception*, Cambridge University Press, New York, 2001.

[6] D. H. Hubel, "Eye, brain, and vision," *Scientific American*, pp. 86–87, 1995.

[7] A.K. Jain, *Fundamentals of Image Processing*, Prentice Hall, Englewood Cliffs, NJ, 1989.

[8] G. Kanizsa, *Organization in Vision: Essays on Gestalt Perception,* Praeger, Westport, CT, 1979.

[9] J.S. Lim, *Two-Dimensional Signal and Image Processing*, Prentice Hall, Englewood Cliffs, NJ, 1990.

[10] S. Mallat, *A Wavelet Tour of Signal Processing*, Academic Press, New York, 1999.

[11] D. Marr, *Vision, A Computational Investigation into the Human Representation and Processing of Visual Information*, W. H. Freeman, 1982.

[12] J.M.S. Prewitt, "Object enhancement and extraction," *Picture Processing and Psychopictorics*, pp. 75–149, 1970.

[13] L.G. Roberts, "Machine perception of three-dimensional solids," *Optical and Electro-optical Information Processing*, pp. 159–197, 1965.

[14] I. Sobel, "Neighbourhood coding of binary images for fast contour following and general array binary processing," *Computer Vision, Graphics and Image Processing*, **8**, pp. 127–135, 1978.

CHAPTER 11

Image Compression and Coding

Digital images typically have thousands of pixels each and require a lot of bits for representation. For example, a 500 × 500 image with 8 bits/pixel requires 2 M bits. Therefore, images must be compressed for two reasons: (1) to save bandwidth during transmission; and (2) to save data space for storage. Many tools and techniques for compressing data are available, some of which have been discussed in Chapter 9 in the context of speech coding. Some of these techniques can be extended to images.

There are many aspects of image compression and coding, not all of which are used in every application. A typical lossy image compression scheme is shown in Fig. 11.1(*a*). The process of representing the image pixels such that some compression is achieved is called *image coding*. This can be composed of one or more of the techniques of image transformation, quantization, and entropy coding. The coded image pixels are then processed by the *channel coder*, where the symbols (or bits) are modulated appropriately for transmission. The channel coder also codes the symbols for error correction. Common modulation techniques include amplitude modulation, frequency modulation, and phase modulation. The block diagram of the receiver is shown in Fig. 11.1(*b*). The received signal is demodulated, and error is corrected in the channel decoder. The resulting symbols are then passed to the image decoder. The image decoder performs entropy-decoding, inverse quantization operation (possibly inverse DPCM), and inverse transformation in order to produce the reconstructed image.

The two types of image compression methods are *lossless* and *lossy* compression. In lossless image compression, the image can be recovered (or reconstructed) without any degradation of the image. There is no loss of information at all. This feature is important in a number of applications including information-coded images for military intelligence, spectral images gathered from space exploration missions, and watermarked images for copyright protection. On the other hand, in numerous other applications lossy compression is quite acceptable. For example, in digital television and Internet images, some loss of information is acceptable as long as the reception quality is good. In fact, a certain amount of degradation is not even perceptible to the human eye. Of

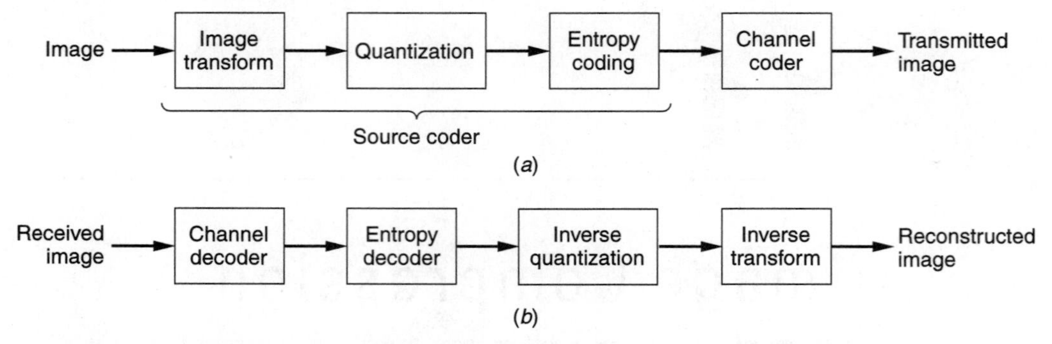

Figure 11.1 (*a*) Image coder; (*b*) image decoder.

course, lossy compression usually yields much higher compression ratios than lossless compression.

One component of image compression is transform coding whereby the image is transformed into another domain, such as the DFT domain. The purpose of this transformation is to take advantage of some characteristics of the image so that the energy of the transformed image is localized into a small number of pixels. In other words, the transformation attempts to decorrelate the pixels of the transformed image. Most images have very uneven frequency distributions, with low frequencies much more common. Therefore, in the DFT domain, the coefficients corresponding to the low frequencies will conceivably be larger in magnitude than those corresponding to high frequencies. The low-frequency pixels can then be coded with a larger number of bits than the others. For lossy compression, all of the coefficients under a specified threshold can be discarded. This is of course one aspect of quantization. Other types of image transforms include the Discrete Cosine Transform (DCT), Discrete Sine Transform (DST), Hadamard Transform (HT), and Karhunen-Loeve Transform (KLT). These transforms have their relative pros and cons.

Further compression can be achieved by quantization methods such as Pulse Code Modulation (PCM) or Differential Pulse Code Modulation (DPCM). In DPCM, the goal is to predict the current pixel from a set of past pixels by using a linear or nonlinear predictor. If the predictor is designed well, then the difference (error) between the predicted pixel and the actual pixel is small. This error signal is encoded in order to achieve compression. The same predictor is used in the decoder in order to reconstruct the image.

The final step in the image compression process is usually entropy coding—the method of assigning codewords to the symbols in such a way that compression is achieved. Typical schemes are Huffman coding and Lempel-Ziv-Welch coding. These lossless coding schemes have become quite popular.

This chapter presents, several different types of image transforms with mathematical derivations and properties. The concepts of image quantization are then briefly reviewed. The technique of 2-D DPCM is then presented in detail. The subsequent section deals with image coding, where we present two popular algorithms, Huffman coding and Lempel-Ziv-Welch coding. The JPEG standards for both lossy and lossless compression are then briefly discussed.

▶ 11.1 IMAGE TRANSFORMS

1-D DFT

The unitary Discrete Fourier Transform (DFT) of a sequence $u(n)$ is defined by

$$V(k) = \frac{1}{\sqrt{N}} \sum_{n=0}^{N-1} u(n) W_N^{kn}, \quad k = 0, 1, \ldots, N-1 \tag{11.1}$$

where $W_N = e^{-j\frac{2\pi}{N}}$. Note that in equation (11.1), we have the factor $\frac{1}{\sqrt{N}}$. This has been introduced to make this transform unitary, as we shall see shortly. Let the sequence $u(n)$ be represented as a vector $\mathbf{u}$ of length-N, and the transformed vector is likewise $\mathbf{v}$. The unitary DFT can then be written as

$$\mathbf{v} = \mathbf{F}\mathbf{u} \tag{11.2}$$

where

$$\mathbf{F} = \frac{1}{\sqrt{N}} \begin{bmatrix} 1 & 1 & 1 & \cdots & 1 \\ 1 & e^{-j\frac{2\pi}{N}} & e^{-j\frac{2\pi}{N}\cdot 2} & \cdots & e^{-j\frac{2\pi}{N}(N-1)} \\ 1 & e^{-j\frac{2\pi}{N}\cdot 2} & e^{-j\frac{2\pi}{N}\cdot 4} & \cdots & e^{-j\frac{2\pi}{N}2(N-1)} \\ \vdots & & & & \\ 1 & e^{-j\frac{2\pi}{N}(N-1)} & e^{-j\frac{2\pi}{N}(N-1)\cdot 2} & \cdots & e^{-j\frac{2\pi}{N}(N-1)(N-1)} \end{bmatrix} \tag{11.3}$$

It turns out that matrix $\mathbf{F}$ is unitary. The concept of unitary matrices was described in detail in Chapter 7; here we repeat the basics for convenience.

Definition 11.1 *A matrix* $\mathbf{A}$ *is said to be unitary if*

$$\mathbf{A}^{\dagger}\mathbf{A} = \mathbf{I} \tag{11.4}$$

where $\dagger$ *represents the conjugate transpose operation and* $\mathbf{I}$ *is the identity matrix.*

Fact 1: A matrix is unitary if and only if its rows (or columns) are orthonormal.

Fact 1 was proved in Chapter 7. It is quite straightforward to show that the DFT kernel matrix $\mathbf{F}$ is unitary. The ith row of $\mathbf{F}$ is

$$\mathbf{q}_i = \frac{1}{\sqrt{N}} \left[1 \quad e^{-j\frac{2\pi}{N}i} \quad e^{-j\frac{2\pi}{N}\cdot 2i} \quad \cdots \quad e^{-j\frac{2\pi}{N}(N-1)i} \right]. \tag{11.5}$$

Therefore,

$$\begin{aligned} \mathbf{q}_i\mathbf{q}_l^{\dagger} &= \frac{1}{N}\left(1 + e^{-j\frac{2\pi}{N}(i-l)} + e^{-j\frac{2\pi}{N}\cdot 2(i-l)} + \cdots + e^{-j\frac{2\pi}{N}(N-1)(i-l)}\right) \\ &= \frac{1}{N} \sum_{m=0}^{N-1} e^{-j\frac{2\pi}{N}mp} \end{aligned} \tag{11.6}$$

where $p = i - l$. By invoking the orthogonality of discrete exponentials, we get

$$\mathbf{q}_i \mathbf{q}_l^\dagger = \frac{1}{N} \begin{cases} N, & p = rN \\ 0, & \text{otherwise.} \end{cases}$$

The above clearly implies that $\mathbf{F}$ is unitary. The inverse DFT is therefore

$$\mathbf{u} = \mathbf{F}^{-1}\mathbf{v} \tag{11.7}$$

$$= \mathbf{F}^{\dagger}\mathbf{v}. \tag{11.8}$$

Since $\mathbf{F}$ is also symmetric,

$$\mathbf{u} = \mathbf{F}^{*}\mathbf{v} \tag{11.9}$$

where $*$ represents the complex conjugate operation. Equation (11.9) can be expanded as

$$u(n) = \frac{1}{\sqrt{N}} \sum_{k=0}^{N-1} v(k) W_N^{-kn}. \tag{11.10}$$

The DFT can be viewed as expanding the input sequence in terms of N basis vectors. This is true for any unitary transformation. Consider the unitary transformation

$$\mathbf{v} = \mathbf{A}\mathbf{u} \tag{11.11}$$

Let

$$\mathbf{A}^T = \begin{bmatrix} \mathbf{a}_0 \, \mathbf{a}_1 & \cdots & \mathbf{a}_{N-1} \end{bmatrix} \tag{11.12}$$

where $\mathbf{a}_i$ is a column vector. Since $\mathbf{A}$ is unitary,

$$\mathbf{u} = \mathbf{A}^{\dagger}\mathbf{v}$$

$$= \begin{bmatrix} \mathbf{a}_0^* \, \mathbf{a}_1^* & \ldots & \mathbf{a}_{N-1}^* \end{bmatrix} \mathbf{v} \tag{11.13}$$

$$= \mathbf{a}_0^* v(0) + \mathbf{a}_1^* v(1) + \cdots + \mathbf{a}_{N-1}^* v(N-1). \tag{11.14}$$

The input is therefore represented in terms of the transform coefficients $v(i)$ and the basis vectors. The basis vectors are the conjugate of the rows of the transform kernel matrix. In other words, the basis vectors $\mathbf{a}_i^*$ are the columns of $\mathbf{A}^{\dagger}$.

▶ **EXAMPLE 11.1** *Consider the 1-D unitary DFT of* $\mathbf{u} = [8\ 1\ 0\ 1]^T$. *The DFT coefficients are given by*

$$\mathbf{v} = \mathbf{F}\mathbf{u}$$

$$= \frac{1}{2} \begin{bmatrix} 1 & 1 & 1 & 1 \\ 1 & -j & -1 & j \\ 1 & -1 & 1 & -1 \\ 1 & j & -1 & -j \end{bmatrix} \begin{bmatrix} 8 \\ 1 \\ 0 \\ 1 \end{bmatrix} = \begin{bmatrix} 5 \\ 4 \\ 3 \\ 4 \end{bmatrix}.$$

The basis vectors are

$$f_0^* = \frac{1}{2}\begin{bmatrix}1\\1\\1\\1\end{bmatrix}, \quad f_1^* = \frac{1}{2}\begin{bmatrix}1\\j\\-1\\-j\end{bmatrix}, \quad f_2^* = \frac{1}{2}\begin{bmatrix}1\\-1\\1\\-1\end{bmatrix}, \quad f_3^* = \frac{1}{2}\begin{bmatrix}1\\-j\\-1\\j\end{bmatrix}.$$

The input can therefore be represented as

$$\begin{bmatrix}8\\1\\0\\1\end{bmatrix} = 5f_0^* + 4f_1^* + 3f_2^* + 4f_3^* \tag{11.15}$$

which can be easily verified by substituting the basis vectors.

Now we address the reason for using a unitary transformation. Clearly, orthnormality is an inherent property of a unitary matrix. This property leads to minimization of the sum of square errors (SSE) if the representation of the input as in (11.14) is truncated.

General Orthonormal Transform

The general orthonormal transform pair is given by

$$v(k) = \sum_{n=0}^{N-1} u(n)a_k(n), \quad 0 \le k \le N-1, \tag{11.16}$$

$$u(n) = \sum_{k=0}^{N-1} v(k)a_k^*(n), \quad 0 \le n \le N-1, \tag{11.17}$$

where

$$\sum_{n=0}^{N-1} a_k(n)a_{k'}^*(n) = \delta(k-k'). \tag{11.18}$$

Let the series expansion of the input given by (11.17) be truncated to P terms as

$$u_P(n) = \sum_{k=0}^{P-1} v(k)a_k^*(n), \quad P \le N. \tag{11.19}$$

In other words, the input is now reconstructed from a subset of the transform coefficients. This is typical in image compression applications, where the coefficients with small magnitude are not transmitted (or coded). Now we define the sum of squares error (SSE) between $u(n)$ and $u_P(n)$ as

$$S = \sum_{n=0}^{N-1} |u(n) - u_P(n)|^2. \tag{11.20}$$

The following result is a powerful motivation for using orthonormal transforms.

Theorem 11.1 *The orthonormality property assures that the sum of squares error, S, is minimized in an orthonormal transform.*

Proof. The SSE is

$$
\begin{aligned}
S &= \sum_{n=0}^{N-1} |(u(n) - u_P(n))|^2 \\
&= \sum_{n=0}^{N-1} \left| (u(n) - \sum_{k=0}^{P-1} v(k) a_k^*(n)) \right|^2 \\
&= \sum_{n=0}^{N-1} \left[u(n) - \sum_{k=0}^{P-1} v(k) a_k^*(n) \right] \left[u^*(n) - \sum_{k'=0}^{P-1} v^*(k') a_{k'}(n) \right] \\
&= \sum_{n=0}^{N-1} \left[|u(n)|^2 - u(n) \sum_{k'=0}^{P-1} v^*(k') a_{k'}(n) - u^*(n) \sum_{k=0}^{P-1} v(k) a_k^*(n) \right] \\
&\quad + \sum_{n=0}^{N-1} \left[\left(\sum_{k=0}^{P-1} v(k) a_k^*(n) \right) \left(\sum_{k'=0}^{P-1} v^*(k') a_{k'}(n) \right) \right].
\end{aligned}
\tag{11.21}
$$

The derivative of the SSE can be found as

$$
\begin{aligned}
\frac{\partial S}{\partial v(k_0)} &= - \sum_{n=0}^{N-1} u(n) a_{k_0}(n) - \sum_{n=0}^{N-1} u^*(n) a_{k_0}^*(n) \\
&\quad + \sum_{n=0}^{N-1} \sum_{k=0}^{P-1} v(k) a_k^*(n) a_{k_0}(n) + \sum_{n=0}^{N-1} \sum_{k'=0}^{P-1} v^*(k') a_{k'}(n) a_{k_0}^*(n).
\end{aligned}
\tag{11.22}
$$

Setting the derivative equal to zero gives the optimum coefficients,

$$
\frac{\partial S}{\partial v(k_0)} = 0.
$$

The above gives

$$
\begin{aligned}
&\sum_{n=0}^{N-1} u(n) a_{k_0}(n) + \sum_{n=0}^{N-1} u^*(n) a_{k_0}^*(n) \\
&= \underbrace{\sum_{k=0}^{P-1} v(k) \sum_{n=0}^{N-1} a_k^*(n) a_{k_0}(n)}_{v(k_0)} + \underbrace{\sum_{k'=0}^{P-1} v^*(k') \sum_{n=0}^{N-1} a_{k'}(n) a_{k_0}^*(n)}_{v^*(k_0)}.
\end{aligned}
\tag{11.23}
$$

Using the orthonormality property, we have

$$\sum_{n=0}^{N-1} u(n)a_{k_0}(n) + \sum_{n=0}^{N-1} u^*(n)a_{k_0}^*(n) = v(k_0) + v^*(k_0). \tag{11.24}$$

Equating the real and the conjugate terms, we get

$$v(k_0) = \sum_{n=0}^{N-1} u(n)a_{k_0}(n), \quad k_0 = 0, 1, \ldots, P-1. \tag{11.25}$$

■

Theorem 11.1 states that no other coefficients can yield a lower SSE than the unitary transform coefficients.

2-D DFT

Now let us revisit the 2-D DFT algorithm we established in Chapter 2. It was shown that given a 2-D signal (image) matrix $\mathbf{U}$, its 2-D DFT can be obtained efficiently by first taking the 1-D DFT of each of its columns and then performing the 1-D DFT of the rows of the resulting matrix. To express this mathematically, let the $N_1 \times N_2$ input matrix be

$$\mathbf{U} = \begin{bmatrix} \mathbf{u}_0 & \mathbf{u}_1 & \cdots & \mathbf{u}_{N_2-1} \end{bmatrix}, \tag{11.26}$$

where $\mathbf{u}_i$ is the ith column of the matrix. Let $\mathbf{F}_{N_1}$ represent the $N_1 \times N_1$, 1-D DFT kernel matrix. The above algorithm suggests that we first perform

$$\begin{aligned} \mathbf{G} &= \begin{bmatrix} \mathbf{F}_{N_1}\mathbf{u}_0 & \mathbf{F}_{N_1}\mathbf{u}_1 & \cdots & \mathbf{F}_{N_1}\mathbf{u}_{N_2-1} \end{bmatrix} \\ &= \mathbf{F}_{N_1}\begin{bmatrix} \mathbf{u}_0 & \mathbf{u}_1 & \cdots & \mathbf{u}_{N_2-1} \end{bmatrix} \\ &\triangleq \begin{bmatrix} \mathbf{g}_0 \\ \mathbf{g}_1 \\ \vdots \\ \mathbf{g}_{N_1-1} \end{bmatrix}, \end{aligned} \tag{11.27}$$

where $\mathbf{g}_i$ is the ith row of $\mathbf{G}$. Next we transform each row of $\mathbf{G}$ with the $N_2 \times N_2$ 1-D DFT kernel matrix $\mathbf{F}_{N_2}$.

$$\mathbf{V} = \begin{bmatrix} (\mathbf{F}_{N_2}\mathbf{g}_0^T)^T \\ (\mathbf{F}_{N_2}\mathbf{g}_1^T)^T \\ \vdots \\ (\mathbf{F}_{N_2}\mathbf{g}_{N_1-1}^T)^T \end{bmatrix} \tag{11.28}$$

$$\mathbf{V} = \begin{bmatrix} \mathbf{g}_0\mathbf{F}_{N_2}^T \\ \mathbf{g}_1\mathbf{F}_{N_2}^T \\ \vdots \\ \mathbf{g}_{N_1-1}\mathbf{F}_{N_2}^T \end{bmatrix}$$

$$= \mathbf{F}_{N_1}\mathbf{U}\mathbf{F}_{N_2}^T. \tag{11.29}$$

We saw in Chapter 2 that this row-column decomposition algorithm works because the 2-D DFT is a separable transform. Indeed, this method works for any 2-D separable transform, not just the DFT.

For any given unitary matrices $\mathbf{A}_{N_1}$ and $\mathbf{A}_{N_2}$, the transform pair is

$$\mathbf{V} = \mathbf{A}_{N_1}\mathbf{U}\mathbf{A}_{N_2}^T \tag{11.30}$$

$$\mathbf{U} = \mathbf{A}_{N_1}^{\dagger}\mathbf{V}\mathbf{A}_{N_2}^{*}. \tag{11.31}$$

Now we proceed to represent an $N \times N$ matrix $\mathbf{U}$ as a linear combination of basis images. There is a set of N^2 basis images, and we now show how to construct them. Define

$$\mathbf{A}^{\dagger} = \begin{bmatrix} \mathbf{a}_0^* & \mathbf{a}_1^* & \cdots & \mathbf{a}_{N-1}^* \end{bmatrix} \tag{11.32}$$

where $\mathbf{a}_i^*$ is the ith column of $\mathbf{A}^{\dagger}$.

From (11.31), we have

$$\mathbf{U} = \begin{bmatrix} \mathbf{a}_0^* & \mathbf{a}_1^* & \cdots & \mathbf{a}_{N-1}^* \end{bmatrix} \mathbf{V} \begin{bmatrix} \mathbf{a}_0^{*T} \\ \mathbf{a}_1^{*T} \\ \vdots \\ \mathbf{a}_{N-1}^{*T} \end{bmatrix} \tag{11.33}$$

$$= \mathbf{a}_0^*\mathbf{V}_{00}\mathbf{a}_0^{*T} + \mathbf{a}_0^{*T}\mathbf{V}_{01}\mathbf{a}_1^{*T} + \cdots + \mathbf{a}_0^*\mathbf{V}_{0,N-1}\mathbf{a}_{N-1}^{*T} + \cdots$$
$$+ \mathbf{a}_{N-1}^*\mathbf{V}_{N-1,0}\mathbf{a}_0^{*T} + \cdots + \mathbf{a}_{N-1}^*\mathbf{V}_{N-1,N-1}\mathbf{a}_{N-1}^{*T}. \tag{11.34}$$

The basis images are therefore given by

$$\mathbf{A}_{ij}^* = \mathbf{a}_i^*\mathbf{a}_j^{*T}. \tag{11.35}$$

▶ **EXAMPLE 11.2** *Consider the $N = 4$ 2-D DFT. The first 4 of the 16 total basis images are*

$$F^*_{00} = f^*_0 f^{*T}_0 = \frac{1}{4}\begin{bmatrix}1\\1\\1\\1\end{bmatrix}\begin{bmatrix}1 & 1 & 1 & 1\end{bmatrix} = \frac{1}{4}\begin{bmatrix}1 & 1 & 1 & 1\\1 & 1 & 1 & 1\\1 & 1 & 1 & 1\\1 & 1 & 1 & 1\end{bmatrix}$$

$$F^*_{01} = f^*_0 f^{*T}_1 = \frac{1}{4}\begin{bmatrix}1\\1\\1\\1\end{bmatrix}\begin{bmatrix}1 & j & -1 & -j\end{bmatrix} = \frac{1}{4}\begin{bmatrix}1 & j & -1 & -j\\1 & j & -1 & -j\\1 & j & -1 & -j\\1 & j & -1 & -j\end{bmatrix}$$

$$F^*_{02} = f^*_0 f^{*T}_2 = \frac{1}{4}\begin{bmatrix}1\\1\\1\\1\end{bmatrix}\begin{bmatrix}1 & -1 & 1 & -1\end{bmatrix} = \frac{1}{4}\begin{bmatrix}1 & -1 & 1 & -1\\1 & -1 & 1 & -1\\1 & -1 & 1 & -1\\1 & -1 & 1 & -1\end{bmatrix}$$

$$F^*_{03} = f^*_0 f^{*T}_3 = \frac{1}{4}\begin{bmatrix}1\\1\\1\\1\end{bmatrix}\begin{bmatrix}1 & -j & -1 & j\end{bmatrix} = \frac{1}{4}\begin{bmatrix}1 & -j & -1 & j\\1 & -j & -1 & j\\1 & -j & -1 & j\\1 & -j & -1 & j\end{bmatrix}$$

Let the image be

$$\mathbf{U} = \begin{bmatrix}0 & 1 & 0 & 0\\0 & 1 & 0 & 0\\0 & 1 & 0 & 0\\0 & 1 & 0 & 0\end{bmatrix}.$$

The 2-D DFT is given by

$$\mathbf{V} = \begin{bmatrix}1 & -j & -1 & j\\0 & 0 & 0 & 0\\0 & 0 & 0 & 0\\0 & 0 & 0 & 0\end{bmatrix}.$$

Since **V** *has four nonzero entries,* **U** *can be expressed in terms of 4 of the 16 basis images as*

$$\mathbf{U} = (1)F^*_{00} + (-j)F^*_{01} + (-1)F^*_{02} + (j)F^*_{03}$$
$$= \begin{bmatrix} 0 & 1 & 0 & 0 \\ 0 & 1 & 0 & 0 \\ 0 & 1 & 0 & 0 \\ 0 & 1 & 0 & 0 \end{bmatrix}.$$

Now consider the 2-D DFT of an image. The Fourier spectrum of an image has a very large dynamic range, typically higher than the display device. Therefore, if we plot this spectrum as an image, only the very bright spots show up on the screen. Because of this difficulty, the DFT spectrum $|F(k_1, k_2)|$ is often plotted in the following scale:

$$D(k_1, k_2) = \log_{10}[1 + |F(k_1, k_2)|]. \tag{11.36}$$

This will be referred to as the log-magnitude spectrum, and it performs the compression of the dynamic range. Consider the "Lena" image of Fig. 11.2(a) as an example. Part (b) of this figure shows the phase of the image. The log-magnitude spectrum of the unitary DFT of the image is shown in part (c). The low frequencies are at the corners of the image. We can use the MATLAB command **fftshift** to move the low frequencies to the center, which gives part (d) of the figure. Very few coefficients have such a large magnitude, and they are all located at the lower frequencies. So, let us investigate what happens if we use only a few of the large coefficients to reconstruct the image using the inverse DFT. This is done in Fig. 11.3. The original image is 512 × 512, and so is its DFT. We now use the central $M \times M$ coefficients of the (centrally) shifted DFT to reconstruct the image. The other coefficients are simply assumed to be zero. The results for $M =$ 60, 80, and 120 are given in parts (b), (c), and (d), respectively, of Fig. 11.3. As we increase the number of DFT coefficients in our reconstruction, the image gets better. For $M = 120$, the reconstructed image is almost as good as the original. If we transmitted the original image, it would require sending 512 × 512 elements. In this case, we are only sending 120 × 120 complex coefficients. This represents a compression ratio of approximately 9. This is a very simple example where we assume that all the coefficients are coded with the same number of bits, which is not usually done in practice. Nevertheless, this example demonstrates the power of transform coding.

1-D DCT

The Discrete Cosine Transform (DCT) is similar to the DFT. There are several motivations for this transform notably: (a) the DCT coefficients are purely real; (b) the DCT has near-optimal property for energy compaction; and (c) the DCT can be computed efficiently using fast algorithms that are computationally comparable to the FFT. As a result of these advantages, the DCT is used extensively in image compression.

The DCT can be formatted in the following way. Consider a length-N sequence $x(n)$, an example of which is shown in Fig. 11.4(a). The first step is to make this sequence symmetric by the following mapping:

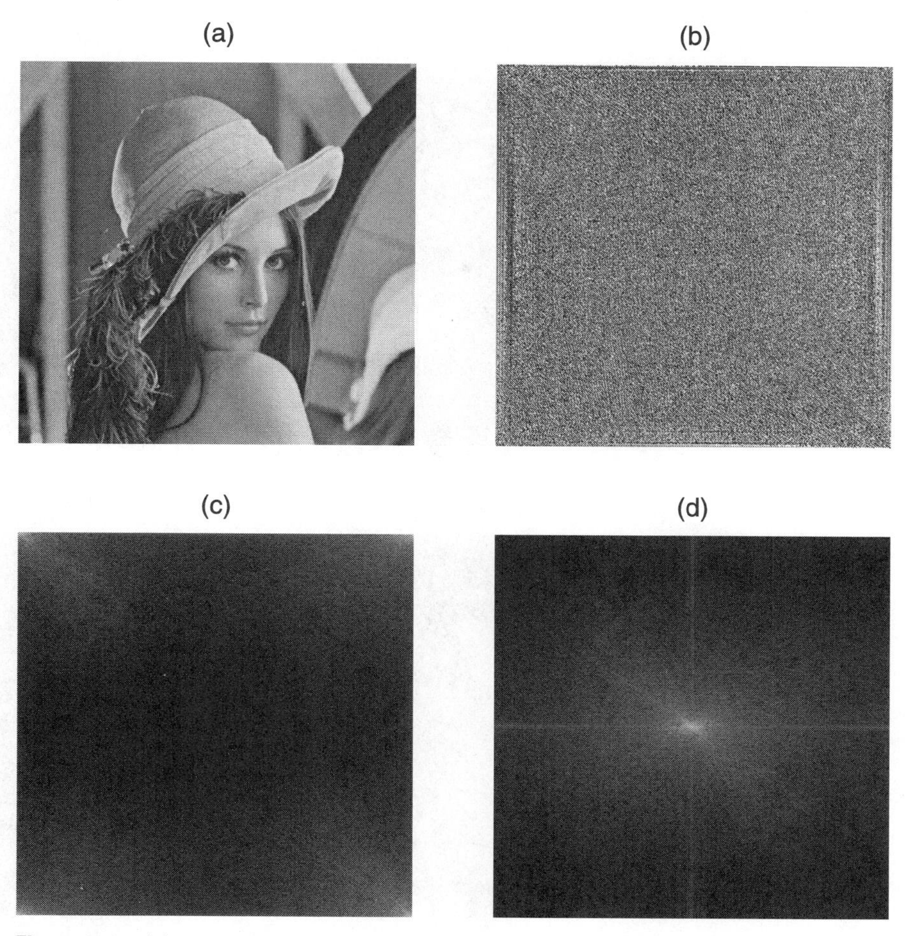

Figure 11.2 (*a*) Original "Lena" image; (*b*) phase spectrum; (*c*) log-magnitude spectrum; (*d*) centrally shifted log-magnitude spectrum.

$$\hat{x}(n) = \begin{cases} x(n), & 0 \leq n \leq N-1 \\ x(2N-1-n), & N \leq n \leq 2N-1 \end{cases} \tag{11.37}$$

For the example at hand, the resulting sequence $\hat{x}(n)$ is shown in Fig. 11.4(*b*). This sequence is of length $2N$ and is symmetric about $n = N - \frac{1}{2}$. The next step is to take the DFT of this symmetric sequence. The $2N$-point DFT of $\hat{x}(n)$ is

$$\hat{X}(k) = \sum_{n=0}^{N-1} x(n) W_{2N}^{kn} + \sum_{n=N}^{2N-1} x(2N-1-n) W_{2N}^{kn}. \tag{11.38}$$

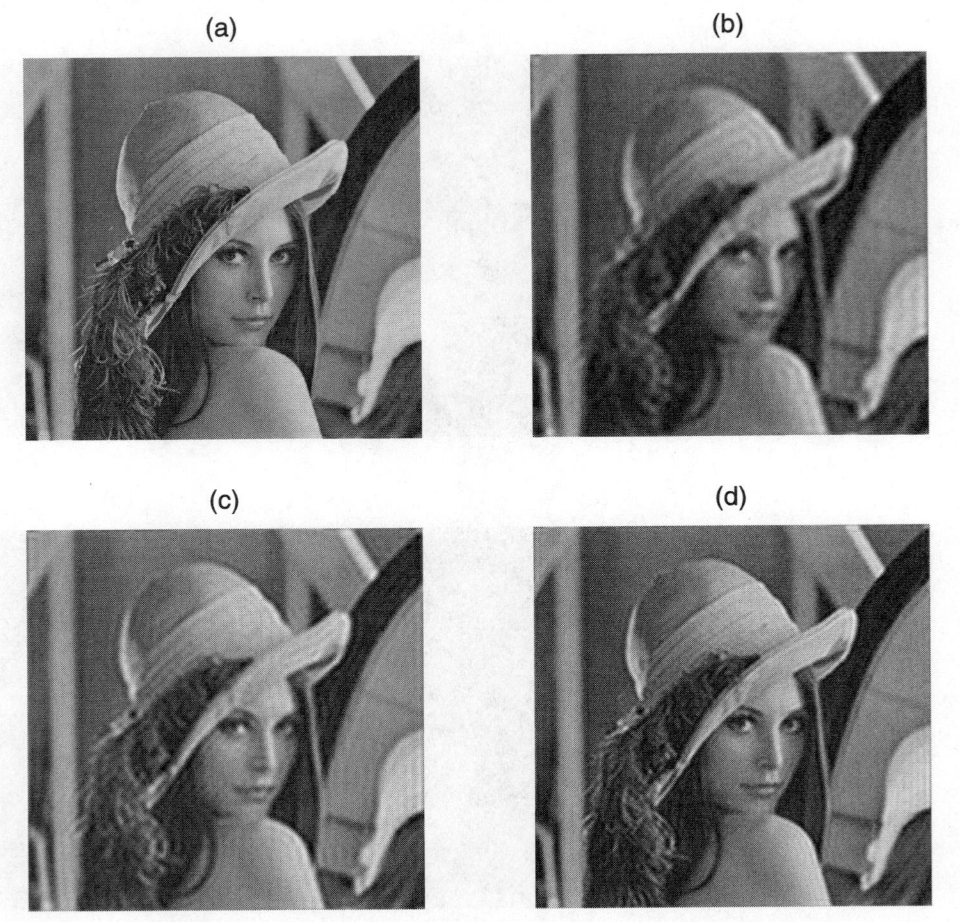

Figure 11.3 (a) Original "Lena" image; (b) Image reconstructed with 60 × 60 DFT coefficients; (c) 80 × 80 DFT coefficients; (d) 120 × 120 DFT coefficients.

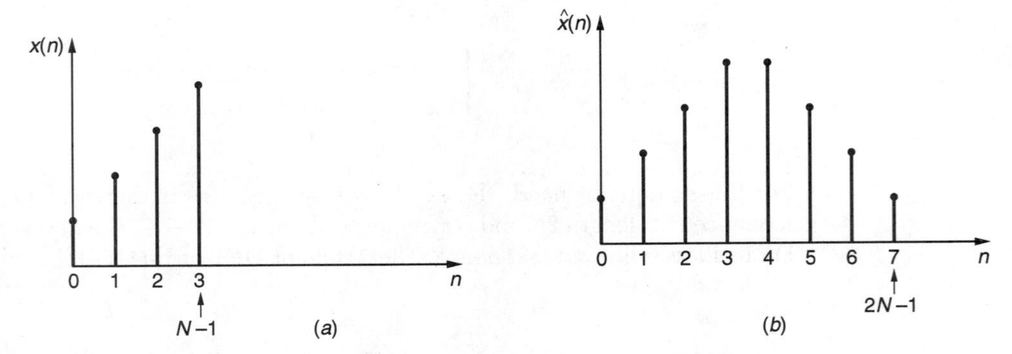

Figure 11.4 (a) An example of $x(n)$; (b) $\hat{x}(n)$ is a symmetric extension of $x(n)$.

Performing the change of variables $p = 2N - 1 - n$, we get

$$\hat{X}(k) = \sum_{n=0}^{N-1} x(n)W_{2N}^{kn} + \sum_{p=0}^{N-1} x(p)W_{2N}^{2kN}W_{2N}^{-k}W_{2N}^{-kp}. \tag{11.39}$$

Using the fact that $W_{2N}^{2kN} = 1$ and letting $p = n$ in the second summation, we get

$$\begin{aligned}\hat{X}(k) &= \sum_{n=0}^{N-1} x(n)W_{2N}^{kn} + \sum_{n=0}^{N-1} x(n)W_{2N}^{-(n+1)k} \\ &= W_{2N}^{-\frac{k}{2}} \sum_{n=0}^{N-1} x(n)\left[W_{2N}^{k(n+\frac{1}{2})} + W_{2N}^{-k(n+\frac{1}{2})}\right] \\ &= W_{2N}^{-\frac{k}{2}} \sum_{n=0}^{N-1} 2x(n)\cos\left[\frac{\pi k}{N}\left(n+\frac{1}{2}\right)\right].\end{aligned} \tag{11.40}$$

Notice that $0 \leq k \leq 2N - 1$ in equation (11.40). Finally, the DCT is defined as $W_{2N}^{\frac{k}{2}}\hat{X}(k)$ in the range $0 \leq k \leq N-1$ and zero otherwise. This will make the DCT purely real. The DCT of $x(n)$ is then

$$C_x(k) = \sum_{n=0}^{N-1} 2x(n)\cos\frac{(2n+1)\pi k}{2N}. \tag{11.41}$$

This is the forward transform. The actual DCT is scaled so that it is unitary. This gives the transform

$$C_x(k) = \alpha(k)\sum_{n=0}^{N-1} x(n)\cos\frac{(2n+1)k\pi}{2N} \tag{11.42}$$

where

$$\alpha(0) = \frac{1}{\sqrt{N}}, \quad \alpha(k) = \sqrt{\frac{2}{N}}, \quad 1 \leq k \leq N-1. \tag{11.43}$$

The inverse unitary DCT is

$$x(n) = \sum_{k=0}^{N-1} \alpha(k)C_x(k)\cos\frac{(2n+1)\pi k}{2N}. \tag{11.44}$$

We need to prove that (11.42) and (11.44) together form a unitary DCT pair. The proof of the fact that the pair is unitary is assigned as Problem 9. We now show that they indeed to form a DCT transform pair. This can be done by substituting (11.44) into the right-hand side of (11.42) and showing that it simplifies to $C_x(k)$. We do this in two parts—first for $k = 0$ and then for $1 \leq k \leq N-1$.

Case 1: $k = 0$

In this case, we need to prove that

$$C_x(0) = \alpha(0) \sum_{n=0}^{N-1} x(n) \cos \frac{(2n+1)\pi k}{2N} \bigg|_{k=0}$$
$$= \frac{1}{\sqrt{N}} \sum_{n=0}^{N-1} x(n). \tag{11.45}$$

With $x(n)$ as in (11.44), the RHS of the above equation is

$$\frac{1}{\sqrt{N}} \sum_{n=0}^{N-1} x(n) = \frac{1}{N} \sum_{n=0}^{N-1} C_x(0) + \frac{\sqrt{2}}{N} \sum_{k=l}^{N-1} C_x(k) \sum_{n=0}^{N-1} \cos \frac{(2n+1)\pi k}{2N}$$
$$= C_x(0) + \frac{\sqrt{2}}{N} \sum_{k=1}^{N-1} C_x(k) \underbrace{\sum_{n=0}^{N-1} \cos \frac{(2n+1)\pi k}{2N}}_{T(k)} \tag{11.46}$$

By using Euler's formula and after some algebraic manipulations, it can be shown (Problem 1) that $T(k) = 0$, which gives the desired result.

Case 2: $1 \le k \le N-1$

The RHS of (11.42) is

$$\sqrt{\frac{2}{N}} \sum_{n=0}^{N-1} x(n) \cos \frac{(2n+1)\pi k}{2N}$$
$$= \frac{2}{N} \sum_{n=0}^{N-1} \sum_{l=1}^{N-1} C_x(l) \cos \frac{(2n+1)l\pi}{2N} \cos \frac{(2n+1)k\pi}{2N}$$
$$= \frac{2}{N} \sum_{l=1}^{N-1} C_x(l) \frac{1}{2} \sum_{n=0}^{N-1} \left[\cos \frac{(2n+1)(l+k)\pi}{2N} + \cos \frac{(2n+1)(l-k)\pi}{2N} \right]. \tag{11.47}$$

It can be shown that (Problem 6)

$$\sum_{n=0}^{N-1} \cos \frac{(2n+1)(l \pm k)\pi}{2N} = 0, \quad l \ne k. \tag{11.48}$$

Then (11.47) reduces to ($l = k$)

$$\sqrt{\frac{2}{N}} \sum_{n=0}^{N-1} x(n) \cos \frac{(2n+1)\pi k}{2N} = \frac{1}{N} C_x(k) \underbrace{\sum_{n=0}^{N-1} \left(\cos \frac{(2n+1)\pi k}{N} + 1 \right)}_{P(k)} \tag{11.49}$$

It is easy to show that

$$P(k) = N. \tag{11.50}$$

Then

$$\sqrt{\frac{2}{N}} \sum_{n=0}^{N-1} x(n) \cos \frac{(2n+1)\pi k}{2N} = C_x(k),$$

which is the desired result.

Properties of the DCT

Define the DCT kernel matrix as **C**, so that for a given vector **x**, the transformed vector is

$$\mathbf{v} = \mathbf{C}\mathbf{x}. \tag{11.51}$$

The transform kernel matrix can easily be found from (11.42) as

$$\mathbf{C} = \begin{bmatrix} \frac{1}{\sqrt{N}} & \frac{1}{\sqrt{N}} & \cdots & \frac{1}{\sqrt{N}} \\ \sqrt{\frac{2}{N}} \cos \frac{\pi}{2N} & \sqrt{\frac{2}{N}} \cos \frac{3\pi}{2N} & \cdots & \sqrt{\frac{2}{N}} \cos \frac{(2(N-1)+1)\pi}{2N} \\ \vdots & & & \vdots \\ \sqrt{\frac{2}{N}} \cos \frac{(N-1)\pi}{2N} & \sqrt{\frac{2}{N}} \cos \frac{3(N-1)\pi}{2N} & \cdots & \sqrt{\frac{2}{N}} \cos \frac{(2(N-1)+1)(N-1)\pi}{2N} \end{bmatrix}. \tag{11.52}$$

Property 1: The DCT is real and orthonormal, that is,

$$\mathbf{C} = \mathbf{C}^* \text{ and } \mathbf{C}\mathbf{C}^T = \mathbf{I}.$$

The first part is clearly trivial. The proof of the second part is left as an exercise (Problem 9).

Property 2: The DCT is a fast transform. That is, the DCT of an N-vector can be computed in $O(N \log_2 N)$ operations. This is shown below.

The derivation is based on taking the even samples, $x(2n)$, and odd samples, $x(2n+1)$, and constructing a new sequence as follows:

$$\begin{aligned} \tilde{x}(n) &= x(2n) \\ \tilde{x}(N-n-1) &= x(2n+1), \quad 0 \le n \le \frac{N}{2} - 1. \end{aligned} \tag{11.53}$$

The DCT of the sequence $x(n)$ is

$$C_x(k) = \alpha(k)\left\{\sum_{n=0}^{\frac{N}{2}-1} x(2n)\cos\frac{\pi(4n+1)k}{2N} + \sum_{n=0}^{\frac{N}{2}-1} x(2n+1)\cos\frac{\pi(4n+3)k}{2N}\right\}$$

$$= \alpha(k)\left\{\sum_{n=0}^{\frac{N}{2}-1} \tilde{x}(n)\cos\frac{\pi(4n+1)k}{2N} + \sum_{n=0}^{\frac{N}{2}-1} \tilde{x}(N-n-1)\cos\frac{\pi(4n+3)k}{2N}\right\}. \tag{11.54}$$

The obvious change of variables is now $m = N - n - 1$. This gives

$$C_x(k) = \alpha(k)\left\{\sum_{n=0}^{\frac{N}{2}-1} \tilde{x}(n)\cos\frac{\pi(4n+1)k}{2N} + \sum_{m=N-1}^{\frac{N}{2}} \tilde{x}(m)\cos\left(\frac{\pi k}{2N}(4(N-m-1)+3)\right)\right\} \tag{11.55}$$

$$C_x(k) = \alpha(k)\left\{\sum_{n=0}^{\frac{N}{2}-1} \tilde{x}(n)\cos\frac{\pi(4n+1)k}{2N} + \sum_{m=N-1}^{\frac{N}{2}} \tilde{x}(m)\cos\left(2\pi k - \frac{\pi k}{2N}(4m+1)\right)\right\} \tag{11.56}$$

$$= \alpha(k)\sum_{n=0}^{N-1} \tilde{x}(n)\cos\frac{\pi(4n+1)k}{2N}$$

$$= \alpha(k)\sum_{n=0}^{N-1} \tilde{x}(n)\mathrm{Re}\left\{e^{-j\left(\frac{2\pi nk}{N}+\frac{\pi k}{2N}\right)}\right\}$$

$$= Re\left\{\alpha(k)e^{-j\frac{\pi k}{2N}}\sum_{n=0}^{N-1} \tilde{x}(n)e^{-j\frac{2\pi nk}{N}}\right\}$$

$$= Re\left\{\alpha(k)e^{-j\frac{\pi k}{2N}} DFT\{\tilde{x}(n)\}\right\}. \tag{11.57}$$

Thus, the DCT of $x(n)$ can be computed via a FFT of the sequence $\tilde{x}(n)$.

Property 3: The DCT has excellent energy compaction for highly correlated data. For a signal $x(n)$ that generates a smooth extension for $x(n) + x(N - n - 1)$, the DCT is well localized; that is, the energy is packed into a small number of coefficients. This is the case for most images, and therefore the 2-D DCT has become very popular for image compression.

2-D DCT

The 2-D unitary DCT of a $N \times N$ sequence $x(n_1, n_2)$ is defined by

$$C_x(k_1, k_2) = \alpha(k_1)\alpha(k_2) \sum_{n_1=0}^{N-1} \sum_{n_2=0}^{N-1} x(n_1, n_2) \cos\frac{(2n_1+1)\pi k_1}{2N} \cos\frac{(2n_2+1)\pi k_2}{2N} \quad (11.58)$$

where $\alpha(k_1)$ and $\alpha(k_2)$ are defined as in (11.43). Notice that this transform is separable. Therefore, the 2-D DCT can be performed in terms of 1-D DCTs using the row-column decomposition method. The inverse unitary DCT is given by

$$x(n_1, n_2) = \sum_{k_1=0}^{N-1} \sum_{k_2=0}^{N-1} \alpha(k_1)\alpha(k_2) C_x(k_1, k_2) \cos\frac{(2n_1+1)\pi k_1}{2N} \cos\frac{(2n_2+1)\pi k_2}{2N}. \quad (11.59)$$

For a given input matrix $\mathbf{X}$, its 2-D unitary DCT can be written as

$$\mathbf{V} = \mathbf{C}\mathbf{X}\mathbf{C}^T. \quad (11.60)$$

This was shown in equation (11.30). Since $\mathbf{C}$ is real and unitary, the 2-D inverse DCT is

$$\mathbf{X} = \mathbf{C}^{\mathrm{T}}\mathbf{V}\mathbf{C}. \quad (11.61)$$

Equations (11.60) and (11.61) are compact representations of the 2-D DCT pair in terms of the 1-D DCT kernel matrix.

An example of a 2-D DCT of a 512×512 image is shown in Fig. 11.5. The log-magnitude of the DCT is plotted as in equation (11.36). The origin of (k_1, k_2) is the upper left corner. The magnitude of the DCT coefficients rapidly decreases as the pair (k_1, k_2) increases. That is, some decorrelation of DCT coefficients has occurred. Now, we use a subset of these coefficients to construct the image using the inverse DCT. The rest of the coefficients are discarded (assumed zero). In particular, we use the first $M \times M$ DCT

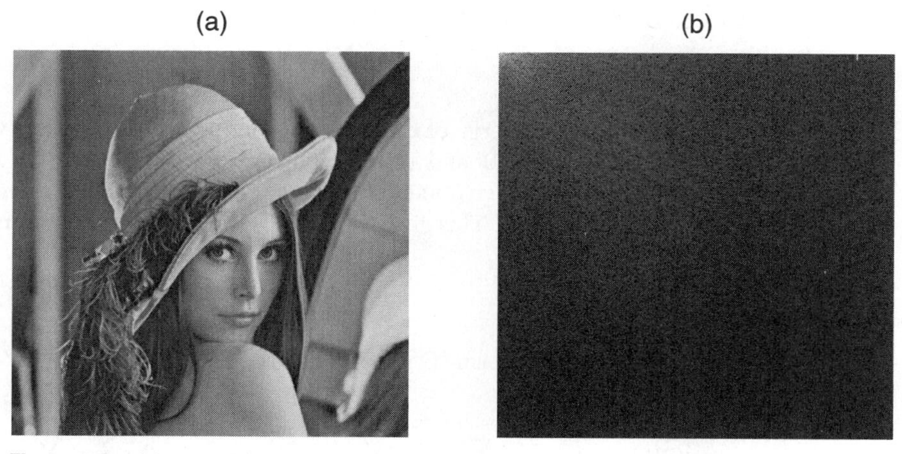

Figure 11.5 (*a*) Original "Lena" image; (*b*) magnitude of DCT.

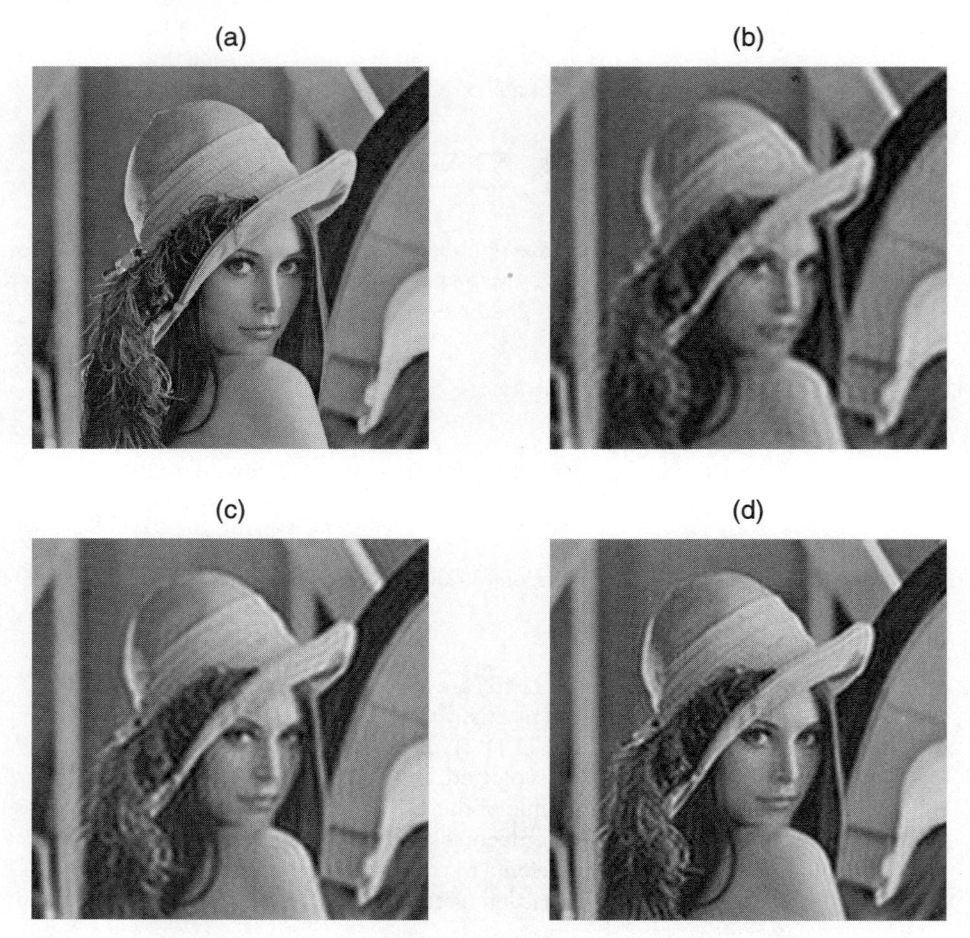

Figure 11.6 (*a*) Original "Lena" image; (*b*) image reconstructed with 60 × 60 DCT coefficients; (*c*) 80 × 80 DCT coefficients; (*d*) 110 × 110 DCT coefficients.

coefficients, since the energy of the image is concentrated in the low values of (k_1, k_2). The results for $M = 60$, 80, and 110 are shown in Fig. 11.6. For $M = 110$, the result is visibly about the same as we obtained using $M = 120$ for the DFT. In addition, all of the DCT coefficients are real. This gives us a compression ratio of approximately 22.

Discrete Sine Transform

The Discrete Sine Transform (DST) of a 1-D sequence $x(n)$ of length N is defined as

$$S_x(k) = \sqrt{\frac{2}{N+1}} \sum_{n=0}^{N-1} x(n) \sin \frac{\pi(k+1)(n+1)}{N+1}, \quad 0 \leq k \leq N-1. \tag{11.62}$$

The inverse DST is

$$x(n) = \sqrt{\frac{2}{N+1}} \sum_{k=0}^{N-1} S_x(k) \sin \frac{\pi(k+1)(n+1)}{N+1}, \quad 0 \le n \le N-1. \tag{11.63}$$

The 1-D DST kernel is therefore given by

$$S(k,n) = \sqrt{\frac{2}{N+1}} \sin \frac{\pi(k+1)(n+1)}{N+1}, \quad 0 \le k,n \le N-1. \tag{11.64}$$

The DST is then

$$\mathbf{S}_x = \mathbf{Sx} \tag{11.65}$$

where $\mathbf{S}_x$ is the $N \times 1$ transformed vector.

The 2-D DST is defined as a separable transform. For an $N \times N$ matrix $\mathbf{X}$, the 2-D DST is an $N \times N$ matrix $\mathbf{V}$, which can be computed as

$$\mathbf{V} = \mathbf{SXS}^T = \mathbf{SXS}. \tag{11.66}$$

The inverse 2-D DST is then

$$\mathbf{X} = \mathbf{S}^{\dagger}\mathbf{VS}^{*} = \mathbf{SVS}. \tag{11.67}$$

Properties of the DST

Property 1: The DST is real and symmetric. It is also orthonormal; that is,

$$\mathbf{S}^* = \mathbf{S} = \mathbf{S}^T = \mathbf{S}^{-1}. \tag{11.68}$$

The proof is assigned as Problem 11.

Property 2: The DST is a fast transform; that is, it can be computed via a FFT. This can be proved in a similar fashion to that of the DCT.

Property 3: The DST has good to excellent energy compaction property for images.

Hadamard Transform

The Hadamard transform is a very unique and interesting transform. The kernel matrix is a Hadamard matrix, whose elements are all equal to either $+1$ or -1. This obviously implies that the Hadamard transform does not require any multiplications. Many interesting theories have been presented on Hadamard matrices. Here we present the very basics of Hadamard matrix theory and then proceed to define this transform.

Definition 11.2 *A Hadamard matrix* $\mathbf{H}$ *of side*[1] N *is an* $N \times N$ *matrix with elements* ± 1 *and the property that*

$$\mathbf{HH}^T = N\mathbf{I}. \tag{11.69}$$

[1] *We use the term* side *to denote the row or column of a square matrix. This is common in the discrete mathematics literature.*

The Hadamard matrix is therefore orthogonal, which is a desirable property of image transforms. The orthogonal property leads to the following result.

Theorem 11.2 *An $N \times N$ matrix with elements h_{ij} is Hadamard if and only if*

$$\sum_{n=1}^{N} h_{in} h_{jn} = 0 \text{ for } i \neq j. \tag{11.70}$$

Proof. *Suppose $h_{ij} = \pm 1$ for matrix $\mathbf{H}$. The (i,j)th element of $\mathbf{HH}^T$ is given by*

$$\sum_{n=1}^{N} h_{in} h_{jn}. \tag{11.71}$$

The hypothesis (11.70) is true if and only if $\mathbf{HH}^T = N\mathbf{I}$. If $i = j$, the defined elements are

$$\sum_{n=1}^{N} h_{in}^2 = N. \tag{11.72}$$

■

Theorem 11.3 *Let $\mathbf{K}$ be a Hadamard matrix of side N and $\mathbf{L}$ be a Hadamard matrix of side R. Then*

$$\mathbf{H} \triangleq \mathbf{K} \otimes \mathbf{L} \tag{11.73}$$

is also a Hadamard matrix.

Proof. Let k_{ij} denote the elements of matrix $\mathbf{K}$. The Kronecker product $\mathbf{K} \otimes \mathbf{L}$ is

$$\mathbf{H} = \begin{bmatrix} k_{11}\mathbf{L} & k_{12}\mathbf{L} & \cdots & k_{1N}\mathbf{L} \\ k_{21}\mathbf{L} & k_{22}\mathbf{L} & \cdots & k_{2N}\mathbf{L} \\ \vdots & & & \vdots \\ k_{N1}\mathbf{L} & k_{N1}\mathbf{L} & \cdots & k_{NN}\mathbf{L} \end{bmatrix}. \tag{11.74}$$

Matrix $\mathbf{H}$ has size $NR \times NR$. It can be also be considered to be an $N \times N$ array of $R \times R$ submatrices. Matrix $\mathbf{HH}^T$ is of the same size. The submatrix in the (i,j) position of $\mathbf{HH}^T$ is

$$\begin{aligned} (\mathbf{HH}^T)_{ij} &= \sum_{r=1}^{N} k_{ir} k_{jr} \mathbf{LL}^T \\ &= \sum_{r=1}^{N} k_{ir} k_{jr} R\mathbf{I}_{R \times R}. \end{aligned} \tag{11.75}$$

Since $\mathbf{K}$ is Hadamard,

$$\sum_{r=1}^{N} k_{ir} k_{jr} = \begin{cases} N, & i = j \\ 0, & i \neq j. \end{cases} \tag{11.76}$$

Therefore

$$(\mathbf{HH}^T)_{ij} = \begin{cases} NR, & i = j \\ 0, & i \neq j. \end{cases},$$

$$\mathbf{HH}^T = NR\mathbf{I}_{NR \times NR} \tag{11.77}$$

which implies that **H** is also Hadamard. ■

In the following, we present two more interesting results on Hadamard matrices. The proofs are omitted and can be found in [1] along with many other interesting results.

Theorem 11.4 *If there is a Hadamard matrix of side N, then N is 1, 2, or a multiple of 4.*

Theorem 11.5 *If* **H** *is a Hadamard matrix, then* **K** *is also Hadamard if it is obtained from* **H** *in one of the following ways:*

(a) *By negating some or all rows.*
(b) *By negating some or all columns.*
(c) *By permuting the rows and permuting the columns.*

Now we consider some examples of Hadamard matrices. Let $\mathbf{H}_n$ denote a Hadamard matrix of side 2^n. Then

$$\mathbf{H}_0 = \begin{bmatrix} 1 \end{bmatrix}$$

$$\mathbf{H}_1 = \begin{bmatrix} 1 & 1 \\ 1 & -1 \end{bmatrix}.$$

By Theorem 11.3, we can find

$$\mathbf{H}_2 = \mathbf{H}_1 \otimes \mathbf{H}_1$$

$$= \left[\begin{array}{cc|cc} 1 & 1 & 1 & 1 \\ 1 & -1 & 1 & -1 \\ \hline 1 & 1 & -1 & -1 \\ 1 & -1 & -1 & 1 \end{array}\right],$$

$$\mathbf{H}_3 = \mathbf{H}_2 \otimes \mathbf{H}_1,$$

and so on.

The 1-D unitary Hadamard transform of a $N \times 1$ vector $\mathbf{x}$ is defined as

$$\mathbf{v} = \frac{1}{\sqrt{N}}\mathbf{Hx} \tag{11.78}$$

where **H** is a Hadamard matrix of side N. The inverse transform is

$$\mathbf{x} = \frac{1}{\sqrt{N}}\mathbf{Hv}. \tag{11.79}$$

The 2-D Hadamard transform for an $N \times N$ matrix $\mathbf{X}$ is

$$\mathbf{V} = \frac{1}{N}\mathbf{H}\mathbf{X}\mathbf{H}^T$$
$$= \frac{1}{N}\mathbf{H}\mathbf{X}\mathbf{H}. \tag{11.80}$$

The inverse transform is

$$\mathbf{X} = \frac{1}{N}\mathbf{H}\mathbf{V}\mathbf{H}. \tag{11.81}$$

Properties of the Hadamard Transform

Property 1: The Hadamard transform kernel $\frac{1}{\sqrt{N}}\mathbf{H}$ is real, symmetric, and orthonormal.

Property 2: The Hadamard transform is a fast transform. The 1-D transform of an N-length vector can be computed in $O(N\log_2 N)$ additions and subtractions.

Property 3: The Hadamard transform has a good to very good energy compaction property for most images.

KL Transform

The Karhunen-Loeve (KL) transform was originally introduced by Karhunen [2] and Loeve [3] and used as a series expansion for random processes. The KL expansion is essentially the same as the principal component analysis introduced by Hotelling. It is therefore also called the Hotelling transform, or principal component transform [4]. It is also known as the eigenvector transform because the transform is based on the eigenvectors of the autocorrelation matrix of the input sequence, as we now define.

Let $\mathbf{R}$ denote the autocorrelation matrix of an $N \times 1$ input vector $\mathbf{x}$. If λ_k and $\boldsymbol{\phi}_k$ are, respectively, the kth eigenvalue and corresponding eigenvector of $\mathbf{R}$, then

$$\mathbf{R}\boldsymbol{\phi}_k = \lambda_k\boldsymbol{\phi}_k \tag{11.82}$$

for $1 \leq k \leq N$. By using the eigenvectors $\boldsymbol{\phi}_k$ as the columns, form a new matrix $\boldsymbol{\Psi}$. The KL transform is defined as

$$\mathbf{v} = \boldsymbol{\Psi}^\dagger\mathbf{x}. \tag{11.83}$$

By using Property 6 in Chapter 7, we have

$$\Psi^\dagger\mathbf{R}\boldsymbol{\Psi} = \boldsymbol{\Lambda} = \text{diag}\left\{\lambda_1, \; \cdots, \; \lambda_N\right\}. \tag{11.84}$$

That is, the autocorrelation matrix is unitarily diagonalizable by a matrix of its eigenvectors.

▶ **EXAMPLE 11.3** *[6] Consider a 2×1 random vector consisting of two realizations of equal probability given by $\mathbf{x}_1 = \begin{bmatrix}2 & 1\end{bmatrix}^T$ and $\mathbf{x}_2 = \begin{bmatrix}3 & 2\end{bmatrix}^T$. The autocorrelation matrix is therefore*

$$R = \frac{1}{2}\left(\mathbf{x}_1\mathbf{x}_1^T + \mathbf{x}_2\mathbf{x}_2^T\right)$$

$$= \begin{bmatrix} 6.5 & 4 \\ 4 & 2.5 \end{bmatrix}.$$

The eigenvalues and eigenvectors can be formed as

$$\lambda_1 = 8.972, \quad \lambda_2 = 0.0279$$

$$\boldsymbol{\phi}_1 = \begin{bmatrix} 0.851 \\ 0.526 \end{bmatrix}, \ \boldsymbol{\phi}_2 = \begin{bmatrix} 0.526 \\ -0.851 \end{bmatrix}.$$

This gives the KL transform kernel matrix

$$\boldsymbol{\Psi}^\dagger = \begin{bmatrix} \boldsymbol{\phi}_1^\dagger \\ \boldsymbol{\phi}_2^\dagger \end{bmatrix} = \begin{bmatrix} 0.851 & 0.526 \\ 0.526 & -0.851 \end{bmatrix}.$$

The transform of each of the input vectors is therefore

$$\mathbf{v}_1 = \Psi^\dagger \mathbf{x}_1 = \begin{bmatrix} 2.228 \\ 0.201 \end{bmatrix}$$

$$\mathbf{v}_2 = \Psi^\dagger \mathbf{x}_2 = \begin{bmatrix} 3.605 \\ -0.124 \end{bmatrix}.$$

In each transform vector above, one coefficient is dominant. That is, most of the energy is packed into that coefficient. The autocorrelation matrix of the transformed vector is

$$\mathbf{R}_v = \frac{1}{2}(\mathbf{v}_1\mathbf{v}_1^T + \mathbf{v}_2\mathbf{v}_2^T)$$

$$= \begin{bmatrix} 8.972 & 0 \\ 0 & 0.0279 \end{bmatrix}.$$

Thus, the KL transform has produced transform coefficients that are uncorrelated. The variances of these transform coefficients are clearly equal to the eigenvalues of $\mathbf{R}$.

Example 11.3 is actually a demonstration of the following important property of the KL transform.

Theorem 11.6 *If input* $\mathbf{x}$ *has zero mean* ($E\{\mathbf{x}\} = 0$), *then the KL transform coefficients have zero mean and are uncorrelated; that is,*

$$E\{\mathbf{v}\} = 0 \tag{11.85}$$

$$E\{v(k)v^*(l)\} = \lambda_k \delta(k - l). \tag{11.86}$$

The proof is straightforward and assigned as a problem. Another important property of the KL transform deals with energy compaction in the first m transform coefficients, which is obviously important for compression. The following theorem presents this important result. The proof is omitted and can be found in [5].

Theorem 11.7 *Among all unitary transforms* $\mathbf{v} = \mathbf{Ax}$, *the KL transform packs the maximum average energy in* $m \leq N$ *samples of* $\mathbf{v}$. *Define the variances of the kth coefficients as*

$$\sigma_k^2 \triangleq E\{|v(k)|^2\}. \tag{11.87}$$

It is assumed that the coefficients are ordered in decreasing values of variance as

$$\sigma_1^2 \geq \sigma_2^2 \geq \cdots \geq \sigma_N^2. \tag{11.88}$$

Define the energy of the first m coefficients as

$$E_m \triangleq \sum_{k=1}^{m} \sigma_k^2.$$

Then for any fixed m,

$$E_m(KLT) \geq E_m(\mathbf{A}).$$

For the 2-D transform definition, we assume separability of the image autocorrelation function. This is done for computational advantages as we now explain. An $N \times N$ image has an autocorrelation function $r\left(m,\ n,\ m',\ n'\right)$. Its correlation matrix therefore has N^4 elements. In general, we have N^2 eigenvectors and eigenvalues. Finding these is not computationally practical. Let $r\left(m,\ n,\ m',\ n'\right)$ be modeled to be separable, that is,

$$r\left(m,\ n,\ m',\ n'\right) = r_1\left(m,\ n,\right) r_2\left(m',\ n'\right). \tag{11.89}$$

Then, the elements of the basis functions can be expressed as

$$\zeta_{k,l}(m,n) = \Psi_1(m,k)\Psi_2(n,l). \tag{11.90}$$

This is again separable. So, the 2-D KL transform is now defined as

$$\mathbf{V} = \mathbf{\Psi}_1^{\dagger}\mathbf{X}\mathbf{\Psi}_2^{*} \tag{11.91}$$

and

$$\mathbf{X} = \mathbf{\Psi}_1\mathbf{V}\mathbf{\Psi}_2^{T}. \tag{11.92}$$

For an $N \times N$ matrix, the separability assumption reduces the computation from $O(N^6)$ to $O(N^3)$ [5].

The obvious disadvantage of the KL transform is its computational complexity. In general, there is no fast KL transform. In addition, the KL transform depends on the statistics of the input, which are usually not known a priori. In many cases, these statistics can be modeled, and the kernel matrix thus obtained can be used for images in the same class. Because of these difficulties, the KL transform is not used in practice for data compression. However, it is commonly used as a benchmark for evaluating the performance of fast transforms.

▶ 11.2 PIXEL CODING

In this section, we discuss some popular methods of coding the individual pixels of an image so that some compression is achieved. There are two types of coding methods, namely, *uniform-length* coding and *variable-length* coding. In uniform-length coding, as the name implies, all pixels are coded with the same number of bits regardless of their probability of occurrence. Although this is a simple coding scheme, it does not offer a compression ratio as high as the variable-length schemes. Besides compression ratio, a designed coding scheme has several other desirable properties, as described next.

1. **Uniquely decodable:** The codes must be uniquely decodable by the receiver. Otherwise, the received data will be distorted. For unique decodability, no codeword can be the prefix of another codeword.
2. **Instantaneous decodability:** The codewords of the receiver must be decodable in real time as the bits arrive. In other words, the receiver should not have to wait for a few bits to decide whether the previous bits were part of the last code or the next code.
3. **Average codeword length:** The smaller the average codeword length, the higher is the compression ratio. Of course, there is a limit to the smallest average codeword length, which is related to the entropy of the source as defined below.

Definition 11.3 *The entropy of a source is the average information given by*

$$H = -\sum_{k=1}^{L} p_k \log_2 p_k \quad \textit{bits/symbol} \tag{11.93}$$

where p_k is the probability of the symbol a_k. If the symbols, a_k, are uniformly distributed, then $p_k = \frac{1}{L}$, and the entropy is

$$H_{\max} = -\sum_{k=1}^{L} \frac{1}{L}\log_2 \frac{1}{L} = -\log_2 \frac{1}{L} \quad \text{bits/symbol} \tag{11.94}$$

This is the maximum possible entropy of a given source. For example, consider two symbols a_1 and a_2 with probabilities p and $1-p$, respectively. The entropy is then

$$H = -p\log_2 p - (1-p)\log_2(1-p). \tag{11.95}$$

The entropy as a function of p is given in Fig. 11.7. If $p = 0$, the entropy is 0, and this is the minimum possible for two symbols. On the other hand, the maximum entropy is unity and occurs at $p = \frac{1}{2}$.

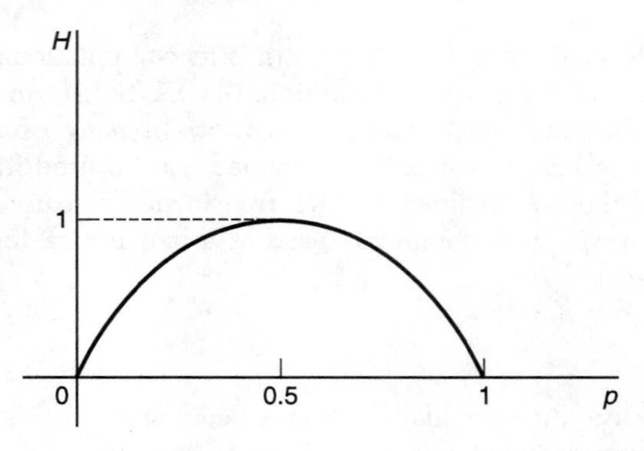

Figure 11.7 Entropy of a two-symbol (binary) source.

Theorem 11.8 *(Shannon's Noiseless Coding Theorem)*
A source with entropy H can be coded without distortion using an average of $H + \epsilon$ bits/symbol, where ϵ is arbitrarily small.

Shannon's theorem is very powerful in that it gives the lower bound for any compression algorithm as the entropy of the source. The entropy is therefore a benchmark for comparing different algorithms. Revisiting the example of Fig. 11.7, if $p = \frac{1}{2}$, that is, if the two symbols are equally probable, then it is possible to code the symbol using 1 bit/message. If $p = 0$, then we need no bits because the output symbol is always known with probability 1. For $p = \frac{1}{8}$, we only need 0.2 bit/symbol average to code the symbols. Now we present some popular coding schemes.

Huffman Coding

Huffman coding is an optimal coding scheme that results in the lowest possible average bit rate. It is very easy to implement, as we now show with an example. Consider a set of five symbols $\{a_1, a_2, a_3, a_4, a_5\}$ to be coded. Let the probabilities (histogram) of these symbols be

$$\left\{p_1, p_2, p_3, p_4, p_5\right\} = \left\{\frac{1}{2}, \frac{1}{4}, \frac{1}{8}, \frac{3}{32}, \frac{1}{32}\right\}.$$

The algorithm is now described using the following steps with reference to Fig. 11.8.

Step 1 Arrange the symbols as nodes in descending order of their probabilities. Consider these nodes as the leaf nodes of a tree.

Step 2 Select two nodes with the lowest probabilities (a_4, a_4) and combine them to form a new node (a_6). The probability of this new node is the sum of the probabilities of the original nodes. Again, we combine two nodes with the lowest probabilities (a_3, a_6) to get a new node (a_7). This process is repeated until we obtain a final node with probability 1. This is called the root node.

Step 3 Arbitrarily assign 1 and 0 to each pair of branches that lead to a node.

Step 4 The code for each symbol is found by reading the 1 or 0 on each branch sequentially from the root node to the leaf node. For example, a_5 has the code 1 1 1 1; a_4 has the code 1 1 1 0; and so on.

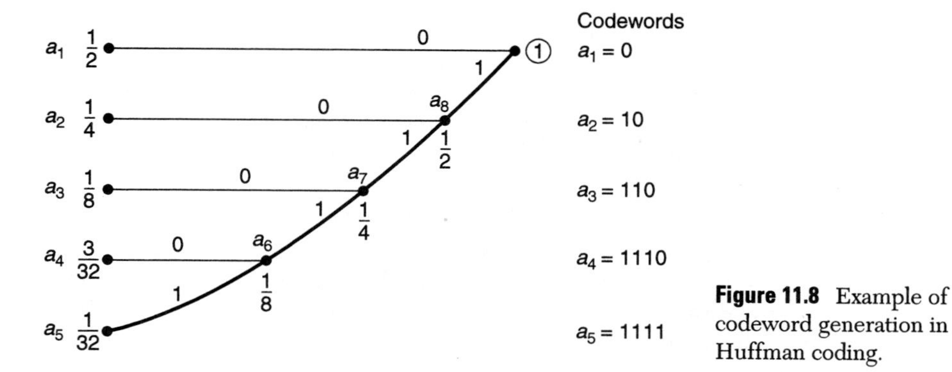

Figure 11.8 Example of codeword generation in Huffman coding.

Now we compare the performance of the Huffman code to uniform-length coding. Since we have five symbols, uniform-length coding requires 3 bits per symbol. For Huffman coding, we find the average length of the codeword as

$$\begin{aligned} H_p &= 1 \cdot \frac{1}{2} + 2 \cdot \frac{1}{4} + 3 \cdot \frac{1}{8} + 4 \cdot \frac{3}{32} + 4 \cdot \frac{1}{32} \\ &= 1.875 \text{ bits/symbol.} \end{aligned}$$

The entropy of the source is

$$\begin{aligned} H &= -\left(\frac{1}{2}\log_2\frac{1}{2} + \frac{1}{4}\log_2\frac{1}{4} + \frac{1}{8}\log_2\frac{1}{8} + \frac{3}{32}\log_2\frac{3}{32} + \frac{1}{32}\log_2\frac{1}{32}\right) \\ &= 1.851 \text{ bits/symbol.} \end{aligned}$$

In this example, Huffman coding yields an average codeword length that is very close to the entropy of the source. This is quite typical of Huffman coding in most cases. However, Huffman coding has the following disadvantages that prohibit its use in some applications. (a) For a large number of symbols, the Huffman tree can be very large, which implies a high computational cost. (b) The probability histogram of the image must be computed, requiring a total of two passes through the data, once for computing the histogram and again for coding.

LZW Coding

Ziv and Lempel [10] invented a universal algorithm for data compression in 1977. A universal algorithm is one that is independent of the symbol statistics. The Lempel-Ziv algorithm was later modified by Welch [9] in 1984, and the resulting algorithm is called the LZW algorithm. This algorithm is based on a dictionary (or translation table), that maps the input symbols into fixed-length codes. The dictionary has a prefix property. This means that for every string in the dictionary, its prefix string is also in the dictionary. For example, let s be a string of symbols and c be a single symbol. If sc is in the dictionary, then s must also be in the dictionary. The dictionary is built by concatenating new characters to previously encountered strings. Therefore, as the dictionary grows, the length of the strings is coded, and compression is achieved. The parser observes the incoming symbols. As long as the new symbol sequence after the last phrase (string) coincides with an existing phrase, no new phrase is introduced and another symbol is considered. As soon as the new phrase is different from anything in the dictionary, it is recognized as a new phrase, encoded, and also saved in the dictionary.

Consider the following sequence of symbols as they arrive from left to right:

$$abaccabcbabcabca\ldots$$

The formation of the dictionary and the encoding process are shown in Fig. 11.9. We start with an empty dictionary, which has a set of indices (or memory locations). Initially, we have null symbol {} in index 0. The parser then observes the first symbol a, which is not in the dictionary. So, it saves a in index 1 and encodes it as $0a$. The next symbol is b, which is also not in the dictionary. So b is saved in index 2 and encoded as $0b$. The next symbol is a, which is already in the dictionary in index 1. So, the parser considers the next symbol, c, and forms the string ac. This string is now in the dictionary and is saved in index 3 and encoded as $1c$, which means that it is a string composed of whatever is in index 1 followed by c. This process is continued, and we have shown up to index-8 in Fig. 11.9. The decoding process is quite straightforward. The decoder also starts with an empty dictionary and with null {} in index 0. When it first receives $0a$, it takes whatever is in index 0 (null), follows it by a, and puts it into index 1. Then it receives $0b$, and so it puts b in index 2. Then it gets $1c$, which means that a new string is created with the contents of index 1 (which is a) followed by c. The new string ac is saved in index 3. This process is continued. The decoder therefore builds the dictionary in the same way as was done in the encoder. The encoding and decoding schemes are quite simple to implement. The size of the dictionary is, of course, finite. Therefore, when it gets full, it must be purged. The rules for purging must be the same for the encoder and the decoder; otherwise there will be errors in decoding. When the dictionary gets full, one option is to start over with a brand-new empty dictionary. This is a simple implementation scheme but is not the most efficient. Another option is to leave the single-symbol strings in the dictionary and purge the rest. Many other variations are possible, and the best possible one would depend on the application and computational complexity that can be tolerated.

Source: *a b a c c a b c b a b c a b c a ...*

Index	Symbol String	Transmitted Code
0	{ }	
1	*a*	0*a*
2	*b*	0*b*
3	*ac*	1*c*
4	*c*	0*c*
5	*ab*	1*b*
6	*cb*	4*b*
7	*abc*	5*c*
8	*abca*	7*a*

Figure 11.9 Example of LZW encoding scheme.

A typical LZW implementation uses 12-bit codes with 8-bit input symbols. This implies a dictionary size of 4096 locations. Observe that each transmitted string consists of an index (12-bit) followed by a single symbol (8-bit). In most coding schemes, transmission is done in bytes. Therefore, this would require 3 bytes of data per code. However, we can do better by a technique called "bit packing." In this method, several codes are used and partitioned into appropriate bytes. In other words, we do not use up 3 bytes for 20 bits (code + symbol), but instead we get the next code and partition the bits appropriately. This increases the efficiency of compression but adds to the complexity of the encoder and decoder.

The LZW algorithm is well suited for data compression for many applications. It adapts to the data being processed and therefore needs no programmer guidance or pre-analysis of data. Its simplicity permits very-high-speed execution. The LZW algorithm is used in the "compress" program used in Unix, in the "zip" program for Windows, and in the "gif" formatting standard.

11.3 DPCM OF IMAGES

One of the most commonly used algorithms in image and speech compression is Differential Pulse Code Modulation (DPCM). This method is popular because it yields a relatively high compression ratio in spite of its simplicity in implementation. The concept of DPCM is to remove mutual redundancy between successive samples in the data. Only new information in the signal is encoded rather than the original signal. DPCM was introduced in Chapter 9 in the context of linear prediction and 1-D adaptive filtering. Here we repeat some of the concepts for convenience and also extend them to two dimensions. DPCM consists of two main blocks: a *quantizer* and a *predictor*. In DPCM, a fixed predictor is employed and designed based on the assumption that signal statistics are stationary and some prior knowledge of the correlation is known. The quantizer can be fixed or adaptive and is used to quantize a prediction error, which is actually encoded for storage or transmission. The receiver (or decoder) can reconstruct the original signal from a quantized prediction error signal.

The design of the predictor dictates the efficiency of the DPCM system. A designer often has to trade off between the computational complexity of the predictor and accuracy of prediction. This chapter presents the design of a predictor for DPCM of images.

The 2-D DPCM system is shown in Fig. 11.10 and is as follows. Let $u(n_1, n_2)$ represent an input sequence and $u_d(n_1, n_2)$ a decoded or reconstructed sequence. For

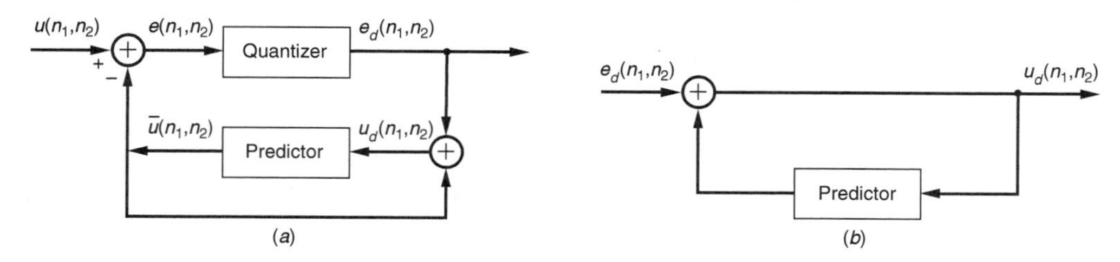

Figure 11.10 2-D DPCM.

a current sample $u(n_1, n_2)$, its predicted value $\overline{u}(n_1, n_2)$ is estimated from previously decoded samples as

$$\overline{u}(n_1, n_2) = \sum_{(k,l)\in R} \sum a(k,l) u_d(n_1 - k, n_2 - l) \tag{11.96}$$

where $a(k, l)$ represent the predictor coefficients and R is a causal prediction window over which $a(k, l)$ is nonzero. Define the prediction error as

$$e(n_1, n_2) = u(n_1, n_2) - \overline{u}(n_1, n_2).$$

This error is quantized, encoded, and then stored or transmitted. Let $e_d(n_1, n_2)$ be a quantized value of $e(n_1, n_2)$; that is,

$$e_d(n_1, n_2) = Q\{e(n_1, n_2)\}$$

where Q{.} is a quantizer. Then the decoded or reproduced sample $u_d(n_1, n_2)$ can be written as

$$u_d(n_1, n_2) = \overline{u}(n_1, n_2) + e_d(n_1, n_2).$$

This process continues recursively.

The technique established in [7] is employed to obtain the predictor coefficients as follows. Assume that an image is scanned line by line starting from the top. Let S_0 represent the current pixel $u(n_1, n_2)$, and let S_i represent the previously scanned pixels given by $u(n_1 - k, n_2 - l)$ for $(k,\ l) \in R$, as shown below.

n_2 →

	$\otimes S_{19}$	$\otimes S_{18}$	$\otimes S_{17}$	$\times S_{16}$	$\times S_{15}$	
$\otimes S_{20}$	$\otimes S_6$	$\otimes S_7$	$\otimes S_2$	$\times S_{10}$	$\times S_{11}$	$\times S_{14}$.
$\otimes S_{21}$	$\otimes S_8$	$\otimes S_3$	$\otimes S_1$	$\times S_9$	$\times S_{12}$	$\times S_{13}$
$\otimes S_{22}$	$\otimes S_5$	$\otimes S_4$	$\times S_0$			

n_1 ↓

Pixels used to predict S_0.

The pixels denoted by $\otimes$ are used to predict S_0. By using notations S_i, as shown above, the predictor equation (7.56) can be written as

$$\overline{S}_0 = \sum_{i=1}^{n} b_i S_i \tag{11.97}$$

where b_i are the prediction coefficients corresponding to $a(k, l)$, that is, $b_1 = a(1,\ 0), b_4 = a(0,\ 1), b_3 = a(1, 1)$, and so on. The prediction coefficients b_i are determined so that a mean-squared estimation error, $E[u(n_1, n_2) - \overline{u}(n_1, n_2)]^2 = E[S_0 - \overline{S}_0]^2$, is minimized. A method to find the prediction coefficients is as follows. An image can be represented using a real random process. Let $R\ (m, n, p, q)$ be an autocorrelation function defined as

$$R(m,\ n,\ p,\ q) = E\{u(m, n) u(p, q)\}.$$

Consider a simple predictor given by

$$\bar{u}(n_1,n_2) = a(1,0)u(n_1-1,n_2) + a(1,1)u(n_1-1,n_2-1) + a(0,1)u(n_1,n_2-1). \quad (11.98)$$

Then

$$E[u(n_1,n_2)-\bar{u}(n_1,n_2)]^2 = E\left[u(n_1,n_2) - \left\{\begin{array}{c} a(1,0)\,u(n_1-1,n_2) \\ +a(1,1)\,u(n_1-1,n_2-1) \\ +a(0,1)\,u(n_1,n_2-1) \end{array}\right\}\right]^2.$$

Differentiating with respect to $a(1{,}0)$, $a(1{,}1)$, and $a(0{,}1)$, and setting each equation equal to zero, the following is obtained.

$$2E\left[\left(u(n_1,n_2) - \left\{\begin{array}{c} a(1,0)\,u(n_1-1,n_2)+ \\ a(1,1)\,u(n_1-1,n_2-1) \\ +a(0,1)\,u(n_1,n_2-1) \end{array}\right\}\right) u(n_1-1,n_2)\right] = 0,$$

$$2E\left[\left(u(n_1,n_2) - \left\{\begin{array}{c} a(1,0)\,u(n_1-1,n_2)+ \\ a(1,1)\,u(n_1-1,n_2-1) \\ +a(0,1)\,u(n_1,n_2-1) \end{array}\right\}\right) u(n_1-1,n_2-1)\right] = 0,$$

$$2E\left[\left(u(n_1,n_2) - \left\{\begin{array}{c} a(1,0)\,u(n_1-1,n_2)+ \\ a(1,1)\,u(n_1-1,n_2-1) \\ +a(0,1)\,u(n_1,n_2-1) \end{array}\right\}\right) u(n_1,n_2-1)\right] = 0.$$

Using the definition of $R(m,n,p,q)$, we find that the above equations become

$$\left\{\begin{array}{c} a(1,0)\,R(n_1-1,n_2,n_1-1,n_2)+ \\ a(1,1)\,R(n_1-1,n_2-1,n_1-1,n_2) \\ +a(0,1)\,R(n_1,n_2-1,n_1-1,n_2) \end{array}\right\} = R(n_1,n_2,n_1-1,n_2),$$

$$\left\{\begin{array}{c} a(1,0)\,R(n_1-1,n_2,n_1-1,n_2-1)+ \\ a(1,1)\,R(n_1-1,n_2-1,n_1-1,n_2-1) \\ +a(0,1)\,R(n_1,n_2-1,n_1-1,n_2-1) \end{array}\right\} = R(n_1,n_2,n_1-1,n_2-1),$$

$$\left\{\begin{array}{c} a(1,0)\,R(n_1-1,n_2,n_1,n_2-1)+ \\ a(1,1)\,R(n_1-1,n_2-1,n_1,n_2-1) \\ +a(0,1)\,R(n_1,n_2-1,n_1,n_2-1) \end{array}\right\} = R(n_1,n_2,n_1,n_2-1).$$

By solving the above, the predictor coefficients can be obtained. The above can be simplified further by assuming that an image that belongs to a real random field is homogeneous. That is, its mean is constant and assumed to be zero, and its autocorrelation function is shift-invariant; that is, $R(m,n,p,q) = R(m-p, n-q) = R(p-m, q-n)$.

$$\begin{aligned} a(1,0)\,R(0,0) + a(1,1)R(0,1) + a(0,1)\,R(1,-1) &= R(1,0), \\ a(1,0)\,R(0,1) + a(1,1)R(0,0) + a(0,1)\,R(1,0) &= R(1,1), \\ a(1,0)\,R(1,-1) + a(1,1)R(1,0) + a(0,1)\,R(0,0) &= R(0,1). \end{aligned} \tag{11.99}$$

Observe that (11.98) is equivalent to

$$\overline{S}_0 = b_4 S_4 + b_3 S_3 + b_1 S_1.$$

For further simplicity, let $R_s(i,j) = E\{S_i S_j\}$ and let it be calculated by spatially averaging the products $S_i S_j$ over an entire image. Therefore, (11.99) can be written in terms of $R_s(i,j)$ and b_i as

$$\begin{aligned} b_4 R_s(4,4) + b_3 R_s(3,4) + b_1\,R_s(1,4) &= R_s(0,4), \\ b_4 R_s(4,3) + b_3 R_s(3,3) + b_1\,R_s(1,3) &= R_s(0,3), \\ b_4 R_s(4,1) + b_3 R_s(3,1) + b_1\,R_s(1,1) &= R_s(0,1). \end{aligned}$$

Because of the homogeneous property of the image, the above becomes

$$\begin{aligned} b_4 R_s(0,0) + b_3 R_s(0,1) + b_1\,R_s(0,9) &= R_s(0,4), \\ b_4 R_s(0,1) + b_3 R_s(0,0) + b_1\,R_s(0,4) &= R_s(0,3), \\ b_4 R_s(0,9) + b_3 R_s(0,4) + b_1\,R_s(0,0) &= R_s(0,1). \end{aligned}$$

Note that $R_s(i,j)$ can be replaced by $R_s(0,i)$, that is, $R_s(i,i) = R_s(0,0)$, $R_s(3,4) = R_s(4,3) = R_s(0,1)$, and so on.

Using a similar approach, we can determine the predictor coefficients b_i for $i = 1, 2, \ldots, n$ by solving the following set of n *normal* equations:

$$R_s(0,i) = \sum_{j=1}^{n} b_j R_s(i,j), \quad i = 1, 2, \ldots, n. \tag{11.100}$$

For quantization of the prediction error $e(n_1, n_2)$, an adaptive quantizer called the *Jayant* quantizer [8] is employed. The Jayant quantizer is an adaptive uniform quantizer whose step size is recursively adjusted as

$$q(n+1) = M(n)q(n)$$

where $q(n)$ is a step size and $M(n)$ is a multiplication factor. The value of this factor depends on the quantization level for the sample $e(n_1, n_2)$. Figure 11.11 illustrates a 2-bit Jayant quantizer, which is later used in our computer simulations.

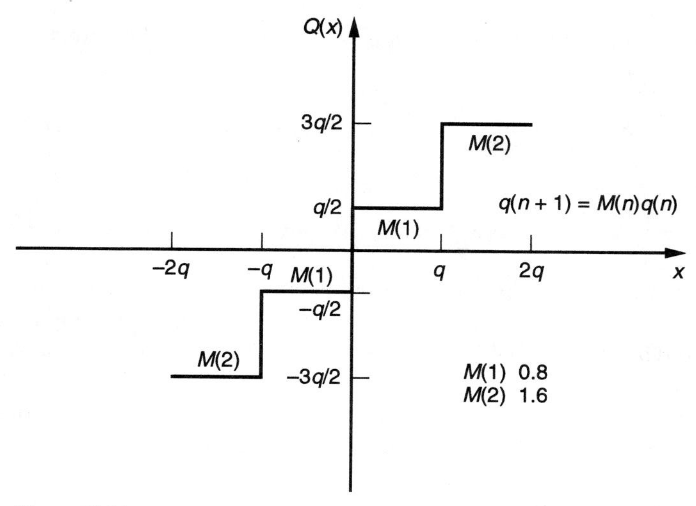

Figure 11.11 Jayant quantizer characteristic.

Simulation Results

Let $u(n_1, n_2)$ be an original image and $u_d(n_1, n_2)$ be the decoded or reproduced image. Note that the predictor processes the decoded or reproduced samples $u_d(n_1, n_2)$ and not the original image. Two cases are considered, as we will explain in the context of the following table:

		n_2
	$+S_7$	$+S_2$
$\otimes S_8$	$\otimes + S_3$	$+S_1$
$\otimes S_5$	$\otimes S_4$	$\times S_0$
n_1		

The pixel S_0 denoted by $\times$ is the pixel to be predicted. The pixels denoted by $\otimes$ are the ones used for prediction in Case 1, and the ones denoted by $+$ are the ones used for prediction in Case 2.

Case 1: $u_d(n_1, n_2) = S_4$ and the predictor uses four previous pixels (S_3, S_4, S_5, S_8) to predict the current sample $S_0 = u(n_1, n_2)$.

Case 2: $u_d(n_1, n_2) = S_1$ and the predictor uses four previous pixels (S_1, S_2, S_3, S_7) to predict the current sample $S_0 = u(n_1, n_2)$.

Observe that for Case 1, the four previous pixels correspond to the pixels of the original and reproduced images, respectively, as

$$\{S_3, S_4, S_5, S_8\} \Rightarrow \{u(n_1 - 1, n_2 - 1), u(n_1, n_2 - 1), u(n_1, n_2 - 2), u(n_1 - 1, n_2 - 2)\}$$
$$\Rightarrow \{u_d(n_1 - 1, n_2), u_d(n_1, n_2), u_d(n_1, n_2 - 1), u_d(n_1 - 1, n_2 - 1)\}.$$

For Case 2, the four previous samples correspond to the pixels of the original and reproduced images, respectively, as

$$\{S_2, S_1, S_3, S_7\} \Rightarrow \{u(n_1-2,n_2), u(n_1-1,n_2), u(n_1-1,n_2-1), u(n_1-2,n_2-1)\}$$
$$\Rightarrow \{u_d(n_1-1,n_2), u_d(n_1,n_2), u_d(n_1,n_2-1), u_d(n_1-1,n_2-1)\}.$$

The above implies that the predictor is causal with respect to the decoded samples $u_d(n_1, n_2)$.

Given an image u, the mean of the image is first removed, that is, $u = u - m$ where m denotes mean(u). Then the image is normalized such that the maximum value of the normalized image is 1, that is, $u = \frac{u}{max(u)}$. This normalized image is used in the prediction algorithm.

For illustration, Case 1 and Case 2 are now used for processing the standard image "Lena" (Fig. 11.12). Let the pixel on the leftmost corner be pixel (1,1).

Case 1 The DPCM starts predicting at pixel (2,3) because at this pixel, the predictor has complete information; that is, the four previous pixels are available. Assume that pixels of the first row and the first two columns of the image are used without quantization; that is, each one is represented by an infinite-precision number. The coefficients (impulse response) of the FIR predictor determined using (11.100) are as follows.

$$h(0,0) = b_4 = 1.2323, \quad h(0,1) = b_5 = -0.4245,$$
$$h(1,0) = b_3 = 0.1605, \quad h(1,1) = b_8 = -0.0046.$$

Case 2 The DPCM starts predicting at pixel (3,2) for the same reason as in Case 1. Assuming that pixels of the first two rows and the first column of the image are used without quantization, the coefficients (impulse response) of the FIR predictor determined using (11.100) are as follows.

$$h(0,0) = b_1 = 1.4620, \quad h(0,1) = b_3 = 0.0799,$$
$$h(1,0) = b_2 = -0.5777, \quad h(1,1) = b_7 = 0.0190.$$

The simulation results with the SNR of the decoded images are shown in Table 11.1. The numbers in the first column represent the SNR when the image is directly quantized to 2 bits using the Jayant quantizer but no DPCM. The second and third columns are for DPCM Case 1 and Case 2, respectively. The resulting images are given in Figs. 11.13–11.18.

In Table 11.1, C.1(b_i) denotes the predictor for Case 1 and C.2(b_i) denotes the predictor for Case 2.

11.4 JPEG STANDARD

The Joint Photographic Experts Group (JPEG) Committee is responsible for developing image compression standards. In 1988, the JPEG Committee's goal was to set a lossless compression standard with a compression ratio of about 2 : 1. The standardization activity

Figure 11.12 Original "Lena" image.

TABLE 11.1 Simulation Results of 2-D DCPM System

SNR(dB)	Quan	C.1(b_i)	C.2(b_i)
Lena	8.810	16.495	20.189

Figure 11.13 2-bit Jayant quantized "Lena."

Figure 11.14 Error image using 2-bit Jayant quantizer.

received several proposals, and the consensus was that the DCT would be used as the image transform. The committee deliberated on the other aspects of the proposal and came up with a draft of the standard in 1990. After several refinements, the international community approved the lossy compression standard in 1992. In the last decade, JPEG has refined the lossy standards and has also developed a lossless compression standard.

JPEG Lossy Image Compression Standard

Block diagrams of the JPEG encoder and decoder are shown in Fig. 11.19. The decoder performs the inverse operations of the encoder and in reverse order.

Figure 11.15 Decoded image using predictor of Case 1.

Figure 11.16 Error image for predictor of Case 1.

DCT

The first step in the encoder is the DCT. The input image is partitioned into 8×8 blocks of data. For each block, an 8×8 2-D DCT is performed. The DCT coefficients corresponding to the low-frequency bins are usually large in magnitude, and they are deemed to be perceptually most significant. The higher frequency coefficients are typically smaller. These features of the DCT coefficients are exploited in designing the quantizer, where the bulk of the compression takes place.

Quantizer

In the quantizer, each DCT coefficient $X(k_1, k_2)$ in the 8×8 block is quantized in a special way depending on its index (k_1, k_2). A predefined 8×8 quantization table is used for this purpose. Let $Q(k_1, k_2)$ denote the entries of this quantization table. Each of these entries

Figure 11.17 Decoded image using predictor of Case 2.

Figure 11.18 Error image using predictor of Case 2.

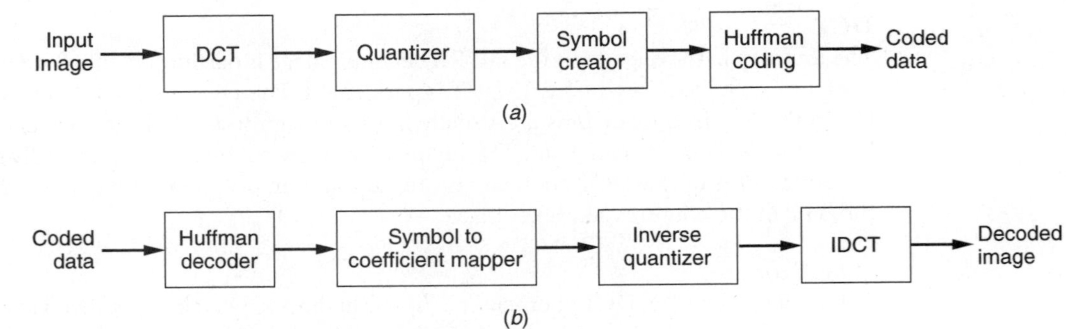

Figure 11.19 (*a*) JPEG encoder, (*b*) JPEG decoder.

is a number between 0 and 255, for 8-bit images. The quantized value for each DCT coefficient is given by

$$X_q(k_1, k_2) = \left[\frac{X(k_1, k_2)}{Q(k_1, k_2)}\right]_{\text{round}}. \tag{11.101}$$

A large value of $Q(k_1, k_2)$ implies a small (in magnitude) quantized value, whereas a small value of $Q(k_1, k_2)$ yields a large quantized value. The quantization tables for the luminance and the chrominance components are given in Fig. 11.20. Notice that the $Q(k_1, k_2)$ entries for lower values of (k_1, k_2) are small, which implies that the quantized values will be large. For higher values of (k_1, k_2), the quantized coefficients will be small. That is, the lower frequency coefficients will require a larger number of bits than the higher frequency components. The quantization tables given here were obtained from numerous psychovisual experiments. These tables have become widely accepted, although the JPEG standard does not specify these particular tables or any others. Notice that the quantization table for the chrominance component has larger values than the luminance. This will yield smaller quantized coefficients for the chrominance. This is done because the human visual system is more insensitive to the chrominance component than the luminance.

As an example, consider the 8×8 block of data shown in Fig. 11.21(*a*). This is a block from the image "earth" that is found in the MATLAB image processing toolbox. The DCT of this block is given in Fig. 11.21(*b*). The DCT matrix is then quantized according to (11.101) with the luminance table given earlier. The quantized coefficients are given in Fig. 11.21(*c*). There is a large number of zeros in this matrix. This makes this array especially suitable for Run Length Coding (RLC), which is explained in the next subsection.

16	11	10	16	24	40	51	61
12	12	14	19	26	58	60	55
14	13	16	24	40	57	69	56
14	17	22	29	51	87	80	62
18	22	37	56	68	109	103	77
24	35	55	64	81	104	113	92
49	64	78	87	103	121	120	101
72	92	95	98	112	100	103	99

(*a*)

17	18	24	47	99	99	99	99
18	21	26	66	99	99	99	99
24	26	56	99	99	99	99	99
47	66	99	99	99	99	99	99
99	99	99	99	99	99	99	99
99	99	99	99	99	99	99	99
99	99	99	99	99	99	99	99
99	99	99	99	99	99	99	99

(*b*)

Figure 11.20 Quantization tables: (*a*) luminance, (*b*) chrominance.

$$
\mathbf{A} = \begin{bmatrix}
6 & 120 & 6 & 120 & 6 & 6 & 6 & 6 \\
6 & 6 & 6 & 6 & 6 & 6 & 6 & 6 \\
6 & 120 & 6 & 120 & 6 & 6 & 6 & 6 \\
6 & 6 & 6 & 6 & 6 & 6 & 6 & 6 \\
6 & 120 & 6 & 120 & 6 & 6 & 6 & 6 \\
6 & 6 & 120 & 6 & 6 & 6 & 6 & 6 \\
6 & 120 & 6 & 6 & 6 & 6 & 6 & 6 \\
6 & 6 & 120 & 6 & 6 & 6 & 6 & 6
\end{bmatrix}
$$

(a)

$$
\text{DCT}\{\mathbf{A}\} = \begin{bmatrix}
16.9706 & 178.1909 & 97.5807 & 137.8858 & 16.9706 & 16.9706 & 16.9706 & 16.9706 \\
0 & 29.0583 & -87.5723 & 76.4521 & 0 & 0 & 0 & 0 \\
0 & 0 & 30.8482 & -21.8130 & 0 & 0 & 0 & 0 \\
0 & 34.2767 & 8.5110 & 23.1565 & 0 & 0 & 0 & 0 \\
0 & 0 & 0 & 40.3051 & 0 & 0 & 0 & 0 \\
0 & 51.2986 & -42.7877 & -4.6061 & 0 & 0 & 0 & 0 \\
0 & 0 & 74.4741 & 52.6611 & 0 & 0 & 0 & 0 \\
0 & 146.0862 & -58.5139 & 114.4187 & 0 & 0 & 0 & 0
\end{bmatrix}
$$

(b)

$$
\mathbf{A}q = \begin{bmatrix}
1 & 16 & 10 & 9 & 1 & 0 & 0 & 0 \\
0 & 2 & -6 & 4 & 0 & 0 & 0 & 0 \\
0 & 0 & 2 & -1 & 0 & 0 & 0 & 0 \\
0 & 2 & 0 & 1 & 0 & 0 & 0 & 0 \\
0 & 0 & 0 & 1 & 0 & 0 & 0 & 0 \\
0 & 1 & -1 & 0 & 0 & 0 & 0 & 0 \\
0 & 0 & 1 & 1 & 0 & 0 & 0 & 0 \\
0 & 2 & -1 & 1 & 0 & 0 & 0 & 0
\end{bmatrix}
$$

(c)

Figure 11.21 Example: (*a*) an 8×8 image matrix, A, (*b*) DCT matrix of A, (*c*) quantized DCT matrix.

Symbol Creater and Coder

This block creates symbols for the coefficients of the quantized 8×8 DCT matrix denoted by $X_q(k_1, k_2)$. The symbols are created in a very special way so that compression is achieved. The matrix of quantized DCT coefficients is converted to a vector sequence by traversing the matrix in a zig-zag way as shown in Fig. 11.22. As we know from the last example, the quantized DCT matrix has a large number of zeros in the high-frequency bins. Forming the sequence in this zig-zag way assures that the zeros are mostly toward the end. This will lead to efficient coding by RLC, as we shall see shortly.

The DC coefficient, $X_q(0,0)$, is handled very differently than the rest because a high degree of correlation exists between the DC coefficients in two adjacent 8×8 blocks of data. These coefficients are therefore differentially coded. Let $X_q^{(i)}(0,0)$ denote the quantized DC coefficient in the ith block. Then the difference

$$d_i = X_q^{(i)}(0,0) - X_q^{i-1}(0,0) \tag{11.102}$$

is mapped to a symbol. For an 8-bit image, the DCT coefficients are signed and lie in the range $[-1023, 1023]$. Therefore, d_i lies in the range $[-2046, 2046]$. This is a very large range and will require too much computation and storage to create the Huffman coding tree. Therefore, this range is subdivided into categories, and these categories are

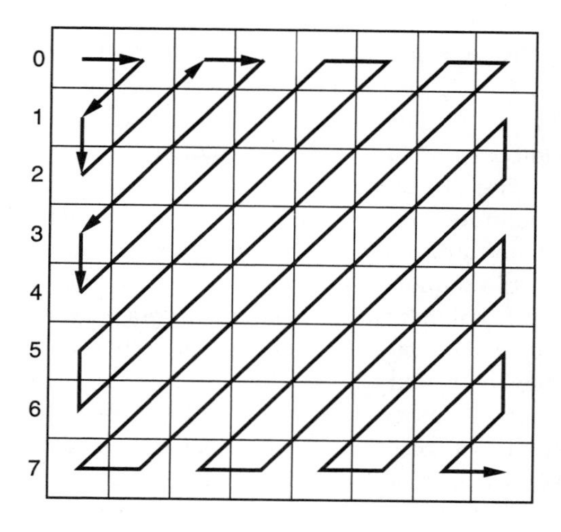

Figure 11.22 Zig-zag procedure to create a vector sequence from the 8 × 8 quantized DCT matrix.

Huffman coded. Each of the positive and negative ranges can be represented by 11 bits. That is, they can be subdivided into 12 different categories. The categories are defined as follows.

$$\text{Category } k\text{:}\quad \text{Element range: } [2^{k-1}, 2^k - 1] \text{ and } [-2^k + 1, -2^{k-1}]$$

We will now consider some example categories. Let $\delta_i^{(+)}$ and $\delta_i^{(-)}$ denote the possible positive and negative values of the difference signal. Then we have the following examples:

$$k = 3: \quad 4 \le \delta_i^{(+)} \le 7, \quad \text{and} \quad -7 \le \delta_i^{(-)} \le -4$$

$$k = 4: \quad 8 \le \delta_i^{(+)} \le 15, \quad \text{and} \quad -15 \le \delta_i^{(-)} \le -8$$

$$k = 5: \quad 16 \le \delta_i^{(+)} \le 31, \quad \text{and} \quad -31 \le \delta_i^{(-)} \le -16, \quad \text{and so on.}$$

The categories are Huffman coded, and the amplitude of the signal within the category is coded in k bits and transmitted as is. At the receiver, the Huffman decoder decodes a string of bits into a category, say k_0. Then the next k_0 bits are interpreted as the amplitude. The subsequent bits are then decoded for a category, and the next corresponding bits are the amplitude, and so on. Positive numbers are coded in binary, whereas the negative numbers are coded in one's complement. As an example, consider the DC values of quantized DCT coefficients of three consecutive blocks as 48,60, and 44. The following symbols are generated:

Difference	Category	Amplitude
12	k = 4	1100
−16	k = 5	01111

Notice that in the case of $k = 5$, if we assume straight binary, the amplitude is 15. But this is not an option for this category. Therefore, it must be interpreted as a negative one's complement number, which yields the desired result.

Now we discuss the JPEG standard for coding the AC coefficients, many of which are zero as we have already seen. Recall that in the case of DC coefficients, we Huffman code each category, which is a symbol between 0 and 11. In this case, the symbol to be Huffman coded is composed of (a) run-length and (b) category. There are 4 bits reserved for the category. The run-length is the number of zero coefficients before a nonzero coefficient is encountered in the sequence. In the begining of the 64-coefficient sequence, there are not many zeros and the run-lengths will be mostly zero. But as we proceed toward the end of the sequence, the run-lengths will be large numbers. There are 4 bits reserved for the run-length for each coefficient, which allows for up to 15 zeros encountered before a nonzero coefficient. If there are more than 15 zeros in a row, then the run-length is 15 and the coefficient will be taken as zero. The symbol to be Huffman coded is composed of 8 bits: 4 bits of run-length followed by 4 bits of category. The category is found as in the case of DC coefficients. Also, if the category is k, then k bits of amplitude are transmitted, which are not Huffman coded. At the receiver, once the 8-bit symbol (run-length + category) is decoded, it tells the receiver how many zeros were encountered prior to the current coefficient and the category value k. Therefore, the receiver will interpret the following k bits as the amplitude and also insert an appropriate number of zeros before the current coefficient, thereby reconstructing the transmitted coefficients exactly.

As a special case, if after a nonzero coefficient, all the rest of the coefficients in the 64-point block are zero, then the symbol of 0 (binary: 0000 0000) is transmitted. This denotes the end of block. As an example of this procedure, consider the example of the quantized DCT coefficient block of Fig. 11.21(c). By using the zig-zag traversing process (Fig. 11.22), we get the first few AC coefficients as $[16, 0, 0, 2,\ 10, 9, -6, 0, 0, 0, 2, \ldots]$. The coding of these coefficients is as follows:

Terminating Value	Run-length/ Category	Symbol to be Huffman Coded	Amplitude Bits
16	0/5	5	10000
2	2/2	34	10
10	0/4	4	1010
9	0/4	4	1001
−6	0/3	3	001
2	3/2	50	10

The coding process in this table is continued until we complete all the 64 coefficients in the block, or we encounter all zeros after a point, in which case the end of block symbol is the last one to be transmitted. The method of generating the Huffman codes is identical to what we have described earlier in this chapter. However, Huffman tables must be transmitted along with the data. More on the generation and transmission of these tables is explained in the context of the JPEG lossless compression standard.

JPEG Lossless Image Compression Standard

The original JPEG lossless compression standard was adopted at about the same time as the lossy compression standard was approved in 1992. The lossless standard had little in common with the DCT-based lossy standard. The lossless standard was supposed to achieve a compression ratio of about 2 : 1 on similar sets of images. The JPEG lossless compression standard is based on predictive coding, followed by Huffman coding. For prediction, as the encoder scans the input image from left to right and from top to bottom, each pixel is predicted as a linear combination of previous pixels. The prediction error is then encoded using Huffman coding. The standard allows the user to choose from eight different predictors. Let a 2×2 block of pixels be given as follows:

C	B
A	Pixel to be predicted

The predictor options are denoted by modes and are as follows.

Mode #	Predicted pixel
0	No prediction
1	A
2	B
3	C
4	$A+B-C$
5	$A+(B-C)/2$
6	$B+(A-C)/2$
7	$(A+B)/2$

The difference between the predicted value and the actual value is Huffman coded. The coding scheme is exactly the same as that used in encoding the DC values in the lossy JPEG scheme. Each symbol is represented by a category and an amplitude. The category symbol is Huffman coded, and the amplitude is simply transmitted as a string of bits. The Huffman code must meet the following conditions:

1. The maximum code length is 16 bits.
2. A canonical Huffman code is used. Let each k codeword be n bits long. The codewords are then given by $x+1, x+2, \ldots, x+k$, where x is obtained by left-shifting the largest magnitude number by an $(n-1)$ bit codeword.

These two conditions are used for fast encoding and decoding of the Huffman table. The JPEG bit stream must also contain the Huffman table, which is coded as follows. Two lists are specified, BITS and HUFFVAL [11]. BITS is a 16-byte array in which the nth byte gives the number of codewords on length n. HUFFVAL is a list of symbol values in increasing order of codeword length. In the case of two symbols having the same codeword length, the symbol with the smaller magnitude is given first. This encoding scheme significantly speeds up the decoding. The decoding process requires two passes through the image.

For lossy compression, the Huffman coding scheme is not strictly specified as above. Many hardware implementations of the lossy standard do not implement this scheme, but simply directly input the Huffman table into the JPEG stream. The informative section of the standard provides example Huffman tables for the luminance and chrominance components. These tables work quite well in practice for a wide variety of images for lossy compression, and so they are often used. With the use of standard tables, the decoder requires only one pass through the image. But this is not possible in the case of lossless compression.

The JPEG lossless standard based on Huffman coding has been adopted widely and is used in some public domain implementations. However, research in the early 1990s showed that performance could be significantly improved by using better predictors and new algorithms. In 1994, JPEG solicited proposals for improved lossless compression standards that could be decoded with one single pass of the image. Several proposals were received, and in 1997, the JPEG-LS standard was approved. Details of the standard can be found in [12] and [13]. The new standard works significantly better than the old one, and several of its extensions are currently under consideration for standardization.

11.5 SUMMARY

In this chapter, we have presented the fundamentals of image compression. We started with a block diagram of the basic elements in a compression scheme. The topic of image transforms was then covered in detail, including the DFT, DCT, DST, Hadamard transform, and KL transform. The section on pixel coding discussed the popular algorithms, Huffman coding, and LZW coding. The topic of quantization was not covered in detail here since it was discussed in Chapters 6 and 9. We briefly discussed the design of a 2-D DPCM. The popular JPEG lossless and lossy standards have also been presented. Sufficient detail has been provided about these standards to allow students to implement the simple versions of the lossy and lossless schemes.

11.6 PROBLEMS

1. Prove that $T(k) = 0$ in equation (11.46).
2. Consider the sequences of Fig. P11.1.

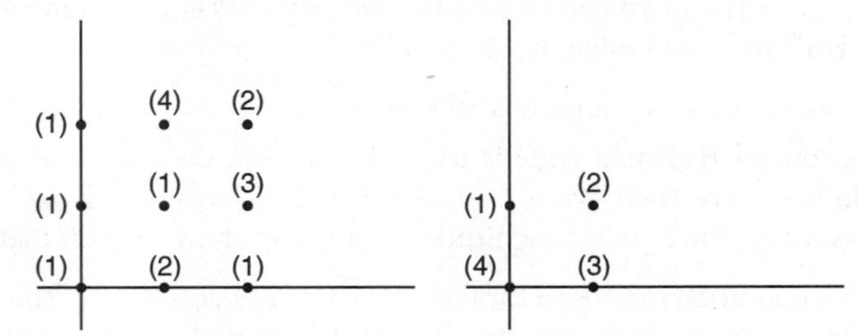

Figure P11.1

(a) Determine the circular convolution of these sequences using the 2-D unitary DFT.
(b) Determine the linear convolution using the 2-D unitary DFT.

3. Let the sequence $x(n_1, n_2)$ be in the domain $[N_1 - 1, N_2 - 1]$. This sequence is also separable, so that $x(n_1, n_2) = x_1(n_1)x_2(n_2)$. Show that the 2-D DCT of $x(n_1, n_2)$ is also separable. Give an example to illustrate this result.

4. Let $C_x(k_1, k_2)$ denote the 2-D unitary DCT of $x(n_1, n_2)$. Show that $C_x^*(k_1, k_2)$ is the unitary DCT of $x^*(n_1, n_2)$. Also show that

$$C_x(k_1, k_2) = C_x(-k_1, k_2) = C_x(k_1, -k_2) = C_x(-k_1, -k_2).$$

5. An image matrix is given by

$$\mathbf{U} = \begin{bmatrix} 2 & 3 \\ 4 & 2 \end{bmatrix}.$$

(a) Find the 2-D DCT and the basis images.
(b) Find the 2-D DFT and the basis images.
(c) Find the 2-D DST and the basis images.

6. Prove equation (11.48).

7. Prove that the magnitude of the determinant of a unitary transform is unity.

8. Show that all eigenvalues of a unitary matrix have unit magnitude.

9. Prove that the 1-D DCT kernel matrix $\mathbf{C}$ is orthonormal, that is, $\mathbf{CC}^T = \mathbf{I}$.

10. Explicitly write down the formulas for the 2-D unitary DST and the inverse 2-D unitary DST. Assume a separable transform.

11. Prove that the 1-D DST kernel matrix $\mathbf{S}$ is symmetric and orthonormal. Verify your proof for $N = 4$.

12. Prove that the 1-D DST is a fast transform. That is, it can be computed via a FFT. Use a technique similar to that used to prove the same for the DCT.

13. Prove that the Hadamard transform is symmetric and orthonormal.

14. Prove that the Hadamard transform is a fast transform; that is, it can be computed in $O(N \log_2 N)$ additions and subtractions.

15. An image matrix is given by

$$\mathbf{U} = \begin{bmatrix} 2 & 3 \\ 4 & 2 \end{bmatrix}.$$

Find the 2-D Hadamard transform and its basis images.

16. An image matrix is given by

$$\mathbf{U} = \begin{bmatrix} 1 & 2 & 4 & 3 \\ 1 & 3 & 7 & 2 \\ 7 & 1 & 2 & 3 \\ 2 & 2 & 1 & 4 \end{bmatrix}.$$

Find the 2-D Hadamard transform and its basis images.

17. Suppose that a $N \times N$ matrix is row ordered into a $N^2 \times 1$ vector. Show that if we take the 1-D Hadamard transform of this vector and then map it into a matrix, it is equivalent to taking the 2-D Hadamard transform of the original matrix. Also, illustrate this with a numerical example.

18. Prove Property 1 of the KL transform; that is, prove equations (11.85) and (11.86).

19. Given a vector $\mathbf{x} = [1 \;\; 3 - 1 \;\; 4]^T$, find its KL transform and the basis vectors.

20. Let $\mathbf{x}$ and $\mathbf{y}$ be random vectors with autocorrelation matrices $\mathbf{R}_x$ and $\mathbf{R}_y$, respectively. Suppose that $\mathbf{R}_x\mathbf{R}_y = \mathbf{R}_y\mathbf{R}_x$. Then show that these vectors have the same KL transform.

21. The message set $A = \{-2, -1, 0, 1, 2, 3\}$ has the probability histogram $P = \left\{\frac{5}{8}, \frac{3}{32}, \frac{1}{32}, \frac{3}{32}, \frac{1}{8}, \frac{1}{32}\right\}$.

(a) Find the entropy of the source.
(b) Let the source be quantized as follows: $Q(-2) = Q(-1) = 4$; $Q(0) = Q(1) = 0$; $Q(2) = Q(3) = -4$. Find the entropy of the quantized source.

22. The output of a source is a random variable X that is uniformly distributed with length-N. Find the entropy of the source.

23. The message set $A = \{a, b, c, d, e, f\}$ has the probability histogram $P = \left\{\frac{5}{8}, \frac{3}{32}, \frac{1}{32}, \frac{3}{32}, \frac{1}{8}, \frac{1}{32}\right\}$. Construct the Huffman tree and find the codes for these symbols. Find the entropy of the source and the average bits/symbol achieved by the Huffman code.

24. The message set $A = \{a, b, c, d, e, f\}$ has the probability histogram $P = \left\{\frac{3}{8}, \frac{3}{16}, \frac{3}{32}, \frac{5}{16}, \frac{1}{32}\right\}$. Construct the Huffman tree and find the codes for these symbols. Find the entropy of the source and the average bits/symbol achieved by the Huffman code.

25. Find the LZW code for the following binary sequence:

$$00101000111101001010101111000110101110001100100010101000001111111.$$

Use two passes through your data and decide on the size of the dictionary.

26. A popular image model is given by the correlation function

$$R(n_1, n_2) = \rho^{|n_1| + |n_2|}, \quad 0 \le \rho \le 1.$$

DPCM is used to quantize the source with the following predictor:

$$f(n_1, n_2) = af(n_1 - 1, n_2) + bf(n_1, n_2 - 1) + cf(n_1 - 1, n_2 - 1),$$

where a, b, and c are the predictor coefficients. Find the coefficient values in terms of ρ.

27. A popular image model is given by the correlation function

$$R(n_1, n_2) = \rho^{|n_1| + |n_2|}, \quad 0 \le \rho \le 1.$$

In a Pulse Code Modulation (PCM) system, the input $f(n_1, n_2)$ is directly quantized. Find the variance of the quantized signal.

28. Let the following vector represent the DC values of blocks of quantized DCT coefficients:

$$Dc = [100 \;\; 90 \;\; 110 \;\; 100 \;\; 105 \;\; 80 \;\; 70].$$

These values need to be differentially coded. The difference sequence will be represented by a category and an amplitude. The category is to be Huffman coded, and the amplitude bits are to be transmitted without coding. Find the coded bit stream for this source.

29. Let the following vector be a quantized DCT vector of a source:

$$x = [54 \;\; 33 \;\; 0 \;\; 0 \;\; 0 \;\; 21 \;\; 0 \;\; 17 \;\; 0 \;\; 0 \;\; 0 \;\; 0 \;\; 0 \;\; 3].$$

Let this sequence be run-length coded with Huffman coding. Let each symbol to be Huffman coded be composed of 11 bits: 3 bits for run-length followed by 8 bits for the category. Find the Huffman codes for these symbols and the amplitude bits for the coded data.

30. Let the following matrix be an 8×8 block of an image:

$$\begin{bmatrix} 100 & 89 & 70 & 70 & 80 & 70 & 64 & 59 \\ 90 & 40 & 90 & 40 & 19 & 20 & 10 & 20 \\ 89 & 80 & 80 & 80 & 70 & 60 & 50 & 50 \\ 70 & 38 & 75 & 65 & 55 & 45 & 35 & 30 \\ 70 & 45 & 60 & 55 & 50 & 45 & 40 & 35 \\ 80 & 20 & 80 & 75 & 70 & 65 & 60 & 65 \\ 70 & 80 & 75 & 77 & 75 & 75 & 75 & 75 \\ 60 & 79 & 60 & 60 & 60 & 60 & 60 & 60 \end{bmatrix}$$

(a) Find the DCT of this block.
(b) Find the quantized DCT block using the luminance quantization table.
(c) Find the quantized DCT block using the chrominance quantization table.
(d) Comment on the results obtained in (*b*) and (*c*).

11.7 COMPUTER PROJECTS

Project 1: Huffman Coding

(a) Write a program in MATLAB or C++ for Huffman coding a sequence of symbols in the range [0,8]. Your program should make two passes through the data. In the first pass, it finds the histogram and creates the Huffman tree. In the second pass, it codes the symbols and finds the bit stream.

(b) Write a program to decode a Huffman-coded bit stream given the Huffman table. Run this program on your coded data in part (*a*) and compare the decoded symbols with the transmitted ones.

Project 2: LZW Coding

(a) Write a computer program in C++ or MATLAB to implement the Lempel-Ziv-Welch algorithm. Assume that the data to be compressed are drawn from a source file with 256 symbols (1 byte). Use a dictionary size of 4096 entries (12-bit index). Your program should read from a file and write the compressed data to a file. The program should account for "bit packing" of indices.

(b) Write a program that does decompression of the data compressed by the Lempel-Ziv-Welch algorithm.

(c) Test the performance of your compression algorithm by recording the compression ratio for each experiment for the files provided by your instructor. One of these files is a text file, and the other is an image file. You must verify that the uncompressed data is identical to the original source file.

(d) Compare your compression results with those of other standard algorithms, such as "compress," "pkzip," and "gzip."

Project 3: 2-D DPCM

(a) Write a program to implement a 2-D FIR DPCM coder. The quantizer should be an adaptive Jayant quantizer. Your program should have the following inputs: (i) the FIR predictor coefficients input as a 3×3 matrix; (ii) the number of bits for quantization; and (iii) an input image as a matrix. The program should create a quantized error matrix for transmission. Your program should compute the predictor coefficients from the given image by using the method of this chapter.

(b) Write a program to implement a 2-D DPCM decoder. The parameters of the decoder should be the same as those of the encoder. The decoder should take the quantized error matrix for (a) as input and recover the original data.

(c) Use three different input images and find the SNR of the reconstructed image for 2b, 3b, and 4b quantization.

(d) Submit the original and reconstructed images and the data in tabular form.

Project 4: DCT

This project performs the transform coding part of JPEG.

(a) Use an 8-bit image of your choice. Find the DCT of the 8×8 blocks of the image. Quantize the DCT matrix using the standard luminance table.

(b) Let the nonzero coefficients be transmitted only. For uniform-length coding, find the maximum number of bits required per symbol. Find the compression ratio.

(c) Reconstruct the image from the quantized DCT matrix. Find the SNR of the reconstructed image.

(d) Submit the original image, reconstructed image, DCT image, quantized DCT image, and all relevant data.

Project 5: 2-D ADPCM

Write a program to perform 2-D ADPCM using the LMS algorithm. Use a 3×3 or 5×5 causal 2-D FIR filter as your predictor. Use the Jayant quantizer. Perform ADPCM on an image of your choice. Use 2b and 3b quantizers.

In your report, submit the following: (a) description of all algorithms; (b) image results and the SNR of the directly quantized image versus that of the ADPCM image; (c) discussion and conclusion (in particular, include an analysis of your results for the different types of quantizers, predictors, and the number of bits in the quantizer); and (d) source code.

▶ BIBLIOGRAPHY

[1] A.P. Street and W.D. Wallis, *Combinatorics: A First Course*, Charles Babbage Research Center, Manitoba, Canada, 1982.

[2] H. Karhunen, "Uber lineare methoden in der Wahrscheinlich- Keitsrechnung," *Ann. Acad. Science Fenn.*, Ser. A.I. 37, Helsinki, 1947.

[3] M. Loeve, "Fonctions aleatoires de seconde ordre," in P. Levy, *Processus Stochastiques et Mouvement Brownien*, Hermann, France, Paris, 1948.

[4] H. HOTELLING, "Analysis of a complex of statistical variables into principal components," *J. Educ. Psychology 24*, pp. 417–441 and pp. 498–520, 1933.

[5] A.K. JAIN, *Digital Image Processing*, Prentice Hall, Englewood Cliffs, NJ, 1989.

[6] J. VILLASENOR, *Digital Image Processing Class Notes*, Short Course, UCLA, 1992.

[7] A. ROSENFELD and A.C. KAK, *Digital Picture Processing*, Academic Press, 1982 (Chapter 5).

[8] V.K. INGLE and J.G. PROAKIS, *Digital Signal Processing Using MATLAB V.4*, PWS Publishing Company, 1997 (Chapter 10).

[9] T.A. WELCH, "A technique for high-performance data compression," *IEEE Computer Magazine*, pp. 8–19, 1984.

[10] J. ZIV and A. LEMPEL, "A universal algorithm for sequential data compression," *IEEE Trans. Information Theory*, IT-23, (3), pp. 337–343, May 1977.

[11] N. MEMON and R. ANSARI, "The JPEG lossless image compression standards," *Handbook of Image & Video Processing*, A. Bovik, ed., Academic Press, San Diego, CA, 2000.

[12] ISO/IEC JTC1/SC29/WG1, "JPEG LS coding system," ISO Working Document ISO/IEC JTC1/SC29/WG1 N399-WD14495, July 1996.

[13] W.B. PENNEBAKER and J.L. MITCHELL, *JPEG Still Image Data Compression Standard*, Van Nostrand Reinhold, New York, 1993.

Appendix

Concepts of Linear Algebra

This appendix reviews some essential topics in linear algebra. For each topic, we present some definitions, basic properties, and numerical examples. We begin with the fundamentals of matrix theory, including basic operations, eigenvalues, eigenvectors, and Kronector products and sums. Then we present the linear independence of vectors, basis vectors, and the rank of a matrix. Next, we discuss vector norms, matrix norms, condition numbers, and similarity transformation of matrices. Special matrices and properties are also presented. In particular, we define and discuss the properties of matrices such as the Normal, Hadamard, Hankel, Vandermonde, Unitary, Circulant, and Jordan matrices. Finally, we briefly describe the Singular Value Decomposition (SVD) and pseudo-inverses.

A.1 NOTATIONS

An $m \times n$ *matrix* $\mathbf{A}$ consists of m rows and n columns and mn elements (real or complex numbers) and is denoted by

$$\mathbf{A} = \begin{bmatrix} a_{11} & a_{12} & \cdots & a_{1n} \\ a_{21} & a_{22} & \cdots & a_{2n} \\ \vdots & \vdots & \ddots & \vdots \\ a_{m1} & a_{m2} & \cdots & a_{mn} \end{bmatrix} = [a_{ij}]_{i,j=1}^{m\,n} = [a_{ij}]_{mn} = [a_{ij}].$$

The element a_{ii} is called the ith *diagonal element* of $\mathbf{A}$, and a_{ij} for $i \neq j$ is called the (i,j)th element of $\mathbf{A}$. We say the *size* of $\mathbf{A}$ is $m \times n$, or the *order* of $\mathbf{A}$ is m when $m = n$. An $m \times 1$ matrix is said to be an m-vector or a *column m-vector*, and an $1 \times n$ matrix is said to be an n-vector or a *row n-vector*. To avoid any confusion, an n-vector means a column vector in this appendix, and a row vector is represented by the transpose (it will be defined shortly) of a column vector. Commonly, R^n and C^n are notations for the sets of real and complex column n-vectors, respectively; and $R^{m \times n}$ and $C^{m \times n}$ are notations for the sets that contain all $m \times n$ real and complex matrices, respectively. If we do not specify the type of a matrix

A, then **A** can be either real or complex. The following are examples of a 2×3 matrix, a column 2-vector, and a row 3-vector:

$$\mathbf{A} = \begin{bmatrix} 1 & 0 & 2-i \\ -2.5 & 3i & -4 \end{bmatrix}, \quad \mathbf{v} = \begin{bmatrix} 1 \\ -2 \end{bmatrix}, \quad \mathbf{w} = \begin{bmatrix} a\,b\,c \end{bmatrix}.$$

A is said to be a *square matrix* if $m = n$, otherwise it is a *rectangular matrix*. **Z** is said to be a *zero matrix*, denoted by $\mathbf{Z} = [0]_{mn} = \mathbf{0}$, if all elements of **Z** are zero. Matrix **D** is said to be an $n \times n$ *diagonal matrix* if all elements of **D** are zero except its diagonal elements and is commonly written as $\mathbf{D} = \text{diag}\{d_1, \ldots, d_n\}$. An $n \times n$ diagonal matrix with all diagonal elements equal to 1 is called the $n \times n$ *identity matrix*, denoted by $\mathbf{I}_n$ or **I**. A matrix **T** is said to be an upper (lower) triangular matrix if all its elements below (above) its diagonal are zero. A matrix **S** is said to be a *submatrix* of **A** if the rows and columns of **S** are consecutive rows and columns of **A**. If the rows and columns start from the first ones, **S** is also called a *leading submatrix* of **A**. For example, $\mathbf{S} = \begin{bmatrix} 1 & 2 \end{bmatrix}$ is a submatrix of $\mathbf{A} = \begin{bmatrix} 3 & -1 & 4 \\ 5 & 1 & 2 \end{bmatrix}$ and $\mathbf{S} = \begin{bmatrix} 1 & -2 \\ -4 & 5 \end{bmatrix}$ is a leading submatrix of $\mathbf{A} = \begin{bmatrix} 1 & -2 & 3 \\ -4 & 5 & -6 \\ 7 & -8 & 9 \end{bmatrix}$.

A.2 BASIC OPERATIONS

Transpose and Hermitian

Given $\mathbf{A} = [a_{ij}]$ in $R^{m \times n}$, the *transpose* of **A**, denoted by $\mathbf{A}^T$, is an $n \times m$ matrix whose rows are columns of **A** and columns are rows of **A**. When **A** is in $C^{m \times n}$, the *Hermitian* of **A**, denoted by $\mathbf{A}^\dagger$, is in $C^{n \times m}$ and its (i,j)th element is a_{ji}^*. For example,

$$\mathbf{A} = \begin{bmatrix} 1 & 2 & 3 \\ 4 & 5 & 6 \end{bmatrix}, \quad \mathbf{A}^T = \begin{bmatrix} 1 & 4 \\ 2 & 5 \\ 3 & 6 \end{bmatrix} \text{ and } \mathbf{B} = \begin{bmatrix} 1+i & -2i \\ 3 & 4-i \end{bmatrix}, \mathbf{B}^\dagger = \begin{bmatrix} 1-i & 3 \\ 2i & 4+i \end{bmatrix}.$$

Trace of a Square Matrix

The *trace* of an $n \times n$ real square matrix $\mathbf{A} = [a_{ij}]$ is defined by the sum of the diagonal elements of **A**; that is, $\text{tr}(\mathbf{A}) = \sum_{k=1}^{n} a_{kk}$.

EXAMPLE A.1 *Let* $\mathbf{A} = \begin{bmatrix} 1 & 2 \\ 3 & 4 \end{bmatrix}$. *Then,* $tr(\mathbf{A}) = 1 + 4 = 5$.

It is not difficult to show that $\text{tr}(\mathbf{AB}) = \text{tr}(\mathbf{BA})$ provided that both **AB** and **BA** exist.

Dot Product (or Inner Product) and Orthogonality

Given two vectors $\mathbf{u} = \begin{bmatrix} u_1 \\ \vdots \\ u_n \end{bmatrix}$ and $\mathbf{v} = \begin{bmatrix} v_1 \\ \vdots \\ v_n \end{bmatrix}$ in C^n, the *dot product* or *inner product* of $\mathbf{u}$ and $\mathbf{v}$ is a scalar α and is defined as

$$\alpha = \mathbf{u}^\dagger \mathbf{v} = \begin{bmatrix} u_1^* & \dots & u_n^* \end{bmatrix} \begin{bmatrix} v_1 \\ \vdots \\ v_n \end{bmatrix} = \sum_{k=1}^{n} u_k^* v_k.$$

▶ **EXAMPLE A.2** *Let* $\mathbf{u} = \begin{bmatrix} 1 \\ 2-3i \end{bmatrix}$, $\mathbf{v} = \begin{bmatrix} -4+i \\ 5-6i \end{bmatrix}$ *and* $\mathbf{w} = \begin{bmatrix} -3 \\ 2 \end{bmatrix}$. *Then*

$$\mathbf{u}^\dagger \mathbf{v} = \begin{bmatrix} 1 & 2+3i \end{bmatrix} \begin{bmatrix} -4+i \\ 5-6i \end{bmatrix} = -24+4i, \textit{ and } \mathbf{w}^T \mathbf{w} = \begin{bmatrix} -3 & 2 \end{bmatrix} \begin{bmatrix} -3 \\ 2 \end{bmatrix} = 13.$$

Vectors $\mathbf{u}$ and $\mathbf{v}$ are said to be *orthogonal* if $\mathbf{u}^\dagger \mathbf{v} = 0$. A set of vectors $\{\mathbf{v}_1, \dots, \mathbf{v}_m\}$ is said to be orthogonal if $\mathbf{v}_i^\dagger \mathbf{v}_j = 0$ for all $i \neq j$, and is said to be orthonormal if in addition $\mathbf{v}_i^\dagger \mathbf{v}_i = 1$ for all $i = 1, \dots, m$. Consider vectors

$$\mathbf{u}_1 = \begin{bmatrix} 1 \\ -1 \end{bmatrix}, \quad \mathbf{u}_2 = \begin{bmatrix} 2 \\ 2 \end{bmatrix}, \quad \mathbf{v}_1 = \frac{1}{\sqrt{2}} \begin{bmatrix} 1 \\ -1 \end{bmatrix} \quad \text{and } \mathbf{v}_2 = \frac{1}{\sqrt{2}} \begin{bmatrix} 1 \\ 1 \end{bmatrix}.$$

The set $\{\mathbf{u}_1, \mathbf{u}_2\}$ is orthogonal and the set $\{\mathbf{v}_1, \mathbf{v}_2\}$ is orthonormal.

The dot product satisfies the *Cauchy-Schwarz Inequality:*

$$(\mathbf{x}^\dagger \mathbf{y})^2 \leq (\mathbf{x}^\dagger \mathbf{x})(\mathbf{y}^\dagger \mathbf{y})$$

for any vectors $\mathbf{x}$ and $\mathbf{y}$ in C^n.

Matrix Addition and Scalar Multiplication

Two matrices with the same size can be added or subtracted elementwise, and a matrix can be multiplied by a scalar (real or complex) elementwise. Let $\mathbf{A} = [a_{ij}]_{mn}$, $\mathbf{B} = [b_{ij}]_{mn}$ and α, β be scalars. Then

$$\mathbf{A} + \mathbf{B} = [a_{ij} + b_{ij}]_{mn}, \quad \alpha \mathbf{A} = [\alpha a_{ij}]_{mn}, \quad \text{and } \alpha \mathbf{A} + \beta \mathbf{B} = [\alpha a_{ij} + \beta b_{ij}]_{mn}.$$

▶ **EXAMPLE A.3** *Let* $\mathbf{A} = \begin{bmatrix} 1 & 2 & 3 \\ 4 & 5 & 6 \end{bmatrix}$, $\mathbf{B} = \begin{bmatrix} 7 & 8 & 9 \\ 10 & 11 & 12 \end{bmatrix}$ *and* $\alpha = -2j$.

Then

$$\mathbf{A} + \mathbf{B} = \begin{bmatrix} 8 & 10 & 12 \\ 14 & 16 & 18 \end{bmatrix}, \quad \alpha \mathbf{A} = \begin{bmatrix} -2j & -4j & -6j \\ -8j & -10j & -12j \end{bmatrix}, \quad \textit{and } 3\mathbf{A} - 2\mathbf{B} = \begin{bmatrix} -11 & -10 & -9 \\ -8 & -7 & -6 \end{bmatrix}.$$

Matrix addition and scalar multiplication have the following properties:

1. $\mathbf{A} + \mathbf{B} = \mathbf{B} + \mathbf{A}$;
2. $\mathbf{A} + (\mathbf{B} + \mathbf{C}) = (\mathbf{A} + \mathbf{B}) + \mathbf{C}$;
3. $(\alpha\beta)\mathbf{A} = \alpha(\beta\mathbf{A}) = \beta(\alpha\mathbf{A})$;
4. $(\mathbf{A} + \mathbf{B})^T = \mathbf{A}^T + \mathbf{B}^T$.

Matrix Multiplication

Given two matrices $\mathbf{A} = [a_{ij}]$ and $\mathbf{B} = [b_{kl}]$ with sizes $m \times r$ and $r \times n$, the product $\mathbf{C} = \mathbf{AB} = [c_{ij}]$ is an $m \times n$ matrix, and its (i,j)th element is defined as

$$c_{ij} = \sum_{k=1}^{r} a_{ik} b_{kj} = \begin{bmatrix} a_{i1} & a_{i2} & \cdots & a_{ir} \end{bmatrix} \begin{bmatrix} b_{1j} \\ b_{2j} \\ \vdots \\ b_{rj} \end{bmatrix}$$

$$= \begin{array}{l} \text{the dot product of the } i\text{th row of } \mathbf{A} \\ \quad \text{and the } j\text{th column of } \mathbf{B}. \end{array}$$

EXAMPLE A.4 *Let* $\mathbf{A} = \begin{bmatrix} 1 & 2 \\ 3 & 4 \\ 5 & 6 \end{bmatrix}$, *and* $\mathbf{B} = \begin{bmatrix} 1 & -1 \\ -1 & 2 \end{bmatrix}$. *Then*

$$\mathbf{AB} = \begin{bmatrix} -1 & 3 \\ -1 & 5 \\ -1 & 7 \end{bmatrix}, \quad \mathbf{BA}^T = \begin{bmatrix} -1 & -1 & -1 \\ 3 & 5 & 7 \end{bmatrix}, \quad \mathbf{BB} = \begin{bmatrix} 2 & -3 \\ -3 & 5 \end{bmatrix}, \quad \textit{and}$$

$$\mathbf{AA}^T = \begin{bmatrix} 5 & 11 & 17 \\ 11 & 25 & 39 \\ 17 & 39 & 61 \end{bmatrix}.$$

For a square matrix $\mathbf{A}$, the notation $\mathbf{A}^n$ for a positive integer n stands for the product $\mathbf{AA} \ldots \mathbf{A}$ (n times) and $\mathbf{A}^0 \equiv \mathbf{I}$. Matrix multiplication has the following properties:

1. $\mathbf{ABC} = \mathbf{A}(\mathbf{BC}) = (\mathbf{AB})\mathbf{C}$;
2. $(\mathbf{A} + \mathbf{B})\,\mathbf{C} = \mathbf{AC} + \mathbf{BC}$;
3. $\mathbf{A}(\mathbf{B} + \mathbf{C}) = \mathbf{AB} + \mathbf{AC}$;
4. $(\mathbf{AB})^T = \mathbf{B}^T\mathbf{A}^T$; if $\mathbf{A}$ and $\mathbf{B}$ are real, and $(\mathbf{AB})^\dagger = \mathbf{B}^\dagger\mathbf{A}^\dagger$ if $\mathbf{A}$ and $\mathbf{B}$ are complex.

In general, matrix multiplication is not commutative; that is, $\mathbf{AB} \neq \mathbf{BA}$ even if both $\mathbf{AB}$ and $\mathbf{BA}$ are well defined and have the same size.

When $\mathbf{A}$ is a matrix and $\mathbf{B}$ is a vector, we can write $\mathbf{AB}$ in terms of the columns of $\mathbf{A}$ and elements of $\mathbf{B}$, or the rows of $\mathbf{A}$ and vector $\mathbf{B}$. Let $\mathbf{A}$ be an $m \times n$ matrix and

$$\mathbf{A} = \begin{bmatrix} \mathbf{C}_1 & \ldots & \mathbf{C}_n \end{bmatrix} = \begin{bmatrix} \mathbf{R}_1 \\ \vdots \\ \mathbf{R}_m \end{bmatrix}$$

where $\mathbf{C}_i's$ and $\mathbf{R}_i's$ are columns and rows of $\mathbf{A}$, respectively. Let $\mathbf{B} = \begin{bmatrix} b_1 \\ \vdots \\ b_n \end{bmatrix}$.

Then

$$\mathbf{AB} = b_1\mathbf{C}_1 + \cdots + b_n\mathbf{C}_m = \begin{bmatrix} \mathbf{R}_1\mathbf{B} \\ \vdots \\ \mathbf{R}_m\mathbf{B} \end{bmatrix}.$$

Partitioned Matrices

In many applications, it is convenient to partition a matrix into blocks (submatrices). For example, the matrix $\mathbf{A} = \begin{bmatrix} 1 & 2 & 3 \\ 4 & 5 & 6 \\ 7 & 8 & 9 \end{bmatrix}$ can be partitioned as $\mathbf{A} = \begin{bmatrix} \mathbf{A}_{11} & \mathbf{A}_{12} \\ \mathbf{A}_{21} & \mathbf{A}_{22} \end{bmatrix}$ where $\mathbf{A}_{11} = \begin{bmatrix} 1 & 2 \\ 4 & 5 \end{bmatrix}$, $\mathbf{A}_{12} = \begin{bmatrix} 3 \\ 6 \end{bmatrix}$, $\mathbf{A}_{21} = \begin{bmatrix} 7 & 8 \end{bmatrix}$, and $\mathbf{A}_{22} = \begin{bmatrix} 9 \end{bmatrix}$; or $\mathbf{A}_{11} = [1 \;\; 2]$, $\mathbf{A}_{12} = [3]$, $\mathbf{A}_{21} = \begin{bmatrix} 4 \\ 7 \end{bmatrix}$, and $\mathbf{A}_{22} = \begin{bmatrix} 5 & 6 \\ 8 & 9 \end{bmatrix}$. Operations on partitioned matrices work as if the blocks were scalars. For example,

$$\begin{bmatrix} \mathbf{A}_{11} & \mathbf{A}_{12} & \mathbf{A}_{13} \\ \mathbf{A}_{21} & \mathbf{A}_{22} & \mathbf{A}_{23} \end{bmatrix} + \begin{bmatrix} \mathbf{B}_{11} & \mathbf{B}_{12} & \mathbf{B}_{13} \\ \mathbf{B}_{21} & \mathbf{B}_{22} & \mathbf{B}_{23} \end{bmatrix} = \begin{bmatrix} \mathbf{A}_{11} + \mathbf{B}_{11} & \mathbf{A}_{12} + \mathbf{B}_{12} & \mathbf{A}_{13} + \mathbf{B}_{13} \\ \mathbf{A}_{21} + \mathbf{B}_{21} & \mathbf{A}_{22} + \mathbf{B}_{22} & \mathbf{A}_{23} + \mathbf{B}_{23} \end{bmatrix},$$

$$\begin{bmatrix} \mathbf{A}_{11} & \mathbf{A}_{12} \\ \mathbf{A}_{21} & \mathbf{A}_{22} \\ \mathbf{A}_{31} & \mathbf{A}_{32} \end{bmatrix} \begin{bmatrix} \mathbf{B}_{11} & \mathbf{B}_{12} \\ \mathbf{B}_{21} & \mathbf{B}_{22} \end{bmatrix} = \begin{bmatrix} \mathbf{A}_{11}\mathbf{B}_{11} + \mathbf{A}_{12}\mathbf{B}_{21} & \mathbf{A}_{11}\mathbf{B}_{12} + \mathbf{A}_{12}\mathbf{B}_{22} \\ \mathbf{A}_{21}\mathbf{B}_{11} + \mathbf{A}_{22}\mathbf{B}_{21} & \mathbf{A}_{21}\mathbf{B}_{12} + \mathbf{A}_{22}\mathbf{B}_{22} \\ \mathbf{A}_{31}\mathbf{B}_{11} + \mathbf{A}_{32}\mathbf{B}_{21} & \mathbf{A}_{31}\mathbf{B}_{12} + \mathbf{A}_{32}\mathbf{B}_{22} \end{bmatrix}$$

provided that all the block products are well defined.

▶ A.3 DETERMINANT OF A SQUARE MATRIX

Determinant

The *determinant* of a square matrix **A**, denoted by det (**A**), is a scalar that provides some useful information about **A**. The determinants of 2 × 2 and 3 × 3 matrices are defined, respectively, as

$$\det\left(\begin{bmatrix} a_{11} & a_{12} \\ a_{21} & a_{22} \end{bmatrix}\right) = a_{11}a_{22} - a_{12}a_{21},$$

$$\det\left(\begin{bmatrix} a_{11} & a_{12} & a_{13} \\ a_{21} & a_{22} & a_{23} \\ a_{31} & a_{32} & a_{33} \end{bmatrix}\right) = \begin{matrix} a_{11}a_{22}a_{33} + a_{21}a_{13}a_{32} + a_{31}a_{12}a_{23} \\ -a_{11}a_{23}a_{32} - a_{21}a_{12}a_{33} - a_{31}a_{13}a_{22} \end{matrix}.$$

For a general $n \times n$ matrix $\mathbf{A} = [a_{ij}]$, the determinant is defined as

$$\det(\mathbf{A}) = \sum_{k=1}^{n} (-1)^{i+k} a_{ik} \det(\mathbf{A}_{ik}) = \sum_{k=1}^{n} (-1)^{k+j} a_{kj} \det(\mathbf{A}_{kj})$$

for any $1 \le i, j \le n$ where $\mathbf{A}_{pq}$ is the $(n-1) \times (n-1)$ matrix resulting from the deletion of the pth row and the qth column of **A**. For example,

$$\det\begin{pmatrix} 1 & -2 & 3 \\ -4 & 5 & -6 \\ 7 & -8 & 9 \end{pmatrix} \overset{\text{let } i=1}{=} (-1)^{1+1}(1)\det\left(\begin{bmatrix} 5 & -6 \\ -8 & 9 \end{bmatrix}\right)$$

$$+(-1)^{1+2}(-2)\det\left(\begin{bmatrix} -4 & -6 \\ 7 & 9 \end{bmatrix}\right) + (-1)^{1+3}(3)\det\left(\begin{bmatrix} -4 & 5 \\ 7 & -8 \end{bmatrix}\right)$$

$$= (-3) - (-2)(6) + (3)(-3) = 0$$

$$\det\begin{pmatrix} 1 & -2 & 3 \\ -4 & 5 & -6 \\ 7 & -8 & 9 \end{pmatrix} \overset{\text{let } j=2}{=} (-1)^{1+2}(-2)\det\left(\begin{bmatrix} -4 & -6 \\ 7 & 9 \end{bmatrix}\right)$$

$$+(-1)^{2+2}(5)\det\left(\begin{bmatrix} 1 & 3 \\ 7 & 9 \end{bmatrix}\right) + (-1)^{3+2}(-8)\det\left(\begin{bmatrix} 1 & 3 \\ -4 & -6 \end{bmatrix}\right)$$

$$= -(-2)(6) + (5)(-12) - (-8)(6) = 0.$$

The determinant of $\mathbf{A}_{pq}$, $\det(\mathbf{A}_{pq})$, is called the (p,q)th *minor* of **A**, and $(-1)^{p+q}\det(\mathbf{A}_{pq})$ is called the *cofactor* of a_{pq}. Directly from the definition, the determinant of a diagonal matrix is the product of its diagonal elements, and the determinant of an upper or lower triangular matrix is also the product of its diagonal elements. Determinants have the following properties:

1. $\det(\mathbf{AB}) = \det(\mathbf{A})\det(\mathbf{B})$;
2. $\det(\alpha\mathbf{A}) = \alpha^n \det(\mathbf{A})$ for any scalar α and $n \times n$ matrix $\mathbf{A}$;
3. $\det(\mathbf{A}^T) = \det(\mathbf{A})$;
4. $\det(\mathbf{A}^k) = (\det(\mathbf{A}))^k$;
5. $\det(\mathbf{A}) = 0$ if any row (or column) of $\mathbf{A}$ is a scalar multiple of another row (or column);
6. $\det(\mathbf{A}) = 0$ if any row (or any column) of $\mathbf{A}$ is zero;
7. If $\mathbf{B}$ is obtained from $\mathbf{A}$ by interchanging two rows (or two columns), then $\det(\mathbf{B}) = -\det(\mathbf{A})$.

Singular and Nonsingular Matrices

A square matrix $\mathbf{A}$ is said to be *nonsingular* if $\det(\mathbf{A}) \neq 0$ and is *singular* if $\det(\mathbf{A}) = 0$. Since $\det(\mathbf{AB}) = \det(\mathbf{A})\det(\mathbf{B})$, the matrix $\mathbf{AB}$ is singular if and only if either $\mathbf{A}$ is singular or $\mathbf{B}$ is singular, and is nonsingular if and only if both $\mathbf{A}$ and $\mathbf{B}$ are nonsingular.

▶ A.4 LINEAR INDEPENDENCE OF VECTORS, BASIS VECTORS, AND RANK OF A MATRIX

Linear Independence of Vectors

Let $\{\mathbf{v}_1, \ldots, \mathbf{v}_m\}$ be a set of vectors in R^n or C^n. A vector $\mathbf{u}$ in R^n or C^n is said to be a *linear combination* of $\mathbf{v}_1, \ldots, \mathbf{v}_m$ if there are scalars $\alpha_1, \ldots, \alpha_m$ such that $\mathbf{u} = \sum_{i=1}^{m} \alpha_i \mathbf{v}_i$. The vector $\mathbf{u} = [-1 \ \ 2]^T$ is a linear combination of $\mathbf{v}_1 = [2 \ \ 3]^T$ and $\mathbf{v}_2 = [-3 \ \ 5]^T$ since $\mathbf{u} = \frac{1}{19}\mathbf{v}_1 + \frac{7}{19}\mathbf{v}_2$. The vector $= [\alpha_1 \ \ \ldots \ \ \alpha_n]^T$ is a solution of the following linear system:

$$[\mathbf{v}_1 \ldots \mathbf{v}_m] \begin{bmatrix} \alpha_1 \\ \vdots \\ \alpha_m \end{bmatrix} = \mathbf{u}.$$

If the solution does not exist, then $\mathbf{u}$ is not a linear combination of $\mathbf{v}_1, \ldots, \mathbf{v}_m$. When $m = n$, the solution $\begin{bmatrix} \alpha_1 & \ldots & \alpha_n \end{bmatrix}^T$ is unique if it exists. The set S containing all linear combinations of $\mathbf{v}_1, \ldots, \mathbf{v}_m$ is called the *spanning set* of $\mathbf{v}_1, \ldots \mathbf{v}_m$ and is denoted by $S = \text{span}(\mathbf{v}_1, \ldots \mathbf{v}_m)$. A set of vectors $\mathbf{v}_1, \ldots, \mathbf{v}_m$ is said to be *linearly independent* if $\alpha_1\mathbf{v}_1 + \cdots + \alpha_m v_m = \mathbf{0}$ implies $\alpha_1 = 0, \ldots, \alpha_m = 0$. A set of vectors is *linearly dependent* if it is not linearly independent; that is, there are some $\alpha_i \neq 0$ such that $\alpha_1\mathbf{v}_1 + \cdots + \alpha_m\mathbf{v}_m = \mathbf{0}$. Among the vectors

$$\mathbf{v}_1 = \begin{bmatrix} 2 \\ 3 \end{bmatrix}, \quad \mathbf{v}_2 = \begin{bmatrix} -3 \\ 5 \end{bmatrix}, \quad \mathbf{u}_1 = \begin{bmatrix} 1 \\ 2 \end{bmatrix}, \quad \mathbf{u}_2 = \begin{bmatrix} -3 \\ -6 \end{bmatrix},$$

the set $\{\mathbf{v}_1, \mathbf{v}_2\}$ is linearly independent and the set $\{\mathbf{u}_1, \mathbf{u}_2\}$ is linearly dependent since $(-3)\mathbf{u}_1 + \mathbf{u}_2 = 0$. A set of two vectors is linearly dependent if one vector is a scalar multiple of the other, that is, $\mathbf{v}_2 = \alpha\mathbf{v}_1$ for some nonzero scalar α. A set of m vectors is linearly dependent if one vector is a linear combination of others. Independent vectors have the following properties.

1. A set of orthogonal vectors is linear independent.
2. Let $\mathbf{v}_1, \ldots, \mathbf{v}_n$ be in R^n and $\mathbf{A} = \begin{bmatrix}\mathbf{v}_1 \ldots \mathbf{v}_n\end{bmatrix}$. The set of $\mathbf{v}_1, \ldots, \mathbf{v}_n$ is linearly independent if and only if $\det(\mathbf{A}) \neq 0$. So, if $\det(\mathbf{A}) = 0$, the set of $\mathbf{v}_1, \ldots \mathbf{v}_n$ is linearly dependent.

Consider the above vectors $\mathbf{v}_1, \mathbf{v}_2, \mathbf{u}_1$ and $\mathbf{u}_2$. Since

$$\det\left(\begin{bmatrix}\mathbf{v}_1 & \mathbf{v}_2\end{bmatrix}\right) = \det\left(\begin{bmatrix}2 & -3\\ 3 & 5\end{bmatrix}\right) = 19 \neq 0 \text{ and}$$

$$\det\left(\begin{bmatrix}\mathbf{u}_1 & \mathbf{u}_2\end{bmatrix}\right) = \det\left(\begin{bmatrix}1 & -3\\ 2 & -6\end{bmatrix}\right) = 0,$$

we can conclude that $\{\mathbf{v}_1, \mathbf{v}_2\}$ is linearly independent and $\{\mathbf{u}_1, \mathbf{u}_2\}$ is linearly dependent.

Basis Vectors

Let $S = \text{span}(\mathbf{v}_1, \ldots, \mathbf{v}_m)$ where $\mathbf{v}_i$'s are in R^n. If $\{\mathbf{v}_1, \ldots, \mathbf{v}_m\}$ is linearly independent, then $\{\mathbf{v}_1, \ldots, \mathbf{v}_m\}$ is called a set of *basis vectors* for S and the dimension of S is m. In this case, S is also called an m-dimensional subspace of the space R^n. Let $\mathbf{e}_i$ be in R^n with all elements zero except the ith element which is 1; that is, $\mathbf{e}_i = \begin{bmatrix}0 & \ldots & 0 & 1 & 0 & \ldots & 0\end{bmatrix}^T$ for $i = 1, \ldots, n$. A vector $\mathbf{u} = \begin{bmatrix}u_1 & \ldots & u_n\end{bmatrix}^T$ in R^n can be written as

$$\mathbf{u} = u_1\mathbf{e}_1 + u_2\mathbf{e}_2 + \cdots + u_n\mathbf{e}_n,$$

that is, a linear combination of $\mathbf{e}_1, \ldots, \mathbf{e}_n$. So, $\{\mathbf{e}_1, \ldots, \mathbf{e}_n\}$ is a set of basis vectors for R^n, and the dimension for R^n is n. Hence, any n linearly independent vectors in R^n form a set of basis vectors for R^n, and m linearly independent vectors in R^n for $m < n$ cannot form a set of basis vectors for R^n.

If $\mathbf{v}_1, \ldots, \mathbf{v}_m$ form a set of basis vectors for an m-dimensional subspace S of R^n, then each element $\mathbf{u}$ in S can be written as

$$\mathbf{u} = \alpha_1\mathbf{v}_1 + \cdots + \alpha_m\mathbf{v}_m = \begin{bmatrix}\mathbf{v}_1 & \ldots & \mathbf{v}_m\end{bmatrix}\begin{bmatrix}\alpha_1\\ \vdots\\ \alpha_m\end{bmatrix}.$$

Values of $\alpha_1, \ldots, \alpha_m$ can be obtained by solving the linear system:

$$\begin{bmatrix} \mathbf{v}_1 & \ldots & \mathbf{v}_m \end{bmatrix} \begin{bmatrix} \alpha_1 \\ \vdots \\ \alpha_m \end{bmatrix} = \mathbf{u}.$$

Rank of a Matrix

The rank of an $m \times n$ matrix $\mathbf{A}$, denoted by rank ($\mathbf{A}$), is the largest number of columns (or rows) of $\mathbf{A}$ that form a set of linearly independent vectors.

▶ **EXAMPLE A.5** *Let* $\mathbf{A} = \begin{bmatrix} 1 & -2 & 3 \\ -4 & 5 & -6 \\ 7 & -8 & 9 \end{bmatrix}$. *The rank of* $\mathbf{A}$ *is 2 since the third column of* $\mathbf{A}$ *is a linear combination of the first two columns and the first two columns are linearly independent:*

$$\begin{bmatrix} 3 \\ -6 \\ 9 \end{bmatrix} = (-1) \begin{bmatrix} 1 \\ -4 \\ 7 \end{bmatrix} + (-2) \begin{bmatrix} -2 \\ 5 \\ -8 \end{bmatrix} \text{ and } \begin{bmatrix} -2 \\ 5 \\ -8 \end{bmatrix} \neq \alpha \begin{bmatrix} 1 \\ -4 \\ 7 \end{bmatrix}.$$

The rank of a matrix has the following properties:

1. An $n \times n$ matrix $\mathbf{A}$ is nonsigular if and only if rank($\mathbf{A}$) $= n$;
2. Rank ($\mathbf{AB}$) $\leq$ min {rank($\mathbf{A}$), rank ($\mathbf{B}$)};
3. Rank ($\mathbf{A} + \mathbf{B}$) $\leq$ rank ($\mathbf{A}$)+ rank ($\mathbf{B}$);
4. The rank of a zero matrix is 0;
5. If rank ($\mathbf{A}$) $= k$, then there is a $k \times k$ submatrix of $\mathbf{A}$ with nonzero determinant, but all $(k+1) \times (k+1)$ submatrices of $\mathbf{A}$ have determinant 0.

▶ **EXAMPLE A.6** *Let* $\mathbf{A} = \begin{bmatrix} 1 & -2 & 3 \\ -4 & 5 & -6 \\ 7 & -8 & 9 \end{bmatrix}$. det ($\mathbf{A}$) $= 0$, *but* det $\left(\begin{bmatrix} 1 & -2 \\ -4 & 5 \end{bmatrix}\right) \neq 0$. *So, rank*($\mathbf{A}$)$=2$ *by Property 5.*

▶ A.5 INVERSE OF A SQUARE MATRIX

The *inverse matrix* of an $n \times n$ matrix $\mathbf{A}$, denoted by $\mathbf{A}^{-1}$, is an $n \times n$ matrix such that

$$\mathbf{AA}^{-1} = \mathbf{A}^{-1}\mathbf{A} = \mathbf{I}_n.$$

A matrix is said to be *invertible* if its inverse exists. The inverse matrix of a square matrix is unique if it exists. The inverse of a 2×2 matrix $\mathbf{A} = \begin{bmatrix} a & b \\ c & d \end{bmatrix}$ is $\mathbf{A}^{-1} =$

$\frac{1}{ad-bc}\begin{bmatrix} d & -b \\ -c & a \end{bmatrix}$, which exists if and only if $ad - bc \neq 0$. For $n \geq 3$, the inverse matrix $\mathbf{A}^{-1}$ can be obtained by solving n linear systems: $\mathbf{A}\mathbf{A}^{-1} = \mathbf{I}_n$; that is,

$$\mathbf{A}\begin{bmatrix}\mathbf{B}_1 & \cdots & \mathbf{B}_n\end{bmatrix} = \begin{bmatrix}\mathbf{e}_1 & \cdots & \mathbf{e}_n\end{bmatrix} \text{ or } \mathbf{A}\mathbf{B}_i = \mathbf{e}_i \text{ for } i = 1, \cdots n \text{ and } \mathbf{A}^{-1} = \begin{bmatrix}\mathbf{B}_1 & \ldots & \mathbf{B}_n\end{bmatrix}.$$

A linear system $\mathbf{Ax} = \mathbf{b}$ has a unique solution if and only if $\mathbf{A}$ is nonsingular. Hence, the inverse of $\mathbf{A}$ exists if and only if $\mathbf{A}$ is nonsingular or $\det(\mathbf{A}) \neq 0$, or $\text{rank}(\mathbf{A}) = n$.

EXAMPLE A.7 *Let* $\mathbf{A} = \begin{bmatrix} 1 & -2 & 3 \\ -4 & 5 & -6 \\ 7 & -8 & 10 \end{bmatrix}$. *Find* $\mathbf{A}^{-1}$ *if it exists.*

Since $\det(\mathbf{A}) = -3 \neq 0$, $\mathbf{A}^{-1}$ *exists.* $\mathbf{A}^{-1}$ *can be computed as follows.*

Solve

$$\begin{bmatrix} 1 & -2 & 3 \\ -4 & 5 & -6 \\ 7 & -8 & 10 \end{bmatrix}\mathbf{B}_1 = \begin{bmatrix} 1 \\ 0 \\ 0 \end{bmatrix}, \quad \mathbf{B}_1 = \begin{bmatrix} -\frac{2}{3} \\ \frac{2}{3} \\ 1 \end{bmatrix},$$

$$\begin{bmatrix} 1 & -2 & 3 \\ -4 & 5 & -6 \\ 7 & -8 & 10 \end{bmatrix}\mathbf{B}_2 = \begin{bmatrix} 0 \\ 1 \\ 0 \end{bmatrix}, \quad \mathbf{B}_2 = \begin{bmatrix} \frac{4}{3} \\ \frac{11}{3} \\ 2 \end{bmatrix}$$

$$\begin{bmatrix} 1 & -2 & 3 \\ -4 & 5 & -6 \\ 7 & -8 & 10 \end{bmatrix}\mathbf{B}_3 = \begin{bmatrix} 0 \\ 0 \\ 1 \end{bmatrix}, \quad \mathbf{B}_3 = \begin{bmatrix} 1 \\ 2 \\ 1 \end{bmatrix}$$

$$\mathbf{A}^{-1} = [\mathbf{B}_1\ \mathbf{B}_2\ \mathbf{B}_3] = \begin{bmatrix} -\frac{2}{3} & \frac{4}{3} & 1 \\ \frac{2}{3} & \frac{11}{3} & 2 \\ 1 & 2 & 1 \end{bmatrix}.$$

There is a closed form for the inverse of a nonsingular square matrix $\mathbf{A} = [a_{ij}]$, given by

$$\mathbf{A}^{-1} = \frac{1}{\det(\mathbf{A})}\text{Adj}(\mathbf{A}); \text{ where } \text{Adj}(\mathbf{A}) = [\text{cofactor of } a_{ji}] = \left[(-1)^{j+i}\det(\mathbf{A}_{ji})\right].$$

Inverse matrices have the following properties:

1. $\det(\mathbf{A}^{-1}) = \dfrac{1}{\det(\mathbf{A})}$.
2. The matrix $\mathbf{AB}$ is invertible if and only if both $\mathbf{A}$ and $\mathbf{B}$ are invertible and $(\mathbf{AB})^{-1} = \mathbf{B}^{-1}\mathbf{A}^{-1}$.
3. If $\mathbf{A}$ is invertible, then the linear system $\mathbf{Ax} = \mathbf{b}$ has a unique solution $\mathbf{x} = \mathbf{A}^{-1}\mathbf{b}$.
4. Let $\mathbf{A}$ be an $n \times n$ invertible matrix, and $\mathbf{u}$ and $\mathbf{v}$ be in C^n. Then

$$(\mathbf{A} + \mathbf{uv}^{\dagger})^{-1} = \mathbf{A}^{-1} - \frac{1}{1 + \mathbf{v}^{\dagger}\mathbf{A}^{-1}\mathbf{u}}\mathbf{A}^{-1}\mathbf{uv}^{\dagger}\mathbf{A}^{-1}.$$

   This is known as the *Matrix Inversion Lemma*. The RLS algorithm is based on this formula.

5. Let $\mathbf{A}$ and $\mathbf{R}$ be $n \times n$ and $m \times m$ invertible matrices, respectively. Let $\mathbf{U}$ be in $C^{n \times m}$ and $\mathbf{V}$ be in $C^{m \times n}$. Then

$$(\mathbf{A} + \mathbf{URV}^{\dagger})^{-1} = \mathbf{A}^{-1} - \mathbf{A}^{-1}\mathbf{U}(\mathbf{R}^{-1} + \mathbf{V}^{\dagger}\mathbf{A}^{-1}\mathbf{U})^{-1}\mathbf{VA}^{-1}.$$

   This is known as the *Sherman-Morrison formula*. The block RLS algorithm is based on this formula.

▸ A.6 EIGENVALUES AND EIGENVECTORS OF A SQUARE MATRIX AND SPECTRAL RADIUS

Eigenvalues and Eigenvectors of a Square Matrix

Let $\mathbf{A}$ be an $n \times n$ matrix, λ a scalar and x a nonzero vector in C^n. The pair $(\lambda, \mathbf{x})$ is said to be an *eigenpair* of $\mathbf{A}$ if the equation $\mathbf{Ax} = \lambda\mathbf{x}$ holds. In this case, λ is called an *eigenvalue* of $\mathbf{A}$ and $\mathbf{x}$ is called an *eigenvector* of $\mathbf{A}$ associated with λ.

▸ EXAMPLE A.8 *Let*

$$\mathbf{A} = \begin{bmatrix} 2 & -1 \\ -1 & 2 \end{bmatrix}, \lambda_1 = 1,\ \lambda_2 = 3,\ \mathbf{x}_1 = \begin{bmatrix} 1 \\ 1 \end{bmatrix} \textit{ and } \mathbf{x}_2 = \begin{bmatrix} 1 \\ -1 \end{bmatrix}.$$

With a quick check,

$$\begin{bmatrix} 2 & -1 \\ -1 & 2 \end{bmatrix}\begin{bmatrix} 1 \\ 1 \end{bmatrix} = (1)\begin{bmatrix} 1 \\ 1 \end{bmatrix} \textit{ and } \begin{bmatrix} 2 & -1 \\ -1 & 2 \end{bmatrix}\begin{bmatrix} 1 \\ -1 \end{bmatrix} = (3)\begin{bmatrix} 1 \\ -1 \end{bmatrix}$$

$(\lambda_1, \mathbf{x}_1)$ *and* $(\lambda_2, \mathbf{x}_2)$ *are eigenpairs of* $\mathbf{A}$.

An eigenpair $(\lambda, \mathbf{x})$ of $\mathbf{A}$ satisfies the equation $\mathbf{Ax} - \lambda\mathbf{x} = 0$, which implies $\det(\mathbf{A} - \lambda\mathbf{I}_n) = 0$. Define $P(\lambda) = \det(\mathbf{A} - \lambda\mathbf{I}_n)$. $P(\lambda)$ is an nth degree polynomial in λ and is called the characteristic polynomial of $\mathbf{A}$. Eigenvalues of $\mathbf{A}$ are zeros of the polynomial $P(\lambda)$ or, equivalently, roots of the polynomial equation $P(\lambda) = 0$. $\mathbf{A}$ has an eigenvalue λ_i

with multiplicity k if and only if $P(\lambda)$ has a factor $(\lambda - \lambda_i)^k$. The following are the steps needed to compute all eigenpairs of $\mathbf{A}$:

Step I. Solve the polynomial equation $P(\lambda) = 0$ for λ.

Step II. For each solution λ obtained in Step I, solve $(\mathbf{A} - \lambda \mathbf{I}_n)\mathbf{x} = \mathbf{0}$ for all $\mathbf{x}$ that are linearly independent.

EXAMPLE A.9 *Let* $\mathbf{A} = \begin{bmatrix} 1 & -2 \\ -3 & 4 \end{bmatrix}$. *Then*

$$P(\lambda) = \det(\mathbf{A} - \lambda \mathbf{I}_2) = \det\left(\begin{bmatrix} 1-\lambda & -2 \\ -3 & 4-\lambda \end{bmatrix}\right) = \lambda^2 - 5\lambda - 2.$$

Step I. *Solving* $P(\lambda) = 0$ *gives*

$$\lambda_1 = \frac{5}{2} + \frac{1}{2}\sqrt{33}, \quad \textit{and} \quad \lambda_2 = \frac{5}{2} - \frac{1}{2}\sqrt{33}.$$

Step II. *Solve for* $\mathbf{x}$. For $\lambda_1 = \frac{5}{2} + \frac{1}{2}\sqrt{33}$, solve $\left(\mathbf{A} - \left(\frac{5}{2} + \frac{1}{2}\sqrt{33}\right)\mathbf{I}_2\right)\mathbf{x}_1 = \mathbf{0}$ for $\mathbf{x}_1$.

$$\begin{bmatrix} 1 - \left(\frac{5}{2} + \frac{1}{2}\sqrt{33}\right) & -2 \\ -3 & 4 - \left(\frac{5}{2} + \frac{1}{2}\sqrt{33}\right) \end{bmatrix} \begin{bmatrix} x_{11} \\ x_{21} \end{bmatrix} = \begin{bmatrix} 0 \\ 0 \end{bmatrix}$$

$$\begin{bmatrix} x_{11} \\ x_{21} \end{bmatrix} = \begin{bmatrix} t \\ -\frac{3}{4}t - \frac{1}{4}t\sqrt{33} \end{bmatrix} \textit{for any nonzero scalar } t.$$

Let $t = 1$. Then $\mathbf{x}_1 = \begin{bmatrix} 1 \\ -\frac{3}{4} - \frac{1}{4}\sqrt{33} \end{bmatrix}$.

For $\lambda_2 = \frac{5}{2} - \frac{1}{2}\sqrt{33}$, solve $\left(\mathbf{A} - \left(\frac{5}{2} - \frac{1}{2}\sqrt{33}\right)\mathbf{I}_2\right)\mathbf{x}_2 = 0$ for $\mathbf{x}_2 = 0$ for $\mathbf{x}_2$. In a similar way, we have $\mathbf{x}_2 = \begin{bmatrix} \frac{1}{2} + \frac{1}{6}\sqrt{33} \\ 1 \end{bmatrix}$. Hence,

$$\left(\frac{5}{2} + \frac{1}{2}\sqrt{33}, \begin{bmatrix} 1 \\ -\frac{3}{4} - \frac{1}{4}\sqrt{33} \end{bmatrix}\right) \textit{ and } \left(\frac{5}{2} - \frac{1}{2}\sqrt{33}, \begin{bmatrix} \frac{1}{2} + \frac{1}{6}\sqrt{33} \\ 1 \end{bmatrix}\right)$$

are eigenpairs of $\mathbf{A}$.

When $n \geq 3$, it may not be possible to find the solutions of the equation $P(\lambda) = 0$ analytically. In general, eigenpairs $(\lambda, \mathbf{x})$ are solved numerically. Eigenvalues and eigenvectors have the following properties.

1. If $(\lambda, \mathbf{x})$ is an eigenpair of $\mathbf{A}$, then $(\lambda, \alpha\mathbf{x})$ is also an eigenpair of $\mathbf{A}$ for any nonzero scalar α.
2. Let $(\lambda_1, \mathbf{x}_1), \ldots, (\lambda_k, \mathbf{x}_k)$ be eigenpairs of $\mathbf{A}$. If $\lambda_1, \ldots, \lambda_k$ are distinct, then $\mathbf{x}_1, \ldots, \mathbf{x}_k$ are linearly independent.
3. $\mathbf{A}$ has a zero eigenvalue if and only if $\mathbf{A}$ is singular.
4. Eigenvalues of $\mathbf{A}$ and $\mathbf{A}^T$ are the same.
5. Eigenvalues of a diagonal matrix, an upper triangular matrix, or a lower triangular matrix are the diagonal elements of the matrix.
6. Let $\mathbf{A}$ be invertible. If $(\lambda, \mathbf{x})$ is an eigenpair of $\mathbf{A}$, then $\left(\frac{1}{\lambda}, \mathbf{x}\right)$ is an eigenpair of $\mathbf{A}^{-1}$.
7. Let $\mathbf{A}$ be an $n \times n$ matrix with eigenvalues $\lambda_1, \ldots, \lambda_n$. Then the determinant and the trace of $\mathbf{A}$ can be expressed respectively, as

$$\det(\mathbf{A}) = \prod_{i=1}^{n} \lambda_i = \lambda_1 \ldots \lambda_n, \quad \text{and} \quad \mathrm{tr}(\mathbf{A}) = \sum_{i=1}^{n} \lambda_i = \lambda_1 + \cdots + \lambda_n.$$

8. If $(\lambda, \mathbf{x})$ is an eigenpair of $\mathbf{A}$, then $(\lambda + \mu, \mathbf{x})$ is an eigenpair of the matrix $\mathbf{A} + \mu\mathbf{I}$.

Spectral Radius of a Square Matrix

For an $n \times n$ matrix $\mathbf{A}$ with eigenvalues $\lambda_1, \ldots, \lambda_n$, the *spectral radius* of $\mathbf{A}$, denoted by $\rho(\mathbf{A})$, is defined as

$$\rho(\mathbf{A}) = \max_{1 \le i \le n} \{|\lambda_i|\}$$

where $|a + ib| = \sqrt{a^2 + b^2}$. A matrix $\mathbf{A}$ is said to be *convergent* if $\lim_{k\to\infty} \mathbf{A}^k = \mathbf{0}$. Let $\mathbf{A}$ be an $n \times n$ complex matrix. We have the following two results.

Theorem 12.1 $\mathbf{A}$ *is convergent if and only if* $\rho(\mathbf{A}) < 1$.

Theorem 12.2 *If* $\rho(\mathbf{A}) < 1$, *then the matrix* $\mathbf{I} - \mathbf{A}$ is invertible and

$$(\mathbf{I} - \mathbf{A})^{-1} = \mathbf{I} + \mathbf{A} + \mathbf{A}^2 + \cdots + \mathbf{A}^n + \cdots.$$

▶ A.7 KRONECKER PRODUCT AND KRONECKER SUM

Kronecker Product

Let $\mathbf{A} = [a_{ij}]_{mn}$ and $\mathbf{B} = [b_{kl}]_{pq}$. The Kronecker product of $\mathbf{A}$ and $\mathbf{B}$ denoted by $\mathbf{A} \otimes \mathbf{B}$ is an $mp \times nq$ matrix defined by $\mathbf{A} \otimes \mathbf{B} = [a_{ij}\mathbf{B}]$.

▶ **EXAMPLE A.10** *Let* $\mathbf{A} = \begin{bmatrix} 1 & 2 & 3 \\ 4 & 5 & 6 \end{bmatrix}$ *and* $\mathbf{B} = \begin{bmatrix} 1 & -1 \\ 1 & 2 \end{bmatrix}$. Then

$$\mathbf{A} \otimes \mathbf{B} = \begin{bmatrix} 1 & -1 & 2 & -2 & 3 & -3 \\ 1 & 2 & 2 & -4 & 3 & 6 \\ 4 & -4 & 5 & -5 & 6 & -6 \\ 4 & 8 & 5 & 10 & 6 & 12 \end{bmatrix}.$$

The Kronecker product is not commutative in general. It has the following properties:

1. $(\alpha\mathbf{A}) \otimes \mathbf{B} = \mathbf{A} \otimes (\alpha\mathbf{B})$;
2. $(\mathbf{A} \otimes \mathbf{B}) \otimes \mathbf{C} = \mathbf{A} \otimes (\mathbf{B} \otimes \mathbf{C})$;
3. $(\mathbf{A} + \mathbf{B}) \otimes \mathbf{C} = \mathbf{A} \otimes \mathbf{C} + \mathbf{B} \otimes \mathbf{C}$;
4. $\mathbf{A} \otimes (\mathbf{B} + \mathbf{C}) = \mathbf{A} \otimes \mathbf{B} + \mathbf{A} \otimes \mathbf{C}$;
5. $(\mathbf{A} \otimes \mathbf{B})^T = \mathbf{A}^T \otimes \mathbf{B}^T$ and $(\mathbf{A} \otimes \mathbf{B})^\dagger = \mathbf{A}^\dagger \otimes \mathbf{B}^\dagger$;
6. $\mathrm{tr}(\mathbf{A} \otimes \mathbf{B}) = \mathrm{tr}(\mathbf{A})\mathrm{tr}(\mathbf{B})$;
7. Let $\mathbf{A}$ be $m \times m$ and $\mathbf{B}$ be $n \times n$. $\det(\mathbf{A} \otimes \mathbf{B}) = [\det(\mathbf{A})]^m[\det(\mathbf{B})]^n$;
8. $(\mathbf{A} \otimes \mathbf{B})(\mathbf{C} \otimes \mathbf{D}) = (\mathbf{AC}) \otimes (\mathbf{BD})$;
9. Let $\mathbf{A}$ and $\mathbf{B}$ be nonsingular. Then $\mathbf{A} \otimes \mathbf{B}$ is nonsingular and $(\mathbf{A} \otimes \mathbf{B})^{-1} = \mathbf{A}^{-1} \otimes \mathbf{B}^{-1}$;
10. $\mathbf{A} \otimes \mathbf{B} = \mathbf{0}$ if and only if $\mathbf{A} = \mathbf{0}$ or $\mathbf{B} = \mathbf{0}$;
11. If $(\lambda_i, \mathbf{x}_i)$ is an eigenpair of $\mathbf{A}$ and $(\mu_j, \mathbf{y}_j)$ is an eigenpair of $\mathbf{B}$, then $(\lambda_i\mu_j, \mathbf{x}_i \otimes \mathbf{y}_j)$ is an eigenpair of $\mathbf{A} \otimes \mathbf{B}$.

Kronecker Sum

Let $\mathbf{A}$ and $\mathbf{B}$ be $m \times m$ and $n \times n$ matrices, respectively. The Kronecker sum of $\mathbf{A}$ and $\mathbf{B}$ denoted by $\mathbf{A} \oplus \mathbf{B}$ is an mn matrix defined by $\mathbf{A} \oplus \mathbf{B} = (\mathbf{I}_n \otimes \mathbf{A}) + (\mathbf{I}_m \otimes \mathbf{B})$.

EXAMPLE A.11 Let $\mathbf{A} = \begin{bmatrix} 1 & 2 & 3 \\ 4 & 5 & 6 \\ 7 & 8 & 9 \end{bmatrix}$ and $\mathbf{B} = \begin{bmatrix} 1 & -1 \\ 1 & 2 \end{bmatrix}$. Then

$$\mathbf{A} \oplus \mathbf{B} = (\mathbf{I}_2 \otimes \mathbf{A}) + (\mathbf{I}_3 \otimes \mathbf{B}) = \begin{bmatrix} \mathbf{A} & 0 \\ 0 & \mathbf{A} \end{bmatrix} + \begin{bmatrix} \mathbf{B} & 0 & 0 \\ 0 & \mathbf{B} & 0 \\ 0 & 0 & \mathbf{B} \end{bmatrix}$$

$$= \begin{bmatrix} 1 & 2 & 3 & 0 & 0 & 0 \\ 4 & 5 & 6 & 0 & 0 & 0 \\ 7 & 8 & 9 & 0 & 0 & 0 \\ 0 & 0 & 0 & 1 & 2 & 3 \\ 0 & 0 & 0 & 4 & 5 & 6 \\ 0 & 0 & 0 & 7 & 8 & 9 \end{bmatrix} + \begin{bmatrix} 1 & -1 & 0 & 0 & 0 & 0 \\ 1 & 2 & 0 & 0 & 0 & 0 \\ 0 & 0 & 1 & -1 & 0 & 0 \\ 0 & 0 & 1 & 2 & 0 & 0 \\ 0 & 0 & 0 & 0 & 1 & -1 \\ 0 & 0 & 0 & 0 & 1 & 2 \end{bmatrix} = \begin{bmatrix} 2 & 1 & 3 & 0 & 0 & 0 \\ 5 & 7 & 6 & 0 & 0 & 0 \\ 7 & 8 & 10 & -1 & 0 & 0 \\ 0 & 0 & 1 & 3 & 2 & 3 \\ 0 & 0 & 0 & 4 & 6 & 5 \\ 0 & 0 & 0 & 7 & 9 & 11 \end{bmatrix}.$$

If $(\lambda_i,\ \mathbf{x}_i)$ is an eigenpair of $\mathbf{A}$ and $(\mu_j,\ \mathbf{y}_j)$ is an eigenpair of $\mathbf{B}$, then $(\lambda_i + \mu_j,\ \mathbf{y}_j \otimes \mathbf{x}_i)$ is an eigenpair of $\mathbf{A} \oplus \mathbf{B}$.

A.8 VECTOR NORMS, MATRIX NORMS, AND CONDITION NUMBERS

Vector Norms

Let S be a set in R^n or C^n. A real valued function $||\cdot||$ defined on S is said to be a *norm* or *vector norm* if $||\cdot||$ satisfies the following properties.

P1 $||\mathbf{x}|| \geq 0$ for any $\mathbf{x}$ in S, and $||\mathbf{x}|| = 0$ if and only if $\mathbf{x} = \mathbf{0}$, a zero vector.

P2 $||\alpha\mathbf{x}|| = |\alpha|\ ||\mathbf{x}||$ for any $\mathbf{x}$ in S where α is an arbitrary scalar.

P3 $||\mathbf{x} + \mathbf{y}|| \leq ||\mathbf{x}|| + ||\mathbf{y}||$ for any $\mathbf{x}$ and $\mathbf{y}$ in S. (triangular inequality)

Commonly used vector norms: let $\mathbf{x} = [x_1\ \dots\ x_n]^T$.

E1 The l_1- norm: $\quad ||\mathbf{x}||_1 \equiv \sum_{k=1}^{n} |x_k|$.

E2 The l_p- norm: $\quad ||\mathbf{x}||_p \equiv (\sum_{k=1}^{n} |x_k|^p)^{1/p}$ for p an integer and $1 < p < \infty$.

In particular, $\quad ||\mathbf{x}||_2 = \sqrt{x_1^2 + \cdots + x_n^2}$.

E3 The l_∞- norm: $\quad ||\mathbf{x}||_\infty \equiv \max_{1 \leq k \leq n} \{|x_k|\}$.

A vector $\mathbf{u}$ is a *unit vector* if $||\mathbf{u}|| = 1$. A given vector $\mathbf{x}$ can be normalized as a unit vector: $\mathbf{u} = \dfrac{\mathbf{x}}{||\mathbf{x}||}$. Vector norms satisfy the following inequalities. Let $\mathbf{x}$ and $\mathbf{y}$ be in R^n or C^n.

1. *The Cauchy-Schwarz Inequality*. $(\mathbf{x}^\dagger\mathbf{y})^2 \leq ||\mathbf{x}||_2\ ||\mathbf{y}||_2$ with equality if and only if $\mathbf{y} = \alpha\mathbf{x}$ for some scalar α.
2. $|\,||\mathbf{x}|| - ||\mathbf{y}||\,| \leq ||\mathbf{x} - \mathbf{y}||$.
3. $\lim_{k\to\infty}\mathbf{x}^{(k)} = \mathbf{x}$ if any only if $\lim_{k\to\infty}||\mathbf{x}^{(k)} - \mathbf{x}|| = 0$.

Matrix Norms

Let S be a set in $R^{m\times n}$ or $C^{m\times n}$. A real valued function $|||\cdot|||$ defined on S is said to be a *norm* or *matrix norm* if $|||\cdot|||$ satisfies the following properties.

P1 $|||\mathbf{A}||| \geq 0$ for any $\mathbf{A}$ in S, and $|||\mathbf{A}||| = 0$ if and only if $\mathbf{A} = \mathbf{0}$, a zero matrix.

P2 $|||\alpha\mathbf{A}||| = |\alpha|\ |||\mathbf{A}|||$ for any $\mathbf{A}$ in S where α is an arbitrary scalar.

P3 $|||\mathbf{A} + \mathbf{B}||| \leq |||\mathbf{A}||| + |||\mathbf{B}|||$ for any $\mathbf{A}$ and $\mathbf{B}$ in S. (triangular inequality)

P4 $||\mathbf{AB}|| \leq |||\mathbf{A}|||\ |||\mathbf{B}|||$ for any $\mathbf{A}$ and $\mathbf{B}$ in S. (submultiplicative)

Commonly used matrix norms: let $\mathbf{A} = [a_{ij}]_{i,j=1}^{mn}$.

E1 The l_1 matrix norm : $||\mathbf{A}||_1 \equiv \sum_{i=1}^{m} \sum_{j=1}^{n} |a_{ij}|$.

E2 The l_p matrix norm : $||\mathbf{A}||_p \equiv \left(\sum_{i=1}^{m} \sum_{j=1}^{n} |a_{ij}|^p\right)^{1/p}$ for p an integer and $1 < p < \infty$. In particular, the l_2 matrix norm or the Euclidean norm : $||\mathbf{A}||_2 \equiv \sqrt{\sum_{i=1}^{m} \sum_{j=1}^{n} |a_{ij}|^2}$.

E3 The maximum column sum matrix norm: $|||\mathbf{A}|||_1 \equiv \max_{1 \le j \le n} \sum_{k=1}^{m} |a_{kj}|$.

E4 The maximum row sum matrix norm: $|||\mathbf{A}|||_\infty \equiv \max_{1 \le i \le m} \sum_{k=1}^{n} |a_{ik}|$.

Note that when $\mathbf{A}$ is real $|||\mathbf{A}|||_\infty = |||\mathbf{A}^T|||_1$.

E5 The spectral norm: $|||\mathbf{A}|||_2 \equiv \max\left\{\sqrt{\lambda} : \lambda \text{ is an eigenvalue of } \mathbf{A}^\dagger \mathbf{A}\right\}$.

Note that $|||\mathbf{A}|||_2 = \sqrt{\rho(\mathbf{A}^\dagger \mathbf{A})}$.

F6 The Frobenius norm: $|||\mathbf{A}|||_F \equiv \sqrt{\sum_{i=1}^{m} \sum_{j=1}^{n} a_{ij}^2}$.

Note that $|||\mathbf{A}|||_F = \sqrt{\text{tr}\,(\mathbf{A}^\dagger \mathbf{A})} = ||\mathbf{A}||_2$.

EXAMPLE A.12 *Let* $\mathbf{A} = \begin{bmatrix} -1 & 2 & -3 \\ 4 & -5 & 6 \end{bmatrix}$. Then

$$||\mathbf{A}||_1 = 21, ||\mathbf{A}||_2 = \sqrt{91}, |||\mathbf{A}|||_1 = 9, |||\mathbf{A}|||_\infty = 15,$$

$$|||\mathbf{A}|||_2 = \sqrt{\rho\,(\mathbf{A}^T\mathbf{A})} = \sqrt{\frac{91}{2} + \frac{1}{2}\sqrt{8065}}, |||\mathbf{A}|||_F = \sqrt{91}.$$

Notice that $||\mathbf{A}||_\infty \equiv \max_{1 \le ij \le n} |a_{ij}|$ is not a matrix norm since it does not satisfy the submultiplicative property.

EXAMPLE A.13 $\mathbf{A} = \begin{bmatrix} 1 & 1 \\ 1 & 1 \end{bmatrix}, \mathbf{A}^2 = \begin{bmatrix} 2 & 2 \\ 2 & 2 \end{bmatrix}$. Then $||\mathbf{A}||_\infty = 1$, and

$$||\mathbf{A}^2||_\infty = 2 \ge ||\mathbf{A}||_\infty ||\mathbf{A}||_\infty.$$

Notice also that the spectral radius $\rho\,(\cdot)$ is not a matrix norm.

EXAMPLE A.14 *Let*

$$\mathbf{A}_1 = \begin{bmatrix} 0 & 1 \\ 0 & 0 \end{bmatrix}, \textit{ and } \mathbf{A}_2 = \begin{bmatrix} 0 & 0 \\ 1 & 0 \end{bmatrix}.$$

It is easy to verify that properties P1, P3, and P4 do not hold for these two matrices:

$P1: \quad \rho(\mathbf{A}_1) = \rho(\mathbf{A}_2) = 0 \text{ but } \mathbf{A}_1 \neq \mathbf{0} \text{ and } \mathbf{A}_2 \neq \mathbf{0}.$

$P3: \quad \rho(\mathbf{A}_1 + \mathbf{A}_2) = 1 > 0 = \rho(\mathbf{A}_1) + \rho(\mathbf{A}_2).$

$P4: \quad \rho(\mathbf{A}_1\mathbf{A}_2) = 1 > 0 = \rho(\mathbf{A}_1)\rho(\mathbf{A}_2).$

The following inequalities hold for an $m \times n$ matrix $\mathbf{A}$:

1. $|||\mathbf{A}|||_2 \leq |||\mathbf{A}|||_F \leq \sqrt{n}|||\mathbf{A}|||_2$;
2. $\max_{1\leq i\leq m,\, 1\leq j\leq n}|a_{ij}| \leq |||\mathbf{A}|||_2 \leq \sqrt{mn}\max_{1\leq i\leq m, 1\leq j\leq n}|a_{ij}|$;
3. $\frac{1}{\sqrt{n}}|||\mathbf{A}|||_\infty \leq |||\mathbf{A}|||_2 \leq \sqrt{m}|||\mathbf{A}|||_\infty$;
4. $\frac{1}{\sqrt{m}}|||\mathbf{A}|||_1 \leq |||\mathbf{A}|||_2 \leq \sqrt{n}|||\mathbf{A}|||_1$;
5. $|||\mathbf{A}|||_2 \leq \sqrt{|||\mathbf{A}|||_1|||\mathbf{A}|||_\infty}$ when $\mathbf{A}$ is a square matrix;
6. $\rho(\mathbf{A}) \leq |||\mathbf{A}|||$ for any square matrix and any matrix norm.

Condition Numbers

For a square matrix $\mathbf{A}$, the quantity

$$\kappa(\mathbf{A}) \equiv \begin{cases} |||\mathbf{A}|||\;|||\mathbf{A}^{-1}||| & \text{if } \mathbf{A} \text{ is nonsingular} \\ \infty & \text{if } \mathbf{A} \text{ is singular} \end{cases}$$

is called the *condition number* of $\mathbf{A}$ with respect to the matrix norm $|||\cdot|||$. Commonly, κ_1, κ_2, and κ_∞, respectively, denote the condition numbers of $\mathbf{A}$ with respect to the matrix norms $|||\mathbf{A}|||_1$, $|||\mathbf{A}|||_2$, and $|||\mathbf{A}|||_\infty$.

▶ **EXAMPLE A.15** *Let* $\mathbf{A} = \begin{bmatrix} 1 & -2 \\ -3 & 4 \end{bmatrix}$. *Then* $\mathbf{A}^{-1} = \begin{bmatrix} -2 & -1 \\ -\frac{3}{2} & -\frac{1}{2} \end{bmatrix}$ *and*

$$|||\mathbf{A}|||_1 = 21, \; |||\mathbf{A}|||_2 = 14.93303, \textit{ and } |||\mathbf{A}|||_\infty = 21.$$

Condition numbers have the following properties.

1. $\kappa(\mathbf{I}) = 1$ for any matrix norm.
2. If $\mathbf{A}$ is invertible, then $\kappa(\mathbf{A}^{-1}) = \kappa(\mathbf{A})$.
3. If $\mathbf{A}$ is invertible with eigenvalues $\lambda_1, \ldots, \lambda_n$, then $\kappa(\mathbf{A}) \geq \left|\frac{\lambda_{\max}(\mathbf{A})}{\lambda_{\min}(\mathbf{A})}\right|$ for any matrix norm where
$$\lambda_{\max}(\mathbf{A}) = \max_{1\leq i\leq n}\{|\lambda_i|\}, \text{ and } \lambda_{\min}(\mathbf{A}) = \min_{1\leq i\leq n}\{|\lambda_i|\}.$$
4. $\kappa(\mathbf{AB}) \leq \kappa(\mathbf{A})\kappa(\mathbf{B})$.

▶ A.9 SIMILARITY TRANSFORMATION

An $n \times n$ matrix $\mathbf{A}$ is said to be *similar* to an $n \times n$ matrix $\mathbf{B}$, denoted by $\mathbf{A} \sim \mathbf{B}$, if there exists a nonsingular $n \times n$ matrix $\mathbf{S}$ such that $\mathbf{A} = \mathbf{S}^{-1}\mathbf{BS}$. The transformation from $\mathbf{B}$

to $\mathbf{S}^{-1}\mathbf{BS}$ is called a *similarity transformation* by the similarity matrix $\mathbf{S}$. Similarity is an equivalence relation on square matrices; that is, similarity is

(a) reflective: $\mathbf{A} \sim \mathbf{A}$;

(b) symmetric: $\mathbf{B} \sim \mathbf{A}$ implies $\mathbf{A} \sim \mathbf{B}$; and

(c) transitive: $\mathbf{C} \sim \mathbf{B}$ and $\mathbf{B} \sim \mathbf{A}$ imply $\mathbf{C} \sim \mathbf{A}$.

Similar matrices have the following properties.

1. If $\mathbf{A} \sim \mathbf{B}$, then $\mathbf{A}$ and $\mathbf{B}$ have the same eigenvalues, counting multiplicity.
2. If $\mathbf{A} \sim \mathbf{B}$, then $\operatorname{rank}(\mathbf{A}) = \operatorname{rank}(\mathbf{B})$.
3. If $\mathbf{A} \sim \mathbf{C}$ and $\mathbf{B} \sim \mathbf{D}$, then $\mathbf{A} + \mathbf{B} \sim \mathbf{C} + \mathbf{D}$.

A.10 SPECIAL MATRICES AND PROPERTIES

Symmetric and Hermitian Matrices

A real (complex) square matrix $\mathbf{A}$ is said to be *symmetric* (*Hermitian*) if $\mathbf{A}^T = \mathbf{A}$ ($\mathbf{A}^\dagger = \mathbf{A}$), and *skew-symmetric* (*skew-Hermitian*) if $\mathbf{A}^T = -\mathbf{A}$ ($\mathbf{A}^\dagger = -\mathbf{A}$). Any real (complex) square matrix $\mathbf{A}$ can be written as $\mathbf{A} = \mathbf{A}_h + \mathbf{A}_s$ where

$$\mathbf{A}_h = \frac{1}{2}(\mathbf{A} + \mathbf{A}^T) \ \left(\mathbf{A}_h = \frac{1}{2}(\mathbf{A} + \mathbf{A}^\dagger)\right)$$

a symmetric (Hermitian) matrix and

$$\mathbf{A}_s = \frac{1}{2}(\mathbf{A} - \mathbf{A}^T) \ \left(\mathbf{A}_s = \frac{1}{2}(\mathbf{A} - \mathbf{A}^\dagger)\right)$$

a skew-symmetric (skew-Hermitian) matrix. Symmetric and Hermitian matrices have the following basic properties.

1. $\mathbf{A} + \mathbf{A}^T, \mathbf{A}\mathbf{A}^T$, and $\mathbf{A}^T\mathbf{A}$ are symmetric matrices, and $\mathbf{A} + \mathbf{A}^\dagger$, $\mathbf{A}\mathbf{A}^\dagger$, and $\mathbf{A}^\dagger\mathbf{A}$ are Hermitian matrices.
2. If $\mathbf{A}$ is symmetric (Hermitian), then $\mathbf{A}^k$ is symmetric (Hermitian) for all $k = 1, 2, \ldots$. If $\mathbf{A}$ is also nonsingular, then $\mathbf{A}^{-1}$ is symmetric (Hermitian).
3. If $\mathbf{A}$ and $\mathbf{B}$ are symmetric (Hermitian), then $\alpha\mathbf{A} + \beta\mathbf{B}$ is symmetric (Hermitian) for all real scalars α and β.
4. If $\mathbf{A}$ is an $n \times n$ symmetric (Hermitian) matrix, then $\mathbf{x}^T\mathbf{A}\mathbf{x}(\mathbf{x}^\dagger\mathbf{A}\mathbf{x})$ is real for all $\mathbf{x}$ in C^n and therefore, all eigenvalues of $\mathbf{A}$ are real.
5. If $\mathbf{A}$ is an $n \times n$ symmetric (Hermitian) matrix, then $\mathbf{S}^T\mathbf{AS}$ ($\mathbf{S}^\dagger\mathbf{AS}$) is symmetric (Hermitian) for all $\mathbf{S}$ in $R^{n\times n}$ ($C^{n\times n}$).
6. If $\mathbf{A}$ is symmetric (Hermitian) and invertible, then $\rho(\mathbf{A}) = |||\mathbf{A}|||_2$ and $\kappa_2(\mathbf{A}) = \left|\frac{\lambda_{\max}(\mathbf{A})}{\lambda_{\min}(\mathbf{A})}\right|$ where

$$\lambda_{\max}(\mathbf{A}) = \max_{1\le i\le n}\{|\lambda_i|\}, \text{ and } \lambda_{\min}(\mathbf{A}) = \min_{1\le i\le n}\{|\lambda_i|\},$$

$\lambda_i's$ are eigenvalues of $\mathbf{A}$.

Normal Matrices

An $n \times n$ matrix $\mathbf{A}$ is said to be *normal* if $\mathbf{A}^{\dagger}\mathbf{A} = \mathbf{A}\mathbf{A}^{\dagger}$. If a matrix is symmetric or Hermitian, then it is also normal. If $\mathbf{A} \otimes \mathbf{B}$ is normal, then $\mathbf{B} \otimes \mathbf{A}$ is also normal.

Orthogonal and Unitary Matrices

An $n \times n$ real (complex) matrix $\mathbf{U}$ is said to be *orthogonal* (*unitary*) if $\mathbf{U}^T\mathbf{U} = \mathbf{I}_n$ ($\mathbf{U}^{\dagger}\mathbf{U} = \mathbf{I}_n$). The determinant of an orthogonal matrix is either 1 or -1 since $\det(\mathbf{U}^2) = (\det(\mathbf{U}))^2 = 1$.

▶ **EXAMPLE A.16** *Let*

$$\mathbf{U}_1 = \begin{bmatrix} \frac{1}{\sqrt{2}} & -\frac{1}{\sqrt{2}} \\ \frac{1}{\sqrt{2}} & \frac{1}{\sqrt{2}} \end{bmatrix}, \text{ and } \mathbf{U}_2 = \begin{bmatrix} -\frac{1}{\sqrt{2}} & -\frac{1}{\sqrt{2}} \\ -\frac{1}{\sqrt{2}} & \frac{1}{\sqrt{2}} \end{bmatrix}.$$

Both $\mathbf{U}_1$ *and* $\mathbf{U}_2$ *are orthogonal matrices since* $\mathbf{U}_1^T\mathbf{U}_1 = \mathbf{I}_2$ *and* $\mathbf{U}_2^T\mathbf{U}_2 = \mathbf{I}_2$. *And* $\det(\mathbf{U}_1) = 1$ *and* $\det(\mathbf{U}_2) = -1$.

The following are equivalent:

1. $\mathbf{U}$ is unitary;
2. $\mathbf{U}$ is nonsingular and $\mathbf{U}^{\dagger} = \mathbf{U}^{-1}$.
3. $\mathbf{U}\mathbf{U}^{\dagger} = \mathbf{I}$.
4. The columns $\mathbf{u}_i$ for $i = 1, \ldots, n$ of $\mathbf{U}$ are orthonormal, that is, $\mathbf{u}_i^{\dagger}\mathbf{u}_j = \begin{cases} 0 & \text{if } i \neq j \\ 1 & \text{if } i = j \end{cases}$.
5. The rows of $\mathbf{U}$ are orthogonal.
6. For all $\mathbf{x}$ in C^n, the 2-norm of $\mathbf{y} = \mathbf{U}\mathbf{x}$ is the same as the one of $\mathbf{x}$, that is, $\mathbf{y}^{\dagger}\mathbf{y} = \mathbf{x}^{\dagger}\mathbf{x}$.

Since $\mathbf{U}\mathbf{U}^{\dagger} = \mathbf{I} = \mathbf{U}^{\dagger}\mathbf{U}$, $\mathbf{U}$ is also normal. If $\mathbf{U}_1$ and $\mathbf{U}_2$ are unitary, then $\mathbf{U}_1\mathbf{U}_2$ and $\mathbf{U}_2\mathbf{U}_1$ are also unitary. If $\mathbf{A} \otimes \mathbf{B}$ is unitary, then $\mathbf{B} \otimes \mathbf{A}$ is also unitary.

Positive Definite and Semidefinite Matrices

An $n \times n$ matrix $\mathbf{A}$ is said to be *positive definite* (*semidefinite*) if $\mathbf{A}$ is a Hermitian matrix and $\mathbf{x}^{\dagger}\mathbf{A}\mathbf{x} > 0$ ($\mathbf{x}^{\dagger}\mathbf{A}\mathbf{x} \geq 0$) for all nonzero $\mathbf{x}$ in C^n. Note that $\mathbf{x}^{\dagger}\mathbf{A}\mathbf{x}$ is always a nonnegative real number. For a positive definite (semidefinite) matrix $\mathbf{A}$, commonly it is indicated as $\mathbf{A} > 0$ ($\mathbf{A} \geq 0$). Positive definite (semidefinite) matrices have the following properties.

1. All eigenvalues of a positive definite (semidefinite) matrix are positive (nonnegative).
2. All diagonal elements of a positive definite (semidefinite) matrix are positive (nonnegative).
3. If $\mathbf{A}$ and $\mathbf{B}$ are positive definite (semidefinite), then $\mathbf{A} + \mathbf{B}$ is positive definite (semidefinite).

4. Let $\mathbf{A} = [a_{ij}]_{i,j=1}^{n}$ be positive definite. Then

$$\det(\mathbf{A}) \leq \prod_{k=1}^{n} a_{kk}$$

with equality if and only if $\mathbf{A}$ is diagonal.

5. If $\mathbf{A}$ and $\mathbf{B}$ are positive definite (semidefinite), then $\mathbf{A} \otimes \mathbf{B}$ is positive definite (semidefinite).

Vandermonde Matrices

An $n \times n$ *Vandermonde matrix* $\mathbf{V}$ is a matrix of the form

$$\mathbf{V} = \begin{bmatrix} 1 & x_1 & \cdots & x_1^{n-2} & x_1^{n-1} \\ 1 & x_2 & \cdots & x_2^{n-2} & x_2^{n-1} \\ \vdots & \vdots & \ddots & \vdots & \vdots \\ 1 & x_n & \cdots & x_n^{n-2} & x_n^{n-1} \end{bmatrix}.$$

The matrix $\mathbf{V}$ depends on n elements $x_1, \ldots, x_n$. The transpose of a Vandermonde matrix is also called a Vandermonde matrix. Note that the DFT matrix is Vandermonde. It is a fact that

$$\det(\mathbf{V}) = \prod_{i,j=1, i>j}^{n} (x_i - x_j).$$

So, a Vandermonde matrix is nonsingular if and only if $x_1, x_2, \ldots, x_n$ are distinct.

EXAMPLE A.17 Let $\mathbf{V} = \begin{bmatrix} 1 & 2 & 4 \\ 1 & -3 & 9 \\ 1 & 4 & 16 \end{bmatrix}$. Here $x_1 = 2$, $x_2 = -3$ and $x_4 = 4$. Since $x_i's$ are distinct, $\det(\mathbf{V}) \neq 0$.

Actually,

$$\det(\mathbf{V}) = (-3-2)(4-2)(4-(-3)) = -70.$$

Circulant Matrices

An $n \times n$ matrix $\mathbf{C}$ of the form

$$\mathbf{C} = \begin{bmatrix} c_1 & c_2 & \cdots & c_{n-1} & c_n \\ c_n & c_1 & \cdots & c_{n-2} & c_{n-1} \\ c_{n-1} & c_n & \cdots & c_{n-3} & c_{n-2} \\ \vdots & \vdots & \ddots & \vdots & \vdots \\ c_2 & c_3 & \cdots & c_n & c_1 \end{bmatrix} = \text{circ}(c_1, c_2, \ldots, c_n)$$

is called a *circulant matrix*. Each row is obtained by cyclically shifting to the right the previous row.

▶ **EXAMPLE A.18** $\mathbf{C} = \text{circ}(1,2,3,4) = \begin{bmatrix} 1 & 2 & 3 & 4 \\ 4 & 1 & 2 & 3 \\ 3 & 4 & 1 & 2 \\ 2 & 3 & 4 & 1 \end{bmatrix}.$

A circulant matrix $\mathbf{C} = \text{circ}(c_1, \ldots, c_n)$ can also be written in the form

$$\mathbf{C} = \sum_{k=0}^{n-1} c_{k+1}\mathbf{P}^k, \text{ where } \mathbf{P} = \begin{bmatrix} 0 & 1 & 0 & \ldots & 0 \\ 0 & 0 & 1 & \ldots & 0 \\ \vdots & \vdots & \ddots & \vdots & \vdots \\ 0 & 0 & 0 & 0 & 1 \\ 1 & 0 & 0 & 0 & 0 \end{bmatrix}.$$

Matrix $\mathbf{P}$ is called the basic circulant permutation matrix and satisfies the equation: $\mathbf{P}^n = \mathbf{I}_n$. Circulant matrices have the following properties.

1. $\mathbf{C}$ is circulant if and only if $\mathbf{C}^\dagger$ is circulant.
2. If $\mathbf{C}_1$ and $\mathbf{C}_2$ are circulant, then $\mathbf{C}_1\mathbf{C}_2$ is circulant.
3. Circulant matrices are normal matrices. That is, $\mathbf{C}^T\mathbf{C} = \mathbf{C}\mathbf{C}^T$ or $\mathbf{C}^\dagger\mathbf{C} = \mathbf{C}\mathbf{C}^\dagger$.
4. Let $(\lambda_l, \mathbf{x}_l)$ for $l = 1, \ldots, n$ be eigenpairs of the circulant matrix $\mathbf{C} = \text{circ}(c_1, \ldots, c_n)$.

Then

$$\lambda_l = \sum_{k=1}^{n} c_k e^{-j(2\pi((k-1)l)/n)}, \text{ and } \mathbf{x}_l = \frac{1}{\sqrt{n}} \begin{bmatrix} 1 \\ e^{-j(2\pi l/n)} \\ e^{-j(4\pi l/n)} \\ \vdots \\ e^{-j(2(n-1)\pi l/n)} \end{bmatrix} \quad \text{for } l = 1, \ldots, n.$$

Note that the eigenvectors of a circulant matrix do not depend on the c_i's. So, all $n \times n$ circulant matrices for a given n have the same set of eigenvectors.

Toeplitz Matrices

An $n \times n$ matrix $\mathbf{T}$ of the form

$$\mathbf{T} = \begin{bmatrix} t_0 & t_1 & \ldots & t_{n-2} & t_{n-1} \\ t_{-1} & t_0 & \ldots & t_{n-3} & t_{n-2} \\ t_{-2} & t_{-1} & \ldots & t_{n-4} & t_{n-3} \\ \vdots & \vdots & \ddots & \vdots & \vdots \\ t_{-(n-1)} & t_{-(n-2)} & \ldots & t_{-1} & t_0 \end{bmatrix}$$

is called a *Toeplitz matrix*. A Toeplitz $\mathbf{T}$ depends on $2n-1$ elements $t_{-(n-1)}$, $t_{-(n-2)}$, $\ldots, t_{-1}, t_0, t_1, \ldots, t_{n-2}, t_{n-1}$.

EXAMPLE A.19 $\mathbf{T} = \begin{bmatrix} 1 & 2 & 3 & 4 \\ 5 & 1 & 2 & 3 \\ 6 & 5 & 1 & 2 \\ 7 & 6 & 5 & 1 \end{bmatrix}$ is a 4×4 Toeplitz matrix.

Circulant matrices are also Toeplitz matrices.

Hankel Matrices

An $n \times n$ matrix $\mathbf{H}$ of the form

$$\mathbf{H} = \begin{bmatrix} h_0 & h_1 & \ldots & h_{n-2} & h_{n-1} \\ h_1 & h_2 & \ldots & h_{n-1} & h_n \\ h_2 & h_3 & \ldots & h_n & h_{n+1} \\ \vdots & \vdots & \ddots & \vdots & \vdots \\ h_{n-1} & h_n & \ldots & h_{2n-2} & h_{2n-1} \end{bmatrix}$$

is called a *Hankel matrix*. A Hankel matrix $\mathbf{H}$ depends on $2n-1$ elements h_0, $h_1, \ldots,$ h_{2n-1}, h_{2n-1}, and its elements are the same along the diagonals perpendicular to the main diagonal.

EXAMPLE A.20 $\mathbf{H} = \begin{bmatrix} 1 & 2 & 3 & 4 \\ 2 & 3 & 4 & 5 \\ 3 & 4 & 5 & 6 \\ 4 & 5 & 6 & 7 \end{bmatrix}$ is a 4×4 Hankel matrix.

Note that a real square Hankel matrix is symmetric.

Hadamard Matrices

An $n \times n$ matrix $\mathbf{H}$ whose elements are ± 1 such that $\mathbf{H}\mathbf{H}^T = n\mathbf{I}_n$ is called a *Hadamard matrix*. An $n \times n$ Hadamard matrix with $n > 2$ exists only if n is divisible by 4. The 2×2 and 4×4 Hadamard matrices $\mathbf{H}_2$ and $\mathbf{H}_4$ are

$$\mathbf{H}_2 = \begin{bmatrix} 1 & 1 \\ 1 & -1 \end{bmatrix}, \text{and } \mathbf{H}_4 = \begin{bmatrix} 1 & 1 & 1 & 1 \\ 1 & -1 & 1 & -1 \\ 1 & 1 & -1 & -1 \\ 1 & -1 & -1 & 1 \end{bmatrix}.$$

A $2n \times 2n$ Hadamard matrix can be obtained by

$$\mathbf{H}_{2n} = \mathbf{H}_2 \otimes \mathbf{H}_n.$$

Note that a Hadamard matrix after normalization is unitary.

▶ A.11 DIAGONALIZATION, UNITARY TRIANGULARIZATION, AND JORDAN FORM

Diagonalization

A square matrix $\mathbf{A}$ is said to be *diagonalizable* if it is similar to a diagonal matrix $\mathbf{D}$, that is, there exists a nonsingular matrix $\mathbf{S}$ such that $\mathbf{A} = \mathbf{S}^{-1}\mathbf{D}\mathbf{S}$.

▶ **EXAMPLE A.21** *Let* $\mathbf{A} = \begin{bmatrix} 2 & -1 \\ -1 & 2 \end{bmatrix}$ *and* $\mathbf{B} = \begin{bmatrix} 1 & 1 \\ 0 & 1 \end{bmatrix}$. *Then*

$$\mathbf{A} = \begin{bmatrix} 1 & -1 \\ 1 & 1 \end{bmatrix}\begin{bmatrix} 1 & 0 \\ 0 & 3 \end{bmatrix}\begin{bmatrix} \frac{1}{2} & \frac{1}{2} \\ -\frac{1}{2} & \frac{1}{2} \end{bmatrix} = \mathbf{SDS}^{-1}$$

is diagonalizable and $\mathbf{B}$ *is not diagonalizable.*

Since $\mathbf{A}$ and $\mathbf{D}$ are similar, they have the same eigenvalues. Eigenvalues of $\mathbf{D}$ are diagonal elements of $\mathbf{D}$. So eigenvalues of $\mathbf{A}$ are known if we know $\mathbf{D}$. The process of finding matrices $\mathbf{D}$ and $\mathbf{S}$ is called *diagonalization*. The following results identify diagonalizable matrices: let $\mathbf{A}$ be an $n \times n$ matrix.

1. If $\mathbf{A}$ has a set of n linearly independent eigenvectors $\mathbf{v}_1, \ldots, \mathbf{v}_n$, then $\mathbf{A}$ is diagonalizable and $\mathbf{A} = \mathbf{S}^{-1}\,\mathbf{DS}$ where $\mathbf{S} = [\mathbf{v}_1 \ldots \mathbf{v}_n]$ and $\mathbf{D} = \text{diag}\{\lambda_1, \ldots, \lambda_n\}$, λ_i, is the eigenvalue corresponding to the eigenvector $\mathbf{v}_i$. Recall that eigenvectors corresponding to distinct eigenvalues are linearly independent. So, if eigenvalues of $\mathbf{A}$ are distinct, then $\mathbf{A}$ is diagonalizable.
2. Every symmetric (Hermitian) matrix can be diagonalized by an orthogonal (unitary) matrix $\mathbf{U}$: $\mathbf{A} = \mathbf{UDU}^T (\mathbf{A} = \mathbf{UDU}^\dagger)$. Recall that if $\mathbf{A}$ is positive definite (semidefinite), then all eigenvalues $\lambda_i's$ of $\mathbf{A}$ are positive (nonnegative). Then $\mathbf{D} = \text{diag}\{\lambda_1, \ldots, \lambda_n\}$, can be written as $\mathbf{D} = \sqrt{\mathbf{D}}\sqrt{\mathbf{D}}$ where $\sqrt{\mathbf{D}} \equiv \text{diag}\{\sqrt{\lambda_1}, \ldots, \sqrt{\lambda_n}\}$, and $\mathbf{A} = \mathbf{U}\sqrt{\mathbf{D}}\sqrt{\mathbf{D}}\mathbf{U}^T = \mathbf{VV}^T (\mathbf{A} = \mathbf{VV}^\dagger)$ where $\mathbf{V} = \mathbf{U}\sqrt{\mathbf{D}}$.
3. An $n \times n$ matrix $\mathbf{A}$ is normal if and only if there exists unitary matrix $\mathbf{U}$ such that $\mathbf{A} = \mathbf{UDU}^\dagger$ where $\mathbf{D} = \text{diag}\{\lambda_1, \ldots, \lambda_n\}$ and $\lambda_i's$ are eigenvalues of $\mathbf{A}$.
4. An $n \times n$ circulant matrix $\mathbf{C} = \text{circ}(c_1, \ldots, c_n)$ can be diagonalized by $\frac{1}{\sqrt{n}}\mathbf{F}$; that is, $\mathbf{C} = \frac{1}{n}\mathbf{F}\mathbf{\Lambda}\mathbf{F}^\dagger$, where

$$\mathbf{F} = \begin{bmatrix} 1 & 1 & 1 & \ldots & 1 \\ 1 & \omega & \omega^2 & \ldots & \omega^{m-1} \\ 1 & \omega^2 & \omega^4 & \ldots & \omega^{2(m-1)} \\ \vdots & \vdots & \vdots & \ddots & \vdots \\ 1 & \omega^{m-1} & \omega^{2(m-1)} & \ldots & \omega^{(m-1)^2} \end{bmatrix}, \omega = e^{-j2\pi/m} \text{and } j = \sqrt{-1},$$

$$\mathbf{\Lambda} = \text{diag}\{\lambda_1, \ldots, \lambda_n\} \text{ where } \lambda_l = \sum_{k=1}^{n} c_k e^{-j(2\pi((k-1)l)/n)}.$$

5. If **A** and **B** are diagonalizable simultaneously—that is, there exists a nonsingular matrix **S** such that $\mathbf{A} = \mathbf{S}^{-1}\mathbf{D}_\mathbf{A}\mathbf{S}$ and $\mathbf{B} = \mathbf{S}^{-1}\mathbf{D}_\mathbf{B}\mathbf{S}$ where $\mathbf{D}_\mathbf{A}$ and $\mathbf{D}_\mathbf{B}$ are diagonal matrices—then **A** and **B** commute, that is, $\mathbf{AB} = \mathbf{BA}$.

Unitary Triangularization

For any square matrix **A**, there is a unitary matrix **U** such that $\mathbf{A} = \mathbf{UTU}^{\dagger}$ where **T** is an upper triangular matrix. So, every square matrix is similar to an upper triangular matrix. Note that eigenvalues of **A** are diagonal elements of **T**.

Jordan Canonical Form

For any square matrix **A**, there is a nonsingular matrix **X** such that

$$\mathbf{A} = \mathbf{X}\ \text{diag}\ \{\mathbf{J}_{k_1}(\lambda_1), \mathbf{J}_{k_2}(\lambda_2), \ldots, \mathbf{J}_{k_1}(\lambda_l)\}\mathbf{X}^{-1}$$

where $\mathbf{J}_k(\lambda)$, called a *Jordan block*, in $C^{k \times k}$ is of the form

$$\mathbf{J}_k(\lambda) = \begin{bmatrix} \lambda & 1 & 0 & \ldots & 0 & 0 \\ 0 & \lambda & 1 & \ldots & 0 & 0 \\ 0 & 0 & \lambda & \ldots & 0 & 0 \\ \vdots & \vdots & \vdots & \ddots & \vdots & \vdots \\ 0 & 0 & 0 & \ldots & \lambda & 1 \\ 0 & 0 & 0 & \ldots & 0 & \lambda \end{bmatrix}$$

and $\lambda_1, \ldots, \lambda_l$ are eigenvalues of **A** with multiplicity $k_1, \ldots, k_l$, respectively. The block diagonal matrix

$$\text{diag}\ \{\mathbf{J}_{k_1}(\lambda_1), \mathbf{J}_{k_2}(\lambda_2), \ldots, \mathbf{J}_{k_l}(\lambda_l)\} = \begin{bmatrix} \mathbf{J}_{k_1}(\lambda_1) & 0 & \ldots & 0 \\ 0 & \mathbf{J}_{k_2}(\lambda_2) & \ldots & 0 \\ \vdots & \vdots & \ddots & \vdots \\ 0 & 0 & \ldots & \mathbf{J}_{k_l}(\lambda_2) \end{bmatrix}$$

is called a *Jordan matrix*. The *Jordan canonical form* is a set of Jordan matrices. Observe that a Jordan matrix is "almost diagonal" and is diagonal when all k_i's are 1 (then **A** is diagonalizable).

▶ A.12 SINGULAR VALUES, SINGULAR VALUE DECOMPOSITION, AND PSEUDO-INVERSES

Singular Values

Singular values of an $m \times n$ matrix **A**, denoted by $\sigma_i\ (\mathbf{A})$, are the square root of eigenvalues of the matrix $\mathbf{AA}^{\dagger}$.

▶ **EXAMPLE A.22** *Let* $\mathbf{A} = \begin{bmatrix} 1 & -2 \\ -3 & 4 \end{bmatrix}$. *Then* $\mathbf{AA}^T = \begin{bmatrix} 5 & -11 \\ -11 & 25 \end{bmatrix}$, *and its eigenvalues are* $\lambda_1 = 15 + \sqrt{221}$ *and* $\lambda_2 = 15 - \sqrt{221}$. *So, singular values of* **A** *are:* $\sigma_1 = \sqrt{15 + \sqrt{221}} = 5.464986$ *and* $\sigma_2 = \sqrt{15 - \sqrt{221}}$.

Notice that the spectral norm of a square matrix **A** is $\max\{\sigma_i\}$, that is $|||\mathbf{A}|||_2 = \max\{\sigma_i\}$.

Singular Value Decomposition

Let **A** be an $m \times n$ matrix. Then **A** can be written as $\mathbf{A} = \mathbf{U\Sigma V}^\dagger$ where **U** is an $m \times m$ unitary matrix, **V** is an $n \times n$ unitary matrix, and

$$\mathbf{\Sigma} = \begin{bmatrix} \mathbf{\Sigma}_p \\ 0 \end{bmatrix} \text{ or } \mathbf{\Sigma} = \begin{bmatrix} \mathbf{\Sigma}_p & 0 \end{bmatrix}$$

is an $m \times n$ matrix where $\mathbf{\Sigma_A} = \text{diag}\{\sigma_1, \ldots, \sigma_p\}, p = \min\{m, n\}$, and $\sigma_1 \geq \ldots \geq \sigma_p$. $\mathbf{U\Sigma U}^\dagger$ is called the *singular value decomposition*, SVD in short, of **A**. Columns of **U** are eigenvectors of $\mathbf{AA}^\dagger$, and columns of **V** are eigenvectors of $\mathbf{A}^\dagger\mathbf{A}$. If **A** is a real matrix, then **U** and **V** are orthogonal matrices.

▶ **EXAMPLE A.23** *Let* $\mathbf{A} = \begin{bmatrix} 1 & -2 & 3 \\ -4 & 5 & -6 \end{bmatrix}$ *and* $\mathbf{B} = \begin{bmatrix} 1 & -2 \\ -3 & 4 \\ 5 & -6 \end{bmatrix}$. *The numerical SVD of* **A** *and* **B** *are:*

$$\mathbf{A} = \begin{bmatrix} -0.38632 & 0.92237 \\ 0.92237 & 0.38632 \end{bmatrix} \begin{bmatrix} 9.508 & 0 & 0 \\ 0 & 0.77287 & 0 \end{bmatrix} \begin{bmatrix} -0.42867 & 0.56631 & -0.70395 \\ -0.80596 & 0.11238 & 0.58120 \\ 0.40825 & 0.81650 & 0.40825 \end{bmatrix}$$

and

$$\mathbf{B} = \begin{bmatrix} -0.22985 & -0.88346 & 0.40825 \\ 0.52474 & 0.24078 & 0.81650 \\ -0.81964 & 0.40190 & 0.40825 \end{bmatrix} \begin{bmatrix} 9.5255 & 0 \\ 0 & 0.5143 \\ 0 & 0 \end{bmatrix} \begin{bmatrix} -0.61963 & 0.78489 \\ 0.78489 & 0.61963 \end{bmatrix}$$

If the rank of **A** is r, then $\sigma_1 \geq \ldots \geq \sigma_r > 0$ and $\sigma_{r+1} = 0, \ldots, \sigma_p = 0$.

Pseudo-inverses

Let **A** be an $m \times n$ matrix with rank r and $\mathbf{A} = \mathbf{U\Sigma V}^\dagger$. Then

$$\mathbf{\Sigma} = \begin{bmatrix} \mathbf{\Sigma}_r & 0 \\ 0 & 0 \end{bmatrix} \text{ where} \mathbf{\Sigma}_r = \text{diag}\{\sigma_1, \ldots, \sigma_r\}.$$

Define $\mathbf{\Sigma}^{-1} = \begin{bmatrix} (\mathbf{\Sigma}_r)^{-1} & 0 \\ 0 & 0 \end{bmatrix}$. Then$\mathbf{A}^{\dagger} = \mathbf{V}\mathbf{\Sigma}^{-1}\mathbf{U}^{\dagger}$ is called the *pesudo-inverse* of $\mathbf{A}$, or the Moore-Penrose generalized inverse of $\mathbf{A}$. It can be shown that $\mathbf{A}^{\dagger}$ satisfies the following properties:

1. $\mathbf{A}\mathbf{A}^{\dagger}and\mathbf{A}^{\dagger}\mathbf{A}$ are Hermitian;
2. $\mathbf{A}\mathbf{A}^{\dagger}\mathbf{A} = \mathbf{A}$; and
3. $\mathbf{A}^{\dagger}\mathbf{A}\mathbf{A}^{\dagger} = \mathbf{A}$.

One of the applications is to solve the linear system $\mathbf{Ax} = \mathbf{b}$ where $\mathbf{A} = \mathbf{U}\mathbf{\Sigma}\mathbf{V}^{\dagger}$ is an $m \times n$ matrix with rank r. A least-squares solution $\mathbf{x}_{ls}$ to the linear system $\mathbf{Ax} = \mathbf{b}$ is

$$\mathbf{x}_{ls} = \mathbf{A}^{\dagger}\mathbf{b} = (\mathbf{V}\mathbf{\Sigma}^{-1}\mathbf{U}^{\dagger})\mathbf{b}.$$

A.13 SUMMARY

In this appendix, we presented several topics in linear algebra that are useful in signal processing. Most of these concepts are used in this text, while others are given for useful reference. There are several excellent books on linear algebra, including [1], [2], and [3].

BIBLIOGRAPHY

[1] G. GOLUB and C. VAN LOAN, *Matrix Computation*, Johns Hopkins University Press, Baltimore, MD, 1983.

[2] R. HORN and C. JOHNSON, *Matrix Analysis*, Cambridge University Press, 1985.

[3] G. STRANG, *Introduction to Linear Algebra*, Wellesley-Cambridge Press, 1998.

Index